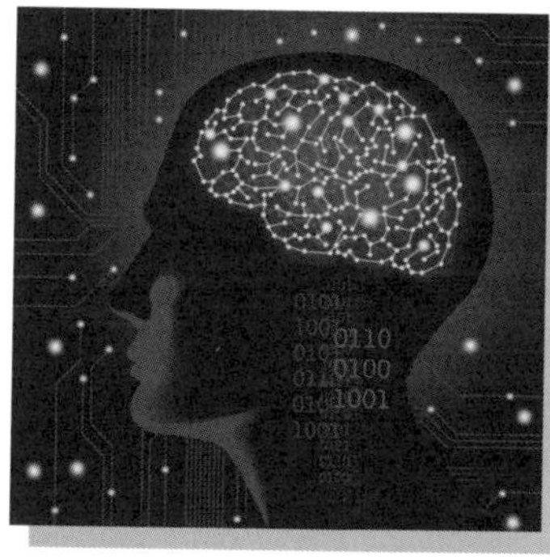

THE HANDBOOK ON REASONING-BASED INTELLIGENT SYSTEMS

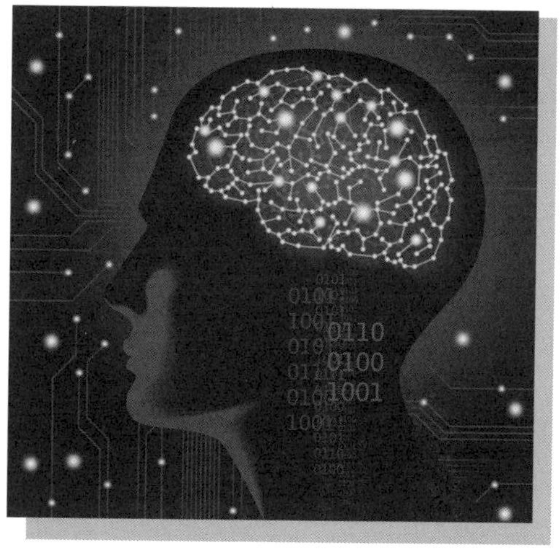

THE HANDBOOK ON REASONING-BASED INTELLIGENT SYSTEMS

edited by

Kazumi Nakamatsu
University of Hyogo, Japan

Lakhmi C Jain
University of South Australia, Australia

World Scientific

NEW JERSEY • LONDON • SINGAPORE • BEIJING • SHANGHAI • HONG KONG • TAIPEI • CHENNAI

Published by

World Scientific Publishing Co. Pte. Ltd.
5 Toh Tuck Link, Singapore 596224
USA office: 27 Warren Street, Suite 401-402, Hackensack, NJ 07601
UK office: 57 Shelton Street, Covent Garden, London WC2H 9HE

British Library Cataloguing-in-Publication Data
A catalogue record for this book is available from the British Library.

THE HANDBOOK ON REASONING-BASED INTELLIGENT SYSTEMS

ISBN 978-981-4329-47-7

Typeset by Stallion Press
Email: enquiries@stallionpress.com

Printed in Singapore by Mainland Press Pte Ltd.

CONTENTS

PREFACE

Recently, artificial systems have become increasingly more intelligent and consequently more complicated. Various kinds of intelligent reasoning are employed in these systems. We therefore present this handbook on reasoning-based intelligent systems. It covers both the fundamental theories and practical applications of reasoning-based intelligent systems. It is hoped that researchers, scientists and engineers who wish to develop reasoning-based intelligent systems and students wishing to learn the foundations and the applications of reasoning-based intelligent systems will find it of use. Consequently it is our hope that this book would contribute to academicians and practitioners.

It is a great honor and pleasure to publish this Handbook on Reasoning-Based Intelligent Systems. This book contains 23 chapters written by prominent researchers from countries all over the world. We remain grateful to the authors. We express our profound gratitude to the reviewers for their constructive feedback during the review process of this book. All staff members of the publisher involved in publishing the handbook are gratefully acknowledged. We wish to express our appreciation to the University of Hyogo, Japan and the University of South Australia, Australia for supporting this project.

Kazumi Nakamatsu Prof. Dr.
University of Hyogo
Japan
nakamatu@shs.u-hyogo.ac.jp

Lakhmi C. Jain Prof. Dr.
University of South Australia
Australia
Lakhmi.jain@unisa.ed.au

Chapter 1

ADVANCES IN INTELLIGENT SYSTEMS

Lakhmi C. Jain
University of South Australia
South Australia
Australia

Kazumi Nakamatsu
University of Hyogo
Japan

This chapter presents a brief overview of the recent advances in intelligent systems. A brief summary of each chapter included in the book is presented. A list of resources is included to explore this field further.

1.1. Introduction

Intelligent systems are finding their use in every field such as aviation, computer science, engineering, healthcare, management and so on. Researchers are keen to see intelligent machines which can to some extent mimic the human behavior [1–4]. A number of techniques are reported in the literature to fuse intelligence in machines. These include Knowledge-Based Systems [5], Neural Networks [6–10], Fuzzy Systems [11–13], Evolutionary Systems [14–17], Intelligent Agents [18–23], fusion of intelligent paradigms [24–28], and so on. Knowledge-Based Systems (KBS) also called expert systems were successfully used for performing tasks such as diagnosis, design, planning and so on. However, KBS are not very popular due to their inability to learn similar to humans. Neural Networks attempt to mimic biological problem solving mechanism that can learn, by itself, to solve a problem through a process of training.

Neural networks are successfully used in many applications due to their ability to generalize. It means that neural networks usually come up with sensible solutions to problems for which they have not seen examples in training. Evolutionary systems use genetic selection of the biological world in solving problems. A number of variations of ES are proposed by researchers to apply it widely in a number of areas. Fuzzy systems provide an approximate but an effective means of describing behavior of systems that are too complex to manage mathematically. FS are used successfully in many applications where the dominant role is played by humans.

There are many definitions of agents. Some researchers define agents as the software systems for performing the tasks delegated to them. Researchers have implemented agents which are flexible and autonomous and can work in dynamic and unpredictable environments. Agents can perform a task that has been assigned to them. Multiple agents are used in many applications to interact with each other and environment through cooperation, coordination and learning. Researchers are predicting that the agents will be established firmly in every field in this decade.

Reasoning-Based Techniques [29] are important to design intelligent machines which can mimic human behavior in a limited but impressive way. With this view in mind Professor Nakamatsu initiated an International Journal on Reasoning-Based Intelligent Systems in 2009 [30].

It is a usual practice to fuse various intelligent paradigms for offsetting the demerits of one technique by the merits of another technique. A number of researchers have demonstrated the merits the fusion technology [26].

This book presents the theory which can be used to develop reasoning-based intelligent systems. A number of applications in various fields are also included.

1.2. Chapters Included in the Book

This book contains 23 chapters. Each chapter was written by prominent researchers from countries all over the world. The chapters have been peer reviewed by experts in the fields and have been enhanced by their welcome efforts. The book is organized as follows. Chapters 2 to 10 show the basic theories used by intelligent reasoning methods. Present chapter introduces intelligent systems and sets a scene for rest of the chapters included in the book. Chapter 2 deals with a model of cognitive reasoning. Chapter 3 deals with an intelligent method for clustering. Chapter 4 considers a Non-Deterministic Information System from a view point of Rough Set. Chapter 5 considers the mathematical limits of computation. Chapter 6 treats the building of hypothesis using intelligent agent systems. Chapter 7 considers a heuristic algorithm procedure for solving Allocation Problems. Chapter 8 considers non-classical logic and the use of its reasoning-based system. Chapter 9 introduces a paraconsistent logic program bf-EVALPSN and its applications. Chapter 10 discusses intelligent abductive models.

Chapters 11 to 13 are devoted to both the theories and the practical applications of intelligent reasoning methods. Chapter 11 deals with supervised neural network learning. Chapters 12 and 13 introduce applications of paraconsistent logic to an artificial neural network, automation and robotics.

Chapters 14 to 23 are devoted to the practical application of intelligent reasoning methods. Chapter 14 proposes a new intelligent learning method for online learning. Chapter 15 considers an automatic test for embedded peripherals. Chapter 16 is devoted to an intelligent reasoning-based discovery of communication patterns.

Chapter 17 treats the quality enhancement and storage of signatures and finger print images. In Chapter 18 the knowledge representation method of electronic circuits in logic programming is proposed.

Chapter 19 deals with an intelligent case-based reasoning model for predicting the changes of tropical cyclone intensities, Chapter 20 analyzes sequential data related to tool manufacturing. Chapter 21 investigates reasoning-based adaptive economic agents. Chapter 22 deals with intelligent reasoning and knowledge acquisition obtained from medical databases. Chapter 23 considers the approximate processing used in medical diagnosis by the use of deductive agents.

1.3. Conclusion

This chapter presents a brief overview of the advances in intelligent systems. A brief abstract of all chapters is included. The interested readers are welcome to use resources listed in Section 1.5.

1.4. References

[1] Teodorescu, H.N., Watada, J. and Jain, L.C. (Editors), Intelligent Systems and Technologies, Springer-Verlag, Germany, 2009.
[2] Abraham, A., Jain, L.C. and van der van Zwagg, B. (Editors), Innovations in Intelligent Systems, Springer-Verlag, Germany, 2004.
[3] Abraham, A., Jain, L.C. and Kacprzyk, J. (Editors), Recent Advances in Intelligent Paradigms and Applications, Springer-Verlag, Germany, 2003.
[4] Jain, L.C. and De Wilde, P. (Editors), Practical Applications of Computational Intelligence Techniques, Kluwer Academic Publishers, USA, 2001.
[5] Jain, L.C., Johnson, R.P., Takefuji, Y. and Zadeh, L.A. (Editors), Knowledge-Based Intelligent Techniques in Industry, CRC Press USA, 1998.
[6] Jain, L.C. and Fanelli, A.M. (Editors), Recent Advances in Artificial Neural Networks: Design and Applications, CRC Press, USA, 2000.
[7] Medsker, L. and Jain, L.C. (Editors), Recurrent Neural Networks: Design and Applications, CRC Press, USA, 2000.
[8] Jain, L.C. (Editor), Soft Computing Techniques in Knowledge-Based Intelligent Engineering Systems, Springer-Verlag, Germany, 1997.
[9] Seiffert, U. and Jain, L.C. (Editors), Self-Organising Neural Networks, Springer-Verlag, Germany, 2002.
[10] Jain, L.C. and Vemuri, R. (Editors), Industrial Applications of Neural Networks, CRC Press USA, 1998.
[11] Sato, M. and Jain, L.C., Innovations in Fuzzy Clustering, Springer-Verlag, 2006.
[12] Dumitrescu, D., Lazzerini, B., and Jain, L.C., Fuzzy Sets and their Applications to Clustering and Training, CRC Press USA, 2000.
[13] Sato, M., Sato, Y. and Jain, L.C., Fuzzy Clustering Models and Applications, Springer-Verlag, Germany, 1997.
[14] Jain, L.C., Srinivasan, D. and Palade, V. (Editors), Advances in Evolutionary Computing for system Design, Springer-Verlag, 2007.
[15] Ghosh, A. and Jain, L.C. (Editors), Evolutionary Computation in Data Mining, Springer-Verlag, Germany, 2005.

[16] Jain, L.C. (Editor), Evolution of Engineering and Information Systems, CRC Press USA, 2000.
[17] Dumitrescu, D., Lazzerini, B., Jain, L.C. and Dumitrescu, A., Evolutionary Computing and Applications, CRC Press USA, 2000.
[18] Srinivasan, D. and Jain, L.C. (Editors), Advances in Multi-Agent Systems and Applications, Springer-Verlag, 2010.
[19] Jarvis, J., Ronnquist, R, Jarvis, D. and Jain, L.C., Holonic Execution: A BDI Approach, Springer-Verlag, 2008.
[20] Nguyen, N.T. and Jain, L.C. (Editors), Intelligent Agents in the Evolution of Web and Applications, Springer-Verlag, Germany, 2009.
[21] Resconi, G. and Jain, L.C., *Intelligent Agents: Theory and Applications*, Springer-Verlag, 2004.
[22] Khosla, R., Ichalkaranje, N. and Jain, L.C. (Editors), Design of Intelligent Multi-Agent Systems, Springer-Verlag, Germany, 2005.
[23] Jain, L.C., Chen. Z. and Ichalkaranje, N. (Editors), Intelligent Agents and their Applications, Springer-Verlag, Germany, 2002.
[24] Mumford, C. and Jain, L.C. (Editors), Computational Intelligence: Collaboration, Fusion and Emergence, Springer-Verlag, 2009.
[25] Jain, L.C. and Martin, N.M. (Editors), Fusion of Neural Networks, Fuzzy Logic and Evolutionary Computing and their Applications, CRC Press USA, 1999.
[26] Jain, L.C. and Jain, R.K. (Editors), Hybrid Intelligent Engineering Systems, World Scientific Publishing Company, Singapore, 1997.
[27] Van Rooij, A., Jain, L.C. and Johnson, R.P., Neural Network Training Using Genetic Algorithms, World Scientific Publishing Company, Singapore, December 1996.
[28] Vonk, E., Jain, L.C. and Johnson, R.P., Automatic Generation of Neural Networks Architecture Using Evolutionary Computing, World Scientific Publishing Company, Singapore, 1997.
[29] Holmes, D. (Guest Editor), Features, The Reasoner, Volume 2, Number 3, March 2008, pp. 1–6.
[30] Nakamatsu, K., Editor-in-Chief, An International Journal of Reasoning-Based Intelligent Systems, *Inderscience*, 2009–2011.

1.5. Resources

The following resources are recommended to explore the field of intelligent systems further. This list is neither complete, nor exclusive. It can be seen as a starting point to explore the field further.

1. Journals

Advanced Engineering Informatics, Elsevier, The Netherlands.
http://www.elsevier.com/wps/find/journaldescription.cws_home/622240/description

AI Magazine, AAAI Press, USA.
http://www.aaai.org/Magazine/magazine.php

Fuzzy Optimization and Decision Making, Springer, USA.
http://www.springer.com/math/journal/10700

IEEE Intelligent Systems, IEEE Press, USA.
http://www.computer.org/portal/site/intelligent/

IEEE Transactions on Neural Networks, IEEE Press, USA.
http://ieee-cis.org/pubs/tnn/

IEEE Transactions on Evolutionary Computing, IEEE Press, USA.
http://www.ieee-cis.org/pubs/tec/

IEEE Transactions on Fuzzy Systems, IEEE Press, USA.
http://ieeexplore.ieee.org/xpl/RecentIssue.jsp?punumber=91

IEEE Computational Intelligence Magazine, IEEE Press, USA.
http://ieeexplore.ieee.org/xpl/RecentIssue.jsp?punumber=10207

Intelligent Decision Technologies: An International Journal, IOS Press, The Netherlands.
http://www.iospress.nl/loadtop/load.php?isbn=18724981

International Journal of Hybrid Intelligent Systems, IOS Press, The Netherlands.
http://ijhis.hybridsystem.com/

International Journal of Intelligent and Fuzzy Systems, IOS Press, The Netherlands.
http://www.iospress.nl/loadtop/load.php?isbn=10641246

International Journal of Knowledge-Based and Intelligent Engineering Systems, IOS Press, The Netherlands.
http://www.iospress.nl/loadtop/load.php?isbn=13272314

Machine Learning, Springer, USA.
http://www.springer.com/computer/artificial/journal/10994

Neural Computing and Applications, Springer, London.
http://www.springer.com/computer/mathematics/journal/521

Neurocomputing, Elsevier, The Netherlands.
http://www.elsevier.com/locate/neucom

International Journal of Intelligent Defence Systems
http://www.inderscience.com/IJIDSS

International Journal of Knowledge and web intelligence
http://www.inderscience.com/ijkwi

International Journal of Advanced Intelligence Paradigms
http://www.inderscience.com/IJAIP

International Journal of Reasoning-based Intelligent Systems
www.inderscience.com/ijris

International Journal of Computational Intelligence Studies
http://www.inderscience.com/ijcistudies

International Journal of Knowledge Engineering and Soft Data Paradigms
http://www.inderscience.com/ijkesdp

International Journal of Artificial Intelligence and Soft Computing
http://www.inderscience.com/ijaisc

International Journal of Granular Computing, Rough Sets and Intelligent Systems
http://www.inderscience.com/sample.php?id=315

International Journal of Innovative Computing, Information and Control
http://www.ijicic.org/

Interactive Technology and Smart Education, Troubador Publishing
http://www.troubador.co.uk/itse/

International Journal of Computational Intelligence Systems
http://www.atlantis-press.com

International Journal of Biomedical Soft Computing and Human Sciences, Biomedical Fuzzy Systems Association, Japan
http://www.f.waseda.jp/watada/BMFSA/journal-IJ/

Multiagents and Grid Systems: An International Journal

Journal of Universal Computer Science

International Journal of Artificial Intelligence, USA

Machine Intelligence and Robotic Control, An International Journal, Cyber Scientific, Japan

Applied Soft Computing, Elsevier Science

International Transactions on Systems Science and Applications

IEEE Transactions on Systems, Man and Cybernetics, Part B, IEEE Press, USA

International Journal of Pattern Recognition and Artificial Intelligence

Computer and Information Systems, Europe

Neural Computing and Applications, Springer-Verlag, Germany

2. Conferences

- KES International Conference Series
  www.kesinternational.org/
- AAAI Conference on Artificial Intelligence
  www.aaai.org/aaai08.php

3. Books

Some books listed in this section may also appear in Section 7 due to their association with the book series.

- Kwaśnicka, H. and Jain, L.C., Innovations in Intelligent Image Analysis, Springer-Verlag, 2011.
- Tolk, A. and Jain, L.C., Intelligence-Based Systems Engineering, Springer-Verlag, 2011.
- Brahnam, S. and Jain, L.C., Advanced Computational Intelligence Paradigms in Healthcare 6: Springer-Verlag, 2011.
- Brahnam, S. and Jain, L.C., Advanced Computational Intelligence Paradigms in Healthcare 5: Springer-Verlag, 2011.
- Montani, S. and Jain, L.C., Successful case-Based Reasoning Applications 1, Springer-Verlag, 2010.
- Nishida, T., Jain, L.C. and Faucher, C. (Editors), Modelling Machine Emotions for Realizing Intelligence, Springer-Verlag, 2010.
- Jain, L.C. and Lim, C.P. (Editors), Handbook on Decision Making: Techniques and Applications, Springer-Verlag, 2010.
- Tsihrintzis, G., Virvou, M. and Jain, L.C. (Editors), Multimedia Services in Intelligent Environments: Software Development Challenges and Solutions, Springer-Verlag, 2010.
- Tsihrintzis, G. and Jain, L.C. (Editors), Multimedia Services in Intelligent Environments: Integrated Systems, Springer-Verlag, 2010.
- Bichindaritz, I., Vaidya, S., Jain, S. and Jain, L.C. (Editors), Advanced Computational Intelligence in Healthcare-4, Springer-Verlag, 2010.
- Pratihar, D. and Jain, L.C. (Editors), Intelligent Autonomous Systems: Foundations and Applications, Springer-Verlag, 2010.
- Brahman, S. and Jain, L.C. (Editors), Advanced Computational Intelligence in healthcare 5, Springer-Verlag, 2010.
- Velasquez and Jain, L.C., Advanced Techniques in Web Intelligence 1, Springer-Verlag, 2010.
- Bianchini, M., Maggini, M., Scarselli, F. and Jain, L.C., Innovations in Neural Information Processing Paradigms, Springer-Velag, Germany, 2009.
- Lim, C.P., Jain, L.C. and Satchidananda, D., Innovations in Swarm Intelligence, Springer-Velag, Germany, 2009.

- Tolk, A. and Jain, L.C. (Editors), Complex Systems in Knowledge-based Environments, Springer-Verlag, Germany, 2009.
- Jain, L.C. and Nguyen, N.T. (Editors), Knowledge Processing and Decision Making in Agent-Based Systems, Springer-Verlag, Germany, 2009.
- Nicoletti, M.D.C. and Jain, L.C. (Editors), Computational Intelligence Techniques for Bioprocess Modelling, Supervision and Control, Springer-Verlag, Germany, 2009.
- Castellano, G., Jain, L.C. and Fanelli, A.M. (Editors), Web Personalization in Intelligent Environments, Springer-Verlag, Germany, 2009.
- Pan, J.S., Huang, H.C. and Jain, L.C. (Editors), Information Hiding and Applications, Springer-Verlag, Germany, 2009.
- Damiani, E., Jeong, J., Howlett, R.J. and Jain, L.C. (Editors), New Directions in Intelligent Interactive Multimedia Systems and Services — 2, Springer-Verlag, Germany, 2009.
- Rakus-Andersson, E., Yager, R.R., Ichalkaranje, N. and Jain, L.C. (Editors), Recent Advances in Decision Making, Springer-Verlag, Germany, 2009.
- Wang, F.H., Pan, J.S. and Jain, L.C., Innovations in Digital Watermarking Techniques, Springer-Verlag, Germany, 2009.
- Jain, L.C., Sato, M., Virvou, M., Tsihrintzis, G., Balas, V. and Abeynayake, C. (Editors), Computational Intelligence Paradigms: Volume 1 — Innovative Applications, Springer-Verlag, 2008.
- Phillips-Wren, G., Ichalkaranje, N. And Jain, L.C. (Editors), Intelligent Decision Making-An AI-Based Approach, Springer-Verlag, 2008.
- Fulcher, J. and Jain, L.C. (Editors), Computational Intelligence: A Compendium, Springer-Verlag, 2008.
- Sordo, M., Vaidya, S. and Jain, L.C. (Editors), Advanced Computational Intelligence Paradigms in Healthcare 3, Springer-Verlag, 2008.
- Virvou, M. And Jain, L.C. (Editors), Intelligent Interactive Systems in Knowledge-Based Environments, Springer-Verlag, 2008.
- Sommerer, C., Jain, L.C. and Mignonneau, L. (Editors), The Art and Science of Interface and Interaction Design, Volume 1, Springer-Verlag, 2008.
- Nayak, R., Ichalkaranje, N. and Jain, L.C. (Editors), Evolution of the Web in Artificial Intelligence Environments, Springer-Verlag, 2008.
- Tsihrintzis, G. and Jain, L.C. (Editors), Multimedia Services in Intelligent Environments, Springer-Verlag, 2008.
- Holmes, D. and Jain, L.C. (Editors), Innovations in Bayesian Networks, Springer-Verlag, 2008.
- Magnenat-Thalmann, N., Jain, L.C. and Ichalkaranje, N., New Advances in Virtual Humans, Springer-Verlag, 2008.
- Tsihrintzis, G., Virvou, M., Howlett, R.J. and Jain, L.C. (Editors), New Directions in Intelligent Interactive Multimedia, Springer-Verlag, 2008.
- Baba, N., Jain, L.C. and Handa, H. (Editors), Advanced Intelligent Paradigms in Computer Games, Springer-Verlag, 2007.

- Chahl, J.S., Jain, L.C., Mizutani, A. and Sato-Ilic, M. (Editors), Innovations in Intelligent Machines 1, Springer-Verlag, 2007.
- Jain, L.C, Tedman, R. and Tedman, D. (Editors), Evolution of Teaching and Learning in Intelligent Environment, Springer-Verlag, 2007.
- Zharkova, V. and Jain, L.C. (Editors), Artificial Intelligence in Recognition and Classification of Astrophysical and Medical Images, Springer-Verlag, 2007.
- Pan, J-S. Huang, H-C, Jain, L.C. and Fang, W-C (Editors), Intelligent Multimedia Data Hiding, Springer-Verlag, 2007.
- Yoshida, H., Jain, A., Ichalkaranje, A. Jain, L.C. and Ichalkaranje, N. (Editors), Advanced Computational Intelligence Paradigms in Healthcare 1, Springer-Verlag, 2007.
- Vaidya, S., Jain, L.C. and Yoshida, H. (Editors), Advanced Computational Intelligence Paradigms in Healthcare 2, Springer-Verlag, 2007.
- Ichalkaranje, N., Ichalkaranje, A. and Jain, L.C., Intelligent Paradigms for Assistive and Preventive Healthcare, Sprinter-Verlag, 2006.
- Patnaik, S., Jain, L.C., Tzafestas, S.G., Resconi, G. and Konar, A. (Editors), Innovations in Robot Mobility and Control, Springer-Verlag, 2006.
- Apolloni, B., Ghosh, A., Alpaslan, F, Jain, L.C. and Patnaik, S. (Editors), Machine Learning and Robot Perception, Springer-Verlag, 2006.
- Palade, V., Bocaniala, C.D. and Jain, L.C. (Editors), Computational Intelligence in Fault Diagnosis, Springer-Verlag, 2006.
- Holmes, D. and Jain, L.C. (Editors), Innovations in Machine Learning, Springer-Verlag, 2006.
- Chen, S.H., Jain, L.C. and Tai, A.C. (Editors), Computational Economics, IGP, USA, 2006.
- Seiffert, U., Jain, L.C. and Schweizer, P. (Editors), Bioinformatics Using Computational Intelligence Paradigms, Springer-Verlag, ISBN 3-54022-901-9, 2005.
- Konar, A. and Jain, L.C., Cognitive Engineering, Springer-Verlag, ISBN 1-85233-975-6, 2005.
- Phillips-Wren, G. and Jain, L.C. (Editors), Intelligent Decision Support Systems in Agent-Mediated Environments, IOS Press, The Netherlands, 2005.
- Silvermann, B., Jain, A., Ichalkaranje, A. and Jain, L.C. (Editors), Intelligent Paradigms in Healthcare Enterprises, Springer-Verlag, Germany, 2005.
- Ghaoui, C. Bannore, V. Jain, M. and Jain, L.C. (Editors), Knowledge-Based Virtual Education, Springer-Verlag, Germany, 2005.
- Abraham, A., Jain, L.C. and Goldberg, R. (Editors), Evolutionary Multiobjective Optimization, Springer-Verlag, London, 2005.
- Pal, N. and Jain, L.C. (Editors), Advanced Techniques in Knowledge Discovery and Data Mining, Springer-Verlag, London, 2005.
- Tonfoni, G. and Jain, L.C., Visualizing Document Processing, Mouton De Gruyter, Germany, 2004.
- Fulcher, J. and Jain, L.C. (Editors), Applied Intelligent Systems, Springer-Verlag, Germany, 2004.

- Damiani, E., Jain, L.C. and Madravio, M. (Editors), Soft Computing in Software Engineering, Springer-Verlag, Germany, 2004.
- Pan, J-S., Huang, H.C. and Jain, L.C. (Editors), Intelligent Watermarking Techniques, World Scientific Publishing Company Singapore, 2004.
- Tonfoni, G. and Jain, L.C. (Editors), Innovations in Decision Support Systems, AKI, 2003.
- Shapiro, A. and Jain, L.C., Intelligent and Other Computational Techniques in Insurance, World Scientific Publishing Company Singapore, 2003.
- Tonfoni, G. and Jain, L.C., The Art and Science of Documentation Management, Intellect, UK, 2003.
- Howlett, R., Ichalkaranje, N., Jain, L.C. and Tonfoni, G. (Editors), Internet-Based Intelligent Information Processing, World Scientific Publishing Company Singapore, 2002.
- Jain, L.C. and Kacprzyk, J. (Editors), New Learning Paradigms in Soft Computing, Springer-Verlag, Germany, 2002.
- Jain, L.C., Ichalkaranje, N. and Tonfoni, G. (Editors), Advances in Intelligent Systems for Defence, World Scientific Publishing Company Singapore, 2002.
- Jain, L.C., Howlett, R.J., Ichalkaranje, N., and Tonfoni, G. (Editors), Virtual Environments for Teaching and Learning, World Scientific Publishing Company Singapore, 2002.
- Schmitt, M., Teodorescu, H.N., Jain, A, Jain, A., Jain, S. and Jain, L.C. (Editors). Computational Intelligence Processing in Medical Diagnosis, Springer-Verlag, 2002.
- Howlett, R.J. and Jain, L.C. (Editors), Radial Basis Function Networks 1, Springer-Verlag, Germany, 2001.
- Howlett, R.J. and Jain, L.C. (Editors), Radial Basis Function Networks 2, Springer-Verlag, Germany, 2001.
- Teodorescu, H.N., Jain, L.C. and Kandel, A. (Editors), Hardware Implementation of Intelligent Systems, Springer-Verlag, Germany, 2001.
- Baba, N. and Jain, L.C. (Editors), Computational Intelligence in Games, Springer-Verlag, 2001.
- Jain, L.C. (Editor), Innovative Teaching and Learning, Springer-Verlag, Germany, 2000.
- Jain, L.C., Lazzerini, B. and Halici, U. (Editors), Innovations in ART Neural Networks, Springer-Verlag, Germany, 2000.
- Jain, A., *et al.* (Editors), Artificial Intelligence Techniques in Breast Cancer Diagnosis and Prognosis, World Scientific Publishing Company, Singapore, 2000.
- Jain, L.C., Halici, U., Hayashi, I., Lee, S.B. and Tsutsui, S. (Editors) Intelligent Biometric Techniques in Fingerprint and Face Recognition, CRC Press, USA, 2000.
- Jain, L.C. and De Silva, C.W. (Editors), Intelligent Adaptive Control, CRC Press USA, 1999.

- Jain, L.C. and Lazzerini, B., (Editors), Knowledge-Based Intelligent Techniques in Character Recognition, CRC Press USA, 1999.
- Teodorescu, H.N., Kandel, A. and Jain, L.C. (Editors), Soft Computing Techniques in Human Related Science, CRC Press USA, 1999.
- Teodorescu, H.N., Kandel, A. and Jain, L.C. (Editors), Fuzzy and Neuro-fuzzy Systems in Medicine, CRC Press USA, 1998.
- Jain, L.C. and Fukuda, T. (Editors), Soft Computing for Intelligent Robotic Systems, Springer-Verlag, Germany, 1998.

4. Conference Proceedings

- Phillips-Wren, G., Jain, L.C., Howlett, R.J. and Nakamatsu, K., Further Advances in Decision Support Systems, Springer-Verlag, 2010.
- Jedrzejowicz, P., Nguyen, N.T., Howlett, R.J. and Jain, L.C., Agent and Multi-Agent Systems: Technologies and Applications, Lecture Notes in Artificial Intelligence, LNAI 6070, Springer-Verlag, 2010.
- Jedrzejowicz, P., Nguyen, N.T., Howlett, R.J. and Jain, L.C., Agent and Multi-Agent Systems: Technologies and Applications, Lecture Notes in Artificial Intelligence, LNAI 6071, Springer-Verlag, 2010.
- Tsihrintzis, G.A., Damiani, E., Virvou, M., Iain, L.C. and Howlett, R.J., Intelligent Interactive Multimedia Systems and Services, Springer-Verlag, 2010.
- Setchi, R., Jordanov, I., Howlett, R.J. and Jain, L.C., Knowledge-Based and Intelligent Information and Engineering Systems, LNAI 6276, Springer-Verlag, 2010.
- Setchi, R., Jordanov, I., Howlett, R.J. and Jain, L.C., Knowledge-Based and Intelligent Information and Engineering Systems, LNAI 6277, Springer-Verlag, 2010.
- Setchi, R., Jordanov, I., Howlett, R.J. and Jain, L.C., Knowledge-Based and Intelligent Information and Engineering Systems, LNAI 6278, Springer-Verlag, 2010.
- Setchi, R., Jordanov, I., Howlett, R.J. and Jain, L.C., Knowledge-Based and Intelligent Information and Engineering Systems, LNAI 6279, Springer-Verlag, 2010.
- Echizen, I. Pan, J., Fellner, D., Nouak, A., Kuijper, A., and Jain, L.C., 2010 Sixth International Conference on Intelligent Information Hiding and Multimedia Signal Processing, IEEE Press, USA, 2010.
- Håkansson, A., Nguyen, N.T., Hartung, R., Howlett, R.J. and Jain, L.C. (Editors), Agents and Multi-Agents Systems: Technologies and Applications, Lecture Notes in Artificial Intelligence, Springer-Verlag, Germany, 2009.
- Velasquez, J.D., Rios, S.A., Howlett, R.J. and Jain, L.C. (Editors), Knowledge-Based and Intelligent Information and Engineering Systems, Lecture Notes in Artificial Intelligence, LNAI 5711, Volume 1, Springer-Verlag, Germany, 2009.

- Velasquez, J.D., Rios, S.A., Howlett, R.J. and Jain, L.C. (Editors), Knowledge-Based and Intelligent Information and Engineering Systems, Lecture Notes in Artificial Intelligence, LNAI 5712, Volume 2, Springer-Verlag, Germany, 2009.
- Pan, J.S., Chen, Y.W. and Jain, L.C. (Editors), Intelligent Information Hiding and Multimedia Signal Processing, IEEE Computer Society Press, USA, 2009.
- Nakamatsu, K., Phillips-Wren, G., Jain, L.C. and Howlett, R.J. (Editors), New Advances in Intelligent Decision Technologies, Springer-Verlag, 2009.
- Nguyen, N.T., Jo, G.S., Howlett, R.J. and Jain, L.C. (Editors), Agents and Multi-Agents Systems: Technologies and Applications, Lecture Notes in Artificial Intelligence, Springer-Verlag, Germany, 2008.
- Lovrek, I., Howlett, R.J. and Jain, L.C. (Editors), Knowledge-Based and Intelligent Information and Engineering Systems, Lecture Notes in Artificial Intelligence, LNAI 5177, Volume 1, KES 2007, Springer-Verlag, Germany, 2008.
- Lovrek, I., Howlett, R.J. and Jain, L.C. (Editors), Knowledge-Based and Intelligent Information and Engineering Systems, Lecture Notes in Artificial Intelligence, LNAI 5178, Volume 2, KES 2007, Springer-Verlag, Germany, 2008.
- Lovrek, I., Howlett, R.J. and Jain, L.C. (Editors), Knowledge-Based and Intelligent Information and Engineering Systems, Lecture Notes in Artificial Intelligence, LNAI 5179, Volume 3, KES 2007, Springer-Verlag, Germany, 2008.
- Pan, J.S., Niu, X.M., Huang, H.C. and Jain, L.C. (Editors), Intelligent Information Hiding and Multimedia Signal Processing, IEEE Computer Society Press, USA, 2008.
- Jain, L.C., Lingras, P., Klusch, M., Lu, J., Zhang, C., Cercone, N. and Cao, L. (Editors), Proceedings of the 2008 IEEE/WIC/ACM International Conference on Web Intelligence, WI 2008, IEEE Computer Society, USA, 2008.
- Jain, L.C., Gini, M., Faltings, B.B., Terano, T., Zhang, C., Cercone, N. and Cao, L. (Editors), Proceedings of the 2008 IEEE/WIC/ACM International Conference on Intelligent Agent Technology, IAT 2008, IEEE Computer Society, USA, 2008.
- Apolloni, B., Howlett, R.J. and Jain, L.C. (Editors), Knowledge-Based Intelligent Information and Engineering Systems, Lecture Notes in Artificial Intelligence, Volume 1, LNAI 4692, KES 2007, Springer-Verlag, Germany, 2007.
- Apolloni, B., Howlett, R.J. and Jain, L.C. (Editors), Knowledge-Based Intelligent Information and Engineering Systems, Lecture Notes in Artificial Intelligence, Volume 2, LNAI 4693, KES 2007, Springer-Verlag, Germany, 2007.
- Apolloni, B., Howlett, R.J. and Jain, L.C. (Editors), Knowledge-Based Intelligent Information and Engineering Systems, Lecture Notes in Artificial Intelligence, Volume 3, LNAI 4694, KES 2007, Springer-Verlag, Germany, 2007.
- Nguyen, N.T., Grzech, A., Howlett, R.J. and Jain, L.C., Agents and Multi-Agents Systems: Technologies and Applications, Lecture Notes in Artificial Intelligence, Springer-Verlag, Germany, 2007, 2007, 1046 pp., ISBN: 978-3-540-72829-0.

- Liao, B.-H., Pan, J.-S., Jain, L.C., Liao, M., Noda, H. and Ho, A.T.S., Intelligent Information Hiding and Multimedia Signal Processing, Volume 1, IEEE Computer Society Press, USA, 2007. ISBN: 0-7695-2994-1.
- Liao, B.-H., Pan, J.-S., Jain, L.C., Liao, M., Noda, H. and Ho, A.T.S., Intelligent Information Hiding and Multimedia Signal Processing, Volume 2, IEEE Computer Society Press, USA, 2007. ISBN: 0-7695-2994-1.
- Howlett, R.P., Gabrys, B. and Jain, L.C. (Editors), Knowledge-Based Intelligent Information and Engineering Systems, Lecture Notes in Artificial Intelligence, KES 2006, Springer-Verlag, Germany, Vol. 4251, 2006.
- Howlett, R.P., Gabrys, B. and Jain, L.C. (Editors), Knowledge-Based Intelligent Information and Engineering Systems, Lecture Notes in Artificial Intelligence, KES 2006, Springer-Verlag, Germany, Vol. 4252, 2006.
- Howlett, R.P., Gabrys, B. and Jain, L.C. (Editors), Knowledge-Based Intelligent Information and Engineering Systems, Lecture Notes in Artificial Intelligence, KES 2006, Springer-Verlag, Germany, Vol. 4253, 2006.
- Khosla, R., Howlett, R.P., and Jain, L.C. (Editors), Knowledge-Based Intelligent Information and Engineering Systems, Lecture Notes in Artificial Intelligence, KES 2005, Springer-Verlag, Germany, Vol. 3682, 2005.
- Skowron, A., Barthes, P., Jain, L.C., Sun, R., Mahoudeaux, P., Liu, J. and Zhong, N. (Editors), Proceedings of the 2005 IEEE/WIC/ACM International Conference on Intelligent Agent Technology, Compiegne, France, IEEE Computer Society Press, USA, 2005.
- Khosla, R., Howlett, R.P., and Jain, L.C. (Editors), Knowledge-Based Intelligent Information and Engineering Systems, Lecture Notes in Artificial Intelligence, KES 2005, Springer-Verlag, Germany, Vol. 3683, 2005.
- Khosla, R., Howlett, R.P., and Jain, L.C. (Editors), Knowledge-Based Intelligent Information and Engineering Systems, Lecture Notes in Artificial Intelligence, KES 2005, Springer-Verlag, Germany, Vol. 3684, 2005.
- Khosla, R., Howlett, R.P., and Jain, L.C. (Editors), Knowledge-Based Intelligent Information and Engineering Systems, Lecture Notes in Artificial Intelligence, KES 2005, Springer-Verlag, Germany, Vol. 3685, 2005.
- Negoita, M., Howlett, R.P., and Jain, L.C. (Editors), Knowledge-Based Intelligent Engineering Systems, KES 2004, Lecture Notes in Artificial Intelligence, Vol. 3213, Springer, 2004.
- Negoita, M., Howlett, R.P., and Jain, L.C. (Editors), Knowledge-Based Intelligent Engineering Systems, KES 2004, Lecture Notes in Artificial Intelligence, Vol. 3214, Springer-Verlag, 2004.
- Negoita, M., Howlett, R.P., and Jain, L.C. (Editors), Knowledge-Based Intelligent Engineering Systems, KES 2004, Lecture Notes in Artificial Intelligence, Vol. 3215, Springer-Verlag, 2004.
- Murase, K., Jain, L.C., Sekiyama, K. and Asakura, T. (Editors), Proceedings of the Fourth International Symposium on Human and Artificial Intelligence Systems, University of Fukui, Japan, 2004.

- Palade, V., Howlett, R.P., and Jain, L.C. (Editors), Knowledge-Based Intelligent Engineering Systems, Lecture Notes in Artificial Intelligence, Vol. 2773, Springer-Verlag, 2003.
- Palade, V., Howlett, R.P., and Jain, L.C. (Editors), Knowledge-Based Intelligent Engineering Systems, Lecture Notes in Artificial Intelligence, Vol. 2774, Springer-Verlag, 2003.
- Damiani, E., Howlett, R.P., Jain, L.C. and Ichalkaranje, N. (Editors), Proceedings of the Fifth International Conference on Knowledge-Based Intelligent Engineering Systems, Volume 1, IOS Press, The Netherlands, 2002.
- Damiani, E., Howlett, R.P., Jain, L.C. and Ichalkaranje, N. (Editors), Proceedings of the Fifth International Conference on Knowledge-Based Intelligent Engineering Systems, Volume 2, IOS Press, The Netherlands, 2002.
- Baba, N., Jain, L.C. and Howlett, R.P. (Editors), Proceedings of the Fifth International Conference on Knowledge-Based Intelligent Engineering Systems (KES'2001), Volume 1, IOS Press, The Netherlands, 2001.
- Baba, N., Jain, L.C. and Howlett, R.P. (Editors), Proceedings of the Fifth International Conference on Knowledge-Based Intelligent Engineering Systems (KES'2001), Volume 2, IOS Press, The Netherlands, 2001.
- Howlett, R.P. and Jain, L.C. (Editors), Proceedings of the Fourth International Conference on Knowledge-Based Intelligent Engineering Systems, IEEE Press, USA, 2000. Part 1.
- Howlett, R.P. and Jain, L.C. (Editors), Proceedings of the Fourth International Conference on Knowledge-Based Intelligent Engineering Systems, IEEE Press, USA, 2000. Part 2.
- Jain, L.C. (Editor), Proceedings of the Third International Conference on Knowledge-Based Intelligent Engineering Systems, IEEE Press, USA, 1999.
- Jain, L.C. and Jain, R.K. (Editors), Proceedings of the Second International Conference on Knowledge-Based Intelligent Engineering Systems, Volume 1, IEEE Press, USA, 1998.
- Jain, L.C. and Jain, R.K. (Editors), Proceedings of the Second International Conference on Knowledge-Based Intelligent Engineering Systems, Volume 2, IEEE Press, USA, 1998.
- Jain, L.C. and Jain, R.K. (Editors), Proceedings of the Second International Conference on Knowledge-Based Intelligent Engineering Systems, Volume 3, IEEE Press, USA, 1998.
- Jain, L.C. (Editor), Proceedings of the First International Conference on Knowledge-Based Intelligent Engineering Systems, Volume 1, IEEE Press, USA, 1997.
- Jain, L.C. (Editor), Proceedings of the First International Conference on Knowledge-Based Intelligent Engineering Systems, Volume 2, IEEE Press, USA, 1997.
- Narasimhan, V.L. and Jain, L.C. (Editors), The Proceedings of the Australian and New Zealand Conference on Intelligent Information Systems, IEEE Press, USA, 1996.

- Jain, L.C. (Editor), Electronic Technology Directions Towards 2000, ETD2000, Volume 1, IEEE Computer Society Press, USA, May 1995.
- Jain, L.C. (Editor), Electronic Technology Directions Towards 2000, ETD2000, Volume 2, IEEE Computer Society Press, USA, May 1995.

5. Special Issue of Journals

- Lim, C.P., Abeynayake, C. and Jain, L.C. (Guest Editors), Special Issue on Innovations in Intelligent Techniques and Their Applications, International Journal of Knowledge Engineering and Soft Data Paradigms, Volume 2, Number 1, 2010.
- Srinivasan, D., Howlett, R.J., Lovrek, I., Jain, L.C. and Lim, C.P. (Guest Editors), Design and Application of Neural Networks and Intelligent Learning Systems, Neurocomputing, Elsevier, Volume 73, Issue 4, 2010.
- Howlett, R., Lovrek, I., Jain, L.C., Lim, C.P., and Gabrys, B. (Guest Editors), Advances in Design and Application of Neural Networks, Neural Computing and Applications, Springer, 2010.
- Loverek, I., Howlett, R.J., Lim, C.P., Jain, L.C. and Phillips-Wren, G. (Guest Editors), Knowledge-based Intelligent Systems and Their Applications, Journal of Intelligent and Fuzzy Systems, Volume 21, Number 1-2, 2010.
- Lim, C.P., Howlett, R.J., Jain, L.C. and Sato-Ilic, M. (Guest Editors), Advances and Applications of Knowledge-Based Intelligent Systems, International Journal of Advanced Intelligence Paradigms, Volume 2, Number 1, 2010.
- Nguyen, N.T., Lim, C.P., Jain, L.C. and Balas, V.E. (Guest Editors), Theoretical Advances and Applications of Intelligent Paradigms, Journal of Intelligent and Fuzzy Systems, IOS Press, Volume 20, Number 1&2, 2009.
- Lovrek, I., Howlett, R.J., Lim, C.P. and Jain, L.C. (Guest Editors), Knowledge-based Intelligent Systems and Their Applications, Journal of Intelligent & Fuzzy Systems, 2009.
- Lim, C.P., Jain, L.C., Tsihrintzis, G.A., Virvou, M. and Balas, V.E. (Guest Editors), Intelligent Modelling and Data Analysis Techniques, International Journal of Intelligent Defence Support Systems, Volume 2, Number 2, 2009.
- Abeynayake, C., Lim, C.P., Castellano, G., Jain, L.C., Tsihrintzis, G.A., Virvou, M. and Balas, V.E. (Guest Editors), Special Issue on Selected Papers from the KES2008 Conference, International Journal of Intelligent Defence Support Systems, Volume 2, Number 1, 2009.
- Lim, C.P., Sunde, J. and Jain, L.C. (Guest Editors), Special Issue on Intelligent Knowledge Processing and Decision Making Techniques, International Conference, Intelligent Decision Technologies: An International Journal, Volume 3, Number 2, 2009.
- Lim, C.P., Howlett, R.J., Jain, L.C. and Sato-Ilic (Guest Editors), Special Issue on Advances and Applications of Knowledge-Based Systems, International Journal of Advanced Intelligence Paradigms, Volume 2, Number 1, 2009.

- Jain, L.C., Lim, C.P. and Nguyen, N.T. (Guest Editors), Recent Advances in Intelligent Paradigms Fusion and Their Applications, International Journal of Hybrid Intelligent Systems, Volume 5, Issue 3, 2008.
- Lim, C.P., Jain, L.C., Nguyen, N.T. and Balas, V. (Guest Editors), Advances in Computational Intelligence Paradigms and Applications, An International Journal on Fuzzy Optimization and Decision Making, Springer, Volume 7, Number 3, 2008.
- Abraham, A., Jarvis, D., Jarvis, J. and Jain, L.C. (Guest Editors), Special issue on Innovations in agents: An International Journal on Multiagent and Grid Systems, IOS Press, Volume 4, Issue 4, 2008.
- Abraham, A., Smith, K., Jain, R. and Jain, L.C. (Guest Editors), Special issue on Network and Information Security: A Computational Intelligence Approach, Journal of Network and Computer Applications, Elsevier Publishers, Volume 30, Issue 1, 2007.
- Ghosh, A., Seiffert, U. and Jain, L.C., (Guest Editors), Special Issue on Evolutionary Computation in Bioinformatics, Journal of Intelligent and Fuzzy Systems, IOS Press, The Netherlands, Volume 18, Number 6, 2007.
- Palade, V. and Jain, L.C. (Guest Editors), Practical Applications of Neural Networks, Journal of "Neural Computing and Applications", Springer-Verlag, Germany, Volume 14, No. 2, 2005.
- Abraham, A. and Jain, L.C. (Guest Editors), Computational Intelligence on the Internet, Journal of Network and Computer Applications, Elsevier Publishers, Volume 28, Number 2, 2005.
- Abraham, A., Thomas, J., Sanyal, S. and Jain, L.C. (Guest Editors), Information Assurance and Security, Journal of Universal Computer Science, Volume 11, Issue 1, 2005.
- Abraham, A. and Jain, L.C. (Guest Editors), Special issue on Optimal Knowledge Mining, Journal of Fuzzy Optimization and Decision Making, Kluwer Academic Publishers, Volume 3, Number 2, 2004.
- Palade, V., Ghaoui, C. and Jain, L.C. (Guest Editors), Special issue on Intelligent Instructional Environments, Journal of Interactive Technology and Smart Education, Troubador Publishing Ltd, UK, Volume 1, Issue 3, August 2004.
- Palade, V., Ghaoui, C. and Jain, L.C. (Guest Editors), Special issue on Engineering Applications of Computational Intelligence, Journal of Intelligent and Fuzzy systems, IOS Press, Volume 15, Number 3, 2004.
- Alahakoon, D., Abraham, A. and Jain, L.C. (Guest Editors), Special issue on Neural Networks for Enhanced Intelligence, Neural Computing and Applications, Springer-Verlag, UK, Volume 13, No. 2, June 2004.
- Abraham, A., Jonkar, I., Barakova, E., Jain, R. and Jain, L.C. (Guest Editors), Special issue on Hybrid Neurocomputing, Neurocomputing, Elsevier, The Netherlands, Volume 13, No. 2, June 2004.

- Abraham, A. and Jain, L.C., (Guest Editors), Special issue on Knowledge Engineering, Journal of Intelligent and Fuzzy Systems, The IOS Press, The Netherlands, Volume 14, Number 3, 2003.
- Jain, L.C. (Guest Editor), Special issue on Fusion of Neural Nets, Fuzzy Systems and Genetic Algorithms in Industrial Applications, IEEE Transactions on Industrial Electronics, USA, December 1999.
- De Silva, C. and Jain, L.C. (Guest Editors), Special Issue on Intelligent Electronic Systems, Engineering Applications of Artificial Intelligence, an international journal, USA, January 1998.
- Jain, L.C. (Guest Editor), Special issue on Intelligent Systems: Design and Applications, Journal of Network and Computer Applications (An International Journal published by Academic Press, England). **Vol. 2**, April 1996.
- Jain, L.C. (Guest Editor), Special issue on Intelligent Systems: Design and Applications, Journal of Network and Computer Applications (An International Journal published by Academic Press, England). **Vol. 1**, January, 1996.

6. Book Chapters

- Kwaśnicka, H. and Jain, L.C., Advances in Intelligent Image Analysis, Chapter 1, Springer-Verlag, 2011, pp. 1–6.
- Jain, L.C. and Aidman, E., Advances in Defence Support Systems, Chapter 1, Springer-Verlag, 2011, pp. 1–9.
- Lim, C.P. and Jain, L.C., Advances in Decision Making, Springer-Verlag, 2010, pp. 3–28.
- Tsihrintzis, G., Virvou, M. and Jain, L.C. (Editors), Advances in Multimedia Services in Intelligent Environments: Software Development Challenges and Solutions, Springer-Verlag, 2010, pp. 1–3.
- Tsihrintzis, G. and Jain, L.C. (Editors), Advances in Multimedia Services in Intelligent Environments: Integrated Systems, Springer-Verlag, 2010, pp. 1–3.
- Bichindaritz, I. and Jain, L.C., Advances in Computational Intelligence in Healthcare, Springer-Verlag, 2010, pp. 1–7.
- Pratihar, D. and Jain, L.C., Towards Intelligent Autonomous Systems, Springer-Verlag, 2010, pp. 1–6.
- Tran, M.D.J., Abeynayake, C., Jain, L.C. and Lim, C.P., An Automated Decision System for Landmine Detection and Classification Using Metal Detector Signals, Springer-Verlag, 2010, pp. 175–200.
- Franklin, S., Finn, A., Pattison, J. and Jain, L.C., Motion Optimization Scheme for Cooperative Mobile Robot Robots, Springer-Verlag, 2010, pp. 139–173.
- Brahman, S. and Jain, L.C., Intelligent Decision Support Systems in Healthcare, Springer-Verlag, 2010.
- Jain, L.C. and Lim, C.P., Advances in Decision Making, Chapter 1, Springer-Verlag, 2009, pp. 1–6.

- Castellano, G., Fanelli, A.M., Torsello, M.A. and Jain, L.C., Innovations in Web Personalization, Chapter 1, Springer-Verlag, 2009, pp. 1–26.
- Khazab, M., Tweedale, J. and Jain, L.C., Dynamic Applications using Multi-agent Systems, Castellano, G., Fanelli, A.M., Torsello, M.A. and Jain, L.C., Innovations in Web Personalization, Chapter 4, Springer-Verlag, 2009, pp. 67–79.
- Rakus-Andersson, E. and Jain, L.C., Computational Intelligence in Medical Decision Making, Chapter 9, Springer-Verlag, 2009, pp. 145–159.
- Jain, L.C., Lim, C.P. and Nguyen, N.T., Innovations in Knowledge Processing and Decision Making in Agent-Based Systems, Springer-Verlag, Germany, 2009, Chapter 1, pp. 1–18.
- Tweedale, J. and Jain, L.C., The Evolution of Intelligent Agents Within the World Wide Web, Springer-Verlag, Germany, 2009, Chapter 1, pp. 1–9.
- Tolk, A. and Jain, L.C., An Introduction to Complex Systems in Knowledge-based Environments, Volume 168, SCI Series, Springer-Verlag, Germany, 2009, pp. 1–5.
- Tweedale, J., Bollenbeck, F., Jain, L.C. and Urlings, P., Agent Transport Layer Adaptation System, Chapter 11, SCI 170, Springer-Verlag, Germany, pp. 247–273, 2009.
- Lim, C.P. and Jain, L.C., Advances in Intelligent Methodologies and Techniques, Chapter 1, Springer-Verlag, Germany, pp. 3–28, 2009.
- Nicoletti, M.D.C., Jain, L.C. and Giordano, R.C., Computational Intelligence Techniques as Tools for Bioprocess Modelling, Optimization, Supervision and Control, Chapter 1, Springer-Verlag, Germany, pp. 1–23, 2009.
- Pedrycz, W., Ichalkaranje, N., Phillips-Wren, G., and Jain, L.C., Introduction to Computational Intelligence for Decision Making, Springer-Verlag, 2008, pp. 75–93, Chapter 3.
- Tweedale, J., Ichalkaranje, N., Sioutis, C., Urlings, P. and Jain, L.C., Future Directions: Building a Decision Making Framework using Agent Teams, Springer-Verlag, 2008, pp. 381–402, Chapter 14.
- Virvou, M. and Jain, L.C., Intelligent Interactive Systems in Knowledge-Based Environments: An Introduction, Springer-Verlag, 2008, pp. 1–8, Chapter 1.
- Tsihrintzis, G. and Jain, L.C., An Introduction to Multimedia Services in Intelligent Environments, Springer-Verlag, pp. 1–10, 2008, Chapter 1.
- Jain, L.C. and Lim, C.P., An Introduction to Computational Intelligence Paradigms, Springer-Verlag, pp. 1–15, 2008 Chapter 1.
- Nayak, R. and Jain, L.C., An Introduction to the Evolution of the Web in an Artificial Intelligence Environment, pp. 1–15, 2008, Chapter 1.
- Nayak, R. and Jain, L.C., Innovations in Web Applications using Artificial Intelligence Paradigms, pp. 17–40, 2008, Chapter 2.
- Zharkova, V.V. and Jain, L.C., Introduction to Paper Recognition and Classification in Medical and Astrophysical Images, Springer-Verlag, 2007, pp. 1–18, Chapter 1, ISBN: 10-3-540-47511-7.

- Yoshida, H., Vaidya, S. and Jain, L.C., Introduction to Computational Intelligence in Healthcare, Springer-Verlag, 2007, pp. 1–4, Chapter 1, ISBN: 978-3-540-72374-5.
- Huang, H.C., Pan, J.S., Fang, W.C. and Jain, L.C., An Introduction to Intelligent Multimedia Data Hiding, Springer-Verlag, 2007, Chapter 1, ISBN: 10-3-540-711686.
- Jain, L.C., Quteishat, A. and Lim, C.P., Intelligent Machines :An Introduction, Springer-Verlag, 2007, pp. 1–9, ISBN : 978-3-540-72695-1.
- Jain, L.C., Tan, S.C. and Lim, C.P., Introduction to Evolutionary Computing in System Design, Springer-Verlag, 2007, pp. 1–9, ISBN : 978-3-540-72376-9.
- Castellano, G., Castiello, C., Fanelli, M. and Jain, L.C., Evolutionary Neuro-Fuzzy Systems and Applications, Springer-Verlag, 2007, pp. 11–45, ISBN : 978-3-540-72376-9.
- Do, Q.V, Lozo, P. and Jain, L.C., Vision-Based Autonomous Robot Navigation, in Innovations in Robot Mobility and Control, Springer-Verlag, 2006, pp. 65–103, Chapter 2.
- Tran, C., Abraham, A. and Jain, L., Soft Computing Paradigms and Regression Trees in Decision Support Systems, in Advances in Applied Artificial Intelligence, Idea Group Publishing, 2006, pp. 1–28, Chapter 1.
- Jarvis, B., Jarvis, D. and Jain, L., Teams in Multi-Agent Systems, in IFIP International Federation for Information Processing, Vol. 228, Intelligent Information Processing III, Springer-Verlag, 2006, pp. 1–10, Chapter 1.
- Consoli, A., Tweedale, J. and Jain, L., The Link Between Agent Coordination and Cooperation, in IFIP International Federation for Information Processing, Vol. 228, Intelligent Information Processing III, Springer-Verlag, 2006, pp. 11–19, Chapter 2.
- Bannore, B. and Swierkowski, L. An Iterative Approach to Super Resolution, in IFIP International Federation for Information Processing, Vol. 228, Intelligent Information Processing III, Springer-Verlag, 2006, pp. 473–482.
- Abraham, A. and Jain, L.C., Industry, Evolutionary Multiobjective Optimization, in Springer-Verlag's book, ISBN 1-85233-787-7, 2005, pp. 1–6.
- Wang, F., Jain, L.C. and Pan, J., Genetic Watermarking on Spatial Domain, in Intelligent Watermarking Techniques, World Scientific, 2004, pp. 481–514, Chapter 17.
- Wang, F., Jain, L.C. and Pan, J., Watermark Embedding System based on Visual Cryptography, in Intelligent Watermarking Techniques, World Scientific, 2004, pp. 377-394, Chapter 13.
- Jain, L.C. and Chen, Z., Industry, Artificial Intelligence In, in Encyclopedia of Information Systems, Elsevier Science, USA, 2003, pp. 583–597.
- Teodorescu, H.N. and Jain, L.C., An Introduction to Computational Intelligence in Medical Diagnosis, Springer-Verlag, 2002, Chapter 1, pp. 1–11.

- Konar, A. and Jain, L.C., An Introduction to Computational Intelligence Paradigms, in Practical Applications of Computational Intelligence Techniques, Springer-Verlag, 2001, pp. 1–38, Chapter 1.
- Kuncheva, L.I., Wrench, J., Jain, L.C. and Al-zaidan, A., A Fuzzy Model of Heavy Metal Loadings in Marine Environment, in Soft Computing for Risk Evaluation and Management, Springer-Verlag, 2001, pp. 355–371.
- Tedman, D. and Jain, L.C., An Introduction to Innovative Teaching and Learning, in Teaching and Learning, Springer-Verlag, 2000, pp. 1–30, Chapter 1.
- Filippidis, A., Russo, M. and Jain, L.C., Novel Extension of ART2 in Surface Landmine Detection, Springer-Verlag, 2000, pp. 1–25, Chapter 1.
- Jain, L.C. and Lazzerini, B., An Introduction to Handwritten Character and Word Recognition, in Knowledge-Based Intelligent Techniques in Character Recognition, CRC Press, 1999, 3–16.
- Filippidis, A., Jain, L.C. and Martin, N.N., "Computational Intelligence Techniques in Landmine Detection," in Computing with Words in Information/Intelligent Systems 2, Edited by Zadeh, L. and Kacprzyk, J., Springer-Verlag, Germany, 1999, pp. 586–609.
- Halici, U., Jain, L.C. and Erol, A., Introduction to Fingerprint Recognition, in Intelligent Biometric Techniques in Fingerprint and Face Recognition, CRC Press, 1999, pp. 3–34.
- Teodorescu, H.N., Kandel, A. and Jain, L., Fuzzy Logic and Neuro-Fuzzy Systems in Medicine: A historical Perspective, in Fuzzy and Neuro-Fuzzy Systems in Medicine, CRC Press, 1999, pp. 3–16.
- Jain, L.C. and Vemuri, R., An Introduction to Intelligent Systems, in Hybrid Intelligent Engineering Systems, World Scientific, 1997, pp. 1–10, Chapter 1.
- Karr, C. and Jain, L.C., Genetic Learning in Fuzzy Control, in Hybrid Intelligent Engineering Systems, World Scientific, 1997, pp. 69–101, Chapter 4.
- Karr, C. and Jain, L.C., Cases in Geno-Fuzzy Control, in Hybrid Intelligent Engineering Systems, World Scientific, 1997, pp. 103–132, Chapter 5.
- Katayama, R., Kuwata, K. and Jain, L.C., Fusion Technology of Neuro, Fuzzy, GA and Chaos Theory and Applications, in Hybrid Intelligent Engineering Systems, World Scientific, 1997, pp. 167–186, Chapter 7.
- Jain, L.C., Medsker, L.R. and Carr, C., Knowledge-Based Intelligent Systems, in Soft Computing Techniques in Knowledge-Based Intelligent Systems, Springer-Verlag, 1997, pp. 3–14, Chapter 1.
- Babri, H., Chen, L., Saratchandran, P., Mital, D.P., Jain, R.K., Johnson, R.P. and Jain, L.C., Neural Networks Paradigms, in Soft Computing Techniques in Knowledge-Based Intelligent Systems, Springer-Verlag, 1997, pp. 15–43, Chapter 2.
- Jain, L.C., Tikk, D. and Koczy, L.T., Fuzzy Logic in Engineering, in Soft Computing Techniques in Knowledge-Based Intelligent Systems, Springer-Verlag, 1997, pp. 44–70, Chapter 3.

- Tanaka, T. and Jain, L.C., Analogue/Digital Circuit Representation for Design and Trouble Shooting in Intelligent Environment, in Soft Computing Techniques in Knowledge-Based Intelligent Systems, Springer-Verlag, 1997, pp. 227–258, Chapter 7.
- Jain, L.C., Hybrid Intelligent System Design Using Neural Network, Fuzzy Logic and Genetic Algorithms — Part I, Cognizant Communication Corporation USA, 1996, Chapter 9.
- Jain, L.C., Hybrid Intelligent System Applications in Engineering using Neural Network and Fuzzy Logic — Part II, Cognizant Communication Corporation USA,1996, Chapter 10.
- Jain, L.C., Introduction to Knowledge-Based Systems, Electronic Technology Directions to the Year 2000, IEEE Computer Society Press USA, 1995, pp. 17–27, Chapter 1.
- Jain, L.C. and Allen, G.N., Introduction to Artificial Neural Networks, Electronic Technology Directions to the Year 2000, IEEE Computer Society Press USA, 1995, pp. 36–62, Chapter 2.
- Jain, L.C. and Karr, C.L., Introduction to Fuzzy Systems, Electronic Technology Directions to the Year 2000, IEEE Computer Society Press USA, 1995, pp. 93–103, Chapter 3.
- Jain, L.C. and Karr, C.L., Introduction to Evolutionary Computing Techniques, Electronic Technology Directions to the Year 2000, IEEE Computer Society Press USA, 1995, pp. 121–127, Chapter 4.
- Marwood, W. and Jain, L.C., Microelectronic Engineering, Electronic Technology Directions to the Year 2000, IEEE Computer Society Press USA, 1995, pp. 227–241, Chapter 6.
- Devanathan, R., Mital, D.P. and Jain, L.C., Control Instrumentation — A survey of Current Practice and Some Future Trends, Electronic Technology Directions to the Year 2000, IEEE Computer Society Press USA, 1995, pp. 277–293, Chapter 7.
- Marwood, W. and Jain, L.C., Electronic Design and Automation, Electronic Technology Directions to the Year 2000, IEEE Computer Society Press USA, 1995, pp. 335–341, Chapter 8.
- Sato, M., Jain, L.C. and Takagi, H., Electronic Design and Automation, Electronic Technology Directions to the Year 2000, IEEE Computer Society Press USA, 1995, pp. 402–430, Chapter 9.
- Furuhashi, T., Takagi, H. and Jain, L.C., Intelligent Systems using Artificial Neural Networks, fuzzy Logic and Genetic Algorithms in Industry, Electronic Technology Directions to the Year 2000, IEEE Computer Society Press USA, 1995, pp. 485–495, Chapter 10.
- Tanaka-Yamawaki, M. and Jain, L.C., Multimedia, Computer Simulations and Earth station, Electronic Technology Directions to the Year 2000, IEEE Computer Society Press USA, 1995, pp. 462–463, Chapter 11.

7. Book Series

Note: Some books reported in this section are also listed in the previous sections.

1.7.1. *Advanced Intelligence and Knowledge Processing, Springer-Verlag, Germany*

http://www.springer.com/series/4738

Series Editor: L.C. Jain and X. Wu

- Liu, H. *et al.* (Editors), Robot Intelligence, Springer-Verlag, London, 2010.
- Mehandjiev, N. and Grefen, P. (Editors), Dynamic Business Process Formation for Instant Virtual Enterprises, Springer-Verlag, London, 2010.
- Adhikari, A., *et al.*, Developing Multi-Database Mining Applications, Springer-Verlag, London, 2010.
- Nin, J. and Herranz, J. (Editors), Privacy and Anonymity in Information Management Systems, Springer-Verlag, London, 2010.
- Tao, J.T. (Editor), Web-based Support Systems, Springer-Verlag, London, 2010.
- Chbeir, R. *et al.* (Editors), Emergent Web Intelligence: Advanced Information Retrieval, Springer-Verlag, London, 2010.
- Badr, Y., *et al.* (Editors), Emergent Web Intelligence: Advanced Semantic Technologies, Springer-Verlag, London, 2010.
- Mulvenna, M. and Nugent, C. (Editors), Supporting People with Dementia using Pervasive Technologies, Springer-Verlag, London, 2010.
- Chekanov, S., Scientific Data Analysis using Jython Scripting and Java, Springer-Verlag, London, 2010.
- Chen, M. and Mugnier, M.L., Graph-based Knowledge Representation, Springer-Verlag, London, 2009.
- Chli, M. and de Wilde, P., Convergence and Knowledge Processing in Multi-agent Systems, Springer-Verlag, London, 2009.
- Monekosso, D., *et al.* (Editors), Intelligent Environments, Springer-Verlag, London, 2009.
- Narahari, Y., *et al.*, Game Theoretic Problems in Network Economics and Mechanism Design Solutions, Springer-Verlag, London, 2009.
- Zarri, G.P., Representation and Management of Narrative Information, Springer-Verlag, London, 2009.
- Zudilova-Seinstra, E. *et al.* (Editors), Trends in Interactive Visualization, Springer-Verlag, London, 2009.
- Camastra, F. and Vinciarelli, A., Machine Learning for audio, Image, and Video Analysis, Springer-Verlag, London, 2008.
- Hu, C., *et al.*, Knowledge Processing with Interval and Soft Computing, Springer-Verlag, London, 2008.
- Kornai, A., Mathematical Linguistics, Springer-Verlag, London, 2008.

- Meisels, A., Distributed Search by Constrained Agents, Springer-Verlag, London, 2008.
- Nguyen, N.T., Advanced Methods for Inconsistent Knowledge Management, Springer-Verlag, London, 2008.
- Okada, A., *et al.* (Editors), Knowledge Cartography, Springer-Verlag, London, 2008.
- Prokopenko, M. (Editor), Advances in Applied Self-Organising Systems, Springer-Verlag, London, 2008.
- Simovici, D.A. and Djeraba, C., Mathematical Tools for Data Mining, Springer-Verlag, London, 2008.
- Pierre, S. (Editor), E-Learning Networked Environments and Architectures, Springer-Verlag, London, 2007.
- Scharl, A., Environmental Online Communication, Springer-Verlag, London, 2007.
- Karny, M. (Editor), Optimized Bayesian Dynamic Advising, Springer-Verlag, London, 2006.
- Liu, S. and Lin, Y., Grey Information: Theory and Practical Applications, Springer-Verlag, London, 2006.
- Maloof, M.A. (Editor), Machine Learning and Data Mining for Computer Security, Springer-Verlag, London, 2006.
- Manolopoulos, Y., *et al.*, R-Trees: Theory and Applications, Springer-Verlag, London, 2006.
- Abraham, A., Jain, L.C. and Goldberg, R. (Editors), Evolutionary Multiobjective Optimization, Springer-Verlag, London, 2005.
- Bandyopadhyay, S., *et al.* (Editors), Advanced Methods for Knowledge Discovery from Complex Data, Springer-Verlag, London, 2005.
- Chen-Burger, Y. and Robertson, D., Automatic Business Modelling, Springer-Verlag, London, 2005.
- Fyfe, C., Hebbian Learning and Negative Feedback Networks, Springer-Verlag, London, 2005.
- Grana, M., *et al.* (Editors), Information Processing with Evolutionary Algorithms, Springer-Verlag, London, 2005.
- Husmeier, D., *et al.* (Editors), Probabilistic Modelling in Bioinformatics and Medical Informatics, Springer-Verlag, London, 2005.
- Konar, A. and Jain, L.C., Cognitive Engineering, Springer-Verlag, London, 2005.
- Pal, N. and Jain, L., Advanced Techniques in Knowledge Discovery and Data Mining, Springer-Verlag, London, 2005.
- Stuckenschmidt, H. and Harmelen, F.V., Information Sharing on the Semantic Web, Springer-Verlag, London, 2005.
- Tan, K.C., *et al.*, Multiobjective Evolutionary Algorithms and Applications, Springer-Verlag, London, 2005.
- Wang, J.T.L. *et al.* (Editors), Data Mining in Bioinformatics, Springer-Verlag, London, 2005.

- Wang, L. and Fu, X., Data Mining with Computational Intelligence, Springer-Verlag, London, 2005.
- Gomez-Perez, *et al.*, Ontological Engineering, Springer-Verlag, London, 2004.
- Ko, C.C., Creating Web-based Laboratories, Springer-Verlag, London, 2004.
- Scharl, A., Environmental Online Communication, Springer-Verlag, London, 2004.
- Zhang, S., *et al.*, Knowledge Discovery in Multiple Databases, Springer-Verlag, London, 2004.
- Mentzas, G., *et al.*, Knowledge Asset Management, Springer-Verlag, London, 2003.
- Vazirgiannis, M., *et al.*, Uncertainty Handling and Quality Assessment in Data Mining, Springer-Verlag, London, 2003.

1.7.2. *Advanced Information Processing, Springer-Verlag, Germany*

http://www.springeronline.com/series/4785

Series Editor: L.C. Jain

Books Published

- Ishibuchi, H., Nakashima, T. and Nii, M., Classification and Modeling with Linguistic Information Granules, Springer-Verlag, Germany, 2005.
- Chen, S.H. and Wang, P.P. (Editors), Computational Intelligence in Economics and Finance, Springer-Verlag, Germany, 2004.
- Liu, J. and Daneshmend, L., Spatial Reasoning and Planning, Springer-Verlag, Germany, 2004.
- Ishida, Y., Immunity-Based Systems, Springer-Verlag, 2004.
- Ohsawa, Y. and McBurney, P. (Editors), Chance Discovery, Springer-Verlag, Germany, 2003.
- Deen, S.M. (Editor), Agent-Based Manufacturing, Springer-Verlag, Germany, 2003.
- Gasós J. and Hirsch B., e-Business Applications, Springer-Verlag, Germany, 2003.
- Harris, C., Hong, X. and Gan, Q., Adaptive Modelling, Estimation and Fusion from Data, Springer-Verlag, Germany, 2002.

1.7.3. *Computational Intelligence and Its Applications Series, IGI Publishing, USA*

http://www.igi-pub.com/bookseries/details.asp?ID=5&v=bs_releases

Series Editors: L.C. Jain, C. Fyfe, N.T. Nguyen

Books Published

- Chen, S.H., *et al.* (Editors), Computational Economics: A Perspectice from Computational Intelligence, 2006.
- Begg, R. and Palaniswami, M., Computational Intelligence for Movement Sciences: Neural Networks and Other Emerging Techniques, 2006.
- Zhang, D., *et al.* Biometric Image Discrimination Technologies, 2006.
- Fulcher, J. (Editor), Advances in Applied Artificial Intelligence, 2006.

1.7.4. *Knowledge-Based Intelligent Engineering Systems Series, IOS Press, The Netherlands*

http://www.kesinternational.org/bookseries.php

Subseries Editor: L.C. Jain and R.J. Howlett

Books Published

- Da Silva Filho, Lambert-Torres, G. and Abe, J.M., Uncertainty Treatment Using Paraconsistent Logic, IOS Press, The Netherlands, Vol. 211, 2010.
- Masulli, M. *et al.* (Editors), Computational Intelligence and Bioengineering, IOS Press, The Netherlands, Vol. 196, 2009.
- Apolloni, B., Bassis, S. and Marinaro, M. (Editors), New Directions in Neural Networks, IOS Press, The Netherlands, Vol. 193, 2009.
- Virvou, M. and Nakamura, T., Knowledge-Based Software Engineering, IOS Press, The Netherlands, Vol. 180, 2008.
- Velasquez, J.D. and Palade, V., Adaptive Web Sites, IOS Press, The Netherlands, Vol. 170, 2008.
- Zha, X.F. and Howlett, R.J., Integrated Intelligent System for Engineering Design, IOS Press, Vol. 149, 2006.
- Nakamatsu, K. and Abe, J.M. (Editors), Advances in Logic Based Intelligent Systems, IOS Press, Vol. 132, 2005.
- Phillips-Wren, G. and Jain, L. (Editors), Intelligent Decision Support systems in Agent-Mediated Environments, IOS Press, The Netherlands, Vol. 115, 2005.
- Turchetti, C., Stochastic Models of Neural Networks, IOS Press, The Netherlands, Vol. 102, 2004.
- Wang, K., Intelligent Condition Monitoring, IOS Press, Vol. 93, 2003.
- Lee, R.S.T. and Liu, J.H.K., Invariant Object Recognition Based on Elastic Graph Matching, IOS Press, The Netherlands, Vol. 86, 2003.
- Abraham, A., Koppen, M. and Franke, K. (Editors), Design and Applications of Hybrid Intelligent Systems, IOS Press, The Netherlands, Vol. 104, 2002.
- Abraham, A., *et al.* (Editors), Soft Computing Systems, IOS Press, The Netherlands, Vol. 87, 2002.
- Loia, V. (Editor), Soft Computing Agents, IOS Press, The Netherlands, Vol. 83, 2002.

- Damiani, E., *et al.* (Editors), Proceedings of the KES2002 Conference, IOS Press, The Netherlands, Vol. 82, Part 1 and Part 2, 2002.
- Motoda, H., Active Mining, IOS Press, The Netherlands, Vol. 79, 2002.
- Namatame, A., *et al.* (Editors), Agent-Based Approaches in Economic and Social Complex Systems, IOS Press, The Netherlands, Vol. 72, 2002.
- Baba, N., Howlett, R.J. and Jain L.C., Proceedings of the KES2001 Conference, IOS Press, The Netherlands, Vol. 69, Part 1 and Part 2, 2001.

1.7.5. *The CRC Press International Series on Computational Intelligence, The CRC Press, USA*

http://www.crcpress.com/shopping_cart/products/product_series.asp?id=&series=747975& parent_id=&sku=1965&isbn=9780849319655&pc

Series Editor: L.C. Jain

Books Published

- Jiming Liu, Hong Kong Baptist University, Multiagent Robot Systems. ISBN 0-8493-2288-X, 2001
- Marco Russo, University of Messina; and Lakhmi Jain (Editors), Fuzzy Learning and Applications. ISBN 0-8493-2269-3, 2001
- Horia-Nicolai Teodorecu, University of South Florida; and L.C. Jain (Editors), Intelligent Systems and Techniques in Rehabilitation Engineering. ISBN 0-8493-0339-7, 2001
- Iluminada Baturone, A. Barriga, C. Jimenez, D. Lopez and S. Sanchez, Instituto de Microelectronica de Sevilla, Microelectronic Design of Fuzzy Logic-Based Systems. ISBN 0-8493-0633-7, 2000
- Lakhmi Jain; and Anna Maria Fanelli, University of Bari (Editors), Recent Advances in Artificial Neural Networks: Design and Applications. ISBN 0-8493-2268-5, 2000
- Lakhmi Jain, Ugur Halici, Middle East Technical University; I. Hayashi, Hannan University; Sang-Bae Lee, Korea Maritime University; and Shigeyoshi Tsutsui, Hannan University (Editors), Intelligent Biometric Techniques in Fingerprint and Face Recognition. ISBN 0-8493-2055-0, 2000
- Beatrice Lazzerini, University of Pisa; D. Dumitrescu, University of Pisa; and Lakhmi Jain, Fuzzy Sets and their Application to Clustering and Training. ISBN 0-8493-0589-6, 2000
- Beatrice Lazzerini, University of Pisa; D. Dumitrescu, University of Pisa; L.C. Jain; and A. Dumitrescu, University de Savoie, Evolutionary Computing and Applications. ISBN 0-8493-0588-8, 2000
- Larry Medsker, American University Washington; and Lakhmi Jain (Editors), Recurrent Neural Networks: Design and Applications. ISBN 0-8493-7181-3, 2000

- Toyoaki Nishida, Nara Institute of Science and Technology, Dynamic Knowledge Interaction. ISBN 0-8493-0314-1, 2000
- Lakhmi Jain; and Clarance de Silva, University of British Columbia (Editors), Intelligent Adaptive Control: Industrial Applications. ISBN 0-8493-9805-3, 1999.
- Lakhmi Jain; and Beatrice Lazzerini, University of Pisa (Editors), Knowledge-Based Intelligent Techniques in Character Recognition. ISBN 0-8493-9807-X, 1999.
- Lakhmi Jain; and Noel Martin, University of South Australia (Editors), Fusion of Neural Networks, Fuzzy Systems, and Genetic Algorithms: Industrial Applications. ISBN 0-8493-9804-5, 1999.
- Charles Karr, Practical Applications of Computational Intelligence for Adaptive Conrol, University of Alabama (Editors), Industrial Applications of Genetic Algorithms. ISBN 0-8493-9801-0, 1999.
- Charles Karr and L.M. Freeman, University of Alabama (Editors), Industrial Applications of Genetic Algorithms. ISBN 0-8493-9801-0, 1999.
- Horia-Nicolai Teodorescu, Abraham Kandel, University of South Florida; and Lakhmi Jain (Editors), Soft Computing Techniques in Human-Related Science. ISBN 0-8493-1635-9, 1999.
- Lakhmi Jain; Ray Johnson, Defence Science and Technology Organisation; Yoshiharu Takefuji, Keio University; and Lotfi Zadeh, University of California-Berkeley (Editors), Knowledge-Based Intelligent Techniques in Industry. ISBN 0-8493-9803-7, 1998.
- Lakhmi Jain; and Rao Vemuri, University of California-Davis (Editors), Industrial Applications of Neural Networks. ISBN 0-8493-9802-9, 1998.
- Horia-Nicolai Teodorescu, Abraham Kandel, University of South Florida; and Lakhmi Jain (Editors), Fuzzy and Neuro-Fuzzy Systems in Medicine. ISBN 0-8493-9806-1, 1998.
- Lakhmi Jain (Editor), Evolution of Engineering and Information Systems and their Applications, 1999. ISBN 0-8493-1965-X, 1997.
- Zhengxin Chen, University of Nebraska at Omaha, Computational Intelligence for Decision Support. ISBN 0-8493-1799-1.
- Horia-Nicolai Teodorescu, Abraham Kandel, University of South Florida, Dynamic Fuzzy Systems and Chaos Applications. ISBN 0-8493-2079-8.

1.7.6. *Series on Innovative Intelligence, World Scientific, Singapore*

http://www.worldscientific.com.sg/books/series/sii_series.shtml

Series Editor: L.C. Jain

Books Published

- Vol. 1: Virtual Environments for Teaching and Learning, 2002, Edited by L C Jain, R J Howlett, N S Ichalkaranje and G Tonfoni

- Vol. 2: Advances in Intelligent Systems for Defence, 2002, Edited by L C Jain, N S Ichalkaranje and G Tonfoni
- Vol. 3: Internet-Based Intelligent Information Processing Systems, 2003, Edited by R J Howlett, N S Ichalkaranje, L C Jain and G Tonfoni
- Vol. 4: Neural Networks for Intelligent Signal Processing, 2003, By A Zaknich
- Vol. 5: Complex-Valued Neural Networks, Theories and Applications, 2003, Edited by A Hirose
- Vol. 6: Intelligent and Other Computational Techniques in Insurance, 2003, Edited by A F Shapiro and L C Jain
- Vol. 7: Intelligent Watermarking Techniques (With CD-Rom), 2004, Edited by J-S Pan, H-C Huang and L C Jain
- Vol. 8: Learning Support Systems for Organizational Learning, 2004, by Joachim P Hasebrook and Hermann A Maurer

1.7.7. *Intelligent Systems Reference Library, Springer-Verlag, Germany*

http://www.springer.com/series/8578

Series Editors: J. Kacprzyk and L.C. Jain

Books Published

- Vol 1: Computational Intelligence: Collaboration, Fusion and Emergence, 2009, Edited by C. Mumford and L.C. Jain
- Vol 2: Tree-Structure based Hybrid Computational Intelligence, 2010, By Y. Chen and A. Abraham
- Vol 3: Developments and Challenges for Autonomous Unmanned Vehicles, 2010, By A. Finn and S. Scheding
- Vol 4: Handbook on Decision Making, Springer, 2010, Edited by L.C. Jain and C.P. Lim
- Vol 5: Intelligent Mathematics: Computational Analysis, 2010, By G.A. Anastassiou
- Vol. 6: Soft Computing in Economics and Finance, 2011 By L. Dymowa
- Vol. 7: Modelling and Control for Intelligent Industrial systems, 2011, By G.G. Rigatos
- Vol. 8: Knowledge Seeker, 2011 by E.H.Y. Lim, J.N.K. Liu, R.S.T. Lee

1.7.8. *Smart Innovations, Systems and Technologies, Springer-Verlag, Germany*

www.springer.com/series/8767

Series Editors: R.J. Howlett and L.C. Jain

Books Published

- Vol. 1: Nishida, T., Jain, L.C. and Faucher, C. (Editors), Modelling Machine Emotions for Realizing Intelligence, Springer-Verlag, 2010.
- Vol. 2: Tsihrintzis, G., Virvou, M. and Jain, L.C. (Editors), Multimedia Services in Intelligent Environments: Software Development Challenges and Solutions, Springer-Verlag, 2010.
- Vol. 3: Tsihrintzis, G. and Jain, L.C. (Editors), Multimedia Services in Intelligent Environments: Integrated Systems, Springer-Verlag, 2010.
- Vol. 4: Phillips-Wren, G., Jain, L.C., Howlett, R.J. and Nakamatsu, K., Advances in Decision Support Systems, Springer-Verlag, 2010.
- Vol. 5: Howlett, R.J., Innovation Through Knowledge Transfer, Springer-Verlag, 2010.
- Vol. 6: Tsihrintzis, G.A., Damiani, E., Virvou, M., Jain, L.C. and Howlett, R.J., Intelligent Interactive Multimedia Systems and Services, Springer-Verlag, 2010.
- Vol. 7: Howlett, R.J., Jain, L.C. and Lee, S.H., Sustainability in Energy and Buildings, Springer-Verlag, 2010.

Chapter 2

STABILITY, CHAOS AND LIMIT CYCLES IN RECURRENT COGNITIVE REASONING SYSTEMS

Aruna Chakraborty[1], Amit Konar[2], Pavel Bhowmik[3] and Atulya K. Nagar[4]
[1]*St. Thomas' College of Engineering and Technology, Khidderpur, Calcutta-23,*
[2,3]*Dept. of Electronics and Tele-Communication Engineering,*
Jadavpur University, Calcutta–700 032
[4]*Intelligence, and Distributed System Lab, Liverpool Hope University,*
Liverpool L169JD.
aruna.stcet@gmail.com; konaramit@yahoo.co.in;
bpavel88@gmail.com; nagara@hope.ac.uk

Cognitive reasoning involves human short term memory to update the contextual information about a phenomenon. Recurrent models are often used to represent the dynamics of short term memory. These models often suffer from chaos and limit cycles. This chapter aims at designing a new approach to analyze the dynamical behavior of a cognitive reasoning system. The chapter addresses three distinct recurrent models for cognitive reasoning. The first model refers to a recurrent dynamics of a propositional temporal reasoning system. The principle of analysis to determine stable points for this model is illustrated. The second model represents a specialized cognitive map with an encoding model for weight adaptation, and a recall model for deriving new inferences. The last model considers a cognitive structure of an emotional system, the dynamics of which may suffer from limit cycles and chaos for certain parametric conditions. A method of stabilization of the system is given for the generation of stable inferences.

2.1. Introduction

A cognitive system deals with the psychological processes involved in various intelligent activities of the human beings [3, 24, 29]. Several models of cognitive systems representing various components of human intelligence are available in the literature [20, 21, 23–27, 30–35]. Usually the information extracted from the real world is stored in the short term memory (STM). The information and their attributes undergo adaptation in STM until a steady-state situation is reached, when they are transferred to long term memory for permanent storage using the hippocampus, located in the human brain [10, 38, 39, 41]. Biologists reveal that the most important module in the brain responsible for cognitive reasoning, called cognitive map [15] is located in the hippocampus. The cognitive map is responsible for automatic encoding and recall and also for spatial reasoning. Researchers in artificial intelligence, however, have loosely used the term cognitive map to represent a structure capable of autonomous reasoning and learning [41]. The pioneering contribution on cognitive map is due to Axelord [1]. Kosko later extended this basic model of cognitive map [36, 37]. Following Kosko several researchers including Zhang *et al.* [42–45] developed various models of cognitive maps for automated reasoning with a power of self learning and adaptability. This chapter attempts to model cognitive map, and particularly human STM. One simple way to model cognitive map/STM is to employ recurrent dynamics. It is a common observation that people while taking a complex decision often gets trapped under limit cycles or chaotic behavior of their psychological states [22]. This chapter attempts to model cognitive map and STM capable of demonstrating stable, chaotic and limit cyclic behavior in their temporal response. Conditions for chaos and limit cycles have been derived, and a solution to arrive at stable decisions even when confronted with limit cycles or chaos is demonstrated.

The chapter will begin with a basic formulation of propositional recurrent model [15] to describe temporal dynamics of a propositional statement [4]. Both continuous and discrete models are presumed, and the analysis demonstrates that the models may either converge to a stable point [5] or oscillate around certain fixed values.

A model of fuzzy cognitive map is introduced next. The objective of this cognitive map is to acquire knowledge about connectivity among concepts through an unsupervised learning mechanism [16–18]. Once the stable connectivity weight between any two concepts is determined, the structure of the map is used for cognitive reasoning [33].

A special type of non-linearity in the model offers a chaotic behavior [11, 28] in the system time response. Duffing oscillator, for example, is one nonlinear oscillator, whose response demonstrates a chaotic fluctuation over time [7, 8]. In Duffing oscillator, nonlinearity is introduced to model non-ideal spring-mass-load system. In many real world problems, nonlinearity is inherent part of system modeling, and studying the dynamic behavior of the system requires plotting phase-trajectories [7, 8]. In this chapter, we consider an illustrative emotional system model, represented by a recurrent nonlinear cognitive dynamics [13] to study the significance of stability, chaos and limit cyclic behavior of the emotional system [12, 14]. It may be added here that in [8], the authors employed Duffing oscillator for emotion classification from EEG [19]. The methodology of cognitive system stability and chaos analysis presented here is different from [7, 8].

The chapter is divided into six sections. In section 2, we outline the basic propositional temporal dynamics, and proposed a method to analyze the stable interpretation of the given system. A fuzzy extension of the propositional temporal dynamics is introduced in section 3, and a Lyapunov based stability analysis is included to study the dynamic behavior of the system. Section 4 includes a simple cognitive neural net capable of adapting its weights by a Hebbian learning algorithm. After the encoding is over, this neural net is capable of automatic reasoning. An illustration to weather forecasting system using the proposed theory is also given in this section. In section 5, we provide an emotional system of a subject, and demonstrate the dynamic behavior of the system through a Lyapunov analysis. Conclusions are listed in section 6.

2.2. Stable Points in Propositional Temporal Dynamics

In a propositional temporal logic, the truth-value of propositions [4] changes over time [9]. The dynamics of a propositional temporal system on a continuous framework can be represented by time-differential equation. The solution of time differential equation in the steady state yields the stable points of the propositional temporal statements.

For the sake of implementation on digital computers, people prefer time-difference equations to time differential equation. A time difference equation has the advantage of evaluating the next value of a proposition, if its current value and its previous values are known. A recurrent model of time difference equation is generally used in the literature of dynamic system to model many practical problems on a discrete framework. Example 1 below provides the construction process of a time difference model of propositional temporal statements, and the procedure to determine stable points of the logical dynamics.

Example 1. Let $p(t)$ and $q(t)$ be two propositions, signifying "I am here at time t" and "you are here at time t" respectively. Consider two propositional temporal rules. The first rule states that I shall be here at time $(t+1)$ if you are here at time t, and the condition: if you are here at time t then I am here at time t is true. Thus $p(t+1)$ is true if $p(t) \wedge q(t) \to p(t)$ is true. More formally,

$$\begin{aligned} p(t+1) &= p(t) \wedge (q(t) \to p(t) \\ &= p(t) \wedge (p(t) \vee \neg q(t) \end{aligned} \tag{1}$$

Similarly, suppose that we a have a second rule like the first rule where you is replaced by I and vice-versa. The resulting logical statement corresponding to this is given by

$$q(t+1) = q(t) \wedge (q(t) \vee \neg p(t)) \tag{2}$$

In this example, we want to determine the local stable points of the discrete propositional temporal given in (1) and (2).

Let $t = t^*$ be the equilibrium time, i.e., the minimum time after which the dynamical system will attain equilibrium. The condition of equilibrium is given by

$$p(t^*+1) = p(t^*) = p^*, \ say \tag{3}$$

$$q(t^*+1) = q(t^*) = q^*, \ say. \tag{4}$$

The equilibrium values of p and q can be determined by setting conditions (3) and (4) in (1) and (2) respectively, which results in

$$p^* = p^* \wedge (p^* \vee \neg q^*) \tag{5}$$

and

$$q^* = q^* \wedge (q^* \vee \neg p^*) \tag{6}$$

Since p^* in the right hand side has to be equal to p^* in the left hand counterpart, (5) holds only if

$$p^* \leq p^* \vee \neg q^* \tag{7}$$

or,

$$p^* \leq \text{Max}(p^*, \neg q^*) \tag{8}$$

Therefore,

$$\begin{aligned} p^* &= \text{Max}(p^*, \neg q^*) \quad \text{if } p^* > \neg q^* \\ &= p^*, \quad \text{if } p^* > \neg q^*, \end{aligned}$$

i.e.,

$$(p^*, q^*) = (1, 1). \tag{9}$$

Again,

$$p^* < \text{Max}(p^*, \neg q^*) \quad \text{if } \neg q^* > p^*,$$

i.e.,

$$(p^*, q^*) = (0, 0). \tag{10}$$

$\therefore$ The stable points of the propositional temporal dynamic system are $(0, 0)$ and $(1, 1)$.

Interested readers may find a number of interesting formulations of propositional temporal dynamics and their stable points in [15]. Now we show an alternative method to analyze the stability issue of propositional temporal dynamics using Lyapunov energy function [40].

2.2.1. *Stability of propositional temporal system using Lyapunov energy function*

The local stability analysis we introduced in the last section attempts to determine all the stable points of a dynamical system. Among the stable points, some are relatively more stable than the rest. What does relative stability mean? It means that for a small variation in the stable point, the dynamics may lose its stability and may have a transition to another state.

For determining the stability in the asymptotic sense, we need to identify stable points that allow larger disturbances without violating the stability of the system. In this system, we briefly outline the Lyapunov energy function and demonstrate as to how to use this function for determining the asymptotic stability of a dynamical system.

2.2.1.1. *The Lyapunov energy function*

A Lyapunov energy function [40] represents the energy contour of a particle when subjected to a set of forcing functions. Usually, the differential or the difference equation for which the stable points need to be investigated, act as forcing functions on a point mass particle residing over the Lyapunov energy surface. To analyze the stability of the dynamical systems, we have to determine a suitable energy surface that includes the trajectory of motion of the particle by the forcing functions. The dynamics is said to be stable if the particle loses energy over time.

Let $L(p(t), q(t))$ be a Lyapunov energy function, then it should satisfy the following criteria:

$$\left.\begin{array}{l} \text{i)}\quad L(0,0) = 0 \text{ (zero at origin)} \\ \text{ii)}\quad L(p,q) > 0 \quad \text{for} \quad (p,q) \neq (0,0) \text{ (positive definite)} \\ \text{iii)}\quad \dfrac{\partial L}{\partial p} \quad \text{and} \quad \dfrac{\partial L}{\partial q} \text{ drop should exist (smooth surface).} \end{array}\right\} \tag{11}$$

For a selected Lyapunov energy function L, the given dynamics is asymptotically stable if $dL/dt < 0$ for continuous system dynamics and $\Delta L < 0$ for discrete system dynamics.

2.2.1.2. *Asymptotic stability analysis of the propositional temporal system*

The smoothness criteria in (11), i.e. $\partial L/\partial p$ and $\partial L/\partial q$ should exist, require that L must be a continuous and derivable function. Since $\wedge$ and $\vee$ are discontinuous functions, we consider t and s norm instead of $\wedge$ and $\vee$ respectively. Thus let

$$p\,t\,q = p \cdot q \tag{12}$$

and

$$p\,s\,q = p + q - pq. \tag{13}$$

Example 2. Consider the equations given in (14) and (15). Let us determine the asymptotic stability of the dynamics, if it exists, by Lyapunov energy function.

$$p(t+1) = q(t) \rightarrow p(t) \tag{14}$$

and

$$q(t+1) = p(t) \rightarrow q(t) \tag{15}$$

Simplifying, we have:

$$\begin{aligned} p(t+1) &= \neg q(t)\ s\ p(t) \\ &= \neg q(t) + p(t) - (\neg q(t))p(t)) \end{aligned} \tag{16}$$

and

$$q(t+1) = \neg p(t) + q(t) - (\neg p(t))(q(t)) \tag{17}$$

For determining the global stable points of the above dynamics, let

$$L(p,q) = p + q \tag{18}$$

be the Lyapunov energy function. It is to be noted that the above function satisfies the Lyapunov energy criteria, i.e.,

$$\left.\begin{aligned} &L(0,0) = 0 \\ &L(p,q) > 0 \quad \text{for } (p,q) \neq (0,0) \\ &\frac{\partial L}{\partial p} \quad \text{and} \quad \frac{\partial L}{\partial q} \ \text{should exist.} \end{aligned}\right\} \tag{19}$$

Since the given system is discrete, we compute

$$\begin{aligned} \Delta L &= L(p(t+1), q(t+1)) - L(p(t), q(t)) \\ &= \{p(t+1) + q(t+1)\} - \{p(t) + q(t)\} \end{aligned} \tag{20}$$

Substituting (16) and (17) in (20) we have

$$\begin{aligned} \Delta L &= \{(1-q(t)) + p(t) - (1-q(t)) \cdot p(t)\} \\ &\quad + \{(1-p(t)) + q(t) - (1-p(t)) \cdot q(t)\} - p(t) - q(t) \\ &= \{(1-q(t)) - (1-q(t)) \cdot p(t)\} + \{(1-p(t)) - (1-p(t)) \cdot q(t) \\ &= (1-q(t))(1-p(t)) + (1-p(t))(1-q(t)) \\ &= 2(1-p(t))(1-q(t)) \\ &\geq 0, \quad \text{for } p(t), q(t) \in \{0,1\}. \end{aligned} \tag{21}$$

$\therefore$ the given dynamics have no global stable points.

2.3. Stability Analysis of Fuzzy Temporal Dynamics

In this section, we consider stability analysis of fuzzy temporal dynamics. Consider for example the following two fuzzy temporal rules.

Rule 1: If $x_1(t)$ is LARGE and $x_1(t)$ is SMALLER than k
and the contact between $x_1(t)$ and $x_2(t)$ is SMALL
Then $x_1(t+1)$ is LARGE

Rule 2: If $x_2(t)$ is LARGE
and the contact between $x_1(t)$ and $x_2(t)$ is SMALL
Then $x_2(t+1)$ is LARGE.

The fuzzy qualification of SMALL and LARGE may be changed to consider different fuzzy sets to take into account of the entire possibility space of all sets in the universe of x_1 and x_2.
Let

$\mu_{LARGE}(x_1, t)$ be the membership of x_1 to be large at time t,
$\mu_{NEGATIVE}((x_1(t) - k), t)$ be the membership of $x_1(t)$ to be smaller than k at time t
$\mu_{SMALL}((x_1(t) \cdot x_2(t)), t)$ be the membership of the product x_1, x_2 to be small at time t.

The Rule 1 thus can be stated in the following form of fuzzy temporal equation (22).

$$\begin{aligned}\mu_{LARGE}(x_1, t+1) = a\mu_{LARGE}(x_1, t) \times \mu_{NEGATIVE}((x_1(t) - k), t) \\ - b\mu_{SMALL}((x_1(t) \cdot x_2(t)), t),\end{aligned} \tag{22}$$

where a and b are two constants representing relative strengths of the two antecedents of Rule 1.

Representing $\mu_{SMALL}((x_1(t) \cdot x_2(t)), t)$ by $\mu_{SMALL}(x_1(t), t) \cdot \mu_{SMALL}(x_2(t), t)$.

we finally obtain (23).

$$\begin{aligned}\mu_{LARGE}(x_1, t+1) = a\mu_{LARGE}(x_1, t) \times \mu_{NEGATIVE}(x_1 - k, t) \\ - b\mu_{SMALL}(x_1, t).\mu_{SMALL}(x_2, t)\end{aligned} \tag{23}$$

The Rule 2 can analogously be represented in the following form of fuzzy temporal equation:

$$\mu_{LARGE}(x_2, t+1) = a\mu_{LARGE}(x_2, t) - c\mu_{SMALL}(x_1, t) \cdot \mu_{SMALL}(x_2, t) \tag{24}$$

The coefficients of the two terms in (24) could be any real numbers. We just selected them arbitrarily as a and b respectively.

Stability Analysis

Abbreviating L for LARGE, N for NEGATIVE, and S for SMALL, we consider the following Lyapunov energy function for the asymptotic stability analysis of the dynamics (22) and (24).

Let V be a Lyapunov energy function given by

$$\begin{aligned}&V(\mu_L(x_1, t+1), \mu_L(x_2, t+1)) \\ &\quad = (b\mu_L(x_2, t+1) - c\mu_L(x_1, t+1))^2 \\ &\quad = (-ac\mu_L(x_1, t) \times \mu_{NEG}(x_1 - k, t) + a\, b(\mu_L(x_2, t))^2\end{aligned} \tag{25}$$

$$\begin{aligned}\Delta V &= V(\mu_L(x_1, t+1), \mu_L(x_2, t+1)) - V(\mu_L(x_1, t), \mu_L(x_2, t)) \\ &= (ab\mu_L(x_2, t) - ac\mu_L(x_1, t) \times \mu_{NEG}(x_1 - k, t))^2 \\ &\quad - (b\mu_L(x_2, t) - c\mu_L(x_1, t))^2 \\ &< 0, \text{ if}\end{aligned} \tag{26}$$

$$\begin{aligned}&ab\mu_L(x_2, t) - ac\mu_L(x_1, t) \times \mu_{NEG}(x_1 - k, t) < b\mu_L(x_2, t) - c\mu_L(x_1, t) \\ &\Rightarrow (a-1)b\ \mu_L(x_2, t) < c\mu_L(x_1, t)(a\mu_{NEG}(x_1 - k, t) - 1)\end{aligned} \tag{27}$$

Or,

$$\frac{\mu_L(x_2, t)}{\mu_L(x_1, t)} < \left(\frac{c}{b(a-1)}\right)(a\mu_{NEG}(x_1 - k, t) - 1) \tag{28}$$

If $\mu_{NEG}(x_1 - k, t) = 1$ (*i.e.* $x_1(t) - k < 0$ *is certain*) then

$$\frac{\mu_L(x_2, t)}{\mu_L(x_1, t)} < c\frac{(a-1)}{b(a-1)} = \frac{c}{b}. \tag{29}$$

The principle of stability analysis illustrated above can be applied to any fuzzy recurrent system like the one given in (22) and (24).

It is indeed important to note that reasoning in fuzzy temporal systems can be performed by classical Mamdani based fuzzy reasoning [15]. This, however, goes beyond the scope of this chapter, as the major focus of the chapter is on stability analysis of fuzzy temporal dynamics.

2.4. Reasoning with Fuzzy Cognitive Map

Pal and Konar presented an alternative model [41] of fuzzy cognitive map with a slightly different nomenclature. According to them a cognitive map is an associative structure consisting of nodes and directed arcs where the nodes carry fuzzy beliefs and the arcs or edges carry connectivity strength.

Let $n_i(t)$ and $n_j(t)$ be the fuzzy beliefs of node N_i and N_j respectively and $\delta w_{ij}(t)$ be the change in connectivity strength (CS) w_{ij} of an edge E_{ij} connected from node N_i to N_j.

Then following the principles of Hebbian learning [34] and considering self-mortality of w_{ij}, we can write

$$\delta w_{ij}(t) = -\alpha w_{ij}(t) + S(n_i(t)).S(n_j(t)) \tag{30}$$

where

$$S(n_k) = \frac{1}{1 + \exp(-n_k)} \quad \text{for } k \in (i, j) \tag{31}$$

and α represents the forgetfulness (mortality) rate.

It may be noted that that the encoding model (32) is similar with Kosko's model [37]. But there is a significant difference between the recall models of Pal & Konar with that of Kosko. The recall model of Pal & Konar [41] in point-wise notation is given below.

$$n_i(t+1) = \max\left[n_i(t), \max_{\forall k}\{(\min(n_k, w_{ki})\}\right]. \tag{32}$$

The above equation states that the cognitive system will attempt to restore the last value of N_i through a memory refresh cycle. The fuzzy belief of node N_i thus will assume either of i) the last value or ii) the supremum of (n_k, w_{ki}), whichever is larger.

To study the stability of the cognitive map Pal and Konar [41] derived some interesting results presented below.

Property 1: *For all* $t, n_i(t+1)$ *in expression* (32) *is bounded between* 0 *and* 1.

Proof. Let us consider three possible ranges of w_{ki}: (i) $w_{ki} < 0$, (ii) $0 < w_{ki} \leq 1$, (iii) $w_{ki} > 1$.

Case
I: When $w_{ki}(t) < 0$, for $\exists(\forall)k$, let us assume that $\max_k\{\min(n_k(t), w_{ki}(t))\} < 0$. However, since $0 < n_i(0) < 1$, recursive use of expression (6.18) reveals that $\forall t$, $0 < n_i\ (t+1) < 1$.

Case II: When $0 < w_{ki}(t) < 1$, $\exists(\forall)k$, the proof of property 1 is obvious.

Case III: When $w_{ki}(t) > 1$, $\exists(\forall)k$, since $0 < n_k(0) < 1$, $0 < \min(n_k(0), w_{ki}(0)) < 1$. So $\max_k\ \{(\min(n_k(0), w_{ki}(0))\} < 1$. Again since $0 < n_i(0) < 1$ by expression (6.18), $0 < n_i(1) < 1$. Property 1 thus can easily be proved by the method of induction.

Property 1, therefore, holds for all the above cases. □

Theorem 1. *The recall model given by expression (32) is unconditionally stable.*

Proof. It is clear from expression (32) that $n_i(t+1) \geq n_i(t)$ for all t. Since an oscillation requires both an increase and decrease in value and $n_i(t+1)$ is never less than $n_i(t)$, for any t, therefore, $n_i(t+1)$ cannot exhibit oscillatory time-response. Further, since $0 < n_i(t+1) < 1$ for all t, vide property 1, and $n_i(t)$ does not exhibit oscillation, so $n_i(t)$ must reach steady state with value between 0 and 1. Since the model represented by expression (32) need not require to satisfy any condition for its stability, the statement of the theorem follows.

Corollary 1 follows from the above theorem. In this corollary, we consider a node vector $\mathbf{N}$, the i-th component of which represents the fuzzy belief of node $\mathbf{N}_i$. □

Corollary 1. *Given a node vector* N *such that its i-th component corresponds to the belief of node* Ni *for all* i. *Also given an edge connectivity matrix* W *whose* (i, j)*-th component* w_{ij} *denotes connectivity strength from node* N_i *to node* N_i. *The condition for steady-state for the node vector* N *then satisfies the following inequality:*

$$N^* \geq W^{*T} \circ N^* \tag{33}$$

where N^* *and* W^* *represent the steady-state values of* N *and* W *respectively.*

Proof. The vector-matrix form corresponding to expression (32) is given by

$$N(t+1) = N(t) \vee (W^T \circ N(t)) \tag{34}$$

where $\vee$ and o denote fuzzy or (max) and max-min composition operators respectively. The $\vee$ operator between two vectors computes the component-wise maximum of the two vectors like addition of two column vectors.

Now to satisfy $N(t+1) = N(t) = N^*$ at time $t = t^*$ we require to satisfy

$$N^* \geq W^{*T} \circ N^*.$$

Thus the result follows. □

Theorem 2. *The condition for stability, limit cycles and instability of the dynamics (30) depends on the choice of α as prescribed below:*

Stable: when $0 < \alpha < 2$,
Limit cyclic: when $\alpha = 2$, and
Unstable: when $\alpha > 2$.

Proof. Replacing δ by $(E-1)$in expression (30) we obtain:

$$(E - 1 + \alpha)w_{ij} = X \tag{35}$$

where

$$X = \left(\frac{1}{1+\exp(-n_i)}\right)\left(\frac{1}{1+\exp(-n_j)}\right). \tag{36}$$

The auxiliary equation for expression (35) is given by

$$(E - 1 + \alpha)w_{ij} = 0 \tag{37}$$

or,

$$E = 1 - \alpha. \tag{38}$$

Therefore the complementary function is given by

$$w_{ij}(t) = A(1-\alpha)^t \tag{39}$$

where A is a constant to be determined from boundary conditions. Let steady-state value of w_{ij} and X be denoted by w^*_{ij} and X^* respectively. Now, for computing w^*_{ij} we first need to compute X^*. The steady-state value of w_{ij} can now be obtained from the particular integral

$$w_{ij}(t) = \frac{X^*}{(E-1+\alpha)}. \tag{40}$$

Since X^* is constant, therefore, we substitute $E = 1$ in expression (40) to determine w_{ij}^*. Thus,

$$w_{ij}^* = \frac{X^*}{\alpha}. \tag{41}$$

The complete solution for (37) is given by

$$w_{ij}(t) = A(1-\alpha)^t + \frac{X^*}{\alpha} \tag{42}$$

Substituting $t = 0$ in (44) we obtain:

$$A = w_{ij}(0) - \frac{X^*}{\alpha}. \tag{43}$$

Combining (42) and (43) we obtain the complete solution for $w_{ij}(t)$ given below:

$$w_{ij}(t) = \left(w_{ij}(0) - \frac{X^*}{\alpha}\right)(1-\alpha)^t + \frac{X^*}{\alpha}. \tag{44}$$

The condition for stability, limit cycles and instability as stated in the theorem follows directly from (44). □

Theorem 3. *The steady-state value of X always satisfies the following inequality:*

$$0.25 \le X^* \le 0.48. \tag{45}$$

Proof. From (38) we have

$$X^* = \left(\frac{1}{1+\exp(-n_i^*)}\right)\left(\frac{1}{1+\exp(-n_j^*)}\right), \tag{46}$$

where n_i^* and n_j^* denote the steady-state values of n_i and n_j respectively.

Since right hand side of (48) is monotonic for $0 \le n_i^*, n_j^* \le 1$, we determine the range of X^* by substituting $n_i^* = n_j^* = 0$ and $n_i^* = n_j^* = 1$ respectively in (46). Hence, the result follows. □

Corollary 2. *The steady-state value of w_{ij} always satisfies the following inequality: $\frac{0.25}{\alpha} \le w_{ij}^* \le \frac{0.48}{\alpha}$.*

Proof. Proof of the corollary directly follows from (41) and (46). □

Corollary 3. *$w_{ij}^* > 0$ for all weights $w_{ij}(0) \ne 0$ follows directly from Theorem 3 and (47).*

The significance of the above results is that the cognitive learning system never destroys a weight of a link permanently. In other words, the cognitive learning model maintains stability of the structure for the cognitive map.

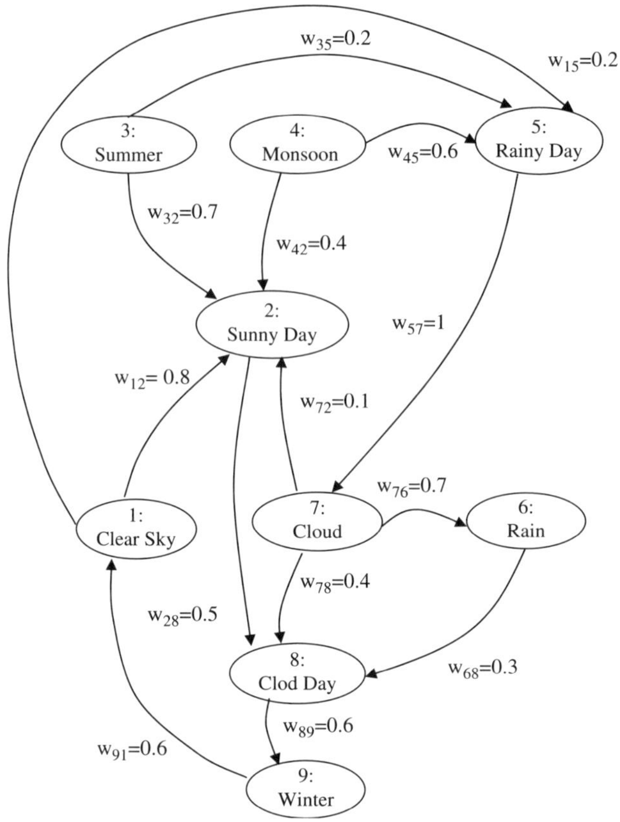

Fig. 2.1. Cognitive map of a typical weather forecasting system.

Computer Simulation: Konar and Pal [33] illustrated the behavior of their cognitive model with reference to a simulated weather forecasting system (Fig. 2.1). To describe the construction of their cognitive map, let us consider one typical rule embedded in the map using nodes and arcs.

Rule 1: An increase in winter causes an increase in clear sky by an amount 0.8 with a CS = 0.6.

The above rule has been encoded by nodes 1 and 9 and the weight w_{91}. After the construction of the cognitive map is over, initial fuzzy beliefs and CS are mapped at appropriate nodes and arcs of the network.

The CS of the arcs are then updated in parallel and with the new CS the fuzzy beliefs at all nodes are updated in parallel. This may be termed as one complete "encode-recall" cycle. A number of cycles are repeated until the steady-state condition is reached. The steady-state beliefs thus obtained represent a permanent non-volatile image of cognitive system's memory.

The cognitive map presented in Fig. 2.1 has been simulated on a computer and the temporal variations of beliefs and CS of nodes and arcs are obtained for a

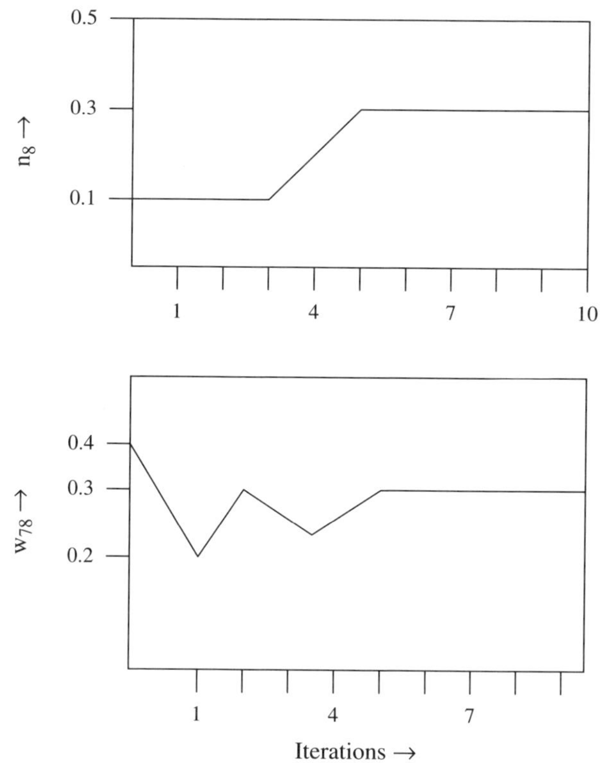

Fig. 2.2. Typical responses of the cognitive map representing the weather forecasting problem.

selected value of α. Plots of belief for one typical node (say node 9) and one weight (say w_{78}) are given in Fig. 2.2 for a value of $\alpha(= 1.6)$ in the stable range.

The cognitive model presented by Pal and Konar [41] thus is found to be stable for $0 < \alpha < 2$. Further, the steady-state values of weights are never zero indicating the stability of the structure.

2.5. Chaos and Limit Cycles in Emotion Based Cognitive Reasoning System

The encoding dynamics of fuzzy cognitive map introduced in section 2.4 exhibits stable and limit cyclic response, but is unable to represent chaotic behavior. Chaos, however, is a common phenomenon in natural reasoning. This section proposes a new model of cognitive reasoning system, whose temporal response includes chaos, limit cycles and stable behavior, depending on the choice of its parameter set. The model is designed for an emotional system, whose nodes represent emotional states, and connectivity between any two nodes represenst a natural phenomenon.

Let x_i, x_j and x_k be three representative emotional states that describe the concentration of individual emotions. Suppose x_j for some j co-operates with x_i, while x_k, for some k, competes with x_i. In other words, the growth rate of x_i will be accelerated with an increase in x_j, but will be decelerated with an increase in x_k. Assuming that there exist m number of co-operative emotional states like x_j and n number of competitive emotional states like x_k, we can represent the dynamic behavior of emotional state x_i by the following differential equation.

$$\frac{dx_i}{dt} = a_{ii}x_i\left(1 - \frac{x_i}{k}\right) + \sum_{\exists j} b_{ji}x_i(1 - \exp(-\beta_{ji}x_j)) - \sum_{\exists k} c_{ki}x_i(1 - \exp(-\lambda_{ki}x_k)) \tag{47}$$

The 1st term in the R.H.S. of (47) corresponds to self-growth of emotional state x_i. Here, a_{ii} denotes the inertial co-efficient that regulates the self-growth of x_i. The factor $(1 - x_i/k)$ is a controlling term that selects the sign of intrinsic growth rate a_{ii}. For instance, when $x_i < k$, the first term in the right hand side is positive, when $x_i = k$ it becomes zero, and when $x_i > k$, it becomes negative. In other words, x_i is allowed to increase up to a level of k, and a fall-off in the growth rate in x_i starts once it exceeds k. The second term represents the co-operation between emotion x_j and x_i for some j. It is indeed important to note that the second term takes into account the co-operation of x_i with a growing x_j. The third term on the other hand represents competition of x_i with growing x_k for some k. The parameters: β_{ji} and λ_{ki} control the growth of x_j and x_k respectively in the growth dynamics of x_i.

To illustrate the use of the proposed dynamics in emotion modeling, we consider the emotional states of a thalasemia patient's mother (Fig. 2.3) [14]. The dynamics

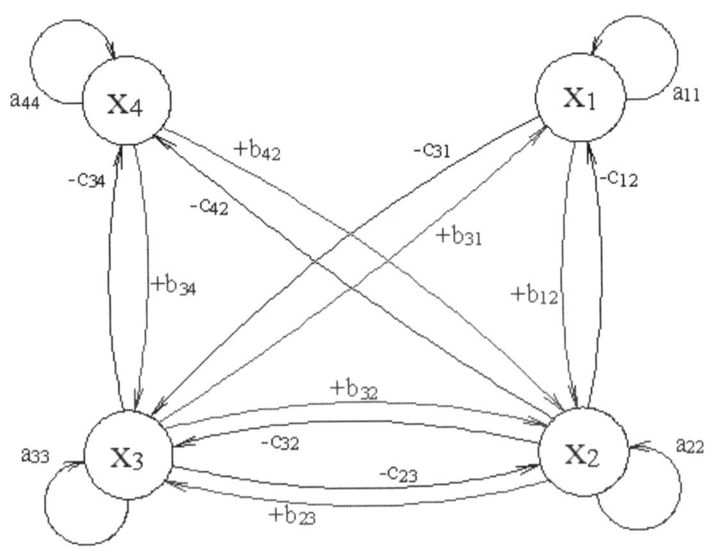

Fig. 2.3. The state transition diagram of emotional dynamics of a thalasemia patient.

of the system shown in Fig. 2.3 is presented here by equations (48)–(51).

$$\frac{dx_1}{dt} = a_{11}x_1\left(1 - \frac{x_1}{k}\right) + b_{31}x_1(1 - \exp(-\beta_{31}x_3)) - c_{21}x_1(1 - \exp(-\lambda_{21}x_2)) \tag{48}$$

$$\frac{dx_2}{dt} = a_{22}x_2\left(1 - \frac{x_2}{k}\right) + b_{12}x_2(1 - \exp(-\beta_{12}x_1)) + b_{32}x_2(1 - \exp(-\beta_{32}x_3)) + b_{42}x_2(1 - \exp(-\beta_{42}x_4)) - c_{32}x_2(1 - \exp(-\lambda_{32}x_3) \tag{49}$$

$$\frac{dx_3}{dt} = a_{33}x_3\left(1 - \frac{x_3}{k}\right) + b_{23}x_3(1 - \exp(-\beta_{23}x_2)) - c_{23}x_3(1 - \exp(-\lambda_{23}x_2)) - c_{13}x_3(1 - \exp(-\lambda_{13}x_1)) - c_{43}x_3(1 - \exp(\lambda_{43}x_4)) \tag{50}$$

$$\frac{dx_4}{dt} = a_{44}x_4\left(1 - \frac{x_4}{k}\right) + b_{34}x_4(1 - \exp(-\beta_{34}x_3)) - c_{24}x_4(1 - \exp(-\lambda_{24}x_2)) \tag{51}$$

where the parameters have their usual meaning as discussed above.

2.5.1. *Effect of parameter variation on the response of the cognitive dynamics of emotion*

In this section, we study the effect of variation in the parameters of the dynamics in the emotional response. We fix the parameters as listed in Tables 1 through 5, and report the effect of changing the parameters in the appropriate sections.

Table 2.1. List of parameters of equation (48).

a_{11}	b_{31}	c_{12}	β_1	λ_1
0.33	0.5	0.55	0.006	0.0033

Table 2.2. List of parameters of equation (49).

a_{22}	b_{12}	b_{32}	b_{42}	c_{32}	β_{12}	β_{32}	β_{42}	λ_{32}
0.002	0.2	0.2	0.3	0.4	0.001	0.001	0.001	0.005

Table 2.3. List of parameters of equation (50).

a_{33}	b_{23}	c_{23}	c_{13}	c_{43}	β_{23}	λ_{23}	λ_{14}	λ_{15}
0.2	0.4	0.5	0.45	0.3	0.005	0.005	0.005	0.005

Table 2.4. List of parameters of equation (51).

a_{44}	b_{34}	c_{24}	β_{34}	λ_{24}
0.1	0.4	0.35	0.001	0.001

Table 2.5. Initial values of the parameters.

$X_1(0)$	$X_2(0)$	$X_3(0)$	$X_4(0)$
500	800	900	700

Variation in a_{ii}

The a_{ii} plays an important role in the stabilization of the said emotional dynamics. It has been noted that keeping the other parameters of the dynamics constant, a large variation in a_{11} forces the system to a limit cyclic or chaotic situations. This observation, of course, is apparent from the dynamics. It is to be noted that an increase in a_{11} causes a positive feedback in x_1 through the first dynamics (48). The rapid growth in x_1 induces a growth in x_2 following equation (49). When c_{13} in Eqn. (50) is small, the growth in x_1 cannot influence x_3 significantly. Thus a large a_{11} pushes the dynamics towards chaos/limit cycles. On the other hand, when a_{22} or a_{33} increases, stabilization of the system is hampered. Further an increase in a_{44} increases the stability of the overall dynamics. The effect of changes in a_{22}, a_{33} and a_{44} on the dynamics, as discussed above, can be explained in a manner similar to the explanation of a_{11}. Fig. 2.4(a), (b), and, (c) illustrate the phenomena of chaotic, limit cyclic, stable behavior of the dynamics due to variation in a_{11}.

Variation in c_{ij}

Decreasing the value of c_{21} from 0.55 reduces oscillatory nature of the dynamics. When c_{21} attains 0.35, there exists an overshoot in the under-damped case, but oscillations die out. Apparently, this seems to be counter-intuitive, as a decrease in c_{21} causes a relative increase in the growth rate of x_1, causing over-arousal of emotion x_1. In fact, the same thing happens in the transient phase, but in the steady-state phase, the increased growth rate of x_1 suppresses the arousal of x_2, x_3 and x_4 states. Consequently, decreasing c_{21} stabilizes the behavior of all the emotional states. It has been noted that when c_{21} is decreased to 0.01, all the states return to the equilibrium state.

An increase in c_{21} pushes the dynamics towards limit cyclic behavior. An over-increase in c_{21} causes a large negative feedback to the growth rate in x_1, causing x_1 to gradually diminishing towards zero. It is, however, noteworthy that even when x_1 stabilizes to zero level, x_2, x_3, x_3 and x_4 continue to demonstrate limit cyclic behavior (Fig. 2.5).

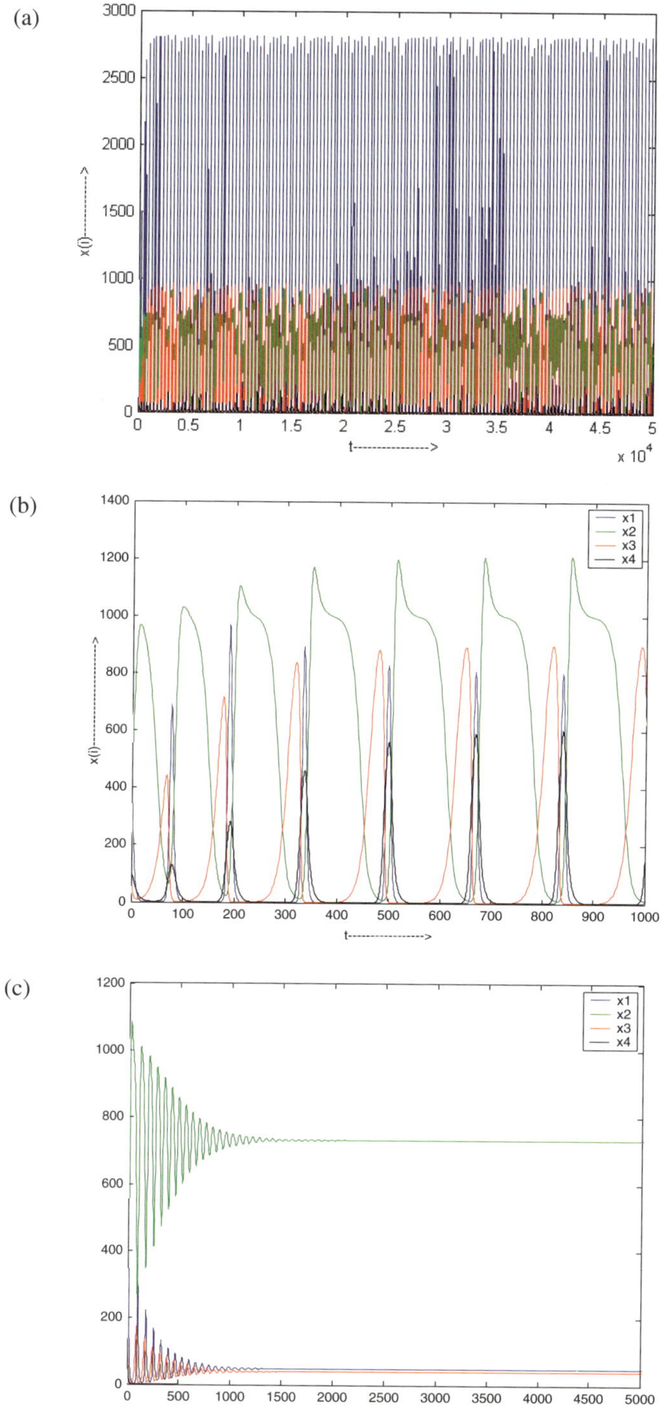

Fig. 2.4. Chaotic, limit cyclic and stable behavior in the emotional dynamics (48–51) for varying a_{11}.

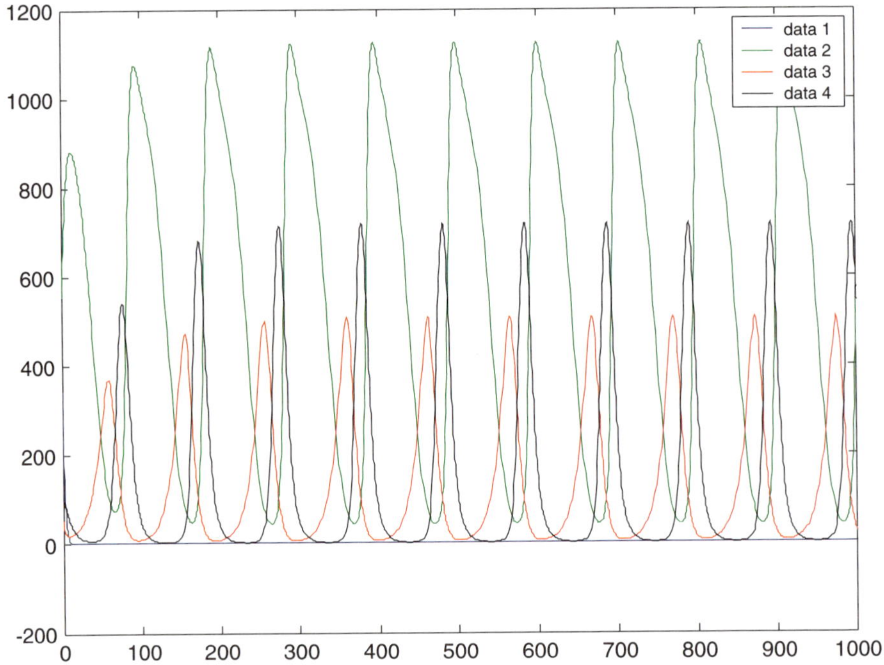

Fig. 2.5. Chaotic variation of all the emotional states due to increase in c_{12}.

Variation of b_{ij}

Computer simulation reveals that a decrease in b_{31} below 0.5, keeping all other parameters fixed, retains the oscillatory nature of the dynamics. When b_{31} is increased up to 0.9, the oscillation in all the emotional states, x_1, x_2, x_3, x_4 are sustained. When b_{31} is increased further to 1.0, the oscillations die out, and all the emotional states attain equilibrium. The phenomena for sustaining oscillation at small values of b_{ij} and maintaining equilibrium at higher value than b_{ij} are explained as follows.

When b_{31} is small (≤ 0.9), the positive feedback provided by the emotional states x_3 to x_1 is inadequate to maintain large amplitude in x_1 because of a large negative feedback from other emotional states. It is apparent that with a large initial value in x_2, the emotion level of x_1 has a significant phase delay of around 180°, causing x_1 to oscillate with low amplitude (around 500 unit). Naturally, the dynamics of x_3 and x_4, which are strongly influenced by x_2 have similar behavior like x_2. When b_{31} approaches 1, the dynamics of x_2 settles down with small amplitude of oscillation around 830 units, causing x_1, x_3 and x_4 to maintain low equilibrium levels depending on their feedback co-efficient from x_2 to x_1, x_3 and x_4 (Fig. 2.6). The effect of variations of a_{ii}, b_{ij}, c_{ij} for other values of i and j can be envisaged in a similar manner as discussed above.

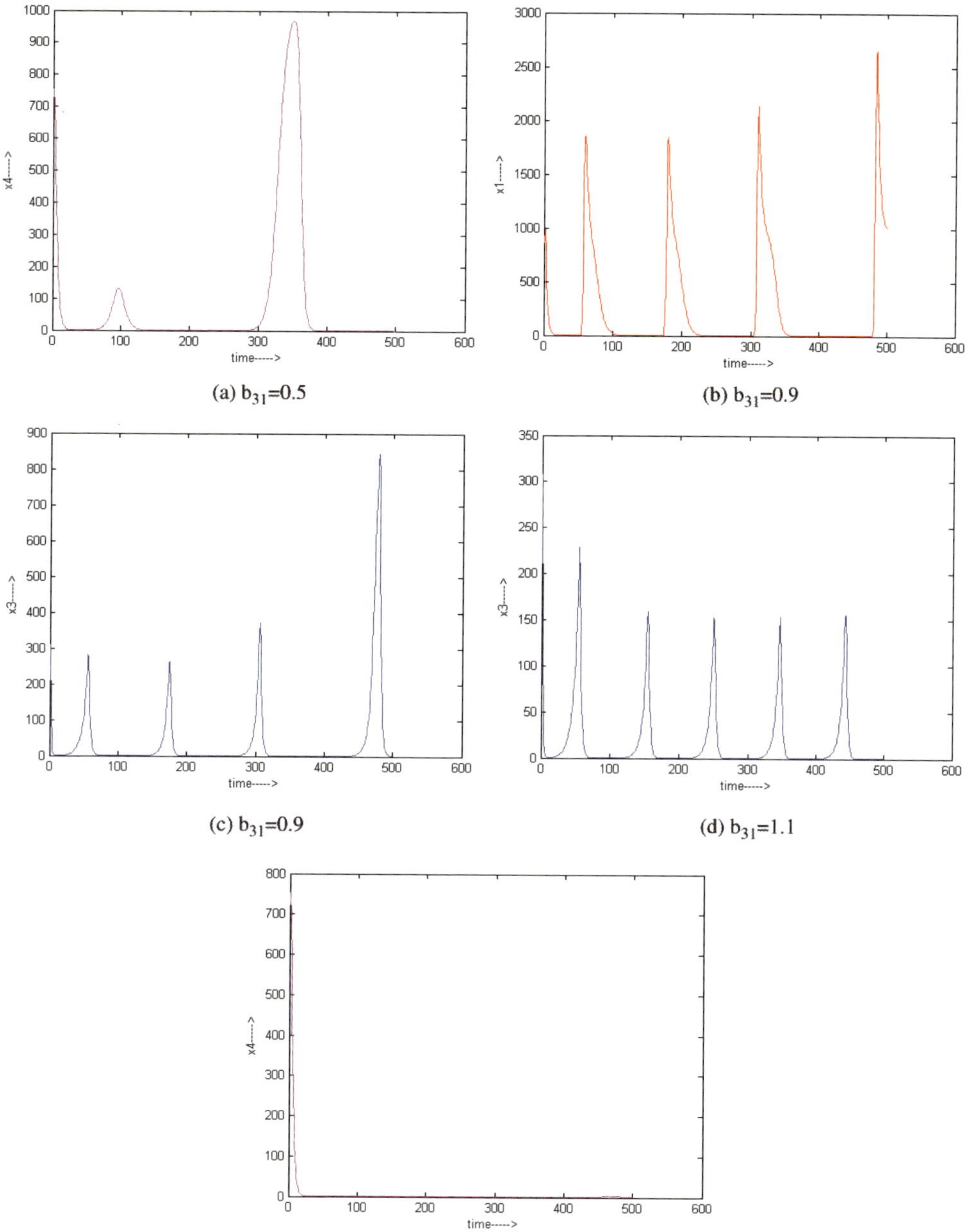

(a) b_{31}=0.5

(b) b_{31}=0.9

(c) b_{31}=0.9

(d) b_{31}=1.1

(e) b_{31}=1.1

Fig. 2.6. The effect of variation of b_{31} on the dynamics.

Chaotic Fluctuation in Emotional State

The dynamics of emotion exhibit stable, limit cyclic and chaotic behavior depending on the parameter set of the dynamics. When the intrinsic growth rate for state x_2, i.e. a_{22}, is decreased significantly, the dynamics describes a chaotic behavior in all the emotional states. Figure 2.7 (a), (b), (c), (d), describe the chaotic

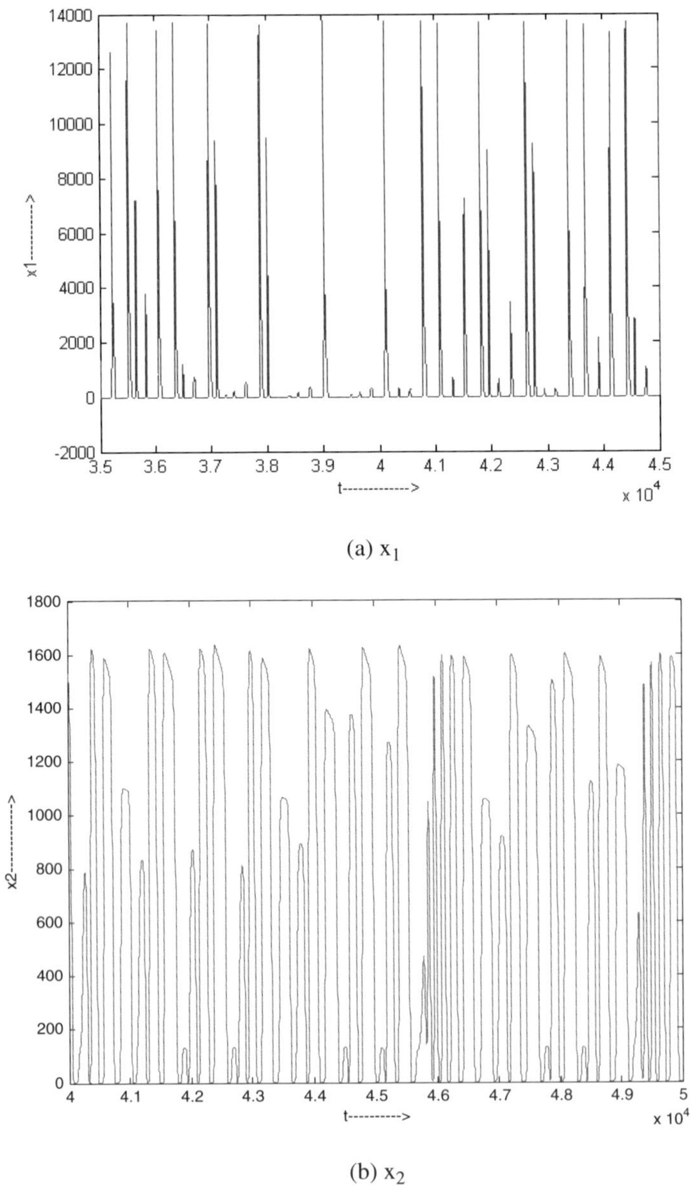

(a) x_1

(b) x_2

Fig. 2.7. Chaotic temporal variation of all the emotional states.

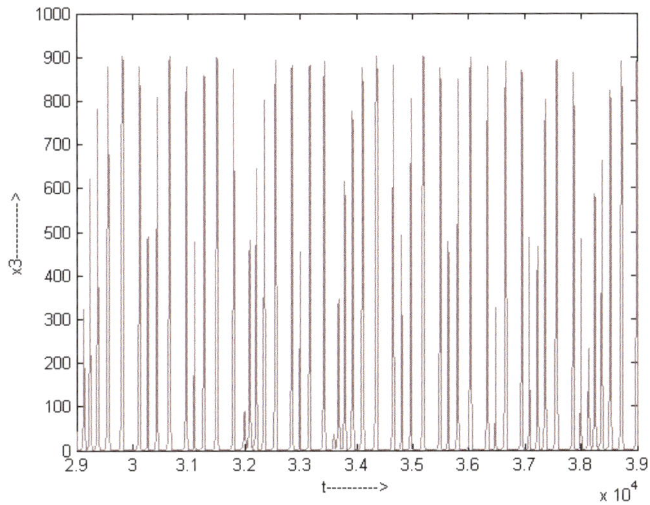

(c) x_3

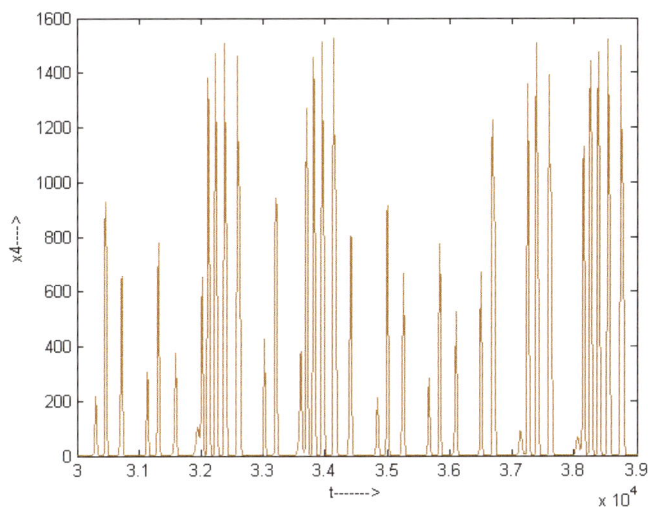

(d) x_4

Fig. 2.7. (*Continued*)

variations of individual emotional states. One simple way to represent chaotic dynamics is to draw a phase trajectory of 2 or 3 variables, when all of which have a chaotic variation over time. Figure 2.8 describes a phase trajectory of 2 emotional states. The phase portrait in the present context exhibits several annular rings of gradually changing dimensions. Such diagram usually provides a clear indication of chaos.

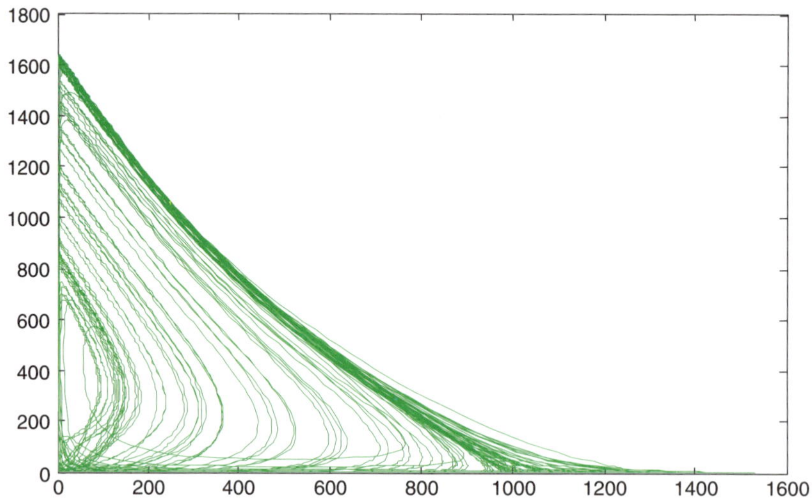

Fig. 2.8. Phase trajectory for chaotic variation of states: Sad (x_2) versus Anxious (x_2).

2.5.2. *Stability analysis of the proposed emotional dynamics by Lyapunov energy function*

Theorem 4. *The dynamics (52) is unconditionally stable and L (xi, xj) is the Lyapunov function to prove its asymptotic stability.*

$$\frac{dx_i}{dt} = a_{ii}x_i\left(1 - \frac{x_j}{k}\right) + \sum_{j=1}^{n} b_{ji}x_i(1 - e^{-\beta_{ji}x_j}) \quad \text{for } i = 1 \text{ to } n. \tag{52}$$

$$L(x_i, x_j) = -\left[\left(a_{ii}\frac{x_i^2}{2} - a_{ii}\frac{x_i^3}{3K}\right) + \sum_{j=1}^{n}\int_0^{x_i} b_{ji}x_i(1 - e^{-\beta_{ji}x_j})dx_j\right] \tag{53}$$

Proof. To prove that $L\ (x_i, x_j)$ is a Lyapunov energy function, we verify that

1. $L(0,0) = 0$,
2. $[L(x_i, x_j)]_{\substack{x_i \neq 0 \\ x_j \neq 0}} > 0$
3. $\frac{\partial L}{\partial x_i}$ and $\frac{\partial L}{\partial x_j}$ both exist,

Consequently, $L(x_i, x_j)$ is a Lyapunov energy function. Now, to show that dL/dt is unconditionally negative we evaluate

$$\begin{aligned} \frac{dL}{dt} &= \sum_{i=1}^{n} \frac{\partial L}{dx_i}\frac{dx_i}{dt} \\ &= -\sum_{i=1}^{n}\left\{a_{ij}x_i\left(1 - \frac{x_i}{k}\right) + \sum_{j=1}^{n} b_{ij}x_i(1 - e^{-\beta_j x_j})\right\}^2 < 0 \end{aligned} \tag{54}$$

Since $\frac{dL}{dt}$ is negative, the dynamics is unconditionally stable. □

To study the chaos in a dynamics $\frac{dx_i}{dt} = f(x_i, x_j, x_k)$, we define Lyapunov exponent

$$\lambda = \underset{N\to\infty}{Lt} \frac{1}{N}\sum_{n=1}^{N} \log_2 \frac{dx_{n+1}}{dx_n},$$

The dynamics is chaotic if $\lambda > 0$. It is apparent that if $\frac{dx_{n+1}}{dx_n} > 0\,\forall n$, then λ is essentially greater than 0.

Theorem 5. *The dynamics (55) is chaotic if*

$$a_{ii} + \sum_{j=1}^{n} b_{ji}\beta_{ji}x_j > 2a_{ii}\frac{x_i}{k} + \sum_{k=1}^{m} c_{ki}\lambda_{ki}x_k.$$

$$\frac{dx_i}{dt} = a_{ii}x_i(1 - \frac{x_i}{k}) + \sum_{j=1}^{n} b_{ji}x_i(1 - e^{-\beta_{ji}x_j}) - \sum c_{ki}x_i(1 - \exp(-\lambda_{ki}x_k)) \quad (55)$$

Proof. Representing

$$\begin{aligned}\frac{dx_i}{dt} &= \frac{x_i(t+1) - x_i(t)}{(t+1) - t}\\ &= x_i(t+1) - x_i(t),\end{aligned}$$

Substituting it in (55), we have:

$$x_i(t+1) = x_i(a_{ii} - \frac{a_{ii}x_i}{k} + 1) + \sum_{j=1}^{n} b_{ji}x_i(1 - e^{-\beta_{ji}x_j}) - \sum_{k=1}^{m} c_{ki}x_i(1 - e^{-\lambda_{ki}x_k})$$

$$\frac{dx_i(t+1)}{dx_i(t)} = a_{ii} - 2a_{ii}\frac{x_i}{k} + 1 + \sum_{j=1}^{n} b_{ji}x_i(1 - e^{-\beta_{ji}x_j}) - \sum_{k=1}^{m} c_{ki}x_i(1 - e^{-\lambda_{ki}x_k}) \quad (56)$$

Now expanding

$$\begin{aligned}1 - e^{-\beta_{ji}x_j} &\approx 1 - (1 + \beta_{ji}x_j)\\ &= \beta_{ji}x_j,\end{aligned} \quad (57)$$

and

$$\begin{aligned}1 - e^{-\lambda_{ki}x_k} &\approx 1 - (1 + \lambda_{ki}x_k)\\ &= \lambda_{ki}x_k,\end{aligned} \quad (58)$$

and substituting (57) and (58) in (56) we have

$$\frac{dx_i(t+1)}{dx_i(t)} = \left(a_{ii} - 2a_{ii}\frac{x_i}{k} + 1\right) + \sum_{j=1}^{n} b_{ji}\beta_{ji}x_j - \sum_{k=1}^{m} c_{ki}\lambda_{ki}x_k \quad (59)$$

Now, for chaos $\frac{dx_i(t+1)}{dx_i(t)} > 1$, which requires:

$$a_{ii} - 2a_{ii}\frac{x_i}{k} + \sum_{j=1}^{n} b_{ji}\beta_{ji}x_j - \sum_{k=1}^{m} c_{ki}\lambda_{ki}x_k > 0$$

or,

$$a_{ii} + \sum_{j=1}^{n} b_{ji}\beta_{ji}x_j > 2a_{ii}\frac{x_i}{k} + \sum_{k=1}^{m} c_{ki}\lambda_{ki}x_k. \tag{60}$$

□

It may be noted that the dynamics (55) will be limit cyclic if $\frac{dx_i(t+1)}{dx_i(t)} = 1$, which yields

$$a_{ii} + \sum_{j=1}^{n} b_{ji}\beta_{ji}x_j = 2a_{ii}\frac{x_i}{k} + \sum_{k=1}^{m} c_{ki}\lambda_{ki}x_k. \tag{61}$$

and stable when the equality in (61) is replaced by less than operator. Several parameter sets are identified to study the chaotic and stable behavior of the dynamics (55). Selection of suitable parameters can be done by satisfying the necessary conditions for stability, chaos and limit cycles as derived above.

Reasoning with the proposed dynamics calls for solving the equations (48–51). But a non-stabilized dynamics needs to be stabilized first for reasoning applications. This can be performed by the methodology introduced in the next section.

2.5.3. *A stabilization scheme for the mixed emotional dynamics*

The mixed emotional dynamics model presented by Equation (55) exhibits chaotic, limit cyclic and stable behavior for different parameter sets of the dynamics. In this section, we present a scheme for controlling damping (Fig. 2.9) for the proposed non-linear model of the emotional dynamics. Let $x_1(t)$, $x_2(t)$, $x_3(t)$ and $x_4(t)$

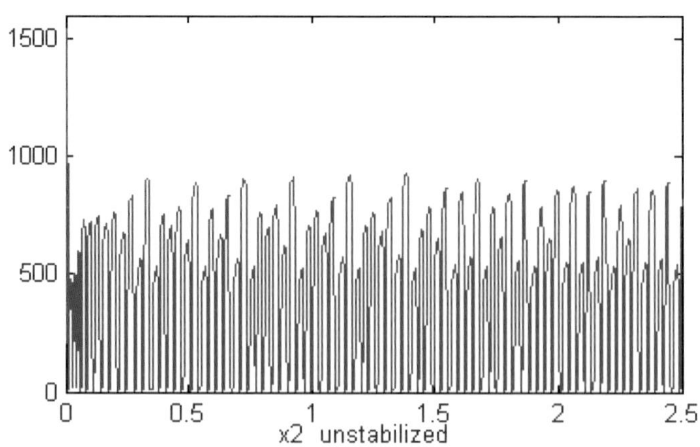

Fig. 2.9. The un-stabilized (chaotic) response of x_2.

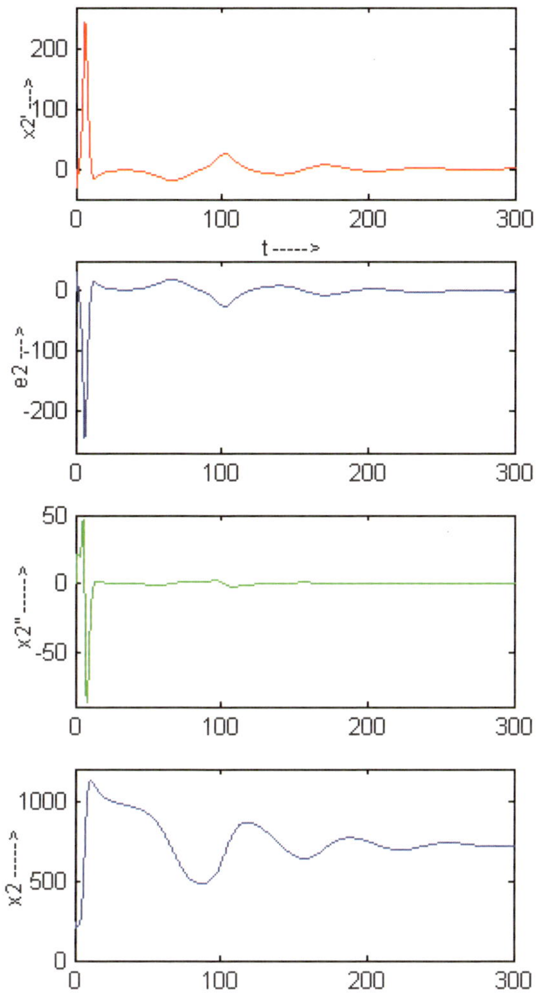

Fig. 2.10. The stabilized response of the emotion controller.

be four emotional states, exhibiting chaotic or limit cyclic behavior. Under this circumstance, we determine the damping level of the dynamics from its transient response. If the predicted damping level is low, we call it under-damped, and we attempt to increase the intrinsic growth rate a_{ii} of the dynamics. On the other hand, if the dynamics is found over-damped, we need not generate any control signal.

Prediction of the damping level from the response of the dynamics, indeed, is an open research problem until this date. A classical model for damping level prediction proposed by Konar and Roy [35] in population control of a predator-prey system, however, seems to be useful in the present context. The damping level predicted here by examining the sign constraint is presented next.

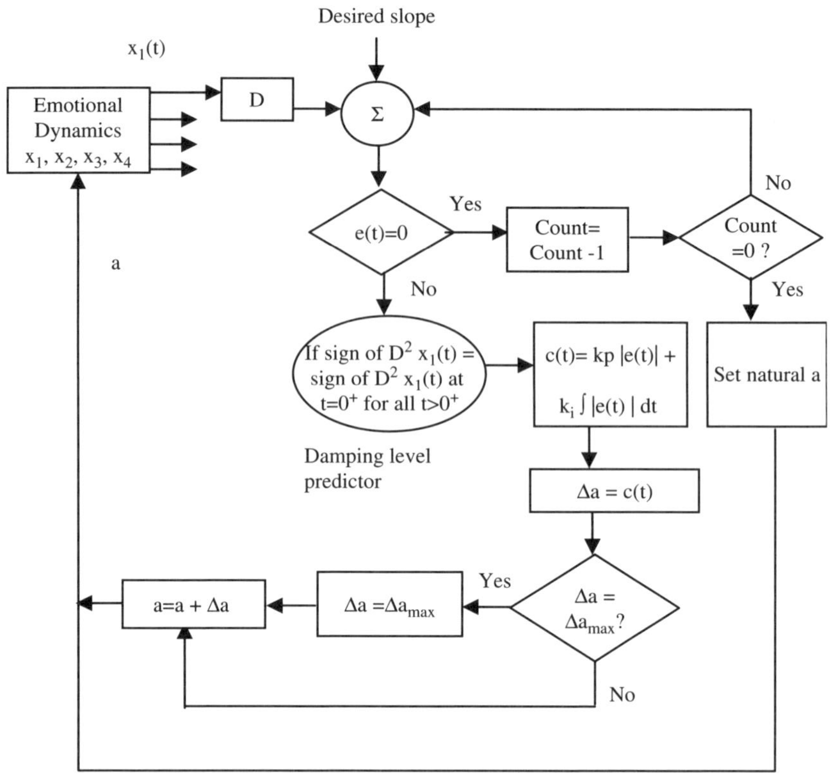

Fig. 2.11. A scheme of controlling damping in emotional dynamics for stabilization.

Sign constraint: if sign of $x_i(t)$ = sign of $x_i(t = 0)$, $\forall t \geq 0$ then the dynamics is said to be under-damped, else it is over-damped.

The above damping level prediction rule, as proposed by Konar and Roy, has been employed here to determine the damping level of the non-linear emotional dynamics. In case it is found under-damped, a proportional plus integral (PI) type controller is employed to control the parameter a_{ii} of the emotional dynamics. In the process of generating control signal, special emphasis is given to keep the dynamics *minimally over-damped*. The phrase *minimally over-damped* refers to the minimum level of damping that causes an over-damped response of the dynamics.

In Fig. 2.11 we have 3 major modules:

i) an emotional dynamics,
ii) a damping level prediction, and
iii) a PI type controller used for enrichment of the intrinsic growth rate a_{ii}, so as to control the damping level of the dynamics.

A computer simulation of the proposed damping level prediction and stabilization scheme of emotional system is undertaken. Simulation results reveal that an under-damped chaotic emotional dynamics (Fig. 2.9) can be stabilized by artificially increasing the intrinsic growth rate a_{11} of the proposed system. The resulting response (Fig. 2.10) exhibits a significant decrease in percentage of peak overshoot.

2.6. Conclusions

The chapter introduced three different types of recurrent dynamics to describe cognitive reasoning system. The first system considers a temporal dynamics of a propositional reasoning system, and demonstrated a method to analyze the stable points of the system. The second problem addressed an encoding and recall paradigm to represent a cognitive map. A detailed analysis of the stability in the encoding phase is given to determine the parametric condition for convergence of the encoding cycle. The recall model is used for automatic reasoning with the encoded set of weights of the cognitive map.

The third problem undertaken here is concerned with a specialized cognitive structure that connects the states of emotion of a subject through incidental information. The dynamics of the proposed cognitive structure of emotion is analyzed to determine the condition of chaos and stability of the dynamical system. Numerical illustrations have been given to demonstrate the stable, limit cyclic and chaotic behavior of the emotional dynamics. The chapter also proposed a novel method to stabilize the emotional dynamics, when it suffers from chaotic or limit cyclic behavior. Computer simulation to control the dynamic behavior of the emotional cognitive structure has also been studied by adapting a parameter of the dynamics.

The principles of analysis given here can be employed in the next generation reasoning systems for electronic commerce, where the buyer's emotional states can be predicted from his previous transactions and current dealings. The predicted status of the buyer can be used for future negotiation and price fixation. The proposed method of analysis can also be used for predicting the opponent's attitude in electronic/computer games.

References

1. Axelrod, R., *Structure of Decision: the Cognitive Maps of Political Elites*, Princeton University Press, Princeton, NJ.
2. Baddeley, A. D., "The fractionation of human memory," *Psychological Medicine*, vol.14, pp. 259–264, 1984.
3. Beauregard, M., Levesque, J., and Paquette V., "Neural Basis of Conscious and Voluntary Self-Regualtion of Emotion," in *Consciousness, Emotional Self-regulation and the Brain*, Beauregard, M. (Ed.), Benjamins Pub., 2004.
4. Ben- Ari, M., *Mathematical Logic for Computer Science*, Prentice – Hall, Englewood Cliffs, NJ, pp. 200–241, 1993.
5. Besnard, P., *An Introduction to Default Logic*, Springer Verlag, Berlin, pp. 27–35, pp. 163–177, 1989.
6. Bharick, H. P., "Semantic memory content in permastore: Fifty years of memory for Spanish learned in school," *Journal of Experimental Psychology: General*, vol. 120, pp. 20–33, 1984.

7. Bhowmik, P., Das, S., Nandi, D., Chakraborty, A., Konar, A., & Nagar, A. K., "Electroencephalographic Signal Based Clustering of Stimulated Emotion Using Duffing Oscillator", *The 3rd International Multi-Conference on Engineering and Technological Innovation, IMETI 2010*, June 29th–July 2nd, Orlando, Florida, USA.
8. Bhowmik, P., Das, S., Nandi, D., Chakraborty, A., Konar, A., & Nagar, A. K., "Emotion Clustering from Stimulated Electroencephalographic Signals Using Duffing Oscillator", *The International Journal of Computers in Healthcare*, Volume 1, Issue 1, pp. 66–85.
9. Bolc, L. and Szalas, A. (Eds.), *Time and Logic: A Computational Approach*, UCL Press, London, pp. 1–50, 1995.
10. Buchel, C., Dolan, R. J., Armony, J.L., and Friston, K.J. "Amygdala-Hippocampal Involvement in human Aversive Trace Conditioning Revealed through Event-Related fMRI," *The J. of Neuroscience*, 19(24), 1999.
11. Cambel, A. B., *Applied Chaos Theory: A Paradigm for Complexity*, Academic Press, San Diego, CA, 1923.
12. Chakraborty, A. and Konar, A., "Chaotic Emotional Dynamics Realized with Recurrent Neural Network," Organized by Computer Society of India, Calcutta- 2005.
13. Chakraborty, A., Konar, A., Chakraborty, U. K., and Chatterjee, A. (2009). Emotion recognition from facial expressions and its control using fuzzy logic, *IEEE Transactions on Systems, Man, and Cybernetics, Part A: Systems and Humans*, v.39 n.4, p.726–743
14. Chakraborty, A., Ghosh, M., Acharya, A., Konar, A., and Nagar, A. (2009), "A recurrent neural dynamics for parameter estimation of mixed emotions from facial expressions of the subject," *Proc. of the Int. Joint Conf. on Neural Networks*, Atlanta.
15. Chakraborty, A. and Konar, A., *Emotional Intelligence: A Cybernetic Approach, Springer*, 2009.
16. Chang, T. M., "Semantic memory: facts and models," *Psychological Bulletin*, vol. 99, pp. 199–220, 1986.
17. Downs, R. M. and Davis, S., *Cognitive Maps and Spatial Behavior: Process and Products*, Aldine Publishing Co., 1973.
18. Duncan, E. M. and Bourg, T., "An examination of the effects of encoding and decision processes on the rate of mental rotation," *Journal of Mental Imagery*, vol. 7, pp. 33–56, 1983.
19. Durka, P., *Matching Pursuit and Unification in EEG Analysis*, Artech House, Norwood, MA, 2007.
20. Farah, M. J., "Is visual imagery really visual? Overlooked evidence from neuropsychology," *Psychological Review*, vol. 95, pp. 307–317, 1988.
21. Frackowiak, R. S. J., Friston, K. J., Frith, C. D., Dolan, R. J., Price, C. J., Zeki, S., Ashburner, J., and Penny, W., *Human Brain Function*, Elsevier, 2005.
22. Ghosh, M., Chakraborty, A., Konar, A. and Nagar, A., Detection of Chaos and Limit Cycles in Emotional Dynamics from the Facial Expressions of the Stimulated Subjects, In *Computational Intelligence, Control and Computer Vision in Robotics and Automation*, Subudhi, B. (Ed.), Narosa Publisher, New Delhi, 2009.
23. Goldenberg, G., Podreka, I., Steiner, M., Suess, E., Deecke, L. and Willmes, K., Pattern of regional cerebral blood flow related to visual and motor imagery, Results of Emission Computerized Tomography, In *Cognitive and Neuropsychological Approaches to Mental Imagery*, Denis, M, Engelkamp, J. and Richardson, J. T. E. (Eds.), Martinus Nijhoff Publishers, Dordrecht, The Netherlands, pp. 363–373, 1988.
24. Greeno, J. G., Process of understanding in problem solving, in *Cognitive Theory*, Castellan, N. J., Pisoni, Jr. D. B. and Potts, G. R. (Eds.), Hillsdale, Erlbaum, NJ, vol. 2, pp. 43–84,1976.

25. Gross, C. G. and Zeigler, H. P., *Readings in Physiological Psychology: Learning and Memory*, Harper & Row, NY, 1969.
26. Grossberg, S. and Seidman, D., Neural dynamics of autistic behaviors: Cognitive, emotional and timing substrates, *Psychological Review*, vol. 113, no. 3, pp. 483–525, July 2006.
27. Guyton, B. C. and Hall, J. E., *Textbook of Medical Physiology*,
28. Harter, D. and Kozma, R., Chaotic neurodynamics for autonomous agents, *IEEE Trans. Neural Networks*, vol. 16, no. 3, May 2005.
29. Jolicoeur, P., The time to name disoriented natural objects, *Memory and Cognition*, vol. 13, pp. 289–303, 1985.
30. Kintsch, W., The role of knowledge in discourse comprehension: A construction-integration model, *Psychological Review*, vol. 95, pp. 163–182, 1988.
31. Kintsch, W. and Buschke, H., Homophones and synonyms in short term memory, Journal of Experimental Psychology, vol. 80, pp. 403–407, 1985.
32. Konar, A., *Artificial Intelligence and Soft Computing: Behavioral and Cognitive Modeling of the Human Brain*, RC Press, Boca Raton, Florida, 1999.
33. Konar, A. and Pal, S., Modeling cognition with fuzzy neural nets, In *Fuzzy Logic Theory: Techniques and Applications*, Leondes, C. T. (Ed.), Academic Press, NY, 1999.
34. Konar, A. and Jain, L., *Cognitive Engineering*, Springer-Verlag, Heidelberg, 2005.
35. Konar, A. and Roy, A. B., Population drift and control of damping in a predator-prey system, *IMA J. of Mathematics Applied in Medicine and Biology*, vol. 7, pp. 245–259, 1990.
36. Kosko, B., *Fuzzy Engineering*, Prentice-Hall, NJ, ch. 15, pp. 499–528, 1996.
37. Kosko, B., "Fuzzy cognitive maps," *Int. J. of Man-Machine Studies*, vol. 24, pp. 65–75, 1986.
38. Kosslyn, S. M., "Aspects of cognitive neuroscience of mental imagery," *Science*, vol. 240, pp. 1621–1626, 1988.
39. Kosslyn, S. M., Mental imagery, In *Visual Cognition and Action: An invitation to Cognitive Science*, Osherson, D. N. and Hollerback, J. M. (Eds.), vol. 2, pp. 73–97, 1990.
40. Kuo, B. C., *Digital Control Systems*, Holt-Saunders International Editions, Japan, Ch-5, pp. 267–302, 1981.
41. Pal, S. and Konar, A., "Cognitive reasoning with fuzzy neural nets," *IEEE Trans. on Systems, Man and Cybernetics*, Part - B, August, 1996.
42. Zhang, W. R., "NPN fuzzy sets and NPN qualitative algebra: a computational network for bipolar cognitive modeling and multi-agent decision analysis," *IEEE Trans. Systems, Man and Cybernetics*, vol. 26, no. 4, 1996.
43. Zhang, W. R., Chen, S. S. and Bezdek, J. C., "Pool2: a generic system for cognitive map development and decision analysis," *IEEE Trans. on Systems, Man and Cybernetics*, vol. 19, no. 1, pp. 31–39, 1989.
44. Zhang, W. R. "Pool- a semantic model for approximate reasoning and its application in decision support," *Management Information System*, vol. 3, no. 41, pp. 65–78, 1986.
45. Zhang, W. R., Chen, S. S., Wang, W. and King, R., "A cognitive map based approach to the co-ordination of distributed co-operative agents," *IEEE Trans. on Systems, Man and Cybernetics*, vol. 22, no. 1, pp. 103–114, 1992.

Chapter 3

SOME STUDIES ON DATA MINING

Dilip Kumar Pratihar
Department of Mechanical Engineering
Indian Institute of Technology,
Kharagpur-721302, West Bengal, India
dkpra@mech.iitkgp.ernet.in

This chapter deals with data mining, whose aim is to extract useful information from a data set pertaining to input–output relationships of a process. A few dimensionality reduction techniques are discussed, which aim to map higher dimensional data into either 2-D or 3-D for visualization. The principles of some clustering techniques have been explained, which work based on the concept of similarity between two data points. The method of conventional regression analysis has been discussed, which can tackle the problem of forward mapping efficiently. However, it may not be always possible to carry out reverse mapping with the help of the derived regression equations. It has been felt that fuzzy logic- and neural network-based approaches can solve the problems of both forward as well as reverse mappings efficiently. Moreover, the need for design and development of an intelligent data miner has been realized recently.

3.1. Introduction

We, human beings, have a natural thirst to know input–output relationships of a process. Data mining aims to extract useful information from a data set containing input–output relationships of any process. Classification and prediction are two forms of data analysis, which are used to extract useful information from a dataset. Classification deals with discrete response, whereas predictive model works on continuous response. Classification models (also known as classifiers), such as decision-tree classifier (Quinlan, 1986), Bayesian classifier (Domingos and Pazzani, 1997), rule-based classifier (Lawrence and Wright, 2001), neural networks (Lacher *et al.*, 1995), support vector machines (Boser *et al.*, 1992), k-nearest-neighbor classifier (Fix and Hodges, 1951), genetic algorithms (Holland, 1975), fuzzy sets (Zadeh, 1965), and others, help us to categorize the data into some groups. On the other hand, predictive models developed based on regression analysis aim to determine input–output relationships of the data set. It is necessary to establish the said relationships in both forward as well as reverse directions to automate the process. The prime goal of data mining is to determine the above relationships in both the said directions as accurately as possible. In forward mapping, output variables of the process (also known as the responses) are expressed as the functions of input variables (also called the process parameters). On the other hand, the process parameters are represented as the functions of the responses in reverse mapping. To automate any process, it may be required to know its input–output relationships in both forward as well as reverse directions, on-line.

Regression analysis can be carried out response-wise using input–output data of a process collected in some fashions, such as full-factorial design, fractional-factorial design, central composite design, and others (Montgomery, 2003). In forward mapping, both linear as well as non-linear response equations can be obtained using the principle of a least square error technique (Montgomery, 2003). As the regression analysis is conducted response-wise, it may not be possible to capture the dynamics of the process accurately. Moreover, the derived transformation matrix to express the responses as the functions of input process parameters may not be invertible always and consequently, it becomes impossible to carry out the reverse mapping. Realizing the above difficulties, soft computing-based approaches (Pratihar, 2008) have also been tried recently to solve the said problems (Parappagoudar *et al.*, 2008). Soft computing is a family consisting of fuzzy logic (Zadeh, 1965), neural networks (McCulloach and Pitts, 1943), genetic algorithms (Holland, 1975) and others, and their various possible combinations.

Visualization of input–output space of a process may sometimes be useful to extract necessary information from the data set. However, we, human beings, cannot visualize more than 3-D space. Higher dimensional (that is, more than 3-D) data may be mapped into either 2-D or 3-D for visualization. Several methods had been proposed by various researchers for the said purpose. Those methods include Sammon's non-linear mapping (Sammon, 1969), VISOR algorithm (Konig *et al.*, 1994), Self-Organizing Map (SOM) (Kohonen, 1995), and others.

Some sort of similarities may exist among the data points (representing input–output relationships of a process) present in a set. These data can be clustered based on their similarity values for deriving useful information from them. The clusters may be either hard or fuzzy in nature. There exist a large number of clustering techniques, namely fuzzy C-means algorithm (Bezdek, 1973), entropy-based algorithm (Yao *et al.*, 2000), and others.

Most of the real-world problems are non-linear in nature. However, the degree of nonlinearity may not be the same over the entire range of the variables. Input–output relationships of a process may sometimes be local in nature. Thus, a particular set of input–output relationships may not hold good for the entire range of the variables. The data may be clustered into a number of groups (also called the clusters) depending on their similarities among them and the input–output relationships can be developed cluster-wise (Ganjigatti *et al.*, 2007).

The performance of the above data miner depends on a number of parameters related to clustering and reasoning. Thus, an intelligent data miner should be evolved to establish the said relationships efficiently in both the directions. It may be treated as one of the key steps towards automating a process. Soft computing (Pratihar, 2008) could be a potential tool to be used to reach the said milestone, as it can handle imprecision and uncertainty inherent to input–output data of a process.

The present chapter deals with statistical regression analysis of the data, study on dimensionality reduction techniques, clustering of the data, and others. Moreover, the need for the design and development of an intelligent data miner has been realized at the end.

3.2. Classification Tools

Classification is done based on supervised learning, for which the training set is to be known before-hand. The working principles of some classifiers are discussed below in brief.

A decision tree consists of a starting (also known as root) node, some internal nodes and a few branches, and some terminal nodes, which are put in the form of a flowchart-like structure. Each internal node offers a test on an attribute and each branch coming out from an internal node represents the outcome of the test. The terminal nodes carry information of the class labels.

Bayesian classifier works based on Bayes' theorem. Its aim is to determine a probability that a given training data set belongs to a class.

A rule-based classifier uses a set of IF-THEN rules. An IF-THEN rule is written as follows:

IF condition THEN action

The IF and THEN parts are called antecedent and consequent, respectively. It might have a better interpretability compared to the decision tree has.

A multilayer feed-forward neural network trained by a back-propagation algorithm can also be used as a classifier. It consists of three layers, namely input, hidden and output; and each layer contains a number of artificial neurons. During training, the connecting weights between input and hidden layers and those between hidden and output layers are updated. It has the ability to classify inherent patterns in a data set. Its interpretability is poor compared to that of a rule-based classifier.

A support vector machine is a powerful classifier. It uses a non-linear mapping first to transform the data to be classified into a higher dimension. It then searches for linear optimal separating hyperplane, which separates the data from one class to another. This algorithm is computationally expensive.

K-nearest-neighbor classifier works based on a comparison of test data with the training data. In order to determine k-nearest neighbors of the test data, their Euclidean distance values from the training data are considered. This algorithm is computationally expensive. Moreover, it may have poor accuracy for a noisy data.

Genetic algorithms (GAs) are population-based search and optimization tools, which work based on the mechanics of natural genetics and Darwin's principle of natural selection. This algorithm starts with a population of initial solutions selected at random and these solutions are updated iteratively using the operators like reproduction, crossover, mutation, and others. A binary-coded GA may be used to solve the classification problems, in which each rule can be represented by a string of bits. In this representation, the leftmost bits represent attributes and the rightmost bit denotes the class. The GA will be able to solve the classification problem through a number of iterations (also known as generations).

A fuzzy set is a set with vague or imprecise boundaries. It is a powerful tool for dealing with imprecision and uncertainties. The concept of fuzzy sets has been used to develop rule-based systems for classification.

3.3. Statistical Regression Analysis

Conventional regression analysis aims to determine input–output relationships of a process involving multiple variables. It is carried out using the experimental data of a process collected according to a well-planned statistical design of experiments (DOE), so that the information related to whole range of the variables can be captured with the help of a minimum number of experiments. Regression analysis may be either linear or non-linear in nature. The following steps are used to establish input–output relationships of a process:

1. Identification of input process parameters along with their feasible ranges and the responses (that is, output variables).
2. Selection of a suitable design of experiments.
3. Experimental data collection according to the design of experiments.
4. Regression analysis to establish input–output relationships of the process.
5. Checking adequacy of the model using analysis of variance.
6. Studying the effects of input variables on the responses.
7. Testing the performance of the developed model on some new cases.

Several designs of experiments are available in the literature (Montgomery, 2003) and some of those are discussed below.

3.3.1. *Design of experiments*

Statistical design of experiments is nothing but an appropriate planning of experiments, so that an adequate amount of data can be collected for conducting the regression analysis. A few commonly used designs, such as full-factorial, fractional-factorial, central composite designs are explained below.

3.3.1.1. *Full-factorial design of experiments*

Let us consider three factors (that is, input process parameters), such as A, B and C; each having two levels, namely high and low denoted by + and −, respectively. As there are three factors, the space of interest can be represented by the volume of

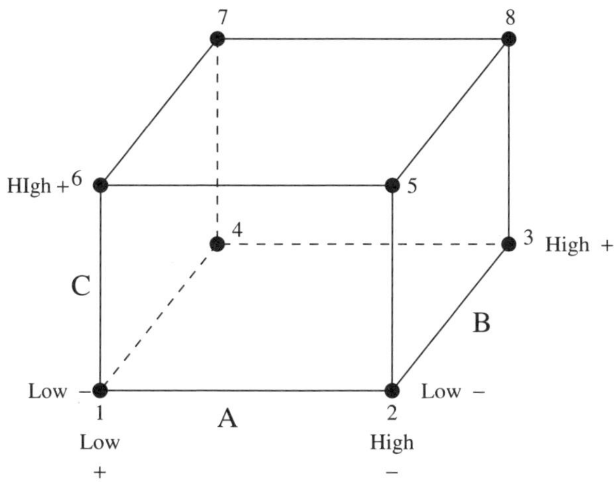

Corner points	Symbols
1	(–, –, –)
2	(+, –, –)
3	(+, +, –)
4	(–, +, –)
5	(+, –, +)
6	(–, –, +)
7	(–, +, +)
8	(+, +, +)

Fig. 3.1. Full-factorial design of experiments.

a cube, as shown in Fig. 3.1. According to this full-factorial design, experiments are to be carried out for $2^3 = 8$ different combinations of input process parameters represented by eight corners of the cube. For example, corner 1 of the cube indicates a particular combination of the factors, each set at its low level $(-)$. Thus, the first corner of the cube shown in the said figure represents a combination $(-,-,-)$. Similarly, the second, third, fourth, fifth, sixth, seventh and eight corners are denoted by the symbols $(+,-,-)$, $(+,+,-)$, $(-,+,-)$, $(+,-,+)$, $(-,-,+)$, $(-,+,+)$ and $(+,+,+)$, respectively.

Experimental data collected according to the above two-level full-factorial design can be utilized to carry out linear regression analysis involving both main as well as interaction terms. Thus, the response will be represented with the help of the main factors and their interaction terms. However, it will not be able to capture the curvature effect, if any, as it is a two-level design.

3.3.1.2. *Central composite design of experiments*

Let us consider central composite design of experiments for a process involving three input process parameters (also known as factors) and one response. Here, three levels are considered for each factor, such as high, low and middle denoted by $+$, $-$ and 0,

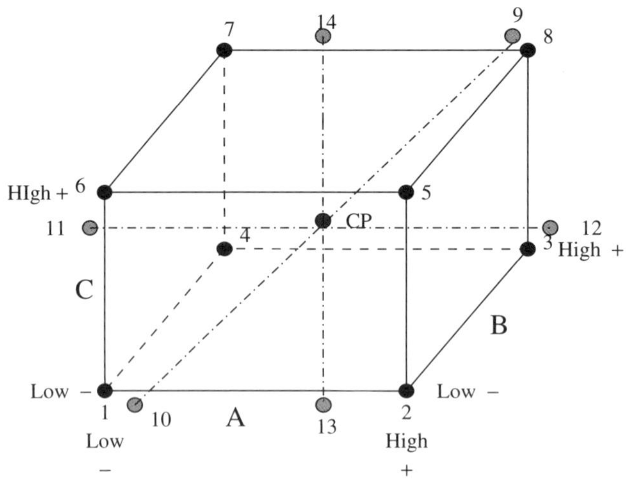

Corner points	Symbols
1	(−, −, −)
2	(+, −, −)
3	(+, +, −)
4	(−, +, −)
5	(+, −, +)
6	(−, −, +)
7	(−, +, +)
8	(+, +, +)

Star points	Symbols
9	(0, +, 0)
10	(0, −, 0)
11	(−, 0, 0)
12	(+, 0, 0)
13	(0, 0, −)
14	(0, 0, +)
Center point, CP	(0, 0, 0)

Fig. 3.2. Central composite design of experiments.

respectively. The design matrix consists of corner points, star points and center point (generally considered for a number of times), which can be represented with the help of a cube also, as shown in Fig. 3.2. Eight corner points of the cube representing the full-factorial design, such as $(-,-,-)$, $(+,-,-)$, $(+,+,-)$, $(-,+,-)$, $(+,-,+)$, $(-,-,+)$, $(-,+,+)$ and $(+,+,+)$ have been considered. Star points are assumed to be located at a distance α_d from the center point (0,0,0). The value of α_d generally varies in the range of $(1,K^{0.5})$, where K represents the number of factors. Assuming $\alpha_d = 1.0$, six star points, such as 9-th, 10-th, 11-th, 12-th, 13-th and 14-th can be represented as (0,+,0), (0,−,0), (−,0,0), (+,0,0), (0,0,−) and (0,0,+), respectively. Moreover, as a thumb rule, three to five runs are taken at the center point (0,0,0). If three center points are considered, the design matrix will consist of $8+6+3=17$ combinations of the input parameters.

3.3.2. *Regression analysis*

Both linear as well as non-linear regression analyses may be conducted to establish input–output relationships of a process, as explained below.

3.3.2.1. *Linear regression analysis*

It is carried out to represent the response of a process in terms of its input process parameters. For simplicity, let us consider three input process parameters expressed in their coded terms (X_1, X_2, X_3) and one response, that is, Y. The input–output relationships can be expressed as follows:

$$Y = \beta_0 + \beta_1 X_1 + \beta_2 X_2 + \beta_3 X_3 + \beta_4 X_1 X_2 + \beta_5 X_1 X_3 + \beta_6 X_2 X_3 + \beta_7 X_1 X_2 X_3 + \varepsilon, \tag{1}$$

where ε denotes the error in fitting; β values represent the coefficients to be determined utilizing the principle of least square error in fitting; X_1X_2, X_1X_3 and X_2X_3 are two-factor interaction terms; $X_1X_2X_3$ indicates the three-factor interaction term.

3.3.2.2. *Non-linear regression analysis*

Non-linear relationship of the response with the input process parameters can also be represented as given below.

$$Y = \beta_0 + \sum_{i=1}^{k} \beta_i X_i + \sum_{i=1}^{k} \beta_{ii} X_i^2 + \sum\sum_{i<j} \beta_{ij} X_i X_j + \varepsilon, \tag{2}$$

where ε represents the error in fitting. The above equation contains linear terms like $X_1, X_2, \ldots, X_k$; square terms, such as $X_1^2, X_2^2, \ldots, X_k^2$; interaction terms like $X_1X_2, X_1X_3, \ldots, X_{k-1}X_k$. The coefficients, that is, β values can be determined using the principle of least square error.

3.3.3. *Adequacy of the model*

Once the model has been built, its adequacy can be checked through the Analysis of Variance (ANOVA) (Montgomery, 2003). The health of the developed model is checked in terms of correlation co-efficient. The model is said to be adequate, if its correlation coefficient turns out to be close to 1.0. Significance test is carried out to identify the significant terms for a response. Moreover, the contributions of various factors towards a particular response can be estimated through either linear plots (in case of linear regression model) or surface plots (in case of non-linear regression model).

3.3.4. *Drawbacks*

Forward mapping, in which the responses of a process are expressed in terms of input process parameters, can be done efficiently using the principle of above regression analysis. However, it may not be always possible to tackle the problem of reverse mapping, as the transformation matrix may become singular. In regression analysis, input–output relationship is determined response-wise. Thus, it may not be able to capture the dynamics of the process completely.

3.4. Dimensionality Reduction Techniques

Higher dimensional data can be mapped into $2-/3-D$ for visualization as explained below.

3.4.1. *Sammon's non-linear mapping (Sammon, 1969)*

It is a distance preserving technique of non-linear mapping, where the error in mapping is minimized using a steepest descent method iteratively.

Let us consider N points in an L-dimensional ($L > 3$) space denoted by X_i, where $i = 1, 2, 3, \ldots, N$. These N-points are to be mapped from $L - D$ space to $2 - D$ plane or $3 - D$ space, represented by Y_i, where $i = 1, 2, 3, \ldots, N$. The scheme is shown in Fig. 3.3.

The N points in $L - D$ space are expressed as follows:

$$X_1 = \begin{bmatrix} x_{11} \\ x_{12} \\ \cdot \\ \cdot \\ x_{1L} \end{bmatrix}, \quad X_2 = \begin{bmatrix} x_{21} \\ x_{22} \\ \cdot \\ \cdot \\ x_{2L} \end{bmatrix}, \ldots, \quad X_N = \begin{bmatrix} x_{N1} \\ x_{N2} \\ \cdot \\ \cdot \\ x_{NL} \end{bmatrix}.$$

Similarly, the N-points in $2 - D$ plane or $3 - D$ space can be represented like the following.

$$Y_1 = \begin{bmatrix} y_{11} \\ \cdot \\ \cdot \\ \cdot \\ y_{1D} \end{bmatrix}, \quad Y_2 = \begin{bmatrix} y_{21} \\ \cdot \\ \cdot \\ \cdot \\ y_{2D} \end{bmatrix}, \ldots, \quad Y_N = \begin{bmatrix} y_{N1} \\ \cdot \\ \cdot \\ \cdot \\ y_{ND} \end{bmatrix}.$$

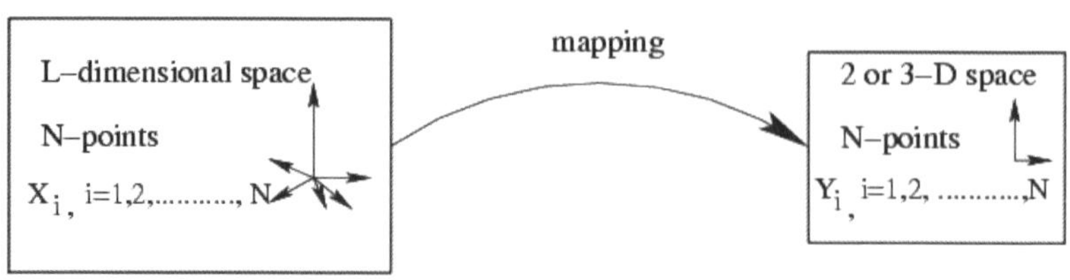

Fig. 3.3. Mapping from L-dimensional space to $2 - D$ plane or $3 - D$ space using NLM (Patihar, 2008).

This method consists of the following steps:

- Generate N points in $2-D$ plane at random.
- Calculate the mapping error as follows: Let d^*_{ij} and d_{ij} be the Euclidean distance values between two points i and j in $L-D$ space and $2-D$ plane, respectively. For an error-free mapping, d^*_{ij} should be equal to d_{ij}.

The mapping error at m^{th} iteration E(m) can be determined mathematically as follows:

$$E(m) = \frac{1}{C}\sum_{i=1}^{N}\sum_{j=1(i<j)}^{N}\frac{[d^*_{ij} - d_{ij}(m)]^2}{d^*_{ij}}, \tag{3}$$

where

$$C = \sum_{i=1}^{N}\sum_{j=1(i<j)}^{N} d^*_{ij} \quad \text{and} \quad d_{ij}(m) = \sqrt{\sum_{k=1}^{D}[y_{ik}(m) - y_{jk}(m)]^2}.$$

- The mapping error has been minimized using the steepest descent method, as given below.

$$y_{pq}(m+1) = y_{pq}(m) - (MF)\Delta_{pq}(m), \tag{4}$$

where MF is a magic factor representing the step length in an iteration, which varies in the range of 0.0 to 1.0, and

$$\Delta_{pq}(m) = \frac{\partial E(m)/\partial y_{pq}(m)}{\left|\partial^2 E(m)/\partial y^2_{pq}(m)\right|}.$$

As a steepest descent method is utilized to minimize the mapping error, there is a chance of its solutions for being trapped into the local minima.

3.4.2. *VISOR algorithm (Konig et al., 1994)*

It is basically a geometrical method of data projection by preserving the distance information, whose principle is explained below.

Let us suppose that N points of $L-D$ space are to be mapped into a $2-D$ plane as accurately as possible.

Figure 3.4 displays the scheme of VISOR algorithm consisting of the following steps:

- The pivot vectors V_1, V_2, V_3 are located in $L-D$ space using the procedure given below.
  1. Compute centroid M of all N data points in $L-D$ space.

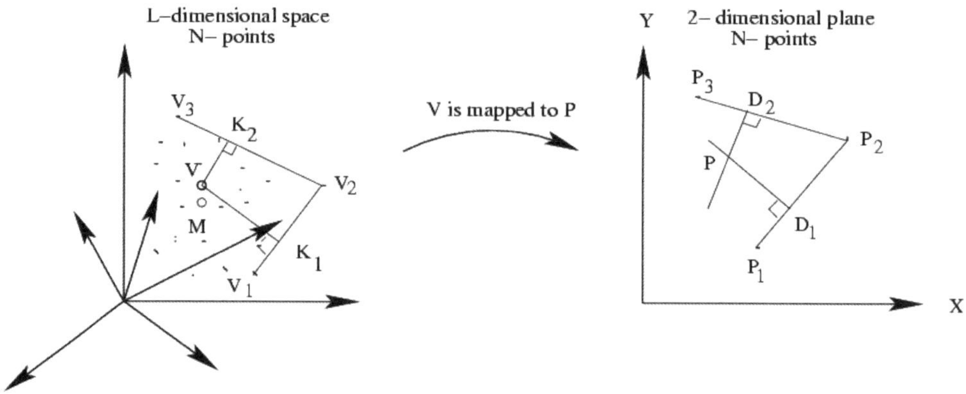

Fig. 3.4. VISOR Algorithm (Pratihar, 2008).

2. Identify the pivot-vector V_1, so that the Euclidean distance

$$d(V_1, M) = \max_{i=1}^{N}(d(v_i, M)).$$

3. Determine the pivot-vector V_2, such that the Euclidean distance

$$d(V_2, V_1) = \max_{i=1}^{N-1}(d(v_i, V_1)).$$

4. Locate the pivot-vector V_3, so that the Euclidean distance

$$d(V_3, V_2) = \max_{i=1}^{N-2}(d(v_i, V_2)) \quad \text{and}$$

$$d(V_3, V_1) = \max_{i=1}^{N-2}(d(v_i, V_1)).$$

- Obtain the pivot-vectors - P_1, P_2, P_3 in $2-D$ plane corresponding to the vectors V_1, V_2, V_3, respectively.
- Map the remaining $(N-3)$ points as discussed below.

1. Let us assume that a particular point V of $L-D$ space (refer to Fig. 3.4) is to be mapped to a corresponding point in $2-D$ plane. The straight lines V_1V_2 and V_2V_3 are considered first and then two perpendicular lines are drawn from the point V to them. Thus, the points K_1 and K_2 are obtained on the lines V_1V_2 and V_2V_3, respectively.
2. In $2-D$, locate D_1 on the line P_1P_2, such that D_1 divides the line P_1P_2 in the same proportion as K_1 has divided the line V_1V_2 (in $L-D$ space). Similarly, the point D_2 is identified on the line P_2P_3.
3. The perpendiculars are drawn at the points D_1 and D_2 to the lines P_1P_2 and P_2P_3, respectively, which intersect at the point P. Thus, the point V in $L-D$ space is mapped to a corresponding point P in $2-D$ plane.

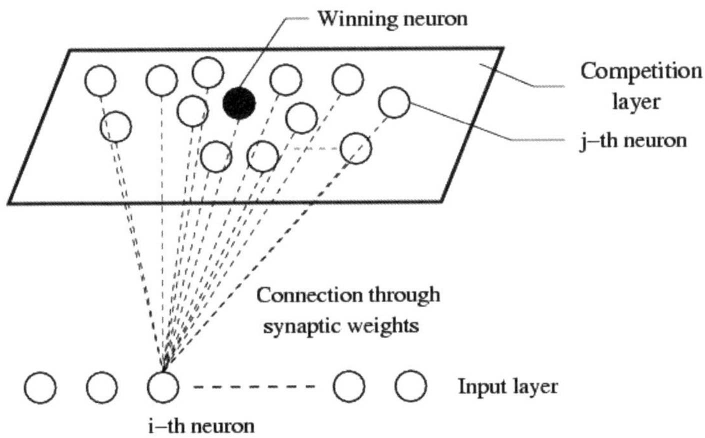

Fig. 3.5. A schematic representing self-organizing map (Pratihar, 2008).

This algorithm is easy to understand and implement. As the mapping is done in one iteration only, it is expected to be faster than other dimensionality reduction techniques.

3.4.3. *Self-organizing map (Kohenen, 1995)*

It is a topology preserving dimensionality reduction technique, which produces self-organizing feature maps similar to those occur in our brain. It works based on the principle of unsupervised and competitive learning. It consists of two layers, namely input and competition layers (refer to Fig. 3.5). The input layer contains higher dimensional data to be mapped to a lower dimensional plane/space. The number of neurons present in the competition layer is assumed to be equal to the number of data points present in the input layer. Each multivariate data of the input layer is connected to all neurons of the competition layer through some synaptic weights. The neurons in the competition layer undergo three basic operations, namely competition, cooperation and updating. In the competition stage, neurons in the competition layer compete among them to be identified as a winner. A neighborhood surrounding the winning neuron is then identified in the cooperation stage. The winning neuron along with its neighbors is updated in the third stage. The said stages are explained below in detail.

(1) *Competition* — Let us consider L-dimensional N points (neurons) in the input layer. Thus, a particular input neuron i can be represented as follows:

$$X_i = [x_{i1}, x_{i2}, \ldots, x_{iL}]^T, \quad \text{where } i = 1, 2, \ldots, N.$$

Let us also assume that the synaptic weight vector between input neuron i and neuron j lying in the competition layer is indicated by

$$W_j^i = [w_{j1}^i, w_{j2}^i, \ldots, w_{jL}^i]^T, \quad \text{where } j = 1, 2, \ldots, N.$$

Let us denote the neuron lying in the competition layer that has the best match with the input vector X_i by n. Thus, Euclidean distance between n and X_i can be represented as follows:

$$n(X_i) = \text{Minimum of } \sqrt{(X_i - W_j^i)^2}, \quad \text{where } j = 1, 2, \ldots, N. \tag{5}$$

(2) *Cooperation* — The winning neuron decides its topological neighborhood of excited neurons for cooperation. The neighborhood function is assumed to follow a Gaussian distribution for the said purpose as given below.

$$h_{j,n(x_i)}(t) = \exp\left(-\frac{d^2_{j,n(X_i)}}{2\sigma_t^2}\right), \tag{6}$$

where $t = 0, 1, 2, \ldots$ and $d_{j,n(x_i)}$ represents the lateral distance between the winning neuron n and excited neuron j, σ_t indicates the value of standard deviation at t^{th} iteration, which is determined as $\sigma_t = \sigma_0 \exp(-(t/\tau))$, where σ_0 indicates the initial value of standard deviation and τ represents the maximum number of iterations. The topological neighborhood is assumed to shrink with the increase of iteration number.

(3) *Updating* — The synaptic weights of the winning neuron and all excited neurons lying in its neighborhood are updated by following the rule given below.

$$W_j^i(t+1) = W_j^i(t) + \eta(t) h_{j,n(X_i)}(t)[X_i - W_j^i(t)], \tag{7}$$

where $\eta(t)$ represents the learning rate lying between 0.0 and 1.0.

As it is a topology-preserving tool, the neighbourhood information of higher dimensional data will remain the same in the lower dimension.

3.4.4. *GA-like approach (Dutta and Pratihar, 2006)*

Genetic algorithm (GA) is a non-traditional tool for optimization, where various operators of natural genetics have been modeled artificially and the principle of natural selection has been considered (Holland, 1975). A GA-like approach was proposed by Dutta and Pratihar (2006) to map the higher dimensional data into a lower dimension, after preserving the distance information. The scheme is shown in Fig. 3.6.

It consists of the following steps:

- Generate a set of N data points lying within the ranges of the variables in a higher dimensional space and determine their Euclidean distance values d^*_{ij}, where i and j represent two different points.
- Create at random an initial population of solutions of size P. Some bits (consisting of 1 and/or 0) are assigned to represent each solution. The decoded value of each solution will be used as a seed for another random number generator. Thus, the initial population of solutions will contain P seed values. Using a particular seed value, N data points are generated in $2-D$, at random. The Euclidean distance

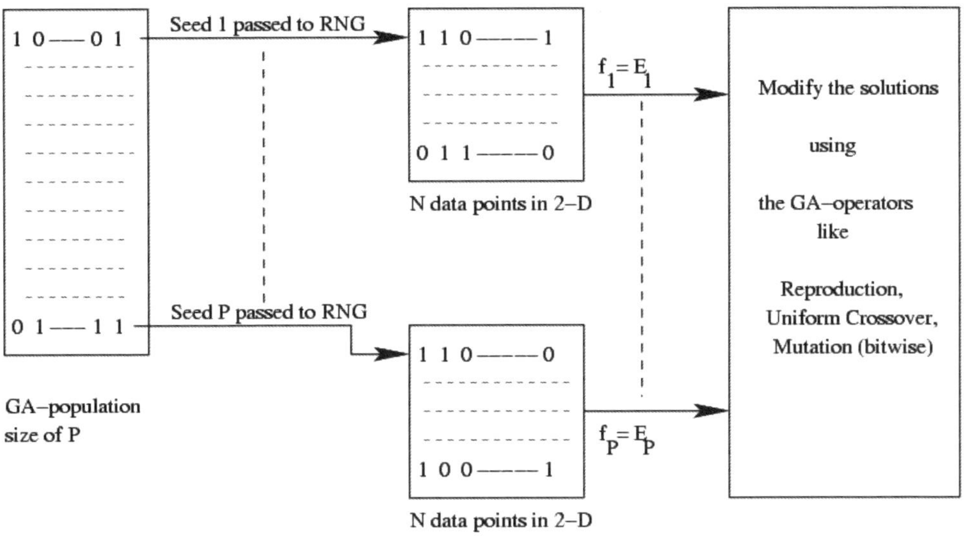

Fig. 3.6. GA-like dimensionality reduction technique (Dutta and Pratihar, 2006).

between the points i and j is then calculated, in $2-D$. Thus, this GA-like approach carries out its search in a wider space, compared to the conventional GA does.

- Determine the error in mapping E_p (for p^{th} solution) utilizing the expression given below.

$$E_p = \frac{1}{C} \sum_{i=1}^{N} \sum_{j=1(i<j)}^{N} \frac{[d_{ij}^{*} - d_{ij}]^2}{d_{ij}^{*}} \tag{8}$$

The symbols have the same meaning as they have in Sammon's NLM.

- The fitness of pth solution is made equal to E_p, which is to be minimized. The fitness values for all solutions of the population are calculated by following the same procedure.
- Modify the population of solutions utilizing the operators, namely tournament selection, uniform crossover and bit-wise mutation.
- Continue the process until the termination criterion is reached.

As it is a population-based search and optimization technique, the chance of its solutions for being trapped into the local minima is less.

3.4.5. *Comparisons*

The performances of the above four dimensionality reduction techniques have been compared to map higher dimensional data of Schaffer's F1 (test function) into $2-D$ for visualization (refer to Fig. 3.7). VISOR algorithm fails to yield well-distributed data in the lower dimension, whereas the remaining three approaches have

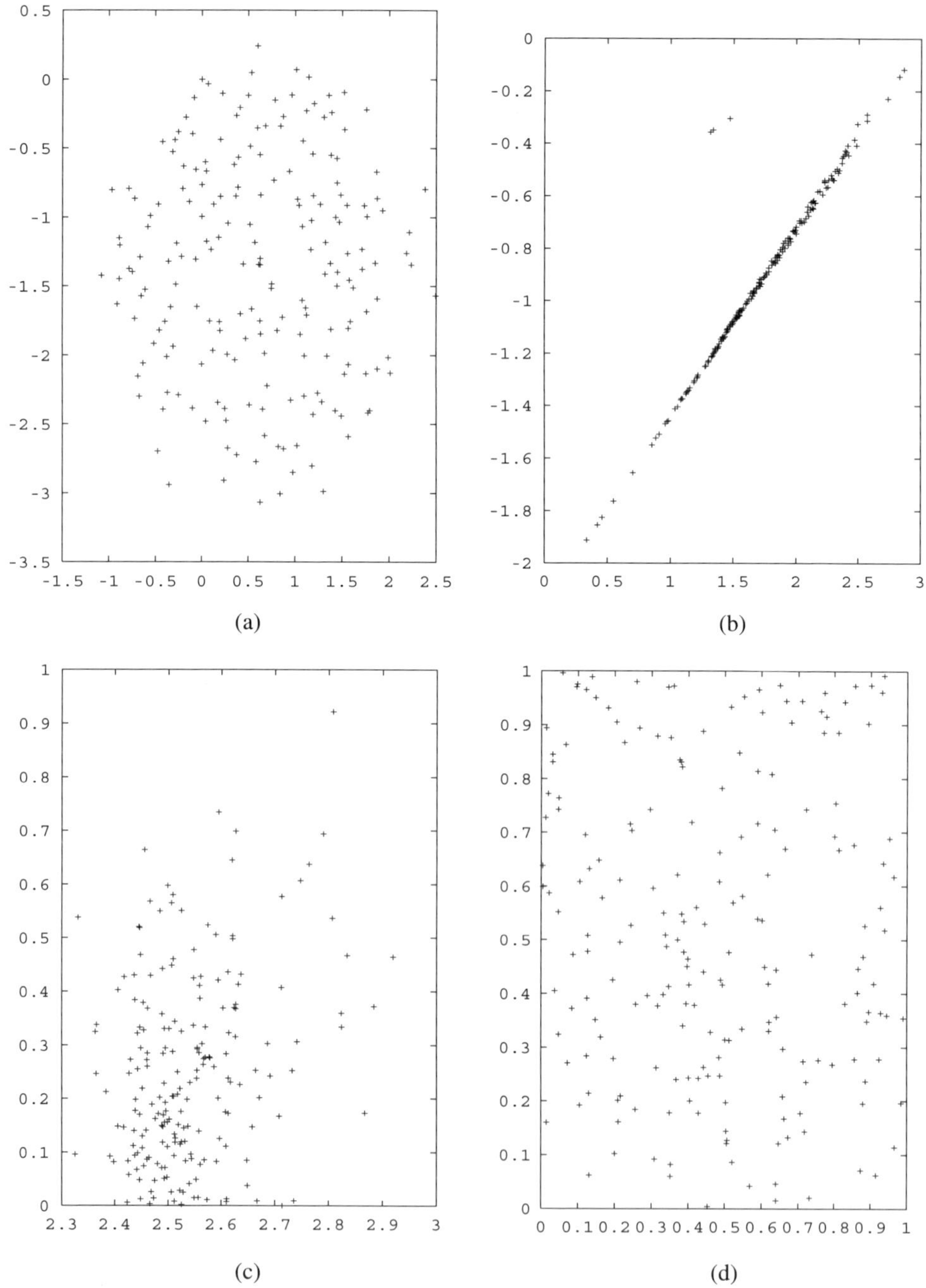

Fig. 3.7. Comparison of four dimensionality reduction techniques (Pratihar, 2008) — (a) Sammon's NLM, (b) VISOR algorithm, (c) Self-organizing Map, (d) GA-like approach.

successfully done it. However, VISOR algorithm is seen to be the fastest of all approaches.

3.4.6. *Dimensionality reduction approaches for large data sets*

High dimensional data sets consisting of a large number of features may contain a considerable amount of irrelevant and redundant information. It may affect the performances of machine learning algorithms. Feature extraction is the process of selecting a subset of original features based on some optimization criteria. It may help the machine learning algorithms to improve their performances in learning, and consequently, increase accuracy in predictions and enhance comprehensibility of learned results. There are basically two models for feature extraction of large data sets, namely filter model (Yu and Liu, 2003) and wrapper model (Ruiz *et al.*, 2006). The filter model concentrates on the features of training data to choose some important ones without using a learning algorithm. On the other hand, the wrapper model utilizes a learning algorithm to select some important features of the high dimensional data. The wrapper model is able to provide better results compared to the filter model, as expected but at the cost of more computations. Thus, the filter model may be chosen for feature extraction of very large data sets due to its less computational complexity. In order to get the advantages of both the above models, a hybrid model has also been proposed to extract useful features from the large data sets (Das, 2001).

3.5. Clustering Techniques

Clustering aims to put similar points into one group and dissimilar points into different groups. Clusters may be either fuzzy or hard in nature. Hard clusters have well-defined boundaries, whereas the fuzzy clusters have vague boundaries among them. The present chapter deals with the fuzzy clusters only. The working principle of two clustering techniques, namely Fuzzy C-Means (FCM) algorithm and Entropy-based Fuzzy Clustering (EFC) algorithm are explained below, in detail.

3.5.1. *Fuzzy C-means algorithm (Bezdek, 1973)*

Fuzzy C-Means (FCM) algorithm is one of the most popular fuzzy clustering techniques, in which membership values of the data points with the cluster centers will be updated iteratively. Let us consider M-dimensional N data points represented by $X_i (i = 1, 2, \ldots, N)$, which are to be clustered. The FCM algorithm consists of the following steps:

Step 1: Assume the number of clusters to be made, i.e., C, where $2 \leq C \leq N$.
Step 2: Select an appropriate level of cluster fuzziness $f > 1.0$.

Step 3: Initialize the $N \times C$ sized membership matrix $[U]$, at random, such that $U_{ij} \in [0, 1]$ and $\sum_{j=1}^{N} U_{ij} = 1.0$, for each i.

Step 4: Calculate k^{th} dimension of jth cluster center CC_{jk} using the expression given below.

$$CC_{jk} = \frac{\sum_{i=1}^{N} U_{ij}^{f} x_{ik}}{\sum_{i=1}^{N} U_{ij}^{f}} \tag{9}$$

Step 5: Determine the Euclidean distance between i^{th} data point and j^{th} cluster center like the following.

$$D_{ij} = \|(X_i - CC_j)\| \tag{10}$$

Step 6: Update fuzzy membership matrix $[U]$ according to D_{ij}. If $D_{ij} > 0$, then

$$U_{ij} = \frac{1}{\sum_{c=1}^{C} \left(\frac{D_{ij}}{D_{ic}}\right)^{\frac{2}{f-1}}} \tag{11}$$

If $D_{ij} = 0$, then the data point coincides with j^{th} cluster center CC_j and it will have the full membership value, that is, $U_{ij} = 1.0$.

Step 7: Repeat Steps 4 through 6, until the changes in $[U]$ come out to be less than some pre-specified values.

A parameter γ (in percent) is introduced to check the validity of the clusters. If the number of data points contained in a cluster comes out to be either more than or equal to $\gamma N/100$, it will be declared a valid cluster, otherwise the said points will be called the outliers.

It is important to mention that the performance of this algorithm is dependent on the selection of number of clusters to be made C, level of fuzziness f and threshold used to determine the outliers, that is, γ.

3.5.2. *Entropy-based fuzzy clustering (Yao* et al.*, 2000)*

In this approach, entropy values of the data points are determined first and then the data point with the minimum entropy value is selected as the cluster center. Here, the data points are clustered using the threshold value of similarity. Its working principle is explained below.

Let us consider N data points in M-dimensional $[T]$ hyperspace, where each data point $X_i (i = 1, 2, 3, \ldots, N)$ is represented by a set of M values (i.e., $x_{i1}, x_{i2}, \ldots, x_{iM}$). The Euclidean distance between any two data points (that is, i and j) is determined as follows:

$$D_{ij} = \sqrt{\sum_{k=1}^{M} (x_{ik} - x_{jk})^2} \tag{12}$$

Now, similarity between the two points (that is, i and j) can be calculated like the following.

$$S_{ij} = e^{-\alpha D_{ij}} \tag{13}$$

where α is a numerical constant. Thus, the similarity value between any two points lies in the range of 0.0 to 1.0. The value of α is decided based on the assumption that the similarity S_{ij} becomes equal to 0.5, when the distance between them (i.e., D_{ij}) turns out to be equal to the mean distance $\overline{D}$, which can be determined as follows:

$$\overline{D} = \frac{1}{^{N}C_2} \sum_{i=1}^{N} \sum_{j=1}^{N,j<i} D_{ij} \tag{14}$$

From equation (13), α can be calculated as follows:

$$\alpha = -\frac{\ln 0.5}{\overline{D}}$$

Now, the entropy of each data point (E_i) is determined with respect to other data points like the following.

$$E_i = -\sum_{j \in X}^{j \neq i} (S_{ij} \log_2 S_{ij}) + (1 - S_{ij}) \log_2(1 - S_{ij}) \tag{15}$$

This algorithm consists of the following steps:

Step 1: Calculate $E_i (i = 1, 2, 3, \ldots, N)$ for each X_i lying in $[T]$ hyperspace.
Step 2: Determine minimum E_i and declare $X_{i,Min}$ as the cluster center.
Step 3: Put $X_{i,Min}$ and the data points having similarity with $X_{i,Min}$ greater than β (threshold value for similarity) into a cluster and remove them from $[T]$.
Step 4: Check whether $[T]$ hyperspace is empty. If yes, terminate the program, else go to Step 2.

In this algorithm, E_i is calculated in such a way that a data point, which is far away from rest of the data points, may also be selected as a cluster center. To prevent such an odd situation, a threshold value γ (in %) is utilized. If the number of data points present in a cluster turns out to be greater than or equal to $\gamma N/100$, the cluster is declared to be a valid one. Otherwise, these data points will be called the outliers.

It is interesting to note that the performance of this algorithm depends on a number of parameters, such as α, β and γ.

3.5.3. *Comparisons*

Multi-dimensional TIG welding data (Juang *et al.*, 1998) have been clustered using the above two methods of clustering and the higher dimensional clustered data are mapped into $2-D$ using the self-organizing map (Kohenen, 1995) for visualization.

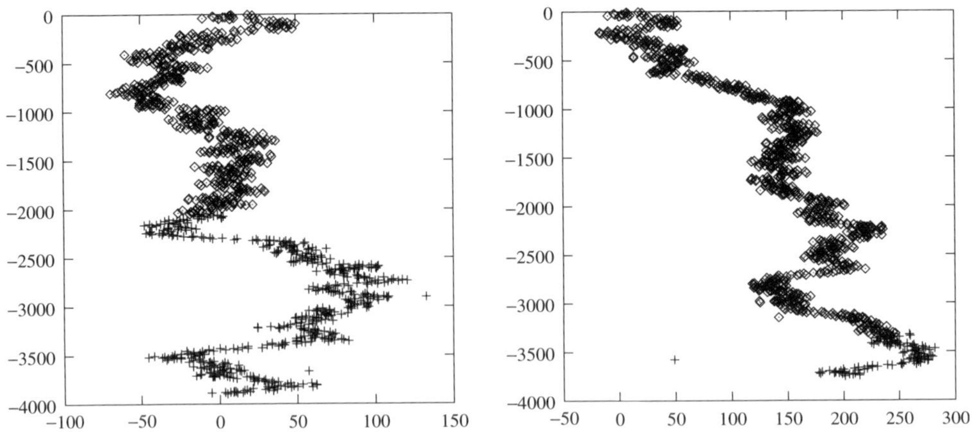

(a) (b)

Fig. 3.8. Comparison of the clusters obtained using two methods — (a) FCM algorithm, (b) EFC algorithm.

Figure 3.8 compares the clusters obtained using the said methods. The FCM algorithm is found to perform a slightly better than the EFC algorithm in terms of the quality of obtained clusters. The clusters should ideally be both distinct as well as compact in nature.

3.5.4. *Clustering of large spatial data sets*

The above clustering algorithms may not be suitable for handling large spatial data sets. In order to perform clustering of the large data sets, the algorithms are expected to have the following properties: minimal requirements of domain knowledge to determine the input parameters; ability to discover the clusters with arbitrary shapes, such as spherical, drawn-out, linear, elongated and others. Ng and Han (1994) developed a crisp clustering algorithm for large data sets using the concept of k representative objects (that is, medoids). This algorithm is known as CLARANS (Clustering Large Applications based on RANdomized Search). The medoids are determined from the data set in such a way that the sum of the within cluster dissimilarities is minimized. This algorithm tries to find a new set of k medoids by considering a candidate set of k medoids in the neighborhood of the current set of k medoids. Ester *et al.* (1996) proposed a density-based clustering algorithm known as DBSCAN (Density-Based Spatial Clustering of Applications with Noise) for the large data sets. A cluster can be identified because its density of points is higher than that of its outside. Moreover, the density of noise-region is less than that of any cluster. In order to declare a valid cluster, each point of it must have a minimum number of points (*MinPts*) in its *Eps-neighborhood*. The *Eps-neighborhood* of a point p is denoted by $N_{Eps}(p) = \{q\}$, such that $dist(p, q) \leq Eps$, where $dist(p, q)$ is the distance between the points: p and q. It is important to mention that DBSCAN could outperform CLARANS algorithm in terms of its ability to discover clusters of arbitrary shapes and efficiency.

3.6. Cluster-wise Regression Analysis

Let us assume that the input–output relationship of a process is non-linear in nature. Each response of this process is represented as a non-linear function of input process parameters. Thus, a non-linear response equation is derived to capture the said relationship for the entire range of input process parameters. However, the said non-linear relationship may be local in nature and consequently, the degree of non-linearity may not be the same for the entire range of the variables. Under these circumstances, the same non-linear response equation may not be able to make the predictions accurately for the whole range of the variables. To overcome this difficulty, input–output space of the process may be clustered using the concept of similarity as discussed above. Once the clusters are obtained, regression analysis can be carried out cluster-wise to determine the input–output relationships more accurately (Ganjigatti *et al.*, 2007).

3.7. Intelligent Data Mining

An intelligent data miner is yet to be developed, which can carry out both forward as well as reverse mappings accurately, after realizing inherent input–output relationships of the data set. Either a fuzzy logic or a neural network could be a potential tool for the said purpose. A genetic algorithm can also be used to evolve a set of suitable clusters, based on which either a fuzzy logic or neural network-based data miner will be developed to solve the said problem efficiently. Thus, an intelligent data miner will have to be evolved to establish input–output relationships of the process accurately on-line, which may be required to automate the same.

3.8. Summary

To automate any process, we need to know its input–output relationships on-line, both in forward and reverse directions accurately. A data miner can provide the above necessary information. It may also be required sometimes to visualize the said input–output hyper-space of the process to be automated. It has been felt that clustering can also help in extracting useful information from a data set. Conventional statistical regression analysis can solve the problems of forward mapping efficiently. However, it may not be able to provide the information related to reverse mapping always. Either a fuzzy logic-based or neural network-based approach may be able to tackle the problems of both forward as well as reverse mappings efficiently. Moreover, it has been felt that an intelligent data miner may be required, which will be able to understand the dynamics of the said input–output relationships of the data set in an adaptive way. A genetic algorithm may be able to evolve an intelligent fuzzy logic-based or neural network-based data miner to solve this problem efficiently.

Acknowledgement

The author acknowledges financial support of the Department of Science and Technology, Government of India, New Delhi, for carrying out research on intelligent data mining.

References

Bezdek J. C., (1973) *Fuzzy mathematics in pattern classification*, (Ph.D. thesis, Cornell University, Ithaca).

Boser B., Guyon I., V. Vapnik N., (1992). *Proc. of Fifth Annual Workshop on Computational Learning Theory*, p. 144.

Das S., (2001) *Proc. of 18-th International Conference on Machine Learning*, p. 74–81.

Domingos P., Pazzani M., (1997). *Machine Learning*, **29**, 103.

Dutta P., Pratihar D.K., *Intl. Jl. of Business Intelligence and Data Mining*, **1**, 347.

Ester M., Kriegel H. P., Sander J., Xu X., (1996). *Proc. of KDD Conference, AAAI*, pp. 226–231.

Fix E., Hodges J. L., (1951).*in Technical Report 4, UASF School of Aviation Medicine, Randolph Field, Texas.*

Ganijigatti J. P., Pratihar D. K. and RoyChoudhury A., (2007) *Jl. of Materials Processing Technology*, **189**, 352.

Holland J. H., (1975). *Adaptation in Natural and Artificial Systems*, (The University of Michigan Press, Ann Arbor, MI, USA).

Juang S. C., Tarng Y. S., Lii H. R., (1998). *Jl. of Materials Processing Technology*, **75**, 54.

Konig A., Bulmahn O. and Glessner M., (1994) *Proc. of Fifth British Machine Vision Conference*, p. 195.

Lacher R. C., Coats P. K., Sharma S. C., L. F. Fant, (1995). *Eur. J. Oper. Res.*, **85**, 53.

Lawrence R. L., Wright A., (2001). *Photogrammetric Engineering and Remote Sensing*, **67**, 1137.

McCulloach W. S., Pitts W., (1943). *Bulletin of Mathematical Biophysics*, **5**, 115.

Montgomery D. C., (2003) *Design and Analysis of Experiments*, (John Wiley & Sons (Asia) Pte. Ltd., Singapore, Fifth edition).

Ng R. T., Han J., (1994) *in Proc. of 20th International Conference on Very Large Data Bases*, p. 144–155.

Parappagoudar M. B., Pratihar D. K. and Datta G. L., (2008) *Applied Soft Computing*, **8**, 239.

Pratihar D. K., (2008) *Soft Computing*, (Narosa Publishing House, New Delhi, India).

Quinlan J. R., (1986). *Machine Theory*, **1**, 81.

Ruiz R., Riquelme J. C., Aguilar-Ruiz J. S., (2006). *Pattern Recognition*, **39**, 2383.

Sammon J. W., (1969). *IEEE Trans. on Computers*, **C-18(5)**, 401.

Yao J., Dash M., Tan S. T. and Liu H., (2000) *Fuzzy Sets and Systems,* **113**, 381.

Yu L., Liu H., (2003). *Proc. of Twentieth International Conference on Machine Learning.*

Zadeh L.A., (1965). *Information and Control*, **8**, 338.

Chapter 4

ROUGH NON-DETERMINISTIC INFORMATION ANALYSIS FOR UNCERTAIN INFORMATION

Hiroshi Sakai*, Hitomi Okuma[†], Mao Wu* and Michinori Nakata[‡]
**Mathematical Sciences Section, Department of Basic Sciences*
Faculty of Engineering, Kyushu Institute of Technology
Tobata, Kitakyushu 804, Japan
sakai@mns.kyutech.ac.jp
wumogaku@yahoo.co.jp

[†]Faculty of Education and Welfare Science,
Oita University
Dannoharu, Oita 870, Japan
okuma@oita-u.ac.jp

[‡]Faculty of Management and Information Science
Josai International University
Gumyo, Togane, Chiba 283, Japan
nakatam@ieee.org

Non-deterministic Information Systems (*NISs*) have been recognized to be the most important framework for handling the information incompleteness in tables, and some theoretical work about *NISs* has been investigated. We follow this robust framework, and we have been developing algorithms and utility programs for handling *NISs*. We are simply calling our work *Rough Non-deterministic Information Analysis* (*RNIA*). In this chapter, a framework of *RNIA* is briefly surveyed at first, then two significant issues on rule generation are examined. In the first issue, rule generation characterized by the consistency is picked up, and *minimal certain rules* and *discernibility functions* are introduced into *NISs*. In the second issue, rule generation characterized by *support* and *accuracy* criteria in *NISs* is investigated. An algorithm which is an extension of *Apriori* algorithm is also proposed. The solution of these two issues will enhance the framework of *RNIA*. Finally, rule generation on numerical data is considered. Here, *numerical patterns* with '@' and '#' are newly introduced into numerical values, and the equivalence classes depending upon numerical patterns are examined. The introduction of numerical patterns enhances the framework of *RNIA*.

4.1. Introduction

Rough set theory is recognized as a mathematical foundation of soft computing. This theory usually handles tables with deterministic information. Many applications of

*This work was partly supported by the Grant-in-Aid for Scientific Research (C) (No.16500176, No.18500214), Japan Society for the Promotion of Science.

this theory to rule generation, machine learning and knowledge discovery have been presented.[1–10]

We follow rule generation in *Deterministic Information Systems* (*DISs*),[1–5] and propose rule generation in *NISs*. *NISs* were proposed by Pawlak, Orłowska and Lipski in order to handle the information incompleteness in *DISs*, like null values, unknown values, missing values.[11–14] From the beginning of the research on incomplete information, *NISs* have been recognized to be the most important framework for handling the information incompleteness. Therefore, rule generation in *NISs* will also be an important framework for rule generation from incomplete information.

The following shows some important research on rule generation from incomplete information. Lipski showed a question-answering system besides an axiomatization of logic.[11,12] Orłowska established rough set analysis for non-deterministic information,[13–16] and Grzymała-Busse developed a system named *LERS*, which depends upon *LEM1* and *LEM2* algorithms.[9,17,18] Stefanowski and Tsoukias defined non symmetric similarity relations and valued tolerance relations for analyzing incomplete information.[19,20] Kryszkiewicz proposed a framework of rules in incomplete information systems.[21–23] According to the authors' knowledge, these are the most important research on incomplete information. We have also investigated incomplete information and rule generation. Nakata has been coping with the semantic aspect for incomplete information.[24–26] Sakai has focused on *NISs*, and investigated several problems.[27–36] We are simply calling this work *Rough Non-deterministic Information Analysis* (*RNIA*).

In this chapter, we briefly survey a framework of *RNIA*, and we study significant issues related to rule generation in *NISs*. Namely, we consider the following two issues with respect to *RNIA*:

(Issue 1) Rule generation on the basis of the consistency in *NISs* (Certain and Possible Rule Generation).
(Issue 2) Rule generation on the basis of the criterion values in *NISs*.

The solution of these significant issues will enhance the framework of *RNIA*. Furthermore, we also consider rule generation from *DISs* with numerical values. In *NISs*, we implicitly handle discrete data sets, but it will be necessary to consider continuous or numerical data sets for handling more general cases of information incompleteness. This rule generation in *DISs* with numerical values is the first step to rule generation in *NISs* with numerical values.

4.2. An Overview of *RNIA*

This section summarizes a framework of *RNIA*.

4.2.1. *Basic Definitions*

A *Deterministic Information System* (*DIS*) is a quadruplet

$$(OB, AT, \{VAL_A | A \in AT\}, f).^{1,3}$$

Here, *OB*, *AT* and VAL_A are finite sets, and we sequentially call every element an *object*, an *attribute* and an *attribute value*, respectively. Furthermore, f is a mapping such that $f : OB \times AT \rightarrow \cup_{A \in AT} VAL_A$.

We usually define two sets $CON \subseteq AT$ which we call *condition attributes* and $DEC \subseteq AT$ ($CON \cap DEC = \{\}$) which we call *decision attributes*. An object $x \in OB$ is *consistent* (with any distinct object $y \in OB$), if $f(x, A) = f(y, A)$ for every $A \in CON$ implies $f(x, B) = f(y, B)$ for every $B \in DEC$. We call a pair $[A, val_A]$ ($A \in AT, val_A \in VAL_A$) a *descriptor*. For $CON = \{A_1, \ldots, A_n\}$ and $DEC = \{B_1, \ldots, B_m\}$, we usually consider an implication

$$\wedge_i[A_i, f(x, A_i)] \Rightarrow \wedge_j[B_j, f(x, B_j)].$$

If x is consistent, we say this implication is *consistent*.

A *Non-deterministic Information System* (*NIS*) is also a quadruplet

$$(OB, AT, \{VAL_A | A \in AT\}, g),$$

where $g : OB \times AT \rightarrow P(\cup_{A \in AT} VAL_A)$ (a power set of $\cup_{A \in AT} VAL_A$).[11, 13, 14] Every set $g(x, A)$ is interpreted as that there is an actual value in this set but this value is not known. For a $NIS = (OB, AT, \{VAL_A | A \in AT\}, g)$ and a set $ATR \subseteq AT$, we name a $DIS = (OB, ATR, \{VAL_A | A \in ATR\}, h)$ satisfying $h(x, A) \in g(x, A)$ a *derived DIS* (*for ATR*) *from a NIS*.

Let us consider Table 4.1, which expresses an exemplary NIS_1. In NIS_1, every attribute value is not a value but a set. We interpret $g(1, Head) = \{yes, no\}$ as that the value of the first person's headache is either *yes* or *no*, and the real value is uncertain. This NIS_1 is an artificial data set. It is a very small size data set, but it will be enough to show several concepts.

The derived *DIS* in Table 4.2 implies a possible case in NIS_1. In our previous research, we have considered rough sets based concepts depending upon all possible cases. Of course, it is possible to apply the typical rough sets based methods to every

Table 4.1. A table of NIS_1.

OB	*Temp(erature)*	*Head(ache)*	*Nausea*	*Flu*
1	{*high*}	{*yes, no*}	{*no*}	{*yes*}
2	{*high, very_high*}	{*yes*}	{*yes*}	{*yes*}
3	{*normal, high, very_high*}	{*no*}	{*no*}	{*no*}
4	{*high*}	{*yes*}	{*yes, no*}	{*yes*}
5	{*high*}	{*yes, no*}	{*yes*}	{*no*}
6	{*normal*}	{*yes*}	{*yes, no*}	{*no*}
7	{*normal*}	{*no*}	{*yes*}	{*no*}
8	{*normal, high, very_high*}	{*yes*}	{*yes, no*}	{*yes*}

Table 4.2. A table of a derived *DIS from* NIS_1.

OB	*Temp(erature)*	*Head(ache)*	*Nausea*	*Flu*
1	*high*	*yes*	*no*	*yes*
2	*high*	*yes*	*yes*	*yes*
3	*normal*	*no*	*no*	*no*
4	*high*	*yes*	*yes*	*yes*
5	*high*	*yes*	*yes*	*no*
6	*normal*	*yes*	*yes*	*no*
7	*normal*	*no*	*yes*	*no*
8	*normal*	*yes*	*yes*	*yes*

Table 4.3. A table with incomplete information $*$. Each $*$ symbol means that the information is uncertain.

OB	*Temp(erature)*	*Head(ache)*	*Nausea*	*Flu*
1	*high*	*	*no*	*yes*
2	*	*yes*	*yes*	*yes*
3	*	*no*	*no*	*no*
4	*high*	*yes*	*	*yes*
5	*high*	*	*yes*	*no*
6	*normal*	*yes*	*	*no*
7	*normal*	*no*	*yes*	*no*
8	*	*yes*	*	*yes*

derived *DIS*. However, the number of all derived *DISs* increases in exponentially. In NIS_1, there are 576 ($2^6 \times 3^2$) derived *DISs*. For this reason, most of the research investigated Table 4.3 instead of NIS_1.

In tables with incomplete information, it is necessary to define the meaning of the $*$ symbol or some assumptions to the $*$ symbol. According to these assumptions, some different frameworks are defined.[18] On the other hand, every definition in *NISs* is very simple, and there is no doubt in every definition. Furthermore, rough sets based concepts defined in *NISs* are different from the concepts defined in tables with incomplete information.[29] For this reason, we have been investigating algorithms which do not depend upon the number of all derived *DISs*.[27–36]

4.2.2. *Two Modalities in RNIA*

Due to the definition of $g(x, A)$, we suppose that there exists a derived DIS^{real} with real information. In order to handle information based on DIS^{real}, two modalities *certainty* and *possibility* are usually defined.

(Certainty) If a formula ψ holds in every derived *DIS* from a *NIS*, ψ also holds in DIS^{real}. In this case, we say ψ *certainly holds* (in DIS^{real}).

(Possibility) If a formula ψ holds in some derived *DISs* from a *NIS*, there exists such a possibility that ψ holds in DIS^{real}. In this case, we say ψ *possibly holds* (in DIS^{real}).

In order to handle these two modalities, we extended concepts in *DISs* to *NISs*, for example, concepts like *the definability of a set, the consistency of an object, dependencies among attributes, rules and criteria including support, accuracy and coverage, reduction of attributes* and so on. In Appendix 1, we show an example NIS_1. As for the algorithms, see Refs. 27–30 for more details.

4.2.3. *Properties and Obtained Results in RNIA*

For a set $ATR = \{A_1, \ldots, A_n\} \subseteq AT$ and any $x \in OB$, let $PT(x, ATR)$ denote the Cartesian product $g(x, A_1) \times \cdots \times g(x, A_n)$. We name every element a *possible tuple (for ATR) of x*. For a possible tuple $\zeta = (\zeta_1, \ldots, \zeta_n) \in PT(x, ATR)$, let $[ATR, \zeta]$ denote a formula $\bigwedge_{1 \leq i \leq n}[A_i, \zeta_i]$. For condition attributes $CON \subseteq AT$ and decision attributes $DEC \subseteq AT$ $(CON \cap DEC = \{\})$, let $PI(x, CON, DEC)$ $(x \in OB)$ denote a set

$$\{[CON, \zeta] \Rightarrow [DEC, \eta] | \zeta \in PT(x, CON),\ \eta \in PT(x, DEC)\}.$$

We name an element of $PI(x, CON, DEC)$ a *possible implication (from CON to DEC) of x*. A possible implication τ may belong to $PI(x, CON, DEC) \cap PI(y, CON, DEC)$ $(x \neq y)$, therefore we may employ the notation τ^x (from $x \in OB$) instead of τ.

Now, we define six classes of possible implications, certain rules and possible rules. For any $\tau^x \in PI(x, CON, DEC)$, let $DD(\tau^x, x, CON, DEC)$ denote a set

$$\left\{\varphi \,\middle|\, \begin{array}{l}\varphi \text{ is a derived } DIS \text{ for } CON \cup DEC \text{ such that} \\ \text{an implication from } x \text{ in } \varphi \text{ is equal to } \tau^x\end{array}\right\}.$$

If $PI(x, CON, DEC)$ is a singleton set $\{\tau^x\}$, we say τ^x is *definite*. Otherwise, we say τ^x is *indefinite*. If the following equation holds

$$\begin{aligned}&\{\varphi \in DD(\tau^x, x, CON, DEC) \mid x \text{ is consistent in } \varphi\} \\ &\quad = DD(\tau^x, x, CON, DEC),\end{aligned}$$

we say τ^x is *globally consistent* (*GC*). If this set is equal to $\{\}$, we say τ^x is *globally inconsistent* (*GI*). Otherwise, we say τ^x is *marginal* (*MA*). By combining two cases, i.e., '*D(efinite)* or *I(ndefinite)*' and '*GC, MA* or *GI*', we define six classes, *DGC, DMA, DGI, IGC, IMA, IGI* in Table 4.4. A possible implication τ^x belonging to *DGC* class is consistent in all derived *DISs*, and this τ^x is not influenced by the information incompleteness, therefore we name this τ^x a *certain rule*. A possible implication τ^x belonging to either *IGC, DMA* or *IMA* class is consistent in some $\varphi \in DD(\tau^x, x, CON, DEC)$. Therefore, we name such τ^x a *possible rule*.

Now, we give the necessary and sufficient conditions for characterizing *GC, MA* and *GI* classes. For any $\zeta \in PT(x, ATR)$, we define two sets as follows:

(1) $inf(x, ATR, \zeta) = \{y \in OB | PT(y, ATR) = \{\zeta\}\} \cup \{x\}$,
(2) $sup(x, ATR, \zeta) = \{y \in OB | \zeta \in PT(y, ATR)\}$.

Table 4.4. Six classes of possible implications in *NISs*.

	GC	*MA*	*GI*
Definite	*DGC*	*DMA*	*DGI*
Indefinite	*IGC*	*IMA*	*IGI*

Intuitively, $inf(x, ATR, \zeta)$ implies a set of objects whose tuples are ζ and definite. A set $sup(x, ATR, \zeta)$ implies a set of objects whose tuples may be ζ. In *DISs*, $inf(x, ATR, \zeta) = sup(x, ATR, \zeta)$ holds, and $\{x\} \subseteq inf(x, ATR, \zeta) \subseteq sup(x, ATR, \zeta)$ holds in *NISs*. By using these two sets *inf* and *sup*, we characterize each class.

Theorem 4.1 (Sakai and Okuma, 2004). *For a NIS, let us consider a possible implication* $\tau^x : [CON, \zeta] \Rightarrow [DEC, \eta] \in PI(x, CON, DEC)$. *Then, the following holds.*

(1) τ^x *belongs to GC class, if and only if* $sup(x, CON, \zeta) \subseteq inf(x, DEC, \eta)$.
(2) τ^x *belongs to MA class, if and only if* $inf(x, CON, \zeta) \subseteq sup(x, DEC, \eta)$.
(3) τ^x *belongs to GI class, if and only if* $inf(x, CON, \zeta) \not\subseteq sup(x, DEC, \eta)$.

Proposition 4.1 (Sakai and Okuma, 2004). *For a NIS, let* $ATR \subseteq AT$ *be* $\{A_1, \ldots, A_n\}$, *and let a possible tuple* $\zeta \in PT(x, ATR)$ *be* $(\zeta_1, \ldots, \zeta_n)$. *Then, the following holds.*

(1) $inf(x, ATR, \zeta) = \cap_i inf(x, \{A_i\}, (\zeta_i))$.
(2) $sup(x, ATR, \zeta) = \cap_i sup(x, \{A_i\}, (\zeta_i))$.

For generating *inf* and *sup* information for ATR, we first prepare two arrays *arrayinf[A,val]* and *arraysup[A,val]* for each attribute A and $val \in VAL_A$. If $g(x, A) = \{val\}$, we add x to *arrayinf[A,val]* and *arraysup[A,val]*. If $val \in g(x, A)$, we add x to *arraysup[A,val]*. Thus, the complexity of generating *inf* and *sup* for attribute A is $o(|OB|)$. Then, we apply Proposition 4.1 to the generation of *inf* and *sup* for the ATR.

For an implication $\tau^x \in PI(x, CON, DEC)$ in a *DIS*, the following criteria may be employed for defining rules.

(1) $support(\tau^x)$ = (the amount of τ)/(the amount of objects),
(2) $accuracy(\tau^x)$ = (the amount of τ)/(the amount of condition part in τ).

In a *NIS*, they are extended to the minimum support and the maximum support as follows:

(1) $minsupp(\tau^x) = Min_{\psi \in DD(\tau^x, x, CON, DEC)}\{support(\tau^x) \text{ in } \psi\}$,
(2) $maxsupp(\tau^x) = Max_{\psi \in DD(\tau^x, x, CON, DEC)}\{support(\tau^x) \text{ in } \psi\}$.

Similarly, the minimum accuracy $minacc(\tau^x)$ and the maximum accuracy $maxacc(\tau^x)$ are defined. These criteria depend upon all derived *DISs*, however we developed an effective method, which does not depend upon the number of derived *DISs*.

Theorem 4.2 (Sakai and Okuma, 2004). *For a NIS, let us consider a possible implication* $\tau^x : [CON, \zeta] \Rightarrow [DEC, \eta] \in PI(x, CON, DEC)$. *We also define the next two sets.*

$$INACC = [sup(x, CON, \zeta) \setminus inf(x, CON, \zeta)] \cap sup(x, DEC, \eta),$$
$$OUTACC = [sup(x, CON, \zeta) \setminus inf(x, CON, \zeta)] \setminus inf(x, DEC, \eta).$$

Then, the following holds.

(1) $minsupp(\tau^x) = |inf(x, CON, \zeta) \cap inf(x, DEC, \eta)|/|OB|$.
(2) $maxsupp(\tau^x) = |sup(x, CON, \zeta) \cap sup(x, DEC, \eta)|/|OB|$.
(3) $minacc(\tau^x) = Numerator/Denominator$,
Here, $Numerator = |inf(x, CON, \zeta) \cap inf(x, DEC, \eta)|$,
$Denominator = |inf(x, CON, \zeta)| + |OUTACC|$.
(4) $maxacc(\tau^x) = Numerator/Denominator$,
Here, $Numerator = |inf(x, CON, \zeta) \cap sup(x, DEC, \eta)| + |INACC|$,
$Denominator = |inf(x, CON, \zeta)| + |INACC|$.

In $minacc(\tau^x)$, the numerator $inf(x, CON, \zeta) \cap inf(x, DEC, \eta)$ and the denominator $inf(x, CON, \zeta)$ are fixed. We adjust every

$$y \in [sup(x, CON, \zeta) \setminus inf(x, CON, \zeta)]$$

for minimizing *accuracy*. For every

$$y \in [sup(x, CON, \zeta) \setminus inf(x, CON, \zeta)] \setminus inf(x, DEC, \eta),$$

we can obtain a possible implication $[CON, \zeta] \Rightarrow [DEC, \eta']$ $(\eta' \neq \eta)$. This implication is just counted in the denominator; it is not counted in the numerator. As for $maxacc(\tau^x)$, it is possible to obtain τ^x from every $y \in [sup(x, CON, \zeta) \setminus inf(x, CON, \zeta)] \cap sup(x, DEC, \eta)$. This implication is counted both in the denominator and the numerator. Since $m/n \leq (m+k)/(n+k)$ $(0 \leq m \leq n,\ n \neq 0,\ k > 0)$ holds, we obtain (4) in Theorem 4.2.

Example 4.1. Let us consider the above Theorem 4.1, 4.2 and Proposition 4.1 by using NIS_1. There are 576 derived *DISs* for all attributes. For $CON = \{Temp, Head\}$ and $DEC = \{Flu\}$, there are $72(=2^3 \times 3^2)$ derived *DISs*. Here, $PT(2, \{Temp, Head\}) = \{(high, yes), (very\_high, yes)\}$ and $PT(2, \{Flu\}) = \{(yes)\}$ hold, and $PI(2, \{Temp, Head\}, \{Flu\})$ consists of two possible implications

$$\tau_1^2 : [Temp, high] \wedge [Head, yes] \Rightarrow [Flu, yes],$$
$$\tau_2^2 : [Temp, very\_high] \wedge [Head, yes] \Rightarrow [Flu, yes].$$

According to Proposition 4.1, the following holds.

$$\begin{aligned}&sup(2, \{Temp, Head\}, (high, yes))\\&\quad = sup(2, \{Temp\}, (high)) \cap sup(2, \{Head\}, (yes))\\&\quad = \{1,2,3,4,5,8\} \cap \{1,2,4,5,6,8\} = \{1,2,4,5,8\}.\end{aligned}$$

Since $inf(2, \{Flu\}, (yes)) = \{1,2,4,8\}$ holds, the following is derived.

$$sup(2, \{Temp, Head\}, (high, yes)) \not\subseteq inf(2, \{Flu\}, (yes)).$$

Due to (1) in Theorem 4.1, we know τ_1^2 does not belong to *IGC* class. However,

$$\begin{aligned}&inf(2, \{Temp, Head\}, (high, yes))\\&\quad = inf(2, \{Temp\}, (high)) \cap inf(2, \{Head\}, (yes))\\&\quad = \{1,2,4,5\} \cap \{2,4,6,8\} = \{2,4\},\end{aligned}$$

and $sup(2, \{Flu\}, (yes)) = \{1,2,4,8\}$ holds. According to (2) in Theorem 4.1, we know τ_1^2 belongs to *IMA* class. Similarly,

$$\begin{aligned}&sup(2, \{Temp, Head\}, (very\_high, yes)) = \{2,8\} \text{ and}\\&inf(2, \{Flu\}, (yes)) = \{1,2,4,8\}\end{aligned}$$

conclude

$$sup(2, \{Temp, Head\}, (very\_high, yes)) \subseteq inf(2, \{Flu\}, (yes)).$$

According to (1) in Theorem 4.1, we know τ_2^2 belongs to *IGC* class.

Now, let us consider criterion values of τ_2^2 according to Theorem 4.2.

$$\begin{aligned}&INACC\\&\quad = [sup(2, \{Temp, Head\}, (high, yes)) \backslash inf(2, \{Temp, Head\}, (high, yes))]\\&\qquad \cap sup(2, \{Flu\}, (yes)) = [\{1,2,4,5,8\} \backslash \{2,4\}] \cap \{1,2,4,8\} = \{1,8\},\\&OUTACC = \{1,5,8\} \backslash inf(2, \{Flu\}, (yes)) = \{5\}.\end{aligned}$$

Each element in *INACC* causes an increase in *accuracy*, and each element in *OUTACC* causes a decrease in *accuracy*. Thus, we obtain the following:

$$\begin{aligned}&minsupp(\tau_2^2) = |\{2,4\} \cap \{1,2,4,8\}|/8 = 1/4,\\&minacc(\tau_2^2) = |\{2,4\} \cap \{1,2,4,8\}|/(|\{2,4\}| + |\{5\}|) = 2/3,\\&maxsupp(\tau_2^2) = |\{1,2,4,5,8\} \cap \{1,2,4,8\}|/8 = 1/2,\\&maxacc(\tau_2^2) = (|\{2,4\} \cap \{1,2,4,8\}| + |\{1,8\}|)/(|\{2,4\}| + |\{1,8\}|) = 1.\end{aligned}$$

By using criteria *minacc* and *maxacc*, we can also characterize certain and possible rules.

Proposition 4.2. *For a NIS, let us consider a possible implication* τ^x. *Then, the following holds.*

Table 4.5. For an implication τ^1 : [*Temperature,high*] => [*Flu*, *yes*] in NIS_1, the following is a DIS_{worst} and a DIS_{best} to the right. $minsupp(\tau^1) = 0.25$, $minacc(\tau^1) = 0.5$, $maxsupp(\tau^1) = 0.5$, $maxacc(\tau^1) = 0.8$ hold.

OB	*Temperature*	*Flu*
1	*high*	*yes*
2	*very_high*	*yes*
3	*high*	*no*
4	*high*	*yes*
5	*high*	*no*
6	*normal*	*no*
7	*normal*	*no*
8	*normal*	*yes*

OB	*Temperature*	*Flu*
1	*high*	*yes*
2	*high*	*yes*
3	*normal*	*no*
4	*high*	*yes*
5	*high*	*no*
6	*normal*	*no*
7	*normal*	*no*
8	*high*	*yes*

(1) τ^x *is a certain rule, if and only if* τ^x *is definite and* $minacc(\tau^x) = 1$.
(2) τ^x *is a possible rule, if and only if* $maxacc(\tau^x) = 1$.

We have obtained another important property in the following.

Theorem 4.3 (Sakai *et al.*, 2007). *For each* τ^x *there is a derived* DIS_{worst}, *where both support*(τ^x) *and accuracy*(τ^x) *are minimum. There is also a derived* DIS_{best}, *where both support*(τ^x) *and accuracy*(τ^x) *are maximum.*

We call DIS_{worst} a *derived DIS* with the *worst condition* for τ^x. We also call DIS_{best} a *derived DIS* with the *best condition* for τ^x. Both DIS_{worst} nor DIS_{best} may not be unique. An example is given in Table 4.5.

4.3. Issue 1: Rule Generation on the Basis of the Consistency in *NISs* (Certain and Possible Rule Generation)

We follow rule generation on the basis of the consistency in *DISs*, and extend it to rule generation in *NISs*. This section copes with the following.

(Issue 1-1) For a *NIS*, let *DEC* be decision attributes and let η be a tuple of decision attribute values. Then, find certain rules, i.e., implications $[CON, \zeta] \Rightarrow [DEC, \eta]$ in *DGC*.

(Issue 1-2) In the same condition as in Issue 1-1, find minimal[17] certain rules.

(Issue 1-3) In the same condition as in Issue 1-1, find possible rules i.e., implications $[CON, \zeta] \Rightarrow [DEC, \eta]$ in *DMA*, *IGC* or *IMA* classes.

4.3.1. *Certain Rule Generation by the Order of Attributes*

At first, we characterize every object, from which certain rules can be generated.

Proposition 4.3. *In a NIS, let us consider an object* $x \in OB$ *satisfying* $PT(x, DEC) = \{\eta\}$. *Let* $(AT \backslash DEC)^*$ *denote a set* $\{A \in AT \setminus DEC \mid PT(x, \{A\})$ *is a singleton set* $\{\zeta_A\}\}$, *and let* ζ^* *denote this unique element of* $PT(x, (AT \backslash DEC)^*)$.

A certain rule whose decision part is $[DEC, \eta]$ *can be generated from* x, *if and only if* $sup(x, (AT \setminus DEC)^*, \zeta^*) \subseteq inf(x, DEC, \eta)$ *holds.*

Proof. If the condition holds, it is possible to generate a certain rule

$$[(AT \setminus DEC)^*, \zeta^*] \Rightarrow [DEC, \eta] \text{ from an object } x$$

according to Theorem 4.1. On the other hand, if a certain rule is generated from an object x, there is a subset $CON \subseteq (AT \setminus DEC)^*$ satisfying

$$\text{(A) } sup(x, CON, \zeta_{CON}) \subseteq inf(x, DEC, \eta).$$

Due to the definition of *CON*, we have the following.

$$\text{(B) } sup(x, (AT \setminus DEC)^*, \zeta^*) = \cap_{A \in (AT \setminus DEC)^*} sup(x, \{A\}, (\zeta_A))$$
$$\subseteq \cap_{A \in CON} sup(x, \{A\}, (\zeta_A)) = sup(x, CON, \zeta_{CON}).$$

Thus, we obtain the inclusion relation from (A) and (B). □

Let $OB(DEC, \eta)$ denote a set

$$\{x \in OB \mid x \text{ generates a certain rule whose decision part is } [DEC, \eta]\}.$$

If $OB(DEC, \eta) = \{\}$, it is impossible to generate any certain rule. Proposition 4.3 is applicable to the generation of $OB(DEC, \eta)$. Namely, Proposition 4.1 is applied to generating $sup(x, (AT \setminus DEC)^*, \zeta^*)$, then (1) in Theorem 4.1 is examined. Thus, the complexity to obtain $OB(DEC, \eta)$ is $o(|OB| \times |AT|)$.

Example 4.2. In NIS_1, let us consider possible implications in the form of $[CON, \zeta] \Rightarrow [Flu, yes]$. Here, $DEC = \{Flu\}$ and $g(x, Flu) = \{yes\}$ for (x = 1,2,4,8). In object 1, $(\{Temp, Head, Nausea, Flu\} \setminus \{Flu\})^* = \{Temp, Nausea\}$ holds. Since $sup(1, \{Temp, Nausea\}, (high, no)) \not\subseteq inf(1, \{Flu\}, (yes))$, $1 \notin OB(\{Flu\}, (yes))$ is derived. Like this, we obtain $OB(\{Flu\}, (yes)) = \{\}$. Namely, we cannot obtain any certain rules whose decision part is $[Flu, yes]$. Then, let us consider possible implications in the form of $[CON, \zeta] \Rightarrow [Flu, no]$. In object 7, $sup(7, \{Temp, Head, Nausea\}, (normal, no, yes)) \subseteq inf(1, \{Flu\}, (no))$ holds, and $7 \in OB(\{Flu\}, (no))$ is derived. Therefore, $\tau_1^7 : [Temp, normal] \wedge [Head, no] \wedge [Nausea, yes] \Rightarrow [Flu, no]$ is a certain rule.

Because of (1) in Theorem 4.1, Issue 1-1 is reduced to find $[CON, \zeta]$ satisfying $sup(x, CON, \zeta) \subseteq inf(x, DEC, \eta)$. In Example 4.2, we dealt with a case that $CON = (AT \setminus DEC)^*$. However, we may deal with a case that $CON \subset (AT \setminus DEC)^*$. In reality, $\tau_2^7 : [Temp, normal] \wedge [Head, no] \Rightarrow [Flu, no]$ is also a certain rule. If we employ a total order which is defined by the significance of attributes over *AT*, it is easy to define *CON*. Namely, *CON* is sequentially $\{A_1\}, \{A_1, A_2\}, \ldots, \{A_1, \ldots, A_n\}$ for the order $A_1, A_2, \ldots, A_n$ of attributes. However, this method generates a certain rule depending upon the significance of attributes, and some other certain rules may not be generated. Furthermore, the minimal condition part in the next subsection is not assured. In reality, we have realized utility programs depending upon a total order of attributes, but we are much more interested in Issue 1-2.

4.3.2. *Minimal Certain Rules*

Now, let us consider minimal certain rules in *NISs*. We follow the usual definition in *DISs*, and give the definition of minimal certain rules in *NISs*.

Definition 4.1. Let us consider a possible implication $\tau^x : [CON, \zeta] \Rightarrow [DEC, \eta]$, which belongs to DGC class. We say τ^x is a *minimal certain rule* from x, if there is no possible implication $(\tau')^x : [CON', \zeta'] \Rightarrow [DEC, \eta]$ such that CON' is a proper subset of CON and it belongs to DGC class.

Issue 1-2 has already been investigated in *DISs*. To find a minimal reduct in a *DIS* is proved to be NP-hard.[37] This means that it is necessary to examine each possible implication $[CON', \zeta'] \Rightarrow [DEC, \eta]$ for $(CON' \subseteq CON)$. Therefore, it will be time-consuming for handling large scale *DISs*.

For this problem, *discernibility functions* in *DISs* have been proposed.[37] These functions are effective to find *core descriptors*, which are useful to obtain minimal reducts. We also introduce discernibility functions in *NISs*, and investigate a method to generate minimal certain rules in *NISs*.

4.3.3. *Discernibility Functions and Minimal Certain Rule Generation*

Issue 1-2 is reduced to find such a minimal conjunction $[CON, \zeta]$ that $sup(x, CON, \zeta) \subseteq inf(x, DEC, \eta)$ and $PT(x, CON) = \{\zeta\}$ hold. Intuitively, a minimal set of descriptors, which discriminate every object in $OB \setminus inf(x, DEC, \eta)$ from $inf(x, DEC, \eta)$, becomes a minimal condition part.

Definition 4.2. Let us consider an object $x \in OB(DEC, \eta)$. Any distinct $y \in OB$ is *discriminated from* x by a definite descriptor $[A_i, \zeta_i]$ $(A_i \in AT \setminus DEC)$ in x, if $y \notin sup(x, \{A_i\}, (\zeta_i))$ and $PT(x, \{A_i\}) = \{\zeta_i\}$. Let $DISC(x, y)$ denote a disjunction of such descriptors in x. We say $[A_i, \zeta_i]$ is a *core* descriptor, if $DISC(x, y) = [A_i, \zeta_i]$.

Definition 4.3. For every $x \in OB(DEC, \eta)$, we define

$$DF(x) = \wedge_{y \in OB \setminus inf(x, DEC, \eta)} DISC(x, y),$$

and we name $DF(x)$ a *discernibility function* of x.

Definition 4.4. For a discernibility function $DF(x)$, let us identify every descriptor in $DF(x)$ with a propositional variable. If a set SOL of descriptors assigns *true* to $DF(x)$, we say *SOL satisfies* $DF(x)$. Especially, if $OB \setminus inf(x, DEC, \eta) = \{\}$ holds, every descriptor satisfies $DF(x)$.

Example 4.3. Let us consider the decision $[Flu, no]$ in Table 4.1 and Appendix 2. Here, $OB(\{Flu\},(no)) = \{3,5,6,7\}$, and we have the next $DF(7)$.

$$DF(7) = [1,[Temp,normal] \vee [Nausea,yes]] \wedge [2,[Temp, normal] \vee [Head,no]] \\ \wedge [4,[Temp,normal] \vee [Head,no]] \wedge [8,[Head,no]].$$

If we assign *true* to both $[Temp,normal]$ and $[Head, no]$, the truth value of $DF(7)$ is *true*. Thus, a set $\{[Temp,normal],[Head,no]\}$ satisfies $DF(7)$, and it certainly discriminates any object in $OB \setminus inf(7,\{Flu\},(no)) = \{1,2,4,8\}$ from object 7.

Lemma 4.1. *If $\tau^x : [CON,\zeta] \Rightarrow [DEC,\eta]$ is minimal and τ^x belongs to DGC class, there is no possible implication $(\tau')^y : [CON',\zeta'] \Rightarrow [DEC,\eta]$ $(x \neq y)$ such that CON' is a proper subset of CON and it belongs to DGC class. Namely, τ^x is a minimal certain rule.*

Proof. Let us suppose the following (A) and (B) according to Theorem 4.1.

(A) $sup(x, CON,\zeta) \subseteq inf(x, DEC,\eta)$,
(B) $sup(y, CON',\zeta') \subseteq inf(y, DEC,\eta)$ and $CON' \subsetneq CON$.

Since both τ^x and $(\tau')^y$ are definite due to the condition of DGC class,

$$\text{(C)}\ inf(y, DEC,\eta) = inf(x, DEC,\eta)$$

holds. Due to the definition of *sup*,

$$\text{(D)}\ sup(y, CON',\zeta') = sup(x, CON',\zeta')$$

clearly holds. We substitute (C) and (D) into (B), then we have the following.

$$\text{(E)}\ sup(x, CON',\zeta') \subseteq inf(x, DEC,\eta).$$

According to the minimal condition of CON, (E) contradicts (A). Therefore, the condition (B) does not hold. □

Theorem 4.4. *For a minimal set SOL satisfying $DF(x)$ in a NIS,*

$$\tau : \wedge_{[A_i,\zeta_i]\in SOL}[A_i,\zeta_i] \Rightarrow [DEC,\eta]$$

is a minimal certain rule.

Proof. SOL is assured to be minimal in the condition, so we show the inclusion relation $sup(x, CON,\zeta) \subseteq inf(x, DEC,\eta)$. Let SOL be a set of descriptors $\{[A_1,\zeta_1],\ldots,[A_n,\zeta_n]\}$. Since SOL satisfies $DF(x)$, SOL assigns *true* to $DISC(x, y)$ for each y, and there is a descriptor

$$[A_k^y,\zeta_k^y] \in SOL \quad \text{and} \quad y \notin sup(x,\{A_k^y\},(\zeta_k^y))$$

according to the definition of $DISC(x, y)$. Therefore,

$$y \notin \cap_{[A_i,\zeta_i]\in SOL} sup(x,\{A_i\},(\zeta_i)) = sup(x, CON,\zeta)$$
$$\text{for any } y \in OB \setminus inf(x, DEC,\eta).$$

is derived. This implies

$$(OB \backslash inf(x, DEC, \eta)) \cap sup(x, CON, \zeta) = \{\}.$$

For any $y \in sup(x, CON, \zeta)$, we have

$$y \notin OB \backslash inf(x, DEC, \eta) \quad \text{and} \quad y \in inf(x, DEC, \eta)$$

due to the above inclusion. For any $y \in sup(x, CON, \zeta)$, it is possible to derive $y \in inf(x, DEC, \eta)$. According to Lemma 4.1, τ is a minimal certain rule. □

According to Theorem 4.4, Issue 1-2 is reduced to find minimal solutions of $DF(x)$. In the subsequent subsections, we examine some methods to obtain minimal solutions of $DF(x)$.

4.3.4. *Enumeration Method for Obtaining Minimal Solutions*

Even though we employ $DF(x)$, we need a procedure of enumeration. This is because, to find a minimal reduct in a *DIS* is proved to be NP-hard.[37]

For every subset SUB of all descriptors in $DF(x)$, a partial order is defined by the set inclusion. We sequentially check whether SUB satisfies $DF(x)$ or not. If so, SUB is a minimal solution, and we remove SUB' satisfying $SUB \subseteq SUB'$, and we repeat this procedure. We name this method the *enumeration method* (*E-method*). This is a simple method, however the number of subsets increases in the power of 2. Therefore, this E-method works just for small size *NISs*.

4.3.5. *Interactive Selection Method for Obtaining Minimal Solutions*

Now, let us consider the second method. We sequentially select a descriptor in $DF(x)$, and we reduce $DF(x)$ to a new $DF_{red}(x)$. By repeating this procedure, it is possible to obtain a set of descriptors satisfying $DF(x)$. We name this method an *interactive selection method* (*IS-method*). This IS-method can obtain solutions of $DF(x)$, which reflect the user's selection.

Generally, a solution of $DF(x)$ in the IS-method may not be minimal. Therefore, it is necessary to apply the E-method to every solution.

Example 4.4. Let a, b, c, d be descriptors in a *NIS*, and let us suppose $DF(1) = (a \vee b) \wedge (b \vee c) \wedge (c \vee d)$. If we select descriptor a in $DF(1)$, $DF(1)$ is revised to $DF_{red}(1) = (b \vee c) \wedge (c \vee d)$. This absorption law is the key procedure to reduce every discernibility function.[22] Similarly, if we select descriptor b in $DF_{red}(1)$, $DF_{red}(1)$ is revised to $DF_{red}(1) = (c \vee d)$. Finally, we select descriptor c and we obtain a set $\{a, b, c\}$. This set satisfies $DF(1)$, but this set is not minimal, since both sets $\{a, c\}$ and $\{b, c\}$ satisfy $DF(1)$. In order to escape from such cases, we apply the E-method to $\{a, b, c\}$. Every subset is examined in the E-method, and we obtain both $\{a, c\}$ and $\{b, c\}$.

4.3.6. *Interactive Selection and Enumeration Method with a Threshold Value for Obtaining Minimal Solutions*

We combine the IS-method with the E-method, and name this method an *interactive selection* and *enumeration method* with a *threshold value* (*ISETV-method*). In the *ISETV*-method, at first we employ the IS-method, and we apply the E-method to a set of descriptors.

Let *SEL* be a set of descriptors, which are selected by the IS-method. When $(|SEL|+|\{[A,\zeta_A]|[A,\zeta_A]$ is in $DF_{red}(x)\}|) \leq K$ for a natural number K, we employ the E-method. A threshold value K allows us to obtain minimal solutions in $DF(x)$. When we select a descriptor $[A,\zeta_A]$, *SEL* is revised to $SEL \cup \{[A,\zeta_A]\}$. However, absorption law often causes drastic reduction of a set $\{[A,\zeta_A]|[A,\zeta_A]$ is in $DF_{red}(x)\}$. The relation between K and the *ISETV*-method is as follows:

(Property 1) For large K, fewer selections are necessary and most of all minimal solutions are obtained. However, it takes much execution time in the E-method, since it is applied to a set $SEL \cup \{[A,\zeta_A]|[A,\zeta_A]$ is in $DF_{red}(x)\}$.

(Property 2) For small K, more selections are necessary and minimal solutions depending upon selections are obtained. In this case, it takes less execution time in the E-method.

4.3.7. *Programs for ISETV-method*

We have realized some programs on two computers PC_1 and PC_2. The *ISETV*-method is implemented on PC_1 with Pentium 4 (3.4 GHz) in prolog. Table 4.6 shows the examples of three *NISs*, which are artificial data generated by using a random number generation program. In reality, the number of derived *DISs* from NIS_3 is 701633053354829913752725347618521088. There is no restriction on data size in prolog, therefore we can obtain this number by using multiplication.

In Table 4.6, *Derived_DISs* means the number of all derived *DISs* for attributes AT. In Table 4.7, program *trans* generates *inf* and *sup* information for $[DEC, \eta]$. Program *allsub* first generates a discernibility function $DF(x)$ for an object $x \in OB(\{J\},(1))$, and generates all minimal certain rules from this object by using the E-method (the threshold value $K = 9$). According to Table 4.7, we may employ $K = 9$ or 10 for *NISs* with a large number of attributes. Namely, we set $K = 9$ or 10, and interactively select descriptors in $DF(x)$. If the condition of descriptors

Table 4.6. Definitions of three *NISs*.

NIS	$\|OB\|$	$\|AT\|$	$\|VAL_A\|$	*Derived_DISs*
NIS_2	50	10	10	1.57×10^{18}
NIS_3	100	10	10	7.01×10^{35}
NIS_4	300	10	10	6.74×10^{86}

Table 4.7. Execution time of programs.

NIS	$trans(sec)$	$\|OB(\{J\},(1))\|$	$allsub(sec)$	$Number\_of\_rules$
NIS_2	0.0	9	0.0	16
NIS_3	1.0	16	1.0	19
NIS_4	8.0	21	1.0	22

is satisfied, the E-method is automatically invoked. In Appendix 2, we show a real execution.

We gave some solutions to Issue 1-1 and 1-2. According to these solutions, we realized utility programs to obtain minimal certain rules in every *NIS*. Most of the discussion for Issue 1-1 and 1-2 including real execution of the $ISETV$-method is in Ref. 31, 33, 34.

4.3.8. *Possible Rule Generation*

For rule generation in IGC class, it is necessary to consider every possible tuple in $PT(x,(AT\backslash DEC))$, whose number of elements is $|\prod_{A\in AT\backslash DEC} g(x,A)|$. There exists the same number of discernibility functions. Therefore, it seems hard to solve Issue 1-3 in the same manner as Issue 1-2.

For rule generation in MA classes, there will be a large number of $[CON,\zeta]$ satisfying $inf(x,CON,\zeta)\subseteq sup(x,DEC,\eta)$, which is (2) in Theorem 4.1. This condition implies there is a derived *DIS*, where $\tau^x : [CON,\zeta] \Rightarrow [DEC,\eta]$ is consistent. In reality, *CON* may be a singleton set.

In any case, there will be lots of possible rules, and we need some other constraints for specifying possible rules. In the next Issue 2, we consider another framework including possible rule generation.

4.4. Issue 2: Rule Generation on the Basis of the Criterion Values in *NISs*

A rule in a *DIS* is often defined by an implication τ^x such that both $support(\tau^x)\geq\alpha$ and $accuracy(\tau^x)\geq\beta$ hold for the threshold values α and β. In a *NIS*, there are derived *DISs* due to the information incompleteness. We extend the definition of a rule in a *DIS* to the *lower* and *upper approximations* of a rule in a *NIS*. We also extend *Apriori* algorithm[38,39] in *DISs* to an algorithm in *NISs*, and realize utility programs for handling lower and upper approximations.

4.4.1. *Some Definitions and the Second Issue*

Definition 4.5. We define the following:

(1) *Lower approximation of a set of rules (from x) in* a *NIS*:
$Rule(x,\alpha,\beta,LA)=\{\tau^x|support(\tau^x)\geq\alpha$ and $accuracy(\tau^x)\geq\beta$ hold for each derived *DIS* in $DD(\tau^x,x,CON,DEC)\}$.

(2) *Lower approximation of a set of rules in* a *NIS*:
$Rule(\alpha, \beta, LA) = \cup_{x \in OB} Rule(x, \alpha, \beta, LA)$.
(3) *Upper approximation of a set of rules (from x) in* a *NIS*:
$Rule(x, \alpha, \beta, UA) = \{\tau^x | support(\tau^x) \geq \alpha$ and $accuracy(\tau^x) \geq \beta$ hold for some derived *DISs* in $DD(\tau^x, x, CON, DEC)\}$.
(4) *Upper approximation of a set of rules in* a *NIS*:
$Rule(\alpha, \beta, UA) = \cup_{x \in OB} Rule(x, \alpha, \beta, UA)$.
(5) *Lower approximation of a set of definite rules in* a *NIS*:
$Rule(\alpha, \beta, LA, def) = \{\tau \in Rule(\alpha, \beta, LA) | \tau$ is definite$\}$.
(6) *Lower approximation of a set of indefinite rules in* a *NIS*:
$Rule(\alpha, \beta, LA, indef) = \{\tau \in Rule(\alpha, \beta, LA) | \tau$ is indefinite$\}$.

Intuitively, every $\tau \in Rule(\alpha, \beta, LA, def)$ is a possible implication, which is not influenced by the information incompleteness. Similarly, every $\tau \in Rule(\alpha, \beta, UA)$ is a possible implication, which satisfies the conditions in some derived *DISs*.

(Issue 2-1) Realize a method to obtain $Rule(\alpha, \beta, LA)$ and $Rule(\alpha, \beta, UA)$.

We have the following propositions according to Theorem 4.3.

Proposition 4.4. *For a NIS, let us consider a possible implication τ. Then, the following are equivalent.*

(1) $\tau \in Rule(\alpha, \beta, LA)$.
(2) $minsupp(\tau^x) \geq \alpha$ *and* $minacc(\tau^x) \geq \beta$ *(for an object x).*
(3) *There exists τ^x such that $support(\tau^x) \geq \alpha$ and $accuracy(\tau^x) \geq \beta$ in a DIS_{worst}.*

Proposition 4.5. *For a NIS, let us consider a possible implication τ. Then, the following are equivalent.*

(1) $\tau \in Rule(\alpha, \beta, UA)$.
(2) $maxsupp(\tau^x) \geq \alpha$ *and* $maxacc(\tau^x) \geq \beta$ *(for an object x).*
(3) *There exists τ^x such that $support(\tau^x) \geq \alpha$ and $accuracy(\tau^x) \geq \beta$ in a DIS_{best}.*

Proposition 4.6. *For a NIS, let us consider a possible implication τ. Then, the following holds.*

(1) *τ is a certain rule, if and only if $\tau \in Rule(1/|OB|, 1.0, LA, def)$.*
(2) *τ is a possible rule, if and only if $\tau \in Rule(1/|OB|, 1.0, UA)$.*

4.4.2. *Definitions of descinf and descsup Instead of inf and sup*

According to Theorem 4.2, the calculation of $minsupp(\tau^x)$, $minacc(\tau^x)$, $maxsupp(\tau^x)$ and $maxacc(\tau^x)$ for a fixed τ^x is easy. However, τ^x depends upon the number of derived *DISs* and condition attributes *CON* ($CON \subseteq 2^{AT \setminus DEC}$), furthermore τ^x depends upon an object x. Therefore, it is hard to apply Theorem 4.2 to every τ^x sequentially.

To solve this, we focus on descriptors $[A, \zeta_A]$. The number of all descriptors is pretty small.

Definition 4.6. For every descriptor $[A, \zeta_A]$ in a *NIS*, we define the following:

(1) $descinf([A, \zeta_A]) = \{x \in OB | g(x, A) = \{\zeta_A\}\}$.
(2) $descinf(\wedge_i [A_i, \zeta_i]) = \cap_i descinf([A_i, \zeta_i])$.
(3) $descsup([A, \zeta_A]) = \{x \in OB | \zeta_A \in g(x, A)\}$.
(4) $descsup(\wedge_i [A_i, \zeta_i]) = \cap_i descsup([A_i, \zeta_i])$.

The definition of $inf(x, CON, \zeta)$ and $descinf([CON, \zeta])$ is almost, but not quite, the same, since $inf(x, CON, \zeta)$ depends upon an object x, but $descinf([CON, \zeta])$ just depends upon the descriptor $[CON, \zeta]$. Therefore, the manipulation on $descinf([CON, \zeta])$ is much simpler than that of $inf(x, CON, \zeta)$. As for $sup(x, CON, \zeta)$ and $descsup([CON, \zeta])$, they define the same set for every x. In correspondence with $descinf([CON, \zeta])$, we define $descsup([CON, \zeta])$.

4.4.3. *Possible Implication and Minsupp, Minacc Values*

In this subsection, we clarify the relation between each possible implication and both *minsupp* and *minacc*. The details are in Ref. 35.

Proposition 4.7. *Let* $[CON, \zeta]$ *denote* $\wedge_{A \in CON}[A, \zeta_A]$ *and* $[DEC, \eta]$ *denote* $\wedge_{B \in DEC}[B, \eta_B]$, *and let us suppose* $x \in descsup([CON, \zeta] \wedge [DEC, \eta])$.

(1) *If* $x \in descinf([CON, \zeta] \wedge [DEC, \eta])$, *we can obtain a possible implication*

$$\tau^x : [CON, \zeta] \Rightarrow [DEC, \eta] \text{ from object } x, \text{ which satisfies}$$
$$minsupp(\tau^x) = |descinf([CON, \zeta] \wedge [DEC, \eta])| / |OB|.$$

(2) *If* $x \in descsup([CON, \zeta] \wedge [DEC, \eta]) \backslash descinf([CON, \zeta] \wedge [DEC, \eta])$, *we can obtain a possible implication* $\tau^x : [CON, \zeta] \Rightarrow [DEC, \eta]$ *from object* x, *which satisfies*

$$minsupp(\tau^x) = (|descinf([CON, \zeta] \wedge [DEC, \eta])| + 1) / |OB|.$$

Proposition 4.8. *Let us suppose the same condition as in Proposition* 4.7, *and let* $OUTMINACC$ *denote a set* $[descsup([CON, \zeta]) \setminus descinf([CON, \zeta])] \setminus descinf([DEC, \eta])$.

(1) *If* $x \in descinf([CON, \zeta] \wedge [DEC, \eta])$, *we can obtain a possible implication*

$$\tau^x : [CON, \zeta] \Rightarrow [DEC, \eta] \text{ from object } x, \text{ which satisfies}$$
$$minacc(\tau^x) = Numerator/Denominator,$$
$$Here, Numerator = |descinf([CON, \zeta] \wedge [DEC, \eta])|,$$
$$Denominator = |descinf([CON, \zeta])| + |OUTMINACC|.$$

(2) *If* $x \in descsup([CON,\zeta] \wedge [DEC,\eta]) \backslash descinf([CON,\zeta] \wedge [DEC,\eta])$, *we can obtain a possible implication* $\tau^x : [CON,\zeta] \Rightarrow [DEC,\eta]$ *from object* x, *which satisfies*

$$minacc(\tau^x) = Numerator/Denominator,$$
$$Here, Numerator = |descinf([CON,\zeta] \wedge [DEC,\eta])| + 1,$$
$$Denominator = |descinf([CON,\zeta]) \cup \{x\}| + |OUTMINACC \backslash \{x\}|.$$

4.4.4. *Possible Implications and Maxsupp, Maxacc Values*

In this subsection, we clarify the relation between each possible implication and both *maxsupp* and *maxacc*. The details are also in Ref. 35.

Proposition 4.9. *Let us suppose the same condition as in Proposition* 4.8. *We can obtain a possible implication* $\tau^x : [CON,\zeta] \Rightarrow [DEC,\eta]$ *from object* x, *which satisfies*

$$maxsupp(\tau^x) = |descsup([CON,\zeta] \wedge [DEC,\eta])|/|OB|.$$

Proposition 4.10. *Let us suppose the same condition as in Proposition* 4.8, *and let INMAXACC denote a set* $[descsup([CON,\zeta]) \setminus descinf([CON,\zeta])] \cap descsup([DEC,\eta])$.

(1) *If* $x \in descinf([CON,\zeta] \wedge [DEC,\eta])$, *we can obtain a possible implication*

$$\tau^x : [CON,\zeta] \Rightarrow [DEC,\eta] \text{ from object } x, \text{ which satisfies}$$
$$maxacc(\tau^x) = Numerator/Denominator,$$
$$Here, Numerator = |descinf([CON,\zeta]) \cap descsup([DEC,\eta])| + |INMAXACC|,$$
$$Denominator = |descinf([CON,\zeta])| + |INMAXACC|.$$

(2) *If* $x \in descsup([CON,\zeta] \wedge [DEC,\eta]) \backslash descinf([CON,\zeta] \wedge [DEC,\eta])$, *it is possible to obtain a possible implication* $\tau^x : [CON,\zeta] \Rightarrow [DEC,\eta]$ *from object* x, *which satisfies*

$$maxacc(\tau^x) = Numerator/Denominator,$$
$$Here, Numerator = |(descinf([CON,\zeta]) \cap descsup([DEC,\eta])) \backslash \{x\}| + |INMAXACC \backslash \{x\}| + 1,$$
$$Denominator = |descinf([CON,\zeta]) \cup \{x\}| + |INMAXACC \backslash \{x\}|.$$

4.4.5. *An Example of Rule Generation on the Basis of the Criterion Values*

This subsection extends *Apriori* algorithm,[38,39] and show an example for generating *Rule*(α, β, LA).

Table 4.8. descinf and descsup information in NIS_1.

	$[Temp, high]$	$[Temp, normal]$	$[Head, no]$	$[Head, yes]$
descinf	$\{1, 4, 5\}$	$\{6, 7\}$	$\{3, 7\}$	$\{2, 4, 6, 8\}$
descsup	$\{1, 2, 3, 4, 5, 8\}$	$\{3, 6, 7, 8\}$	$\{1, 3, 5, 7\}$	$\{1, 2, 4, 5, 6, 8\}$
	$[Nausea, no]$	$[Nausea, yes]$	$[Flu, no]$	$[Flu, yes]$
descinf	$\{1, 3\}$	$\{2, 5, 7\}$	$\{3, 5, 6, 7\}$	$\{1, 2, 4, 8\}$
descsup	$\{1, 3, 4, 6, 8\}$	$\{2, 4, 5, 6, 7, 8\}$	$\{3, 5, 6, 7\}$	$\{1, 2, 4, 8\}$

Table 4.9. descinf and descsup information satisfying either CASE 1 or CASE 2.

	$[Temp, normal] \wedge [Flu, no]$	$[Head, no] \wedge [Flu, no]$	$[Nausea, yes] \wedge [Flu, no]$
descinf	$\{6, 7\}$	$\{3, 7\}$	$\{5, 7\}$
descsup	$\{3, 6, 7\}$	$\{3, 5, 7\}$	$\{5, 6, 7\}$

	$[Temp, high] \wedge [Flu, yes]$	$[Head, yes] \wedge [Flu, yes]$
descinf	$\{1, 4\}$	$\{2, 4, 8\}$
descsup	$\{1, 2, 4, 8\}$	$\{1, 2, 4, 8\}$

Example 4.5. Let us consider *descinf* and *descsup*, which are obtained in NIS_1. We set $\alpha = 0.3$, $\beta = 0.6$, condition attribute $CON \subseteq \{Temp, Head, Nausea\}$ and decision attribute $DEC = \{Flu\}$. Since $|OB|$=8, $minsupp(\tau) = |SET|/8 \geq 0.3$ must hold. Thus, $|SET| \geq 3$ must hold.

(STEP 1) In the first step, we generate Table 4.8 satisfying either of the following.
(CASE A) $|descinf([A, \zeta_A])| \geq 3$ ($A \in \{Temp, Head, Nausea, Flu\}$),
(CASE B) $|descinf([A, \zeta_A])|$=2 and $|descsup([A, \zeta_A])| \geq 3$.

In CASE A, we may obtain a definite possible implication, and in CASE B we may obtain an indefinite possible implication. Other implications do not satisfy the condition $|SET| \geq 3$.

(STEP 2) In the next step, we sequentially generate conjunctions $[A, \zeta_A] \wedge [Flu, val]$ ($A \in \{Temp, Head, Nausea\}$). For these conjunctions, we calculate *descinf* and *descsup* according to Definition 4.6. Then, we examine the condition of CASE A and CASE B. Table 4.9 shows conjunctions satisfying either CASE A or CASE B.
The conjunction $[Temp, normal] \wedge [Flu, no]$ in Table 4.9 means that a possible implication τ^3 : $[Temp, normal] \Rightarrow [Flu, no]$ satisfies the condition $|SET| \geq 3$. For τ^6, object 3 works to decrease both $support(\tau^6)$ and $accuracy(\tau^6)$. Namely, $(\tau')^3$ ($\tau' \neq \tau$) is employed from object 3. Thus, τ^6 nor τ^7 do not satisfy the condition $|SET| \geq 3$. Then, we employ (2) in Proposition 4.8, and obtain $minacc(\tau^3) = 0.75$. In this way, we conclude $\tau^3 \in Rule(0.3.0.6, LA)$, and we remove this conjunction from Table 4.9. For other conjunctions, the same procedure is applied, and we obtain some rules

and the revised table. For this revised table, we recursively repeat STEP 2 until this table is empty. We show the real execution of this example in Appendix 3.

In the above procedure, every calculation just depends upon *descinf* and *descsup*, and the calculation is very simple. The most time-consuming part will be to generate conjunctions of descriptors in Table 4.9.

4.4.6. *Algorithms for Rule Generation on the Basis of the Criterion Values*

Now, we show two algorithms for $Rule(\alpha, \beta, LA)$ and $Rule(\alpha, \beta, UA)$.

Algorithm 4.1 (An Algorithm for Lower Approximation: $Rule(\alpha, \beta, LA)$).
Input: A *NIS*, a decision attribute *DEC*, threshold value α and β.
Output: Every rule defined by $Rule(\alpha, \beta, LA)$.

1. For the condition $minsupp(\tau) = |SET|/|OB| \geq \alpha$, obtain the number *NUM* of elements in *SET*;
2. Generate a set *CANDIDATE*(1), which consists of descriptors $[A, \zeta_A]$ satisfying either (*CASE* A) or (*CASE* B) in the following;
   (*CASE* A) $|descinf([A, \zeta_A])| \geq NUM$,
   (*CASE* B) $|descinf([A, \zeta_A])| = NUM - 1$ and $|(descsup([A, \zeta_A])| \geq NUM$.
3. Generate a set *CANDIDATE*(2) according to the following procedures;
   (Proc 3-1) For every $[A, \zeta_A]$ and $[DEC, \zeta_{DEC}]$ ($A \neq DEC$) in $CANDIDATE(1)$, generate a new descriptor $[\{A, DEC\}, (\zeta_A, \zeta_{DEC})]$;
   (Proc 3-2) Examine condition (*CASE* A) and (*CASE* B) for each $[\{A, DEC\}, (\zeta_A, \zeta_{DEC})]$ and each object x;
   If either (*CASE* A) or (*CASE* B) holds and $minacc(\tau^x) \geq \beta$
   display $\tau^x : [A, \zeta_A] \Rightarrow [DEC, \zeta_{DEC}]$ as a rule;
   If either (*CASE* A) or (*CASE* B) holds and $minacc(\tau^x) < \beta$,
   add this descriptor to $CANDIDATE(2)$;
4. Assign 2 to n;
5. while $CANDIDATE(n) \neq \{\}$, generate $CANDIDATE(n+1)$ according to the following procedures;
   (Proc 5-1) For $DESC_1$ and $DESC_2$ ($[DEC, \zeta_{DEC}] \in DESC_1 \cap DESC_2$) in $CANDIDATE(n)$, generate a new descriptor by using a conjunction of $DESC_1 \wedge DESC_2$;
   (Proc 5-2) Examine the same procedure as (Proc 3-2).
   Assign $n + 1$ to n;
6. end.

Algorithm 4.2 (An Algorithm for Upper Approximation: $Rule(\alpha, \beta, UA)$). Input: A *NIS*, a decision attribute *DEC*, threshold value α and β.
Output: Every rule defined by $Rule(\alpha, \beta, UA)$.

Algorithm 4.2 is proposed as Algorithm 4.1 with the following two revisions:

1. (*CASE* A) and (*CASE* B) in Algorithm 4.1 are replaced with (*CASE* C).
   (*CASE* C) $|descsup([A, \zeta_A])| \geq NUM$.
2. $minacc(\tau^x)$ in Algorithm 4.1 is replaced with $maxacc(\tau^x)$.

We follow *Apriori* algorithm, and extend it to algorithms in *NISs*. *Apriori* algorithm employs a large item set, which corresponds to an equivalence class for a descriptor. On the other hand, Algorithm 4.1 and 4.2 employ two classes, i.e., *descinf* and *descsup*. Due to some theorems, propositions and the manipulation of *descinf* and *descsup*, Algorithm 4.1 and 4.2 can pick up a DIS_{worst} and a DIS_{best} for τ^x, and two algorithms calculate four values, i.e., $minsupp(\tau^x)$, $minacc(\tau^x)$, $maxsupp(\tau^x)$ and $maxacc(\tau^x)$. In this way, Algorithm 4.1 and 4.2 handle non-deterministic information as well as deterministic information. In Algorithm 4.1 and 4.2, it takes two steps of *Apriori* algorithm for manipulating *descinf* and *descsup*. Since the rest is the same, the complexity of Algorithm 4.1 and 4.2 is almost the same as *Apriori* algorithm.

4.4.7. *An Attempt of Applying Utility Programs to Data in UCI Machine Learning Repository*

We are planning to apply our utility programs to data in the UCI Machine Learning Repository.[40] Here, we briefly show the application to the Mammographic data. This original data consists of 961 objects and 6 attributes, i.e., *BI-RADS assessment, Age, Shape, Margin, Density, Severity.* We obtained 150 objects from the top of the data. We name this data set *mammo.dat*. In *mammo.dat*, the ? symbol is marked for every 76 missing value. The number of objects without missing values is 91. For handling a total of 150 objects, we revised *mammo.dat* to the following syntax.

```
150                              /* the number of objects */
6                                /* the number of attributes */
5 60 3 5 3 1                     /* table data */
4 40 1 1 [1,2,3,4] 1             /* ? is replaced with [1,2,3,4] */
5 50 4 5 3 1
4 20 1 1 3 0
5 70 1 5 [1,2,3,4] 1
4 60 1 [1,2,3,4,5] 3 0
4 70 [1,2,3,4] [1,2,3,4,5] 3 0
       :           :           :
```

There are $4^{55} \times 5^{21}$ derived *DISs* for these 150 objects. Probably, it seems hard to enumerate all derived *DISs* sequentially. In the first attribute (*BI-RADS assessment*), 1 (definitely benign) to 5 (highly suggestive of malignancy) were assigned in a double-review process by physicians. We omitted this attribute and employed the remaining four attributes *Age, Shape, Margin, Density* to defining the *Severity*. For this data, we obtained *Rule*$(0.2, 0.5, LA)$ and *Rule*$(0.2, 0.5, UA)$ in the following:

$$
\begin{aligned}
&Rule(0.2, 0.5, LA, def) = Rule(0.2, 0.5, LA)\\
&=\{[Shape, 1] \Rightarrow [Severity, 0], [Shape, 2] \Rightarrow [Severity, 0],\\
&\quad [Margin, 1] \Rightarrow [Severity, 0], [Shape, 4] \Rightarrow [Severity, 1]\},\\
&Rule(0.2, 0.5, UA)\backslash Rule(0.2, 0.5, LA)\\
&=\{[Density, 3] \Rightarrow [Severity, 0], [Density, 3] \Rightarrow [Severity, 1]\}.
\end{aligned}
$$

If we employ a lower approximation strategy to this data, we may miss two implications on *Density* in *Rule*$(0.2, 0.5, UA)\backslash Rule(0.2, 0.5, LA)$. We may be confused by these two implications, because these implications are inconsistent. However, both *maxacc* values are more than 0.6. This comes from the interpretation of DIS_{best}. In the attribute *Density*, 47 attribute values are non-deterministic information, namely every ? symbol is replaced with [1,2,3,4]. The selections from these attribute values cause *maxacc* values.

4.5. Issue 3: Rule Generation in Tables with Numerical Values

In most of *DISs* and *NISs*, we implicitly handle tables with categorical data. As for tables with numerical values, multivariate analysis may often be employed. In order to enhance *RNIA*, it is necessary to investigate rough sets based information analysis for numerical data, too. We introduce numerical patterns into numerical values, and define equivalence classes based on these patterns. Due to this introduction, it is possible to handle the precision of information, namely it is possible to define *fine* information and *coarse* information. These fine and coarse concepts cause more flexible information analysis, including rule generation, for numerical data. The details of this section are in Refs. 32, 36.

4.5.1. *An Exemplary Data with Numerical Values*

Let us consider Table 4.10. There are four persons, and Table 4.10 stores a relation between *Sex, Height* and *Weight*.

For such numerical data, multivariate analysis may usually be applied. According to regression analysis in Microsoft Excel, we easily obtained a regression line

Table 4.10. Exemplary deterministic information system with numerical values.

Person	*Sex*	*Height (cm)*	*Weight (kg)*
1	*male*	162	64.3
2	*female*	162	64.5
3	*female*	164	65.8
4	*male*	175	72.8

from *Height* to *Weight*

$$\text{Weight} = 0.6445 \times \text{Height} - 39.982.$$

Furthermore, we may soon observe such an implication that

$$[Height, 160s] \Rightarrow [Weight, 60s].$$

In some cases, this implication may be more informative than the regression line.

However, we may not obtain this implication according to the typical rough set theory, because such descriptors [Height,160s] nor [Weight,60s] have not been employed yet. In reality, in Table 4.10, there exists a set of equivalence classes on each attribute in the following:

$$\text{eq}(\{\text{Sex}\}) = \{\{1,4\},\{2,3\}\},\ \text{eq}(\{\text{Height}\}) = \{\{1,2\},\{3\},\{4\}\},$$
$$\text{eq}(\{\text{Weight}\}) = \{\{1\},\{2\},\{3\},\{4\}\}.$$

In this case, $eq(\{Sex\}) \not\subseteq eq(\{Weight\})$ and $eq(\{Height\}) \not\subseteq eq(\{Weight\})$ hold, so we do not recognize the consistency from *Sex* to *Weight* nor *Height* to *Weight*. In order to find the consistency, the combination of attributes is usually considered. Namely, $eq(\{Sex,Height\}) = \{\{1\},\{2\},\{3\},\{4\}\}$ is generated, and $eq(\{Sex,Height\}) \subseteq eq(\{Weight\})$ is examined. After this, four consistent implications including the following are recognized.

$$[Sex,male] \wedge [Height, 162] \Rightarrow [Weight, 64.3] \text{ from object 1}$$

However, this implication just represents object 1.

4.5.2. *A Proposal of Meaningful Figures in Numerical Values*

Now in this subsection, we propose a concept of *meaningful figures* in numerical values. Let us consider an irrational number $\pi = 3.14\cdots$. For students in elementary schools, $\pi = 3.14$ will be sufficient for calculating the area of a circle. On the other hand, $\pi = 3.14$ may be insufficient for researchers of numerical analysis. Furthermore, let us suppose the real total price is \$14976.5 for buying a car. In this case, we may see \$15000 is enough to buy this car. The \$0.5 in \$14976.5 must be meaningless. Because of these two examples, it seems important to consider the concept of meaningful figures in numerical data. In the subsequent sections, we introduce this concept of meaningful figures into rough set theory.

4.5.3. *Numerical Patterns and Equivalence Relations*

In order to handle the concept of meaningful figures, we introduce two symbols @ and #, which represent numeric values from 0 to 9. The @ symbol means a significant figure, and the # symbol means a figure, which we do not care about. We implicitly employ 10-adic numbers for expressing numerical data, however we may employ 2-adic or 8-adic numbers.

A *numerical pattern* is a sequence of @ and #, for example @@@, @@#, @##, @@.@ and @#.#. Here, '.' denotes a decimal point, and @ does not occur after #. We see @@@, @@#, @## and ### have the same *type*. Three patterns @@.@, @@.# @#.# have the same type, too. For example, two values of weight 64.3 Kg and 64.5 Kg are different for a numerical pattern @@.@. However, two values are the same for a pattern @@.#. We define the meaningful figures in numerical values based on such numerical patterns.

Clearly, it is possible to introduce an *order* ($<_p$) on *precision* into a set of patterns with the same type. For example, in the attribute *Height*

$$@@@ <_p @@\# <_p @\#\# <_p \#\#\#,$$

and in the attribute *Weight*

$$@@.@ <_p @@.\# <_p @\#.\# <_p \#\#.\#.$$

The pattern @@@ is the most precise, and the pattern ### is the least precise in the attribute *Height*. Such numerical patterns cause different kinds of equivalence relations for the same data. For a set *ATR* of attributes and a numerical pattern *pat*, let *eq(ATR,pat)* denote a set of equivalence classes based on *pat*. In Table 4.10, we have the following:

$$\begin{aligned}
&\text{eq(Height, \#\#\#)} = \{\{1,2,3,4\}\} && \text{eq(Weight, \#\#.\#)} = \{\{1,2,3,4\}\}\\
&\text{eq(Height, @\#\#)} = \{\{1,2,3,4\}\} && \text{eq(Weight, @\#.\#)} = \{\{1,2,3\},\{4\}\}\\
&\text{eq(Height, @@\#)} = \{\{1,2,3\},\{4\}\} && \text{eq(Weight, @@.\#)} = \{\{1,2\},\{3\},\{4\}\}\\
&\text{eq(Height, @@@)} = \{\{1,2\},\{3\},\{4\}\} && \text{eq(Weight, @@.@)} = \{\{1\},\{2\},\{3\},\{4\}\}
\end{aligned}$$

Proposition 4.11. *For numerical patterns pat_1 and pat_2 with the same type, let us suppose $pat_1 <_p pat_2$. Then, there exists $M \in eq(ATR, pat_2)$ such that $N \subseteq M$ for any $N \in eq(ATR, pat_1)$.*

The concept of precision has already been introduced into rough set theory.[1] According to the definition in Ref. 1, we say $eq(ATR, pat_1)$ is *more fine*, and we say $eq(ATR, pat_2)$ is *more coarse*. Recently such fine and coarse information is picked up in Refs. 41, 42. In Ref. 41, *zoom in* and *zoom out* operations are proposed. For a numerical pattern @@#, @@@ takes the role of the *zoom in* operation, and @## takes the role of the *zoom out* operation. Equivalence classes *eq(ATR, pat)*, which depend upon the numerical pattern *pat*, give a real instance of the framework defined in Refs. 1, 41, 42.

The following example clarifies the new property on rule generation with numerical patterns.

Example 4.6. Let us consider Table 4.10, again. According to the typical rough sets based method, an implication

$$[Height, 162] \Rightarrow [Weight, 64.3] \text{ from object 1}$$

is inconsistent with

$$[Height, 162] \Rightarrow [Weight, 64.5] \text{ from object 2.}$$

For obtaining a consistent rule from object 1, it is necessary to add another descriptor to the condition part, and we obtain

$$[Sex,male] \wedge [Height, 162] \Rightarrow [Weight, 64.3] \text{ from object 1.}$$

However, there exist other two cases, which depend upon the numerical patterns, in the following:

(1) Let us consider a pattern @@.# for *Weight*. In this case, $eq(Height, @@@) \subseteq eq(Weight, @@.\#)$ holds, so every implication in the form of $[Height, @@@] \Rightarrow [Weight, @@.\#]$ is consistent. Thus, the next implication is consistent.

$$[Height, 162] \Rightarrow [Weight, 64.\#] \text{ from objects 1 and 2.}$$

(2) Let us consider a pattern @#.# for *Weight*. In this case, $eq(Height, @@\#) \subseteq eq(Weight, @\#.\#)$ holds, and an implication

$$[Height, 16\#] \Rightarrow [Weight, 6\#.\#] \text{ from objects 1, 2 and 3}$$

is also consistent. In this way, we can define consistent implications which have not been defined yet.

4.5.4. *Rule Generation in Numerical Data*

Now, we show the third issue in this chapter. For solving Issue 3-1, we extend *Apriori* algorithm, again.

(Issue 3-1) In a *DIS* with numerical values, realize a method to obtain a set $\{\tau | \tau$ is an implication with numerical patterns, $support(\tau) \geq \alpha$ and $accuracy(\tau) \geq \beta\}$.

Algorithm 4.3. (An Overview of Rule Generation in a *DIS* with Numerical Values)

1. Fix a threshold α and a decision attribute *DEC*.
2. Let *OB* denote the total set of objects. Find candidates of descriptors with numerical patterns,

$$CAN_1 = \{[A, val_{A,pat}] \,|\, |[x]_{A,val_{A,pat}}| \geq \alpha \times |OB| \text{ holds for an object } x\}.$$

3. Generate CAN_2;

$$CAN_2 = \{[A, val_{A,pat1}] \wedge [DEC, val_{DEC,pat2}] \mid$$
$$|[x]_{\{A,DEC\},\{val_{A,pat1},val_{DEC,pat2}\}}| \geq \alpha \times |OB| \text{ holds for an object } x,$$
$$[A, val_{A,pat1}] \in CAN_1, A \neq DEC\},$$

and examine $accuracy(\tau)$ in every implication τ

$$\tau : [A, val_{A,pat1}] \Rightarrow [DEC, val_{DEC,pat2}].$$

If $accuracy(\tau) \geq \beta$, this implication satisfies the condition. We store this implication, and remove $[A, val_{A,pat1}] \wedge [DEC, val_{DEC,pat2}]$ from CAN_2. We also remove the conjunction, which implies a redundant implication previously obtained.
4. Generate CAN_n recursively and obtain rules until $CAN_n = \{\}$.

4.5.5. *An Application of Utility Programs*

The utility program with the user interface is also implemented in C# on the second computer PC_2 with Pentium 4 (3.4GHz). We refer to the application of utility programs to *bunpa.csv* in the UCI Repository.[40] This *bunpa.csv* stores diseases and symptoms of the liver. In order to understand the meaning of every attribute, we need some special knowledge. In this application, we omit the details of such special knowledge. We just deal with *bunpa.csv* as data with numerical values. This data consists of 345 objects and 7 attributes. There is no missing value. The following are the attributes with attribute values and real data.

```
1.MCV:Continuous, 2.ALP:Continuous, 3.SGPT: Continuous,
4.AST:Continuous, 5.GGT:Continuous, 6.Drinks: Continuous,
7.Class:1,2

85,92,45,27,31,0,1
85,64,59,32,23,0,2
86,54,33,16,54,0,2
91,78,34,24,36,0,2
  :    :    :
```

In Appendix 4, we show two examples of the execution. In the first example, in which we did not employ numerical patterns, we could not obtain any rules for *support* = 0.5 and *accuracy* = 0.5. However in the second example, in which we employed numerical patterns, we obtained 11 rules. These examples show us the merits of handling numerical patterns.

4.5.6. *Comparison with Previous Research Results*

There is a lot of research about the discretization of numerical values, like interval theory and cluster analysis. As for rough sets based rule generation, discretization is also investigated in Refs. 43, 44. In Refs. 43, 44, the issue of how to divide an interval into subintervals is mainly considered, and entropy function based discretization is proposed. This method is also employed in decision tree generation, like C4.5. Rules are generated from such discretized data.

In our research, the purpose is to handle the coarse and fine concepts, which are defined by descriptors with numerical patterns, in rule generation. For example, $[Weight, 6\#.\#]$ in Table 4.10 implies that the weight is 60 s Kg. Such information is familiar in our life. In order to handle such coarse information explicitly in rough set theory, we introduced numerical patterns into rough sets based rule generation. The purpose of our research is not to discretize numerical values, but every numerical pattern defines a discretization over attribute values as a result.

4.6. Concluding Remarks

This chapter focused on three issues related to rule generation in *RNIA*, and examined every issue according to Refs. 27–36. For every issue, we implemented utility programs and applied them to some data sets. Due to these utility programs, we can explicitly handle not only deterministic information but also non-deterministic information. We think these utility programs are quite unique.

However, we have not analyzed the real meaning of the obtained rules yet. Because, we have focused on the framework of *RNIA* and the algorithms for implementation. From now on, we will apply these utility programs to real data with non-deterministic information, incomplete information and missing values, and we want to obtain meaningful rules in *NISs*.

Three issues in this chapter are investigated in parallel, and every utility program has been developed independently. It will be necessary to combine these utility programs to a total system for *NISs*. We will enhance the framework of *RNIA* through these issues.

Acknowledgments

The authors would like to thank anonymous referees and Professor John Alan Fulcher for their useful comments. We are grateful to Professor Kazumi Nakamatsu. We also thank Mr.Kazuhiro Koba, Mr.Ryuji Ishibashi and Mr.Kohei Hayashi for their efforts in implementing the utility programs.

More than four years have passed from the preparation of this paper. Therefore, the contents of RNIA are energetically developed. We hope that readers also see a chapter on RNIA in a new book "Emerging Paradigms in Machine Learning" by Springer (Edited by S. Ramanna, R.J. Howlett and L.C. Jain, SIST Series, Vol. 13, pp. 215–247) including the latest work on RNIA.

References

1. Z. Pawlak: Rough Sets, Kluwer Academic Publisher, 1991.
2. Z. Pawlak: Some Issues on Rough Sets, Transactions on Rough Sets, Int'l. Rough Set Society, Vol. 1, pp. 1–58, 2004.

3. J. Komorowski, Z. Pawlak, L. Polkowski and A. Skowron: Rough Sets: a tutorial, Rough Fuzzy Hybridization, Springer, pp. 3–98, 1999.
4. L. Polkowski and A. Skowron (eds.): Rough Sets in Knowledge Discovery 1, Studies in Fuzziness and Soft Computing, Vol. 18, Physica-Verlag, 1998.
5. L. Polkowski and A. Skowron (eds.): Rough Sets in Knowledge Discovery 2, Studies in Fuzziness and Soft Computing, Vol. 19, Physica-Verlag, 1998.
6. S. Tsumoto: Knowledge Discovery in Clinical Databases and Evaluation of Discovered Knowledge in Outpatient Clinic, Information Sciences, Vol. 124, pp. 125–137, 2000.
7. A. Nakamura, S. Tsumoto, H. Tanaka and S. Kobayashi: Rough Set Theory and Its Applications, Journal of Japanese Society for AI, Vol. 11, No. 2, pp. 209–215, 1996.
8. W. Ziarko: Variable Precision Rough Set Model, Journal of Computer and System Sciences, Vol. 46, pp. 39–59, 1993.
9. J. Grzymała-Busse: A New Version of the Rule Induction System LERS, Fundamenta Informaticae, Vol. 31, pp. 27–39, 1997.
10. Rough Set Software, Bulletin of Int'l. Rough Set Society, Vol. 2, pp. 15–46, 1998.
11. W. Lipski: On Semantic Issues Connected with Incomplete Information Data Base, ACM Trans. DBS, Vol. 4, pp. 269–296, 1979.
12. W. Lipski: On Databases with Incomplete Information, Journal of the ACM, Vol. 28, pp. 41–70, 1981.
13. E. Orłowska(ed.): Incomplete Information: Rough Set Analysis, Physica-Verlag, 1998.
14. E. Orłowska and Z. Pawlak: Representation of Nondeterministic Information, Theoretical Computer Science, Vol. 29, pp. 27–39, 1984.
15. S. Demri and E. Orłowska: Incomplete Information: Structure, Inference, Complexity, Monographs in Theoretical Computer Science, Springer, 2002.
16. E. Orłowska: A Roadmap of Information Logics and Information Algebras Inspired by Rough Sets, Plenary Workshop in Rough Sets, Fuzzy Sets, Data Mining and Granular Computing, 2005.
17. J. Grzymała-Busse and P. Werbrouck: On the Best Search Method in the LEM1 and LEM2 Algorithms, Incomplete Information: Rough Set Analysis, Phisica-Verlag, pp. 75–91, 1998.
18. J. Grzymała-Busse: Data with Missing Attribute Values: Generalization of Indiscernibility Relation and Rule Induction, Transactions on Rough Sets, Int'l. Rough Set Society, Vol. 1, pp. 78–95, 2004.
19. J. Stefanowski and A. Tsoukias: On the Extension of Rough Sets under Incomplete Information, Lecture Notes in AI, Springer-Verlag, Vol. 1711, pp. 73–81, 1999.
20. J. Stefanowski and A. Tsoukias: Incomplete Information Tables and Rough Classification, Computational Intelligence, Vol. 7, pp. 212–219, 2001.
21. M. Kryszkiewicz: Rules in Incomplete Information Systems, Information Sciences, Vol. 113, pp. 271–292, 1999.
22. M. Kryszkiewicz and H. Rybinski: Computation of Reducts of Composed Information Systems, Fundamenta Informaticae, Vol. 27, pp. 183–195, 1996.
23. M. Kryszkiewicz: Maintenance of Reducts in the Variable Precision Rough Sets Model, ICS Research Report 31/94, Warsaw University of Technology, 1994.
24. M. Nakata and S. Miyamoto: Databases with Non-deterministic Information, Bulletin of Int'l. Rough Set Society, Vol. 7, pp. 15–21, 2003.
25. M. Nakata and H. Sakai: Rough-set-based Approaches to Data Containing Incomplete Information: Possibility-based Cases, In: Nakamatsu, K. and Abe, J. (Eds.), Advances in Logic Based Intelligent Systems, Frontiers in Artificial Intelligence and Applications, Vol. 132, IOS Press, pp. 234–241, 2005.
26. M. Nakata and H. Sakai: Lower and Upper Approximations in Data Tables Containing Possibilistic Information, Transactions on Rough Sets, Vol. 7, pp. 170–189, 2007.

27. H. Sakai: Effective Procedures for Handling Possible Equivalence Relations in Non-deterministic Information Systems, Fundamenta Informaticae, Vol. 48, pp. 343–362, 2001.
28. H. Sakai: Effective Procedures for Data Dependencies in Information Systems, Rough Set Theory and Granular Computing, Studies in Fuzziness and Soft Computing, Vol. 125, Springer-Verlag, pp. 167–176, 2003.
29. H. Sakai and A. Okuma: Basic Algorithms and Tools for Rough Non-deterministic Information Analysis, Transactions on Rough Sets, Int'l. Rough Set Society, Vol. 1, pp. 209–231, 2004.
30. H. Sakai: Possible Equivalence Relations and Their Application to Hypothesis Generation in Non-deterministic Information Systems, Transaction on Rough Sets, Int'l. Rough Set Society, Vol. 2, pp. 86–112, 2004.
31. H. Sakai and M. Nakata: Discernibility Functions and Minimal Rules in Non-deterministic Information Systems, Lecture Notes in AI, Springer-Verlag, Vol. 3641, pp. 254–264, 2005.
32. H. Sakai, T. Murai and M. Nakata: On a Tool for Rough Non-deterministic Information Analysis and Its Perspective for Handling Numerical Data, Lecture Notes in AI, Springer-Verlag, Vol. 3558, pp. 203–214, 2005.
33. H. Sakai: On a Rough Sets Based Data Mining Tool in Prolog: An Overview, Declarative Programming for Knowledge Management, Revised Selected Papers, LNAI Vol. 4369, pp. 48–65, 2006.
34. H. Sakai and M. Nakata: An Application of Discernibility Functions to Generating Minimal Rules in Non-deterministic Information Systems, Journal of Advanced Computational Intelligence and Intelligent Informatics, Vol. 10, No. 5, pp. 695–702, 2006.
35. H. Sakai, R. Ishibashi, K. Koba and M. Nakata: On Possible Rules and Apriori Algorithm in Non-deterministic Information Systems 2, Lecture Notes in AI, Springer-Verlag, Vol. 4482, pp. 280–288, 2007.
36. H. Sakai, K. Koba and M. Nakata: Rough Sets Based Rule Generation from Data with Categorical and Numerical Values, Journal of Advanced Computational Intelligence and Intelligent Informatics Vol. 12, No. 5, pp. 426–434, 2008.
37. A. Skowron and C. Rauszer: The Discernibility Matrices and Functions in Information Systems, In Intelligent Decision Support — Handbook of Advances and Applications of the Rough Set Theory, Kluwer Academic Publishers, pp. 331–362, 1992.
38. R. Agrawal and R. Srikant: Fast Algorithms for Mining Association Rules, Proc. 20th Very Large Data Base, pp. 487–499, 1994.
39. R. Agrawal, H. Mannila, R. Srikant, H. Toivonen and A. Verkamo: Fast Discovery of Association Rules, Advances in Knowledge Discovery and Data Mining, pp. 307–328, 1996.
40. UCI Machine Learning Repository: `http://mlearn.ics.uci.edu/MLRepository.html`
41. T. Murai, G. Resconi, M. Nakata and Y. Sato: Operations of Zooming In and Out on Possible Worlds for Semantic Fields, E. Damiani et al.(eds), Knowledge-Based Intelligent Information Engineering Systems and Allied Technologies, IOS Press, pp. 1083–1087, 2002.
42. Y. Yao, C. Liau and N. Zhong: Granular Computing Based on Rough Sets, Quotient Space Theory, and Belief Functions, Lecture Notes in AI, Springer-Verlag, Vol. 2871, pp. 152–159, 2003.
43. M. Chmielewski and J. Grzymała-Busse: Global Discretization of Continuous Attributes as Preprocessing for Machine Learning, Int'l. Journal of Approximate Reasoning, Vol. 15, pp. 319–331, 1996.
44. J. Grzymała-Busse and J. Stefanowski: Three Discretization Methods for Rule Induction, Int'l. Journal of Intelligent Systems, Vol. 16, pp. 29–38, 2001.

Appendix 1

Let us show an example of utility programs on PC_1 with Pentium 4 (3.4GHz) in prolog and C. The following is the data file *flu.pl*, which is NIS_1 in Table 4.1. Our utility programs handle every data in this syntax.

```
object(8,4).
data(1,[high,[yes,no],no,yes]).
data(2,[[high,very_high],yes,yes,yes]).
data(3,[[normal,high,very_high],no,no,no]).
data(4,[high,yes,[yes,no],yes]).
data(5,[high,[yes,no],yes,no]).
data(6,[normal,yes,[yes,no],no]).
data(7,[normal,no,yes,no]).
data(8,[[normal,high,very_high],yes,[yes,no],yes]).
```

In the following, every possible equivalence relation in each attribute is generated. Every possible equivalence relation is obtained as a side effect of defining a total set OB.[27,33] Every underline specifies the user's inputs.

```
?-go.
Original File Name:'flu.pl'.
---------- An Attribute Temperature -----     /* Data is stored in 1.pe */
[1] [[1,2,4,5],[3,6,7,8]] 1                   /* An equivalence relation */
[2] [[1,4,5],[2],[3,6,7,8]] 1
[3] [[1,2,4,5,8],[3,6,7]] 1
   :       :      :
[17] [[1,4,5,8],[2,3],[6,7]] 1
[18] [[1,4,5],[2,3,8],[6,7]] 1
POSSIBLE CASES 18
---------- An Attribute Headache -----        /* Data is stored in 2.pe */
[1] [[1,2,4,5,6,8],[3,7]] 1
[2] [[1,2,4,6,8],[3,5,7]] 1
[3] [[1,3,5,7],[2,4,6,8]] 1
[4] [[1,3,7],[2,4,5,6,8]] 1
POSSIBLE CASES 4
---------- An Attribute Nausea -----          /* Data is stored in 3.pe */
[1] [[1,3,4,6,8],[2,5,7]] 1
[2] [[1,3,4,6],[2,5,7,8]] 1
[3] [[1,3,4,8],[2,5,6,7]] 1
[4] [[1,3,4],[2,5,6,7,8]] 1
[5] [[1,3,6,8],[2,4,5,7]] 1
[6] [[1,3,6],[2,4,5,7,8]] 1
[7] [[1,3,8],[2,4,5,6,7]] 1
[8] [[1,3],[2,4,5,6,7,8]] 1
POSSIBLE CASES 8
---------- An Attribute Flu -----             /* Data is stored in 4.pe */
[1] [[1,2,4,8],[3,5,6,7]] 1
POSSIBLE CASES 1
EXEC_TIME=0.0(sec)
yes
```

In *DISs*, every concept in rough set theory is defined by an equivalence relation. In *NISs*, we can obtain possible equivalence relations, and we can calculate every rough sets based concept in *NISs*. We can generate possible equivalence relations for a set $ATR \subseteq AT$ by using program *merge.exe* in C. In the following, possible equivalence relations for $ATR = \{Temp, Head, Nausea\}$ are generated.

```
% ./merge
Merging 1.pe ...
Merging 2.pe ...
Merging 3.pe ...
EXEC_TIME=0.000(sec)
```

After merging three files, we obtain a file 123.*pe*. We can calculate the data dependency from $CON = \{Temp, Head, Nausea\}$ to $DEC = \{Flu\}$ by using program *depratio.exe* in C.

```
% ./depratio
File Name for Condition:123.pe
File Name for Decision:4.pe
----- Dependency Check ------------------------
CRITERION 1(Num_of_Consistent_DISs / Num_of_All_DISs)
   Number of Derived DISs:576
   Number of Derived Consistent DISs:240
   Degree of Consistent DISs:0.417
CRITERION 2(Total_Min_and_Max_Degree)
   Minimum Degree of Dependency:0.125
   Maximum Degree of Dependency:1.000
----- Consistency Ratio for Every Object ------------------------
Consistent ratio of the object 1:0.833(=480/576)
Consistent ratio of the object 2:0.750(=432/576)
Consistent ratio of the object 3:0.833(=480/576)
Consistent ratio of the object 4:0.750(=432/576)
Consistent ratio of the object 5:0.604(=348/576)
Consistent ratio of the object 6:0.833(=480/576)
Consistent ratio of the object 7:1.000(=576/576)
Consistent ratio of the object 8:0.750(=432/576)
EXEC_TIME=0.000(sec)
```

In the above execution, we know there exist 576 derived *DISs* from NIS_1 and 240 derived *DISs* are consistent. In the worst case, the degree of dependency is 0.125, namely only object 7 is consistent and other objects are inconsistent. In the best case, i.e., in 240 consistent derived *DISs*, every object is consistent.

Appendix 2

Let us show two examples of the *ISETV*-method on PC_1 in prolog. In the first example, minimal certain rules $[CON, val] \Rightarrow [Flu,no]$ in NIS_1 are generated. An attribute file *attrib.pl* in the following defines conditions of rules.

```
decision([4]).      /* a definition of the decision attribute */
decval([no]).       /* a definition of the decision attribute value */
condition([1,2,3]). /* candidates of condition attributes */
```

In the preliminary step, program *trans* translates two files to another file. Every attribute is identified with its ordinal number. Namely, $[1, normal]$ means $[Temp,normal]$ and $[2, no]$ means $[Head,no]$.

```
?-trans.
File Name for Read Open:'flu.pl'.
Decision Definition File:'attrib.pl'.
File Name for Write Open:'flu.rs'.
EXEC_TIME=0.0(sec)
yes
?-init.
Rs File:'flu.rs'.
DECLIST:<inf=[3,5,6,7]>
Certain Rules come from [7]
EXEC_TIME=0.0(sec)
yes
?-disfunc(7,M).
Descriptor [2,no] is a core for object 8
M=[[1,[1,normal],[3,yes]],[2,[1,normal],[2,no]],
[4,[1,normal],[2,no]],[8,[2,no]]]
yes
?-minimal.
<<Minimal Certain Rules from object 7>>
Descriptor [2,no] is a core for object 8
[1,normal]&[2,no]=>[4,no] [72/72(=72/72,1/1),DGC]
   This rule covers objects [7], Coverage=1.0
   [(0.125,0.25),(1.0,1.0),(0.25,0.5)]
[2,no]&[3,yes]=>[4,no] [32/32(=32/32,1/1),DGC]
   This rule covers objects [7], Coverage=1.0
   [(0.125,0.25),(1.0,1.0),(0.25,0.5)]
EXEC_TIME=0.0(sec)
yes
```

In the application of program *init*, four objects 3,5,6 and 7 satisfy $g(x, Flu) = \{no\}$. According to Proposition 4.3, we first obtain a set of objects, from which a certain rule can be generated. In this execution, we know certain rules can be generated from object 7. Program *disfunc*$(7, M)$ displays the function $DF(7)$. Since a descriptor $[2, no]$ is a core, it is easy to solve $DF(7)$. The solutions are $\{[2, no], [1, normal]\}$

and $\{[2, no], [3, yes]\}$. Program *minimal* handles such simple cases, and *minimal* does not call the E-method. The first implication τ^7 : $[Temp,normal] \wedge [Head,no] \Rightarrow [Flu, no]$ appears in 72 derived *DISs*, and τ^7 is always consistent in 72 derived *DISs*. Furthermore, we see $minsupp(\tau^7) = 0.125$, $maxsupp(\tau^7) = 0.25$, $minacc(\tau^7) = maxacc(\tau^7) = 1.0$.

Now, we show an example of NIS_4 in Table 4.6. Minimal certain rules in the form of $[CON,val] \Rightarrow [10,2]$ from NIS_4 are generated.

```
?-init.
Rs File:'data300.rs'.
DECLIST:<inf=[23,25,26,68,69,106,128,129,137,173,176,185,187,189,198,245,
   253,269,271,279,293]>
Certain Rules come from [23,25,26,68,69,106,128,129,137,173,176,185,187,
   189,198,245,253,269,271,279,293]
EXEC_TIME=0.0(sec)
yes
```

In the following execution, we sequentially specify descriptors, and interactively reduce the discernibility function. Finally, we obtained 5 minimal certain rules from object 23. In the interactive mode, we specify a value 6 for the threshold number of descriptors. There are three loops. In the first loop, a discernibility function $DF(23)$ is generated, and a descriptor [3,10] is specified. In the second loop, $DF_{red}(23)$ is generated. In $DF_{red}(23)$, both [5,3] and [6,1] are elements in $DISC(23, y)$ (y = 11,16,18,$\cdots$,291,299), therefore a set $\{[3, 10], [4, 6], [5, 3]\}$ assigns *true* to $DF(23)$. In this manner, we obtained two minimal certain rules in Loop 2. In Loop 3, the number of rest descriptors is less than 6, so the E-method is called and three rules are generated.

```
?-isetv(23).
<<=====================================================================>>
      Minimal Certain Rules from object 23
<<=====================================================================>>
====================== NON-INTERACTIVE MODE ============================
Core Descriptors:[]
EXEC_TIME=0.0(sec)
====================== INTERACTIVE MODE ================================
Input a number of Related Descriptors to Start Exhaustive Search:6.
Exhaustive Search for less than 64 Cases !!
[Loop:1]
   Discernibility Function without Core Descriptors:
      [[1,[2,2],[3,10],[4,6],[5,3],[6,1],[7,10],[8,5]],
      [2,[2,2],[3,10],[4,6],[5,3],[6,1],[7,10],[8,5]],
      [3,[2,2],[3,10],[4,6],[5,3],[6,1],[7,10],[8,5]],
      [4,[2,2],[3,10],[4,6],[5,3],[6,1],[7,10],[8,5]],
```

```
      :          :          :
    [296,[2,2],[3,10],[4,6],[5,3],[6,1],[7,10],[8,5]],
    [297,[2,2],[3,10],[4,6],[5,3],[8,5]],
    [298,[2,2],[3,10],[4,6],[6,1],[7,10],[8,5]],
    [299,[2,2],[5,3],[6,1],[7,10],[8,5]],
    [300,[2,2],[3,10],[4,6],[5,3],[7,10],[8,5]]]
  Currently Selected Descriptors:[]
  Descriptors in the Current Discernibility Function:
    [[2,2],[3,10],[4,6],[5,3],[6,1],[7,10],[8,5]]
  Select a Descriptor:[3,10].
  Currently Selected Descriptors:[[3,10]]
[Loop:2]
  Discernibility Function without Core Descriptors:
    [[11,[2,2],[4,6],[5,3],[6,1],[8,5]],
    [16,[4,6],[5,3],[6,1],[7,10],[8,5]],
    [18,[2,2],[4,6],[5,3],[6,1],[8,5]],
      :          :          :
    [291,[2,2],[4,6],[5,3],[6,1],[7,10],[8,5]],
    [299,[2,2],[5,3],[6,1],[7,10],[8,5]]]
  Currently Selected Descriptors:[[3,10]]
  Descriptors in the Current Discernibility Function:
    [[2,2],[4,6],[5,3],[6,1],[7,10],[8,5]]
  Select a Descriptor:[4,6].
  Currently Selected Descriptors:[[3,10],[4,6]]
  Common Descriptors in the Current Discernibility Function:
    [[5,3],[6,1]]
  == Execution of E-method ==
  [3,10]&[4,6]&[5,3]=>[10,2]
    [3609223525487808198316488413675552/3609223525487808198316488413675552
    (=119379916649196013265879904/119379916649196013265879904,
    30233088/30233088),DGC]
    This rule covers objects [23], Coverage=0.047
    [(0.003,0.003),(1.0,1.0),(0.035,0.0476)]
  [3,10]&[4,6]&[6,1]=>[10,2]
    [8120752932347568446212098930769992/8120752932347568446212098930769992
    (=268604812460691029848227784/268604812460691029848227784,
    30233088/30233088),DGC]
    This rule covers objects [23], Coverage=0.047
    [(0.003,0.003),(1.0,1.0),(0.035,0.047)]
[Loop:3]
  Discernibility Function without Core Descriptors:
    [[37,[2,2],[7,10],[8,5]],
```

```
      [57,[2,2],[7,10],[8,5]],
      [136,[2,2],[7,10],[8,5]],
      [210,[7,10]],
      [272,[2,2],[8,5]],
      [299,[2,2],[7,10],[8,5]]]
   Currently Selected Descriptors:[[3,10],[4,6]]
   Descriptors in the Current Discernibility Function:
      [[2,2],[7,10],[8,5]]
   Threshold Value Condition is satisfied.
   == Execution of E-method ==
   [2,2]&[7,10]&[8,5]=>[10,2]
      [160409934466124808814066151718912/160409934466124808814066151718912
      (=5305774073297600589594624/5305774073297600589594624,
      30233088/30233088),DGC]
      This rule covers objects [23], Coverage=0.047
      [(0.003,0.003),(1.0,1.0),(0.035,0.047)]
         :          :          :
   [3,10]&[7,10]&[8,5]=>[10,2]
      [120307450849593606610549613789184/120307450849593606610549613789184
      (=3979330554973200442195968/3979330554973200442195968,
      30233088/30233088),DGC]
      This rule covers objects [23], Coverage=0.047
      [(0.003,0.003),(1.0,1.0),(0.035,0.047)]
yes
```

Appendix 3

Let us show an example of execution on NIS_1. A utility program *nisapri.exe* is implemented on PC_1 in C. We employ Excel for preparing data. The following is the data file *flu.prn* in Excel, and our utility programs handle every data in this syntax.

```
8
4
high [yes,no] no yes
[high,very_high] yes yes yes
[normal,high,very_high] no no no
high yes [yes,no] yes
high [yes,no] yes no
normal yes [yes,no] no
normal no yes no
[normal,high,very_high] yes [yes,no] yes
```

We first translate this data file to another data, then we specify the condition of rule generation in an attribute file. Here, $DEC = \{Flu\}$, $\alpha = 0.3$ and $\beta = 0.6$ are specified in this attribute file. After this preparation, we can execute *nisapri.exe*.

```
% ./nisapri
version 1.2.8
File Name:flu
=====================================================================
      Lower Approximation Strategy
=====================================================================
CAN(1)=[Temperature,high],[Temperature,normal],[Headache,no],
[Headache,yes],[Nausea,no],[Nausea,yes],[Flu,no],[Flu,yes](8)
CAN(2)=[Temperature,normal][Flu,no](<DEF>0.667,<INDEF>0.750),
[Headache,no][Flu,no](<DEF>0.667,<INDEF>0.750),[Nausea,yes][Flu,no]
(<DEF>0.400,<INDEF>0.500),[Temperature,high][Flu,yes](<DEF>0.500,
<INDEF>0.600),[Headache,yes][Flu,yes](<DEF>0.600,<INDEF>0.667)(5)
------- OBTAINED RULE -------
[Temperature,normal]=>[Flu,no]
   (minsupp<DEF>=0.250,minsupp<INDEF>=0.375,minacc<DEF>=0.667,
   minacc<INDEF>=0.750) (<DEF>from )(<INDEF>from 3)
[Headache,no]=>[Flu,no]
   (minsupp<DEF>=0.250,minsupp<INDEF>=0.375,minacc<DEF>=0.667,
   minacc<INDEF>=0.750) (<DEF>from )(<INDEF>from 5)
[Temperature,high]=>[Flu,yes]
   (minsupp<DEF>=0.250,minsupp<INDEF>=0.375,minacc<DEF>=0.500,
   minacc<INDEF>=0.600) (<DEF>from )(<INDEF>from 2,8)
[Headache,yes]=>[Flu,yes]
   (minsupp<DEF>=0.375,minsupp<INDEF>=0.500,minacc<DEF>=0.600,
   minacc<INDEF>=0.667) (<DEF>from 2,4,8)(<INDEF>from 1)
EXEC_TIME=0.000(sec)

=====================================================================
      Upper Approximation Strategy
=====================================================================
CAN(1)=[Temperature,high],[Temperature,normal],[Headache,no],
[Headache,yes],[Nausea,no],[Nausea,yes],[Flu,no],[Flu,yes](8)
CAN(2)=[Temperature,normal][Flu,no](<DEF>1.000,<INDEF>1.000),
[Headache,no][Flu,no](<DEF>1.000,<INDEF>1.000),[Nausea,yes][Flu,no]
(<DEF>0.750,<INDEF>0.750),[Temperature,high][Flu,yes](<DEF>0.800,
<INDEF>0.800),[Headache,yes][Flu,yes](<DEF>0.800,<INDEF>0.800)(5)
------- OBTAINED RULE -------
[Temperature,normal]=>[Flu,no]
   (maxsupp<DEF>=0.375,maxsupp<INDEF>=0.375,maxacc<DEF>=1.000,
   maxacc<INDEF>=1.000) (<DEF>from 6,7)(<INDEF>from 3)
[Headache,no]=>[Flu,no]
   (maxsupp<DEF>=0.375,maxsupp<INDEF>=0.375,maxacc<DEF>=1.000,
   maxacc<INDEF>=1.000) (<DEF>from 3,7)(<INDEF>from 5)
[Nausea,yes]=>[Flu,no]
   (maxsupp<DEF>=0.375,maxsupp<INDEF>=0.375,maxacc<DEF>=0.750,
   maxacc<INDEF>=0.750) (<DEF>from 5,7)(<INDEF>from 6)
```

```
[Temperature,high]=>[Flu,yes]
   (maxsupp<DEF>=0.500,maxsupp<INDEF>=0.500,maxacc<DEF>=0.800,
   maxacc<INDEF>=0.800) (<DEF>from 1,4)(<INDEF>from 2,8)
[Headache,yes]=>[Flu,yes]
   (maxsupp<DEF>=0.500,maxsupp<INDEF>=0.500,maxacc<DEF>=0.800,
   maxacc<INDEF>=0.800) (<DEF>from 2,4,8)(<INDEF>from 1)
EXEC_TIME=0.000(sec)
```

According to this execution, we know the most reliable implication is

$$[Headache,yes] \Rightarrow [Flu,yes] \in Rule(0.3, 0.6, LA, def).$$

Furthermore,

$$[Nausea,yes] => [Flu,no] \in Rule(0.3, 0.6, UA) - Rule(0.3, 0.6, LA)$$

is influenced by the information incompleteness.

Appendix 4

Let us show two examples of the execution of *bunpa.csv*. A utility program with a user interface is implemented in C# on PC_2 with Pentium 4 (3.4GHz). We specify the usage of numerical patterns in the buttons of a user interface. In Fig A.1, the Standard button is selected, and the usual equivalence classes are employed. It took 0.003(sec) for rule generation. In this case, there is no implication satisfying

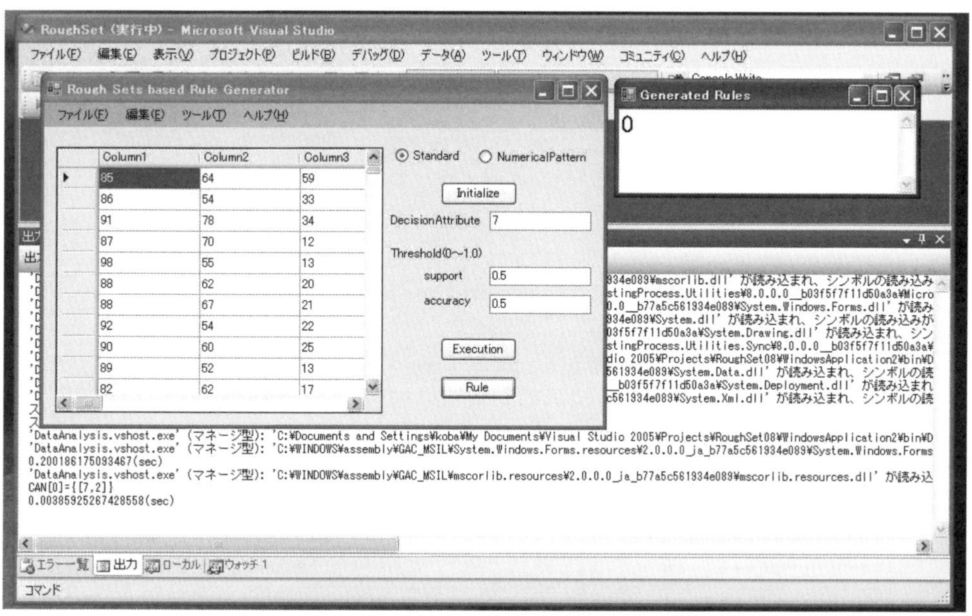

Fig. A.1. Rule generation without using numerical patterns.

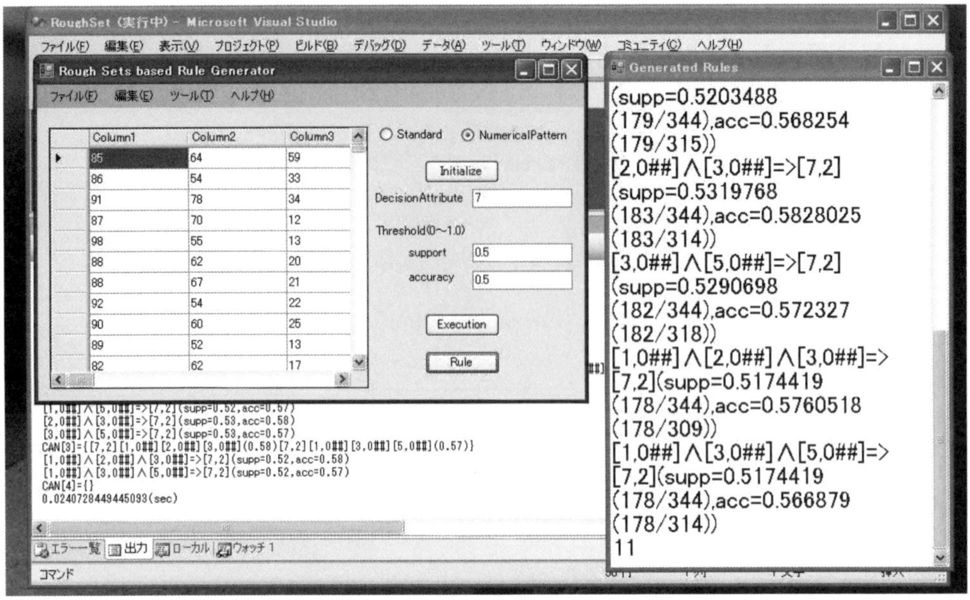

Fig. A.2. Rule generation by using numerical patterns.

support $\geq$ 0.5 and *accuracy* $\geq$ 0.5. In Fig A.2, the Numerical Pattern button is selected, and the equivalence classes with numerical values are employed. It took 0.020(sec) for rule generation. In this case, there are 11 implications satisfying *support* $\geq$ 0.5 and *accuracy* $\geq$ 0.5.

Chapter 5

METAMATHEMATICAL LIMITS TO COMPUTATION

N. C. A. da Costa and F. A. Doria
Advanced Studies Research Group and Fuzzy Sets Laboratory
PIT, Production Engineering Program
COPPE, UFRJ
P.O. Box 68507
21945–972 Rio RJ Brazil.
ncacosta@usp.br
fadoria@gmail.com
fadoria@pep.ufrj.br

We discuss the axiomatization of (theoretic) computer science within theories that include Zermelo–Fraenkel set theory. We then formulate several undecidability and incompleteness results that bear on practical situations within axiomatized versions of computer science.

To appear in K. Nakamatsu, ed., *Handbook of Reasoning–Based Intelligent Systems*, World Scientific.

5.1. Prologue

This text reviews ideas and results that have appeared in print since 1991 and which show the wide–ranging formal limits that exist and which constrain the everyday practice of mathematics. The m ain blueprint for those results is the following (which is stated waving hands):

> *Let P be any predicate in an adequate formal language so that, for two terms x and y, $x \neq y$, one can prove $P(x)$ and $\neg P(y)$. Then there is a term z so that one can neither prove $P(z)$ nor $\neg P(z)$.*

$\neg$ is "not," and we clarify below the meaning of "adequate formal language." We mainly think of axiomatic systems like PA (Peano Arithmetic) or preferably ZFC (Zermelo–Fraenkel set theory plus the axiom of choice), in order to ensure enough elbow room for our arguments. The blueprint for the result stated above is Rice's Theorem [30] in the case of arithmetic; Rice's Theorem was extended by the authors to the whole of ZFC and even beyond in 1991 [6] (see also [10]).

Our main goal here is: suppose that we wish to axiomatize computer science within some well–known axiomatic framework such as ZFC. What do we gain when we formalize it? What do we lose?

The results we present here have been discussed and commented by several authors. See [2; 4; 33; 34; 35; 37; 38; 39].

A cautionary note

Here when we talk about some axiomatics for (theoretic) computer science we mean a formalization of the theory of Turing machines within ZFC or adequate extensions of it. We use in our arguments a basic system S which is essentially ZFC or an adequate extension; however we are aware that several of our arguments can be formulated within PA.

Main concepts

For concepts from logic that we require see [28]; for concepts in computer science see [26; 32]. We use: $\neg$, "not," $\vee$, "or," $\wedge$, "and," $\rightarrow$, "if... then...," $\leftrightarrow$, "if and only if," $\exists x$, "there is a x," $\forall x$, "for every x." $P(x)$ is a formula with x free; it roughly means "x has property P." $T \vdash \xi$ means T proves ξ, or ξ is a theorem of T. ω is the set of natural numbers, $\omega = \{0, 1, 2, \ldots\}$.

We deal with algorithmic functions. These are given by their programs coded in Gödel numbers e [32]. We will sometimes use Turing machines (noted by sans–serif letters with the Gödel number as index M_e) or partial recursive functions, noted $\{e\}$. $\mathsf{M}_e = \{e\}$.

We list here a few concepts that will frequently appear in what follows; we give naïve explanations for them:

- *Algorithm.* A procedure to compute a computable function. A Turing machine table is an algorithm.
- *Computable function.* A function from the integers on the integers that can be computed by a Turing machine.
- *Consistency.* A formal system based on classical logic is consistent if we cannot prove both ξ and $\neg\xi$ in it.
- *Effective procedure.* The same as algorithm.
- *Formal system.* We will use here formal systems with the following characteristics:
  - The alphabet of the formal system is finite, or denumerably infinite;
  - The set of well–formed formulae is recursive;
  - The set of theorems is recursively enumerable.
  - Logical framework for our formal systems is the first–order classical predicate calculus.
  - We also ask that the formal systems we use here include ZFC, that is, axiomatic set theory, and moreover ask that S have a model with standard arithmetic. (We say that S is arithmetically sound.)

    So S is supposed consistent.

We write S for theories with those properties.

- *Gödel number.* An algorithm, or program, can be written as a finite string of symbols, and as such can be coded by an integer. That integer is the algorithm's Gödel number. We can choose a Gödel numbering in such a way that every integer codes some algorithm.
- *Halting Problem.* The Halting Problem asks: given an arbitrary Turing machine of Gödel number m, and an arbitrary integer n given as its input, is there an algorithm that settles whether machine coded by m over n stops and produces an output? The answer is, no.
- *Incompleteness.* A formal system S is incomplete if there is a sentence ξ so that S neither proves ξ nor proves its negation $\neg\xi$.
- *Partial recursive function.* Another formalization of the concept of computability, equivalent to the Turing machine picture. We use both terms interchangeably.
- *Primitive recursive function.* A subset of the set of partial recursive functions [26; 32]. The primitive recursive functions are all total functions, and are bounded in its growth by Ackermann's Function [32] F_ω.
- *Program.* The same as algorithm, or Turing machine table, or partial recursive function.
- *Recursive subset of the set of the integers.* A subset $A \subset \omega$, $\omega = 0, 1, 2, \ldots$, is recursive if there is an algorithm A so that $\mathsf{A}(n) = 1$ if and only if $n \in A$, and $\mathsf{A}(n) = 0$, if and only if $n \notin A$.
- *Recursively enumerable subset of the integers.* A subset $B \subset \omega$ is recursively enumerable if the elements of B are enumerated by some algorithm B, repetitions being allowed, and without the need of the enumeration being in some prescribed order. Notice that there are recursively enumerable sets that are not recursive.
- *Total function, total Turing machine.* A total function is a function whose domain is the whole set of integers, ω. A total Turing machine is an algorithm for that function (note that one such algorithm only exists if the function is computable).
- *Turing machine.* A device conceived by Alan Turing in 1937 to emulate computation procedures. A Turing machine inputs a string of binary digits (a binary word) and, if it stops, it outputs a string of binary digits. A Turing machine is composed of a head, and a tape divided into squares. The head runs over the tape for a finite number of squares; it starts over a given square, reads what is printed on the square, and given the inner state of the head, prints a new symbol on the square, and moves to the right square, to the left square, or stays over the same square. Turing machines [32] are introduced via their tables which specify elementary operations like moving the head to the left or to the right, erasing or printing a symbol on the tape's square under the head, and so on.
- *Undecidability.* A set $C \subset \omega$ is undecidable if there is no algorithm C that makes set C recursive.

5.2. Preliminary Results

(For this review see [3] on which we base it.) We start from a very simple theory of arithmetic, noted $A1$. Its language includes variables $x, y, \ldots$, two constants, $\mathbf{0}$ and $\mathbf{1}$, the equality sign $=$, and two operation signs, $+, \times$. Basically $A1$ has axioms for the operations $+$ and $\times$, the behavior of constants $\mathbf{0}$ and $\mathbf{1}$, and the trichotomy axiom, that is, given two natural numbers x and y, either $x < y$ or $x = y$ or $x > y$. $A1$ contains no induction axiom. $P(x)$ is a formula where x appears as a free variable.

We will also require Russell's ι symbol [25]. Its meaning translates as follows: if $P(x)$ is a formula where x appears free, then $\iota_x P(x)$ is, roughly, the x such that P.

The *standard interpretation* for $A1$ is: the variables $x, y, \ldots$ range over the natural numbers, and $\mathbf{0}$ and $\mathbf{1}$ are seen as, respectively, zero and one.

Definition 5.2.1. A Turing machine of Gödel number e operating on x with output y, $\{e\}(x) = y$ is **representable** in theory $A1$ if there is a formula $F_e(x, y)$ in the language of $A1$ so that:

1. $A1 \vdash \{F_e(x,y) \wedge F_e(x,z) \rightarrow [y = z]\}$, and
2. For natural numbers a, b, if $\{e\}(a) = b$, then $A1 \vdash F_e(a, b)$.

Then:

Proposition 5.2.2. *Every Turing machine is representable in $A1$. Moreover there is an effective procedure that allows us to obtain F_e from the Gödel number e.*

Remark 5.2.3. We consider here theories S that are *arithmetically sound*, that is, which have a model with standard arithmetic for its arithmetical segment.

A first example of generalized incompleteness; Rice's Theorem

The example we now give shows that Gödel incompleteness is a pervasive phenomenon, from an arithmetic theory like $A1$ upwards, that is, it affects all theories that contain $A1$, are based on a first–order classical language, have a model where arithmetic is standard and have a recursively enumerable set of theorems. This also shows that we actually require very little in order to argue that undecidability and incompleteness are widespread phenomena.

Suppose that the theory S we deal with has also the ι symbol. Let $P(x)$ be a formula with a single free variable so that for terms $\xi \neq \zeta$, $S \vdash P(\xi)$ and $S \vdash \neg P(\zeta)$ (we call such P, nontrivial predicates). Then, for the term:

$$\eta = \iota_x[(x = \xi \wedge \alpha) \vee (x = \zeta \wedge \neg\alpha)],$$

where α is an undecidable sentence in S, that is, $S \nvdash \alpha$ and $S \nvdash \neg\alpha$:

Proposition 5.2.4. *$S \nvdash P(\eta)$ and $S \nvdash \neg P(\eta)$.*

This shows that incompleteness is found everywhere within theories like S.

Remark 5.2.5. Notice that this implies Rice's Theorem [26; 30; 32] in computer science. With some hand–waving: suppose that there is an algorithm that settles $P(n)$, for each $n \in \omega$. Then by the representation theorem we may internalize that algorithm into S, and obtain a proof of $P(n)$ for arbitrary $n \in \omega$, which is a contradiction, given Proposition 5.2.4.

Richardson's map

Our main tool will be an explicit expression for the Halting Function, that is, the function that settles the halting problem [32]. We will show that it can be constructed within the language of classical analysis. We use the Richardson transforms [10; 31]. Our formal background, if required, is $S \supseteq$ ZFC. We start from a strengthening of Proposition 5.2.2:

Proposition 5.2.6. *If $\{e\}(a) = b$, for natural numbers a, b, then we can algorithmically construct a polynomial p_e over the natural numbers so that $[\{e\}(a) = b] \leftrightarrow [\exists x_1, x_2, \ldots, x_k \in \omega\, p_e(a, b, x_1, x_2, \ldots, x_k) = 0]$.*

Then:

Proposition 5.2.7. *$a \in R_e$, where R_e is a recursively enumerable set, if and only if there are e and p so that $\exists x_1, x_2, \ldots, x_k \in \omega\, [p_e(a, x_1, x_2, \ldots, x_k) = 0]$.*

Richardson's map [10; 31] allows us to obtain in an algorithmic way, given any $p_e(a, \ldots)$, a polynomial over the natural numbers, a real–defined and real–valued function $f_e(a, \ldots)$ that has roots if and only if $p_e(a, \ldots)$ has roots as a Diophantine equation.

Richardson's map: multidimensional version

Let $\mathcal{A}$ be the algebra of subelementary functions (polynomials over the reals, sines, cosines; everything closed under $+, -$, products by real numbers and by the functions that generate the algebra, to which we add function composition). Let R denote the real line.

If $\mathcal{P}$ is the set of all finite–length polynomials over ω:

Proposition 5.2.8. (Richardson's Map, I) *There is an injection $\kappa_{\mathcal{P}} : \mathcal{P} \to \mathcal{A}$, where $\mathcal{P}$ denotes the algebra of ω–valued polynomials in a finite number of variables, and $\mathcal{A}$ is the algebra of subelementary functions described above, such that:*

1. *Map $\kappa_{\mathcal{P}}$ is constructive, that is, given the expression for p in A1, there is an effective procedure so that we can obtain the corresponding expression for $F = \kappa_{\mathcal{P}}(p)$ in S.*
2. *$\kappa_{\mathcal{P}}$ is 1–1.*
3. *For $\vec{x} = (x_1, \ldots, x_n)$, $\exists \vec{x} \in \omega^n\, p(m, \vec{x}) = 0$ if and only if $\exists \vec{x} \in \mathsf{R}^n\, F(m, \vec{x}) = 0$ if and only if $\exists \vec{x} \in \mathsf{R}^n\, F(m, \vec{x}) \leq 1$, for $p \in \mathcal{P}$ and $F \in \mathcal{A}$.*
4. *The injection $\kappa_{\mathcal{P}}$ is proper.*

Step 3 allows us to translate the existence of roots for Diophantine equations into roots of the corresponding transformed real–defined and real–valued function, with some extras.

Richardson's map: one–dimensional version

Proposition 5.2.9 (Richardson's Map, II) *Let $\mathcal{A}_1$ be the algebra of subelementary functions over a single real variable x. Then there is a map $\kappa' : \mathcal{P} \to \mathcal{A}_1$ such that:*

1. *Map κ' is constructive.*
2. *κ' is 1–1.*
3. *The inclusion $\kappa'(\mathcal{P}) \subset \mathcal{A}_1$ is proper.*
4. *$\exists \vec{x} \in \omega^n\ p(m, \vec{x}) = 0$ if and only if $\exists x \in \mathsf{R}\, L(m, x) = 0$ if and only if $\exists x \in \mathsf{R}\, G(m, x) \leq 1$.*

This is enough to imply the algorithmic unsolvability of the following problem: *given the interior D of a square, there is no general algorithm to decide whether a given curve described by elementary functions meets some straight line within D.* It suffices to compactify the previous function over the interval $[-1, +1]$ and consider its restriction to $(-1, +1)$.

The Halting Function

The key result in Alan Turing's great 1937 paper, "On computable numbers, with an application to the Entscheidungsproblem" [40], is a proof of the algorithmic unsolvability of a version of the halting problem:

> *Given an arbitrary Turing machine of Gödel number e, for input x, there is no algorithm that decides whether $\{e\}(x)$ stops and outputs something, or enters an infinite loop.*

Let $\mathsf{M}_m(a) \downarrow$ mean: "Turing machine of Gödel number m stops over input a and gives some output." Similarly $\mathsf{M}_m(a) \uparrow$ means, "Turing machine of Gödel number m enters an infinite loop over input a." Then we can define the halting function θ:

- $\theta(m, a) = 1$ if and only if $\mathsf{M}_m(a) \downarrow$.
- $\theta(m, a) = 0$ if and only if $\mathsf{M}_m(a) \uparrow$.

$\theta(m, a)$ is the halting function for M_m over input a. θ isn't an algorithmic function, of course [32; 40]. Yet we can explicitly write an expression for a function in the language of classical analysis that settles the halting problem:

- Given Turing machine $\mathsf{M}_m(a) = b$, for natural numbers a, b, we can algorithmically obtain [15] a polynomial $p_m(\langle a, b\rangle, x_1, \ldots, x_k)$ so that:

$$\mathsf{M}_m(a) = b \leftrightarrow \exists x_1, \ldots, x_2 \in \omega\, [p_m(\langle a, b\rangle, x_1, \ldots, x_k) = 0].$$

- Given F_m, real–defined and real–valued, we have that:

$$\exists x_1, \ldots, x_2 \in \omega \, [p_m(\langle a, b \rangle, x_1, \ldots, x_k) = 0] \leftrightarrow$$
$$\exists x_1, \ldots, x_k \in \mathsf{R} \, F_m(\langle a, b \rangle, x_1, \ldots, x_k) \leq 1.$$

and

$$\forall x_1, \ldots, x_2 \in \omega \, [p_m(\langle a, b \rangle, x_1, \ldots, x_k) \neq 0]$$
$$\leftrightarrow \forall x_1, \ldots, x_k \in \mathsf{R} \, F_m(\langle a, b \rangle, x_1, \ldots, x_k) > 1.$$

- That is to say: $\mathsf{M}_m(a) \downarrow$ if and only if $F_m(a, \ldots)$ goes below 1, and $\mathsf{M}_m(a) \uparrow$ if and only if $F_m(a, \ldots)$ stays above 1.
This is the property we use in order to construct the halting function θ_m.

We now need the concept of an *universal Diophantine polynomial.* Davis [15] describes an algorithmic procedure out of which, given a Turing machine with input a $\mathsf{M}_m(a)$, we obtain a polynomial $p_m(a, x_1, \ldots)$ so that it has roots if and only if $\mathsf{M}_m(a)$ converges (outputs some result). Now, if $\mathsf{U}(m, a)$ is an universal Turing machine [32; 40], we can similarly obtain a polynomial $p(\langle m, a \rangle, \ldots)$ which stands for $p_m(a, \ldots)$.

If $[\exists x_1, \ldots, x_k \, p_m(\langle a, b \rangle, x_1, \ldots, x_k) = 0] \leftrightarrow [\mathsf{M}_m(a) = b]$, then, for the universal polynomial $p(\langle m, a, b \rangle, \ldots)$:

$$[\exists x_1, \ldots, x_r \, p(\langle m, a, b \rangle, x_1, \ldots, x_r) = 0]$$
$$\leftrightarrow [\exists x_1, \ldots, x_k \, p_m(\langle a, b \rangle, x_1, \ldots, x_k) = 0].$$

Now, if σ is the sign function, $\sigma(\pm x) = \pm 1$ and $\sigma(0) = 0$:

Proposition 5.2.10 (The Halting Function.) *The halting function $\theta(n, q)$ is explicitly given by:*

$$\theta(n, q) = \sigma(G_{n,q}),$$
$$G_{n,q} = \int_{-\infty}^{+\infty} C_{n,q}(x) e^{-x^2} dx,$$
$$C_{m,q}(x) = |F_{m,q}(x) - 1| - (F_{m,q}(x) - 1).$$
$$F_{n,q}(x) = \kappa_P p_{n,q}.$$

Here $p_{n,q}$ is the two–parameter universal Diophantine polynomial

$$p(\langle n, q \rangle, x_1, x_2, \ldots, x_r)$$

and κ_P is as in Proposition 5.2.8. It is clear that there are infinitely many alternative explicit expressions for the halting function θ [10].

General undecidability and incompleteness results

Everything proceeds within S, so that we can obtain all the maps given by Richardson's functor into $\mathcal{A}$ and extensions. Let also $\mathcal{B} \supseteq \mathcal{A}$ be any similar extension; let P be a predicate (a formula with a free variable $P(x)$) defined for $\mathcal{B}$ such that P is nontrivial, that is, for terms a, b, $a \neq b$, so that $S \vdash P(a)$ and $S \vdash \neg P(b)$,

If ξ is any word in that language, we write $\|\xi\|$ for its complexity, as measured by the number of letters from the alphabet of S in ξ. Also we define the *complexity of a proof* $C_S(\xi)$ of ξ in the language of S to be the minimum length that a deduction of ξ from the axioms of S can have, as measured by the total number of letters in the expressions that belong to the proof.

Then:

Proposition 5.2.11. *If S is arithmetically sound, then:*

1. *There is an $h \in \mathcal{B}$ so that neither $S \nvdash \neg P(h)$ nor $S \nvdash P(h)$, but model* $\mathbf{M}$ $\models P(h)$*, where* $\mathbf{M}$ *makes S arithmetically sound.*
2. *There is a denumerable set of functions $h_m(x) \in \mathcal{B}$, $m \in \omega$, such that there is no general decision procedure to ascertain, for an arbitrary m, whether $P(h_m)$ or $\neg P(h_m)$ is provable in* ZFC.
3. *Given the set $K = \{m : S \vdash \phi(\widehat{m})\}$, and given an arbitrary total recursive function $g : \omega \to \omega$, there is an infinite number of values for m so that $C_S(P(\widehat{m})) > g(\|P(\widehat{m})\|)$.*

Proof. Let θ be as above. Let f_0, g_0 satisfy our conditions on P, that is, $S \vdash P(f_0)$ and $S \vdash \neg P(g_0)$. Then define:

$$h(m, x) = \theta(m, x) f_0 + (1 - \theta(m, x)) g_0.$$

This settles (2). Now let us specify β so that the corresponding Diophantine equation $p = 0$ is never solvable in the standard model for arithmetic, while that fact cannot be proved in ZFC. We then form, for such an indicator function,

$$h = \beta f_0 + (1 - \beta) g_0.$$

This settles (1). Finally, for (3), we notice that as K is recursively enumerable but not recursive, it satisfies the conditions in the Gödel–Ehrenfeucht–Mycielski theorem [19] about the length of proofs. □

Proposition 5.2.12. *Let $\forall x\, R(x)$ be the formula that formalizes Riemann's Hypothesis as a Π_1 arithmetic sentence ([32], p. 322). Then:*

1. *Given any nontrivial P in S, there is a term ξ such that $S \vdash P(\xi)$ if and only if $S \vdash \forall x\, R(x)$.*
2. *There is a term ξ such that for a nontrivial P, $P(\xi)$ is S-arithmetically expressible as a Π_{m+1} problem, but not as any Σ_k problem, $k \leq m$.*
3. *There is a term ξ so that $P(\xi)$ isn't arithmetically expressible.*

(Recall that sentence ϕ is arithmetically expressible if there is an arithmetic sentence ψ so that $S \vdash \phi \leftrightarrow \psi$.)

We have what amounts to an extension of the preceding result together with the theorems that give expressions for the halting function. We need here the concepts of arithmetic degree, arithmetic hierarchy, complete arithmetic degree and set in the arithmetic hierarchy; for a review see [32]:

Proposition 5.2.13. *For all $p \in \omega$ expressions $\theta^{(p)}(m)$ can be explicitly constructed for characteristic functions of sets in the complete arithmetic degrees $\mathbf{0}^{(p}$.*

Higher–level intractability

The next results give an idea of how intractable those problems can be (we repeat the first one). For function F see below Definition 5.4.4.

Proposition 5.2.14.

1. *Given theory S, then there is a partial recursive function F so that S neither proves nor disproves it to be total, and for some arbitrarily chosen constructive ordinal α in the Turing–Feferman hierarchy over S, $S_\alpha \vdash [\mathsf{F}$ is total$]$, but for no $\beta < \alpha$ does S proves that F is total.*
2. *Given axiomatic set theory* ZFC *plus some large cardinal hypothesis, there is a partial recursive function F so that* ZFC *thus enlarged neither proves nor disproves F to be total.*

Comments on the proof: See Section 5.7 for a detailed proof.

5.3. More Comments About Undecidability and Incompleteness in Strong Theories

The next propositions just add to the preceding results.

Proposition 5.3.1. *There is a real number s whose decimal expansion is proved by* ZFC *to be generated by an algorithmic procedure and yet such that* ZFC *cannot compute a single digit of it.*

Proof. Let r and r' be real numbers whose decimal expansions are infinite and algorithmically generated. Impose also that, for all i, the i–th digits $r_i \neq r'_i$. Then for β so that $\beta = 0$ holds in a model for ZFC with standard arithmetic, and $\beta = 1$ in nonstandard models, as in [10], $s = \beta r + (1 - \beta) r'$. □

A second result:

Proposition 5.3.2. *There is a real number Ω' so that* ZFC *cannot compute a single digit of it.*

Sketch of proof: List (this isn't a recursive enumeration!) for some partial recursive algorithms all instances of the halting problem that cannot be proved to diverge within ZFC. Those instances appear as Π_1 sentences. Order them by their respective Gödel numbers. This is a noncomputable function, which can be coded as a real number in several ways.

ZFC cannot prove by construction a single digit of that number.

Another construction:

Proposition 5.3.3. *There are infinitely many real numbers in* ZFC *so that none of its decimal places are computable.*

Proof. Loosely speaking, recall that no well–orderings of the real numbers are expressible in ZFC, but there are infinitely many such orderings.

Let Γ be one such ordering, and let r be the first real number in Γ so that none of its decimal places are definable — there is one such r, as the set of numbers which have definable decimal places is denumerable. □

Remark 5.3.4. Still another simple but interesting result about the decidability of Π_1 sentences and hierarchies of theories goes as follows. Π_1 arithmetic sentences code several interesting facts, e.g. Riemann's Hypothesis. If we are able to ascertain by some analog procedure [10] that Riemann's Hypothesis is true of the standard model for arithmetic, then given the Fefermann hierarchy [21] of theories $S_0 = S$, $S_1 = S + \mathrm{Con}\, S$, $S_2 = S + \mathrm{Con}\, S + \mathrm{Con}(S + \mathrm{Con} S), \ldots$, there will be a S_j so that $S_j \vdash \forall x\, R(x)$.

Another interesting series of results about undecidability and incompleteness in the theory of partial differential equations was recently presented by E. Reyes [29].

5.4. An Axiomatization for (Theoretical) Computer Science

The present section is based on four papers by the authors, namely [5; 8; 11; 14].

Is there a "reasonable" or "natural" way of axiomatizing computer science? We answer that question in the affirmative, and exhibit that axiomatization here. However such an axiom system presents difficulties that will be considered in this section.

Suppose that there is a Turing machine that enumerates all theorems in a theory S (actually there is one such machine, sketched below). Since S includes Peano Arithmetic, its set of theorems will be recursively enumerable but not recursive, due to Gödel's incompleteness. Start our enumerating machine and gradually build up a list of theorems of S. Separate those that formally assert "recursive function f_e, which has Gödel number e, is total." Place them in a second list. These are the S–provably total recursive functions.

More precisely:

- A Turing machine M_S that enumerates all theorems of S can operate as follows. First, start a listing of all well–formed formulae in the language of S. Then consider each integer, $1, 2, 3, \ldots, k, \ldots$ and see if each such integer is the Gödel number of the proof of some well–formed formula in the language of S.
- If so, pick it up from the previous listing and place it in a new list, which will be the enumeration of theorems of S.
- Therefore we can compute a Gödel number e_S for such an enumerating machine.

Remark 5.4.1. Notice that it is possible to have a well–defined function which however isn't recursive. The best example is the Busy Beaver function; another example is the counterexample function to the $P = NP$ hypothesis [13; 14].

Now construct by the usual diagonalization procedure a function F over that second list. That function is immediately seen to be recursive (the diagonal procedure over that second listing gives a simple algorithm for it) and total. Yet it cannot be S–provably total.

That it to say, no theorem of S has the form "F is total." Also, as S is sound — a fact that is a consequence of its arithmetic portion having an interpretation in the standard model — then the negation of that sentence, "F isn't total" cannot appear among the theorems of S. Therefore such a sentence is independent of the axioms of S.

(We give a more rigorous construction for a function like F below.)

"Reasonable" axiomatics

We start from a well–established result which can be paraphrased as:

Turing machine theory can be seen as arithmetic under disguise.

Therefore in order to axiomatize theoretical computer science we use Diophantine equations within ZFC set theory.

Specifically, it can be seen as a privileged domain within arithmetic: the theory of Diophantine equations. So, we "translate" Turing machine theory into the theory of Diophantine equations [15] in order to axiomatize it.

Let $p_U(k, x_0, x_1, x_2, \ldots, x_k)$ be an universal Diophantine polynomial [15; 36] which we suppose to be fixed. We define the partial recursive function $\{e\}$ of Gödel number e that acts on natural number m as its input and has natural number n as its output as:

Definition 5.4.2.

$$[\{e\}(m) = n] \leftrightarrow_{\text{Def}} [\exists x_0, \ldots, x_k \in \omega \, p_U(\langle e, m, n\rangle, x_0, x_1, \ldots, x_k) = 0].$$

$\langle x, y, z\rangle = \langle x, \langle y, z\rangle\rangle$ and $\langle \ldots \rangle$ is the usual pairing function; for the computation of e and the construction of p_U see [15; 32; 36].

Turing machine $\{e\}$ is given by the preceding definition. Of course there is a relation between (the abstract objects) Turing machines and concrete objects of our real world such as computers (which can be best seen as realizations of Turing machines with a bounded tape) but we restrict our attention to the mathematical object characterized above.

The next discussion and examples [8; 14] ponder well–known phenomena from the theory of fast–growing recursive functions in the light of our main theme, that is, axiomatization of CS (computer science) within ZFC or any other standard axiomatics for set theory, based on the classical predicate calculus, with a recursively enumerable set of theorems, and which extend enough of arithmetic to adequately handle Definition 5.4.2.

The main techniques we use are the trick presented in [8] together with a version of the construction of the extension of Rice's Theorem (originally in [6]).

The first example we now give starts from the well–known property of intuitively total recursive functions described at the beginning of this section when framed within usual axiomatic systems:

Not every intuitively total recursive function can be proved so within a given axiomatic framework.

That is to say, axiomatic systems like the ones we consider here aren't able to "see" as total functions some recursive functions that are intuitively total. This has been already discussed above in an informal way. We now give a more rigorous discussion for that phenomenon [9; 17]:

Remark 5.4.3. Suppose that we are given a prescription so that, for any integer n, we can compute a finite set of numbers S_n. Then put: $\mathsf{F}^*(n) = \max S_n + 1$. Most professional mathematicians would immediately agree that F^* is both computable and total. But is that really so?

Can we construct that function within ZFC? We surely can give a Turing program, or a recursive–function definition for it relative to the recursivity of S_n — but we won't then be able to prove that similar functions are total, in the general case, as it is well–known.

Let's take a closer look at a function F as the ones naïvely described above:

Definition 5.4.4. For each n, $\mathsf{F}(n)$ is the sup of those $\{e\}(k)$ such that:

1. $k \le n$.
2. $\lceil \mathrm{Pr}_{\mathrm{ZFC}}(\lceil \forall x\, \exists z\, T(e, x, z) \rceil) \rceil \le n$.
   We explain the notation:
   - $\mathrm{Pr}_S(\phi)$ means "the proof of ϕ in axiom system S."
   - $\lceil \xi \rceil$ denotes the Gödel number of ξ.
   - $T(e, x, z)$ means, Turing machine $\{e\}$ of Gödel number e, given input x, stops in z computing steps.

So the second condition means that there is a proof of [$\{e\}$ is total] in ZFC whose Gödel number is $\leq n$.

F^* in Remark 5.4.3 is always intuitively total, and so is the particular version F in Definition 5.4.4. For some particular value of n we will be able to construct a model for a collection of cardinality $\leq n$ of sentences of ZFC, and so to compute $\mathsf{F}(n)$. The analogy is: we can *individually* check that, for each natural number n, $\mathsf{F}(n)$ exists and can be computed. However we cannot "join" all those results together to show that F is total.

Granted those remarks:

Proposition 5.4.5. *We can explicitly compute a Gödel number* e_{F} *so that* $\{e_{\mathsf{F}}\} = \mathsf{F}$.

F is such that:

Proposition 5.4.6. *The formal sentence* $\forall x\, \exists z\, T(e_{\mathsf{F}}, x, z)$ *cannot be proved or disproved from the axioms of* ZFC*, supposed consistent, that is,*

$$\forall x\, \exists z\, T(e_{\mathsf{F}}, x, z)$$

is independent of the ZFC *axioms.*

So F cannot be proved to be total in ZFC. In other words: this means that even if F is intuitively total (that is to say, it holds of the standard model for arithmetic), there must exist a model for ZFC with a nonstandard part where [F isn't total] is verified. (This result originates in Kleene's 1936 paper [23]; see also [24], p. 257.)

A brief remark on Σ_1–soundness

We now briefly mention Feferman's Σ_1–soundness reflection principle. Naïvely a system S is Σ_1–sound if, given a proof of $\exists x\, P(x)$, where P is a predicate with values in the natural numbers, then there is some natural number n so that $P(n)$ is true. That concept was formalized by Feferman as one of its reflection principles [20]. Then there is an important result that connects F to Feferman's Σ_1–reflection principles [20; 21].

Recall that a formaized reflection principle in Feferman's sense [20; 21] has (roughly) two aspects:

- If we prove in S that there is the Gödel number of a proof of ϕ, then ϕ.
- If we obtain for each n, $\mathrm{Pr}_S\ulcorner\phi(n)\urcorner \to \phi(n)$, then we can collect all those under an universal quantifier, that is, there is a form of the ω–rule at work here.

Σ_1–soundness is such a principle restricted to Σ_1 sentences ϕ. The chief result is:

Proposition 5.4.7. $S \vdash$ [F *is total*] $\leftrightarrow$ [S *is* Σ_1*–sound*].

Follows that:

Corollary 5.4.8. $S \vdash [\mathsf{F}\ \textit{is total}] \to \operatorname{Consis} S$.

$\operatorname{Consis} S$ translates as "S is consistent." For a more precise characterization see before remark 5.6.1. Proofs of both facts appear in [5].

5.5. Can We Handle Arbitrary Infinite Sets of Poly Machines in ZFC?

The next example was first discussed in [8; 14]. We may argue that the results about F deal with fast–growing functions, which are objects quite far from the everyday realm of programs and concrete computers. So we move on to another example, that bears on concepts related to the $Pvs.NP$ question. (See the last section of this text.)

The main point here is: the concept of an *infinite set of poly machines* may be quite difficult to handle within an axiomatic system, even a strong one like ZFC.

A *poly machine*, or a polynomially time–bounded Turing machine is a total Turing machine M_m that on a binary input x of length $|x|$ outputs a binary word y after less than $p_m(|x|)$ operation steps, p_m a polynomial kept fixed for M_m. Consider the following example, which is an application of the previous results (the trick described is due to [1]):

Example 5.5.1. Let $\langle m, a, b\rangle$ denote a Turing machine M_m coupled to a clock $\mathsf{C}_{(a,b)}$ — another Turing machine — that stops M_m after it executes $|x|^a + b$ cycles over input x of length $|x|$. We note that pair $\langle \mathsf{M}_m, \mathsf{C}_{(a,b)}\rangle$. We agree that if clock $\mathsf{C}_{(a,b)}$ interrupts the operation of M_m then M_m outputs 0 and stops.

Consider the set $A = \{(m, \mathsf{F}(a), \mathsf{F}(b)) : m, a, b \in \omega\}$, F as in Definition 5.4.4. Each individual machine $\langle m, \mathsf{F}(a), \mathsf{F}(b)\rangle$, m, a, b integers like $0, 1, 2, \ldots$, in it is certainly a poly machine, but we cannot prove in ZFC that the whole set only contains poly machines. It is in fact undecidable: "A is a set of poly machines" holds of the standard integers, but doesn't hold in some models for ZFC with nonstandard arithmetic part.

We have shown for, say, ZFC:

Proposition 5.5.2. *The sentence "A is a set of poly machines" is independent of* ZFC, *supposed consistent.*

Now we may ask: does it help if we add some strong axiom — here noted X — to ZFC, so that the resulting theory ZFC + X, supposed consistent, proves the consistency of ZFC itself? (Think of X, say, as some large cardinal axiom, for instance.)

Proposition 5.5.3. *There is a set A' so that the sentence "A' is a set of poly machines" is independent of* ZFC + X, *supposed consistent.*

Proof. Suffices to obtain the function F' for ZFC $+ X$. □

This example is crucial, because among other consequences, any (tentative) proof of the hypothesis $P < NP$ must include a step that says, "for *every* poly machine, any poly machine will give a wrong answer at least once," and a sentence like "A is a set of poly machines" is undecidable within consistent ZFC, even if we can pick up infinitely many of its elements and individually show each of them to be a poly machine. This raises the following question: will any system like the S we have been considering here be able to prove $P < NP$? We believe that the answer is negative [13].

5.6. More Examples of Incompleteness for Computer Science in S

Hartmanis and Hopcroft presented in 1976 some examples of undecidable sentences in formal theories that include the theory of Turing machines [14; 22]. They start from a formal theory that:

- Includes set theory (more precisely, they ask that the theory be of "sufficient power to prove the basic theorems of set theory").
  Also the theory must allow for predicate symbols $P, Q, \ldots$.
- It has a recursively enumerable set of theorems.
- Its theorems are "intuitively true."
  This is too strong and also vague for the whole of set theory with the axiom of choice — for instance, is the Banach–Tarski theorem intuitively true?. So, we take this third requirement to be the *arithmetically soundness* condition, that is, S must have a model with standard arithmetic.

S adequately fits the framework for the Hartmanis and Hopcroft results. We now endow S with the ι–symbol [25]. Then:

- The first undecidability result in [22] has to do with the BGS relativization result [1]. The BGS result says that there are recursive oracles $A, B, A \neq B$, so that one has (for the relativized versions) $P^A = NP^A$ and $P^B < NP^B$. Hartmanis and Hopcroft show that there is an oracle C so that the assertion $P^C = NP^C$ is undecidable with respect to the axioms of S. Their proof is by a diagonal argument; we give below an alternative, quite general argument.
- Then they show that there is an algorithm A (a Turing machine) of which it is true that for input x it runs in time x^2, but so that the formal version of the sentence "$\mathsf{A}(x)$ runs in time $t_{\mathsf{A}} < 2^x$" is undecidable in S.

We present here a general argument for these results which is based on the version of Rice's theorem in fragments of set theory with the ι symbol (which we suppose to be available) already stated and used here [6; 10].

Recall that the formal sentence Consis S can be naïvely "translated" as: "there is no proof in S of the sentence $0 = 1$."

Remark 5.6.1. For consistent S, let $\mathrm{Consis}\, S$ be the usual formal sentence that asserts the consistency of S; $S \nvdash \mathrm{Consis}\, S$ and $S \nvdash \neg\mathrm{Consis}\, S$. Let ξ, ζ be terms in the language of S, so that for some predicate P in the language of S, $S \vdash P(\xi)$ while $S \vdash \neg P(\zeta)$. Then:

$$\lambda = \iota_x\{[\mathrm{Consis}\, S \wedge x = \xi] \vee [\neg\mathrm{Consis}\, S \wedge x = \zeta]\}.$$

$S \nvdash \lambda = \xi$ and $S \nvdash \lambda = \zeta$, but if $\mathbf{N} \models S$ and $\mathbf{N}$ has a standard arithmetic part, then $\mathbf{N} \models \lambda = \xi$. Moreover, $S \nvdash P(\lambda)$ and $S \nvdash \neg P(\lambda)$, while $\mathbf{N} \models P(\lambda)$.

- For the first result, put oracles A, B as $\xi = A$ and $\zeta = B$. Then C:

  $$C = \iota_x\{[\mathrm{Consis}\, S \wedge x = A] \vee [\neg\mathrm{Consis}\, S \wedge x = B]\}$$

  is proved to be a recursive oracle in S, but $S \nvdash C = A$ and $S \nvdash C = B$. So, $S \nvdash P^C = NP^C$ and $S \nvdash P^C < NP^C$.
- For the second result, if P is a polynomial Turing machine, and E is an exponential Turing machine, then:

  $$\mathsf{M} = \iota_x\{[\mathrm{Consis}\, S \wedge x = \mathsf{P}] \vee [\neg\mathrm{Consis}\, S \wedge x = \mathsf{E}]\}$$

  is such that S proves M to be a total Turing machine which has an exponential time bound which cannot be improved in S, but such that it is true of $\mathbf{N}$ that it is time–polynomial.

We may also use the term:

$$\lambda = \iota_x[(x = \xi \wedge \beta = 0] \vee [x = \zeta \wedge \beta = 1].$$

β [10] is an algebraic expression already used here. It can be explicitly constructed and is such that $S \nvdash \beta = 0$ and $S \nvdash \beta = 1$, while $\beta = 0$ holds of the standard model for arithmetic; see the references for details.

Remark 5.6.2. The second result can be easily extended to a result by Loo [27]: that there is a Turing machine of arbitrarily large complexity when "seen" from within the standard model, but which is only polynomial in an adequate nonstandard model. An analogous example was discussed in 2002 by one of the authors [16]. An alternative argument that clarifies the above discussion stems from the following observation: consider,

$$p(x) = 1 + x + \frac{x^2}{2!} + \frac{x^3}{3!} + \cdots + \frac{x^K}{K!},$$

where K is a nonstandard positive integer.

If $\mathbf{N} \models S$, where $\mathbf{N}$ has a standard arithmetic part, while also $\mathbf{M} \models S$, and $\mathbf{M}$ has a nonstandard arithmetic part, with $K \in \mathbf{M}$, K nonstandard, then we notice that the restriction $p(x)|\mathbf{N}$ is an exponential, while $p(x) \in \mathbf{M}$ is a polynomial.

5.7. Function F and Function G

Now recall that a consistent system S is ω–consistent if S doesn't simultaneously prove $\exists x\, P(x)$, and $\neg P(0), \neg P(1), \neg P(2), \ldots$. ω–consistency implies consistency, while the converse isn't true. Consider PA which is supposed consistent, and add a new symbol ζ to its alphabet, with the (new) predicate $N(x)$ which should intuitively mean, "x is a natural number."

Then by the usual compacity argument the system $\mathrm{PA} + \zeta \neq 0 + \zeta \neq 1 + \cdots + N(\zeta)$ is consistent. Also, from $N(\zeta) \to \exists x\, N(x)$ one proves $\exists x\, N(x)$. We then get the ω–inconsistent system $\mathrm{PA} + \zeta \neq 0 + \zeta \neq 1 + \cdots + \exists x\, N(x)$, which however is consistent. F gives rise to one such situation:

Remark 5.7.1. Notice that given function F, we can see that [F is total] is independent of S, but the structure of that function for the nonstandard models where $\neg$[F is total] holds isn't clear at all. We immediately get an ω–inconsistency result, for

$$S + \neg[\mathsf{F} \text{ is total}] + \exists y\, (\mathsf{F}(0) = y) + \exists y\, (\mathsf{F}(1) = y) + \cdots$$

is a consistent theory. Thus Σ_1–unsoundness (or equivalently $\neg$[F is total] implies ω–inconsistency.

We can say that for very large values of its argument, F ceases to be defined in such ω–inconsistent systems.

Construction of function G

But nevertheless we can ask: is there a partial recursive function G with an infinite domain and yet with a behavior similar to that of F? We answer this question in the affirmative through the following:

Proposition 5.7.2. *There is a partial recursive function* G *so that:*

1. *If* $\mathbf{N} \models S$ *and has standard arithmetic, then* $\mathbf{N} \models \mathsf{F} = \mathsf{G}$.
2. $S + [\mathsf{F}$ *is total*$] \vdash$ [Domain G *is infinite and* G *is increasing*].
3. $S + [\mathsf{F}$ *is total*$] \nvdash [\mathsf{G}$ *is total*$]$ *and* $S + [\mathsf{F}$ *is total*$] \nvdash \neg[\mathsf{G}$ *is total*$]$.
4. *If* $S_{(\alpha)}$, $\alpha > 1$ *an ordinal, is in the Turing–Feferman hierarchy over* S *extended by* Σ_1*–soundness reflection principles, then we can choose* α *as high as we wish in that hierarchy* $\leq \omega$, *so that* $S_{(\alpha)} \vdash [\mathsf{G}$ *is total*$]$, *but such that* $S_{(\beta)} \nvdash [\mathsf{G}$ *is total*$]$, $\beta < \alpha$.

(For the Turing–Feferman hierarchy see [20; 21].)

Proof. As we have supposed that S has a model $\mathbf{N}$ with standard arithmetic, then so does theory $S' = S + [\mathsf{F}$ is total$]$:

- We can explicitly obtain a Diophantine polynomial $p(x_1, \ldots, x_k)$ so that:
  1. $S' \nvdash [\forall x_1, \ldots p(x_1, \ldots) > 0]$ while $\forall x_1, \ldots p(x_1, \ldots) > 0$ holds of $\mathbf{N}$.

2. $S' \not\vdash [\exists x_1, \ldots p(x_1, \ldots) = 0]$ and for some model $\mathbf{M}$ with nonstandard arithmetic, $\mathbf{M} \models \exists x_1, \ldots p(x_1, \ldots) = 0$.

- Define: $\zeta = a_1$, given that for some (necessarily nonstandard) model $\mathbf{M}$,

$$\mathbf{M} \models [p(a_1, a_2, \ldots, a_k) = 0].$$

(We must impose some uniqueness condition on $\langle a_1, \ldots, a_k \rangle$.)

By construction, for all models of S', [F is total]. □

Remark 5.7.3. We now informally describe an algorithm for the function we are looking for. Let h be such that S proves h to be total and strictly increasing. Put, for G:

- If $m < \zeta$, $\mathsf{G}(m) = \mathsf{F}(m)$.
- If $m > \zeta$ and $n = \mathsf{h}(m)$, then $\mathsf{G}(n)$ is undefined.
- If $m > \zeta$ and $n \neq \mathsf{h}(m)$, then $\mathsf{G}(n) = \mathsf{F}(n)$.

So G will always have an infinite domain whenever [F is total] holds. And due to the dependence of G on ζ, it cannot be proved total even in a strong theory such as $S +$ [F is total], that is, $S + [S$ is Σ_1–sound]. For the last assertion, given the hierarchy over S plus Σ_1–sound reflection principles, it suffices to choose an adequate p so that $\forall x, \ldots\ p(x, \ldots) > 0$ is only proved by $S_{(\alpha)}$, but not by any $S_{(\beta)}$, $\beta < \alpha \leq \omega$. (For the Turing–Feferman theorem see [20; 21].)

We can give a formal expression for G with the help of the ι symbol, which we then add to our formal background — S and the required extensions.

- We can write down the algorithm for a Turing machine that never stops over any input n. Let $\{e_0\}$ be that machine.
- Consider h as in Remark 5.7.3. h is a S–provably total recursive function. Then write:

$$\begin{aligned}\mathsf{H}(n) = \iota_x\{&[(x = \mathsf{F}(n)) \wedge (n \notin \text{Image}(\mathsf{h}))]\\ &\vee [(x = \{e_0\}(n)) \wedge (n \in \text{Image}(\mathsf{h}))]\}.\end{aligned}$$

  Therefore S proves that such a function equals F (which is total for all models of S), but for the values of h, where it is undefined. As the image of h doesn't exhaust all of ω, its complement is infinite, and H will have an infinite domain.
- Now consider the first Diophantine polynomial p above together with $\zeta = a_1$. If we write $\forall x\, p > 0$ for $[\forall x_1, \ldots p(x_1, \ldots) > 0]$, and similarly $\exists x\, p = 0$, then consider the next expression that can be seen to be in the language of S (possibly

extended for definitions):

$$\begin{aligned}\mathsf{G}(n) = \iota_x \{&[(\forall x\, p > 0) \wedge (x = \mathsf{F}(n))] \\ &\vee[(\exists x\, p = 0) \wedge [(n < \zeta) \wedge (x = \mathsf{F}(n))] \\ &\vee[(n \geq \zeta) \wedge (x = \mathsf{H}(n))]]\}.\end{aligned}$$

ζ is as above. This construction originates in the (symbolic) form $(1 - \beta)\mathsf{X} + \beta\mathsf{X}'$, which was used in [6]; for the expression above see the construction in [10], p. 34. It is also akin to the construction in the result known as Kreisel's Lemma [14].

5.8. The *P vs. NP* Question

We give here a brief summary of our work on the P vs. NP question, which suggests that the formal sentences $P < NP$ and $P = NP$ are independent of axioms systems like PA or ZFC. For details see [9; 13; 14].

The P vs. NP question deals with problems where it is difficult to find a solution, but once we get it, it is easy to test it to see whether it actually settles the problem. Hard to find, easy to test, is the motto of the problem. By hard we mean: time–exponential on the length of the input; by easy, time–polynomial on the length of the input. The input is always coded in a binary alphabet.

Remark 5.8.1. There are several such problems; best known is the traveling salesman problem[a]. We consider here the Boolean satisfiability problem. Given a Boolean expression, find (if any) a set of truth–values for its variables that will give value "true" for the whole expression. We restrict our attention to expressions in conjunctive normal form (cnf), that is, Boolean expressions that are conjunctions of disjunctions of Boolean variables or of its negations.

The set of all satisfiable boolean expressions in cnf — after we have excluded the unsatisfiable, totally false expressions — can be coded in several easy ways by the whole of ω through a primitive recursive map. We not that set SAT.

We now give the formal version of the sentence $P = NP$, noted $[P = NP]$, as it is well–known it can be naïvely translated as, "there is a fast, time–polynomial on the input, algorithm that settles all instances of the Boolean satisfiability problem."

The formal version goes as follows:

[P = NP] and [P < NP]

Definition 5.8.2.

- The standard formalization for $[P = NP]$ is:

$$[P = NP] \leftrightarrow \exists m \in \omega, a\, \forall x \in \omega\, [(t_m(x) \leq |x|^a + a) \wedge R(x, m)].$$

---

[a]See http://en.wikipedia.org/wiki/Traveling-salesman

- $[P < NP] \leftrightarrow_{\mathrm{Def}} \neg[P = NP]$.
- R is a poly predicate, that is to say, $\langle x, m\rangle$ satisfies R if and only if there is a poly machine $\mathsf{M}_R(x, y) = 1$. Otherwise $\mathsf{M}_R(x, y) = 0$.

The exotic formalization

The exotic formalization is a simple trick that allows us to easily derive undecidable sentences for the question. Alas! they are not the real thing, they are, we may say, one step away from it:

For the next definitions and results let f be in general a (possibly partial) recursive function which is strictly increasing over its domain, and let e_{f} be the Gödel number of an algorithm that computes f. Let $p(\langle e_{\mathsf{f}}, b, c\rangle, x_1, x_2, \ldots, x_k)$ be [15] an universal Diophantine polynomial with parameters e_{f}, b, c; that polynomial has integer roots if and only if $\{e_f\}(b) = c$. We may if needed suppose that polynomial to be ≥ 0. We omit the "$\in \omega$" in the quantifiers, since they all refer to natural numbers.

Definition 5.8.3.

$$M_{\mathsf{f}}(x, y) \leftrightarrow_{\mathrm{Def}}$$

$$\exists x_1, \ldots, x_k\, [p(\langle e_{\mathsf{f}}, x, y\rangle, x_1, \ldots, x_k) = 0].$$

Actually $M_{\mathsf{f}}(x, y)$ stands for $M_{e_{\mathsf{f}}}(x, y)$, or better,

$$M(e_{\mathsf{f}}, x, y),$$

as dependence is on the Gödel number e_{f}.

Definition 5.8.4. $\neg Q(m, a, x) \leftrightarrow_{\mathrm{Def}} [(t_m(x) \leq |x|^a + a) \rightarrow \neg R(x, m)]$.

Proposition 5.8.5. $[P < NP] \leftrightarrow \forall m, a\, \exists x\, \neg Q(m, a, x)$.

Definition 5.8.6. $\neg Q_{\mathsf{f}}(m, a, x) \leftrightarrow_{\mathrm{Def}} \exists a'\, [M_{\mathsf{f}}(a, a') \wedge \neg Q(m, a', x)]$.

We will sometimes write $\neg Q(m, \mathsf{f}(a), x)$ for $\neg Q_{\mathsf{f}}(m, a, x)$, whenever f is total.

Definition 5.8.7 (Exotic formalization.)

$$[P < NP]^{\mathsf{f}} \leftrightarrow_{\mathrm{Def}} \forall m, a\, \exists x\, \neg Q_{\mathsf{f}}(m, a, x).$$

Definition 5.8.8. $[P = NP]^{\mathsf{f}} \leftrightarrow_{\mathrm{Def}} \neg[P < NP]^{\mathsf{f}}$.

We now use function F as above (see Definition 5.4.4):

Proposition 5.8.9. $S \vdash [P < NP]^{\mathsf{F}} \leftrightarrow \{[\mathsf{F}\ \textit{is total}] \wedge [P < NP]\}$.

Lemma 5.8.10.

1. $S \vdash [P < NP]^{\mathsf{F}} \rightarrow [\mathsf{F}\ \textit{is total.}]$.
2. $S \vdash [P < NP] \rightarrow \mathrm{Consis}\, S$.

Proposition 5.8.11. *If S is consistent, then S doesn't prove $[P < NP]^{\mathsf{F}}$.*

Corollary 5.8.12. *$[P = NP]^{\mathsf{F}}$ is consistent with S.*

If $\mathbf{N}$ is a model for S and makes it arithmetically sound, that is, $\mathbf{N}$ has a standard arithmetic part for the arithmetic in S:

Proposition 5.8.13. $\mathbf{N} \models S + [P < NP] \leftrightarrow [P < NP]^{\mathsf{F}}$.

Proposition 5.8.14. *If S is consistent, then $[P < NP] \leftrightarrow [P < NP]^{\mathsf{F}}$ is independent of S.*

The two hypotheses

We now add two hypotheses to our argument:

Hypothesis 5.8.15. There is a recursive set P so that:

1. It is a primitive recursive transformation of the BGS set.
2. We can show that for the induced transform A of K which is the primitive recursive predicate in the definition $\forall x \, \exists y \, K$ of $P < NP$, is associated to a Skolem function that cannot be proved total in S.

Hypothesis 5.8.15 implies that $S \nvdash [P < NP]$; the exotic formulation has been an effort in the direction of a proof of that hypothesis (for BGS see below).

Hypothesis 5.8.16. If $S \nvdash [P < NP]$ and $S \vdash [P = NP]$, then S proves that the set of BGS poly machines that solve any problem in the NP class is recursively enumerable.

(A BGS machine is an ordered pair $\langle \mathsf{M}_n, p_r(|x|) \rangle|$ where p_r is a degree r polynomial on the length $|x|$ of a binary input x that determines that a clock should shut down the operation of M_n over x when the time bound $p_r(|x|)$ is reached.) For a discussion plus plausibility arguments for those hypotheses see [1; 13]. They lead to the following:

Proposition 5.8.17. *If Hypotheses 5.8.15 and 5.8.16 hold, then $[P = NP]$ and $[P < NP]$ are independent of S. Moreover, Peano Arithmetic strengthened with Shoenfield's ω–rule proves $[P < NP]$.*

For a discussion of these results see also [39].

Acknowledgments

The authors wish to thank Professor K. Nakamatsu for his kind invitation to contribute to this handbook.

The ongoing research program that led to this text has been sponsored by the Advanced Studies Group, Production Engineering Program, COPPE–UFRJ, Rio, Brazil.

The authors also wish to thank the Institute for Advanced Studies at the University of São Paulo for support of the early steps of this research project; we wish to acknowledge support from the Brazilian Academy of Philosophy and its chairman Professor J. R. Moderno. Portions of this work were done during the COBERA March 2005 Workshop at Galway, Ireland; FAD wishes to thank Professor Vela Velupillai for the stimulating and fruitful environment he so kindly sponsored at that meeting. Finally FAD wishes to thank Professors R. Bartholo, C. A. Cosenza and S. Fuks for their invitation to join the Fuzzy Sets Lab at COPPE–UFRJ and the Philosophy of Science Program at the same institution.

The authors acknowledge partial support from CNPq, Philosophy Section. They also wish to thank Professor F. Zamberlan, chairman of the Production Engineering Program at COPPE–UFRJ for support.

References

1. T. Baker, J. Gill and R. Solovay, "Relativizations of the $P = ?NP$ question", *SIAM J. Comp.* **4**, 431–442 (1975).
2. J. Barrow, *Impossibility: The Limits of Science and the Science of Limits*, Oxford (1998).
3. R. Bartholo, C. Cosenza, F. A. Doria, C. Lessa, "Can economic systems be seen as computing devices?" *Journal of Economic Behavior and Organization* **70**, 72–80 (2009).
4. L. Blum, F. Cucker, M. Shub, S. Smale, *Complexity and Real Computation*, Springer (1998).
5. W. A. Carnielli and F. A. Doria, "Is computer science logic–dependent?" in C. Dgremont., *et al.* (Eds.), *Dialogues, Logics and other Strange Things: Essays in Honor of Shahid Rahman*, College Publications (2008).
6. N. C. A. da Costa and F. A. Doria, "Undecidability and incompleteness in classical mechanics," *International Journal of Theoretical Physics* **30**, 1041–1073 (1991).
7. N. C. A. da Costa and F. A. Doria, "An undecidable Hopf bifurcation with an undecidable fixed point," *Int. J. Theoretical Physics* **33**, 1885–1903 (1994).
8. N. C. A. da Costa and F. A. Doria, "On set theory as a foundation for computer science," *Bulletin of the Section of Logic, (University of Lodz)* **33**, 33–40 (2004).
9. N. C. A. da Costa and F. A. Doria, "Consequences of an exotic formulation for $P = NP$", *Applied Mathematics and Computation* **145**, 655–665 (2003); also "Addendum", *Applied Mathematics and Computation* **172**, 1364–1367 (2006).
10. N. C. A. da Costa and F. A. Doria, "Computing the Future," in K. V. Velupillai, ed., *Computability, Complexity and Constructivity in Economic Analysis*, Blackwell (2005).
11. N. C. A. da Costa and F. A. Doria, "Some thoughts on hypercomputation," *Applied Mathematics and Computation*, special issue on hypercomputation, **178**, 83–92 (2006).
12. N. C. A. da Costa and F. A. Doria, "Janus–faced physics: on Hilbert's 6th Problem," in C. Calude, ed., *Complexity and Randomness: from Leibniz to Chaitin (Festschrift in Honor of G. Chaitin on his 60th Birthday)*, World Scientific (2007).

13. N. C. A. da Costa and F. A. Doria, "Hypotheses that imply the independence of $P = NP$ from strong axiomatic systems," to appear in S. Zambelli, *Festschrift in Honor of V. Velupillai*, Routledge (2008).
14. N. C. A. da Costa, F. A. Doria and E. Bir, "On the metamathematics of the P *vs* NP question," *Applied Mathematics and Computation* **189**, 1223–1240 (2007).
15. M. Davis, "Hilbert's Tenth Problem is unsolvable" in *Computability and Unsolvability*, Dover (1982).
16. F. A. Doria, "Metamathematics of $Pvs.NP$", talk at the Suppes Fest, Fed. University at Santa Catarina, Florianópolis (Brazil), April (2002).
17. F. A. Doria, "Informal vs. formal mathematics", *Synthèse* **154**, 401–415 (2007).
18. F. A. Doria and J. F. Costa, eds., Special issue on hypercomputation, *Applied Math. Computation* **178** (2006).
19. A. Ehrenfeucht and J. Mycielski, "Abbreviating proofs by adding new axioms," *AMS Bull.* **77**, 366–367 (1971).
20. S. Feferman, "Transfinite recursive progressions of axiomatic theories", *J. Symbolic Logic* **27**, 259–316 (1962).
21. T. Franzen, "Transfinite progressions: a second look at completeness," *Bull. Symbolic Logic* **10**, 367-389 (2004).
22. J. Hartmanis and J. Hopcroft, "Independence results in computer science", *SIGACT News*, 13, Oct. Dec. (1976).
23. S. C. Kleene, "General recursive functions of natural numbers", *Math. Annalen* **112**, 727–742 (1936).
24. S. C. Kleene, *Mathematical Logic*, Wiley (1967).
25. G. T. Kneebone, *Mathematical Logic*, Van Nostrand (1963).
26. M. Machtey and P. Young, *An Introduction to the General Theory of Algorithms*, North–Holland (1979).
27. K. Loo, "Internal Turing machines", arXiv:math-ph/0407056 v2 (2004).
28. E. Mendelson. *Introduction to Mathematical Logic*, 4th ed., Chapman & Hall (1997).
29. E. Reyes, "Undecidability and incompleteness in the theory of partial differential equations," submitted to the *Logic and Time: Einstein and Gödel* meeting in Rio (2007)
30. H. G. Rice, "Classes of recursively enumerable sets and their decision problems," *Trans. Amer. Math. Society* **74**, 358–366 (1953).
31. D. Richardson, "Some Unsolvable Problems Involving Elementary Functions of a Real Variable," *J. Symbolic Logic* **33**, 514–522 (1968).
32. H. Rogers Jr., *Theory of Recursive Functions and Effective Computability*, McGraw–Hill (1967).
33. B. Scarpellini, "Two undecidable problems of analysis" (translation from the 1963 original text), *Minds and Machines* **13**, 49–77 (2003).
34. B. Scarpellini, "Comments on 'Two undecidable problems of analysis,' " *Minds and Machines* **13**, 79–85 (2003).
35. S. Smale, "Mathematical problems for the next century," in V. Arnol'd *et al.*, *Mathematics: Frontiers and Perspectives*, AMS (2000).
36. C. Smorýnski, *Logical Number Theory, I*, Springer (1991).
37. I. Stewart, "Deciding the undecidable," *Nature* **352**, 664–665 (1991).
38. I. Stewart, "Deciding the Undecidable," in *The Problems of Mathematics*, 2nd edition, pp. 308–311, Oxford U. P. (1992).
39. A. Syropoulos, *Hypercomputation: Computing Beyond the Church–Turing Barrier* , Monographs in Computer Science, Springer (2008).
40. A. M. Turing, "On computable numbers, with an application to the Entscheidungsproblem," *Proc. London Math. Society* **50**, 230–251 (1937).

Chapter 6

HYPOTHESIS REFINEMENT: BUILDING HYPOTHESES IN AN INTELLIGENT AGENT SYSTEM

Gauvain Bourgne, Nicolas Maudet and Suzanne Pinson
Université Paris-Dauphine
LAMSADE Laboratory
75775 Paris Cedex 16
{bourgne,maudet,pinson}@lamsade.dauphine.fr

This paper deals with distributed hypothesis formation in an intelligent agent system, where individual agents communicate together to overcome the incompleteness of their own observations. We present a Hypothesis Refinement framework encompassing a variety of individual hypothesis formation processes, and formalize the problem of building adequate hypotheses when data are distributed. Then, some communication protocols are proposed to let agents build proper hypotheses by confronting them to other agents in the system. At last, we give a methodology to adapt this framework to specific applications, and illustrate it with an example application in which agents attempt to flee from a building in which a fire has erupted.

6.1. Introduction

In numerous concrete problems and applications, individual agents in a system must make decisions based on incomplete or uncertain informations. Conclusions that can be directly drawn from their certain knowledge might then be unsufficient. When data are incomplete, one can make some assumptions about unknown facts or rules. These *hypotheses* can thus be used to complete the knowledge of the agents, though they remain revisable. This paper is concerned with hypothesis formation in an intelligent agent system. We propose a generic framework to describe this kind of problems, and some mechanisms to deal with the distribution in a society of agents of individual hypothesis formation processes. This work will be illustrated by two running examples in different settings. We take in this paper a methodological perspective, focusing on the study of hypothesis refinement problems, and communication protocols that can be used for it. For detailed applications using this framework or experimental study of the proposed mechanisms, one can refer to SMILE,[1] an application of this framework to incremental concept learning, and to an abductive instantiation,[2,3] dealing with the dynamic fire simulation abduction problem presented in the last section.

In Section 6.2, we will precisely define our framework, and its different parameters. Section 6.3 will then introduce a learner/critic approach for solving such problems, and give several local or global revision mechanisms to build an hypothesis in a system of two or more agents. Afterwards, Section 6.4 will explain how to instantiate our general framework to specific problems. These methodological considerations will be illustrated by a third example, slightly more complex. Finally, Section 6.5 presents related works, and Section 6.6 concludes.

6.2. Hypothesis Refinement Problem

This section defines the hypothesis refinement problem, a general class of problems concerned with distributed formation of hypotheses. The formalisation of this problem intends to encompass a variety of different hypothesis formation problems, and therefore explicit representation of knowledge is not specified. However, it is supposed that there exist some different kinds of knowledge in the system that are represented in some way. To refer to these and to the relations between them, we will use *abstract concepts*, that will have to be instantiated when dealing with specific problems. We shall see that these abstract notions are sufficient to develop some useful properties to describe a system and characterize the class of problem. Moreover, we will show in the next section that some general communication protocols can be proposed for these problems on the basis of these notions. We will first give our characterisation of the different kinds of knowledge and their possible representation, before introducing the notion of consistency relation, that will be detailed at different levels. Then, we will see how we can abstract away from the specifics of the reasoning used for different kinds of hypothesis formation. Using these notions, we shall finally characterize the hypothesis refinement problem, and its possible variants or parameters. All these notions will be illustrated by two simple examples of abductive or inductive hypothesis formation.

6.2.1. *Knowledge representation*

The first concern is to characterize the knowledge of the agents. One main distinction that is of utmost importance in our framework is to know whether some knowledge is *certain* or *revisable*. Another criterion will be to know whether some konwledge is *common* to all agents in the system or specific to one of them. In most cases, we will assume that agents are initially identical, but will become differentiated by their history (perceiving and acquiring individually different pieces of knowledge). We shall then consider three main notions in the knowledge of an agent a_i:

- *Background theory* $\mathcal{T}$ represents *prior knowledge* common to all agents. It is *certain* knowledge.
- *Observation memory* O_i represents *acquired knowledge*, specific to an agent. It is *certain* knowledge that the agent have usually acquired through perception or communication. It is supposed to be *factual*, meaning that it concerns specific data

(*observations*) rather than general rules. Moreover these observations should be *retrievable*, meaning that it is possible to retrieve some specific observation from the observation memory.
- *Hypothesis* h_i represents some *revisable* knowledge in which the agent *believes*, though it knows it is not certain. It is usually built from its other knowledge, or through interaction with other agents.

Note that the distinction between *revisable* and *certain* knowledge corresponds to a distinction between what the agents *believes* to be true, and what the agent *knows* to be true. We assume that the agent can derive some conclusions from its explicit knowledge in some way, and so we shall denote by $Cn(X)$ the knowledge that can be deduced from X. Then, the *certain knowledge* of an agent a_i (what it knows) will be $Cn(\mathcal{T} \cup O_i)$.[a] The *beliefs* of an agent a_i (what it believes) will thus be $Cn(\mathcal{T} \cup O_i \cup h_i)$. Note that the agent always believes what it knows.

Though we use abstract notions to avoid specifying too early which kind of knowledge representation is used, it can still be useful to define some notions about the possible observations or hypotheses.

- We shall denote by $\mathcal{O}$ the *observation language*. It defines what observations can be made in the framework of a given application, providing syntax and semantic to generate them. $\mathcal{O}^A$, the *actual observations language* is a restriction of $\mathcal{O}$ that specify all the observations that can actually be observed by an agent in an instance of the application.
- We shall denote by Ω the *observation memory language*. It is classically the set of all subsets from $\mathcal{O}$, but it can take other forms depending on the way observation are memorised. Ω^A is then the *actual observation memories language*, that specify all possible observation memories of any agent, given an instance of the application.
- At last, we shall denote by $\mathcal{H}$ the *hypothesis language*, that can restrict semantically or syntaxically the hypotheses that can be generated.

Example 6.1 (Simple distributed medical diagnosis). As a first example to illustrate our framework, we propose a simple abductive distributed medical diagnosis. All agents know the possible diseases and their symptoms, and they must build an hypothesis on some patient, each of them initially knowing only part of the symptoms. More formally, we consider:

- a set of possible disease $\mathcal{D}$ = {angina, bronchitis, flu, hayfever}.
- a set of possible symptome $\mathcal{S}$ = {cough, mucus, fatigue, fever}.
- a causality relation E linking a disease to the symptoms it can cause. We have E = {(angina,fatigue), (angina,fever),(bronchitis, cough), (bronchitis, mucus),

[a]Though we use set notations to assemble two kinds of knowledge, this does know imply that knowledge is represented by sets.

(flu, cough), (flu, fatigue), (flu, fever), (hayfever,mucus)}. We define the function *effect* such that gives us all the symptoms caused by a set of diseases D: $\mathit{effect}(D) = \{s | \exists d \in \mathcal{D}, (d, s) \in E\}$.

In this framework, we can thus have a *static diagnosis problem*, given by a set O_i of observations on the patient for each agent. For instance, we could have 3 agents, a_1, a_2, a_3 with $O_1 = \{\text{fatigue}\}$, $O_2 = \{\text{fever}, \neg\text{cough}\}$ and $O_3 = \emptyset$. We could also define a more dynamic approach, where agent have initially no symptom, but can try to observe in turn the absence or presence of a symptom. In this approach, an instance of *progressive diagnosis problem* would correspond to a set of diseases D^+ affecting the patient (for instance $D^+ = \{\text{angina, hayfever}\}$).

In this problem, our different concepts would then map in the following way:

- *Background theory* $\mathcal{T}$ is the tuple $(\mathcal{D}, \mathcal{S}, E)$, that we shall call the *medical theory*. It is certain, and common to all agent.
- *Observations* are either the presence or absence of a symptom. The *observation language* $\mathcal{O}$ can thus be defined by a set: $\mathcal{O} = \mathcal{S} \cup \{\neg s | s \in \mathcal{S}\}$. On the other hand, the *actual observations language* $\mathcal{O}^A$ depends on a specific diagnosis problem. In a static diagnosis problem with h agents, we just have $\mathcal{O}^A = \bigcup_{i \in \{1,\ldots,n\}} O_i$ since no new observation can be made. If it is a progressive diagnosis problem, $\mathcal{O}^A$ would be the set of possible observation, that is the data about presence or absence of every symptom for the patient. Thus, in this case, $\mathcal{O}^A = \mathit{effect}(D^+) \cup \{\neg s | s \in \mathcal{S} \backslash \mathit{effect}(D^+)\}$. For instance, with $D^+ = \{\text{angina, hayfever}\}$, we would have $\mathcal{O}^A = \{\text{fatigue, fever, mucus, } \neg \text{ cough }\}$.
- *Observation memories* are here just simple sets: $\Omega = 2^{\mathcal{O}}$ and $\Omega^A = 2^{\mathcal{O}^A}$.
- Finally, an *hypothesis* is a *diagnosis*, that is, a set of diseases: $\mathcal{H} = 2^{\mathcal{D}}$.

Example 6.2 (Distributed inductive concept learning). As a second illustrative example, we propose a case of multiagent incremental concept learning. Given a set of examples labeled as positive or negative, a society of agents tries to form an hypothesis on the classification criterion (called the target concept). We shall illustrate by learning the concept of 'good day to play tennis'.[4] Examples are days represented by three descriptors and labeled as being good or bad days for playing tennis. A day will be characterized by its outlook, that can be sunny (O_{Sun}), cloudy (O_{Cloud}) or rainy (O_{Rain}), by the wind, that can be strong (W_{Strong}) or weak (W_{Weak}), its humidity, high (H_{High}) or low (H_{Normal}), and its temperature, hot (T_{Hot}) or cold (T_{Cold}). A day is considered to be a good day to play tennis iff it is sunny with normal humidity, rainy with a weak wind or overcast.

We then have the following mapping:

- *Observations* are called *labeled examples* in this context. A *labeled example* will be represented as a label + or − together with a description, that is, a set of atoms corresponding to the value of its different descriptors. The set of possible atoms will be $\mathcal{A} = \{O_{Sun}, O_{Cloud}, O_{Rain}, W_{Strong}, W_{Weak}, H_{High}, H_{Low}, T_{Hot}, T_{Cold}\}$.

The set of possible descriptions for an example would then be $\mathcal{D} = \{d_O, d_W, d_T, d_H — d_O \in \{O_{Sun}, O_{Cloud}, O_{Rain}\}, d_W \in \{W_{Strong}, W_{Weak}\}, d_T \in \{T_{Hot}, T_{Cold}\}, d_H \in \{H_{High}, H_{Low}\}\}$. Note that $\mathcal{D} \subset 2^{\mathcal{A}}$. Thus:

- The *observation language* $\mathcal{O}$ will contains examples that can be build from $\mathcal{A}$. We have $\mathcal{O} = \mathcal{D} \times \{+, -\}$.
- The *actual observations language* $\mathcal{O}^A$ depends on the target concept. It will contains all examples that are correctly labeled according to the target concept. Thus for the concept of a good day to play tennis, we would have $\mathcal{O}^A = \{e = (d, +) | e \in \mathcal{O} s.t. f_{tennis}(d)\} \cup \{e = (d, -) | e \in \mathcal{O} s.t. \neg f_{tennis}(d)\}$ where $f_{tennis}(d)$ is true iff d is a good day to play tennis. Thus, for instance $e_1 = ((O_{Rain}, W_{Weak}, H_{Normal}, T_{Hot}), +) \in \mathcal{O}^A$ and $e_2 = ((O_{Sun}, W_{Strong}, T_{Hot}, H_{High}), -) \in \mathcal{O}^A$.

- *Observation memories* O_i will be example memories $E_i = (E_i^+, E_i^-)$ where E_i^+ is the positive example memory, the set of all description of positive examples known by a_i, and E_i^- is the set of all description of negative examples known by a_i. Thus, the *observation memory language* is $\Omega = 2^{\mathcal{D}} \times 2^{\mathcal{D}}$, and the *actual observation memories language* is $\Omega^A = 2^{\Delta^+} \times 2^{\Delta^-}$ where $\Delta^+ = \{d \in \mathcal{D} | (d, +) \in \mathcal{O}^A\}$ and $\Delta^- = \{d \in \mathcal{D} | (d, -) \in \mathcal{O}^A\}$.
- An *hypothesis* will a be disjunction of conjunctive statements, where a conjuctive statement is a conjonction of atoms from $\mathcal{A}$, that is a formulae of atoms from $\mathcal{A}$ in disjunctive normal form (DNF). Thus $\mathcal{H} = DNF(\mathcal{A})$. For instance, an hypothesis perfectly matching f_{tennis} would be $h = (O_{Sun} \wedge H_{Normal}) \vee O_{Cloud} \vee (O_{Rain} \wedge W_{Weak})$.

6.2.2. *Consistency relation*

As hypotheses are revisable, they might be contradictory with some new observations. The validity of an hypothesis is something that should be ensured and maintained. However, the condition of this validity can vary according to the kinds of hypotheses and observations. We thus need some abstract relation, the *consistency*, to represent this validity. We will introduced below the consistency of an hypothesis, before extending this notion to agents and groups of agents. Then we will introduce another tool to compare hypotheses of a group of agents through the notion of equivalence and homogeneity.

6.2.2.1. *Consistency of an hypothesis*

To represent the adequation of what an agent believes with what it knows, that is the validity of an hypothesis with respect to other knowledge, we will use an abstract relation linking an hypothesis with certain knowledge : the *consistency relation*. As the background theory is common to all agents, we will incorporate it in the consistency relation. Thus, the consistency relation will link a hypothesis with an observation memory. We shall denote by $Cons(h, O)$ the fact that a hypothesis h

is *consistent* with an observation memory O. Such a relation can take different form according to the reasoning involved, but it would frequently include an aspect of *coherence*, ensuring that beliefs are not contradictory, and an aspect of *completeness*, ensuring that the hypothesis is useful and properly complete the lack of information. We give below two possible examples of consistency relation.

Example 6.3 (Simple distributed medical diagnosis). For the distributed medical diagnosis problem, an hypothesis (diagnosis) is valid if the assumed diseases causes the symptoms that are observed and no more. If the diseases cause some symptoms whose absence have been observed, then it is not coherent: we know that at least one of the assumed disease cannot be present. If the diseases does not explain some of the symptoms, then it is incomplete: we know that we either forgot some disease in the diagnosis or chose wrong ones.

More formally: $\forall h \in \mathcal{H}, \forall O \in \Omega, Cons(h, O)$iff

- $\forall s \in \mathcal{S}, \neg s \notin O \vee s \notin \mathit{effect}(h)$ (coherence)
- $\forall s \in O \cap \mathcal{S}, s \in \mathit{effect}(h)$ (completeness)

Thus, using the background theory defined earlier and $h_1 = \{$angina,hayfever$\}$:

- h_1 is consistent with $\{\neg$ cough, fever, mucus$\}$ (coherent and complete).
- h_1 is inconsistent with $\{$fatigue, mucus, $\neg$ fever$\}$ (incoherence since fever should be an effect of angina).
- h_1 is inconsistent with $\{$cough, fever, mucus$\}$ (incompleteness since cough is not an effect of angina nor hayfever).

Example 6.4 (Distributed inductive concept learning). In the case of multi-agent learning, the hypothesis means that all examples whose description is covered by it should be positive. A description (or an example) is covered by an hypothesis if all its atoms are present in one of its conjunctive statement. We denote the fact that h *covers* (or subsume) an example e by $h \models_s e$. Then a hypothesis is incoherent if it covers a negative example, and it is incomplete if there is a positive example in the example memory that is not covered.

More formally: $\forall h \in \mathcal{H}, \forall E = (E^+, E^-) \in \Omega, Cons(h, E)$ iff

- $\forall e \in E^-, h \not\models_s e$ (coherence)
- $\forall e \in E^+, h \models_s e$ (completeness)

6.2.2.2. *Group Consistency*

Now we extend this notion of consistency to agents and groups of agents:

Definition 6.1 (Group Consistency). An agent a_i is group consistent wrt. the group of agents G ($GCons(a_i, G)$) iff $Cons(h_i, \cup_{i \in G} O_i)$)

A stronger notion of consistency requires any agent within the group to be consistent with the entire group.

Definition 6.2 (Mutual Consistency). A group of agents is mutually consistent ($MCons(G)$) iff $\forall a_i \in G$, it is the case that $GCons(a_i, G)$.

Now for the purpose of our work, we shall mainly be interested in some interesting particular cases which depends on the cardinality of the group G:

- *Internal consistency* — this is the limit case when G is limited to a single agent. In this case, Group and Mutual consistency collapse into a single notion that we shall call *internal consistency* ($ICons(a_i)$).
- *Peer consistency* — when the group of agents we consider contains only two agents. In this case we can distinguish both the *peer consistency* of an agent wrt a fellow agent, and the *mutual peer consistency* of a group of two agents. This is especially important in our context, since our communication protocols only deal locally with bilateral communications.
- *MAS-consistency* — we conclude with the limit case involving all agents in the system. Then, we will refer to the *MAS-consistency* of an agent wrt to the society; and to the *mutual MAS-consistency* of a society of agents.

For the sake of readability, we now introduce some notational sugar. When we refer to the internal consistency of an agent a_i, we shall simply write $ICons(a_i)$. As for the case of peer and MAS consistency, we will put the cardinality of the group involved as an exponent (recall that the society is composed of n agents):

	Group Consistency	**Mutual Consistency**
Internal consistency (single agent)	$ICons(a_i)$	$ICons(a_i)$
Peer consistency (pair of agents)	$GCons^2(a_i, \{a_i, a_j\})$	$MCons^2(\{a_i, a_j\})$
MAS-consistency (society of agents)	$GCons^n(a_i, \{a_1, \ldots, a_n\})$	$MCons^n(\{a_1, \ldots, a_n\})$

Example 6.5 (Distributed inductive concept learning). We illustrate these notions with a small system of three agents and the 'good day to play tennis' learning problem described earlier. We have:

$E_1^+ = \{(O_{Sun}, W_{Strong}, T_{Hot}, H_{Normal})\}$, $E_1^- = \emptyset$, $h_1 = O_{Sun} \wedge T_{Hot}$
$E_2^+ = \{(O_{Cloud}, W_{Weak}, T_{Hot}, H_{Low})\}$, $E_2^- = \emptyset$, $h_2 = O_{Sun} \wedge (H_{Normal} \vee O_{Cloud})$
$E_3^+ = \{(O_{Sun}, W_{Weak}, T_{Hot}, H_{Normal})\}$, $E_3^- = \{(O_{Sun}, W_{Strong}, T_{Hot}, H_{High})\}$, $h_3 = (O_{Sun} \wedge W_{Weak}) \vee (O_{Cloud} \wedge H_{Low})$

Then, h_1 is consistent with E_1 but neither with E_2 nor E_3; h_2 is consistent with E_1, E_2 and E_3; h_3 is consistent with E_3 and E_2, but not with E_1. Thus, all agents are *internally consistent.* Moreover, a_3 is *peer-consistent* with a_2 but not with a_1, As a_2 is also *peer-consistent* with a_3, a_2 and a_3 are *mutually peer-consistent.* Finally, a_2 is *peer-consistent* with a_1, and a_3. Thus a_2 is *mas-consistent.* However, a_2 and a_1 are not *mutually peer-consistent* and the system is not *mutually mas-consistent.*

6.2.2.3. *Equivalence and homogeneity*

Note that mutual group consistency does not implies that each agent has the same hypothesis, but only that each agent has an hypothesis that is consistent with all the observations of the group of agents. We introduce here some notions to ensure more similarities between the hypotheses of a group of agent. First, we need some tool to compare hypotheses. If the hypothesis is to be used for prediction of future observations, equivalent hypotheses should have the same domain of consistency.

Definition 6.3 (Equivalence wrt *Cons*.). Given a consistency relation $Cons$, two hypotheses h_1 and h_2 are *equivalent* wrt $Cons$ iff for all observation set $O \in \Omega^A$, $Cons(h_1, O)$ implies $Cons(h_2, O)$ and $Cons(h_2, O)$ implies $Cons(h_1, O)$.

In other words, if we denote by $\equiv_A$ this relation:

$$h_1 \equiv_A h_2 \text{ iff } \forall O \in \Omega^A, (Cons(h_1, O) \Leftrightarrow Cons(h_2, O))$$

This relation is an equivalence relation (reflexive, transitive and symmetric).

Then, a group will be said to be *homogeneous* if all the hypotheses of its agents are in the same equivalence class. Since the similarity criterion might not always be the same for all application, homogeneity is defined wrt an equivalence relation.

Definition 6.4 (Homogeneity). A group of agents $G = a_1, \ldots, a_k$ is *homogeneous* wrt an equivalence relation $\equiv$ if and only if for all pair of agent a_i and a_j in G, the hypotheses h_i and h_j of these agents are equivalent according to $\equiv$, that is, $\forall (a_i, a_j) \in G \times G, h_i \equiv h_j$.

We shall denote this property by $Hom_{\equiv}(G)$.

If this property is not verified, the group will be said to be *heterogeneous*.

Thus, if the purpose is to build a single hypothesis, the system should be homogeneous with respect to equality, whereas if the purpose is to predict the same observations, homogeneity with respect to $\equiv_A$ is enough.

Example 6.6 (Simple distributed medical diagnosis). We consider a system of 3 agents a_1, a_2, a_3 with hypotheses $h_1 = \{$flu, hayfever$\}$, $h_2 = \{$angina, bronchitis$\}$ and $h_3 = \{$flu, hayfever$\}$. We have *effect*$(h_1) = \{$cough, fatigue, fever, mucus$\} =$ *effect*(h_2). Consequently, considering the definition of $Cons$ in this context, we have $h_1 \equiv_A h_2$. Moreover, we have $h_1 = h_3$. Thus $\{a_1, a_2, a_3\}$ is *homogeneous wrt* $\equiv_A$, and $\{a_1, a_3\}$ is *homogeneous wrt equality.*

6.2.3. *Internal hypothesis formation*

Consistency captures the adequacy of an hypothesis, but does not make assumption on how to build such a hypothesis. There exists a number of different algorithms to produce hypotheses with a single agent, that can be used to model agents' internal reasoning. To abstract away from the specificity of different algorithms we will introduce *hypothesis formation functions* and *internal revision mechanisms.*

A *hypothesis formation function* $\mathcal{E}_h$ is a function that takes an agent internal states and returns an hypothesis that is consistent with these internal states. Usually, we will restrict the internal states that can have impact on the result. The internal states that might be used are current observation memory O, last hypothesis h', and last observation memory that was consistent with h': O'. Background theory $\mathcal{T}$ may intervene but, being constant and common, is not explicitly mentioned in the notation. Thus, in the general case, $\mathcal{E}_h(a_i)$ can be written as $\mathcal{E}_h(O, h', O')$ which can be define as a function from $\Omega^A \times \mathcal{H} \times \Omega^A$ to $\mathcal{H}$.

In dynamic situations where observations are received one at a time, we will rather use the procedural notion of *internal revision mechanism.* An *internal revision mechanism* μ is a process triggered by an agent with hypothesis $h \in \mathcal{H}$ and observation memory $O \in \Omega^A$ that receives a new observation $o \in \mathcal{O}^A$ to restore its internal consistency. This process updates the observation memory, and revises the hypothesis if necessary (we will denote the resulting hypothesis by $\mu(h, O, o)$).

We will say that agent has an *order-dependant internal reasoning* if $\mathcal{E}_h$ (or μ) is truly dependant of the last hypothesis and/or observation memory, that is, if $\exists h'_1, h'_2 \in \mathcal{H}, O, O'_1, O'_2 \in \Omega^A, \mathcal{E}_h(O, h'_1, O'_1) \neq \mathcal{E}_h(O, h'_2, O'_2)$ (or $\exists h, h', O, o, \mu(h, O, o) \neq \mu(h', O, o)$). On the other hand, an agent's reasoning is *order-independant* if this function will only depends on the current observation memory. We will thus only use this parameter for such functions, denoting them by $\mathcal{E}_h(O)$. Order-independant hypothesis formation functions will consequently be represented as functions from Ω^A to $\mathcal{H}$.

6.2.3.1. *Full Determinism*

The *hypothesis formation function* $\mathcal{E}_h$ has to return a single hypothesis that is consistent with a given observation set. However, there can be a lot of different hypotheses that are consistent with a given observation set, and the hypothesis formation function will have to select a *single* one among these. It will thus be important to know if this choice is always the same in comparable circumstances. We introduce the notion of *full determinism* to guarantee some persistence in this choice. This notion concerns only *order-independant* hypothesis formation functions.

Definition 6.5 (Full determinism). An (order-independant) hypothesis formation function $\mathcal{E}_h$ is *fully deterministic* for $\equiv$ and *Cons* if and only if it can only return hypotheses from different $\equiv$ equivalence class for two different observation sets if one of these two hypotheses is not consistent with the two observation sets. In other words, $\mathcal{E}_h(O)$ is *fully deterministic* for $\equiv$ and Cons iff:

$$\forall O, O', (\mathcal{E}_h(O) \not\equiv \mathcal{E}_h(O') \Leftrightarrow (\neg Cons(\mathcal{E}_h(O), O') \vee \neg Cons(\mathcal{E}_h(O'), O))$$

In order to guide the selection of a single hypothesis, different criteria can be used depending on the application. Examples could be irredundancy, minimal cardinality, minimal generality or specificity. In general, we can use a pre-order $\preceq$, possibly partial, called the *preference relation*, to specify more precisely which hypothesis should be returned. If this preference relation is a complete order rather than a partial pre-order, it will then be easy to build a fully deterministic hypothesis formation function, as illustrated by the following property:

Proposition 6.1. *If there exists a* complete order $\preceq$ *on the equivalence class of hypotheses according to* $\equiv$, *the hypothesis formation function* $\mathcal{E}_h$ *defined by* $\forall O, \mathcal{E}_h(O) = min_{\preceq}\{h|Cons(h,O)\}$ *is* fully deterministic *for* $Cons$ *and* $\equiv$.

6.2.3.2. *Individualism*

At last, we will discuss the ways in which an hypothesis can be changed. Agent reasoning is a way for the agent to ensure its internal consistency, but we could imagine that an agent could revise its hypothesis from external sources (by adopting another agent's hypothesis, for instance). Whereas this would be perfectly sensible for an agent using an order-dependant form of reasoning, it might not be very useful for an agent using a hypothesis formation process that does not depends on previous hypothesis, unless the adopted hypothesis is perfectly accurate. Indeed, if an agent with such a hypothesis formation function adopts an external hypothesis and get an observation that makes it inconsistent afterwards, it will not revise it, but rather replace it by a new one based on its observation memory. Thus it would be more interesting for the agent to get observations, that can be taken into account for new revision, rather than adopting hypothesis that would be forgotten at the first internal revision. We will thus prevent adoption of external hypotheses for agent using *order-independant* reasonings, requiring them to be *individualistic*:

Definition 6.6 (Individualism). An agent is said to be *individualistic* iff its working hypothesis h_i may only be modified as a consequence of an internal revision mechanism μ guaranteeing internal consistency.

This means that no other agent can directly impose a given hypothesis to an individualistic agent. As a consequence, only a new observation (being it a new perception, or data given by a fellow agent) can result in a modification of its working hypothesis h_i. This ensure that an agent will always base its hypothesis on its own observations, and will thus be able to defend it or adapt it on this basis if needed.

6.2.4. *Assumptions*

We will now present the different assumptions governing this hypothesis refinement framework and the different notions introduced in this section. First assumption ensures consistency of certain knowledge, whereas other assumptions deal with the agents, the consistency relation and the observations.

6.2.4.1. *Consistent world*

The main and most important assumption of this framework is that we rely on some existing *certain* knowledge which must be consistent as it is not revisable. This *consistent world assumption* is defined below:

Assumption 6.1 (Consistent world assumption). There exists a consistent theory $\mathcal{T}^0$ such that all certain knowledge in the system belongs to that theory, specifically, the background theory belongs to it ($\mathcal{T} \subseteq \mathcal{T}^0$) as well as any observation made by an agent in the system ($\forall i, O_i \subseteq \mathcal{T}^0$).

6.2.4.2. *Assumptions on agents*

Assumption 6.2. Agents are *cooperative* and *locally efficient.*

The first requirement that we put on the agents is *cooperativeness.* An agent will not lie or refuse to take part in the revision process. Then, we assume that the agents are *locally efficient*, meaning that they can always find some hypothesis consistent with their observation set. As the consistent world assumption ensures that an observation set is always consistent, this requirement is not very strong. At worst, it can be bypassed by allowing the agent to have a special unknown hypothesis.

6.2.4.3. *Compositionality of the consistency relation*

The requirement that we put on the consistency relation is its compositionality:

Assumption 6.3 (Compositionality of the consistency relation). The consistency relation must be *compositional*, meaning that:

$$Cons(h, O) \text{ and } Cons(h, O') \Leftrightarrow Cons(h, O \cup O')$$

To understand the consequences of this assumption, it is useful to distinguish both directions of this equivalence relation. The $\Rightarrow$ direction is often called *additivity.*[5] It basically means that it is possible to consider independently each observation. The $\Leftarrow$ direction is best understood when we read the contrapositive: it says that the inconsistency is monotonic (that is, when h is not consistent with some observation set O, it cannot become consistent again when that set grows monotically). In other words, an hypothesis assessed inconsistent on the basis of an observation set cannot become consistent when that set grows. We refer to this latter property, following Flach,[5] as *incrementality.*

6.2.4.4. *Assumptions on observations*

At last, we remind here some assumptions on observations and observation memory that were presented before.

Assumption 6.4. Observations are *certain* and *factual.*

First property has been presented before. It means that sensors are considered perfect. Second one requires observations to be *factual*, they concern specific data. In effect, it means that an observation is always a positive or negative litteral or grounded predicate, that is, an atomic instantiated fact, without generality and that has no components which could be observations themselves. Agents never gets formulas such as $o_1 \vee o_2$ as observations. Conjunctions of literals or predicates would be considered as being a set of observations rather than a single one.

Finally, we recall our assumption on observation memory:

Assumption 6.5 (Retrievability of observations). Observations stored in an observation memory are *retrievable*, meaning that it is always possible to get (or build) a stored observation from the observation memory it was stored in.

6.2.5. *Problem description*

A hypothesis refinement problem can now be described as the problem of ensuring the *mutual MAS-consistency* of a system of agents through global revision mechanism implying some communication protocol between the agents (with some associated strategies), combined with some internal reasoning at each agent level. This defines a general class of problems for which we will provide in the next section several revision mechanisms based on a learner/critic approach. However, this class of problem can be more specified by precising a number of parameters that we shall now describe.

6.2.5.1. *Reasoning*

One of the main parameters implied is the type of reasoning used to build hypothesis. Are we concerned with explaining ambiguous facts ? do we seek an hypothesis to classify some observations ? Depending on the approach, we will instantiate differently our framework and its underlying notions (hypothesis and observation language, consistency relation, hypothesis formation function and such). More details on these instantiations will be provided in Section 6.4.1.

The choice of an internal revision mechanism (or hypothesis formation function) will be relevant for selecting a global revision mechanism among these presented in the next section. Especially, it will be important to know whether reasoning is *order-dependant* or not, or rather, whether agents are *individualistic* or not. Moreover, in the case of order-independant hypothesis formation function, we have to know whether or not it is *fully deterministic.*

6.2.5.2. *Homogeneity vs heterogeneity*

A second parameter is whether we want to build a single hypothesis (or equivalence class of hypotheses) or a diverse set of hypotheses (one for each agent). In the first case, we will require *homogeneity* wrt to equality or $\equiv_A$ in addition to mutual

MAS consistency (if agent are not individualistic, we just have to build one MAS-consistent hypothesis). Otherwise, an *heterogenous* system would be favored.

6.2.5.3. *Communicational constraints*

Besides, we can have some *communicational constraints* governing the exchanges between agents. These constraints might specify whether broadcast is allowed, or which agents can speak together. We can represent *topological communicational constraints* by a communication graph where the nodes represent agents, and an edge (a_i, a_j) represent the fact that agent a_i can communicate with agent a_j. We will assume that an agent can always answer to a message that is sent to it. If any agent can speak with any other, the communication graph will be fully connected.

Depending on the dynamicity of the system, these constraints can be static or variable. Moreover, communication might be further limited by temporal consideration. We will detail this with the different level of dynamicity.

6.2.5.4. *Dynamicity*

At last, we can have different level of dynamicity, depending on the evolution of the environment (new observations) and of the topological communicational constraints.

Static problem. In such problems, all observations are present from the start, though distributed. Agents communicate to build consistent hypotheses.

Iterated static revision. In this case, time is divided in turn. At each turn, one agent receives a new observation, and we resolve MAS-consistency as if we were in a static problem. Communicational constraints might vary between each turn, but not in the middle of a turn.

Discrete dynamic system. In such systems, time is also divided in turns. At each turn, one or more (possibly all) agents receives new observations. However, communications are restricted. A given agent can only exchange *data* with one other agent in a given turn (though it might exchange communication requests or rejections with other agents). Communicational constraints might also vary between each turn, but not in the middle of a turn. Thus an agent cannot be sure it would be able to exchange data on the next turn with a given agent with whom it can discuss on the current turn. Such dynamicity is adapted to simulations or turn-based games.

Continuous time. In this last case, new observations might be observed at any time, and communicational constraints might also change at any time.

6.3. Learner/Critic Revision Mechanisms

This section presents some revision mechanisms that can be used to reach or maintain different levels of consistency in a system of agents. They are based on a learner and critic approach. An agent taking a role of *learner* builds locally an hypothesis that is proposed to other agents, acting as *critics*. These critics then criticize the hypothesis and answer by giving *counter-examples*, that is, observations that are inconsistent with the proposed hypothesis, or by accepting the hypothesis. Depending on the mechanism, an agent can play different roles during an exchange. We shall first introduce some notions about communication protocols and local or global revision mechanisms, then we shall describe two basic local revisions mechanisms before discussing how to reach mutual MAS-consistency through local exchanges and describing some complete global revision mechanisms.

6.3.1. *Revision mechanisms and protocols*

We first need to clearly establish what we mean by *communication protocol* or *revision mechanism*. To begin with, in modeling interaction between agents, we classically use the terms *communication protocol* and *strategy*. A *protocol* is a number of specifications, known by all agents involved in the communication, that governs exchanges between these agents.[6,7] In our context, it will govern information exchanges between agents. However a protocol often let an agent choose between several options. We call *strategy* the decision process that allows an agent to choose which message it should send in a given situation, and how to parameterize it properly. Different strategies can thus correspond to a given protocol. We require those strategies to verify some *conformance* property,[8] that is, to respect all the specifications given by the protocol, to be *complete* and *determinist*, that is, to be always able to determine one and only one behaviour for an agent in any situation (and always give the same behaviour for that situation).

A *revision mechanism* $\mathcal{M}$ is a process by which an agent a_i receiving a new observation o communicates with a group G of other agents to revise its beliefs, and eventually those of the other agents. We denote this execution (or application) of $\mathcal{M}$ by a_i with G upon reception of o by $\mathcal{M}(a_i, G, o)$. It would correspond to a protocol together with the strategies of all agents involved. It is a rather algorithmic view of these exchanges, behaving like some procedure triggered by an agent and involving a group of agents, whose internal states (for instance hypothesis and observations memory) might be modified. It focuses on the revision process, whereas protocols focus on the autonomy of the agents, and the interactions between them. The difference between these two terminologies is thus rather a matter of perspective, and we will use both of them indifferently according to the situation.

We just saw that a revision mechanism can involve a group of agents. Depending on the cardinality of this group we will distinguish three levels:

Internal revision mechanism. The limit case when $n = 1$ has already been described previously. Internal revision mechanism μ is a way to model agent reasoning procedurally. It corresponds to a hypothesis formation function.

Local revision mechanism. A local revision mechanism $\mathcal{M}^2$ is the process by which an agent a_i that has received an observation o communicates with another agent a_j to revise its beliefs, and possibly those of a_j. It will usually be based on some internal revision mechanism μ by which agents can take into account new informations.

Global revision mechanism. A global revision mechanism $\mathcal{M}^n$ involves all agents in the system. It is a process by which an agent a_i receiving some new observation o triggers a number of communications (usually local revision mechanisms) involving all agents in the system, and revising the internal states of a_i and possibly those of the other agents.

Though all these revision mechanisms are defined here as being triggered by an agent having received some new observation o, we actually consider that this condition is optional. An agent a_i can apply a revision mechanism (be it internal, local or global) without any reception of observation, especially if links between agent are variables. In such case, the trigger might simply be the fact that an agent wants to check the consistency of its hypothesis with other agents (it will then simply be denoted by $\mathcal{M}(a_i, G)$). The aim of a revision mechanism is indeed to preserve or guarantee some property on the internal states of the agents.

Definition 6.7 (Guarantee or preservation of a property). A revision mechanism $\mathcal{M}$ *preserves* a property $P(a_i, G)$ iff, for any observation o reaching a_i, it is the case that the execution of $\mathcal{M}$ by a_i with G will result in a situation where $P(a_i, G)$ holds if this property was verified before the reception of o.

A revision mechanism $\mathcal{M}$ *guarantees* a property $P(a_i, G)$ iff it is the case that any execution of $\mathcal{M}$ by a_i with G will result in a situation where $P(a_i, G)$ holds.

Classically, we would require that a local revision mechanism guarantees or preserves (mutual) peer-consistency or homogeneity of a_i with a_j, and that a global revision mechanism guarantees or preserves MAS-consistency of a_i or mutual MAS-consistency (or homogeneity) of the whole system.

6.3.2. *Local communication protocols*

We will present in this section our local communication protocols, with associated local revision mechanism guaranteeing peer-consistency, mutual peer-consistency or homogeneity according to the parameters and the variant used.

6.3.2.1. *Unilateral hypothesis exchange*

The most basic local protocol is an asymmetric protocol that we call *Unilateral Hypothesis Exchange* (UHE). The agent applying the mechanism is the *learner*, it takes an active role in building and refining an hypothesis. The second agent is a *critic*, that uses its knowledge to acknowledge or invalidate the proposed hypothesis.

Figure 6.1 illustrates the protocol. The associated strategy is as follows. The *learner* agent a_i first updates its hypothesis h_i to h'_i using an internal revision mechanism μ guaranteeing $ICons(a_i)$. Then it *proposes* it to the partner agent

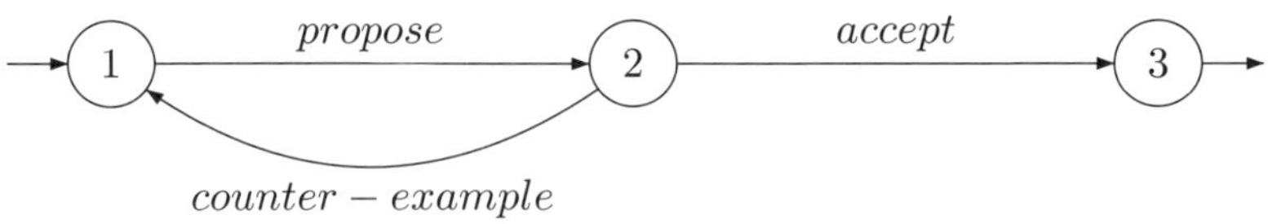

Fig. 6.1. Unilateral Hypothesis Exchange Protocol (UHE).

a_j, called *critic*, and a_j either replies with $accept_{direct}$ and adopts h'_i as its new working hypothesis if $Cons(h'_i, O_j)$, or otherwise sends *counter-example(o')*, where $o' \in O_j$ is such that $Cons(h'_i, \{o'\})$ is false. Upon reception of a counter-example, a_i applies again μ to update its hypothesis with the new observation, and proposes the resulting hypothesis as before, except that an acceptance will now result in a $accept_{indirect}$ message. (The reason justifying the distinction between $accept_{direct}$ and $accept_{indirect}$ will become clear later.)

Example 6.7 (Distributed inductive concept learning). We consider two agents in the 'good day for playing tennis' learning problem. The example memory of the agents are the following:

$E_1^+ = \{(O_{Sun}, W_{Weak}, T_{Hot}, H_{Normal}), (O_{Cloud}, W_{Strong}, T_{Hot}, H_{High})\}$ and $E_1^- = \{(O_{Sun}, W_{Strong}, T_{Cold}, H_{High})\}$
$E_2^+ = \{(O_{Sun}, W_{Strong}, T_{Cold}, H_{Normal}), (O_{Sun}, W_{Strong}, T_{Hot}, H_{Normal})\}$ and $E_2^- = \{(O_{Rain}, W_{Strong}, T_{Hot}, H_{High})\}$

Their hypotheses are $h_1 = T_{Hot}$ and $h_2 = W_{Strong} \wedge H_{Normal}$. If agent a_1 communicates with a_2 using the Unilateral Hypothesis Exchange Protocol, we would get the following dialogue.

a_1 sends to a_2	$propose T_{Hot}$
a_2 sends to a_1	*counter-example*$((O_{Sun}, W_{Strong}, T_{Cold}, H_{Normal}), +)$
a_1 sends to a_2	*propose*$(T_{Hot} \vee (O_{Sun} \wedge W_{Strong} \wedge T_{Cold} \wedge H_{Normal}))$
a_2 sends to a_1	*counter-example*$((O_{Rain}, W_{Strong}, T_{Hot}, H_{High}), -)$
a_1 sends to a_2	*propose*$((O_{Sun} \wedge H_{Normal}) \vee (O_{Cloud} \wedge W_{Strong} \wedge T_{Hot} \wedge H_{High})$
a_2 sends to a_1	*accept*

After this exchange, a_2 would adopt the new hypothesis $h'_1 = (O_{Sun} \wedge H_{Normal}) \vee (O_{Cloud} \wedge W_{Strong} \wedge T_{Hot} \wedge H_{High})$. Thus $h'_2 = h'_1 = (O_{Sun} \wedge H_{Normal}) \vee (O_{Cloud} \wedge W_{Strong} \wedge T_{Hot} \wedge H_{High})$. Moreover, E_1^+ would have been changed to $E_1'^+ = \{(O_{Sun}, W_{Weak}, T_{Hot}, H_{Normal}), (O_{Cloud}, W_{Strong}, T_{Hot}, H_{High}), (O_{Sun}, W_{Strong}, T_{Cold}, H_{Normal})\}$ and E_1^- to $E_1'^- = \{(O_{Sun}, W_{Strong}, T_{Cold}, H_{High}), (O_{Rain}, W_{Strong}, T_{Hot}, H_{High})\}$, while E_2 is unchanged.

We shall denote by $\mathcal{M}_U^2$ the local revision mechanism based on the unilateral hypothesis exchange with both agents using the strategy describe above.

Proposition 6.2. *$\mathcal{M}_U^2$ guarantees mutual peer consistency and homogeneity with respect to equality for any pair of non-individualistic agents.*

A different strategy, using the same protocol and principles, would be that the *critic* agent keeps its own hypothesis instead of adopting the accepted one. Such a strategy can be used by individualistic agents. In this case, the *critic* agent just checks the consistency of the *learner* and provides counter-examples if needed. We will denote by $\mathcal{M}^2_{UwA}$, called unilateral hypothesis exchange revision mechanism without adoption, the corresponding variant local revision mechanism. It only guarantees peer-consistency of the agents applying it with the pair of agents involved.

6.3.2.2. *Bilateral hypothesis exchange*

To get a local revision mechanism that guarantees mutual peer consistency but still preserves the individualism of agents, we symmetrize $\mathcal{M}_{UwA}$. The obtained protocol is called the Bilateral Hypotheses Exchange protocol, and the associated mechanism is denoted by $\mathcal{M}_B$. Here, the *critic* agent becomes *learner* agent once it has validated the other agent's hypothesis. It can be interpreted as two reciprocal applications of the unilateral protocol without adoption of hypothesis. Upon reception of an hypothesis h_i (by *propose*(h_i)), agent a_j ends up in state 2 and can reply either with an *accept*, a *counter-example*, or a *counter-propose*, as specified by the following strategy:

- If $\exists o \in O_j$ s.t. $Cons(h_i, \{o\})$ is false, a *counter-example* consisting of the observation o is sent, as with the UHE mechanism.
- else, we have $Cons(h_i, O_j)$. We now want to check that $Cons(h_j, O_i)$ holds. If $h_i = h_j$, a_j can directly *accept* h_i. Else, it must *counter-propose*(h_j) to invert *critic* and *learner* roles. Then a_i will act as a critic and send counter-examples until consistency is reached.

Proposition 6.3. *$\mathcal{M}^2_B$ guarantees $MCons^2(\{a_i, a_j\})$. If the hypothesis formation function $\mathcal{E}_h$ used in internal revision is* fully deterministic *for $\equiv_A$ (or equality) and* common to both agents, *then $\mathcal{M}^2_B$ guarantees homogeneity for $\equiv_A$ (or equality).*

Example 6.8 (Simple distributed medical diagnosis). We consider the same two agents in the static diagnosis problem. Their observation memory is: $O_1 = \{\text{cough}\}$ and $O_2 = \{\text{fever}, \neg \text{ mucus}\}$. Their hypotheses are $h_1 = \{\text{bronchitis}\}$ and $h_2 = \{\text{angina}\}$. We use a total preference order on hypothesis for selecting favored hypothesis (minimal number of diseases if prefered, and then hayfever is prefered to bronchitis, which is prefered to angine, which is prefered to flu[b]). Now if these two agents were to communicate using the Bilateral Hypothesis Exchange Protocol, we would get the following dialogue.

a_1 sends to a_2	*propose*({bronchitis})
a_2 sends to a_1	*counter-example*(¬ mucus)
a_1 sends to a_2	*propose*({flu})
a_2 sends to a_1	*counter-propose*({angina})
a_1 sends to a_2	*counter-example*(cough)
a_2 sends to a_1	*propose*({flu})
a_1 sends to a_2	*accept*

[b]If we must compare two diagnosis with $k > 1$ diseases, we order them in sequence by preference, and use lexicographic order on the sequence.

After this exchange, we will have: $h'_1 = h'_2 = \{\text{flu}\}$, $E'_1 = \{\text{cough}, \neg \text{ mucus}\}$ and $E'_2 = \{\text{fever}, \neg \text{ mucus}, \text{cough}\}$.

6.3.3. *From local to global*

In dynamic settings, we would like to ensure global mutual MAS-consistency of the system by ensuring locally the mutual peer-consistency of every pair of agents (this is equivalent to MAS-consistency since the consistency relation is compositionnal). However, communicational constraints might prevent some communications. We have different cases according to the structure and dynamicity of the agent network.

6.3.3.1. *Static links: full propagation*

In a fully connected society of agents, where an agent can always communicate with another, it is possible to plan a number of local revisions in order to ensure MAS-consistency. In such case, we will present some complete global revision mechanism that articulates those local revisions in different ways, according to the parameters of the system or the requirements.

If the structure is more complex, but still static and connected, then some agents must act as relay to ensure peer-consistency of two agents that cannot communicate. Through some planned propagation, MAS-consistent hypotheses can be build as long as the network is connected. We will present some complete global revision mechanism for static links in Section 6.3.4.

6.3.3.2. *Rumor-like propagation*

When the communicational links between agents are not static, however, a complete planification is not possible, and we adopt a more reactive approach. We shall indeed try to propagate peer-consistency from neighbour to neighbour in a rumor-like process. At each time step (or as soon as previous communication is finished), an agent tries to engage in a local revision with any neighbour with whom it is not sure to have reach peer-consistency. Still, it is possible that two given agents will never be neighbours. Thus we either need to have agents relaying the hypotheses of other agents in their stead, or to derive peer-consistency through transitiveness. It would thus be convenient to ensure locally a transitive property from which the MAS-consistency can be build. Unfortunately, in most cases, peer-consistency is not transitive. Consequently we will rather use homogeneity with respect to $\equiv_A$, which is transitive, and together with internal consistency, implies peer-consistency.

We detail this for discrete evolution system. In such system, where each agent can only participate in a local revision with *one* agent, we would use dialog initiation protocols. These protocols enables agents to make some requests for communication that can be accepted or not. Several strategies are possible, but the more natural one is to first preselect the agents with whom it would seem beneficial to communicate with, and then attribute a weight to each of the possible requests, in order to compare them with incoming requests and satisfy the request with highest weight. Several requests can be sent until all agents have either no satisfiable request to send, or are engaged in a local revision. Then, we can guarantee that mutual MAS-consistency (and homogeneity) will eventually be reached under a set of conditions.

Theorem 6.1 (Convergence of rumor-like propagation). *Using a rumor-like propagation and some local revision protocol $\mathcal{M}$, agents of a system are bound to reach homogeneity wrt $\equiv_A$ and mutual MAS-consistency of the system under the following conditions:*

- Internal consistency (ICONS) — *agent are equipped with an internal revision mechanism ensuring their consistency at all times;*
- Homogeneity (HOM)— *local revision mechanism has to guarantee the mutual consistency of agents upon termination (which implies termination);*
- Willingness and focus to solve inconsistencies (SOLVE+FOCUS) — *agents want to communicate with any other agents with whom they do not know whether they are mutually peer-consistent, and only with those agents.*
- Willingness to communicate (COMM) — *agents cannot refuse a weighted communication request, unless they have just received or send a request with a greater weight.*
- Commitment to communication request (REQU) — *agents cannot accept a weighted communication request if they have themselves sent a communication request with a greater weight. Therefore, they will not cancel their request unless they have received a communicational request with greater weight.*
- Bounded Perception — *the number of observations coming from the environment that are relevant for building hypotheses is not infinite;*
- Temporal connectivity — *agents must be temporally connected.*[2] *This means that it is always possible to find a chain of agents to transmit an information from an agent to another in a finite time.*

6.3.4. *Complete global communication protocols*

We present here complete global revision mechanisms that can be used when topological communicational constraints are static. We first detailed the basic protocol for fully connected societies, before giving some variants. Then the principles of a mechanism with propagation for structured networks are stated.

6.3.4.1. *Clock-wise hypothesis exchange for fully connected societies*

The general idea is to make repeated uses of a local mechanism guaranteeing mutual peer consistency to eventually get a MAS-consistent hypothesis adopted by all agents. The hypothesis must be validated by all agents in turn without being changed. Any change in the hypothesis forces us to check it again from the beginning. Intuitively, as consistency is additive, the hypothesis should grow precise enough to become consistent with all agents. In more detail, the global revision mechanism $\mathcal{M}_C^n$ can be described as follows. When the *learner* agent a_1 applies it after receiving an observation o, a_1 first applies $\mathcal{M}_U^2$ (unilateral hypothesis exchange local revision mechanism) to reach mutual consistency with agent a_2. Then it does the same with agent a_3. If the local protocol ends with an $accept_{direct}$, then a_1 proceeds to exchange its hypothesis with the next agent (a_4), else ($accept_{indirect}$) it goes back to a_2. This iterates, each $accept_{indirect}$ restarting the process with a new hypothesis submitted to a_2. When a_n sends a $accept_{direct}$, it means that the hypothesis has been accepted and adopted in turn by all agents. In such a case, the mechanism ends, and this common hypothesis is MAS-consistent.

Proposition 6.4. *$\mathcal{M}_C^n$ guarantees mutual MAS-consistency (and homogeneity wrt equality) in any fully connected society of non-individualistic agents.*

6.3.4.2. *Heterogeneous variants*

Similarly to the case of local mechanisms, it is possible to define a simple variant of this mechanism: *clock-like hypothesis exchange without adoption* mechanism $\mathcal{M}_{CwA}^n$ where *critic* agents do not adopt accepted hypotheses, thus respecting the individualism of the agents. However, this variant only revise the hypothesis of the learner agent: It guarantees *MAS-consistency* of the learner agent, but not mutual MAS-consistency of the system. Then, we can make the *clock-like hypothesis exchange without adoption* mechanism $\mathcal{M}_{CwA}^n$ mutually MAS-consistent by making all agents whose hypothesis is inconsistent with o apply it in turn, starting with the agent having received the observation o[c]. We will denote the resulting global revision mechanism, *iterated clock-like hypothesis exchange* by $\mathcal{M}_{C*}^n$. It *preserves mutual MAS-consistency* in any fully connected society of possibly individualistic agents. If the iteration is complete (all agents applies $\mathcal{M}_{CwA}^n$ in turn), then the resulting global mechanism will *guarantee* mutual MAS-consistency in fully connected societies. Note that with all these variants, the resulting system in *heterogeneous*, which can be interesting in learning problem for using voting methods inspired for ensemble learning.[9]

6.3.4.3. *Revision mechanism with propagation*

When agents are topologically constrained in their communications, a single learner cannot directly propose its hypothesis to all others agents, and the above-mentionned global mechanisms cannot be used. Instead, the global mechanism must rely on some kind of propagation. To ensure that propagation does not reach the same agent from two different ways, the first basic idea is to eliminate cycles by

[c]Recall that the agents whose hypothesis is inconsistent with o will be the only ones that will need to change their hypothesis, since hypotheses consistent with o will remain consistent with $\bigcup_{i\in\{1,\dots,n\}} O_i \cup \{o\}$ using the compositionality of the consistency relation.

constructing a spanning tree (a tree-like sub-graph that contains all the nodes of the original communication graph). We describe here a mechanism $\mathcal{M}_P^n$ to construct such a tree while propagating and refining the hypothesis at the same time.

The agent receiving a new observation becomes the root of the tree. It will always act as a learner. It begins the propagation by taking all of its neighbours as children, and then selecting the first one, ensuring mutual consistency with it (through the use of $\mathcal{M}_U^2$, and asking it to propagate this hypothesis). Then this child, who just acted as a critic, will take a role of learner for relaying the hypothesis to its own neighbour. It will first ask its neighbour to become its children (if they do not already belong to the tree), and then use the first one as a *critic* using a local revision mechanism. If the result is $accept_{indirect}$, then it will stop acting as a learner (he is not defending the hypothesis of its parent anymore), and ask its parent to use it as a critic once more since it has learned some counter-examples ($hyp_{changed}$ message). If the result is $accept_{direct}$ it will ask it to propagates in turn the hypothesis. When an agent with no children (and no neighbour that are not in the tree) is asked to propagate its hypothesis, it will just reply with a *strong-accept*, meaning that itself and all its children are consistent with the hypothesis. Its parent would then use its next son to criticize this hypothesis. When all its children have replied with *strong-accept*, it will send in turn a *strong-accept* to its parent. When the root has received *strong-accept* from all its children, the system is *mutually MAS-consistent* and homogeneous wrt equality. Thus:

Proposition 6.5. *The global revision mechanism* $\mathcal{M}_P^n$ *described above guarantees mutual MAS-consistency and homogeneity wrt equality for all connected networks of (non-individualistc) agents.*

6.4. Instantiating the Framework

The purpose of this section is to give guidelines for using our hypothesis framework and learner-critic revision mechanisms, explaining how to instantiate the general problem to specific applications. We will detail a first level of instantiation, depending on the type of reasoning and knowledge representation (in a syntaxical view). Some examples of instantiation to classical types of hypothesis formation will illustrate this part. Afterwards, we shall present the second level of instantiation, the application level, where semantic is defined and other properties of the system are decided. When presenting this level, we shall also mention a possible third level related to specific problem instances in a given application. Thereafter, we will illustrate all these notions by giving a full instantiation of a complex toy problem.

6.4.1. *Reasoning and representation*

The first thing that needs to be specified in hypothesis refinement is the knowledge representation. When instantiating it at reasoning level, *syntaxical* consideration is enough. We need to know whether hypotheses, observations and observation memories are represented as sets, logical theory, propositional or first order formulae, and so on. *Semantic* of the different language is not yet necessary at this level. For instance, it will be enough to know that observation are represented as grounded instances of a specific subset of predicates of a first-order language. The precise list of possible predicates, constants, function, or domains of the parameters of a predicate will be defined when instantiating our reasoning to an application. As a result, we will also refer to this first instantiation as the *syntaxical instantiation.*

Thus, we need to define the kind of knowledge representation that we will use for the background theory $\mathcal{T}$, that is, the kind of *representation language*, in a syntaxical perpective. Besides, the form *hypothesis language* $\mathcal{H}$, observation language $\mathcal{O}$ and observation memory language Ω must also be defined. Once the knowledge representation is decided, we shall instantiate the *consistency relation* $Cons$, defining it for the kind of reasoning and knowledge representation that we shall use. This will specify the validity criterion wrt to observation memories for the hypotheses that agents will build. Then, all that is left is to define the hypothesis formation mechanism. This is done by instantiating the *hypothesis formation function* $\mathcal{E}_h$ or the *internal revision mechanism* μ. We will rather use $\mathcal{E}_h$ for order-independant reasonings, and μ for order-dependant ones. Moreover, when defining $\mathcal{E}_h$, we should check whether this function is *fully deterministic.*

We will now give three classical examples of reasoning instantiation, that can be adapted to cover a good range of hypothesis formation problems.

6.4.1.1. *Logical abduction*

We first present a logical abductive setting, inspired from Poole's framework.[10] The *representation language* will be a first order logic language $\mathcal{L}$, defined over a set of predicates $\mathcal{P}$. Among the set of all litterals (grounded predicates, or negated grounded predicates), we will define some subsets A and Ob, corresponding to litterals that can be used to form respectively, hypotheses and observations. Moreover, we will distinguish some of the litterals as the set of *defaults* Δ. Litterals from Δ are considered true until proven false. They are used for prediction and can contribute to explanations. *Hypotheses* will be conjunctions of grounded abductible predicates (elements of A). Thus the hypothesis language $\mathcal{H}$ will be the set of all possible conjunctions of elements from A (that is, $\{\bigwedge_{e\in E} e | E \subseteq A\}$). *Observations* will be grounded litterals from Ob ($\mathcal{O} = Ob$), and *observation memories* will just be set of observations ($\Omega = 2^{\mathcal{O}}$). Besides, we will define the set $\mathcal{M} \subseteq \mathcal{O}$ of *manifestations*, that is, observations that need to be explained by the hypothesis (usually we do not need to explain normal behaviours). Then *Background theory* will be a logical theory on $\mathcal{L}$. This theory must respect some constraints with respect to the use of the litterals from the different subgroups ($A, Ob, \mathcal{M}, \Delta$). Namely, it must ensure that *manifestations* are the consequences of some of its rules, and that *abducible predicates* are present in the premisses of some of its rules. Moreover, as $\mathcal{T}$ describe *certain knowledge*, it must be consistent (*consistent world assumption*).

Finally, the *consistency relation* can be define as the combination of two properties. We shall have, $\forall h \in \mathcal{H}, \forall O \in \Omega, Cons_{abd}(h, O)$ iff:

- *coherence*: $\mathcal{T} \cup \{h\} \cup O \not\models \Box$, that is, the logical theory formed by adding h and O to $\mathcal{T}$ is coherent (it does not implies a tautology).
- *completeness*: $\forall o \in O \cap \mathcal{M}, \mathcal{T} \cup \{h\} \models o$, that is, every observation in the observation memory that is a manifestation is explained by the hypothesis (together with the background theory).

Note that this notion of consistency is not compositional in the general case, but a proper choice of $\mathcal{T}$, $\mathcal{O}$ and $\mathcal{H}$ can make it compositionnal. In particular, we might require that defaults are *fully specified*, meaning defaults will only be used for compactness of the theory: all defaults that are not proven wrong by $\mathcal{T}$ will never be proven wrong, and can thus be considered true with certainty. In such case, a compositional consistency relation can be defined. This setting will be used in our fire simulation example. We shall not detail the *hypothesis formation function* here. It can be based on Theorist,[10] as described in a previous paper.[11]

6.4.1.2. *Cover-set abduction*

This is another form of abductive reasoning, used in our simple diagnosis running example. We will now describe it in the term of parsimonious covering theory.[12] *Background Theory* is represented by two finite non empty sets D and M of objects respectively called *disorders* and *manifestations* and a relation $C \subseteq D \times M$ called *causation*, or *causality relation*. Then, an *observation* is just a manifestation, that is, $\mathcal{O} = M$, and an *observation memory* is a set of observations (once again, $\Omega = 2^{\mathcal{O}}$). Finally, a *hypothesis* is a set of disorders ($\mathcal{H} = 2^D$). Using this, a hypothesis h is *consistent* with an observation memory O ($Cons_{cov}(h, O)$) iff:

- $\forall m \in O, m \in \mathit{effect}(h)$ where $\mathit{effect}(h) = \{m / \exists d \in h, (d, m) \in C\}$ (completeness, or covering)
- $\forall h' \subset h, \exists m \in O, m \notin \mathit{effect}(h')$ (irredundance)

Here, irredundance is a parsimony criterion. Other parsimony criterion could be use such as minimal cardinality or relevance. Unless two elements of M can be contradictory, or O can contain other elements that can be contradictory with some manifestations, there is no need to check *coherence* (since there would be no possible cause of incoherence).

The *hypothesis formation function* $\mathcal{E}_h$ could use *parsimonious covering*[12] or *hypothesis assembly* algorithms.[13] As these process returns a set of hypothesis rather than a single one, some selection process would have to be added in $\mathcal{E}_h$.

6.4.1.3. *Inductive incremental learning*

At last, we present a form of inductive reasoning through incremental concept learning. It has already been presented briefly in our learning running example.

We summarize it here for other kinds of concepts. We use an atomic knowledge representation, defined over a set of atoms $\mathcal{A}$ representing possible descriptors for an example. There is no explicit negation, but we can use additional atoms like $a^{\neg}$ to take into account the fact that an example does not have descriptor a. Now, there is no explicit *background theory*. *Observations* are called examples, and represented by a description (a set of atoms) and a label + or −. Thus, $\mathcal{O} = \mathcal{D} \times \{+, -\}$ where $\mathcal{D} = 2^{\mathcal{A}}$ is the description language. Observation memories are called example memory E and are constituted of two sets of descriptions corresponding to positive (E^+) and negative (E^-) examples stored. Consequently, $\Omega = 2^{\mathcal{D}} \times 2^{\mathcal{D}}$. At last, *hypotheses* are disjunctions of conjunctive statements over atoms. $\mathcal{H} = \{\bigvee_{c \in C} c | C \subseteq \{\bigwedge_{a \in D} a | D \subseteq \mathcal{A}\}\}$.

For *consistency*, as seen before: $\forall h \in \mathcal{H}, \forall E = (E^+, E^-) \in \Omega, Cons_{ind}(h, E)$ iff

- $\forall e \in E^-, h \not\models_s e$ (coherence)
- $\forall e \in E^+, h \models_s e$ (completeness)

In this context, the reasoning is order-dependant. We will thus rather specify an *internal revision process* μ. One example of such process adapted to this atomic representation is Imind.[14] Another process that would require a slightly different example or hypothesis representation (attribute-value representation for examples and decision trees for hypotheses) could be Id5.[15]

6.4.2. *Application level*

This second level of instantiation is the adaptation of our framework instantiated for a given king of reasoning and knowledge representation to a given application or fully specified problem. As opposed to the first instantiation, it will be called *semantical instantiation*. It is indeed in this step that we shall define the semantics of our languages.

6.4.2.1. *Semantic specification*

As a first step, we need to precise the content of our languages, by detailing their constituting elements (possible elements, predicates, variable domains, propositions, atoms and so on). We will thus precise the semantic of our application, that is, the vocabulary that will be used by the language. It must be done for the representation language as well as for observation, observation memory and hypothesis languages.

We would also have to determine the rules of the system, and specify most of the background theory $\mathcal{T}$, that is, all part of the background theory that will be unchanged in all instances of the application.

6.4.2.2. *Other considerations*

Other considerations depending on the application will then need to be precised. First point is to state the *dynamicity* of the system. Is it a static problem, an iterated revision problem, or some discrete or continuous dynamics problem. We must also precise the *communicational constraints*, and their variability. How do the agents get observations from the environment, if they can, should also be decided. In a dynamic system, we might also have to determine some aspects of the problem that do not directly depends on the hypothesis formation process. For example, agents might be able to act in their environment on the basis of their beliefs, and such things must be modelised.

6.4.2.3. *Instance of a problem*

An application can be a unique one-time problem, but most of the time, it is a class of very similar problems. For example, simple medical diagnosis is an application that can have several instances depending on the diseases that really affect some patients, and 'good day to play tennis' is an instance of concept learning (one could devise other problem instances with same semantic such that 'a good day to have barbecue' whose target concept would be different). Some details depends on given instances of the application. The *actual observations language* $\mathcal{O}^A$ (and consequently also the *actual observation memories language* Ω^A) can only be defined once all parameters that can affect observations have been set. For instance, we cannot know whether or not the observation 'cough' is in $\mathcal{O}^A$ until the diseases really affecting the patient have been chosen. Part of the background theory might also be filled at this time. Then all elements are decided, and the instance of the application can be run.

6.4.3. *Example application: Fire simulation*

This section provides an example toy application of our general framework.

6.4.3.1. *Description*

This experiment involves agents trying to escape from a burning building. The environment is described as a spatial grid with a set of walls and (thankfully) some exits. Time and space are considered discrete. Time is divided in rounds. Agents are localised by their position on the spatial grid. These agents can move and communicate with other agents. In a round, an agent can move of one cell in any of the four cardinal directions, provided it is not blocked by a wall. Agents can also communicate with any other agent that is in view (but only once per turn).

Suddenly, at time t_0, a fire erupts in these premises, from one or more cell called the origins of the fire. From this moment, the fire propagates. Each round, for each cases where there is fire, the fire propagates in the four directions. However, the fire cannot propagate through a wall. If the fire propagates in a case where an agent is positioned, that agent burns and is considered dead. It is removed from the system, and thus can of course no longer move nor communicate. If an agent gets to an exit, it is considered saved, and can no longer be burned.

6.4.3.2. *Syntaxical instantiation*

The described situation implies that our agents are able to cope with partial observations of their environment, to build hypotheses from these observations, to draw conclusions from these hypotheses, and to communicate to refine their explanations. They will use an abductive reasoning for building hypotheses, combined with deductive reasoning to predict evolution from these hypotheses. Poole's framework,[10] which captures explanation and prediction processes in a single axiomatization, is adapted to such situations. We shall thus use the logical abduction instantiation detailed above (6.4.1.1), and require that defaults are fully specified. All the reasoning can be represented logically and a fully determinist *hypothesis formation function* $\mathcal{E}_h$ can be build by combining Theorist with a preference relation $\preceq$ inducing a total order on $\mathcal{H}$. In our practical implementation, an equivalent result will be given by implementing $\mathcal{E}_h$ as a procedural function where background theory and preference relation are included as procedural knowledge.

6.4.3.3. *Semantical instantiation*

Background knowledge $\mathcal{T}$ of the agents concerns the environment and the rules governing its dynamics, that is, agents know the map as well as the rules of fire propagation previously described. They also locally perceive this environment, but cannot see further than 3 cells away, in any direction. Walls also block the line of view, preventing agents from seeing behind them. Within their sight, they can see other agents and whether or not the cells they see are on fire. All these perceptions are memorised in the observation set O_i of each agent a_i. They can be of three different forms: $\mathcal{O} = \{\texttt{Fire}(t_i,\ \texttt{at}\ (x,y)),\ \texttt{Nofire}(t_i,\ \texttt{at}\ (x,y)),\ \texttt{Agent}\ a_j\ (t_i,\ \texttt{at}\ (x,y))\}$. Observation like Agent a_j $(t_i,$ at $(x,y))$ mainly serves to enable an agent to know its neighbours for a given round (so that it knows whom it can request communication with). To decide what they should do, each agent builds a *favorite hypothesis* h_i, chosen among its different possible interpretations of its observations about the fire. The *abducible predicates* A used to define $\mathcal{H}$ are the instantiations of the predicate FireOrigin$(t_i,$ at $(x,y))$. They are used to explain the *manifestations* $\mathcal{M} = \{\texttt{Fire}(t_i,\ \texttt{at}\ (x,y))\}$. Defaults $\Delta = \{\texttt{Free}(d,\ \texttt{at}\ (x,y))\}$ where $d \in \{\texttt{Up, Down, Right, Left}\}$ are here only used to make the representation of prior knowledge more compact. We illustrate it by giving an example of background theory:

$\mathcal{T} = \{$

$\texttt{Fire}(t_i,\texttt{at}(x,y))$ and $\texttt{Free(Up,at}(x,y)) \Rightarrow \texttt{Fire}(t_i+1,\texttt{at}(x,y+1))$,

$\texttt{Fire}(t_i,\texttt{at}(x,y))$ and $\texttt{Free(Down,at}(x,y)) \Rightarrow \texttt{Fire}(t_i+1,\texttt{at}(x,y-1))$,

$\texttt{Fire}(t_i,\texttt{at}(x,y))$ and $\texttt{Free(Right,at}(x,y)) \Rightarrow \texttt{Fire}(t_i+1,\texttt{at}(x+1,y))$,

$\texttt{Fire}(t_i,\texttt{at}(x,y))$ and $\texttt{Free(Left,at}(x,y)) \Rightarrow \texttt{Fire}(t_i+1,\texttt{at}(x-1,y))$,

$\texttt{DownHorizontalWall(at}(x,y)) \Rightarrow \neg\texttt{Free(Down, at}(x,y))$,

```
DownHorizontalWall(at(x,y)) ⇒ ¬Free(Up, at(x,y − 1)),
LeftVerticalWall(at(x,y)) ⇒ ¬Free(Left, at(x,y)),
LeftVerticalWall(at (x,y)) ⇒ ¬Free(Right,at(x − 1,y)),
DownHorizontalWall(at(1,1)),..., DownHorizontalWall(at(9,17)),
LeftVerticalWall(at (1,1)),..., LeftVerticalWall( at (14,15)) }
```

First eight lines represent knowledge about the rules governing fire propagation: rules for propagation in each direction, and rules for blocking the formers when a wall is present. The last two lines represent the knowledge about the map of the premises, expressed in terms of `DownHorizontalWall(at (x,y))` and `LeftVerticalWall(at (x,y))` literals. `DownHorizontalWall(at (x,y))` means that there is a wall between cell (x, y) and the cell below $(x,y - 1)$, whereas `LeftVerticalWall(at (x,y))` indicates a wall between (x, y) and the cell at its left $(x-1,y)$. As each agent fully knows the premises' map, we can get sure of which default will be false or will stay true.

The preference relation $\preceq_p$ used by agents to select their favored hypotheses h_i among possible candidates is the following (note that we only consider hypotheses that are minimal wrt set inclusion): (i) an agent first prefers hypotheses involving as few as possible different fire origins; (ii) when the number of assumed origins is equal, an agent will prefer hypotheses with an origin that is as close as possible to the observed fires. If there are still different candidates for being favorite hypothesis, the hypothesis formation $\mathcal{E}_h$ will introduce a bias by considering first origins that are left and down from the observed fires. Once its favorite hypothesis has been chosen, each agent can predict the future states of the environment, and computes from that an escape plan, selecting the shortest safe way out toward an exit.

6.4.3.4. *Other considerations*

The dynamicity here is clearly a case of discrete evolution. Each turn, fire propagates and agents get an opportunity to do something. As for communicational constraints, we have a limited propagation, since agents can only participate to one conversation in a turn, and topological constraints, since agents can only speak with other agents that they can see. Since the communicational links are highly variable, we will use rumor-like propagation rather than a static complete global protocol. Different kinds of ponderations can be used. We will compare a neutral random heuristics with an heuristic that tries to maximise the effectiveness of local exchange with simple topological consideration (*topological* heuristic) and one that tries to maximise the number of conversations in each turn (*altruist* heuristic). Note that this system is not temporally connected: rumor-like propagation will not ensure homogeneity of the system, as some agents can die before consistency is reached. However, hypothesis refinement can still be used to improve individual accuracy of agents hypotheses. As for the local protocol, since the hypothesis formation function is fully deterministic and order-independant, we will use Bilateral Hypothesis Exchange.

6.4.3.5. *Problem instances*

We defined our settings, but it can be instantiated in different problems corresponding to different initial situations. This situation can vary and have a great impact on the difficulty of the problem or the adequacy of our protocols. For instance, performances of information propagation are highly constrained by the environment topology. With a large number of walls the perceptions of agents are limited, and also the number of possible inter-agent communications, whereas an "open" environment will provide optimal possibilities of perception and information propagation. Moreover, reasoning is more critical in maps where different possible explanations for a given set of observations could imply different escape plans. The *map of the premises* is thus the main aspect to take into account when instantiating this problem (see Fig. 6.2 for a problem instance with subsequent development). Besides, *initial number of agents* is also important: having more agents results in more potential for propagation, but also means that the bilateral exchange restriction will be more crucial. At last, *initial positions of the agents* have a significant

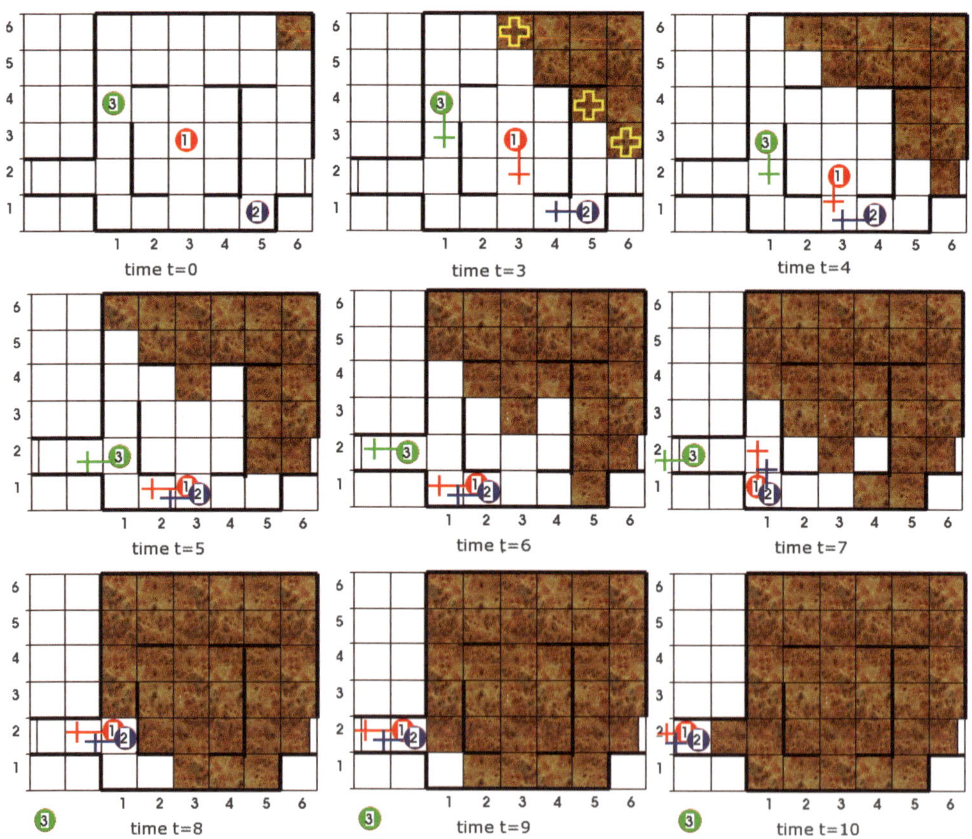

Fig. 6.2. Example of fire simulation problem instance, with full development.

influence on the overall behavior of an instance of our system: being close from an exit will (in general) ease the escape. Restricting the initial positions of the agents can also strengthen the importance of hypothesis accuracy.

6.4.3.6. *A word on experimentations*

This application was studied in detail in previous papers.[2,3] We give here some conclusions from experimental results. Bilateral Hypothesis Exchange (BHE) local protocol was compared with some simpler Observation Exchange protocol, giving a list of all observations deemed interesting: either all unexpected observations (Complete Observation Exchange — COE), or only unexpected fires (Positive Observation Exchange — POE). Rumor-like propagation protocol was used with three different weight attribution strategies. Effectiveness is evaluated through the proportion of surviving agents, while efficiency is related to the mean size of data exchanges.

Local study concluded that Bilateral Hypothesis Exchange is an effective and efficient way to refine hypothesis in complex situations, but if the accuracy of the hypothesis is not crucial, simpler protocol might be as effective and more efficient (such as POE). Indeed, in critical maps, observations about places where there is no fire are more important, and the number of data given by complete observations exchange grows a lot. Bilateral Hypothesis exchange is a good choice for such case, as (i) it ensures mutual peer-consistency and homogeneity, (ii) it gets better effectiveness results and (iii) it is more efficient that COE, and not too much less efficient than POE. POE is indeed still more efficient than BHE. BHE selects only useful example, but as it must also propose hypotheses, more data are exchanged. This is the cost of ensuring mutual peer-consistency. However, for non critical maps, differences in effectiveness disappear, and POE is more efficient.

Experiment on different heuristics for rumor-like propagation confirmed that topological and altruist weight attribution strategies do give better results than neutral random weights. The simple topological heuristic yields better results than the more complex (and much less efficient) altruist one. This hints that it might be better to maximize the impact of each dialog rather than their number. Developing simple weight attribution heuristics adapted to the applications appears as an efficient way to increase the effectiveness of the global process.

6.5. Related Works

A specificity of our approach is to bring together issues of communicational constraints in a context of deliberative agents using hypothesis finding to cope with incomplete knowledge. To the best of our knowledge, this is an original attempt. There are, however, several related trends of work, dealing with specific aspects of our work. We now discuss these related works, dividing them into (i) approaches studying distributed abduction, (ii) approaches studying distributed inductive learning, (iii) other approaches dealing with distributed hypothetical reasoning, and finally (iv) approaches studying information propagation and convergence to consensus in (dynamic) structured networks.

6.5.1. *Distributed abduction*

There is now a solid line of research interested in bringing the power of computational logic into multiagent settings. One of the seminal work in this line of research is that of Kowalski and Sadri,[16] which integrates the reasoning machinery of abductive logic programs within an agent cycle. In particular, agents can communicate with each other in order to exchange informations, and, for instance, collaborate. In fact, the communicative behaviour of agents is elegantly integrated within the agents' reasoning process (communications are seen as actions), and integrity constraints are used to specify protocols.[17] Distributed abduction has been considered in recent years in the ALIAS[18] system. They distribute the abductive programming algorithm of Kakas and Mancarella,[19] using abductive logic program to represent each agents theory. Each agents thus have a local theory (normal logic program) and a set of integrity constraints, as well as some goals (observations to explain). Communications are induced by the inferences rules and fully integrated in the proof procedure. Some other related papers have studied desired properties of some specific protocols, like termination[20] or conformance.[8,21] But it is important to recall that the main objective of these works is to fully integrate communication into an agent's abductive reasoning, with the aim of ultimately using logic-based techniques to prove these properties. This is, in a sense, much more ambitious that what we do here. On the other hand, our objective is to design protocols that can be very widely used (and the context of abductive agents just happens to be one example application), in complex situations of interaction, involving a large number of agents, evolving in dynamic and highly constrained environments. This explains that we focus on different properties, like the efficiency or effectiveness of these protocols.

More recently, DARE[22] addressed a similar problem, but consider possible dynamicity of the system by allowing agents to enter or exit some proof cluster. The algorithm alternates abductive derivations and consistency checks among the agents. A new version based on ASystem,[23] called DAREC,[24] has also been proposed in order to improve parallelization, add constraints and ensure confidentiality (in the sense that internal reasoning is not disclosed to other agents). In none of these works however is the issue of communicational constraints explicitely raised, as they assumed a fully connected society of agents. Recently, the authors have proposed a Multiagent Abductive Reasoning System (MARS)[25] for dealing with abduction of fully distributed clausal theories. This work presents a more complex bilateral protocol to check the consistency of candidate hypotheses, where an explicit context is progressively built to ensure the coherence of hypotheses, and argumentation can be used to justify the completeness of an hypothesis. Indeed, this situation is much more complex than the one studied here since the consistency relation is not assumed to be compositional anymore, and agents must as a result do the critique collaboratively. It makes use of consequence finding technique both to perform abduction and compute validity contexts.

6.5.2. *Distributed inductive learning*

Since 2000,[26] several works dealt with inductive learning in a multiagent setting,[27] but most of them aimed at improving the behaviour of a group of reactive agents, thus favoring reinforcement learning techniques, in which agents try to learn the valuation function of a markov decision process, using feedback given as numerical rewards for some states. Few of those works investigate collaborative supervised learning, and even fewer of them take an inductive, cognitive or logic approach. Other agents are often considered as dynamic elements of the environment rather than learner partners, but there are some examples of works that investigated more direct collaborations. For instance, Ontañon and Plaza[28] have worked on some ensemble learning methods in which agents are lazy learners (using case-based reasoning techniques without explicit representation of the hypothesis) exchanging examples (cases) to improve global learning. In another proposal, by Modi and Shen,[29] each agent observes all the examples, but only perceive a part of their representation (a subset of attributes). Each agent can then evaluate the interest of choosing one of its known attributes to build collaboratively a common decision tree. In another work,[30] agents converge to a common hypothesis (given as a neural network) by using some synchronization procedure. In a more logical setting, A. Guerra-Hernandez investigated in his PhD the learning of a BDI rational agent in a multiagent setting,[31,32] in which agent sharing a library of plans (associated to a trigerring context) tried to improve these contexts by learning from successful or failed executions of those plans. The collaborative aspect of the learning was still quite limited however, as agents shared examples without considering their relevance to the other agent.

As mentioned earlier, our approach has also been instantiated in the domain of inductive reasoning with a multi agent supervised incremental concept learning, SMILE[1] (standing for Sound Multi-agent Incremental LEarning), which is presented in terms of consistency maintenance. This work studied iterated static revisions. In further extensions, variants of the revision mechanism with propagation were used to deal with static or dynamic communicational constraints by propagating the relevant informations (either by using other agents as relays, or by changing the learner when counter-examples are received).[33] An interesting effect of this process is that, for difficult boolean problems (where the difficulty reside in the size of the hypothesis space) the system learns a better hypothesis than a single agent using the same local learning algorithm on all the examples. This improvement was investigated, and a more efficient parallel version preserving this property was proposed.[34] Note that a parallel can be drawn between the proposition of an hypothesis and the equivalence requests used in Query Learning,[35] although these requests are addressed to the system rather than to an oracle. Studying oracle-based system to see when oracle could be replaced by other agents of the system in a multi agent context could thus be an interesting lead.

More recently, Ontañon and Plaza presented some work on logic-based inductive learning.[36] They proposed a method to synchronize hypotheses of two agents, using interactions between the agents similar to the ideas developed here. This method is however not incremental, which might be a problem for online learning in dynamic system. They draw inspiration from argumentative interactions. The idea of augmenting communication between agents by adding arguments has indeed been popular in the multiagent community for a few years.[37] However, although this approach has several advantages (better expressivity and easier relevance verification, by example), its efficiency and effectiveness in terms of speed and impact on the completion of the interaction's goals have not often been tested (though one can find some exceptions[38,39]).

6.5.3. *Other type of distributed hypothetical reasoning*

Beyond the aforementioned approaches which explicitly deal with distributed abduction and induction, there are a couple of other works that touch upon very related concepts. We now discuss them.

6.5.3.1. *Distributed consequence finding*

Using the principles of inverse entailment,[40] it is possible to compute abductive hypotheses deductively, as the negation of an abductive hypothesis is a new consequence of the negated observations with respect to the original theory. Abduction can then be performed by consequence finding (see for instance[41]), and studies of distributed consequence finding are thus relevant for distributed abduction. One of the main approach for consequence finding use a clause splitting method. After computing locally the consequences of their theory, agents propagate to their neighbours the parts of their consequences that belong to their languages. All the consequences of these sub-clauses will later be recombined with the consequences of the rest of the clause. This algorithm is the basis of the SomeWhere plaftorm[42] for supporting peer-to-peer inferences in the propositional case. This algorithm has seen several extensions, in particular for reasoning with inconsistencies.[43,44] Another view is to build a spanning tree over the system, computes consequences in the leaves, and propagates relevant ones to the root. This partition-based approach, first proposed for theorem-proving,[45,46] has been extended to consequence finding.[47] It is designed to improve efficiency when the theories of the agents are individually cohesive, and loosely coupled.

6.5.3.2. *Distributed diagnosis*

In the framework of multiagent diagnosis where distributed entities try to get a satisfying global diagnosis of the system, Roos *et al.*, have shown that the number of messages needed for establishing a complete global diagnosis is bound to be prohibitive if communications are not directed by a proper protocol.[48] Although our agents also try to determine the state of the system from distributed partial observations, the dynamic nature of our problem and the constraints on the interaction between our agents does not allow us to use the same methods.

6.5.4. *Consensus in (dynamic) networks*

The propagation of information in networks has been the subject of intense study, starting with the so-called *gossip problem*. At the beginning, each agent have a distinct piece of information (called a rumor), and the aim is to let all agents know all rumors in as few step as possible.[49] This has also been used to reach consensus.[50] Those approaches study the way to propagate efficiently information in a system, but aims at propagating *all* information, whereas we intend to select those pieces of information that are critical for the system to reach mutual MAS-consistency. The unified framework of *multiagent agreement problem*[51] tackles a related issue, and sets up a "cycle" very similar to the one we use in discrete dynamic system. In particular, a special emphasis is put on the interaction choice to select who to gather information from. The recent work of Glinton *et al.*,[52] a superb illustration of the latest research in this line of work, makes however assumptions closer to ours: agents are to aggregate uncertain information not directly accessible to every agent, but the amount of available information to each agent renders impossible to simply propagate everything. So agents must instead communicate their conclusions only (about some fact). The dynamics of such systems is studied in detail in (non-dynamic) networks. The final key difference lies of course in the reasoning ability of agents: in that work, agents are Bayesian, they essentially update some probabilistic value about some fact.

6.6. Conclusion

We have presented a framework for hypothesis refinement problems, that can be used to build different kinds of hypotheses in a society of intelligent agents. We proposed some revision mechanisms to ensure MAS-consistency of the system based on a learner/critic approach, and detailed the different steps for instantiating this frameworks to specific problems. Though the current framework and solutions proposed are quite general, relaxation of some of the more restrictive assumptions could be interesting. Especially, it would be useful to develop revision mechanism that can be used when the consistency relation is not *compositionnal*. A first study of such a relaxation has been done for cover-set abduction,[53] and some mechanism was proposed for abductive reasoning with distributed clausal theories.[25] Moreover, it is not always possible to assume that observations are certain. Future work will investigate hypotheses refinement mechanisms without relying on this assumption by using an argumentation framework.[54]

References

1. G. Bourgne, A. E. F. Seghrouchni, and H. Soldano. Smile: Sound multi-agent incremental learning ;-). In *Proc. of AAMAS-2007*, pp. 164–171. ACM Press (May, 2007).
2. G. Bourgne, G. Hette, N. Maudet, and S. Pinson. Hypotheses refinement under topological communication constraints. In *Proc. of AAMAS-2007*, pp. 994–1001. ACM Press (May, 2007).
3. G. Bourgne. Protocoles d'échange d'hypothèses sous contraintes communicationnelles: application au raisonnement abductif. In *Actes RJCIA 2007*, pp. 49–66. Cépaduès Éditions (Juillet, 2007).
4. T. Mitchell, *Machine Learning*. (McGraw Hill International Ed., Singapore, 1997).
5. P. A. Flach. Abduction and induction: syllogistic and inferential perspectives. Technical report, University of Bristol, Bristol, UK, (1996).
6. F. Dignum. Agents, markets, institutions, and protocols. In *AgentLink 2001, LNCS 1991*, pp. 98–114, London, UK, (2001). Springer-Verlag.
7. S. van Otterloo. *A Security Analysis of Multi-agent Protocols*. Thesis for the degree of doctor in philosophy, University of Liverpool, (2005).
8. U. Endriss, N. Maudet, F. Sadri, and F. Toni. Aspects of protocol conformance in inter-agent dialogue. In *Proc. of AAMAS 2003*, pp. 982–983. ACM Press (July, 2003).
9. T. G. Dietterich, *Ensemble Learning*, In *The Handbook of Brain Theory and Neural Networks, Second edition*, pp. 405–408. The MIT Press, (2002).
10. D. Poole, Explanation and prediction: An architecture for default and abductive reasoning, *Computational Intelligence*. **5**(2), 97–110, (1989).
11. G. Bourgne, N. Maudet, and S. Pinson. When agents communicate hypotheses in critical situations. In *Declarative Agent Languages and Technologies (DALT–2006)*, vol. 4327, *LNAI*, pp. 89–104. Springer-Verlag (May, 2006).
12. Y. Peng and J. A. Reggia, *Abductive inference models for diagnostic problem-solving*. (Springer-Verlag New York, Inc., New York, NY, USA, 1990).
13. D. Allemang, M. Tanner, T. Bylander, and J. Josephson. Computational complexity of hypothesis assembly. In *Proceedings of the 10th International Joint Conference on Artificial Intelligence*, pp. 1112–1117, (1987).
14. M. Henniche. Mgi: an incremental bottom-up algorithm. In *IEEE Aust. and New Zealand Conference on Intelligent Information Systems*, pp. 347–351, (1994).
15. P. E. Utgoff, Incremental induction of decision trees, *M.L.* **4**(2), 161–186, (1989).
16. R. A. Kowalski and F. Sadri, From logic programming towards multi-agent systems, *Annals of Mathematics and Artificial Intelligence*. **25**(3-4), 391–419, (1999).
17. F. Sadri, F. Toni, and P. Torroni. Dialogues for negotiation: Agent varieties and dialogue sequences. In *Proc. of the International Workshop on Intelligent Agents VIII (ATAL01)*, LNCS 2333, pp. 405–421. Springer, (2001).
18. A. Ciampolini, E. Lamma, P. Mello, F. Toni, and P. Torroni, Cooperation and competition in ALIAS: a logic framework for agents that negotiate, *Annals of Mathematics and Artificial Intelligence*. **37**(1–2), 65–91, (2003). ISSN 1012-2443.
19. A. C. Kakas and P. Mancarella. Database updates through abduction. In *Proc. of VLDB '90*, pp. 650–661. Morgan Kaufmann Pub., (1990). ISBN 1-55860-149-X.
20. P. Torroni. A study on the termination of negotiation dialogues. In *Proc. of the First International Joint Conference on. Autonomous Agents and Multi-Agent Systems (AAMAS'02)*, pp. 1223–1230. ACM Press, (2002).
21. M. Alberti, M. Gavanelli, E. Lamma, F. Chesani, P. Mello, and P. Torroni, Compliance verification of agent interaction: a logic-based software tool, *Applied Artificial Intelligence*. **20**(2-4), 133–157, (2006).

22. J. Ma, A. Russo, K. Broda, and K. Clark, DARE: a system for distributed abductive reasoning, *Journal of Autonomous Agents and Multiagent Systems.* **16**(3), 271–297, (2008). ISSN 1387-2532.
23. A. C. Kakas, B. V. Nuffelen, and M. Denecker. A-system: Problem solving through abduction. In ed. B. Nebel, *Proceedings of the Seventeenth International Joint Conference on Artificial Intelligence (IJCAI 2001)*, pp. 591–596. Morgan Kaufmann, (2001).
24. J. Ma, A. Russo, K. Broda, and E. Lupu. Distributed abductive reasoning with constraints. In *Post-proceedings of Declarative Agents and Language Technologies (DALT'2010)*, LNCS, p. 19pp. Springer, (2011).
25. G. Bourgne, K. Inoue, and N. Maudet. Abduction of distributed theories through local interactions. In *Proceedings of the 19th European Conference on Artificial Intelligence (ECAI-2010)* (August, 2010).
26. G. Weiss, Distributed artificial intelligence meets machine learning: Learning in multiagent environments, *J. Artificial Societies and Social Simulation.* **3**(3), (2000).
27. L. Panait and S. Luke, Cooperative multi-agent learning: The state of the art, *Autonomous Agents and Multi-Agent Systems.* **11**(3), 387–434, (2005). ISSN 1387-2532.
28. S. Ontañón and E. Plaza. Recycling data for multi-agent learning. In *Proceedings of ICML-05*, vol. 119, *ACM International Conference Proceeding Series*, pp. 633–640. ACM, (2005).
29. P. J. Modi and W.-M. Shen. Collaborative multiagent learning for classification tasks. In *Agents*, pp. 37–38, (2001).
30. J. Wang and L. Gasser. Mutual online concept learning for multiple agents. In *Proceedings of AAMAS'02*, pp. 362–369. ACM, (2002).
31. A. Guerra-Hernandez. *Apprentissage d'agents rationnels BDI dans un univers Multi-Agents.* PhD thesis, Université PAris 13 - Institut Galilée, (2003).
32. A. Guerra-Hernández, A. ElFallah-Seghrouchni, and H. Soldano. Learning in BDI multi-agent systems. In *CLIMA IV*, vol. 3259, pp. 218–233. Springer Verlag, (2004).
33. G. Bourgne, A. ElFallah-Seghrouchni, and H. Soldano. Learning in a fixed or evolving network of agents. In *ACM-IAT'09.* IEEE Computer Society, (2009).
34. G. Bourgne, H. Soldano, and A. E. Fallah-Seghrouchni. Learning better together. In *proceedings of ECAI'2010*, vol. 215, *Frontiers in Artificial Intelligence and Applications*, pp. 85–90. IOS Press, (2010).
35. D. Angluin, M. Frazier, and L. Pitt, Learning conjunctions of horn clauses, *Machine Learning.* **9**, 147–164, (1992).
36. S. Ontañón and E. Plaza. Multiagent inductive learning: An argumentation-based approach. In *Proceedings of ICML-10*, pp. 839–846. Omnipress, (2010).
37. S. Parsons, C. Sierra, and N. R. Jennings, Agents that reason and negotiate by arguing, *Journal of Logic and Computation.* **8**(3), 261–292, (1998).
38. H. Jung and M. Tambe. Argumentation as distributed constraint satisfaction: Applications and results. In *Proceedings of the fifth international conference on Autonomous agents (AGENTS01)*, pp. 324–331, (2001).
39. N. C. Karunatillake and N. R. Jennings. Is it worth arguing? In *Proceedings of the First International Workshop on Argumentation in Multi-Agent Systems (ArgMAS04)*, pp. 62–67, (2004).
40. S. Muggleton, Inverse entailment and Progol, *New Generation Comput.* **13**(3&4), 245–286, (1995).
41. K. Inoue, T. Sato, M. Ishihata, Y. Kameya, and H. Nabeshima. Evaluating abductive hypotheses using an EM algorithm on BDDs. In *Proceedings of the Twenty-first International Joint Conference on Artificial Intelligence (IJCAI'09)*, pp. 810–815, (2009).

42. P. Adjiman, P. Chatalic, F. Goasdoué, M.-C. Rousset, and L. Simon, Distributed reasoning in a peer-to-peer setting: Application to the semantic web, *Journal of Artificial Intelligence Research.* **25**, 269–314, (2006).
43. P. Chatalic, G. H. Nguyen, and M.-C. Rousset. Reasoning with inconsistencies in propositional peer-to-peer inference systems. In *Proceedings of the 17th European Conference on Artificial Intelligence (ECAI 2006)*, vol. 141, *Frontiers in Artificial Intelligence and Applications*, pp. 352–356. IOS Press, (2006).
44. A. Binas and S. A. McIlraith. Peer-to-peer query answering with inconsistent knowledge. In *Proceedings of the 11th International Conference on Principles of Knowledge Representation and Reasoning (KR'08)*, pp. 329–339, (2008).
45. E. Amir and S. A. McIlraith, Partition-based logical reasoning for first-order and propositional theories, *Artificial Intelligence.* **162**, 49–88, (2005).
46. S. A. McIlraith and E. Amir. Theorem proving with structured theories. In ed. B. Nebel, *Proceedings of the Seventeenth International Joint Conference on Artificial Intelligence (IJCAI01)*, pp. 624–634. Morgan Kaufmann, (2001).
47. K. Inoue, G. Bourgne, and T. Okamoto. Complete distributed consequence finding with message passing. In *Proceedings of 3rd International Conference on Agents and Artificial Intelligence, ICAART 2011, Rome, Italy*, p. 8pp, (2011).
48. N. Roos, A. ten Tije, and C. Witteveen. Reaching diagnostic agreement in multiagent diagnosis. In *Proceedings of the Third International joint conference on Autonomous Agents and Multi-Agent Systems (AAMAS04)*, pp. 1254–1255, (2004).
49. S.Even and B. Monien. On the number of rounds needed to disseminate information. In *Proc. of the First Annual ACM Symposium on Parallel Algorithms and Architectures*, (1989).
50. B. Chlebus and D. Kowalski. Gossiping to reach consensus. In *Proc., 14th ACM Symp. on Parallel Algorithms and Architectures*, (2002).
51. K. Lakkaraju and L. Gasser. A unified framework for multi-agent agreement. In *Proceedings of the 6th International Joint Conference on Autonomous Agents and Multiagent Systems (AAMAS-2007)*. ACM Press (May, 2007).
52. R. Glinton, P. Scerri, and K. P. Sycara. Exploiting scale invariant dynamics for efficient information propagation in large teams. In *Proceedings of the 9th International Conference on Autonomous Agents and Multiagent Systems (AAMAS 2010)*, pp. 21–30, (2010).
53. G. Bourgne, N. Maudet, and G. Ravilly-Abadie. Raisonnement hypothétique distribué sans connaissance commune. In *Actes de JFSMA2008* (October, 2008).
54. P. M. Dung, On the acceptability of arguments and its fundamental role in nonmonotonic reasoning, logic programming and n-person games, *A.I.* **77**(2), 321–357, (1995).

Chapter 7

A HEURISTIC ALGORITHMIC PROCEDURE TO SOLVE ALLOCATION PROBLEMS WITH FUZZY EVALUATIONS

R. Bartholo*, C. A. N. Cosenza†, F. A. Doria‡ and M. R. Doria§

Professor of Engineering, COPPE–UFRJ
bartholo@pep.ufrj.br

†*Professor of Engineering, Emeritus, at COPPE–UFRJ*
cosenza@pep.ufrj.br

‡*Member of the Brazilian Academy of Philosophy*
Advanced Studies Research Group and Fuzzy Sets Laboratory
PIT, Production Engineering Program, COPPE/UFRJ, P.O. Box 68507
21945–972 Rio RJ Brazil
fadoria@pep.ufrj.br
famadoria@gmail.com

§*Junior researcher, Fuzzy Sets Laboratory, UFRJ*

We use here 'algorithm' in an informal way, to designate a procedure that combines mechanical steps and heuristic oracles. This algorithm is suited to deal with allocation problems. We describe it and then sketch some of its applications.

7.1. Introduction

This text addresses a major theoretical issue with immediate consequences to practical situations. Our concern is:

> *Suppose that we have a problem, or a class of problems, whose analytical solution is intractable (of high computer complexity, for example) in many interesting instances. Can we develop some heuristic procedure to deal with that problem or class of problems in a fast, easy way, in actual, concrete situations?*

We use here 'algorithm' in an informal way, to designate a procedure that combines mechanical steps and heuristic oracles.

We offer here an affirmative answer for several interesting problems which are both of theoretical import and of practical consequences.

Heuristic procedures are in most cases Gordian–knot–like procedures to settle difficult problems, or to get a solution for problems where the usual analytical techniques fail, or are too cumbersome. Such is the case of NP–complete problems, where all known algorithms for the general case turn out to be intractable [7; 8].

The procedure we describe here directly bears on the P vs. NP problem, as we offer a fast algorithm — in the informal sense soon to be clarified — for a problem which is known to belong to the NP class. (For a review of the P vs. NP question from the third author's perspective see [6; 7; 8; 9].)

7.2. Sketch of the Technique

We will first formulate an exact version of the problem, or family of problems, we are going to deal with in the present paper. They are basically allocation problems. We give a concrete example:

- We present a list of industries $\{I_j\}, j = 1, 2, \ldots, k$, that we wish to distribute over a given region.
- Given industry I_j, we list the requirements (factors) $F_{jm}, m = 1, 2, \ldots, p$ for that industry. These are given as exact values, say, 100 tons of ore per month.
- These are the demand side of our problem.
- Now we list all possible locations $\{L_j\}, j = 1, 2, \ldots, q$ for those industries.
- Given each location L_j, we consider the available resources $R_{jm}, m = 1, 2, \ldots, r$. These are again given as exact values and costs, say, a maximum availability of 200 tons of ore each month, at 20 dollars each ton.
- We add the budgetary constraint, usually a maximum value to be expended on the whole project.

The above described problem is a well-known problem in the NP-class, for which there is no known fast algorithm in the general situation. That is to say, it is an intractable problem. (According to the view espoused by the third author, the formalized version of the sentence "there is a fast algorithm for every problem in the NP-class" is independent of a whole family of strong axiomatic theories [8], but that author believes that its negation can be proved in Peano Arithmetic plus Shoenfield's ω-rule.)

The algorithm

We describe here a very simple procedure that combines mechanical steps and heuristic steps originally devised to solve allocation problems and problems that have to do with the distribution of activities over a geographical area. Allocation problems are notoriously difficult to handle, as in the general case they lead to NP-complete problems. The present technique starts out of three considerations:

- We describe what we need and code our requirements in a matrix with linguistic entries.
- We consider the options to fulfill our needs and also code them in a related matrix with linguistic entries.
- We heuristically postulate a map between our needs and the alternatives to settle them. The map is given by a third matrix which may have crisp or fuzzy entries.

More precisely: we formalize the problem through two matrices. The first matrix, A, lists individuals on one side, against data or conditions that we expect to be fulfilled. The second matrix, B, lists the same individuals against actual situations that may satisfy the requirements in matrix A. Finally matrix C is built out of heuristic considerations and allows us to relate A and B and to rank possible solutions to our queries.

The present technique was developed by one of the authors (C. A. Cosenza) in the late 70s as a simpler, low–cost alternative to the Garin–Lowry–Echenique model for the occupation of a territory by some urban system. It has been informally known as an industrial location algorithm [1]. We sometimes call it the COPPE–COSENZA algorithm for short.

It was then noticed that it also provided answers in a wide range of situations ranging from the original allocation problems that motivated it, to applications in diagnostic medicine and even in architecture.

The present paper describes the core ideas of that technique and sketches several of its applications as mentioned above. The technique we exhibit here is by no means an isolated effort: it is akin to procedures discussed in Liang and Wang [10] and in Raj and Kumar [12]. It is however much simpler due to its heuristic part, and can be applied in a straightforward way to large collections of data, since it is of low algorithmic complexity.

A brief note on heuristic procedures

The present technique has a well–known formal structure, that of an algorithm with an oracle [14]. One builds an algorithm so that, at prescribed steps, the computation asks a question that is answered by some external device, the oracle (in our case, the algorithm's operator).

The oracle intervenes in the construction of matrix C which is the evaluation matrix. Given matrix C, an explicit algorithm allows us to build a ranking of solutions for our queries.

An acknowledgment

C. A. Cosenza wishes to stress and to acknowledge his indebtness to Marcial Echenique and to Donald Hay for their support and help in the development of this technique. It was first studied at the Martin Centre, University of Cambridge.

7.3. Main Concepts

The main idea is disarmingly simple: we define two matrices, A and B. The first one, matrix A, tells us what we need the required factors we need. The second matrix, B, exhibits the possible alternatives we have in the real world in order to implement our wishes.

Then there is a third matrix, C, which is a function of A and B, which allows us to compute optimal allocations out of our desiderata and out of the real–world alternatives we have at our disposal. That computation is both simple and fast. The main ideas can be formulated for crisp sets, but a more sensitive algorithm is obtained with the help of fuzzy objects.

Construction of matrix A

Suppose that we have several industries to distribute over a given geographic space, and suppose that we have different potential placements for those industries. The first matrix describes the industries we are interested in, and relates them to requirements for these industries (say, a shoe factory requires a continuous leather supply, water, energy, some chemical inputs and pollution control).

The first matrix, A, with k lines and m columns, has the following structure:

- *Lines* list the industries, $p_1, p_2, \ldots, p_k$.
- *Columns* list the requirements (factors) for these industries, $f_1, f_2, \ldots, f_m$.
- Given matrix A, its entries are linguistic variables A_{ij}, say:
  - — Critical factor.
  - — Decisive factor.
  - — Not decisive factor.
  - — Irrelevant factor.

Construction of matrix B

Matrix B has n lines and k columns and tells us what we have to offer to the demands in matrix A.

Matrix B has the same structure as A but in a transposed way:

- *Lines* list the same requirements $f_1, f_2, \ldots, f_n$ that appear in A. Matrix B tells us what is available in our prospective placements.
- *Columns* list possible placements for our industries, $z_1, z_2, \ldots, z_m$, where in general $n \neq m$.
- Again the entries B_{jk} of B are linguistic variables:
  - — Optimal availability.
  - — Good availability.
  - — Regular availability.
  - — Poor availability.

Construction of matrix C

Matrix C will be the tool we require to do actual computations. Its entries are given as follows, in the crisp case:

1. Suppose that there is demand for factor f_i (1 value of demand), and that region z_j doesn't have that factor (0 value of offer). We put $C_{ik} = 0$.

2. Suppose that there is no demand for f_i (0 value of demand), and yet that region z_j has that factor (value equal to 1). We put $C_{ik} = 1/n$.
3. Suppose that there is no demand for f_i (0 value of demand), and that region z_j doesn't have that factor (value also equal to 0). We put $C_{ik} = 1/n!$.
4. Finally suppose that there is a demand for f_i (1 value of demand), and that region z_j has that factor (value equal to 1). We put $C_{ik} = 1$.

These are simply "marks" we give for the possible alternatives. Cases 1. and 4. are obvious: they correspond to 0% and 100%, respectively. Case 2. gives an intermediate nonzero value because the fact that (momentarily) one doesn't require a factor that is available and which may be required in the future must be taken into account. Finally Case 3 — no demand and absence of a prescribed factor in the region — is given a nonzero mark not to penalize the possibility, as 0 should only be given to a factor that is required and isn't available.

The matrix C_{ij} is then:

$$C = \begin{pmatrix} 1/n! & 1/n \\ 0 & 1 \end{pmatrix}$$

A fuzzy set version

We may also build a fuzzy version of the above matrix C. Here is an example.

Suppose that we have translated the linguistic variables that appear as entries of A and B as fuzzy numbers. Then if μ_A is the usual fuzzy set operator with respect to A that indicates — let's make a paraphrasis — how much of an element x belongs to a set A (that is to say, $\mu_A(x)$ is a number between 0 and 1), then we may have for C:

$$C = \begin{pmatrix} 0^+ & \dots & \dots & 0^{++} \\ C_{11} & 1 & \dots & C_{1m} \\ \dots & \dots & 1 & \dots \\ 0 & \dots & \dots & 1 \end{pmatrix}$$

Here $0^+ = 1/n!$, $0^{++} = 1/n$ and for $i \neq k$, $C_{ik} = 1 + [\mu_B(B_{ik} - A_{ij}]$.

Computation of optimal values

The basic idea behind these constructions is simple: matrix C alows us to give "marks" or "values" to possible allocations given some requirements. So we can for instance select the site with the highest set of marks, or perhaps the site with the highest marks plus the least oscillations among the prescribed values, and so on.

So, matrix C is our main tool, and it is built with the help of a heuristic procedure out of A (matrix of requirements, clad as linguistic variables, as values, as fuzzy numbers, etc) and B (matrix of possible alternatives to satisfy A, similarly expressed).

Ranking techniques

There are several alternative, empirically tests, ranking techniques, which depend on the optimal goals to be attained. We describe here a simple ranking scheme that has given very reliable results in actual optimizing situations:

- Consider the demand for industry j. Form the demand value $D_j = \sum_k D_{kj}$. That is to say:
  — Fix industry j. For site k, sum over all "grades" of the factors required for j. This gives the demand D_j of industry j.
- Examine the offer for site m: $O_m = \sum_i O_{ik}$. The offer is calculated as we do for the demand.
- The rank grade r_{jm} of site m with respect to industry j is given by:

$$r_{jm} = \frac{O_m}{D_j}.$$

Given matrices A, B, C, there are of course many other possibilities to rank locations for industries with respect to the required factors. This is just the first possibility.

We can also have global rankings:

- Compute the global demand, for all industries, $D = \sum_j D_j$.
- Compute the global offer, $O = \sum_j O_j$.
- The rank grade r_m of site m with respect to all industries is: $r_m = O_m/D$.

Of course $r_m < 1$ means that not all requirements are fulfilled, while $r_m \leq 1$ means the fulfilment of the requirements.

Comparison with an analytic solution

Given a fixed budget X, the distribution of F activities among Y locations, and the calculation of an arrangement, if any, that satisfies the prescribed budget is clearly NP–complete. For the obvious algorithm for the computation of an adequate arrangement of activities and locations is exponential, while testing whether some particular arrangement fits the budget can be done in polynomial time on the length of the input data.

7.4. The Proposed Algorithm

The outline of the algorithm is very simple:

- Compute matrix A, the "demand" matrix.
- Compute matrix B, the "offer" matrix.
- Compute matrix C, the grading matrix.
- Compute ranks r_{km} and r_k and order them.
- Pick up the best options.

We can of course consider other ranking possibilities. If the grading matrix C has fuzzy entries, then we can consider those values as probabilities, and define an adequate entropy over them, whose maximization will lead to the desired ranking. However what one has seen out of concrete, real–world situations, is that the simple ranking technique used above provides some very good optimization results.

This algorithm is a fast one (the only more intricate step is the ranking procedure, but if we deal with it in a global way, as sketched before, we have a fast computation).

7.5. Example: the Brazilian Biodiesel Program

The COPPE–COSENZA technique has been recently used [2; 3] to locate biodiesel producing plants in Brazil. Attributes considered were:

- Elements linked to the production cycle.
- Elements related to transportation.
- Services of industrial interest.
- Availability of communication.
- Industrial integration.
- Availability of labor.
- Availability of electric power in regular supply.
- Availability and regular supply of water.
- Sanitary facilities.
- Adequate climatic conditions.
- Soil fertility.
- Other restrictions and facilities related to the existence of industries.
- Capacity to house a population surplus.
- Contribution to the general welfare of the local population.

The technique employed has been as described above. The model predicts that some 15 locations offer the possibility of producing castor oil at costs that range from U.S. $ 1.01 to $ 1.27, which is low enough for competitiveness in foreign markets.

7.6. Example: Diagnosing Temporal Lobe Epilepsy

We follow [4; 5]. The idea is to make a comparative diagnosis of hard to control temporal lobe epilepsy (TLE) out of a sample of 22 patients that were examined by the clinical staff at Rio's research hospital HCCFF.

The problem dealt with here has to do with difficulties in ascertaining a correct diagnosis for hard to control TLE in the clinical services at HCCFF. Since many preliminary diagnoses of that condition were superseded or even contradicted by later tests, the idea was to give a general evaluation of a reasonable number of patients that had submitted to several tests after the preliminary clinical evaluation. Details can be checked in the reference [4], but the test battery included:

- EEG at the first clinical interview.
- Craneal axial computerized tomography.
- MR.
- Video–guided EEG.

Of course diagnoses sometimes changed when going from the first evaluation to the final, post–battery of tests, evaluation. But, could we make a general comparison of those results?

Application to clinical diagnoses

The researchers built matrices A and B according to the following prescriptions:

- Matrix A has as lines the patients, and in the columns we have the several tests that were made, and the corresponding results. The results were given as linguistic variables, eventually interpreted through fuzzy sets.
- Matrix B again lists patients in its lines, and enumerates possible diagnostic results in the columns.

Matrix C is the heuristic step. It basically relates each linguistic evaluation (presence or absence of some symptom, presence or absence of some result, strong or weak symptomatic evidence of some kind) to the possible diagnostics — not relevant, weakly relevant, relevant, strongly relevant. A final calculation, given by a specific algorithm, ranks the possible diagnostics. For details see [4; 5].

7.7. Example: Groundwater Vulnerability

We briefly mention here another recently published application of the Coppe–Cosenza procedure.

The main question here is to exploit groundwater sources with adequate concerns to is environmental impact. The idea has been to compute the intrinsic vulnerability index (IVI) so that one can best evaluate the environmental risks. One considers, for instance, potential contaminant sources such as:

- Gas stations.
- Cemeteries.
- Garbage dumps and landfills.
- Agriculture with chemical fertilizers.
- Polluting industries.
- Non–native vegetation.

For details see [11].

7.8. Comments

The main theoretical and practical interest of the present — widely used, as already stressed — technique is the fact that it allows us to bypass the computational demands of the explicit, fully algorithmic solution of an allocation problem. Namely, given some budget restraint, find the best allocation of resources for some activity, or for a group of activities. It has long been known that oracle–constructions allows us to obtain significant gains in both theoretical and actual computation performances when dealing with high complexity problems [6].

This is the gist of the present technique: a very simple oracle procedure allows us to greatly lower the demands on computational resources that one requires for some allocation problems referred to in this chapter. The best example is the use of this algorithm for the placement of biodiesel plants. One has to do here with thousands of candidate sites, plus thousands of conditions that are given as requirements (the "demand" part) and available factors (the "offer" part). The ranking procedure is then an essentially low time–polynomial procedure, that is to say, the costs of processing raw data are radically curtailed when compared to the extant analytical treatments. Compare that with the huge computational resources required to deal with a 30,000 city arrangement in the Traveling Salesman Problem.

Can we relate the present heuristic approach to the exact algorithmic solution for an allocation problem? Yes. We can proceed along two lines:

- Given some actual allocation problem we can obtain (by computational brute force) an optimal solution, or a family of optimal solutions for it, and then compare that family of solutions to the solutions obtained with our heuristic approach.
- Or we can start from the exact, fully algorithmic, solution, and use it to impose constraints on the entries of matrices A, B and C. We can then ponder the alternatives that result from those constraints and compare them to some real–world heuristic situation for the same problem.

This technique has been used for nearly thirty years with the support of CNPq–Brazil and has consistenly given optimized results in the problems it deals with. Yet its successful applications have remained nearly unnoticed by theoreticians in systems analysis. The present paper is a first, modest attempt, to expose its barebones and to suggest the import of examining it carefully.

A final remark

We notice that we can obtain equivalent algorithms for any problem in the NP–class, out of the preceding computational schemes.

Acknowledgments

C. A. Cosenza wishes to thank Professor P. Arestis and the Department of Land Economy, University of Cambridge, for support.

The authors wish also to acknowledge support from the Fuzzy Sets Laboratory and from the Research Group on Advanced Studies at the Production Engineering Program, COPPE–UFRJ. F. A. Doria is partially funded by CNPq–Brazil, Philosophy Section.

References

1. Cosenza, C. A. N., "Industrial location model: a proposal," preprint, Martin Centre, Cambridge U., Cambridge (1981).
2. Cosenza, C. A. N., "Brazil's biodiesel project," Centre for Brazilian Studies, Oxford University (2005).
3. Cosenza, C. A. N. Neves, C., Rodrigues Lima, F., "A decision–making method for the selection of biodiesel sites in Brazilian regional planning," internal report, PIT–Coppe (2005).
4. Cosenza–Andraus, M. E., Cosenza, C. A. N., Gomes–Nunes, R., Fantezia–Andraus, C., Alves–Leon, S. V., "Monitorización prolongada por videoelectroencefalografía de pacientes con diagnóstico ambulatorio de epilepsia del lóbulo temporal de difícil control: aplicación del modelo de lógica fuzzy" [Long–term monitoring of patients with clinical diagnosis of hard to control frontal–lobe epilepsy through video EEG: application of a fuzzy logic model], *Revista de Neurología*, **43**, 7–14 (2006).
5. Cosenza, C. A. N., Cosenza–Andraus, M. E., Fantezia–Andraus, C., Alves–Leon, S. V., "Fuzzy method for evaluation of spect interictal in patients with difficult control epilepsy," *Proc. of the Second Brazilian Symposium on Mathematical and Computational Biology*, E–Papers, Rio (2003).
6. N. C. A. da Costa and F. A. Doria, "Some thoughts on hypercomputation," *Applied Mathematics and Computation* **178** 83–92 (2006).
7. N. C. A. da Costa and F. A. Doria, "Metamathematical limits to computation," this Handbook (2008).
8. N. C. A. da Costa and F. A. Doria, "Hypotheses that imply the independence of $P = NP$ from strong axiomatic systems," to appear in S. Zambelli, *Festschrift in Honor of V. Velupillai*, Routledge (2010).
9. N. C. A. da Costa, F. A. Doria and E. Bir, "On the metamathematics of the P *vs* NP question," *Applied Mathematics and Computation* **189**, 1223–1240 (2007).
10. Liang, G.–S., Wang, M.–J., "A fuzzy multi–criteria decision–making method for facility site selection," *Int. J. Prod. Res.* **29**, 2313–2330 (1991).
11. Nobre, R. C. M., Rotunno Filho, O. C., Mansur, W. J., Nobre, M. M. M., Cosenza, C. A. N., "Groundwater vulnerability and risk maping using GIS modeling and a fuzzy logic tool," *J. Contaminant Hydrology* **94**, 277–292 (2007).
12. Raj, P. A., Kumar, D. N., "Ranking alternatives with fuzzy weights using maximizing set and minimizing set," *Fuzzy Sets and Systems* **105**, 365–375 (1999).
13. Rheingantz, P. A., "Cosenza's hierarchical analysis model for the performance assessment of office buildings," *Facilities* **21**, 333–346 (2003).
14. H. Rogers Jr., *Theory of Recursive Functions and of Effective Computability*, reprint, MIT Press (1992).
15. Toledo, O. M., Cosenza, C. A. N., "A evolução do modelo Coppetec–Cosenza de modelo de localização industrial para modelo de hierarquia fuzzy e sua aplicabilidade" [Evolution of the Coppetec–Cosenza model from an industrial allocation model to a fuzzy hierarchical model, and applications], COPPE/UFRJ Publications, rio (2000).

Chapter 8

NON-CLASSICAL LOGICS AND INTELLIGENT SYSTEMS

Seiki Akama
C-Republic
1-20-1, Higashi-Yurigaoka, Asao-ku
Kawasaki-shi 215-0012, Japan.
akama@jcom.home.ne.jp

There are many systems of non-classical logics in the literature. Non-classical logics are developed to overcome several defects of classical logic. They can be regarded very important to reasoning-based intelligent systems. In this chapter, we review some important non-classical logics, in particular, modal, many-valued and paraconsistent logics. We also consider their applications to reasoning-based intelligent systems.

8.1. Introduction

By a *reasoning-based intelligent system*, we mean a computational system capable of dealing with several forms of *reasoning*. It is clear that such systems should also equipa mechanism to represent *knowledge*.

Although reasoning-based intelligent systems can perform mathematical and common-sense reasoning, both types of reasoning must be based on logic. So far, many reasoning-based intelligent systems have been investigated and implemented. And they usually use the so-called *classical logic*.

It is known that classical logic is well-established in the sense that its proof and model theory have been fully studied. In addition, a lot of work on *automated theorem-proving* for classical logic has been done. Undoubtedly, automated theorem-proving can be seen as a starting point of study of reasoning-based intelligent systems. The development of theorem-proving has led to *logic programming* like Prolog; see Kowalski (1974).

However, there seems to be no reason to restrict ourselves to classical logic as a foundation for reasoning-based intelligent systems. There are at least two reasons to claim that classical logic is not appropriate to formalize human reasoning. First, classical logic is of no help to represent *intensional concepts* like modality and time. But these notions are prominent in common-sense reasoning. Second, human knowledge may be incomplete and inconsistent. However, classical logic cannot express incomplete and inconsistent information in our world.

Opponents would object to our claims by saying that classical logic can accommodate such problems. Indeed, it is possible to formalize such concepts by using classical logic as a meta-language. One could thus say that these may not be the shortcomings of classical logic. Even if this is so, they cannot be *naturally* handled in classical logic.

Non-classical logic are a family of logics different from classical logic. We can overcome some defects in classical logic, if we replace classical logic by suitable non-classical logic. There are many systems of non-classical logics.

The purpose of this chapter is to review non-classical logics in the context of reasoning-based intelligent systems, because non-classical logics are less known than classical logic.

The rest of this chapter is organized as follows. In Section 8.2, we provide general arguments on non-classical logics. In Section 8.3, we review modal logic which can formalize intensional concepts. In Section 8.4, we discuss intuitionistic logic which rejects the law of excluded middle in classical logic. In Section 8.5, we present a review of many-valued logics which can handle incomplete information. In Section 8.6, we take up paraconsistent logic which can tolerate inconsistent information. In Section 8.7, we discuss how non-classical logics can be applied to reasoning-based intelligent system. Section 8 gives some conclusions.

8.2. Non-Classical Logics

We usually start to study logic as classical logic. As a result, researchers in many fields stick to classical logic as their tool. In fact, classical logic is of use in a number of applications. Unfortunately, this approach does not seem correct.

As discussed in the previous section, classical logic is not appropriate for some issues. In other words, classical logic is not universal. This is not surprising because classical logic has been developed for mathematics. We believe that a way out is to use non-classical logics.

We know many systems of non-classical logics, and they could be applied to corresponding suitable problems. At present, major systems of non-classical logics are as follows:

- modal logic
- intuitionistic logic
- many-valued logic
- paraconsistent logic
- conditional logic
- relevance logic
- substructural logic
- probabilistic logic
- non-monotonic logic

Of course, this list is not exhaustive, and new non-classical logics will emerge in future.

To classify these non-classical logics, we have two types; see Haack (1978). The first type is considered as an extension of classical logic. It extends classical logic with new features. For instance, modal logic adds modal operators to classical logic.

The second type is an alternative to classical logic. It therefore denies some of the features of classical logic. For example, many-valued logic is based on many truth-values, whereas classical logic uses two truth-values, i.e. true and false.

These two types of non-classical logics are conceptually different and their uses heavily depend on applications. In some cases, they can provide more promising results than classical logic.

In the following, we will give a quick review of some non-classical logics from the viewpoint we posed in Section 8.1.

8.3. Modal Logic

Modal logic extends classical logic with modal operators to represent *intensional concepts*. Intensional concepts are beyond the scope of truth and falsity.

So new mechanism for intesionality should be devised. The role can be played by a modal operator. Generally, $\Box$ (necessity) and $\Diamond$ (possibility) are used as modal operators.

A formula of the form $\Box A$ reads "A is necessarily true" and $\Box$ "A is possibly true", respectively. These are dual in the sense that $\Box A \leftrightarrow \neg\Diamond\neg A$.

Reading modal operators differently, we can obtain other intensional logics capable of formalizing some intensional concepts. Currently, many variants of modal logics are known. We list some of them as follows:

- tense logic
- epistemic logic
- doxastic logic
- deontic logic
- dynamic logic
- conditional logic
- intensional logic

Tense logic, also known as *temporal logic*, is a modal logic of tense, originally proposed by Prior (1957). Tense logic aims at modeling tense in natural language. Later, the idea can be extended to deal with the issues related to time and change.

Epistemic logic can model knowledge as a modal logic, thus regarding it as a modal logic of knowledge. Similarly, *doxastic logic* is a modal logic of belief. These modal logics were firstly developed by Hintikka (1962), and also used in Artificial Intelligence.

Deontic logic is a modal logic of obligation, which has been proposed by von Wright (1951). It can thus model obligation and permission as well as various forms of legal reasoning; see Aqvist (2002).

Dynamic logic is a modal logic of program, originally due to Pratt (1976). It can be used for program verification and related topics; see Harel, Kozen and Tiuryn (2000).

Conditional logic is a modal logic of conditional statements in natural language like counterfactual and indicative conditionals; see Stalnaker (1968) and Lewis (1973).

Intensional logic is a modal logic which can handle the distinction of extesion and intension; see Montague (1974). A logical basis of natural language semantics needs intensional logic.

Now, we show basics of proof and model theory for modal logic. The language of the minimal modal logic denoted **K** is the classical propositional logic **CPC** with the necessity operator $\Box$. The name "K" is after Kripke.

A Hilbert system for **K** is formalized as follows:

Modal Logic K
Axiom
(CPC) Axioms of **PC**
(K) $\Box(A \to B) \to (\Box A \to \Box B)$
Rules of Inference
(MP) $\vdash A, \vdash A \to B \Rightarrow \vdash B$
(NEC) $\vdash A \Rightarrow \vdash \Box A$

Here, $\vdash A$ means that A is provable in **K**. (NEC) is called the *necessitation*. The notion of proof is defined as usual.

Systems of *normal modal logic* can be obtained by adding extra axioms to describe properties of modality. Some of the important axioms are listed as follows:

(D) $\Box A \to \Diamond A$
(T) $\Box A \to A$
(B) $A \to \Box\Diamond A$
(4) $\Box A \to \Box\Box A$
(5) $\Diamond A \to \Box\Diamond A$

The name of normal modal logic is systematically given by the combination of axioms. For instance, the extension of **K** with the axiom (D) is called KD. However, some such systems traditionally have the following names:

D = **KD**
T = **KT**
B = **KB**

S4 = **KT4**
S5 = **KT5**

Before the 1960's, the study of modal logic was proof-theoretical due to the lack of model theory. A semantics of modal logic has been developed by Kripke and it is now called *Kripke semantics*; see Kripke (1959), Kripke (1963a), and Kripke (1963b).

Kripke semantics uses a *possible world* to interpret modal operators. Intuitively, the interpretation of $\Box A$ says that A is true in all possible worlds. Possible worlds are linked with the actual world by means of the *accessibility relation*.

A *Kripke model* for the normal modal logic **K** is defined as a triple $M = \langle W, R, V \rangle$, where W is a set of possible worlds, R is an accessibility relation on $W \times W$, and V is a valuation function: $W \times PV \rightarrow \{0,1\}$. We here denote by PV a set of propositional variables. $F = \langle W, R \rangle$ is called a *frame*.

We write $M, w \models A$ to mean that a formula A is true at a world w in the model M. Let p be a propositional variable and $false$ be absurdity. Then, $\models$ can be defined as follows:

$M, w \models p \Leftrightarrow V(w,p) = 1$
$M, w \not\models false$
$M, w \models \neg A \Leftrightarrow M, w \not\models A$
$M, w \models A \wedge B \Leftrightarrow M, w \models A$ and $M, w \models B$
$M, w \models A \vee B \Leftrightarrow M, w \models A$ or $M, w \models B$
$M, w \models A \rightarrow B \Leftrightarrow M, w \models A \;\Rightarrow\; M, w \models B$
$M, w \models \Box A \Leftrightarrow \forall v(wRv \;\Rightarrow\; M, v \models A)$
$M, w \models \Diamond A \Leftrightarrow \forall v(wRv$ and $M, v \models A)$

Here, there are no restrictions on the property of R. We say that a formula A is *valid* in the modal logic S, written $M \models_S A$, just in case $M, w \models A$ for every world w and every model M.

We know that the minimal modal logic **K** is complete.

Theorem 8.1. $\vdash_{\mathbf{K}} A \Leftrightarrow \models_{\mathbf{K}} A$.

By imposing some restrictions on the accessibility relation R, we can give Kripke models for various normal modal logics. The correspondences of axioms and conditions on R are given as follows:

axiom	conditions on R
(K)	no conditions
(D)	$\forall w \exists v(wRv)$ (serial)
(T)	$\forall w(wRw)$ (reflexive)
(4)	$\forall wvu(wRv$ and $vRu \;\Rightarrow\; wRu)$ (transitive)
(5)	$\forall wvu(wRv$ and $wRu \;\Rightarrow\; vRu)$ (euclidean)

For example, the accessibility relation in a Kripke model for modal logic **S4** is reflexive ad transitive since it needs axioms (K), (T) and (4). The completeness results of several modal logics have been established; see Hughes and Cresswell (1996) for details.

If we read modal operators differently, other types of modal logics listed above can be obtained. We can tackle many problems by using such modal logics. In this sense, modal logic is of special importance to applications.

8.4. Intuitionistic Logic

Intuitionistic logic is a rival to classical logic in that it rejects the *law of excluded middle*, i.e. $A \vee \neg A$ in classical logic. Intuitionistic logic is a logic for philosophy of mathematics called *intuitionism* due to Brouwer who claimed that mathematical reasoning should be based on mental activity. Based on Brouwer's philosophy, Heyting formalized intuitionistic logic; see Heyting (1956).

A Hilbert system for intuitionistic logic **Int** is formalized as follows:

Intuitionistic Logic Int
Axiom
(INT1) $A \rightarrow A$
(INT2) $A \rightarrow (B \rightarrow A)$
(INT3) $A \rightarrow (A \rightarrow B) \rightarrow (A \rightarrow B)$
(INT4) $(A \rightarrow (B \rightarrow C)) \rightarrow (B \rightarrow (A \rightarrow C))$
(INT5) $(A \rightarrow B) \rightarrow ((B \rightarrow C) \rightarrow (A \rightarrow C))$
(INT6) $(A \wedge B) \rightarrow A$
(INT7) $(A \wedge B) \rightarrow B$
(INT8) $(A \rightarrow B) \rightarrow ((A \rightarrow C) \rightarrow (A \rightarrow (B \wedge C)))$
(INT9) $A \rightarrow (A \vee B)$
(INT10) $B \rightarrow (A \vee B)$
(INT11) $(A \rightarrow C) \rightarrow ((B \rightarrow C) \rightarrow ((A \vee B) \rightarrow C))$
(INT12) $(A \rightarrow B) \rightarrow (A \rightarrow \neg B) \rightarrow \neg A$
(INT13) $\neg A \rightarrow (A \rightarrow B)$
Rules of Inference
(MP) $\vdash A, \vdash A \rightarrow B \Rightarrow \vdash B$

The logical symbols used here are the same as the ones in classical logic. As in classical logic, intuitionistic negation $\neg A$ can be defined as $A \rightarrow false$. $\vdash_{\mathbf{INT}}$ is also used for provability.

If we add (LEM) or (LDN) to **INT**, we can get classical logic **CPC**.

(LEM) $A \vee \neg A$
(LDN) $\neg\neg A \rightarrow A$

A semantics of intuitionistic logic is also non-truth-functional. Kripke also developed a semantics for **INT**; see Kripke (1965) and Fitting (1969). A Kripke semantics is similar to the one for modal logic **S4** due to the connection that **INT** can be embedded into **S4**.

A *Kripke model* for **INT** is defined as a triple $M = \langle W, \leq, V \rangle$, where W is a set of possible worlds, $\leq$ is binary relation, which is reflexive and transitive, on W, and V is a valuation function: $W \times PV \rightarrow \{0, 1\}$ satisfying the monotonicity condition that for any $w, v \in W$ and propositional variable p, if $V(w, p) = 1$ and wRv then $V(v, p) = 1$.

We write $M, w \models A$ to mean that a formula A is true at a world w in the model M. Then, $\models$ can be defined as follows:

$$M, w \models p \Leftrightarrow V(w, p) = 1$$
$$M, w \not\models false$$
$$M, w \models \neg A \Leftrightarrow \forall v \geq w(M, v \not\models A)$$
$$M, w \models A \wedge B \Leftrightarrow M, w \models A \text{ and } M, w \models B$$
$$M, w \models A \vee B \Leftrightarrow M, w \models A \text{ or } M, w \models B$$
$$M, w \models A \rightarrow B \Leftrightarrow \forall v \geq w(M, v \models A \Rightarrow M, v \models B)$$

We say that A is *valid*, written $\models_{\mathbf{INT}} A$, just in case $M, w \models A$ for every world w and every model M.

Note that the monotonicity of V holds for any formula. The striking feature of Kripke model for **INT** lies in the fact that both implication and negation are interpreted intensionally.

The completeness of **INT** holds.

Theorem 8.2. $\vdash_{\mathbf{INT}} A \Leftrightarrow \models_{\mathbf{INT}} A$.

The reader is invited to consult Fitting (1969) for details. Intuitionistic logic was mainly studied from logical perspectives, but it receives special attention in computer science.

8.5. Many-Valued Logic

Many-valued logic, also known as multiple-valued logic, is a family of logics which have more than three truth-values. Namely, many-valued logics can express other possibilities in addition to truth and falsity.

The idea of many-valued logic is implicit in Aristotle's thinking concerning *future contingents*. Now, many-valued logics are widely used to treat problems in various areas. In particular, three-valued and four-valued logics are well known for applications in computer science. It is also noted that the so-called *fuzzy logic* is classified as a many-valued (infinite-valued) logic.

We start with the exposition of *three-valued logic*. The first serious attempt to formalize a three-valued logic has been done by Lukasiewicz (1920). His system is now known as Lukasiewicz's three-valued logic, denoted $\mathbf{L_3}$, in which the third truth-valued reads "indeterminate" or "possible". Lukasiewicz considered that future contingent propositions should receive the third truth-value denoted by I, which is neither true nor false, although his interpretation is controversial.

The language of $\mathbf{L}_3$ comprises conjunction ($\wedge$), disjunction ($\vee$), implication ($\rightarrow_L$) and negation ($\sim$). The semantics for many-valued logics can be usually given by using the truth-value tables. The truth-value tables for $\mathbf{L}_3$ are as follows:

A	$\sim A$
T	F
I	I
F	T

A	B	$A \wedge B$	$A \vee B$	$A \rightarrow_L B$
T	T	T	T	T
T	F	F	T	F
T	I	I	T	I
F	T	F	F	T
F	F	F	F	T
F	I	F	I	T
I	T	I	T	T
I	F	F	I	I
I	I	I	I	T

Here, we should be careful of the fact that both the law of excluded middle $A \vee \sim A$ and the law of non-contradiction $\sim (A \wedge \sim A)$ do not hold. In fact, these receive I when the truth-values of compound formulas are I.

A Hilbert system for $\mathbf{L}_3$ is as follows:

Lukasiewicz's Three-Valued Logic $\mathbf{L}_3$
Axiom
(L1) $A \rightarrow (B \rightarrow A)$
(L2) $(A \rightarrow B) \rightarrow ((B \rightarrow C) \rightarrow (A \rightarrow C))$
(L3) $((A \rightarrow \sim A) \rightarrow A) \rightarrow A$
(L4) $(\sim A \rightarrow \sim B) \rightarrow (B \rightarrow A)$
Rules of Inference
(MP) $\vdash A, \vdash A \rightarrow B \Rightarrow \vdash B$

HereA$\wedge$ and $\vee$ are defined by means of $\sim$ and $\rightarrow_L$ in the following way.

$A \vee B =_{def} (A \rightarrow B) \rightarrow B$
$A \wedge B =_{def} \sim (\sim A \vee \sim B)$

Kleene also proposed three-valued logic $\mathbf{K_3}$ in connection with recursive function theory; see Kleene (1952). $\mathbf{K_3}$ differs from $\mathbf{L}_3$ in its interpretation of implication $\rightarrow_K$. In $\mathbf{K_3}$, the third truth-value reads "undefined". Consequently, $\mathbf{K_3}$ can be applied to theory of programs. There are no tautologies in $\mathbf{K_3}$, thus implying that we cannot provide a Hilbert system for it.

A	$\sim A$
T	F
I	I
F	T

A	B	$A \wedge B$	$A \vee B$	$A \rightarrow_K B$
T	T	T	T	T
T	F	F	T	F
T	I	I	T	I
F	T	F	F	T
F	F	F	F	T
F	I	F	I	T
I	T	I	T	T
I	F	F	I	I
I	I	I	I	I

K3 is usually called *Kleene's strong three-valued logic*. In the literature, *Kleene's weak three-valued logic* also appears, in which a formula evaluates as I if any compound formula evaluates as I. Kleene's weak three-valued logic is equivalent to Bochvar's three-valued logic.

Four-valued logic is suited as a logic for a computer which must deal with incomplete and inconsistent information. Belnap introduced a four-valued logic which can formalize the internal states of a computer; see Belnap (1977a) and Belnap (1977b).

There are four states, i.e. (T), (F), $(None)$ and $(Both)$, to recognize an input in a computer. Based on these states, a computer can compute suitable outputs.

(T) a proposition is true.
(F) a proposition is false.
(N) a proposition is neither true nor false.
(B) a proposition is both true and false.

Here, (N) and (B) abbreviate $(None)$ and $(Both)$, respectively. From the above, (N) corresponds to incompleteness and (B) inconsistency. Four-valued logic can be thus seen as a natural extension of three-valued logic. In fact, Belnap's four-valued logic can model both incomplete information (N) and inconsistent information (B).

Belnap proposed two four-valued logics **A4** and **L4**. The former can cope only with atomic formulas, whereas the latter can handle compound formulas. **A4** is based on the *approximation lattice* depicted as Figure 8.1.

Here, B is the least upper bound and N is the greatest lower bound with respect to the ordering $\leq$.

Both
/ \
T A4 *F*
\ /
None

Fig. 8.1. Approximation Lattice

L4 has logical symbols; $\sim, \wedge, \vee$, and is based on a set of truth-values $\mathbf{4} = \{T, F, N, B\}$. One of the features of **L4** is the monotonicity of logical symbols. Let f be a logical operation. It is said that f is monotonic iff $a \subseteq b \Rightarrow f(a) \subseteq f(b)$. To guarantee the monotonicity of conjunction and disjunction, they must satisfy the following:

$a \wedge b = a \Leftrightarrow a \vee b = b$
$a \wedge b = b \Leftrightarrow a \vee b = a$

The truth-vaue tables for **L4** are as follows.

	N	F	T	B
$\sim$	B	T	F	N

$\wedge$	N	F	T	B
N	N	F	N	F
F	F	F	F	F
T	N	F	T	B
B	F	F	B	B

$\vee$	N	F	T	B
N	N	N	T	T
F	N	F	T	B
T	T	T	T	T
B	T	B	T	B

Belnap gave a semantics for the language with the above logical symbols. A *setup* is a mapping a set of atomic formulas *Atom* to the set **4**. Then, formulas of **L4** are defined as follows:

$s(A \wedge B) = s(A) \wedge s(B)$
$s(A \vee B) = s(A) \vee s(B)$
$s(\sim A) = \sim s(A)$

Further, Belnap defined an entailment relation $\rightarrow$ as follows:

$A \rightarrow B \Leftrightarrow s(A) \leq s(B)$

for all setups s.

The entailment relation $\rightarrow$ can be axiomatized as follows:

$(A_1 \wedge \ldots \wedge A_m) \rightarrow (B_1 \vee \ldots \vee B_n)$ (A_i shares some B_j)
$(A \vee B) \rightarrow C \leftrightarrow (A \rightarrow C)$ and $(B \rightarrow C)$
$A \rightarrow B \ \Leftrightarrow \sim B \rightarrow \sim A$
$A \vee B \leftrightarrow B \vee A, \ A \wedge B \leftrightarrow B \wedge A$
$A \vee (B \vee C) \leftrightarrow (A \vee B) \vee C$
$A \wedge (B \wedge C) \leftrightarrow (A \wedge B) \wedge C$

$A \wedge (B \vee C) \leftrightarrow (A \wedge B) \vee (A \wedge C)$
$A \vee (B \wedge C) \leftrightarrow (A \vee B) \wedge (A \vee C)$
$(B \vee C) \wedge A \leftrightarrow (B \wedge A) \vee (C \wedge A)$
$(B \wedge C) \vee A \leftrightarrow (B \vee A) \wedge (C \vee A)$
$\sim\sim A \leftrightarrow A$
$\sim (A \wedge B) \leftrightarrow \sim A \vee \sim B, \ \sim (A \vee B) \leftrightarrow \sim A \wedge \sim B$
$A \rightarrow B, B \rightarrow C \ \Leftrightarrow \ A \rightarrow C$
$A \leftrightarrow B, B \leftrightarrow C \ \Leftrightarrow A \leftrightarrow C$
$A \rightarrow B \ \Leftrightarrow \ A \leftrightarrow (A \wedge B) \ \Leftrightarrow \ (A \vee B) \leftrightarrow B$

Note here that $(A \wedge \sim A) \rightarrow B$ and $A \rightarrow (B \vee \sim B)$ cannot be derived in this axiomatization. It can be shown that the logic given above is closely related to the so-called *relevant logic* of Anderson and Belnap (1976). In fact, Belnap's four-valued logic is equivalent to the system of *tautological entailment*.

Fitting (1991) studied *bilattice*, which is the lattice **4** with two kinds of orderings, in connection with the semantics of logic programs. Bilattices introduce non-standard logical connectives.

Other finite-valued logics have been also proposed in the literature. For instance, Garcia and Moussavi (1990) proposed a six-valued logic for incomplete information; also see Akama (1997).

Infinite-valued logic is a many-valued logic having infinite truth-values in $[0, 1]$. *fuzzy logic* and *probabilistic logic* belong to this family. Lukasiewicz introduced infinite-valued logic $\mathbf{L}_\infty$ in 1930; see Lukasiewicz **?**. Its truth-value tables can be generated by the following matrix:

$$
\begin{array}{lcll}
|\sim A| & = & 1-|A| & \\
|A \vee B| & = & \max(|A|,|B|) & \\
|A \wedge B| & = & \min(|A|,|B|) & \\
|A \rightarrow B| & = & 1 & (|A| \leq |B|) \\
 & = & 1-|A|+|B| & (|A| > |B|)
\end{array}
$$

A Hilbert system for $\mathbf{L}_\infty$ is as follows:

Lukasiewicz's Infinite-Valued logic $\mathbf{L}_\infty$
Axioms
(IL1) $A \rightarrow (B \rightarrow A)$
(IL2) $(A \rightarrow B) \rightarrow ((B \rightarrow C) \rightarrow (A \rightarrow C))$
(IL3) $((A \rightarrow B) \rightarrow B) \rightarrow ((B \rightarrow A) \rightarrow A)$
(IL4) $(\sim A \rightarrow \sim B) \rightarrow (B \rightarrow A)$
(IL5) $((A \rightarrow B) \rightarrow (B \rightarrow A)) \rightarrow (B \rightarrow A)$

Rules of Inference
(MP) $\vdash A, \ \vdash A \rightarrow B \ \Rightarrow \vdash B$

Since (IL5) derived from other axioms, it can be deleted.

It is known that $\mathbf{L}_\infty$ was used as the basis of *fuzzy logic*; see Zadeh (1975). Fuzzy logic is a logic of vagueness and is found in many applications. Since 1990's, important work has been done for foundations for fuzzy logic.

8.6. Paraconsistent Logic

Paraconsistent logic is a logical system for developing inconsistent but non-trivial theories. It is therefore possible to tolerate inconsistent information by using a paraconsistent logic. In this sense, paraconsistent logics are very important in knowledge representation.

Let $\vdash$ be a consequence relation. Then, it is called *explosive* if $A, \neg A \vdash B$ for any A and any B. This means that everything can be proved if contradiction $A \wedge \neg A$ is found. Namely, the logic based on $\vdash$ is *trivial* in that $\vdash B$ for any B. A logic is called *paraconsistent* if it is not explosive.

The important point here is that a logic is trivial when the consequence relation is explosive. A trivial system is useless when faced with contradiction. Classical logic and many major logics like intuitionistic logic are not paraconsistent.

The idea of formalizing a paraconsistent logic is not new. In fact, work has been done since the late 1940. There are several systems of paraconsistent logics from different viewpoints. We can list the important systems as follows.

- da Costa logic
- discursive logic
- relevance logic

---

da Costa logic is a family of paraconsistent logics due to da Costa which can be a basis for inconsistent but non-trivial theories; see da Costa (1974). The important feature of da Costa logic is to use novel interpretation of negation avoiding triviality.

Discursive logic was proposed by Jaskowski (1969), which is regarded as a non-adjunctive approach. *Adjunction* is a rule of inference of the form: from $\vdash A$ and $\vdash B$ to $\vdash A \wedge B$. Discursive logic can avoid explosion by prohibiting adjunction.

Relevance logic, also called *relevant logic* is a family of logics based on the notion of relevance in conditionals. Historically, relevance logic was developed to avoid the *paradox of implications*; see Anderson and Belnap (1976) and Anderson, Belnap and Dunn (1992). Not all relevance logics are paraconsistent, but some are considered important.

Here, we review da Costa logic $\mathbf{C_1}$ due to da Costa da Costa (1974). The language of $\mathbf{C_1}$ is based on the logical symbols: $\wedge, \vee, \rightarrow$, and $\neg$. $\leftrightarrow$ is defined as usual. In addition, a formula A°, which is read "A is well-behaved", is shorthand for $\neg(A \wedge \neg A)$. The basic ideas of $\mathbf{C_1}$ contain the following: (1) most valid formulas in the classical logic hold, (2) the law of non-contradiction $\neg(A \wedge \neg A)$ should not be valid, (3) from two contradictory formulas it should not be possible to deduce any formula.

The Hilbert system of $\mathbf{C_1}$ extends the positive intuitionistic logic with the axioms for negation.

da Costa Logic $\mathbf{C_1}$
Axioms
(DC1) $A \to (B \to A)$
(DC2) $(A \to B) \to (A \to (B \to C)) \to (A \to C))$
(DC3) $(A \wedge B) \to A$
(DC4) $(A \wedge B) \to B$
(DC5) $A \to (B \to (A \wedge B))$
(DC6) $A \to (A \vee B)$
(DC7) $B \to (A \vee B)$
(DC8) $(A \to C) \to ((B \to C) \to ((A \vee B) \to C))$
(DC9) $B^\circ \to ((A \to B) \to ((A \to \neg B) \to \neg A))$
(DC10) $(A^\circ \wedge B^\circ) \to (A \wedge B)^\circ \wedge (A \vee B)^\circ \wedge (A \to B)^\circ$
(DC11) $A \vee \neg A$
(DC12) $\neg\neg A \to A$

Rules of Inference
(MP) $\vdash A,\ \vdash A \to B\ \Rightarrow \vdash B$

Here, (DC1)-(DC8) are axioms of the positive intuitionistic logic. (DC9) and (DC10) play a role for the formalization of paraconsistency.

A semantics for $\mathbf{C_1}$ can be given by a two-valued valuation. We denote by $\mathcal{F}$ the set of formulas of $\mathbf{C_1}$. A valuation is a mapping v from $\mathcal{F}$ to $\{0, 1\}$ satisfying the following:

$v(A) = 0\ \Rightarrow\ v(\neg A) = 1$
$v(\neg\neg A) = 1\ \Rightarrow\ v(A) = 1$
$v(B^\circ) = v(A \to B) = v(A \to \neg B) = 1\ \Rightarrow\ v(A) = 0$
$v(A \to B) = 1\ \Leftrightarrow\ v(A) = 0$ or $v(B) = 1$
$v(A \wedge B) = 1\ \Leftrightarrow\ v(A) = v(B) = 1$
$v(A \vee B) = 1\ \Leftrightarrow\ v(A) = 1$ or $v(B) = 1$
$v(A^\circ) = v(B^\circ) = 1 \Rightarrow v((A \wedge B)^\circ) = v((A \vee B)^\circ) = v((A \to B)^\circ) = 1$

A formula A is *valid*, written $\models A$, if $v(A) = 1$ for every valuation v. Completeness holds for $\mathbf{C_1}$.

Theorem 8.3. $\vdash_{\mathbf{C_1}} A \Leftrightarrow\ \models_{\mathbf{C_1}} A$.

The da Costa logic $\mathbf{C_1}$ can be extended to $\mathbf{C_n}$ $(1 \leq n \leq \omega)$. Now, A^1 stands for A^0 and A^n stands for $A^{n-1} \wedge (A^{(n-1)})^\circ, 1 \leq n \leq \omega$.

Then, da Costa logic $\mathbf{C_n}$ $(1 \leq n \leq \omega)$ can be obtained by (DC1)-(DC8), (DC12), (DC13) and the following:

(DC9n) $B^{(n)} \to ((A \to B) \to ((A \to \neg B) \to \neg A))$
(DC10n) $(A^{(n)} \wedge B^{(n)}) \to (A \wedge B)^{(n)} \wedge (A \vee B)^{(n)} \wedge (A \to B)^{(n)}$

Note that da Costa logic $\mathbf{C}_\omega$ has the axioms (DC1)-(DC8), (DC12) and (DC13).

Besides the major paraconsistent logics cited above, there are other approaches. From a philosophical point of view, Priest (1979) proposed the logic of paradox **LP** to give a paraconsistent solution to the *Liar Paradox*.

In connection with reasoning about incomplete and inconsistent information in knowledge representation, Subrahmanian proposed *annotated logics*, which can serve as a foundation for paraconsistent logic programming in Blair and Subrahmanian (1989). Subrahmanian and others studied theoretical aspects of annotated logics; see da Costa, Subrahmanian and Vago (1991), da Costa, Abe and Subrahmanian (1991), Abe (1992).

On the theoretical side, Carnielli and Marcos proposed the *logic of formal inconsistency* (LFI) which is a class of paraconsistent logics in which the concept of inconsistency and consistency can be expressed at object-level; see Carnielli, Coniglio and Marcos (2007).

8.7. How to Use Non-Classical Logics

Now we have many useful non-classical logics. These have own interestig features, but it is not obvious how they are used for different applications. Here, we discuss problems related to their usage.

Logic is universal for various applications. However, the term "logic" here means classical logic. In fact, classical logic is well-established and everyone can use it as a powerful tool. Non-classical logics should be properly used compatible to the applications in question. For instance, if the problem involves intensionality then we have to use modal logic. The problems with vagueness warrant the use of many-valued logic.

Various real problems can be described by certain non-classical logic. The users of non-classical logics rely on formal results like semantics and axiomatization. In this regard, non-classical logics can serve as a *model* for various problems in our world.

Logic is used as a model as well as an inference engine in reasoning-based intelligent systems. This means that proof theory is important. For most non-classical systems, there are corresponding Hilbert systems for their axiomatizations. Unfortunately, Hilbert systems are not suited to automated reasoning. For this reason, we need to work out different proof methods. We can show the following proof methods which could be extended for non-classical logics.

- resolution
- natural deduction
- sequent calculus
- tableau calculus

One of the most popular methods for automated reasoning is *resolution* due to Robinson (1965). It was devised for classical logic and became a theoretical basis for logic programming. However, it is not easy to extend it for non-classical logics.

Natural deduction and *sequent calculus* are classified as the so-called *Gentzen systems*; see Gentzen (1934). In fact, these were proposed for classical and intuitionistic logic. Natural deduction is not always appropriate for the purpose, but sequent calculus is an interesting tool for non-classical logics.

Tableau calculus is seen as a variant of sequent calculus and has computational relevance. There is a rich variety of work on tableau formulations for non-classical logics.

We believe that tableau calculus is the best proof method for automating non-classical logics. But there are no works on tableau for some non-classical logics. In any case, it is very important to investigate a proof method for non-classical logics and non-classical logic programming. These can give a fundamental mechanism for reasoning-based intelligent systems.

8.8. Conclusions

We discussed non-classical logics in the context of reasoning-based intelligent systems. We reviewed major non-classical logics. Of course, there are other interesting non-classical logics.

The non-classical logics we surveyed in this chapter are of particular importance to the issues involving intensionality, incompleteness and inconsistency. It thus means that we could utilize suitable non-classical logics for particular problems.

There are some ongoing issues. We here discuss two issues. The first is the formalization of predicate logic. For example, the representation of common-sense knowledge needs predicate logic to represent concepts like property and relation. It is in fact necessary to study non-classical predicate logics. Little work has been done on the subject because of the difficulties with axiomatization and semantics. In particular, semantics has many possibilities to interpret quantification.

The second issue is to combine non-classical logics. Indeed a non-classical logic is proposed for some purposes, but it is not obvious how two non-classical logics can be combined. There are, however, some approaches in the literature. For example, if we want to deal with time in an incomplete world, *many-valued tense logic* should be developed; see Akama, Nagata and Yamada (2007). There are other intriguing logics like intuitionistic deontic logic. Work is needed for this important problem.

References

Abe, J. M.: (1992). *Foundations of Annotated Logics* (Portuguese), Ph.D. Thesis, University of Sao Paulo.

Akama, S. (1997). A proof method for the six-valued logic for incomplete information, *Proc. of ISMVL'97*, pp. 276–281.

Akama, S. Nagata, Y. and Yamada, C. (2007). A three-valued temporal logic for future contingents, *Logique et Analyse*, **198**, pp. 99–111.

Anderson, A. and Belnap, N. (1976). *Entailment: The Logic of Relevance and Necessity I* (Princeton University Press).

Anderson, A., Belnap, N. and Dunn, J. (1992). *Entailment: The Logic of Relevance and Necessity II* (Princeton University Press).

Aqvist, L. (2002). Deontic logic, D. Gabbay and F. Guenthner (eds.), *Handbook of Philosophical Logic II, 2md Edition*, pp. 147–263 (Kluwer), 2002.
Belnap, N. (1977a). A useful four-valued logic, J.M. Dunn and G. Epstein (eds.), *Modern Used of Multiple-Valued Logic*, pp. 8–37 (Reidel).
Belnap, N. (1977b). How a computer should think, G. Ryle (ed.), *Contemporary Aspects of Philosophy*, pp. 30–55 (Oriel Press).
Blair, H. and Subrahmanian, V. (1989). Paraconsistent logic programming, *Theoretical Computer Science*, **68**, pp.135-154.
Carnielli, W., Coniglio, M. and Marcos, J. (2007). Logics of formal inconsistency, D. Gabbay and F. Guenthner (eds.), *Handbook of Philosophical Logic, vol. 14, Second Edition*, pp. 1–93 (Kulwer).
da Costa, N. (1974). On the theory of inconsistent formal systes, *Notre Dame Journal of Formal Logic*, **15**, pp. 497–510.
da Costa, N., Abe, J. and Subrahmanian, V. (1991). Remarks on annotated logic, *Zeitschrift für mathematische Logik und Grundlagen der Mathematik*, **37**, pp. 561–570.
da Costa, N., Subrahmanian, V. and Vago, C. (1991). The annotated logic $P\tau$, *Zeitschrift für mathematische Logik und Grundlagen der Mathematik*, **37**, pp. 139–148.
Fitting, M. (1969). *Intuitionisic Logic, Model Theory and Forcing* (North-Holland).
Fitting, M. (1991). Bilattice and the semantics of logic programming, *Journal of Logic Programming*, **11**, pp. 91–116.
Garcia, O. and Moussavi, M. (1990). A six-valued logic for representing incomplete information, *Proc. of ISMVL'90*, pp. 110–114.
Gentzen, G. (1934). Untersuchung über das logische Schliessen, *Mathematische Zeitschrift*, **38**, pp. 176–210, 405–430.
Haack, S. (1978). *Philosophy of Logics* (Cambridge University Press).
Harel, D., Kozen, D. and Tiuryn, J. (2000).*Dynamic Logic* (MIT Press).
Heyting, A. (1956). *Intuitionism* (North-Holland).
Hintikka, J. (1962). *Knowledge and Belief* (Cornell University Press).
Hughes, G. and Cresswell, M. (1996). *A New Introduction to Modal Logic* (Routledge).
Jaskowski, S. (1969). Propositional calculus for contradictory deductive systems, *Studia Logica*, **24**, pp.143–157, originally published in 1948.
Kleene, S. (1952). *Introduction to Metamathematics* (NorthHolland).
Kowalski, R. (1974). Predicate logic as a programming language, *Proc. Of IFIPf74*, pp. 569–574.
Kripke, S. (1959). A complete theorem in modal logic, *Journal of Symbolic Logic*, **24**, pp. 1–24.
Kripke, S. (1963a). Semantical considerations on modal logic, *Acta Philosophica Fennica*, **16**, pp. 83–94.
Kripke, S. (1963b). Semantical analysis of modal logic I, *Zeitschrift für mathematische Logik und Grundlagen der Mathematik*, **8**, pp. 67–96.
Kripke, S. (1965). Semantical analysis of intuitionistic logic, J. Crossley ad M. Dummett (eds.), *Formal Systems and Recursve Fnctions* pp. 92–130, North-Holland, Amsteram.
Lewis, D. (1973). *Counterfactuals* (Blackwell).
Lukasiewicz, J. (1920). On 3-valued logic, S. McCall (ed.), *Polish Logic*, 1967 (Oxford University Press).
Lukasiewicz, J. (1930). Many-valued systems of propositional logic, S. McCall (ed.), *Polish Logic*, 1967 (Oxford University Press).
Montague, R. (1974). *Formal Philosophy*, edited by R.H. Thomason (Yale University Press).

Pratt, V (1976). Semantical considerations on Floyd-Hoare logic, *Proc. of the 17th IEEE Symposium on Foundations on Computer Science*, pp. 109–121,1976.
Priest, G. (1979). The logic of paradox, *The Journal of Philosophical Logic*, **8**, pp. 219–241, 1979.
Prior, A. N. (1957). *Time and Modality* (Oxford University Press).
Robinson, J. A. (1965). A machine-oriented logic based on the resolution principle, *Journal of the ACM*, **12**, pp. 23–41, 1965.
Stalnaker, R. (1968). A theory of conditionals, N. Rescsher (ed.), *Studies in Logical Theories*, pp. 98–112, (Oxford University Press).
von Wright, G. (1951). Deontic logic, *Mind*, **60**, pp. 1–15, 1951.
Zadeh, L. (1975). Fuzzy logic and approximate reasoning, *Synthese*, **30**, pp. 407–428.

Chapter 9

A PARACONSISTENT ANNOTATED LOGIC PROGRAM BEFORE-AFTER EVALPSN AND ITS APPLICATION

Kazumi Nakamatsu and Jair Minoro Abe
School of Human Science and Environment, University of Hyogo, Himeji, Hyogo, Japan,
nakamatu@shse.u-hyogo.ac.jp
Graduate Program in Production Engineering, ICET — Paulista University, Sao Paulo, SP Brazil,
jairabe@uol.com.br

A paraconsistent annotated logic program called EVALPSN has been developed for dealing with defeasible deontic reasoning and applied to various kinds of intelligent control and safety verification. Moreover, in order to deal with before-after relation between processes(time intervals), bf(before-after)-EVALPSN has also been developed recently. In this chapter, bf-EVALPSN and its application to real-time process order control based on safety verification are introduced with simple examples. The bf-EVALPSN reasoning system consists of two kinds of inference rules in bf-EVALPSN called basic and transitive bf-inference rules.

9.1. Introduction and Background

One of the main purposes of paraconsistent logic is to deal with inconsistency in a framework of consistent logical systems. It has been more than six decades since the first paraconsistent logical system was proposed by S. Jaskowski [12]. It was four decades later that a family of paraconsistent logic called "annotated logic" was proposed by da Costa *et al.* [8,50], which can deal with inconsistency by introducing many truth values called "annotations" into their syntax as attached information to formulas.

The paraconsistent annotated logic by da Costa *et al.* was developed from the viewpoint of logic programming by Subrahmanian *et al.* [7, 13, 49]. Furthermore, in order to deal with inconsistency and non-monotonic reasoning in a framework of annotated logic programming, ALPSN (Annotated Logic Program with Strong Negation) and its stable model semantics was developed by Nakamatsu and Suzuki [16]. It has been shown that ALPSN can deal with some non-monotonic reasonings such as default logic [47], autoepistemic logic [15] and a non-monotonic Assumption

Based Truth Maintenance System(ATMS) [9] in a framework of annotated logic programming [17, 35, 36]. Even though ALPSN can deal with non-monotonic reasoning such as default reasoning and conflicts can be represented as paraconsistent knowledge in it, it is difficult and complicated to deal with reasoning to resolve conflicts in ALPSN. On the other hand, it is known that defeasible logic can deal with conflict resolving in a logical way [5,42,43], although defeasible logic cannot deal with inconsistency in its syntax and its inference rules are too complicated to implement them easily. In order to deal with conflict resolving and inconsistency in a framework of annotated logic programming, a new version of ALPSN, VALPSN (Vector Annotated Logic Program with Strong Negation) that can deal with defeasible reasoning and inconsistency was also developed by Nakamatsu *et al.* [21]. Moreover, it has been shown that VALPSN can be applied to conflict resolving in various systems [18–20]. It also has been shown that VALPSN provides a computational model of defeasible logic [5, 6]. Later, VALPSN was extended to EVALPSN (Extended VALPSN) by Nakamatsu *et al.* [22, 23] to deal with deontic notions (obligation, permission, forbiddance, etc.) and defeasible deontic reasoning [44, 45]. Recently, EVALPSN has been applied to various kinds of safety verification and intelligent control, for example, railway interlocking safety verification [26], robot action control [24, 27, 28, 38], safety verification for air traffic control [25], traffic signal control [29], discrete event control [30–32] and pipeline valve control [33, 34].

Considering the safety verification for process control, there is an occasion in which the safety verification for process order control is significant. For example, suppose a pipeline network in which two kinds of liquids, nitric acid and caustic soda are used for cleaning the pipelines. If those liquids are processed continuously and mixed in the same pipeline by accident, explosion by neutralization would be caused. In order to avoid such a dangerous accident, the safety for process order control should be strictly verified in a formal way such as EVALPSN. However, it seems to be a little difficult to utilize EVALPSN for verifying process order control as well as the safety verification for each process in process control. We have already proposed a new EVALPSN, bf(before-after)-EVALPSN that can deal with before-after relations between two time intervals [39].

As a related work, an interval temporal logic has been proposed for developing practical planning and natural language understanding systems in Allen [1, 2]. In his logic, a before-after relation between two time intervals is expressed by some special first-order predicates and dealt with in a framework of first order temporal logic. On the other hand, in bf-EVALPSN, a before-after relation between two processes(time intervals) is expressed in the integer vector annotation of a special literal $R(p_i, p_j, t)$ representing a before-after relation between two processes, and dealt with in the framework of annotated paraconsistent logic programming. By representing before-after relations in vector annotations paraconsistently, bf-EVALPSN can easily implement an efficient real-time verification mechanism for before-after relations between processes.

This chapter mainly focuses on introducing bf-EVALPSN and its application to real-time process order control and its safety verification with simple process order control examples. As far as we know there seems to be no other efficient formal computational tool that can deal with the real-time safety verification for process order control than bf-EVALPSN.

This chapter is organized as follows: firstly, the background and overview of the paraconsistent annotated logic program EVALPSN are introduced; in Sec. 9.2, paraconsistent annotated logic as the background knowledge of EVALPSN and EVALPSN itself are formally recapitulated with simple examples; in Sec. 9.3, after bf-EVALPSN is formally defined; in Sec. 9.4, the reasoning system of bf-EVALPSN is introduced in detail by way of simple examples; in Sec. 9.5, how to implement and apply bf-EVALPSN to real-time safety verification for process order control is described with simple practical examples; lastly, a conclusion and remarks are provided.

9.2. Paraconsistent Annotated Logic Program

This section is devoted to clarifying the formal background of the paraconsistent annotated logic program EVALPSN. The details of EVALPSN have been introduced in [39]. We assume that the reader has a basic knowledge of classical logic and logic programming [14]. In order to understand EVLPSN and its reasoning we introduce Paraconsistent Annotated Logics $P\mathcal{T}$ [8] in the following subsection.

9.2.1. *Paraconsistent Annotated Logic $P\mathcal{T}$*

Here we briefly recapitulate the syntax and semantics for propositional paraconsistent annotated logics proposed by da Costa *et al.* [8].

Generally, a truth value called an *annotation* is attached to each atomic formula explicitly in paraconsistent annotated logic, and the set of annotations constitutes a complete lattice. We introduce a paraconsistent annotated logic $P\mathcal{T}$ with the four valued complete lattice $\mathcal{T}$.

Definition 9.1. The primitive symbols of $P\mathcal{T}$ are :

- propositional symbols $p, q, \ldots, p_i, q_i, \ldots$
- each member of $\mathcal{T}$ is an *annotation constant* (we may call it simply an annotation)
- the connectives and parentheses $\wedge, \vee, \rightarrow, \neg, (,)$.

Formulas are defined recursively as follows:

- if p is a propositional symbol and $\mu \in \mathcal{T}$ is an annotation constant, then $p:\mu$ is an *annotated atomic formula (atom)*
- if F, F_1, F_2 are formulas, then $\neg F, F_1 \wedge F_2, F_1 \vee F_2, F_1 \rightarrow F_2$ are formulas.

We suppose that the four-valued lattice in Fig. 9.1 is the complete lattice $\mathcal{T}$, where annotation `t` may be intuitively regarded as the truth value *true* and annotation `f`

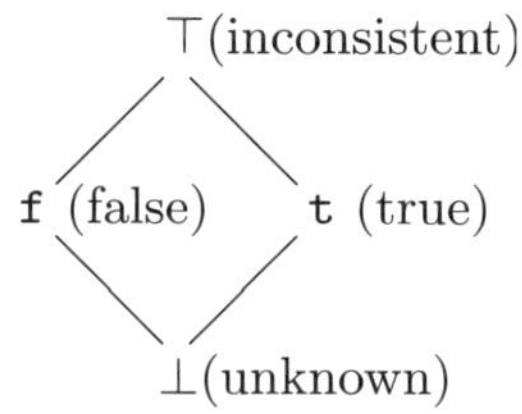

Fig. 9.1. The 4-valued Complete Lattice $\mathcal{T}$.

as the truth value *false*. It may be comprehensible that annotations $\bot, \mathtt{t}, \mathtt{f}$ and $\top$ correspond to the truth values $*, T, F$ and TF in Visser [51] and *None, T, F*, and *Both* in Belnap [4], respectively. Moreover, the complete lattice $\mathcal{T}$ can be viewed as a bi-lattice in which the vertical direction $\overrightarrow{\bot\top}$ indicates *knowledge amount* ordering and the horizontal direction $\overrightarrow{\mathtt{ft}}$ does *truth* ordering [10]. We use the symbol $\leq$ to denote the ordering in terms of knowledge amount (the vertical direction $\overrightarrow{\bot\top}$) over the complete lattice $\mathcal{T}$, and the symbols $\bot$ and $\top$ are used to denote the bottom and top elements, respectively. In the paraconsistent annotated logic $P\mathcal{T}$, each annotated atomic formula can be interpreted epistemically, for example, $p:\mathtt{t}$ may be interpreted epistemically as "the proposition p is known to be true".

There are two kinds of negation in the paraconsistent annotated logic $P\mathcal{T}$, one of them, which is represented by the symbol $\neg$ in Def. 9.1 is called *epistemic negation*, and the epistemic negation in $P\mathcal{T}$ followed by an annotated atomic formula is defined as a mapping between elements of the complete lattice $\mathcal{T}$ as follows:

$$\neg(\bot) = \bot, \quad \neg(\mathtt{t}) = \mathtt{f}, \quad \neg(\mathtt{f}) = \mathtt{t}, \quad \neg(\top) = \top.$$

This definition shows that the epistemic negation maps annotations to themselves without changing the knowledge amounts of the annotations, and the epistemic negation followed by an annotated atomic formula can be eliminated by syntactical operation. For example, the knowledge amount of annotation $\mathtt{t}$ is the same as that of annotation $\mathtt{f}$ as shown in the complete lattice $\mathcal{T}$, and we have the epistemic negation $\neg(p:\mathtt{t}) = p:\neg(\mathtt{t}) = p:\mathtt{f}$.[a] which shows that the knowledge amount in terms of the proposition p cannot be changed by the epistemic negation. There is another negation called *ontological(strong) negation* that is defined by the epistemic negation.

Definition 9.2. (Strong Negation) Let F be any formula,

$$\sim F =_{def} F \rightarrow ((F \rightarrow F) \wedge \neg(F \rightarrow F)).$$

The epistemic negation in the above definition is not interpreted as a mapping between annotations since it is not followed by an annotated atomic formula.

[a] An expression $\neg p:\mu$ is conveniently used for expressing a negative annotated literal instead of $\neg(p:\mu)$ or $p:\neg(\mu)$.

Therefore, the strongly negated formula $\sim F$ is intuitively interpreted so that if the formula F exists, the contradiction $((F \rightarrow F) \wedge \neg(F \rightarrow F))$ is implied. Usually, strong negation is used for denying the existence of the formula following it.

The semantics for the paraconsistent annotated logics $P\mathcal{T}$ is defined as follows.

Definition 9.3. Let ν be the set of all propositional symbols and $\mathcal{F}$ be the set of all formulas. An interpretation I is a function,

$$I : \nu \longrightarrow \mathcal{T}.$$

To each interpretation I, we can associate the valuation function such that

$$v_I : \mathcal{F} \longrightarrow \{0, 1\},$$

which is defined as follows:

- let p be a propositional symbol and μ an annotation,

$$v_I(p:\mu) = 1 \textbf{ iff } \mu \leq I(p), \qquad v_I(p:\mu) = 0 \textbf{ iff } \mu \not\leq I(p)$$

- let A and B be any formulas, and A not an annotated atom,

$$v_I(\neg A) = 1 \textbf{ iff } v_I(A) = 0, \qquad v_I(\sim B) = 1 \textbf{ iff } v_I(B) = 0$$

- other formulas, $A \rightarrow B$, $A \wedge B$, $A \vee B$ are valuated as usual.

We provide an intuitive interpretation for strongly negated annotated atoms with the complete lattice $\mathcal{T}$. For example, the strongly negated literal $\sim (p:\mathtt{t})$ implies the knowledge "p is false($\mathtt{f}$) or unknown($\perp$)" since it denies the existence of the knowledge that "p is true($\mathtt{t}$)". This intuitive interpretation is proved by Def. 9.3 as follows: if $v_I(\sim (p:\mathtt{t})) = 1$, we have $v_I(p:\mathtt{t}) = 0$ and for any annotation $\mu \in \{\perp, \mathtt{f}, \mathtt{t}, \top\} \leq \mathtt{t}$, we have $v_I(p:\mu) = 1$, therefore, we obtain that $\mu = \mathtt{f}$ or $\mu = \perp$.

9.2.2. *EVALPSN(Extended Vector Annotated Logic Program with Strong Negation)*

Generally, an annotation is explicitly attached to each literal in paraconsistent annotated logic programs as well as the paraconsistent annotated logic $P|calT$. For example, let p be a literal, μ an annotation, then $p:\mu$ is called an *annotated literal.* The set of annotations constitutes a complete lattice.

An annotation in EVALPSN has a form of $[(i, j), \mu]$ called an *extended vector annotation.* The first component (i, j) is called a *vector annotation* and the set of vector annotations constitutes a complete lattice,

$$\mathcal{T}_v(n) = \{\ (x, y)|0 \leq x \leq n, 0 \leq y \leq n, x, y \text{ and } n \text{ are integers }\}$$

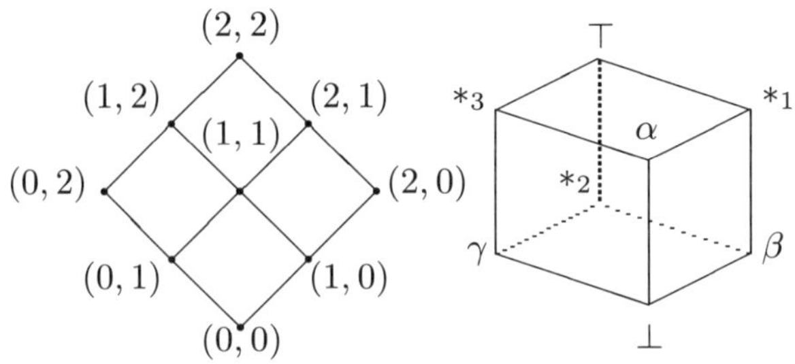

Fig. 9.2. Lattice $\mathcal{T}_v(2)$ and Lattice $\mathcal{T}_d$.

shown by the Hasse's diagram as $n = 2$ in Fig. 9.2. The ordering($\preceq_v$) of the complete lattice $\mathcal{T}_v(n)$ is defined as follows: let (x_1, y_1), $(x_2, y_2) \in \mathcal{T}_v(n)$,

$$(x_1, y_1) \preceq_v (x_2, y_2) \text{ iff } x_1 \leq x_2 \text{ and } y_1 \leq y_2.$$

For each extended vector annotated literal $p : [(i, j), \mu]$, the integer i denotes the amount of positive information to support the literal p and the integer j denotes the amount of negative information. The second component μ is an index of fact and deontic notions such as obligation, and the set of the second components constitutes the following complete lattice,

$$\mathcal{T}_d = \{\bot, \alpha, \beta, \gamma, *_1, *_2, *_3, \top\}.$$

The ordering($\preceq_d$) of the complete lattice $\mathcal{T}_d$ is described by the Hasse's diagram in Fig. 9.2. The intuitive meanings of all members in $\mathcal{T}_d$ are

$\bot$(unknown),
α(fact), β(obligation), γ(non-obligation),
$*_1$(fact and obligation),
$*_2$(obligation and non-obligation),
$*_3$(fact and non-obligation), and
$\top$(inconsistency).

The complete lattice $\mathcal{T}_d$ is a quatro-lattice in which the direction $\overrightarrow{\bot\top}$ measures *knowledge amount*, the direction $\overrightarrow{\gamma\beta}$ *deontic truth*, the direction $\overrightarrow{\bot *_2}$ *deontic knowledge amount* and the direction $\overrightarrow{\bot\alpha}$ *factuality*. For example, the annotation β(obligation) can be intuitively interpreted to be more obligatory than the annotation γ(non-obligation), and the annotations $\bot$(no knowledge) and $*_2$(obligation and non-obligation) are deontically neutral, that is to say, it cannot be said whether they represent obligation or non-obligation.

A complete lattice $\mathcal{T}_e(n)$ of extended vector annotations is defined as the product,

$$\mathcal{T}_v(n) \times \mathcal{T}_d.$$

The ordering($\preceq_e$) of the complete lattice $\mathcal{T}_e(n)$ is also defined as follows: let $[(i_1, j_1), \mu_1]$, $[(i_2, j_2), \mu_2] \in \mathcal{T}_e$,

$$[(i_1, j_1), \mu_1] \preceq_e [(i_2, j_2), \mu_2] \quad \text{iff} \quad (i_1, j_1) \preceq_v (i_2, j_2) \ \text{and} \ \mu_1 \preceq_d \mu_2.$$

There are two kinds of epistemic negation ($\neg_1$ and $\neg_2$) in EVALPSN, which are defined as mappings over the complete lattices $\mathcal{T}_v(n)$ and $\mathcal{T}_d$, respectively.

Definition 9.4. (epistemic negations $\neg_1$ and $\neg_2$ in EVALPSN)

$$\neg_1([(i,j),\mu]) = [(j,i),\mu], \quad \forall \mu \in \mathcal{T}_d$$

$$\neg_2([(i,j),\bot]) = [(i,j),\bot], \quad \neg_2([(i,j),\alpha]) = [(i,j),\alpha],$$

$$\neg_2([(i,j),\beta]) = [(i,j),\gamma], \quad \neg_2([(i,j),\gamma]) = [(i,j),\beta],$$

$$\neg_2([(i,j),*_1]) = [(i,j),*_3], \quad \neg_2([(i,j),*_2]) = [(i,j),*_2],$$

$$\neg_2([(i,j),*_3]) = [(i,j),*_1], \quad \neg_2([(i,j),\top]) = [(i,j),\top].$$

If we regard the epistemic negations in Def. 9.4 as syntactical operations, the epistemic negations followed by literals can be eliminated by the syntactical operations. For example, $\neg_1 p : [(2,0),\alpha] = p : [(0,2),\alpha]$ and $\neg_2 q : [(1,0),\beta] = p : [(1,0),\gamma]$. The strong negation ($\sim$) in EVALPSN is defined as well as the paraconsistent annotated logic $P\mathcal{T}$.

Definition 9.5. (well extended vector annotated literal) Let p be a literal. $p : [(i,0),\mu]$ and $p : [(0,j),\mu]$ are called *weva(well extended vector annotated)-literals*, where $i, j \in \{1, 2, \cdots, n\}$, and $\mu \in \{\ \alpha,\ \beta,\ \gamma\ \}$.

Definition 9.6. (EVALPSN) If $L_0, \cdots, L_n$ are weva-literals,

$$L_1 \wedge \cdots \wedge L_i \wedge \sim L_{i+1} \wedge \cdots \wedge \sim L_n \rightarrow L_0$$

is called an *EVALPSN clause*. An *EVALPSN* is a finite set of EVALPSN clauses.

Fact and deontic notions, "obligation", "forbiddance" and "permission" are represented by extended vector annotations,

$$[(m,0),\alpha],\ [(m,0),\beta],\ [(0,m),\beta],\ \text{and}\ [(0,m),\gamma],$$

respectively, where m is a positive integer. For example,

$p : [(2,0),\alpha]$ is intuitively interpreted as "it is known to be true of strength 2 that p is a fact";
$p : [(1,0),\beta]$ is as "it is known to be true of strength 1 that p is obligatory";
$p : [(0,2),\beta]$ is as "it is known to be false of strength 2 that p is obligatory", that is to say, "it is known to be true of strength 2 that p is forbidden";
$p : [(0,1),\gamma]$ is as "it is known to be false of strength 1 that p is not obligatory", that is to say, "it is known to be true of strength 1 that p is permitted".

Generally, if an EVALPSN contains the strong negation $\sim$, it has stable model semantics [39] as well as other ordinary logic programs with strong negation. However, the stable model semantics may have a problem that some programs may have more than two stable models and others have no stable model. Moreover, computing stable models takes a long time compared to usual logic programming such as PROLOG programming. Therefore, it does not seem to be so appropriate for practical application such as real time processing in general. However, we fortunately have cases to implement EVALPSN practically, if an EVALPSN is a *stratified* program, it has a tractable model called a *perfect model* [46] and the strong negation in the EVALPSN can be treated as the *Negation as Failure* in logic programming with no strong negation. The details of stratified program and some tractable models for normal logic programs can be found in [3, 11, 46, 48], furthermore the details of the stratified EVALPSN are described in [39]. Therefore, inefficient EVALPSN stable model computation does not have to be taken into account in practice since all EVALPSNs that will appear in the subsequent sections are stratified.

9.3. Before-after EVALPSN

In this section, we define bf-EVALPSN that can deal with before-after relations between two processes(time intervals) in detail [40, 41]. The reasoning system of bf-EVALPSN consists of two kinds of inference rules called *basic bf-inference rule* and *transitive bf-inference rule*, which will be introduced with some simple examples of real-time process order control in the following sections.

In bf-EVALPSN, a special annotated literal $R(p_m, p_n, t) : [(i, j), \mu]$ called *bf-literal* whose non-negative integer vector annotation (i, j) represents the before-after relation between processes Pr_m and Pr_n at time t is introduced. The integer components i and j of the vector annotation (i, j) represent the after and before degrees between processes $Pr_m(p_m)$ and $Pr_n(p_n)$, respectively, and before-after relations are represented in vector annotations paraconsistently. In the reasoning system, the basic bf-inference rules are used for determining the vector annotation of a bf-literal in real-time according to the start/finish time information of two processes. On the other hand, the transitive bf-inference rule is used for determining the vector annotation of a bf-literal in real-time based on the vector annotations of two related bf-literals as follows. Suppose that there are three processes, Pr_0, Pr_1 and Pr_2 starting in sequence, then the before-after relation between processes Pr_0 and Pr_2 can be determined from two before-after relations between processes Pr_0 and Pr_1, and between processes Pr_1 and Pr_2. Such process before-after relation reasoning is also formalized as transitive bf-inference rules in bf-EVALPSN. The transitive bf-inference system can contribute to reduce using times of basic bf-inference rules and it is a unique remarkable feature of the reasoning system. Suppose that there is a process order control system dealing with ten processes, Pr_0, Pr_1, $\cdots$ and Pr_9 starting in sequence. Without transitive bf-inference rules, the system has to deal with ${}_{10}C_2 = 45$ before-after relations independently by basic bf-inference rules. However, if we use transitive bf-inference rules, just nine before-after relations between processes Pr_i and Pr_{i+1} $(i = 0, 1, 2, \cdots, 8)$ need to be determined by basic bf-inference rules, and the rest before-after relations could be determined based on the nine before-after relations by using transitive bf-inference rules. For example, the before-after relation of processes Pr_1 and Pr_4 is inferred from two before-after relations between processes Pr_1 and Pr_3, and between processes Pr_3 and Pr_4 by transitive bf-inference rules; moreover, the before-after relation between processes Pr_1 and Pr_3 is inferred from two before-after relations between processes Pr_1 and Pr_2, and between processes Pr_2 and Pr_3 by transitive bf-inference rules.

We introduce bf(before-after)-EVALPSN that can deal with before-after relations between two processes. Hereafter, the word "**b**efore-a**f**ter" is abbreviated as just "bf". A particular literal $R(p_i, p_j, t)$ whose vector annotation represents the bf-relation between processes $Pr_i(p_i)$ and $Pr_j(p_j)$ is introduced, which declares the bf-relation between the processes at time t.

Definition 9.7. (bf-EVALPSN) An extended vector annotated literal

$$R(p_i, p_j, t) : [(i, j), \mu]$$

is called a *bf-EVALP literal* or a *bf-literal* for short, where (i, j) is a vector annotation and $\mu \in \{\alpha, \beta, \gamma\}$. If an EVALPSN clause contains bf-EVALP literals, it is called a *bf-EVALPSN clause* or just a *bf-EVALP clause* if it contains no strong negation. A *bf-EVALPSN* is a finite set of bf-EVALPSN clauses.

We provide a paraconsistent before-after interpretation for vector annotations representing bf-relations in bf-EVALPSN, and such a vector annotation is called *bf-annotations*. Exactly speaking, bf-relations are classified into fifteen meaningful kinds according to bf-relations between each start/finish time of two processes in bf-EVALPSN.

First of all, we define the most basic bf-relation in bf-EVALPSN.

Before (`be`)/After (`af`)

Bf-relations *before/after* are defined according to the bf-relation between each start time of two processes, which are represented by bf-annotations `be`/`af`, respectively. Suppose that there are two processes, Pr_i with its start time x_s and finish time x_f, and Pr_j with its start time y_s and finish time y_f. If one process has started before/after another one starts, then the bf-relations between them are defined as "before(`be`)/after(`af`)", respectively. They are described by the process time chart in Fig. 9.3 with the condition that process Pr_i has started before process Pr_j starts. The bf-relation between their start/finish times is denoted by the inequality $\{x_s < y_s\}$.[b] For example, a fact at time t "process Pr_i has started before process

[b]If time t_1 is earlier than time t_2, we conveniently denote the relation by the inequality $t_1 < t_2$.

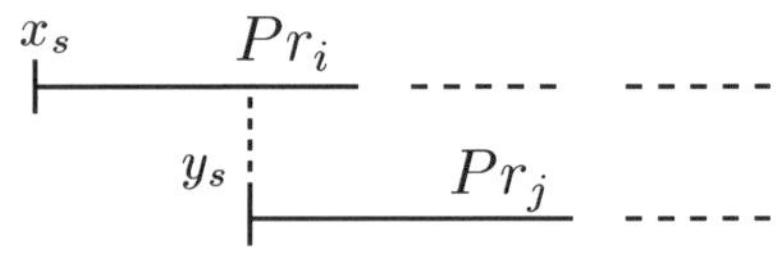

Fig. 9.3. Bf-relations Before(`be`)/After(`af`).

Fig. 9.4. Bf-relations Disjoint Before(`db`)/After(`da`).

Pr_j starts" can be represented in the bf-EVALP clause,

$$R(p_i, p_j, t) : [\texttt{be}, \alpha].$$

We define other kinds of bf-relations as well as before(`be`)/after(`af`).

Disjoint Before (`db`) /After (`da`)

Bf-relations *disjoint before/after* between two processes are represented by bf-annotations `db`/`da`, respectively. The expressions "disjoint before/after" imply that there is a time lag between the earlier process finish time and the later one start time. They also are described by the process time chart in Fig. 9.4 with the condition that process Pr_i had finished before process Pr_j started. The bf-relation between their start/finish times is denoted by the inequality $\{x_f < y_s\}$. For example, an obligation at time t "process Pr_i must start after process Pr_j finishes" can be represented in the bf-EVALP clause

$$R(p_i, p_j, t) : [\texttt{da}, \beta].$$

Immediate Before (`mb`)/After (`ma`)

Bf-relations *immediate before/after* between two processes are represented by bf-annotations `mb`/`ma`, respectively. The expressions "immediate before/after" imply that there is no time lag between the earlier process finish time and the later one start time. The bf-relations are also described by the process time chart in Fig. 9.5 with the condition that process Pr_i had finished immediately before process Pr_j started. The bf-relation between their start/finish times is denoted by the equality $\{x_f = y_s\}$. For example, a fact at time t "process Pr_i has finished immediately

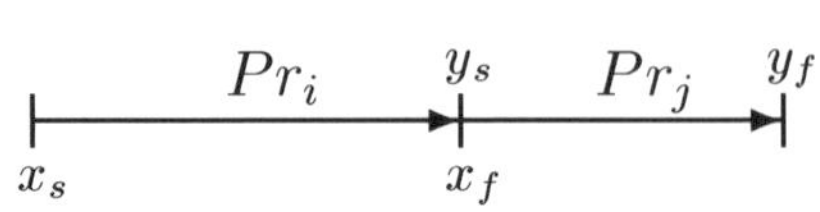

Fig. 9.5. Bf-relations Immediate Before(`mb`)/After(`ma`).

before process Pr_j starts" can be represented in the bf-EVALP clause

$$R(p_i, p_j, t) : [\mathtt{mb}, \alpha]$$

Joint Before (jb)/After (ja)

Bf-relations *joint before/after* between two processes are represented by bf-annotations jb/ja, respectively. The expressions "joint before/after" imply that the two processes overlap and the earlier process had finished before the later one finished. The bf-relations are also described by the process time chart in Fig. 9.6 with the condition that process Pr_i had started before process Pr_j started and process Pr_i had finished before process Pr_j finished. The bf-relation between their start/finish times is denoted by the inequalities $\{x_s < y_s < x_f < y_f\}$. For example, a fact at time t "process Pr_i had started before process Pr_j started and process Pr_i had finished before process Pr_j finished" can be represented in the bf-EVALP clause,

$$R(p_i, p_j, t) : [\mathtt{jb}, \alpha]$$

S-included Before (sb), S-included After (sa)

Bf-relations *s-included before/after* between two processes are represented by bf-annotations sb/sa, respectively. The expressions "s-included before/after" imply that one process had started before another one started and they have finished at the same time. The bf-relations are also described by the process time chart in Fig. 9.7 with the condition that process Pr_i had started before process Pr_j started and they have finished at the same time. The bf-relation between their start/finish times is denoted by the equality and inequalities $\{x_s < y_s < x_f = y_f\}$. For example, a fact at time t "process Pr_i had started before process Pr_j started and they have finished at the same time" can be represented in the bf-EVALP clause,

$$R(p_i, p_j, t) : [\mathtt{sb}, \alpha].$$

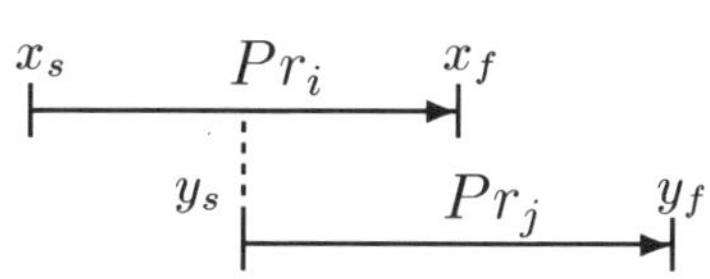

Fig. 9.6. Bf-relations, Joint Before/After.

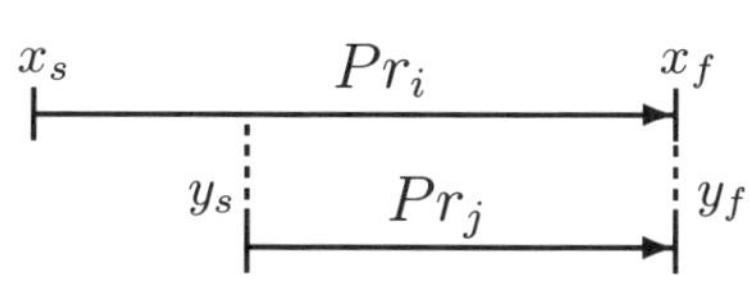

Fig. 9.7. Bf-relations S-included Before(sb)/After(sa).

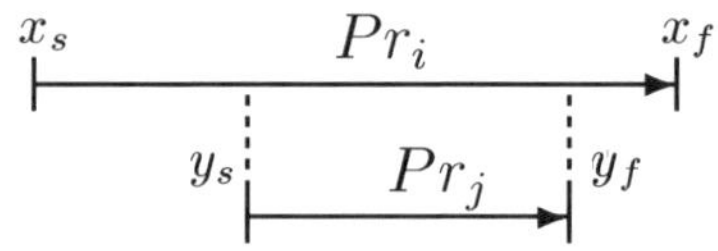

Fig. 9.8. Bf-relations Included Before(ib)/After(ia).

Included Before (ib)/After (ia)

Bf-relations *included before/after* between processes Pr_i and Pr_j are represented by bf-annotations ib/ia, respectively. The expressions "included before/after" imply that one process had started/finished before/after another one started/finished, respectively. The bf-relations are also described by the process time chart in Fig. 9.8 with the condition that process Pr_i had started before process Pr_j started and process Pr_i has finished after process Pr_j finished. The bf-relation between their start/finish times is denoted by the inequailies $\{x_s < y_s,\ y_f < x_f\}$. For example, an obligation at time t "process Pr_i must start before process Pr_j starts and must finish after process Pr_j finishes" can be represented in the bf-EVALP clause,

$$R(p_i, p_j, t) : [\mathtt{ib}, \beta].$$

F-included Before (fb)/After (fa)

bf-relations *f-include before/after* between two processes Pr_i and Pr_j are represented by bf-annotations fb/fa, respectively. The expressions "f-included before/after" imply that the two processes have started at the same time and one process had finished before another one finished. The bf-relations are also described by the process time chart in Fig. 9.9 with the condition that processes Pr_i and Pr_j have started at the same time and process Pr_i has finished after process Pr_j finished. The bf-relation between their start/finish times is denoted by the equality and inequality $\{x_s = y_s,\ y_f < x_f\}$. For example, a fact at time t "processes Pr_i and Pr_j have started at the same time and process Pr_i has finished after the process Pr_j finished" can be represented in the bf-EVALP clause,

$$R(p_i, p_j, t) : [\mathtt{fa}, \alpha].$$

Paraconsistent Before-after (pba)

Bf-relation *paraconsistent before-after* between two processes Pr_i and Pr_j is represented by bf-annotation pba. The expression "paraconsistent before-after" implies that two processes have started at the same time and also finished at the same time.

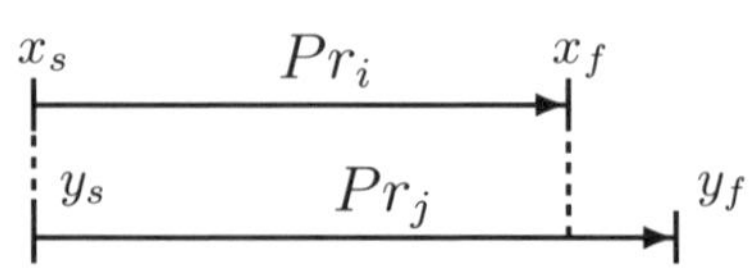

Fig. 9.9. Bf-relations F-included Before(fb)/After(fa).

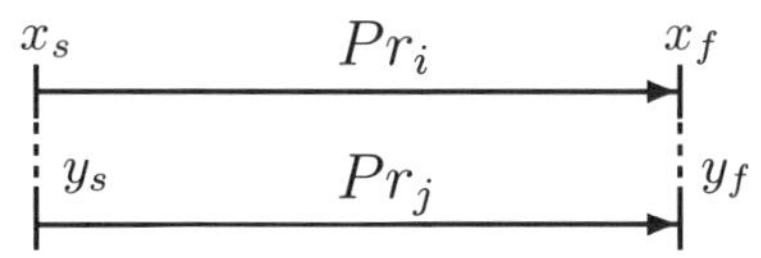

Fig. 9.10. Bf-relation, Paraconsistent Before-after.

The bf-relation is described by the process time chart in Fig. 9.10 with the condition that processes Pr_i and Pr_j have started at the same time and also finished at the same time. The bf-relation between their start/finish times is denoted by the equalities $\{x_s = y_s,\ y_f = x_f\}$. For example, an obligation at time t "processes Pr_i and Pr_j must start at the same time and they also must finish at the same time" can be represented in the bf-EVALP clause,

$$R(p_i, p_j, t) : [\mathtt{pba}, \beta].$$

Now we consider the mapping of the defined bf-relation annotations { be, af, db, da, mb, ma, jb, ja, ib, ia, sb, sa, fb, fa, pba } into vector annotations. If we consider a before-after measure over the 15 bf-annotations, obviously there exists a partial order($<_h$) in the before-after measure, where $\mu_1 <_h \mu_2$ is intuitively interpreted that bf-annotation μ_1 is more "before" than bf-annotation μ_2, and

$$\mu_1, \mu_2 \in \{\mathtt{be}, \mathtt{af}, \mathtt{db}, \mathtt{da}, \mathtt{mb}, \mathtt{ma}, \mathtt{jb}, \mathtt{ja}, \mathtt{ib}, \mathtt{ia}, \mathtt{sb}, \mathtt{sa}, \mathtt{fb}, \mathtt{fa}, \mathtt{pba}\}.$$

If $\mu_1 <_h \mu_2$ and $\mu_2 <_h \mu_1$, then we denote it $\mu_1 \equiv_h \mu_2$. Then we have the following ordering :

$$\mathtt{db} <_h \mathtt{mb} <_h \mathtt{jb} <_h \mathtt{sb} <_h \mathtt{ib} <_h \mathtt{fb} <_h \mathtt{pba} <_h \mathtt{ia} <_h \mathtt{ja} <_h \mathtt{ma} <_h \mathtt{da}$$

and

$$\mathtt{sb} \equiv_h \mathtt{be} <_h \mathtt{af} \equiv_h \mathtt{sa}.$$

On the other hand, if we take the before-after knowledge(information) amount of each bf-relation into account as another measure, obviously there also exists another partial order($<_v$) in terms of the knowledge amount, where $\mu_1 <_v \mu_2$ is intuitively interpreted that the bf-annotation μ_1 has less knowledge amount in terms of before-after relation than the bf-annotation μ_2. If $\mu_1 <_v \mu_2$ and $\mu_2 <_v \mu_1$, we denote it $\mu_1 \equiv_v \mu_2$. Then we have the following ordering :

$$\mathtt{be} <_v \mu_1, \quad \mu_1 \in \{\ \mathtt{db}, \mathtt{mb}, \mathtt{jb}, \mathtt{sb}, \mathtt{ib}, \mathtt{fb}, \mathtt{pba}\ \},$$
$$\mathtt{af} <_v \mu_2, \quad \mu_2 \in \{\ \mathtt{da}, \mathtt{ma}, \mathtt{ja}, \mathtt{sa}, \mathtt{ia}, \mathtt{fa}, \mathtt{pba}\ \},$$
$$\mathtt{db} \equiv_v \mathtt{mb} \equiv_v \mathtt{jb} \equiv_v \mathtt{sb} \equiv_v \mathtt{ib} \equiv_v \mathtt{fb} \equiv_v \mathtt{pba} \equiv_v$$
$$\mathtt{fa} \equiv_v \mathtt{ia} \equiv_v \mathtt{sa} \equiv_v \mathtt{ja} \equiv_v \mathtt{ma} \equiv_v \mathtt{da}$$

and

$$\mathtt{be} \equiv_v \mathtt{af}.$$

If we regard the before-after measure as the horizontal one and the before-after knowledge amount as the vertical one, we obtain the complete bi-lattice $\mathcal{T}_v(12)_{bf}$ of bf-annotations.

$$\begin{aligned}\mathcal{T}_v(12)_{bf} = \{ & \perp_{12}(0,0), \cdots, \mathtt{be}(0,8), \cdots, \mathtt{db}(0,12), \cdots, \mathtt{mb}(1,11), \cdots, \\ & \mathtt{jb}(2,10), \cdots, \mathtt{sb}(3,9), \cdots, \mathtt{ib}(4,8), \cdots, \mathtt{fb}(5,7), \cdots, \\ & \mathtt{pba}(6,6), \cdots, \mathtt{fa}(7,5), \cdots, \mathtt{ia}(8,4), \cdots, \mathtt{sa}(9,3), \cdots, \\ & \mathtt{ja}(10,2), \cdots, \mathtt{ma}(11,1), \cdots, \mathtt{da}(12,0), \cdots, \mathtt{af}(8,0), \cdots, \\ & \top_{12}(12,12) \},\end{aligned}$$

which is described by the Hasse's diagram in Fig. 9.11. Here we can define the epistemic negation $\neg_1$ for bf-annotations, which maps bf-annotations to themselves.

Definition 9.8. (Epistemic Negation $\neg_1$ for Bf-annotations)
The epistemic negation $\neg_1$ over the bf-annotations

$$\{\mathtt{be}, \mathtt{af}, \mathtt{da}, \mathtt{db}, \mathtt{ma}, \mathtt{mb}, \mathtt{ja}, \mathtt{jb}, \mathtt{sa}, \mathtt{sb}, \mathtt{ia}, \mathtt{ib}, \mathtt{fa}, \mathtt{fb}, \mathtt{pba}\}$$

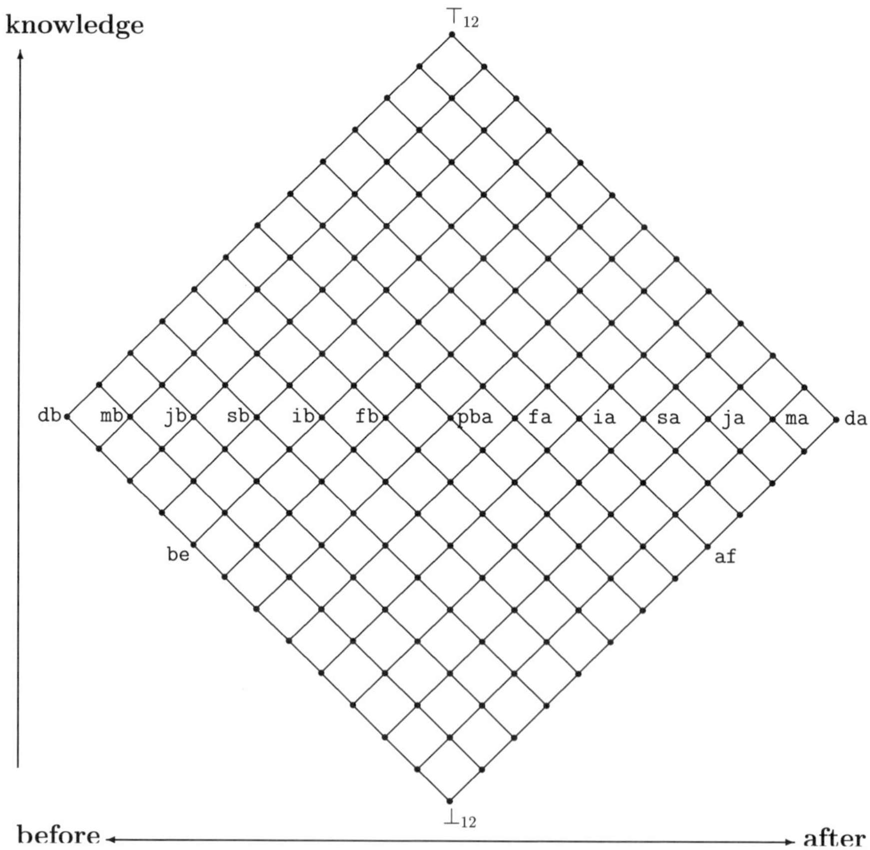

Fig. 9.11. The Complete Lattice $\mathcal{T}_v(12)_{bf}$ of Bf-annotations.

is obviously defined as the following mappings :

$$\neg_1(\texttt{af}) = \texttt{be}, \qquad \neg_1(\texttt{be}) = \texttt{af},$$
$$\neg_1(\texttt{da}) = \texttt{db}, \qquad \neg_1(\texttt{db}) = \texttt{da},$$
$$\neg_1(\texttt{ma}) = \texttt{mb}, \qquad \neg_1(\texttt{mb}) = \texttt{ma},$$
$$\neg_1(\texttt{ja}) = \texttt{jb}, \qquad \neg_1(\texttt{jb}) = \texttt{ja},$$
$$\neg_1(\texttt{sa}) = \texttt{sb}, \qquad \neg_1(\texttt{sb}) = \texttt{sa},$$
$$\neg_1(\texttt{ia}) = \texttt{ib}, \qquad \neg_1(\texttt{ib}) = \texttt{ia},$$
$$\neg_1(\texttt{fa}) = \texttt{fb}, \qquad \neg_1(\texttt{fb}) = \texttt{fa},$$
$$\neg_1(\texttt{pba}) = \texttt{pba}.$$

We note that a bf-EVALP literal

$$R(pi, pj, t) : [\mu_1, \mu_2],$$

where $\mu_1 \in \{\, \texttt{mb}, \texttt{jb}, \texttt{sb}, \texttt{ib}, \texttt{fb}, \texttt{pba}, \texttt{fa}, \texttt{ia}, \texttt{sa}, \texttt{jb}, \texttt{ma} \,\}$ and $\mu_2 \in \{\alpha, \beta, \gamma\}$, would not be well annotated if neither $m = 0$ and $n = 0$, however, since the bf-literal is equivalent to the following two well annotated bf-literals:

$$R(p_i, p_j, t) : [(m, 0), \mu] \qquad \text{and} \qquad R(p_i, p_j, t) : [(0, n), \mu],$$

such a non-well annotated bf-EVALP literal can be dealt with as the conjunction of two well annotated bf-EVALP literals. For example, suppose a non-well annotated bf-EVALP clause

$$R(p_i, p_j, t_0) : [(m, n), \mu_0] \rightarrow R(p_i, p_j, t_1) : [(k, l), \mu_1],$$

where $k \neq 0$, $l \neq 0$, $m \neq 0$ and $n \neq 0$. It can be equivalently transformed into two well annotated bf-EVALP clauses,

$$R(p_i, p_j, t_0) : [(m, 0), \mu_0] \wedge R(p_i, p_j, t_0) : [(0, n), \mu_0] \rightarrow R(p_i, p_j, t_1) : [(k, 0), \mu_1],$$
$$R(p_i, p_j, t_0) : [(m, 0), \mu_0] \wedge R(p_i, p_j, t_0) : [(0, n), \mu_0] \rightarrow R(p_i, p_j, t_1) : [(0, l), \mu_1].$$

9.4. Reasoning System in Bf-EVALPSN

In this section, we introduce the reasoning system in bf-EVALPSN, which consists of the basic inference rule for reasoning a bf-relation by process start/finish times and the transitive inference rule for reasoning one bf-relation from two other bf-relations transitively.

9.4.1. *Examples of Bf-relation Reasoning*

In order to introduce the basic inference rule we show a simple example of bf-relation reasoning. Suppose that processes Pr_0, Pr_1 and Pr_2 are scheduled to be processed according to the time chart in Fig. 9.12, then we show how the bf-relations between those processes are reasoned at each time $t_i (0 \leq i \leq 7)$.

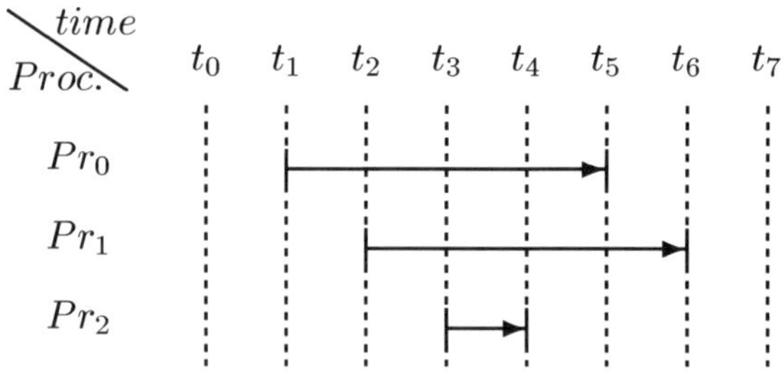

Fig. 9.12. Process Schedule Chart.

At time t_0, no process has started, thus we have no knowledge in terms of the bf-relations between processes $Pr_0(p_0)$, $Pr_1(p_1)$ and $Pr_2(p_2)$. Therefore, we have the bf-EVALP clauses,

$$R(p_0, p_1, t_0) : [(0,0), \alpha],$$
$$R(p_1, p_2, t_0) : [(0,0), \alpha],$$
$$R(p_0, p_2, t_0) : [(0,0), \alpha].$$

At time t_1, only process Pr_0 has started, then it is obviously reasoned that the bf-relation between processes Pr_0 and Pr_1 will be one of the bf-annotations,

$$\{\texttt{db}(0,12), \texttt{mb}(1,11), \texttt{jb}(2,10), \texttt{sb}(3,9), \texttt{ib}(4,8)\}$$

whose greatest lower bound is $(0,8)$ (refer to Fig. 9.11). On the other hand, we still have no knowledge in terms of the bf-relation between processes Pr_1 and Pr_2, thus the vector annotation of bf-literal $R(p_1, p_2, t_1)$ still remains $(0,0)$. Obviously bf-literal $R(p_0, p_2, t_1)$ has the same vector annotation $\texttt{be}(0,8)$ as that of bf-literal $R(p_0, p_1, t_1)$, and we have the bf-EVALP clauses,

$$R(p_0, p_1, t_1) : [(0,8), \alpha],$$
$$R(p_1, p_2, t_1) : [(0,0), \alpha],$$
$$R(p_0, p_2, t_1) : [(0,8), \alpha].$$

At time t_2, process Pr_1 has started before process Pr_0 finishes, then it is obviously reasoned that the bf-relation between processes Pr_0 and Pr_1 will be one of the bf-annotations,

$$\{\texttt{jb}(2,10), \texttt{sb}(3,9), \texttt{ib}(4,8)\}$$

whose greatest lower bound is $(2,8)$ (refer to Fig. 9.11). As process Pr_2 has not started yet, the vector annotation of bf-literal $R(p_1, p_2, t_2)$ turns to $(0,8)$ from $(0,0)$ as well as that of bf-literal $R(p_0, p_2, t_1)$ and the vector annotation of bf-literal

$R(p_0, p_2, t_2)$ still remains $(0, 8)$. Therefore, we have the bf-EVALP clauses,

$$R(p_0, p_1, t_2) : [(2, 8), \alpha],$$
$$R(p_1, p_2, t_2) : [(0, 8), \alpha],$$
$$R(p_0, p_2, t_2) : [(0, 8), \alpha].$$

At time t_3, process Pr_2 has started before processes Pr_0 and Pr_1 finish, then the vector annotation of bf-literal $R(p_0, p_1, t_3)$ still remains $(2, 8)$, and the vector annotation $(0, 8)$ of both bf-literals $R(p_1, p_2, t_3)$ and $R(p_0, p_2, t_3)$ turns to $(2, 8)$ as well as that of bf-literal $R(p_0, p_1, t_2)$. Therefore, we have the bf-EVALP clauses,

$$R(p_0, p_1, t_3) : [(2, 8), \alpha],$$
$$R(p_1, p_2, t_3) : [(2, 8), \alpha],$$
$$R(p_0, p_2, t_3) : [(2, 8), \alpha].$$

At time t_4, only process Pr_2 has finished before processes Pr_0 or Pr_1 finish, then bf-literals $R(p_1, p_2, t_4)$ and $R(p_0, p_2, t_4)$ have the same bf-annotation $\mathtt{ib}(4, 8)$. On the other hand, the vector annotation of bf-literal $R(p_1, p_2, t_4)$ still remains $(2, 8)$. Therefore, we have the bf-EVALP clauses,

$$R(p_0, p_1, t_4) : [(2, 8), \alpha],$$
$$R(p_1, p_2, t_4) : [\mathtt{ib}(4, 8), \alpha],$$
$$R(p_0, p_2, t_4) : [\mathtt{ib}(4, 8), \alpha].$$

At time t_5, process Pr_0 has finished before process Pr_1 finishes, then bf-literal $R(p_0, p_1, t_5)$ has bf-annotation $\mathtt{jb}(2, 10)$. Therefore, even though process Pr_1 has not finished yet, all the bf-relations between processes Pr_0, Pr_1 and Pr_2 have been determined as follows,

$$R(p_0, p_1, t_5) : [\mathtt{jb}(2, 10), \alpha],$$
$$R(p_1, p_2, t_5) : [\mathtt{ib}(4, 8), \alpha],$$
$$R(p_0, p_2, t_5) : [\mathtt{ib}(4, 8), \alpha].$$

9.4.2. *Basic Before-after Inference Rule*

Now we construct basic bf-relation inference rules called *basic bf-inference rules* with referring to the example in Sec. 9.4.1.

In order to represent the basic bf-inference rules in bf-EVALPSN, we newly introduce two more literals:

$st(p_i, t)$, which is intuitively interpreted that process Pr_i starts at time t, and
$fi(p_i, t)$, which is intuitively interpreted that process Pr_i finishes at time t,

which are used for expressing process start/finish information and may have one of the vector annotations, $\{(0, 0), \mathtt{t}(1, 0), \mathtt{f}(0, 1), (1, 1)\}$, where annotations $\mathtt{t}$ and $\mathtt{f}$ can be intuitively interpreted as "true" and "false", respectively.

Firstly, we show a group of basic bf-inference rules to be applied at the initial stage (time t_0) for bf-relation reasoning, which are called $(0,0)$-*rules.*

(0,0)-rules

Suppose that no process has started yet and the vector annotation of bf-literal $R(p_i, p_j, t)$ is $(0,0)$, which shows that there is no knowledge in terms of the bf-relation between processes Pr_i and Pr_j, then the following two basic bf-inference rules are applied at the initial stage.

$(0,0)$-rule-1

If process Pr_i started before process Pr_j starts, then the vector annotation $(0,0)$ of bf-literal $R(p_i, p_j, t)$ should turn to bf-annotation $\texttt{be}(0,8)$, which is the greatest lower bound of the bf-annotations,

$$\{\texttt{db}(0,12),\ \texttt{mb}(1,11),\ \texttt{jb}(2,10),\ \texttt{sb}(3,9),\ \texttt{ib}(4,8)\}.$$

$(0,0)$-rule-2

If both processes Pr_i and Pr_j have started at the same time, then it is reasonably anticipated that the bf-relation between processes Pr_i and Pr_j will be one of the bf-annotations,

$$\{\texttt{fb}(5,7),\ \texttt{pba}(6,6),\ \texttt{fa}(7,5)\}$$

whose greatest lower bound is $(5,5)$ (see Fig. 9.11). Therefore, the vector annotation $(0,0)$ of bf-literal $R(p_i, p_j, t)$ should turn to $(5,5)$.

Basic bf-inference rules $(0,0)$-rule-1 and 2 may be translated into the bf-EVALPSN clauses,

$$R(p_i, p_j, t):[(0,0), \alpha] \wedge st(p_i, t):[\mathtt{t}, \alpha] \wedge \sim st(p_j, t):[\mathtt{t}, \alpha] \\ \rightarrow R(p_i, p_j, t):[(0,8), \alpha], \tag{9.1}$$

$$R(p_i, p_j, t):[(0,0), \alpha] \wedge st(p_i, t):[\mathtt{t}, \alpha] \wedge st(p_j, t):[\mathtt{t}, \alpha] \\ \rightarrow R(p_i, p_j, t):[(5,5), \alpha]. \tag{9.2}$$

Suppose that one of basic bf-inference rules $(0,0)$-rule-1 and 2 has been applied, then the vector annotation of bf-literal $R(p_i, p_j, t)$ should be one of $(0,8)$ or $(5,5)$. Therefore, we have two groups of basic bf-inference rules to be applied immediately after basic bf-inference rules $(0,0)$-rule-1 and 2, which are called $(0,8)$-*rules* and $(5,5)$-*rules* respectively.

(0,8)-rules

Suppose that process Pr_i has started before process Pr_j starts, then the vector annotation of bf-literal $R(p_i, p_j, t)$ should be $(0,8)$. Then we have the following inference rules to be applied immediately after basic bf-inference rule $(0,0)$-rule-1.

<u>$(0,8)$-rule-1</u>

If process Pr_i has finished before process Pr_j starts, and process Pr_j starts immediately after process Pr_i finished, then the vector annotation $(0,8)$ of bf-literal $R(p_i,p_j,t)$ should turn to bf-annotation $\texttt{mb}(1,11)$.

<u>$(0,8)$-rule-2</u>

If process Pr_i has finished before process Pr_j starts, and process Pr_j has not started immediately after process Pr_i finished, then the vector annotation $(0,8)$ of bf-literal $R(p_i,p_j,t)$ should turn to bf-annotation $\texttt{db}(0,12)$.

<u>$(0,8)$-rule-3</u>

If process Pr_j starts before process Pr_i finishes, then the vector annotation $(0,8)$ of bf-literal $R(p_i,p_j,t)$ should turn to $(2,8)$ that is the greatest lower bound of the bf-annotations,

$$\{\texttt{jb}(2,10),\ \texttt{sb}(3,9),\ \texttt{ib}(4,8)\}.$$

Basic bf-inference rules $(0,8)$-rule-1,2 and 3 may be translated into the bf-EVALPSN clauses,

$$R(p_i,p_j,t):[(0,8),\alpha] \wedge fi(p_i,t):[\texttt{t},\alpha] \wedge st(p_j,t):[\texttt{t}] \\ \rightarrow R(p_i,p_j,t):[(1,11),\alpha], \tag{9.3}$$

$$R(p_i,p_j,t):[(0,8),\alpha] \wedge fi(p_i,t):[\texttt{t},\alpha] \wedge \sim st(p_j,t):[\texttt{t}] \\ \rightarrow R(p_i,p_j,t):[(0,12),\alpha], \tag{9.4}$$

$$R(p_i,p_j,t):[(0,8),\alpha] \wedge \sim fi(p_i,t):[\texttt{t},\alpha] \wedge st(p_j,t):[\texttt{t},\alpha] \\ \rightarrow R(p_i,p_j,t):[(2,8),\alpha]. \tag{9.5}$$

(5,5)-rules

Suppose that both processes Pr_i and Pr_j have already started at the same time, then the vector annotation of bf-literal $R(p_i,p_j,t)$ should be $(5,5)$. Then we have the following inference rules to be applied immediately after basic bf-inference rule $(0,0)$-rule-2.

<u>$(5,5)$-rule-1</u>

If process Pr_i has finished before process Pr_j finishes, then the vector annotation $(5,5)$ of bf-literal $R(p_i,p_j,t)$ should turn to bf-annotation $\texttt{sb}(5,7)$.

<u>$(5,5)$-rule-2</u>

If both processes Pr_i and Pr_j have finished at the same time, then the vector annotation $(5,5)$ of bf-literal $R(p_i,p_j,t)$ should turn to bf-annotation $\texttt{pba}(6,6)$.

$(5,5)$-rule-3

If process Pr_j has finished before process Pr_i finishes, then the vector annotation $(5,5)$ of bf-literal $R(p_i,p_j,t)$ should turn to bf-annotation $\mathtt{sa}(7,5)$.

Basic bf-inference rules $(5,5)$-rules-1,2 and 3 may be translated into the bf-EVALPSN clauses,

$$R(p_i,p_j,t):[(5,5),\alpha] \wedge fi(p_i,t):[\mathtt{t},\alpha] \wedge \sim fi(p_j,t):[\mathtt{t},\alpha] \\ \rightarrow R(p_i,p_j,t):[(5,7),\alpha], \tag{9.6}$$

$$R(p_i,p_j,t):[(5,5),\alpha] \wedge fi(p_i,t):[\mathtt{t},\alpha] \wedge fi(p_j,t):[\mathtt{t},\alpha] \\ \rightarrow R(p_i,p_j,t):[(6,6),\alpha], \tag{9.7}$$

$$R(p_i,p_j,t):[(5,5),\alpha] \wedge \sim fi(p_i,t):[\mathtt{t},\alpha] \wedge fi(p_j,t):[\mathtt{t},\alpha] \\ \rightarrow R(p_i,p_j,t):[(7,5),\alpha]. \tag{9.8}$$

If one of basic bf-inference rules, $(5,5)$-rule-1,2 and 3, and $(0,8)$-rule-1 and 2 has been applied, the final bf-relations represented by bf-annotations such as $\mathtt{jb}(2,10)/\mathtt{ja}(10,2)$ between two processes should be derived. On the other hand, even if basic bf-inference rule $(0,8)$-rule-3 has been applied, no bf-annotation could be derived. Therefore, a group of basic bf-inference rules called $(2,8)$-*rules* should be considered after applying basic bf-inference rule $(0,8)$-rule-3.

(2, 8)-rules

Suppose that process Pr_i has started before process Pr_j starts and process Pr_j has started before process Pr_i finishes, then the vector annotation of bf-literal $R(p_i,p_j,t)$ should be $(2,8)$ and the following three rules should be considered.

$(2,8)$-rule-1

If process Pr_i finished before process Pr_j finishes, then the vector annotation $(2,8)$ of bf-literal $R(p_i,p_j,t)$ should turn to bf-annotation $\mathtt{jb}(2,10)$.

$(2,8)$-rule-2

If both processes Pr_i and Pr_j have finished at the same time, then the vector annotation $(2,8)$ of bf-literal $R(p_i,p_j,t)$ should turn to bf-annotation $\mathtt{fb}(3,9)$.

$(2,8)$-rule-3

If process Pr_j has finished before Pr_i finishes, then the vector annotation $(2,8)$ of bf-literal $R(p_i,p_j,t)$ should turn to bf-annotation $\mathtt{ib}(4,8)$.

Basic bf-inference rules $(2,8)$-rule-1,2 and 3 may be translated into the bf-EVALPSN clauses,

$$R(p_i, p_j, t):[(2,8),\alpha] \wedge fi(p_i,t):[\mathtt{t},\alpha] \wedge \sim fi(p_j,t):[\mathtt{t},\alpha] \\ \rightarrow R(p_i,p_j,t):[(2,10),\alpha], \tag{9.9}$$

$$R(p_i, p_j, t):[(2,8),\alpha] \wedge fi(p_i,t):[\mathtt{t},\alpha] \wedge fi(p_j,t):[\mathtt{t},\alpha] \\ \rightarrow R(p_i,p_j,t):[(3,9),\alpha], \tag{9.10}$$

$$R(p_i, p_j, t):[(2,8),\alpha] \wedge \sim fi(p_i,t):[\mathtt{t},\alpha] \wedge fi(p_j,t):[\mathtt{t},\alpha] \\ \rightarrow R(p_i,p_j,t):[(4,8),\alpha]. \tag{9.11}$$

The application orders of all basic bf-inference rules are summarized in Table 9.1.

9.4.3. *Transitive Before-after Inference Rule*

Here we construct another kind of bf-inference rules called *transitive bf-inference rules*, which can reason one vector annotation of bf-literal from two other vector annotations of bf-literals transitively. Bf-relations represented by bf-annotations could be reasoned efficiently by using the transitive bf-inference rules.

Suppose that there are three processes Pr_i,Pr_j and Pr_k starting sequentially, then we consider to reason the vector annotation of bf-literal $R(p_i, p_k, t)$ from those of bf-literals $R(p_i, p_j, t)$ and $R(p_j, p_k, t)$ transitively. Firstly, let us consider the following simple examples for forming some transitive bf-inference rules.

Example 9.9. Suppose that only process Pr_i has started at time t as shown in Fig. 9.13, then we obtain the vector annotation $(0,8)$ of bf-literal $R(p_i,p_j,t)$ by basic bf-inference rule $(0,0)$-rule-1 and the vector annotation $(0,0)$ of bf-literal $R(p_j,p_k,t)$. Then, obviously the vector annotation of bf-literal $R(p_i,p_k,t)$ is reasoned as bf-annotation $\mathtt{be}(0,8)$. Thus, we have the following bf-EVALP clause as a transitive bf-inference rule, which is temporarily named trule-1,

trule-1

$$R(p_i,p_j,t):[(0,8),\alpha] \wedge R(p_j,p_k,t):[(0,0),\alpha] \rightarrow R(p_i,p_k,t):[(0,8),\alpha].$$

Table 9.1. Application Orders of Basic Bf-inference Rules.

vector annotation	rule	vector annotation	rule	vector annotation	rule	vector annotation
$(0,0)$	rule-1	$(0,8)$	rule-1	$(0,12)$		
			rule-2	$(1,11)$		
			rule-3	$(2,8)$	rule-1	$(2,10)$
					rule-2	$(3,9)$
					rule-3	$(4,8)$
	rule-2	$(5,5)$	rule-1	$(5,7)$		
			rule-2	$(6,6)$		
			rule-3	$(7,5)$		

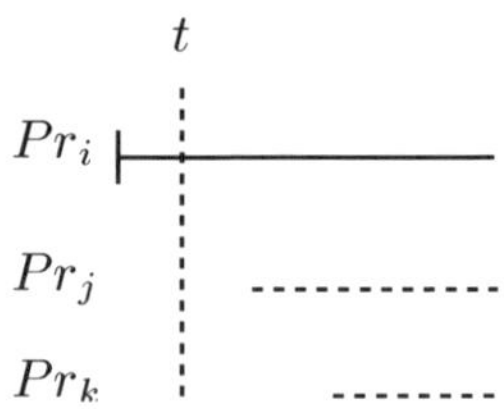

Fig. 9.13. Process Time Chart Ex-1.

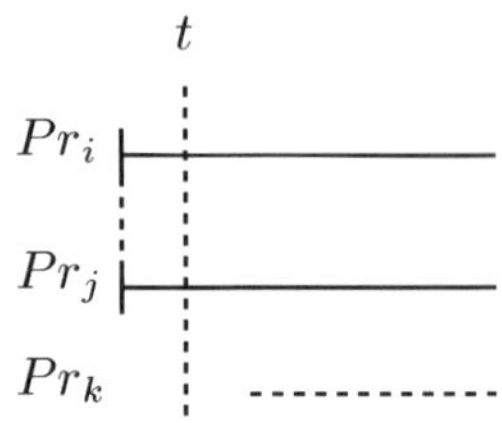

Fig. 9.14. Process Time Chart Ex-2.

Example 9.10. Suppose that processes Pr_i and Pr_j have started at the same time t as shown in Fig. 9.14, then we obtain the vector annotation $(5, 5)$ of bf-literal $R(p_i, p_j, t)$ by basic bf-inference rule $(0, 0)$-rule-2 and the vector annotation $(0, 8)$ of bf-literal $R(p_j, p_k, t)$ by basic bf-inference rule $(0, 0)$-rule-1. Then, the vector annotation of bf-literal $R(p_i, p_k, t)$ is also reasoned as bf-annotation $\texttt{be}(0, 8)$. Thus, we have the following bf-EVALP clause as another transitive bf-inference rule, which is temporarily named trule-2,

trule-2

$$R(p_i, p_j, t) : [(5, 5), \alpha] \wedge R(p_j, p_k, t) : [(0, 8), \alpha] \rightarrow R(p_i, p_k, t) : [(0, 8), \alpha].$$

Example 9.11. Suppose that both processes Pr_i and Pr_j have already started at time t but process Pr_k has not started yet as shown in Fig. 9.15, then we have obtained the vector annotation $(2, 8)$ of bf-literal $R(p_i, p_j, t)$ by basic bf-inference rule $(0, 8)$-rule-3 and the vector annotation $(0, 8)$ of bf-literal $R(p_j, p_k, t)$ by basic bf-inference rule $(0, 0)$-rule-1. Then, obviously the vector annotation of bf-literal

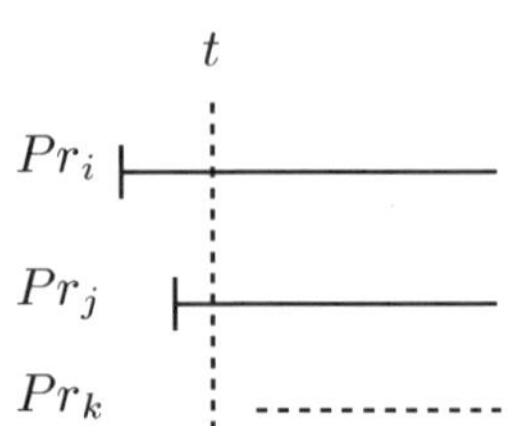

Fig. 9.15. Process Time Chart Ex-3.

$R(p_i, p_k, t)$ is reasoned as bf-annotation $\texttt{be}(0, 8)$. Thus, we also have the following bf-EVALP clause as a transitive bf-inference rule, which is temporarily named trule-3,

trule-3

$$R(p_i, p_j, t) : [(2, 8), \alpha] \wedge R(p_j, p_k, t) : [(0, 8), \alpha] \rightarrow R(p_i, p_k, t) : [(0, 8), \alpha].$$

We could reason bf-relations by using the transitive bf-inference rules in Eq. (9.9), Eq. (9.10) and Eq. (9.11). However, there are exceptional cases in which bf-relations cannot be determined uniquely by transitive bf-inference rules. Such a case is introduced in the following example.

Example 9.12. Here we take another example with three process time charts in Fig. 9.16, Fig. 9.17 and Fig. 9.18. The difference between those process schedules is that process Pr_j starts at different times t_3, t_4 and t_5 in each time chart. The variation of the vector annotations of bf-literals $R(p_i, p_j, t_n)$, $R(p_j, p_k, t_n)$ and $R(p_i, p_k, t_n)$ $(1 \leq n \leq 7)$ in each process schedule are listed in Table 9.2, Table 9.3 and Table 9.4.

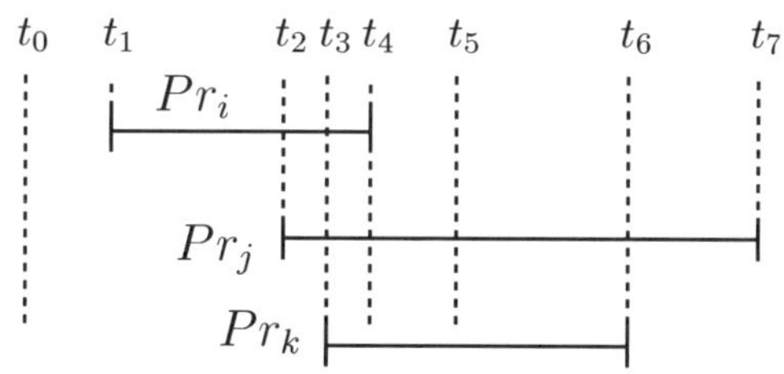

Fig. 9.16. Process Time Chart Ex-4-1.

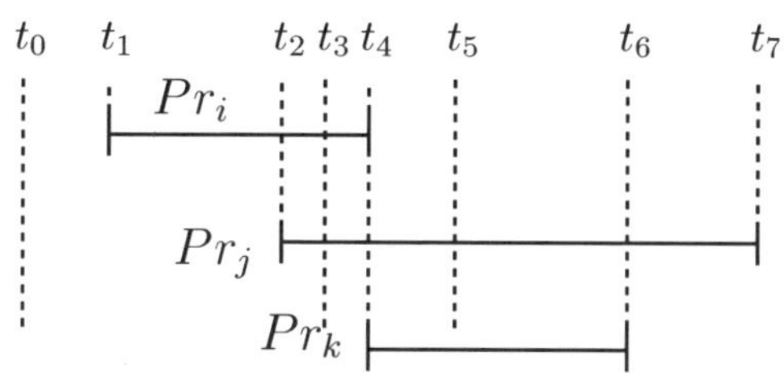

Fig. 9.17. Process Time Chart Ex-4-2.

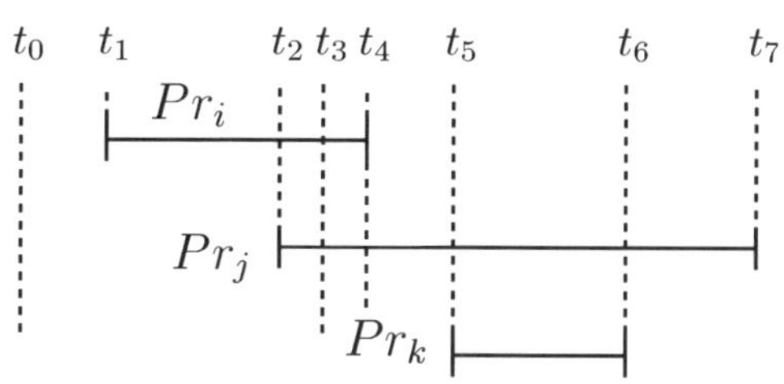

Fig. 9.18. Process Time Chart Ex-4-3.

Table 9.2. Vector Annotations of Process Time Chart Ex-4-1.

bf-literal	t_1	t_2	t_3	t_4	t_5	t_6	t_7
$R(p_i, p_j, t)$	$(0, 8)$	$(2, 8)$	$(2, 8)$	$(2, 10)$	$(2, 10)$	$(2, 10)$	$(2, 10)$
$R(p_j, p_k, t)$	$(0, 0)$	$(0, 8)$	$(2, 8)$	$(2, 8)$	$(2, 8)$	$(4, 8)$	$(4, 8)$
$R(p_i, p_k, t)$	$(0, 8)$	$(0, 8)$	$(2, 8)$	$(2, 10)$	$(2, 10)$	$(2, 10)$	$(2, 10)$

Table 9.3. Vector Annotations of Process Time Chart Ex-4-2.

bf-literal	t_1	t_2	t_3	t_4	t_5	t_6	t_7
$R(p_i, p_j, t)$	$(0, 8)$	$(2, 8)$	$(2, 8)$	$(2, 10)$	$(2, 10)$	$(2, 10)$	$(2, 10)$
$R(p_j, p_k, t)$	$(0, 0)$	$(0, 8)$	$(0, 8)$	$(2, 8)$	$(2, 8)$	$(4, 8)$	$(4, 8)$
$R(p_i, p_k, t)$	$(0, 8)$	$(0, 8)$	$(0, 8)$	$(1, 11)$	$(1, 11)$	$(1, 11)$	$(1, 11)$

Table 9.4. Vector Annotations of Process Time Chart Ex-4-3.

bf-relation	t_1	t_2	t_3	t_4	t_5	t_6	t_7
$R(p_i, p_j, t)$	$(0, 8)$	$(2, 8)$	$(2, 8)$	$(2, 10)$	$(2, 10)$	$(2, 10)$	$(2, 10)$
$R(p_j, p_k, t)$	$(0, 0)$	$(0, 8)$	$(0, 8)$	$(0, 8)$	$(2, 8)$	$(4, 8)$	$(4, 8)$
$R(p_i, p_k, t)$	$(0, 8)$	$(0, 8)$	$(0, 8)$	$(0, 12)$	$(0, 12)$	$(0, 12)$	$(0, 12)$

If we focus on the variation of the vector annotations of the three bf-literals at time t_3 in Table 9.2, Table 9.3 and Table 9.4, the following transitive bf-inference rule could be derived.

trule-4

$$R(p_i, p_j, t) : [(2, 8), \alpha] \wedge R(p_j, p_k, t) : [(2, 8), \alpha] \rightarrow R(p_i, p_j, t) : [(2, 8), \alpha].$$

Moreover, if we consider the variation of the vector annotations at time t_4 in Table 9.2, the following transitive bf-inference rule should be derived.

trule-5

$$R(p_i, p_j, t) : [(2, 10), \alpha] \wedge R(p_j, p_k, t) : [(2, 8), \alpha] \rightarrow R(p_i, p_k, t) : [(2, 10), \alpha].$$

As well as trule-5, the following two transitive bf-inference rules could be derived with focusing on the vector annotations at time t_4 in Table 9.3 and Table 9.4, respectively.

trule-6

$$R(p_i, p_j, t) : [(2, 10), \alpha] \wedge R(p_j, p_k, t) : [(2, 8), \alpha] \rightarrow R(p_i, p_k, t) : [(1, 11), \alpha],$$

trule-7

$$R(p_i, p_j, t) : [(2, 10), \alpha] \wedge R(p_j, p_k, t) : [(0, 8), \alpha] \rightarrow R(p_i, p_k, t) : [(0, 12), \alpha].$$

Here if we consider trule-5 and trule-6, they have the same body part,

$$R(p_i, p_j, t):[(2,10),\alpha] \wedge R(p_j, p_k, t):[(2,8),\alpha], \tag{9.12}$$

and different head parts,

$$R(p_i, p_k, t):[(2,10),\alpha], \quad \text{and} \quad R(p_i, p_k, t):[(1,11),\alpha].$$

This fact implies that such transitive bf-inference rules cannot be deterministically applied. We analyze the reason why there are more than two applicable transitive bf-inference rules at time t_4. If we focus on the application order of transitive bf-inference rules during time t_2 to time t_4, obviously there are two different application orders of transitive bf-inference rules in which trule-5 and trule-6 are applied after trule-4 and trule-3, respectively as follows.

Order-1 trule-3 $\longrightarrow$ trule-4 $\longrightarrow$ trule-5

Order-2 trule-3 $\longrightarrow$ trule-6

Moreover, whenever the same body, the bf-EVALP clause (9.12) of trule-5 and trule-6 is satisfied, the body,

$$R(p_i, p_j, t):[(2,10),\alpha] \wedge R(p_j, p_k, t):[(0,8),\alpha]$$

of trule-7 is also satisfied at time t_4, then its head, the bf-EVALP clause

$$R(p_i, p_k, t):[(0,12),\alpha]$$

can also be derivable. Thus we have another application order.

Order-3 rule-3 $\longrightarrow$ rule-7

Therefore, without considering the application order of transitive bf-inference rules we may not reason bf-relations correctly by transitive bf-inference rules.

Here we list all transitive bf-inference rules with taking their application orders into account. For simplicity, we represent a transitive bf-inference rule,

$$R(p_i, p_j, t):[(n_1,n_2),\alpha] \wedge R(p_j, p_k, t):[(n_3,n_4),\alpha] \rightarrow R(p_i, p_k, t):[(n_5,n_6),\alpha]$$

by vector annotations and logical connectives, $\wedge$ and $\rightarrow$, as follows:

$$(n_1,n_2) \wedge (n_3,n_4) \rightarrow (n_5,n_6)$$

in the list of transitive bf-inference rules.

Transitive Bf-inference Rules

TR0 $(0,0) \wedge (0,0) \rightarrow (0,0)$

TR1 $(0,8) \wedge (0,0) \rightarrow (0,8)$

TR1 − 1 $(0,12) \wedge (0,0) \rightarrow (0,12)$

$$\begin{aligned}
&\textbf{TR1}-\textbf{2} \quad (1,11)\wedge(0,8)\to(0,12)\\
&\textbf{TR1}-\textbf{3} \quad (1,11)\wedge(5,5)\to(1,11)\\
&\textbf{TR1}-\textbf{4} \quad (2,8)\wedge(0,8)\to(0,8)\\
&\quad\textbf{TR1}-\textbf{4}-\textbf{1} \quad (2,10)\wedge(0,8)\to(0,12)\\
&\quad\textbf{TR1}-\textbf{4}-\textbf{2} \quad (4,8)\wedge(0,12)\to(0,8) && (9.13)\\
&\quad\textbf{TR1}-\textbf{4}-\textbf{3} \quad (2,8)\wedge(2,8)\to(2,8)\\
&\qquad\textbf{TR1}-\textbf{4}-\textbf{3}-\textbf{1} \quad (2,10)\wedge(2,8)\to(2,10)\\
&\qquad\textbf{TR1}-\textbf{4}-\textbf{3}-\textbf{2} \quad (4,8)\wedge(2,10)\to(2,8) && (9.14)\\
&\qquad\textbf{TR1}-\textbf{4}-\textbf{3}-\textbf{3} \quad (2,8)\wedge(4,8)\to(4,8)\\
&\qquad\textbf{TR1}-\textbf{4}-\textbf{3}-\textbf{4} \quad (3,9)\wedge(2,10)\to(2,10)\\
&\qquad\textbf{TR1}-\textbf{4}-\textbf{3}-\textbf{5} \quad (2,10)\wedge(4,8)\to(3,9)\\
&\qquad\textbf{TR1}-\textbf{4}-\textbf{3}-\textbf{6} \quad (4,8)\wedge(3,9)\to(4,8)\\
&\qquad\textbf{TR1}-\textbf{4}-\textbf{3}-\textbf{7} \quad (3,9)\wedge(3,9)\to(3,9)\\
&\quad\textbf{TR1}-\textbf{4}-\textbf{4} \quad (3,9)\wedge(0,12)\to(0,12)\\
&\quad\textbf{TR1}-\textbf{4}-\textbf{5} \quad (2,10)\wedge(2,8)\to(1,11)\\
&\quad\textbf{TR1}-\textbf{4}-\textbf{6} \quad (4,8)\wedge(1,11)\to(2,8) && (9.15)\\
&\quad\textbf{TR1}-\textbf{4}-\textbf{7} \quad (3,9)\wedge(1,11)\to(1,11)\\
&\textbf{TR1}-\textbf{5} \quad (2,8)\wedge(5,5)\to(2,8)\\
&\quad\textbf{TR1}-\textbf{5}-\textbf{1} \quad (4,8)\wedge(5,7)\to(2,8) && (9.16)\\
&\quad\textbf{TR1}-\textbf{5}-\textbf{2} \quad (2,8)\wedge(7,5)\to(4,8)\\
&\quad\textbf{TR1}-\textbf{5}-\textbf{3} \quad (3,9)\wedge(5,7)\to(2,10)\\
&\quad\textbf{TR1}-\textbf{5}-\textbf{4} \quad (2,10)\wedge(7,5)\to(3,9)\\
&\textbf{TR2} \quad (5,5)\wedge(0,8)\to(0,8)\\
&\textbf{TR2}-\textbf{1} \quad (5,7)\wedge(0,8)\to(0,12)\\
&\textbf{TR2}-\textbf{2} \quad (7,5)\wedge(0,12)\to(0,8) && (9.17)\\
&\textbf{TR2}-\textbf{3} \quad (5,5)\wedge(2,8)\to(2,8)\\
&\quad\textbf{TR2}-\textbf{3}-\textbf{1} \quad (5,7)\wedge(2,8)\to(2,10)\\
&\quad\textbf{TR2}-\textbf{3}-\textbf{2} \quad (7,5)\wedge(2,10)\to(2,8) && (9.18)\\
&\quad\textbf{TR2}-\textbf{3}-\textbf{3} \quad (5,5)\wedge(4,8)\to(4,8)\\
&\quad\textbf{TR2}-\textbf{3}-\textbf{4} \quad (7,5)\wedge(3,9)\to(4,8)\\
&\textbf{TR2}-\textbf{4} \quad (5,7)\wedge(2,8)\to(1,11)\\
&\textbf{TR2}-\textbf{5} \quad (7,5)\wedge(1,11)\to(2,8) && (9.19)\\
&\textbf{TR3} \quad (5,5)\wedge(5,5)\to(5,5)\\
&\textbf{TR3}-\textbf{1} \quad (7,5)\wedge(5,7)\to(5,5) && (9.20)\\
&\textbf{TR3}-\textbf{2} \quad (5,7)\wedge(7,5)\to(6,6)
\end{aligned}$$

Note. The bottom vector annotation $(0, 0)$ in transitive bf-inference rules shows that any bf-EVALP clause $R(p_j, p_k, t) : [(n, m), \alpha]$ satisfies it.

Here we indicate two important points in terms of the list of transitive bf-inference rules.

(I) Names of transitive bf-inference rules such as TR1-4-3 show their application orders. For example, if transitive bf-inference rule TR1 has been applied, one of transitive bf-inference rules TR1-1,TR1-2,... or TR1-5 should be applied at the following stage; if transitive bf-inference rule TR1-4 has been applied after transitive bf-inference rule TR1, one of transitive bf-inference rules TR1-4-1,TR1-4-2,... or TR1-4-7 should be applied at the following stage; on the other hand, if one of transitive bf-inference rules TR1-1, TR1-2 or TR1-3 has been applied after transitive bf-inference rule TR1, there is no transitive bf-inference rule to be applied at the following stage because bf-annotations $\mathtt{db}(0, 12)$ or $\mathtt{mb}(1, 11)$ between processes Pr_i and Pr_k have already been derived.

(II) The eight transitive bf-inference rules,

TR1-4-2	(9.13),	TR2-2	(9.17),
TR1-4-3-2	(9.14),	TR2-3-2	(9.18),
TR1-4-6	(9.15),	TR2-5	(9.19)
TR1-5-1	(9.16),	TR3-1	(9.20)

have no following rule to be applied at the following stage, even though they cannot derive the final bf-relations between processes represented by bf-annotations such as $\mathtt{jb}(2, 10)/\mathtt{ja}(10, 2)$. For example, suppose that transitive bf-inference rule TR1-4-3-2 has been applied, then the vector annotation $(2, 8)$ of the bf-literal (p_i, p_k, t) just implies that the final bf-relation between processes Pr_i and Pr_k is one of three bf-annotations, $\mathtt{jb}(2, 10)$, $\mathtt{sb}(3, 9)$ and $\mathtt{ib}(4, 8)$. Therefore, if one of the eight transitive bf-inference rules has been applied, one of basic bf-inference rules $(0, 8)$-rule, $(2, 8)$-rule or $(5, 5)$-rule should be applied for deriving the final bf-annotation at the following stage. For instance, if transitive bf-inference rule TR1-4-3-2 has been applied, basic bf-inference rule $(2, 8)$-rule should be applied at the following stage.

9.4.4. *Example of Transitive Bf-relation Reasoning*

Now we show an example of bf-relation reasoning by transitive bf-inference rules taking the process time chart in Fig. 9.18.

At time t_1, transitive bf-inference rule TR1 is applied and we have the bf-EVALP clause,

$$R(p_i, p_k, t_1) : [(0, 8), \alpha].$$

At time t_2, transitive bf-inference rule TR1-2 is applied, however bf-literal $R(p_i, p_k, t_2)$ has the same vector annotation $(0, 8)$ as the previous time t_1. Therefore, we have the bf-EVALP clause,

$$R(p_i, p_k, t_2) : [(0, 8), \alpha].$$

At time t_3, no transitive bf-inference rule can be applied, since the vector annotations of bf-literals $R(p_i, p_j, t_3)$ and $R(p_j, p_k, t_3)$ are the same as the previous time t_2. Therefore, we still have the bf-EVALP clause having the same vector annotation,

$$R(p_i, p_k, t_3) : [(0, 8), \alpha].$$

At time t_4, transitive bf-inference rule TR1-2-1 is applied and we obtain the bf-EVALP clause having bf-annotation $\mathtt{db}(0, 12)$,

$$R(p_i, p_k, t_4) : [(0, 12), \alpha].$$

9.5. Application of Bf-EVALPSN to Process Order Verification

In this section, we present a simple example for application of the bf-relation reasoning system in bf-EVALPSN to process order verification.

Suppose that there is a pipeline system consists of two pipelines PIPELINE-1 and PIPELINE-2, which is also supposed to deal with four pipeline processes Pr_0, Pr_1, Pr_2 and Pr_3 according to the process schedule in Fig. 9.19. Moreover, the process order verification system has four safety properties $SPR - i(i = 0, 1, 2, 3)$ for processes $Pr_j(j = 0, 1, 2, 3)$ to be strictly assured, respectively.

SPR-0 process Pr_0 must start before any other processes, and process Pr_0 must finish before process Pr_2 finishes;
SPR-1 process Pr_1 must start after process Pr_0 starts;
SPR-2 process Pr_2 must start immediately after process Pr_1 finishes;
SPR-3 process Pr_3 must start immediately after processes Pr_0 and Pr_2 finish.

All safety properties $SPR - i(i = 0, 1, 2, 3)$ can be translated into the following bf-EVALPSN clauses.

[SPR-0]

$$\sim R(p_0, p_1, t) : [(0, 8), \alpha] \rightarrow st(p_1, t) : [\mathtt{f}, \beta], \tag{9.21}$$

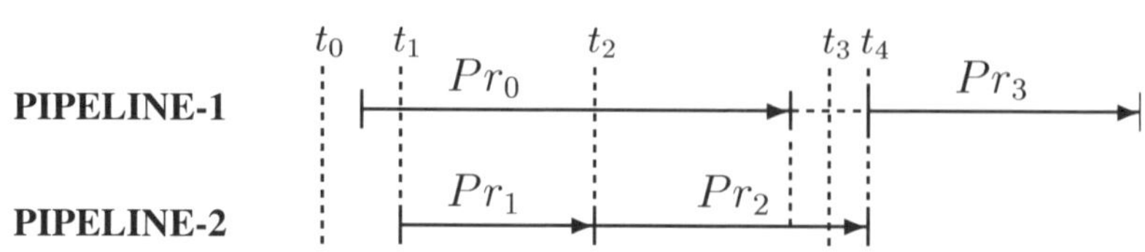

Fig. 9.19. Pipeline Process Schedule.

$$\sim R(p_0, p_2, t):[(0,8), \alpha] \rightarrow st(p_2, t):[\mathtt{f}, \beta], \tag{9.22}$$

$$\sim R(p_0, p_3, t):[(0,8), \alpha] \rightarrow st(p_3, t):[\mathtt{f}, \beta], \tag{9.23}$$

$$st(p_1, t):[\mathtt{f}, \beta] \wedge st(p_2, t):[\mathtt{f}, \beta] \wedge st(p_3, t):[\mathtt{f}, \beta]$$
$$\rightarrow st(p_0, t):[\mathtt{f}, \gamma], \tag{9.24}$$

$$\sim fi(p_0, t):[\mathtt{f}, \beta] \rightarrow fi(p_0, t):[\mathtt{f}, \gamma], \tag{9.25}$$

where the bf-EVALPSN clauses (9.21), (9.22) and (9.23) declare that if process Pr_0 has not started before each process $Pr_i(i = 1, 2, 3)$ starts, it should be forbidden to start each process $Pr_i(i = 1, 2, 3)$, respectively; the bf-EVALPSN clause (9.24) declares that if each process $Pr_i(i = 1, 2, 3)$ is forbidden from starting, it should be permitted to start process Pr_0; and the bf-EVALPSN clause (9.25) declares that if there is no forbiddance from finishing process Pr_0, it should be permitted to finish process Pr_0.

[SPR-1]

$$\sim st(p_1, t):[\mathtt{f}, \beta] \rightarrow st(p_1, t):[\mathtt{f}, \gamma], \tag{9.26}$$

$$\sim fi(p_1, t):[\mathtt{f}, \beta] \rightarrow fi(p_1, t):[\mathtt{f}, \gamma], \tag{9.27}$$

where the bf-EVALPSN clauses (9.26)/(9.27) declare that if there is no forbiddance from starting/finishing process Pr_1 respectively, it should be permitted to start/finish process Pr_1.

[SPR-2]

$$\sim R(p_2, p_1, t):[(11,0), \alpha] \rightarrow st(p_2, t):[\mathtt{f}, \beta], \tag{9.28}$$

$$\sim st(p_2, t):[\mathtt{f}, \beta] \rightarrow st(p_2, t):[\mathtt{f}, \gamma], \tag{9.29}$$

$$\sim R(p_2, p_0, t):[(10,2), \alpha] \rightarrow fi(p_2, t):[\mathtt{f}, \beta], \tag{9.30}$$

$$\sim fi(p_2, t):[\mathtt{f}, \beta] \rightarrow fi(p_2, t):[\mathtt{f}, \gamma], \tag{9.31}$$

where the bf-EVALPSN clause (9.28) declares that if process Pr_1 has not finished before process Pr_2 starts, it should be forbidden to start process Pr_2; the vector annotation $(11, 0)$ of bf-literal $R(p_2, p_1, t)$ is the greatest lower bound of the set $\{\mathtt{da}(12, 0), \mathtt{ma}(11, 1)\}$, which implies that process Pr_1 has finished before process Pr_2 starts in either way; the bf-EVALPSN clauses (9.29)/(9.31) declare that if there is no forbiddance from starting/finishing process Pr_2, it should be permitted to start/finish process Pr_2, respectively; and the bf-EVALPSN clauses (9.30) declares that if process Pr_0 has not finished before process Pr_2 finishes, it should be forbidden to finish process Pr_2.

[SPR-3]

$$\sim R(p_3, p_0, t):[(11,0), \alpha] \rightarrow st(p_3, t):[\mathtt{f}, \beta], \tag{9.32}$$

$$\sim R(p_3, p_1, t):[(11,0), \alpha] \rightarrow st(p_3, t):[\mathtt{f}, \beta], \tag{9.33}$$

$$\sim R(p_3, p_2, t) : [(11, 0), \alpha] \to st(p_3, t) : [\mathtt{f}, \beta], \tag{9.34}$$

$$\sim st(p_3, t) : [\mathtt{f}, \beta] \to st(p_3, t) : [\mathtt{f}, \gamma], \tag{9.35}$$

$$\sim fi(p_3, t) : [\mathtt{f}, \beta] \to fi(p_3, t) : [\mathtt{f}, \gamma], \tag{9.36}$$

where the bf-EVALPSN clauses (9.32),(9.33) and (9.34) declare that if none of processes $Pr_i(i = 0, 1, 2)$ has not finished, it should be forbidden to start process Pr_3; and the bf-EVALPSN clauses (9.35)/(9.36) declares that if there is no forbiddance from starting/finishing process Pr_3, it should be permitted to start/finish process Pr_3, respectively.

Now, we show how the process order verification in bf-EVALPSN is carried out in real-time by considering the process schedule in Fig. 9.19. In the following example, five bf-relations between processes Pr_0, Pr_1, Pr_2 and Pr_3 represented by the vector annotations of bf-literals,

$$R(p_0, p_1, t), \quad R(p_0, p_2, t), \quad R(p_0, p_3, t), \quad R(p_2, p_1, t), \quad R(p_3, p_2, t)$$

are verified in real-time based on safety properties $SPR - 0$, $SPR - 1$, $SPR - 2$ and $SPR - 3$.

Stage 0(at time t_0)

no process has started at time t_0, thus, the bf-EVALP clauses,

$$R(p_0, p_1, t_0) : [(0, 0), \alpha], \tag{9.37}$$

$$R(p_1, p_2, t_0) : [(0, 0), \alpha], \tag{9.38}$$

$$R(p_2, p_3, t_0) : [(0, 0), \alpha] \tag{9.39}$$

are obtained; also, the bf-EVALP clauses,

$$R(p_0, p_2, t_0) : [(0, 0), \alpha], \tag{9.40}$$

$$R(p_0, p_3, t_0) : [(0, 0), \alpha] \tag{9.41}$$

are obtained by transitive bf-inference rule TR0; then, the bf-EVALP clauses (9.37),(9.40) and (9.41) satisfy each body of the bf-EVALPSN clauses (9.21),(9.22) and (9.23) respectively, therefore, the forbiddance from starting each process $Pr_i(i = 1, 2, 3)$,

$$st(p_1, t_0) : [\mathtt{f}, \beta], \tag{9.42}$$

$$st(p_2, t_0) : [\mathtt{f}, \beta], \tag{9.43}$$

$$st(p_3, t_0) : [\mathtt{f}, \beta] \tag{9.44}$$

are derived; moreover, as the bf-EVALP clauses (9.42),(9.43) and (9.44) satisfy the body of the bf-EVALPSN clause (9.24), the permission for starting process Pr_0,

$$st(p_0, t_0) : [\mathtt{f}, \gamma]$$

is derived; therefore, process Pr_0 is permitted to start at this stage.

Stage 1(at time t_1)

process Pr_0 has already started but all other processes $Pr_i (i = 1, 2, 3)$ have not started yet; then the bf-EVALP clauses,

$$R(p_0, p_1, t_1) : [(0, 8), \alpha], \tag{9.45}$$
$$R(p_1, p_2, t_1) : [(0, 0), \alpha], \tag{9.46}$$
$$R(p_2, p_3, t_1) : [(0, 0), \alpha] \tag{9.47}$$

are obtained, where the bf-EVALP clause (9.45) is derived by basic bf-inference rule $(0, 0)$-rule-1; moreover, the bf-EVALP clauses,

$$R(p_0, p_2, t_1) : [(0, 8), \alpha], \tag{9.48}$$
$$R(p_0, p_3, t_1) : [(0, 8), \alpha] \tag{9.49}$$

are obtained by transitive bf-inference rule TR1; as the bf-EVALP clause (9.45) does not satisfy the body of the bf-EVALPSN clause (9.21), the forbiddance from starting process Pr_1,

$$st(p_1, t_1) : [\mathtt{f}, \beta] \tag{9.50}$$

cannot be derived; Then, as there is not the forbiddance (9.50), the body of the bf-EVALPSN clause (9.26) is satisfied, and the permission for starting process Pr_1,

$$st(p_1, t_1) : [\mathtt{f}, \gamma]$$

is derived; on the other hand, as the bf-EVALP clauses (9.48) and (9.49) satisfy the body of the bf-EVALPSN clauses (9.28) and (9.32) respectively, the forbiddance from starting both processes Pr_2 and Pr_3,

$$st(p_2, t_1) : [\mathtt{f}, \beta],$$
$$st(p_3, t_1) : [\mathtt{f}, \beta],$$

are derived; therefore, process Pr_1 is permitted to start.

Stage 2(at time t_2)

process Pr_1 has just finished and process Pr_0 has not finished yet; then, the bf-EVALP clauses,

$$R(p_0, p_1, t_2) : [(4, 8), \alpha],$$
$$R(p_1, p_2, t_2) : [(1, 11), \alpha],$$
$$R(p_2, p_3, t_2) : [(0, 8), \alpha]$$

are derived by basic bf-inference rules $(2, 8)$-rule-3, $(0, 8)$-rule-2 and $(0, 0)$-rule-1, respectively; moreover, the bf-EVALP clauses,

$$R(p_0, p_2, t_2) : [(2, 8), \alpha],$$
$$R(p_0, p_3, t_2) : [(0, 12), \alpha]$$

are obtained by transitive bf-inference rules TR1-4-6 and TR1-2, respectively; then, as the bf-EVALP clause (9.51) does not satisfy the body of the bf-EVALPSN clause (9.28), the forbiddance from starting process Pr_2,

$$st(p_2, t_2):[\mathtt{f}, \beta] \tag{9.51}$$

cannot be derived; as there is not the forbiddance (9.51), it satisfies the body of the bf-EVALPSN clause (9.29), and the permission for starting process Pr_2,

$$st(p_2, t_2):[\mathtt{f}, \gamma]$$

is derived; on the other hand, as the bf-EVALP clause (9.51) satisfies the body of the bf-EVALPSN clause (9.32), the forbiddance from starting process Pr_3,

$$st(p_3, t_2):[\mathtt{f}, \beta]$$

is derived; therefore, process Pr_2 is permitted to start, however process Pr_3 is still forbidden from starting.

Stage 3(at the t_3)

process Pr_0 has finished, process Pr_2 has not finished yet, and process Pr_3 has not started yet; then, the bf-EVALP clauses,

$$R(p_0, p_1, t_3):[(4, 8), \alpha],$$
$$R(p_1, p_2, t_3):[(1, 11), \alpha],$$
$$R(p_2, p_3, t_3):[(0, 8), \alpha],$$

which have the same vector annotations as the previous stage are obtained; moreover, the bf-EVALP clauses,

$$R(p_0, p_2, t_3):[(2, 10), \alpha],$$
$$R(p_0, p_3, t_3):[(0, 12), \alpha]$$

are obtained, where the bf-EVALP clause (9.52) is derived by the basic bf-inference rule $(2, 8)$-rule-1; then, the bf-EVALP clause (9.52) satisfies the body of the bf-EVALP clause (9.34), and the forbiddance from starting process Pr_3,

$$S(p_3, t_3):[\mathtt{f}, \beta]$$

is derived; therefore, process Pr_3 is still forbidden from starting because process Pr_2 has not finished yet at this stage.

Stage 4(at time t_4)

process Pr_2 has just finished and process Pr_3 has not started yet; then, the bf-EVALP clauses,

$$R(p_0, p_1, t_4):[(4,8), \alpha],$$
$$R(p_1, p_2, t_4):[(1,11), \alpha],$$
$$R(p_2, p_3, t_4):[(1,11), \alpha],$$
$$R(p_0, p_2, t_4):[(2,10), \alpha],$$
$$R(p_0, p_3, t_4):[(0,12), \alpha]$$

are obtained; the bf-EVALP clause (9.52) is derived by basic bf-inference rule $(0,8)$-rule-2; moreover, as the bf-EVALP clauses (9.52),(9.52) and (9.52) do not satisfy the bodies of the bf-EVALP clauses (9.32),(9.33) and (9.34), the forbiddance from starting process Pr_3,

$$st(p_3, t_4):[\mathtt{f}, \beta] \tag{9.52}$$

cannot be derived; as there is not the forbiddance (9.52), the body of the bf-EVALPSN clause (9.35) is satisfied, and the permission for starting process Pr_3,

$$st(p_3, t_4):[\mathtt{f}, \gamma]$$

is derived; Therefore, process Pr_3 is permitted to start because processes Pr_0, Pr_1 and Pr_2 have finished.

9.6. Conclusion and Remark

In this chapter, we have introduced a paraconsistent annotated logic program called bf-EVALPSN that can deal with verification for before-after relations between processes(time intervals), its reasoning system with two inference rules, basic bf-inference rules and transitive inference rules, and application to the safety verification for process order control.

Considering that EVALPSN can be implemented as not only software but also hardware, we would like to conclude that the bf-EVALPSN before-after relation verification system introduced in this chapter is very suitable for dealing with process order verification in real-time.

References

1. J. F. Allen, Towards a General Theory of Action and Time, *Artificial Intelligence*, **23**, 123–154, (1984).
2. J. F. Allen, G. Ferguson, Actions and Events in Interval Temporal Logic, *J.Logic and Computation*, **4**, 531–579, (1994).

3. K. R. Apt, H. A. Blair, A. Walker, Towards a theory of declarative knowledge. In J. Minker(ed.) *Foundation of Deductive Database and Logic Programs*, 89–148, (Morgan Kaufmann, 1989).
4. N. D. Belnap, A useful four valued logic. In M. Dunn, G. Epstein(eds.) *Modern Uses of Multiple-Valued Logic*, 8–37, (D. Reidel Publishing,1977).
5. D.Billington,(1993) Defeasible logic is stable, *J.Logic and Computation*, **3**, 379–400, (1993).
6. D. Billington, Conflicting literals and defeasible logic. In A. Nayak, M. Pagnucco(eds.) *Proc. 2nd Australian Workshop Commonsense Reasoning*, 1–15, Perth, Australia, (Dec.,1997).
7. H. A. Blair, V. S. Subrahmanian, Paraconsistent logic programming, *Theoretical Computer Science*, **68**, 135–154, (1989).
8. N. C. A. da Costa, V. S. Subrahmanian, C. Vago, The paraconsistent logics $P\mathcal{T}$, *Zeitschrift für Mathematische Logic und Grundlangen der Mathematik*, **37**, 139–148, (1989).
9. O. Dressler, (1988) An extended basic ATMS. In M. Reinfrank, *et al.* (eds.) *Proc. 2nd Int'l Workshop on Non-monotonic Reasoning*, *LNCS*, **346**, 143–163, Grassau, Germany, (Jun.,1988).
10. M. Fitting, Bilattice and the semantics of logic programming, *J.Logic Programming*, **11**, 91–116, (1991).
11. A. V. Gelder, K. A. Ross, J. S. Schlipf, The well-founded semantics for general logic programs, *J.ACM*, **38**, 620–650, (1991).
12. S. Jaskowski, Propositional calculus for contradictory deductive system (English translation of the original Polish paper), *Studia Logica*, **24**, 143–157, (1948).
13. M. Kifer, V. S. Subrahmanian, Theory of generalized annotated logic programming and its applications, *J.Logic Programming*, **12**, 335–368, (1992).
14. J. W. Lloyd, *Foundations of Logic Programming*, (Springer, 1987) 2nd edition.
15. R. Moore, Semantical considerations on non-monotonic logic, *Artificial Intelligence*, **25**, 75–94, (1985).
16. K. Nakamatsu, A. Suzuki, (1994) Annotated semantics for default reasoning. In R. Dai(ed.) *Proc. 3rd Pacific Rim Intl. Conf. Artificial Intelligence(PRICAI94)*, 180–186, Beijin, China,(Aug.,1994).
17. K. Nakamatsu, A. Suzuki, A nonmonotonic ATMS based on annotated logic programs. In W. Wobcke,*et al.* (eds.) *Agents and Multi-Agents Systems*, *LNAI*, **1441**, 79–93, (1998).
18. K. Nakamatsu, J. M. Abe, Reasonings based on vector annotated logic programs. In M. Mohammadian M(ed.) *Computational Intelligence for Modelling, Control & Automation (CIMCA99), Concurrent Systems Engineering Series*, **55**, 396–403, (IOS Press, Feb.,1999).
19. K. Nakamatsu, J. M. Abe, A. Suzuki, Defeasible reasoning between conflicting agents based on VALPSN. In C. Tessier, L. Chaudron(eds.) *Proc. AAAI Workshop Agents' Conflicts*, 18 July, 20–27, Orlando, FL, USA, (AAAI Press, July, 1999).
20. K. Nakamatsu, J. M. Abe, A. Suzuki, Defeasible reasoning based on VALPSN and its application. In A. Nayak, M. Pagnucco(eds.), *Proc. Third Australian Commonsense Reasoning Workshop*, 114–130, Sydney, Australia, (Dec.,1999).
21. K. Nakamatsu, On the relation between vector annotated logic programs and defeasible theories, *Logic and Logical Philosophy*, **8**, 181–205, (2000).
22. K. Nakamatsu, J. M. Abe, A. Suzuki, A defeasible deontic reasoning system based on annotated logic programming. In D. M. Dubois(ed.) *Proc. 4th Intl. Conf. Computing Anticipatory Systems(CASYS2000), AIP Conference Proceedings*, **573**, 609–620, (2001).

23. K. Nakamatsu, J. M. Abe, A. Suzuki, Annotated semantics for defeasible deontic reasoning. In W. Ziarko, Y. Yao(eds.) *Proc. 2nd Intl. Conf. Rough Sets and Current Trends in Computing(RSCTC2000), LNAI*, **2005**, 432–440, (2001).
24. K. Nakamatsu, J. M. Abe, A. Suzuki, Extended vector annotated logic program and its application to robot action control and safety verification. In A. Abraham, *et al.* (eds.) *Hybrid Information Systems (Advances in Soft Computing Series)*, 665–680, (Physica-Verlag, 2002).
25. K. Nakamatsu, H. Suito, J. M. Abe, A. Suzuki, Paraconsistent logic program based safety verification for air traffic control. In A. El Kamel,*et al.* (eds.) *Proc. IEEE Intl. Conf. System, Man and Cybernetics 02(SMC02)*, Hammamet, Tunisia, (IEEE, 2002), CD-ROM.
26. K. Nakamatsu, J. M. Abe, A. Suzuki, A railway interlocking safety verification system based on abductive paraconsistent logic programming. In A. Abraham,*et al.* (eds.) *Soft Computing Systems(HIS02), Frontiers in Artificial Intelligence and Applications*, **87**, 775–784, (IOS Press, 2002).
27. K. Nakamatsu, J. M. Abe, A. Suzuki, Defeasible deontic robot control based on extended vector annotated logic programming. In D. M. Dubois(ed.) *Proc. 5th Intl. Conf. Computing Anticipatory Systems(CASYS2001), AIP Conference Proceedings*, **627**, 490–500, (2002).
28. K. Nakamatsu, Y. Mita, T. Shibata, Defeasible deontic action control based on paraconsistent logic program and its hardware application. In M. Mohammadian(ed.) *Proc. Intl. Conf. Computational Intelligence for Modelling Control and Automation 2003(CIMCA2003)*, Vienna, Austria, (IOS Press, Feb., 2003), CD-ROM.
29. K. Nakamatsu, T. Seno, J. M. Abe, A. Suzuki, Intelligent real-time traffic signal control based on a paraconsistent logic program EVALPSN. In G. Wang, *et al.* (eds.) *Rough Sets, Fuzzy Sets, Data Mining and Granular Computing(RSFDGrC2003), LNAI*, **2639**, 719–723, (2003).
30. K. Nakamatsu, H. Komaba, A. Suzuki, Defeasible deontic control for discrete events based on EVALPSN. In S. Tsumoto, *et al.* (eds.) *Proc. 4th Intl. Conf. Rough Sets and Current Trends in Computing(RSCTC2004), LNAI*, **3066**, 310–315, (2004).
31. K. Nakamatsu, R. Ishikawa, A. Suzuki, A paraconsistent based control for a discrete event cat and mouse. In M. G. H. Negoita,*et al.* (eds.) *Proc. 8th Intl. Conf. Knowledge-Based Intelligent Information and Engineering Systems(KES2004), LNAI*, **3214**, 954–960, (2004).
32. K. Nakamatsu, S-L, Chung, H. Komaba, A. Suzuki, A discrete event control based on EVALPSN stable model. In D. Slezak,*et al.* (eds.) *Rough Sets, Fuzzy Sets, Data Mining and Granular Computing(RSFDGrC2005), LNAI*, **3641**, 671–681, (2005).
33. K. Nakamatsu, J. M. Abe, S. Akama, An intelligent safety verification based on a paraconsistent logic program. In R. Khosla,*et al.* (eds.) *Proc. 9th Intl. Conf. Knowledge-Based Intelligent Information and Engineering Systems(KES2005), LNAI*, **3682**, 708–715, (2005).
34. K. Nakamatsu, K. Kawasumi, A. Suzuki, Intelligent verification for pipeline based on EVALPSN. In K. Nakamatsu, J. M. Abe(eds.) *Advances in Logic Based Intelligent Systems Frontiers in Artificial Intelligence and Applications*, **132**, 63–70, (IOS Press, 2005).
35. K. Nakamatsu, A. Suzuki, Autoepistemic theory and paraconsistent logic program. In K. Nakamatsu, J. M. Abe(eds.) *Advances in Logic Based Intelligent Systems Frontiers in Artificial Intelligence and Applications*, **132**, 177–184, (IOS Press, 2005).
36. K. Nakamatsu, A. Suzuki, Annotated semantics for non-monotonic reasonings in artificial intelligence – I, II, III, IV. In K. Nakamatsu, J. M. Abe(eds.) *Advances in Logic*

Based Intelligent Systems Frontiers in Artificial Intelligence and Applications, **132**, 185–215, (IOS Press, 2005).
37. K. Nakamatsu, Pipeline valve control based on EVALPSN safety verification, *J.Advanced Computational Intelligence and Intelligent Informatics*, **10**, 647-656, (2006).
38. K. Nakamatsu, Y. Mita, T. Shibata, An intelligent action control system based on extended vector annotated logic program and its hardware implementation, *J.Intelligent Automation and Soft Computing*, **13**, 289-304, (2007).
39. K. Nakamatsu, Paraconsistent annotated logic program EVALPSN and its application. In J. A. Fulcher, L. C. Jain(eds.) *Computational Intelligence: A Compendium Studies in Computational Intelligence*, **115**, 233-306, (Springer, 2008).
40. K. Nakamatsu, S. Akama, J. M. Abe, Transitive reasoning of before-after relation based on bf-EVALPSN. In *Proc. 12th Intl. Conf. Knowledge-based Intelligent Information and Engineering Systems(KES2008), LNCS*, **5178**, 474-482, (2008).
41. K. Nakamatsu, J. M. Abe, The development of Paraconsistent Annotated Logic Program, *Int'l J. Reasoning-based Intelligent Systems*, **1**, 92-112, (2009).
42. D. Nute, Defeasible reasoning. In E. A. Stohr, *et al.* (eds.) *Proc. 20th Hawaii Intl. Conf. System Science(HICSS87)*, **1**, 470–477, Hawaii,USA,(Jan., 1987).
43. D. Nute, Basic defeasible logics. In L. F. del Cerro, M. Penttonen(eds.) *Intensional Logics for Programming*, 125–154, (Oxford University Press, 1992).
44. D. Nute, Defeasible logic. In D. M. Gabbay,*et al.* (eds.) *Handbook of Logic in Artificial Intelligence and Logic Programming*, **3**, 353–396, (Oxford University Press, 1994).
45. D. Nute, Apparent obligatory. In D. Nute(ed.) *Defeasible Deontic Logic, Synthese Library*, **263**, 287–316, (Kluwer Academic Publisher, 1997).
46. T. C. Przymusinski, On the declarative semantics of deductive databases and logic programs. In J. Minker(ed.) *Foundation of Deductive Database and Logic Programs*, 193–216, (Morgan Kaufmann,1988).
47. R. Reiter, A logic for default reasoning, *Artificial Intelligence*, **13**, 81–123, (1980).
48. J. C. Shepherdson, Negation as failure, completion and stratification. In D. M. Gabbay, *et al.* (eds.) *Handbook of Logic in Artificial Intelligence and Logic Programming*, **5**, 356–419, (Oxford University Press, 1998).
49. V. S. Subrahmanian, Amalgamating knowledge bases, *ACM Trans.Database Systems*, **19**, 291–331, (1994).
50. V. S. Subrahmanian, On the semantics of qualitative logic programs. In *Proc. the 1987 Symp. Logic Programming(SLP87)*, 173–182, San Francisco, USA, (IEEE, Sep., 1987).
51. A. Visser, Four valued semantics and the liar, *J.Philosophical Logic*, **13**, 99–112, (1987).

Chapter 10

INSPECTING AND PREFERRING ABDUCTIVE MODELS

Luís Moniz Pereira*, Pierangelo Dell'Acqua†, Alexandre Miguel Pinto‡ and Gonçalo Lopes§

*CENTRIA — Centro de Inteligência Artificial
Universidade Nova de Lisboa, 2829-516 Caparica, Portugal
cmp@fct.unl.pt

†ITN — Department of Science and Technology
Linköping University, 60174 Norrköping, Sweden
pierangelo.dellacqua@liu.se

‡CISUC — Centro de Informática e Sistemas
Universidade de Coimbra, Pólo II, Pinhal de Marrocos
3030-290 Coimbra, Portugal
ampinto@dei.uc.pt

§CENTRIA — Centro de Inteligência Artificial
Universidade Nova de Lisboa, 2829-516 Caparica, Portugal
goncaloclopes@gmail.com

This work proposes the application of preferences over abductive logic programs as an appealing declarative formalism to model choice situations. In particular, both a priori and a posteriori handling of preferences between abductive extensions of a theory are addressed as complementary and essential mechanisms in a broader framework for abductive reasoning. Furthermore, both of these choice mechanisms are combined with other formalisms for decision making, like economic decision theory, resulting in theories containing the best advantages from both qualitative and quantitative formalisms. Several examples are presented throughout to illustrate the enounced methodologies. These have been tested in our implementation, which we explain in detail.

10.1. Introduction

Much work in logic program semantics and procedures has focused on preferences between rules of a theory[1] and among theory literals,[2,3] with or without updates. However, the exploration of the application of preferences to abductive extensions of a theory has still much to progress. In our perspective, handling preferences over abductive logic programs has several advantages, and allows for easier and more concise translation into normal logic programs (NLP) than those prescribed by more general and complex rule preference frameworks.

We argue that preferring among abductive extensions is really a much more appealing formalism than that of hard or soft constraints on program literals, since the latter represent formal conclusions to the program, which are often not defeasible. An abductive extension is, by definition, a defeasible construct, and allows greater flexibility in enforcing preference relations. In,[4] a preliminary theory of revisable preferences between abducible literals was presented, along with a formal semantics based on the definition of *abductive stable models*. In this work we extend the theoretical framework thence proposed, addressing many problems and limitations that remained to be solved.

We also propose to broaden the framework to account for more flexible and powerful means to express preferences between abducibles, besides a priori relevancy rules embedded in a program's theory. In fact, we intend to show that there are many advantages as well to prefer a posteriori, i.e. to enact preferences on the computed models, after the consequences of opting for one or another abducible are known. Furthermore, we combine both of these choice mechanisms with other formalisms for decision making, like economic decision theory, resulting in theories containing the best advantages from both qualitative and quantitative formalisms.

10.2. Abductive Framework

10.2.1. *Basic Abductive Language*

[5]Let $\mathcal{L}$ be a first order propositional language defined as follows. Assume given an alphabet (set) of propositional atoms containing the reserved atom $\bot$ to denote falsity. A literal in $\mathcal{L}$ is an atom A or its default negation $not A$, the latter expressing that the atom is false by default (CWA).

Definition 10.1. A rule in $\mathcal{L}$ takes the form $A \leftarrow L_1, \ldots, L_t$ where A is an atom and $L_1, \ldots, L_t$ $(t \geq 0)$ are literals.

We follow the standard convention and call A the head of the rule, and the conjunction $L_1, \ldots, L_t$ its body. When $t = 0$ we write the rule simply as A, that is without '$\leftarrow$'. An integrity constraint is a rule whose head is $\bot$.

Definition 10.2. A goal or query in $\mathcal{L}$ takes the form $?{-}L_1, \ldots, L_t$, where $L_1, \ldots, L_t$ $(t \geq 1)$ are literals.

A (logic) program P over $\mathcal{L}$ is a finite (countable) set of rules.

Notation 10.1. We adopt the convention of using ';' to separate rules, thus we write a program P as $\{\text{rule}_1; \ldots; \text{rule}_n\}$.

Every program P is associated with a set of abducibles $\mathcal{A}$ consisting of literals which (without loss of generality) do not appear in any rule head of P. Abducibles may be thought of as hypotheses that can be used to extend the current theory in order to provide hypothetical solutions or possible explanations for given queries.

Given an abductive solution, to test whether a certain abducible has been abduced, $\mathcal{L}$ contains the reserved abducible *abduced*(a), for every abducible $a \neq$ *abduced*$(.)$ in $\mathcal{L}$. Thus, *abduced*(a) acts as a constraint that is satisfied in the solution if the abducible a is indeed assumed. It can be construed as meta-abduction in the form of abducing to check (or passively verify) that a certain abduction is adopted.

Example 10.1. Let $P = \{p \leftarrow abduced(a), a; q \leftarrow abduced(b)\}$ with set of abducibles $\mathcal{A}_P = \{a, b, abduced(a), abduced(b)\}$. Then, P has four intended models: $M = \{\}$, $M_2 = \{p, a, abduced(a)\}$, $M_3 = \{q, b, abduced(b)\}$, and $M_4 = \{p, q, a, b, abduced(a), abduced(b)\}$. The set $\{q, abduced(b)\}$ is not an intended model since the assumption of $abduced(b)$ requires the assumption of b.

Given a set of abducibles $\mathcal{A}$, we write $\mathcal{A}^*$ to indicate the subset of $\mathcal{A}$ consisting of all the abducibles in $\mathcal{A}$ distinct from $abduced(.)$, that is:

$$\mathcal{A}^* = \{a : a \neq abduced(.) \text{ and } a \in \mathcal{A}\}.$$

10.2.1.1. *Hypotheses Generation*

The production of alternative explanations for a query is a central problem in abduction, because of the combinatorial explosion of possible explanations. Thus, it is important to generate only those abductive explanations which are relevant for the problem at hand. Several approaches have thus far been proposed, often based on some global criteria, which has the drawback of generally being domain independent and computationally expensive. An alternative to global criteria for competing alternative assumptions is to allow the theory to contain rules encoding domain specific information about which particular assumptions are to be considered in a particular situation.

In our approach, preferences among abducibles can be expressed in order to discard unwanted assumptions. Technically, preferences over alternative abducibles will be coded as constraints over even cycles of default negation, under Stable Model semantics,[6] and triggering one of the preference rules will break the cycle in favor of one abducible or another. The notion of expectation is employed to express preconditions for enabling the assumption of an abducible. An abducible can be assumed only if there is an expectation for it, and there is not an expectation to the contrary. In this case, we say that the abducible is *considered*. These expectations are expressed by the following rules, for any given abducible $a \in \mathcal{A}^*$:

$$expect(a) \leftarrow L_1, \ldots, L_t$$
$$expect\_not(a) \leftarrow L_1, \ldots, L_t$$

Note that $\mathcal{L}$ does not contain atoms of the form $expect(abduced(.))$ and $expect\_not(abduced(.))$.

Example 10.2. Let $P = \{p \leftarrow a; q \leftarrow b; expect(a); expect(b); expect\_not(a) \leftarrow q\}$ with set of abducibles $\mathcal{A}_P = \{a, b, abduced(a), abduced(b)\}$. Then, P has three intended models: $M = \{expect(a), expect(b)\}$, $M_2 = \{p, a, abduced(a), expect(a), expect(b)\}$ and $M_3 = \{q, b, abduced(b), expect(a), expect(b)\}$. It is not possible to assume both a and b because the assumption of b makes q true which in turn makes $expect\_not(a)$ true preventing a to be assumed.

This notion of considered abducible allows us to divide the abductive process into two distinct moments: the generation of hypotheses and the pruning of the unpreferred ones. Computation of preferences between models is problematic when both the generation and comparison get mixed up, as already mentioned in,[3] but in our case we introduce a middle-man condition instead of two distinct computations.

10.2.1.2. *Enforced Abduction*

To express that the assumption of an abducible enforces the assumption of another abducible, $\mathcal{L}$ contains reserved atoms of the form $a \prec b$, for any abducibles $a, b \in \mathcal{A}^*$. The atom $a \prec b$ states that the assumption of b enforces the assumption of a, active abduction behavior. That is, if b is assumed, then $a \prec b$ forces a to be assumed provided that a can be considered. Note that the abducibles a, b are both required to be different from *abduced*(.) since they belong to $\mathcal{A}^*$. Removing this requirement would not add to the expressive power of $\mathcal{L}$.

Example 10.3. Let $P = \{p \leftarrow a; a \prec b; b \prec a; expect(a); expect(b)\}$ with set of abducibles $\mathcal{A}_P = \{a, b, abduced(a), abduced(b)\}$. Then, P has two intended models: $M = \{a \prec b, b \prec a, expect(a), expect(b)\}$ and $M_2 = \{a \prec b, b \prec a, a, b, p, expect(a), expect(b), abduced(a), abduced(b)\}$. This is due to the active abduction behavior of $a \prec b$ and $b \prec a$ that prevents intended models containing either a or b.

10.2.1.3. *Conditional Abduction*

The assumption of an abducible a can be conditional on the assumption of another abducible b. The reserved atom $a \sqsubset b$ in $\mathcal{L}$, for any abducible $a, b \in \mathcal{A}^*$, states that a can be assumed only if b is (without assuming it for the purpose of having a), passive abduction behavior. That is, $a \sqsubset b$ acts as a check passively constraining the assumption of a to the assumption of b (passive abduction behavior). Note that the abducibles a, b are required to be different from *abduced*(.). Removing this requirement would not add to the expressive power of $\mathcal{L}$.

Example 10.4. Let $P = \{p \leftarrow a; q \leftarrow b; a \sqsubset b; expect(a)\}$ with set of abducibles $\mathcal{A}_P = \{a, b, abduced(a), abduced(b)\}$. Then, there exists only one intended model of P: $M = \{a \sqsubset b, expect(a)\}$. Note that $M_2 = \{a \sqsubset b, b, q, expect(a), abduced(b)\}$ and $M_3 = \{a \sqsubset b, a, p, expect(a), abduced(a)\}$ are not intended models. In fact, M_2 is not a model since b cannot be assumed (there is no expectation for it). This fact also prevents the assumption of a (due to $a \sqsubset b$) and consequently M_3 is not a model.

Example 10.5. Let $P = \{p \leftarrow a; q \leftarrow b; a \sqsubset b; expect(a); expect(b)\}$ with abducibles $\mathcal{A}_P = \{a, b, abduced(a), abduced(b)\}$. Then, P has three intended models: $M = \{a \sqsubset b, expect(a), expect(b)\}$, $M_2 = \{a \sqsubset b, b, q, expect(a), expect(b), abduced(b)\}$ and $M_3 = \{a \sqsubset b, a, b, p, q, expect(a), expect(b), abduced(a), abduced(b)\}$.

Mark that the global $a \sqsubset b \leftarrow cond$ can be avoided altogether by replacing every occurrence of the abducible a with $cond, a, abduced(b)$. In fact, when one desires some occurrences of a to be subjected to conditional abduction and others not, then only some occurrences need be replaced as above.

10.2.1.4. *Cardinality Constrained Abduction*

To constrain the number of assumed abducibles, $\mathcal{L}$ contains reserved atoms of the form $L\ \{l_1, \ldots, l_n\}\ U$ where $n \geq 1$, every l_i is an abducible in $\mathcal{A}$, and L and U are natural numbers representing, respectively, the lower and upper bounds on the cardinality of abducibles. $L\ \{l_1, \ldots, l_n\}\ U$ states that at least L and at most U abducibles in $\{l_1, \ldots, l_n\}$ must be assumed (active abduction behavior). Since the abducibles l_i belong to $\mathcal{A}$, they can also take the form $abduced(.)$.

10.2.2. *Declarative Semantics*

The declarative semantics of programs over $\mathcal{L}$ is given in terms of abductive stable models. Before introducing them, we need few definitions. In the following we let P be a program and $\mathcal{A}_P$ the abducibles in P.

A 2-valued interpretation M of $\mathcal{L}$ is any set of literals from $\mathcal{L}$ that satisfies the condition that, for any atom A, precisely one of the literals A or $not\ A$ belongs to M. We say that an interpretation M satisfies a conjunction of literals $L_1, \ldots, L_t$, if every literal L_i in the conjunction belongs to M. We also need to introduce the notion of default assumptions of P with respect to an interpretation M, where a default literal $not\ A$ is considered an atom, and $not\ not\ A \equiv A$.

$$Default(P, M)$$
$$= \{not\ A : \text{there exists no rule } A \leftarrow L_1, \ldots, L_t \text{ in } P \text{ such that } M \models L_1, \ldots, L_t\}$$

Abducibles are false by default since we made the assumption that abducibles are not defined by any rule in P. An interpretation M is a stable model of P iff:

(1) $M \not\models \bot$
(2) $M = least(P \cup Default(P, M))$, where *least* indicates the least model

Let C be $L\,\{l_1, \ldots, l_n\}\,U$. Then, we let $W(C, M)$ be the number of abducibles in $\{l_1, \ldots, l_n\}$ satisfied by an interpretation M:

$$W(C, M) = |\,\{l : l \in \{l_1, \ldots, l_n\} \text{ and } M \models l\}\,|$$

Given a set of abducibles Δ, we write Δ^* to indicate:

$$\Delta^* = \{a\ :\ a \neq abduced(.) \text{ and } a \in \Delta\}$$

Definition 10.3. Let $\Delta \subseteq \mathcal{A}_P$ be a set of abducibles. M is an abductive stable model with hypotheses Δ of P iff:

(1) $M \not\models \bot$
(2) $M = least(Q \cup Default(Q, M))$, where $Q = P \cup \Delta$
(3) $M \models expect(a)$ and $M \not\models expect\_not(a)$, for every $a \in \Delta^*$
(4) for every $a \in \Delta^*$, if $M \models a$ then $M \models abduced(a)$
(5) for every atom $a \prec b$, if $M \models a \prec b$, $M \models expect(a)$, $M \not\models expect\_not(a)$ and $M \models b$, then $M \models a$
(6) for every atom C of the form $L\{l_1, \ldots, l_n\}U$, if $M \models C$ then $L \leq W(C, M) \leq U$
(7) for every $a \in \Delta^*$, if $M \models abduced(a)$ then $M \models a$
(8) for every atom $a \sqsubset b$, if $M \models a \sqsubset b$ and $M \models a$, then $M \models b$

Example 10.6. Let $P = \{p \leftarrow abduced(a); expect(a)\}$ and $\mathcal{A}_P = \{a, abduced(a)\}$. Then, $M = \{p, a, abduced(a), expect(a)\}$ is an abductive stable model with hypotheses $\Delta = \{a, abduced(a)\}$, while $M_2 = \{p, abduced(a), expect(a)\}$ is not since condition (7) of Def. 10.3 is not fulfilled.

Definition 10.4. Let G be a goal. Then, Δ is an abductive explanation for G in P iff:

(1) M is an abductive stable model with hypotheses Δ of P, and
(2) $M \models G$

Definition 10.5. Let G be a goal. Then, Δ is a strict abductive explanation for G in P iff

(1) Δ is a minimal set for which:
  - $M \models G$
  - $M \not\models \bot$
  - $M = least(Q \cup Default(Q, M))$, where $Q = P \cup \Delta$
  - $M \models expect(a)$ and $M \not\models expect\_not(a)$, for every $a \in \Delta^*$
  - for every $a \in \Delta^*$, if $M \models a$ then $M \models abduced(a)$
  - for every atom $a \prec b$, if $M \models a \prec b$, $M \models expect(a)$, $M \not\models expect\_not(a)$ and $M \models b$, then $M \models a$
  - for every atom C of the form $L\{l_1, \ldots, l_n\}U$, if $M \models C$ then $L \leq W(C, M) \leq U$
(2) for every $a \in \Delta^*$, if $M \models abduced(a)$ then $M \models a$
(3) for every atom $a \sqsubset b$, if $M \models a \sqsubset b$ and $M \models a$, then $M \models b$

Note that in Def. 10.5 condition (2) is not subject to minimization. The reason for this is clarified by the next example.

Example 10.7. Reconsider the program P of Example 10.6. Suppose that the goal G is $?{-}p$. It holds that $\Delta = \{a, abduced(a)\}$ is an abductive explanation for G in P, but it is not strict since Δ is not a minimal set satisfying condition (1) of Def. 10.5. Indeed, the minimal set is $\Delta_2 = \{abduced(a)\}$. Hence, there exists no strict abductive explanation for G in P.

The following result relates abductive explanations and strict abductive explanations.

Proposition 10.1. *Let G be a goal and Δ a strict abductive explanation for G in P. Then, Δ is an abductive explanation for G in P.*

Proof. It follows immediately from Definitions 10.4 and 10.5. □

It is easy to see that the converse of Prop 10.1 does not hold since abductive explanations are not subject to the minimality requirement.

10.3. Pragmatics

10.3.1. *Constraining Abduction*

Quite often in domain problems it happens that the assumption of abducibles is subject to the fulfillment of certain conditions, including other assumptions, which must be satisfied. This requirement can be expressed in our framework by exploiting constrained abduction, $a \sqsubset b$. It states that the assumption of the abducible a is subject to the assumption of b with the property that b cannot be assumed in order to assume a.

Example 10.8. Consider a scenario where there is a pub that is open or closed. If the light is on in the pub then it is open or being cleaned. In case it is late night, one can assume that the pub is open if there are people inside. The pub being located in an entertainment district, there is noise around if there is people in the pub or a party nearby. This scenario can be described by the following program P with $\mathcal{A}_P = \{open, cleaning, party, people, abduced(open), abduced(cleaning), abduced(party), abduced(people)\}$.

$$\begin{array}{l} light \leftarrow open, not\ cleaning \\ light \leftarrow cleaning, not\ open, not\ abduced(people) \\ open \sqsubset people \leftarrow late\_night \\ noise \leftarrow party \\ noise \leftarrow people \end{array}$$

Thus, in case it is night (but not late night) and one does observe lights in the pub, then one has two equally plausible explanations for it: $\{open\}$ or $\{cleaning\}$. Otherwise (it is late night), then there is only one explanation for the lights being turned on: $\{cleaning\}$. If instead it is late night and one hears noise also (that is, the query is $?\!-light, noise$), then one will now have three abductive explanations: $\{open, people\}$, $\{cleaning, party\}$ and $\{open, party, people\}$. The last explanation reflects the fact that the pub may be open with late customers simultaneously with a party nearby, both events producing noise.

10.3.2. *Preferring Abducibles*

In this section we illustrate how to express preferences between considered abducibles. To do so, we employ the construct $L\,\langle l_1,\ldots,l_n\rangle\,U$ to constrain the number of abducibles assumed (without having to assume it). Such a construct has a passive abduction behavior and is defined as:

$$L\,\langle l_1,\ldots,l_n\rangle\,U \equiv L\,\{\mathit{abduce}(l_1),\ldots,\mathit{abduce}(l_n)\}\,U$$

for any abducible $l_1,\ldots,l_n$ in $\mathcal{A}^*$. The following two examples illustrate the difference between $L\,\langle l_1,\ldots,l_n\rangle\,U$ and $L\,\{l_1,\ldots,l_n\}\,U$.

Example 10.9. Let $P = \{p \leftarrow \mathit{abduced}(a); 1\,\{a,b\}\,1; \mathit{expect}(a); \mathit{expect}(b)\}$ with set of abducibles $\mathcal{A}_P = \{a, b, \mathit{abduced}(a), \mathit{abduced}(b)\}$. Then, P has two abductive stable models: $M = \{p, a, 1\,\{a,b\}\,1, \mathit{abduced}(a), \mathit{expect}(a), \mathit{expect}(b)\}$ and $M_2 = \{b, 1\,\{a,b\}\,1, \mathit{abduced}(b), \mathit{expect}(a), \mathit{expect}(b)\}$. It holds that $\Delta = \{a, \mathit{abduced}(a)\}$ is a strict abductive explanation for $?\!-p$ in P.

Example 10.10. Let $Q = \{p \leftarrow \mathit{abduced}(a); 1\,\langle a,b\rangle\,1; \mathit{expect}(a); \mathit{expect}(b)\}$ with set of abducibles $\mathcal{A}_Q = \{a, b, \mathit{abduced}(a), \mathit{abduced}(b)\}$. Then, Q has two abductive stable models: $M = \{p, a, 1\,\langle a,b\rangle\,1, \mathit{abduced}(a), \mathit{expect}(a), \mathit{expect}(b)\}$ and $M_2 = \{b, 1\,\langle a,b\rangle\,1, \mathit{abduced}(b), \mathit{expect}(a), \mathit{expect}(b)\}$. In contrast to the program P of Example 10.9, there exists no strict abductive explanation for $?\!-p$ in Q.

Consider for example a situation where there are three alternative abducibles a, b and c can explain an observation p. Suppose that a and c are preferred to b. This situation can be axiomatized by the following program P with abducibles $\mathcal{A}_P = \{a, b, c, \mathit{abduced}(a), \mathit{abduced}(b), \mathit{abduced}(c)\}$:[a]

$$P = \{p \leftarrow a; p \leftarrow b; p \leftarrow c; a \prec b; c \prec b; 0\,\langle a,b,c\rangle\,1; \mathit{expect}(a); \mathit{expect}(b); \mathit{expect}(c)\}$$

P has three abductive stable models:

$$M = \{a \prec b, c \prec b, 0\,\langle a,b,c\rangle\,1, \mathit{expect}(a), \mathit{expect}(b), \mathit{expect}(c)\}$$
$$\Delta = \{\}$$

$$M_2 = \{a \prec b, c \prec b, 0\,\langle a,b,c\rangle\,1, a, p, \mathit{expect}(a), \mathit{expect}(b), \mathit{expect}(c), \mathit{abduced}(a)\}$$
$$\Delta_2 = \{a, \mathit{abduced}(a)\}$$

$$M_3 = \{a \prec b, c \prec b, 0\,\langle a,b,c\rangle\,1, c, p, \mathit{expect}(a), \mathit{expect}(b), \mathit{expect}(c), \mathit{abduced}(c)\}$$
$$\Delta_3 = \{c, \mathit{abduced}(c)\}$$

The strict abductive explanations for $?\!-p$ in P are Δ_2 and Δ_3. It is worth noting that the abducible b cannot be part of any abductive explanation. In fact, if it were the case, the rules $a \prec b$ and $c \prec b$ in P would enforce the abduction of a and c (respectively) violating therefore the cardinality constraint $0\,\langle a,b,c\rangle\,1$.

[a]Note that if the preference relation is a strict partial order, then it must be axiomatized explicitly.

Example 10.11. Consider a situation[b] where Claire drinks either tea or coffee (but not both). Suppose that Claire prefers coffee over tea when sleepy, and doesn't drink coffee when she has high blood pressure. This situation can be represented by a program P over $\mathcal{L}$ with set of abducibles $\mathcal{A}_P = \{tea, coffee, abduced(tea), abduced(coffee)\}$:

$$drink \leftarrow tea$$
$$drink \leftarrow coffee$$

$$expect(tea)$$
$$expect(coffee)$$
$$expect\_not(coffee) \leftarrow blood\_pressure\_high$$

$$0\,\langle tea, coffee\rangle\,1$$
$$coffee \prec tea \leftarrow sleepy$$

Following the abductive stable model semantics, the program above has two models, one with tea and the other with coffee. Also, by adding the literal *sleepy*, the enforced abduction comes into play, defeating the abductive stable model where only *tea* is present (due to the impossibility of simultaneously abducing coffee). However, if later on we add *blood_pressure_high* to the program, coffee is no longer expected, and as such, the transformed preference rule no longer defeats the abduction of *tea* which then becomes the single abductive stable model, despite the presence of *sleepy*.

10.3.3. *Abducible Sets*

In many situations it is desirable not only to include rules about the expectations for single abducibles, but also to express contextual information constraining the powerset of abducibles. For instance, in the previous example we expressed that abducing tea or coffee was mutually exclusive (i.e. only one of them could be abduced), but it is easy to imagine similar choice situations where it would be possible, indeed even desirable, to abduce both, or neither. The behaviour of abducibles over different sets is highly context-dependent, and as such, should also be embedded over rules in the theory.

Overall, the problem is analogous to the ones addressed by *cardinality* and *weight constraint rules* for the Stable Model semantics,[8] and below we present how one can nicely import these results to work with abduction of sets, and also hierarchies of sets.

Example 10.12. Consider a situation where Claire is deciding what to have for a meal from a limited buffet. The menu has appetizers (which Claire doesn't mind skipping, unless she's very hungry), three main dishes, from which one can select

[b] The following example is taken from.[7]

a maximum of two, and drinks, from which she will have a single one. The situation, with all possible choices, can be modelled by the following program P over $\mathcal{L}$ with set of abducibles $\mathcal{A}_P = \{bread, salad, cheese, fish, meat, veggie, wine, juice, water, abduced(bread), abduced(salad), abduced(cheese), abduced(fish), abduced(meat), abduced(veggie), abduced(wine), abduced(juice), abduced(water)\}$:

$$
\begin{aligned}
&0\ \{bread, salad, cheese\}\ 3 \leftarrow appetizers\\
&1\ \{fish, meat, veggie\}\ 2 \leftarrow main\_dishesv\\
&1\ \{wine, juice, water\}\ 1 \leftarrow drinks\\
&2\ \{appetizers, main\_dishes, drinks\}\ 3\\
&main\_dishes \prec appetizers\\
&drinks \prec appetizers\\
&appetizers \leftarrow very\_hungry
\end{aligned}
$$

In this situation we model appetizers as being the least preferred set from those available for the meal. This shows how we can condition sets of abducibles based on the generation of literals from other cardinality constraints along with preferences among such literals.

10.3.4. *Modeling Inspection Points*

When finding an abductive solution for a query, one may want to check whether some other literals become true or false strictly within the abductive solution found, but without performing additional abductions, and without having to produce a complete model to do so. Pereira and Pinto[9] argue that this type of reasoning requires a new mechanism. To achieve it, they introduce the concept of inspection point, and show how one can employ it to investigate side-effects of interest. Procedurally, inspection points can be construed as utilizing a form of meta-abduction, by "meta-abducing" the specific abduction of actually checking (i.e. passively verify) that a certain and corresponding concrete abduction is indeed adopted. That is, one abduces the checking of some abducible A, and the check consists in confirming that A is part of the abductive solution by matching it with the object of the abduced check.

In general, one may want to find one possible set of conditions (literals of the program assumed true) sufficient to entail the query. However, sometimes one may also want to know which are (some of) the consequences (side-effects) of such conditions. That is, one wants to know the truth value of some other literals, not part of the query, whose truth-value may be determined by the abductive conditions found as a solution of the query. In some cases, the focus can be in some specific side-effects of abductions performed. In their approach the side-effects of interest are explicitly indicated by the user by wrapping the corresponding goals within a reserved construct *inspect*/1. Procedurally, *inspect* goals must be solved without abducing regular abducibles, only "meta-abducibles" of the form *abduced*/1. An example will make this concept clear.

Example 10.13. Consider the following program taken from,[9] where *tear_gas*, *fire*, and *water_cannon* are abducibles.

$$\begin{array}{ll} \bot \leftarrow police, riot, not\, contain & \\ contain \leftarrow tear\_gas & contain \leftarrow water\_cannon \\ smoke \leftarrow fire & smoke \leftarrow inspect(tear\_gas) \\ police & riot \end{array}$$

Notice the two rules for *smoke*. The first states that one explanation for smoke is fire, when assuming the hypothesis *fire*. The second states *tear_gas* is also a possible explanation for smoke. However, the presence of tear gas is a much more unlikely situation than the presence of fire; after all, tear gas is only used by police to contain riots and that is truly an exceptional situation. Fires are much more common and spontaneous than riots. For this reason, *fire* is a much more plausible explanation for *smoke* and, therefore, in order to let the explanation for *smoke* be *tear_gas*, there must be a plausible reason — imposed by some other likely phenomenon. This is represented by *inspect*(*tear_gas*) instead of simply *tear_gas*. The *inspect* construct disallows regular abduction — only meta-abduction — to be performed whilst trying to solve *tear_gas*. I.e., if we take tear gas as an abductive solution for fire, this rule imposes that the step where we abduce *tear_gas* is performed elsewhere, not under the derivation tree for *smoke*. Thus, *tear_gas* is an inspection point.

The integrity constraint, since there is *police* and a *riot*, forces *contain* to be true, and hence, *tear_gas* or *water_cannon* or both, must be abduced. *smoke* is only explained if, at the end of the day, *tear_gas* is abduced to enact containment.

Abductive solutions should be plausible. *smoke* is plausibly explained by *tear_gas* if there is a reason, a best explanation, that makes the presence of tear gas plausible; in this case the riot and the police. Plausibility is an important concept in science which lends credibility to hypotheses.

Example 10.14. A gourmet's proper meal goes along with red wine if the main dish is meat, and with white wine if the main dish is fish. Specific wine abduction can be done before main dish abduction, without committing to the kind of main dish. If incompatible, an alternative abduction for the wine can be produced. Consider the following program, with abducibles $\mathcal{A}_P = \{meat, fish, merlot, chardonnay\}$:

$$\begin{array}{ll} \bot \leftarrow not\, proper\_meal & main\_dish \leftarrow meat \\ proper\_meal \leftarrow wine, main\_dish & main\_dish \leftarrow fish \\ & \\ wine \leftarrow red\_wine, inspect(meat) & red\_wine \leftarrow merlot \\ wine \leftarrow white\_wine, inspect(fish) & white\_wine \leftarrow chardonnay \end{array}$$

$$0 \ \langle meat, fish \rangle \ 1$$
$$0 \ \langle merlot, chardonnay \rangle \ 1$$

The alternative combinations for the gourmet's proper meal are thus: {*fish*, *chardonnay*} and {*meat*, *merlot*}. The main dish is not determined by the wine, even if the latter is pre-selected first. This example also illustrates the use of inspection points to express a preference whereby one can only have wine with a main dish.

Inspection points can occur within the proof tree of another inspection point, as the next example shows.

Example 10.15. Consider the following program, where the abducibles are a, b, c, d:

$$x \leftarrow a, \mathit{inspect}(y), b, c, \mathit{not}\, d \qquad y \leftarrow \mathit{inspect}(\mathit{not}\, a)$$
$$z \leftarrow d \qquad y \leftarrow b, \mathit{inspect}(\mathit{not}\, z), c$$

Let the query be $?{-}x$. According to the intended semantics of inspection points, the set $\Delta = \{a, b, c\}$ is an abductive explanation for the query.

The following simple transformation maps programs with inspection points into programs over $\mathcal{L}$. Mark that the abductive stable models of the transformed program clearly correspond to the intended procedural meanings ascribed to the inspection points of the original program.

Definition 10.6. (Transforming Inspection Points) Let P a program containing rules whose body possibly contains inspection points. The program $\Pi(P)$ over $\mathcal{L}$ consists of:

(1) all the rules obtained by the rules in P by replacing:

$$\mathit{inspect}(a) \text{ or } \mathit{inspect}(\mathit{abduced}(a)) \text{ with } \mathit{abduced}(a)$$

if a is an abducible, and keeping $\mathit{inspect}(L)$ otherwise
(2) $\mathit{inspect}(\mathit{not}\, L)$ with $\mathit{not}\ \mathit{inspect}(L)$
(3) for every abducible a, the rule: $\mathit{expect}(a)$
(4) for every rule $A \leftarrow L_1, \ldots, L_t$ in P, the additional rule: $\mathit{inspect}(A) \leftarrow L_1', \ldots, L_t'$ where for every $1 \leq i \leq t$

$$L_i' = \begin{cases} \mathit{abduced}(L_i) & \text{if } L_i \text{ is an abducible} \\ \mathit{inspect}(X) & \text{if } L_i \text{ is } \mathit{inspect}(X) \\ \mathit{inspect}(L_i) & \text{otherwise} \end{cases}$$

Example 10.16. Let P be the program of Example 10.15. Then, $\Pi(P)$ is:

$x \leftarrow a, \mathit{inspect}(y), b, c, \mathit{not}\, d$
$\mathit{inspect}(x) \leftarrow \mathit{abduced}(a), \mathit{inspect}(y), \mathit{abduced}(b), \mathit{abduced}(c), \mathit{abduced}(\mathit{not}\, d)$

$y \leftarrow \mathit{abduced}(\mathit{not}\, a)$
$\mathit{inspect}(y) \leftarrow \mathit{abduced}(\mathit{not}\, a)$

$y \leftarrow b, \mathit{inspect}(\mathit{not}\, z), c$
$\mathit{inspect}(y) \leftarrow \mathit{abduced}(b), \mathit{inspect}(\mathit{not}\, z), \mathit{abduced}(c)$

$z \leftarrow d$
$inspect(z) \leftarrow abduced(d)$

$expect(a).$
$expect(b).$
$expect(c).$
$expect(d).$

The abductive stable model of $\Pi(P)$ respecting the inspection points is:

$$\{x, a, b, c, abduced(a), abduced(b), abduced(c), expect(a), expect(b), expect(c), expect(d)\}.$$

Note that for each $abduced(a)$ the corresponding A is in the model.

10.4. Procedural Semantics

10.4.1. *Framework*

In this section, we consider a first-order propositional language $\mathcal{L}^{\#}$ containing rules as defined in Def. 10.1. Programs over $\mathcal{L}^{\#}$ are constrained to satisfy certain given properties.

Definition 10.7. Let Γ and Σ be sets of rules over $\mathcal{L}^{\#}$. Then (Γ, Σ) is a restricted program.

The basic idea is that Γ contains the rules formalizing the application domain while Σ formalizes the properties that Γ must satisfy. Every restricted program is associated with a set of abducibles $\mathcal{A}_{(\Gamma,\Sigma)}$.

Remark 10.1. Note that the definition of restricted program can be generalized to have Σ be any set of wffs.

In the following let $\Delta \subseteq \mathcal{A}_{(\Gamma,\Sigma)}$ be a set of abducibles.

Definition 10.8. M is a valid stable model with hypotheses Δ of (Γ, Σ) iff:

(1) $M \not\models \bot$
(2) $M = least(\Gamma^+ \cup Default(\Gamma^+, M))$, where $\Gamma^+ = \Gamma \cup \Delta$
(3) $M \models \Sigma$

A valid stable model is an abductive stable model (conditions (1) and (2)) satisfying the wffs in Σ (condition (3)). (explain why not to have strict valid stable models)

Example 10.17. Let (Γ, Σ) be a restricted program where: $\Gamma = \{p \leftarrow a; expect(a)\}$ and $\Sigma = \{q \leftarrow a\}$ with $\mathcal{A}_{(\Gamma,\Sigma)} = \{a, abduced(a)\}$. Then, its unique valid stable model is $M = \{expect(a)\}$. In fact, the only other model could be $M_2 = \{p, a, abduced(a), expect(a), expect(b)\}$ but it is not valid being Σ not satisfied.

Example 10.18. Let (Γ, Σ) be $(\{p \leftarrow a; expect(a); expect(b)\}, \{\bot \leftarrow a, not\, b\})$ with $\mathcal{A}_{(\Gamma,\Sigma)} = \{a, b, abduced(a), abduced(b)\}$. This program has two valid stable models: $M = \{expect(a), expect(b)\}$ and $M_2 = \{p, a, b, abduced(a), abduced(b), expect(a), expect(b)\}$.

Definition 10.9. Let G be a goal. Then, Δ is a valid explanation for G in (Γ, Σ) iff:

(1) M is a valid stable model with hypotheses Δ of (Γ, Σ), and
(2) $M \models G$

Definition 10.10. Let G be a goal. Then, Δ is a strict valid explanation for G in (Γ, Σ) iff

(1) Δ is a minimal set for which:
- $M \not\models \bot$
- $M = least(\Gamma^+ \cup Default(\Gamma^+, M))$, where $\Gamma^+ = \Gamma \cup \Delta$
- $M \models G$

(2) $M \models \Sigma$

Proposition 10.2. *Let G be a goal and Δ a strict valid explanation for G in (Γ, Σ). Then, Δ is an valid explanation for G in (Γ, Σ).*

Proof. It follows immediately from Definitions 10.9 and 10.10. □

It is easy to see that the converse of Prop 10.2 does not hold. Consider the following example.

Example 10.19. Let $\Gamma = \{p \leftarrow a\}$ and $\Sigma = \{\bot \leftarrow not\, b\}$, where $\mathcal{A}_{(\Gamma,\Sigma)} = \{a, b\}$. Then, $M = \{p, a, b\}$ is a valid stable model of (Γ, Σ). Suppose the goal G is $?{-}p$. Then, $\Delta = \{a, b\}$ is a valid explanation for G in P. However, there exists no strict valid explanation for G.

10.4.2. *Program Transformation*

We define a transformation γ mapping programs over $\mathcal{L}$ into restricted programs over $\mathcal{L}^{\#}$.

Definition 10.11. Let P be a program over $\mathcal{L}$ with set of abducibles $\mathcal{A}_P$. The restricted program $\gamma(P) = (\Gamma, \Sigma)$ over $\mathcal{L}^{\#}$ with set of abducibles $\mathcal{A}_{(\Gamma,\Sigma)} = \mathcal{A}_P$ is defined as follows.
Γ consists of:

(1) all the rules in P
(2) $\bot \leftarrow a, not\ expect(a)$
$\bot \leftarrow a, expect\_not(a)$
for every abducible $a \in \mathcal{A}_P^*$

(3) $\bot \leftarrow a, \mathit{not}\ \mathit{abduced}(a)$
for every abducible $a \in \mathcal{A}^*_P$
(4) $\bot \leftarrow a \prec b, \mathit{expect}(a), \mathit{not}\ \mathit{expect\_not}(a), b, \mathit{not}\ a$
for every rule $a \prec b \leftarrow L_1, \ldots, L_t$ in P
(5) $\bot \leftarrow L\,\{l_1, \ldots, l_n\}\,U, \mathit{count}([l_1, \ldots, l_n], N), N \leq L$
$\bot \leftarrow L\,\{l_1, \ldots, l_n\}\,U, \mathit{count}([l_1, \ldots, l_n], N), N \geq U$
for every rule $L\,\{l_1, \ldots, l_n\}\,U \leftarrow L_1, \ldots L_t$ in P

Σ consists of:

(6) $\bot \leftarrow \mathit{abduced}(a), \mathit{not}\ a$
for every abducible $a \in \mathcal{A}^*_P$
(7) $\bot \leftarrow a \sqsubset b, a, \mathit{not}\, b$
for every rule $a \sqsubset b \leftarrow L_1, \ldots, L_t$ in P

Remark 10.2. We assume given the atom $\mathit{count}([l_1, \ldots, l_n], m)$ that holds if m is the number of abducibles belonging to $[l_1, \ldots, l_n]$ that are assumed. That is, if C is the atom $L\,\{l_1, \ldots, l_n\}\,U$, then we have that $M \models \mathit{count}([l_1, \ldots, l_n], m)$ iff $W(C, M) = m$, for any interpretation M.

Remark 10.3. If P contains inspection points, then apply the transformation γ to $\Pi(P)$ (cf. Definition 10.6) instead of P directly.

Example 10.20. Consider again the program P of Example 10.11. The transformed program $\gamma(P) = (\Gamma, \Sigma)$ over $\mathcal{L}^{\#}$ with abducibles $\mathcal{A}_{(\Gamma,\Sigma)} = \{\mathit{tea}, \mathit{coffee}, \mathit{abduced}(\mathit{tea}), \mathit{abduced}(\mathit{coffee})\}$ is:
Γ consists of:

$\mathit{drink} \leftarrow \mathit{tea}$
$\mathit{drink} \leftarrow \mathit{coffee}$

$\mathit{expect}(\mathit{tea})$
$\mathit{expect}(\mathit{coffee})$
$\mathit{expect\_not}(\mathit{coffee}) \leftarrow \mathit{blood\_pressure\_high}$

$0\,\{\mathit{abduce}(\mathit{tea}), \mathit{abduce}(\mathit{coffee})\}\,1$
$\mathit{coffee} \prec \mathit{tea} \leftarrow \mathit{sleepy}$

$\bot \leftarrow \mathit{tea}, \mathit{not}\ \mathit{expect}(\mathit{tea})$
$\bot \leftarrow \mathit{coffee}, \mathit{not}\ \mathit{expect}(\mathit{coffee})$
$\bot \leftarrow \mathit{tea}, \mathit{not}\ \mathit{expect\_not}(\mathit{tea})$
$\bot \leftarrow \mathit{coffee}, \mathit{not}\ \mathit{expect\_not}(\mathit{coffee})$

$\bot \leftarrow \mathit{tea}, \mathit{not}\ \mathit{abduced}(\mathit{tea})$
$\bot \leftarrow \mathit{coffee}, \mathit{not}\ \mathit{abduced}(\mathit{coffee})$

$\bot \leftarrow \mathit{coffee} \prec \mathit{tea} \leftarrow \mathit{expect}(\mathit{coffee}), \mathit{not}\ \mathit{expect\_not}(\mathit{coffee}), \mathit{tea}, \mathit{not}\ \mathit{coffee}$

$\bot \leftarrow 0\,\{\mathit{abduce}(\mathit{tea}), \mathit{abduce}(\mathit{coffee})\}\,1, \mathit{count}([\mathit{abduce}(\mathit{tea}), \mathit{abduce}(\mathit{coffee})], N),$
$N \leq 0$

$\bot \leftarrow 0\,\{abduce(tea), abduce(coffee)\}\,1, count([abduce(tea), abduce(coffee)], N),$
$N \geq 1$

Σ consists of:

$\bot \leftarrow abduced(tea), not\ tea$
$\bot \leftarrow abduced(coffee), not\ coffee$

10.4.3. *Properties*

The correctness of the transformation γ is guaranteed by the following two properties.

Theorem 10.1. *Let P be a program over $\mathcal{L}$ with set of abducibles $\mathcal{A}_P$. M is an abductive stable model with hypotheses $\Delta \subseteq \mathcal{A}_P$ of P iff M is a valid stable model with hypotheses Δ of $\gamma(P)$.*

Proof.

($\Rightarrow$)
Let M be an abductive stable model with hypotheses $\Delta \subseteq \mathcal{A}_P$ of P. Then, we need to show that M is a valid stable model with hypotheses Δ of $\gamma(P) = (\Gamma, \Sigma)$. This is established by proving the three conditions below.

(1) $M \not\models \bot$
Immediate by condition (1) of Definition 10.3.

(2) $M = least(\Gamma^+ \cup Default(\Gamma^+, M))$, where $\Gamma^+ = \Gamma \cup \Delta$
Note first that by Definition 10.11 it holds that $\Delta \subseteq \mathcal{A}_P = \mathcal{A}_{(\Gamma,\Sigma)}$. By condition (2) of Definition 10.3 we have that:

$$M = least(Q \cup Default(Q, M)), \text{ where } Q = P \cup \Delta.$$

Let Q_2 be the program obtained by extending Q with the rules:

$$\bot \leftarrow a, not\ expect(a)$$
$$\bot \leftarrow a, expect\_not(a)$$

for every abducible $a \in \mathcal{A}_P^*$. Then, M is an abductive stable model of Q_2 since if $a \in \Delta$ then the claim follows by condition (3) of Definition 10.3, otherwise ($a \notin \Delta$) the claim is immediate by noting that the body of the rule is false.

Let Q_3 be the program obtained by extending Q_2 with the rules:

$$\bot \leftarrow a, not\ abduced(a)$$

for every abducible $a \in \mathcal{A}_P^*$. By condition (4) of Definition 10.3, it follows immediately that M is an abductive stable model of Q_3.

Let Q_4 be the program obtained by extending Q_3 with the rules:

$$\bot \leftarrow a \prec b, expect(a), not\ expect\_not(a), b, not\ a$$

for every rule $a \prec b \leftarrow L_1, \ldots, L_t$ in P. To establish that M is an abductive stable model of Q_4 consider the following two cases. If it holds that $M \not\models (a \prec b, \mathit{expect}(a), \mathit{not}\ \mathit{expect\_not}(a), b)$ then the claim is immediate. Otherwise, the claim follows by condition (5) of Definition 10.3 stating that $M \models a$.

Let Q_5 be the program obtained by extending Q_4 with the rules:

$$\bot \leftarrow L\,\{l_1, \ldots, l_n\}\,U, \mathit{count}([l_1, \ldots, l_n], N), N \le L$$
$$\bot \leftarrow L\,\{l_1, \ldots, l_n\}\,U, \mathit{count}([l_1, \ldots, l_n], N), N \ge U$$

for every rule $L\,\{l_1, \ldots, l_n\}\,U \leftarrow L_1, \ldots L_t$ in P. M is an abductive stable model of Q_5 by condition (6) of Definition 10.3 under the assumption that $M \models \mathit{count}([l_1, \ldots, l_n], m)$ iff $W(C, M) = m$ with $C = L\,\{l_1, \ldots, l_n\}\,U$.

By noting that $Q_6 = \Gamma^+$, claim (2) follows.

(3) $M \models \Sigma$

Note first that by Definition 10.11 the program Σ consists of two kinds of rules:

$$\bot \leftarrow \mathit{abduced}(a), \mathit{not}\ a$$

for every abducible $a \in \mathcal{A}_P^*$, and

$$\bot \leftarrow a \sqsubset b, a, \mathit{not}\ b$$

for every rule $a \sqsubset b \leftarrow L_1, \ldots, L_t$ in P. Conditions (7) and (8) of Definition 10.3 ensure that the bodies of the rules above is false. Hence, the claim follows.

($\Leftarrow$)

Let M be a valid stable model with hypotheses Δ of $\gamma(P) = (\Gamma, \Sigma)$. We need to show that M is an abductive stable model with hypotheses $\Delta \subseteq \mathcal{A}_P$ of P. The proof of this claim is omitted being quite similar to the proof of the previous case ($\Rightarrow$). □

Lemma 10.1. *Let P be a program over $\mathcal{L}$ and G a goal. Let $\Delta_2 \subseteq \mathcal{A}_P$ and M_2 be a set of abducibles and an interpretation that satisfy condition (1) of Definition 10.10 for G in $\gamma(P)$. Then, Δ_2 and M_2 satisfy condition (1) of Definition 10.5 for G in P.*

Proof. To prove that Δ_2 and M_2 satisfy condition (1) of Definition 10.5 we have to prove the following conditions:

- $M_2 \models G$ and $M_2 \not\models \bot$ since M_2 satisfies condition (1) of Definition 10.10
- $M_2 = \mathit{least}(Q \cup \mathit{Default}(Q, M_2))$, where $Q = P \cup \Delta_2$ by construction of γ, Γ exactly contains all the rules in P plus rules of the form $\bot \leftarrow \mathit{body}$. This claim follows from the facts that:
  - $\mathit{Default}(Q, M_2) = \mathit{Default}(\Gamma^+, M_2)$, and
  - $M_2 = \mathit{least}(\Gamma^+ \cup \mathit{Default}(\Gamma^+, M_2))$

  where $\Gamma^+ = \Gamma \cup \Delta_2$

- $M \models expect(a)$ and $M \not\models expect\_not(a)$, for every $a \in \Delta_2^*$ immediate from case (2) of Definition 10.11 and $M_2 \not\models \bot$
- for every $a \in \Delta_2^*$, if $M_2 \models a$ then $M_2 \models abduced(a)$ immediate from case (3) of Definition 10.11 and $M_2 \not\models \bot$
- for every atom $a \prec b$, if $M_2 \models a \prec b$, $M_2 \models expect(a)$, $M_2 \not\models expect\_not(a)$ and $M_2 \models b$, then $M_2 \models a$ immediate from case (4) of Definition 10.11 and $M_2 \not\models \bot$
- for every atom $C = L\{l_1, \ldots, l_n\}U$, if $M_2 \models C$ then $L \leq W(C, M_2) \leq U$ immediate from case (5) of Definition 10.11 and $M_2 \not\models \bot$

□

Theorem 10.2. *Let P be a program over $\mathcal{L}$ and G a goal. $\Delta \subseteq \mathcal{A}_P$ is a strict abductive explanation for G in P iff Δ is a strict valid explanation for G in $\gamma(P)$.*

Proof.
Let $\gamma(P) = (\Sigma, \Gamma)$.
($\Rightarrow$)
Let $\Delta \subseteq \mathcal{A}_P$ be a strict abductive explanation for G in P. Then, we need to show that Δ is a strict valid explanation for G in $\gamma(P)$.

The proof is by contradiction. Suppose that Δ is not a strict valid explanation for G in $\gamma(P)$. That is, there is no interpretation M for which conditions (1) and (2) of Definition 10.10 hold. Let $\overline{M}$ be an abductive stable model with hypotheses Δ of P. By Theorem 10.1, $\overline{M}$ is a valid stable model with hypotheses Δ of (Γ, Σ), and therefore Δ is a valid explanation for G in (Γ, Σ).

Since Δ is not a strict valid explanation, the two conditions (1) and (2) of Definition 10.10 do not hold for $\overline{M}$. Consider condition (2). Since $\overline{M}$ is a valid stable model of (Γ, Σ), then we have that $\overline{M} \models \Sigma$. Thus, it is condition (1) that makes Δ not strict. Hence, there must exist a set of abducibles $\Delta_2 \subset \Delta$ and an interpretation M_2 that satisfy condition (1) of Definition 10.10:[c]

(1) Δ_2 is a minimal set for which:

- $M_2 \not\models \bot$
- $M_2 = least(\Gamma^+ \cup Default(\Gamma^+, M_2))$, where $\Gamma^+ = \Gamma \cup \Delta_2$
- $M_2 \models G$

The hypotheses Δ_2 satisfy condition (1) of Definition 10.5 by Lemma 10.1, and therefore Δ cannot be a strict abductive explanation for G in P (because Δ would not be a minimal set to satisfy condition (1)).

($\Leftarrow$)
Let $\Delta \subseteq \mathcal{A}_P$ be a strict valid explanation for G in $\gamma(P)$. We need to show that Δ is a strict abductive explanation for G in P.

The proof of this claim is omitted being quite similar to the proof of the previous case. □

[c]Note that M_2 might not satisfy Σ.

10.5. A Posteriori Preferences

While we can indeed place a priori constraints on which abducibles are relevant given contextual knowledge in the situation, more often than not we are only able to enact certain choices after looking at the consequences of adopting one or another abducible. The consequences of each abductive stable model can, and often are, unique to that model, and we cannot model preferences across these consequences during model generation itself. Only after the relevant models are computed can we reason about which consequences, or other features of the models, are determinant for the final choice, i.e. the quality of the model.

One possibility is to consider a quantitative classification of the obtained models, for instance by associating some measure of utility to each choice scenario, like in decision theory. This allows us to consider and integrate many techniques and results from more quantitative decision making frameworks into our own theories, accounting for more elaborate choice models.

When considering abductive logic programs as specifications of choice models of agents, other possibilities come up derived from the capability of the agent to act upon and sense the environment. Namely, if it is the case that the currently available knowledge of the situation is insufficient to commit to any single preferred abductive model, it may be possible for the agent to gather additional information by performing experiments, or consulting an oracle in order to confirm or disconfirm some of the remaining hypotheses. This process may even be nested, so that the available experiments are themselves considered as hypotheses with associated qualitative and quantitative preferences, allowing to express arbitrarily complex context-dependent choices.

We explore these novel approaches in the sequel, along with some telling examples which intend to show the need for, and illustrate, the proposed reasoning schemes.

10.5.1. *The consequences of abduction*

A desirable result of encoding abduction semantics over models of a program (where each abducible literal may be assumed or not) is that we immediately obtain the consequences of committing to any one hypotheses set. Rules which contain abducibles in their bodies can account for the side-effect derivation of certain positive literals in some models, but not others, possibly triggering integrity constraints or indirectly deriving interesting consequences simply as a result of accepting a hypothesis.

Sometimes these computed consequences are relevant to the process of preference handling itself, as we prefer certain consequences to others. However, more often than not it is not possible to simply condition preferences between abducibles based on these consequences, as it may lead to unexpected circular inconsistencies. Also it may be difficult to express more general preferences over what are the preferred literals in a more complete model.

Example 10.21. Consider the simple abductive logic program presented below, with $\mathcal{A} = \{a, b\}$:

$$c \leftarrow a \qquad 1\ \{a, b\}\ 1$$
$$expect(a)$$
$$expect(b)$$

This program has two abductive stable models:

$$M_1 = \{expect(a), expect(b), a, c\}$$
$$M_2 = \{expect(a), expect(b), b\}.$$

Now let us suppose that we want to prefer models where the literal c is not present. We could model this, for this particular program, by stating $b \prec a \leftarrow c$. However, if we introduced another abducible that directly or indirectly resulted in the derivation of c then we would have to shortcut another preference rule for this abducible. Also, if the derivation of c resulted from the combination of more than one abduction, one would have to extensively encode preferences over these sets to all other possible combinations of literals which didn't derive c.

In the long run, the complexity of writing program rules to account for all possible combinations for c would quickly become unsurmountable. Also, if other interesting literals also suggested other types of preferences over the models, and particularly if these other preferences contradicted the previous ones, inconsistencies could easily arise which would destroy entire models of the program. Also, how could we model more general meta-preferences like the one: 'prefer models which have a greater number of abduced literals'.

In these cases, a posteriori reasoning is much more general and powerful to express these kinds of constraints and preference rules which operate on the consequences of the models themselves.

10.5.2. *Utility Theory*

Economic decision theory has been well recognized as a comprehensive and well-founded model for describing ideal rational agents, and much work in the field of AI has been undertaken in order to synthesize models of bounded rationality in computational systems. The logic programming field is no exception, with a number of results being imported into logic models of belief, choice and decision making.[10] A particularly interesting field of study regards modeling agents capable of planning ahead their future choices and actions.

Abduction can also be seen as a mechanism to enable the generation of the possible futures of an agent, with each abductive stable model representing a possibly reachable scenario of interest. Preferring over abducibles in this case is enacting preferences over the imagined future of the agent. In this particular domain, it is unavoidable to deal with uncertainty, a problem that decision theory is ready to address using probability theory coupled with utility functions.

We intend to show that combining the qualitative reasoning addressed in previous sections with the quantitative decision making mechanisms of decision theory is a natural extension to both mechanisms for handling preferences, and neatly accounts for some interesting properties presented by agents in the real world.

In fact, it is possible to associate a quantitative measure of utility to each abductive scenario, by conditioning utility literals on consequences of abducibles. By combining utilities with information regarding the probability of the occurrence of uncertain literals, we end up with an interesting mixture of qualitative and quantitative reasoning, where possible abducibles, constrained by the expectation and preference rules, generate all possible relevant future scenarios which are then associated with a degree of belief to be coupled with the importance that the model be adopted.

Example 10.22. Suppose that agent Claire is spending a day at the beach and she is deciding what means of transportation to adopt. She knows that usually it is faster and more comfortable to go by car, but she also knows that, because it is hot, there is a possibility that there will be a traffic jam. There is also the possibility of using public transportation (by train), but it will take a lot of time, although it meets her wishes of being more environmentally friendly.

This situation can be modeled by the following abductive logic program:

$$go\_to(beach) \leftarrow car$$
$$go\_to(beach) \leftarrow train$$

$$expect(car)$$
$$expect(train)$$
$$1\ \{car, train\}\ 1$$

$$probability(traffic\_jam, 0.7) \leftarrow hot$$
$$probability(not\ traffic\_jam, 0.3) \leftarrow hot$$

$$utility(stuck\_in\_traffic, -8)$$
$$utility(wasting\_time, -4)$$
$$utility(comfort, 10)$$
$$utility(environment\_friendly, 3)$$

$$hot$$

By assuming each of the abductive hypotheses, the general utility of going to the beach can be computed for each particular scenario:

Assume car
Probability of being stuck in traffic $= 0.7$
Probability of a comfortable ride $= 0.3$
Expected utility $= 10 * 0.3 + 0.7 * -8 = -2.6$

Assume train
Expected utility $= -4 + 3 = -1$

It is important to clarify that it wouldn't be possible to condition any kind of comparison or preference between abducibles based on the value of the computed utilities during model computation itself. This results from the fact that the final utilities depend on literals particular to each model, and are not available *a priori*. It should be clear that enacting preferential reasoning over the utilities computed for each model has thus to be performed after the scenarios are available, with an a posteriori meta-reasoning over the models and their respective utilities.

10.5.3. *Oracles*

Performing an experiment can be a critical element in the process of making informed choices. For medical practitioners, for instance, it is a natural extension of abductive reasoning. Preliminary diagnosis (or hypotheses generation) points to expected symptoms and consequences of assuming a certain diagnosis. From these expected symptoms, experiments can be extracted to confirm or disconfirm these consequences. Experiments themselves are abducible choices, and preferences are often applied to specify which experiment should be performed first given the context of a particular patient. Some of them may be more expensive, or more stressing to the patient, or less reliable under certain situations.

New information obtained from performed experiments is incorporated into the original knowledge base in order to refine the preliminary diagnosis. In addition to trimming down on some of the available hypotheses, information from the experiments can actually bring about additional hypotheses for diagnosis that the practitioner was not able to conjecture prematurely. This cycle of refining a diagnosis with additional inquiry is an example of iterated abduction where the dynamics of sensing and acting upon the environment are critical to the very process of opting for the best possible set of hypotheses, and the following diagnosis example illustrates well how it relates to the non-static nature of human preferences.

Example 10.23. A patient shows up at the dentist with signs of pain upon teeth percussion. The expected causes for the observed signs are:

- Periapical lesion
- Horizontal Fracture of the root and/or crown
- Vertical Fracture of the root and/or crown

This setting can be modelled by the following abductive logic program D, representing a partial medical knowledge base of the practitioner:

$$percussion\_pain \leftarrow periapical\_lesion$$
$$percussion\_pain \leftarrow fracture$$

$$radiolucency \leftarrow periapical\_lesion$$

$$fracture \leftarrow horizontal\_fracture$$
$$elliptic\_fracture\_trace \leftarrow horizontal\_fracture$$
$$tooth\_mobility \leftarrow horizontal\_fracture$$

$$fracture \leftarrow vertical\_fracture$$
$$decompression\_pain \leftarrow vertical\_fracture$$

$$0\ \{periapical\_lesion, horizontal\_fracture, vertical\_fracture\}\ 1$$

$$expect(periapical\_lesion)$$
$$expect(horizontal\_fracture)$$
$$expect(vertical\_fracture)$$

$$\bot \leftarrow not\ percussion\_pain$$

The integrity constraint indicates that the practitioner must conclude *percussion_pain* since that is the symptom of the patient that requires explanation. There are three preferred abductive stable models for D corresponding to each of the available hypotheses $\Delta_1 = \{periapical\_lesion\}$, $\Delta_2 = \{horizontal\_fracture\}$ and $\Delta_3 = \{vertical\_fracture\}$. Excluding the expectation domain literals, which are contained in all models, we have:

$M_1 = \{percussion\_pain, periapical\_lesion, radiolucency\}$
with hypotheses Δ_1

$M_2 = \{percussion\_pain, horizontal\_fracture, fracture,$
$tooth\_mobility, elliptic\_fracture\_trace\}$ with hypotheses Δ_2

$M_3 = \{percussion\_pain, vertical\_fracture, fracture,$
$decompression\_pain\}$ with hypotheses Δ_3

Notice that in the collection of abductive stable models we have not only each of the possible diagnosis, but also *all* the expected symptoms of assuming each of the diagnosis.

Following the computation of possible diagnosis scenarios, it is necessary to generate and choose an experiment which can lead to confirmation or disconfirmation of the hypotheses. Let us suppose that the medical practitioner has available an additional knowledge base of possible experiments and rules stating when each experiment is indicated. Consider the following abductive logic program E:

$$1\ \{$$
$$xray, percussion\_test,$$
$$mobility\_test, decompression\_test$$
$$\}\ 1$$

$$expect(percussion\_test) \leftarrow possible(percussion\_pain)$$
$$expect(xray) \leftarrow possible(radiolucency)$$
$$expect(xray) \leftarrow possible(elliptic\_fracture\_trace)$$
$$expect\_not(xray) \leftarrow radiotherapy\_patient$$

$$expect(mobility\_test) \leftarrow possible(tooth\_mobility)$$
$$expect(decompression\_test) \leftarrow$$
$$possible(decompression\_pain)$$

$$mobility\_test \triangleleft xray$$
$$decompression\_test \triangleleft xray$$
$$mobility\_test \triangleleft decompression\_test \leftarrow trauma$$

The literal *possible*/1 indicates that a given symptom is an expected possibility that should be confirmed by an experiment, if one is available. The less invasive experiments are preferred to those which are more invasive (e.g. xray). We also impose the constraint that only one experiment may be executed at any one time.

Let us now take all of the consequences which were extracted from the models of D and were not original symptoms of the patient, and assert them in program E enclosed in distinct *possible*/1 literals. The following rules are then added to E, forming the new abductive logic program E_1:

$$possible(radiolucency)$$
$$possible(elliptic\_fracture\_trace)$$
$$possible(tooth\_mobility)$$
$$possible(decompression\_pain)$$

Computing the models of E_1 yields two preferred abductive hypotheses for conducting a test on the patient: $\Delta_1 = \{mobility\_test\}$ and $\Delta_2 = \{decompression\_test\}$. Both of these hypotheses stand on equal grounds, so it is possible to non-deterministically pick one for execution. Let us assume that Δ_1 is ultimately chosen. Conducting the mobility test on the patient's tooth shows that no significant mobility is present. The new information can now be asserted onto the original diagnosis knowledge base described by D, in the form of the new integrity constraint:

$$\bot \leftarrow tooth\_mobility$$

meaning that we have disproven tooth_mobility so it would be contradictory to conclude it as a consequence of diagnosis.

By recomputing the models for the new abductive logic program D_1, we verify that the previous hypotheses set $\Delta_2 = \{horizontal\_fracture\}$ has been defeated, and now only diagnosis of *periapical_lesion* and *vertical_fracture* remain as possible candidates.

An additional iteration over the abductive logic program $E \cup Ps$, where Ps is the set of *possible*/1 literals derived from the consequences of D_1, reveals that the next experiment to be performed is the non-invasive *decompression_test*. Depending on the result of this experiment, the practitioner will be able to conclude the final diagnosis on a last iteration of the augmented medical knowledge base. This particular example is oversimplified however, since medical knowledge is already structured so as to provide only *crucial literals*. A literal is crucial with respect to two theories if only one of the two theories supports the derivation of that literal. Computing such crucial literals is a non-trivial, but well-known problem, originally addressed in.[11] Its implementation is, nevertheless, outside the scope of this work.

10.6. Sophie's Choice

One of the most discussed cases where the same moral precept gives rise to conflicting obligations is taken from William Styron's Sophie's Choice.[12] Sophie and her two children are at a Nazi concentration camp. A guard confronts Sophie and tells her that one of her children will be allowed to live and one will be killed. But it is Sophie who must decide which child will be killed. Sophie can prevent the death of either of her children, but only by condemning the other to be killed. The guard makes the situation even more excruciating by informing Sophie that if she chooses neither, then both will be killed. With this added factor, Sophie has a morally compelling reason to choose one of her children. But for each child, Sophie has an apparently equally strong reason to save him or her. Thus the same moral precept gives rise to conflicting obligations.

The described scenario can be formalized with the following program P over $\mathcal{L}$ with set of abducibles:

$$\mathcal{A}_P = \{letting\_both\_die, kill(child\_1), kill(child\_2), flip\_a\_coin\}$$

(1) $expect(kill(child\_1))$
$expect(kill(child\_2))$
$expect(flip\_a\_coin)$
$expect(letting\_both\_die)$
(2) $on\_observe(decide) \leftarrow sophie\_choice$
(3) $decide \leftarrow letting\_both\_die$
$decide \leftarrow kill(child\_1)$
$decide \leftarrow kill(child\_2)$
$decide \leftarrow flip\_a\_coin$
(4) $1 \ \langle letting\_both\_die, kill(child\_1), kill(child\_2), flip\_a\_coin\rangle \ 1$
(5) $expect\_not(kill(C)) \leftarrow special\_reason(C)$
(6) $expect\_not(flip\_a\_coin) \leftarrow special\_reason(child\_1), not\ special\_reason(child\_2)$
$expect\_not(flip\_a\_coin) \leftarrow special\_reason(child\_2), not\ special\_reason(child\_1)$
(7) $die(2) \leftarrow abduced(letting\_both\_die)$
$die(1) \leftarrow abduced(kill(child\_1))$
$die(1) \leftarrow abduced(kill(child\_2))$
$die(1) \leftarrow abduced(flip\_a\_coin)$
(8) $pr(feel\_guity, 1) \leftarrow abduced(kill(child\_1))$
$pr(feel\_guity, 1) \leftarrow abduced(kill(child\_2))$
$pr(feel\_guity, 0.5) \leftarrow abduced(flip\_a\_coin)$
(9) $Ai \ll Aj \leftarrow member(die(N), Ai), member(die(K), Aj), N < K$
(10) $Ai \ll Aj \leftarrow member(pr(feel\_guity, Pi), Ai), member(pr(feel\_guity, Pj), Aj),$
$Pi < Pj$

Line 1 says that there is always expectation for every abducible declared in $\mathcal{A}_P$. In addition, if Sophie has no special reason for any child, there is no expectation to the contrary of those abducibles (lines 5–6). Thus, the strict abductive explanations for the goal? — *decide* are:

$$\Delta_1 = \{letting\_both\_die, not\ kill(child\_1), not\ kill(child\_2), not\ flip\_a\_coin\}$$
$$\Delta_2 = \{kill(child\_1), not\ kill(child\_2), not\ flip\_a\_coin\}$$
$$\Delta_3 = \{kill(child\_2), not\ kill(child\_1), not\ flip\_a\_coin\}$$
$$\Delta_4 = \{flip\_a\_coin\}$$

e.g. it is possible for Sophie to decide to let both of her children die, choose one on her own decision, or flip a coin to decide. Then, in the next stage, a posteriori preferences are taken into account to filter out the less preferred abductive solutions. Considering the "a posteriori" preference encoded in line 9, solution that includes *letting_both_die* is ruled out since it leads to the consequence of two children dying, which is less preferred than any of the (equally) preferred remaining solutions (all with the consequence of just one child dying) (line 7). From the three remaining solutions, the ones that kill some child are ruled out since their consequences are the greater probability of Sophie to feel guilty than the one of $\{flip\_a\_coin\}$ (line 8), having taken into account "a posteriori preference" in line 10 where $member(G, A)$ is the standard system predicate meaning in this context to check whether G belongs to the set of literals representing the abductive stable model A. In short, Sophie's final decision is to flip a coin since, according to this, only one child will die and she will feel less guilty about her decision. Next consider the case when some special reason for a single child, e.g. *child_1*, is entered. Then the expectation to the contrary of killing child 1 (line 5) and of flipping a coin (line 6) are held. Therefore, only two strict abductive explanations, one including *letting_both_die* and one including *kill(child_2)* are available for Sophie to choose. Then, as above, by applying the "a posteriori" preference in line 9, the first one is ruled out. In other words, Sophie's final decision is to kill the *child_2*.

10.7. Implementation

10.7.1. *XSB-XASP Interface*

The Prolog language has been for quite some time one of the most accepted means to codify and execute logic programs, and as such has become a useful tool for research and application development in logic programming. Several stable implementations have been developed and refined over the years, with plenty of working solutions to pragmatic issues ranging from efficiency and portability to explorations of language extensions. The XSB Prolog system[d] is one of the most sophisticated, powerful, efficient and versatile among these implementations, with a focus on execution

[d]Both the XSB Logic Programming system and Smodels are freely available at: `http://xsb.sourceforge.net` and `http://www.tcs.hut.fi/Software/smodels`

efficiency and interaction with external systems, implementing program evaluation following the Well-Founded Semantics (WFS) for normal logic programs.

The semantics of Stable Models has become the cornerstone for the definition of some of the most important results in logic programming of the past decade, providing an increase in logic program declarativity and a new paradigm for program evaluation. However, the lack of some important properties of previous language semantics, like relevancy and cumulativity, somewhat reduces its applicability in practice, namely regarding abduction.

The XASP interface[13,16] (standing for XSB Answer Set Programming), is included in XSB Prolog as a practical programming interface to Smodels,[14] one of the most successful and efficient implementations of the Stable Models semantics over generalized logic programs. The XASP system allows one not only to compute the models of a given NLP, but also to effectively combine 3-valued with 2-valued reasoning. The latter is achieved by using Smodels to compute the stable models of the so-called residual program, the one that results from a query evaluated in XSB using tabling.[15] This residual program is represented by delay lists, that is, the set of undefined literals for which the program could not find a complete proof, due to mutual dependencies or loops over default negation for that set of literals, detected by the XSB tabling mechanism. This method allows us to obtain any two-valued semantics in completion to the three-valued semantics the XSB system produces.

This kind of integration allows one to maintain the relevance property for queries over our programs, something that the Stable Model semantics does not originally enjoy. In Stable Models, by the very definition of the semantics, it is necessary to compute all the models for the whole program. In our implementation framework, we sidestep this issue, by using XASP to compute the query relevant residual program on demand, usually after some degree of transformation. Only the resulting program is then sent to Smodels for computation of abductive stable models.

This is one of the main problems which abduction over stable models has been facing, in that it always has to consider all the abducibles in a program and then progressively defeat all those that are irrelevant for the problem at hand. This is not so in our system framework, since we can usually begin evaluation by a top-down derivation of a query, which immediately constrains the set of abducibles that are relevant to the satisfaction and proof of that particular query.

An important consideration of computing consequences, like suggested in Section 10.5.1, is that we could end up having to compute the models of the whole program in order to obtain just a particular relevant subset which will be used to enact a posteriori preferences. This can be easily avoided by performing preliminary computation of the relevant residual program, given the consequences that we expect to observe. This means that the consequences believed significant for model preference can be computed on the XSB side, and their additional residual program sent to Smodels as well. In this phase, we do not allow for additional abduction of literals, but merely enforce that rules for consequences are consumers of considered abducibles which have already been produced.

In this way, we combine a declarative methodology to describe the abductive process, with an efficient and viable implementation of reasoning by complementing a 3-valued well-founded derivation with the computation of the stable models of the residual program, in a natural way to obtain all the possible 2-valued models from the well-founded one.

10.7.2. *Top-Down Proof Procedure*

In our implementation we aim for query-driven evaluation of abductive stable models, so that only the relevant part of the program is considered for computation, as mentioned previously. Computation of such models is performed in two stages.

In the first stage, XSB computes the Well-Founded Model (WFM) of the program w.r.t. the given query, supporting goal derivation on any expected abducibles that are considered by not otherwise being defeated, as detailed in Section 10.2.1.1. Satisfaction of integrity constraints is also enforced at this stage, by always querying for *not* $\bot$ in conjunction with any specified goal.

In this way, we aim to dynamically collect only those abducibles which are necessary to prove the query. However, we cannot assume them either true or false at this stage, so they must come up undefined in the derivation tree. This is achieved by codifying considered abducibles in the following manner:

$$\begin{aligned} consider(A) &\leftarrow expect(A), not\ expect\_not(A), abduce(A). \\ abduce(A) &\leftarrow not\ abduce\_not(A). \\ abduce\_not(A) &\leftarrow not\ abduce(A). \end{aligned}$$

The latter two clauses are defined for every abducible, encoding abductive literals as even-loops over default negation, which guarantees that any considered abducibles come up undefined in the WFM of the program, and hence are contained in the residual program computed by the XSB Prolog-based meta-interpreter as explained in Section 10.7.1.

10.7.3. *Computation of Abductive Stable Models*

In the second stage, Smodels will be used to compute abductive stable models from the residual program obtained from top-down goal derivation. Determination of relevant abducibles is performed by examining the residual program for ground literals which are argument to *consider/1* clauses. Relevant preference rules are pre-evaluated at this stage as well, by querying for any such rules envolving any pair of the relevant abducible set.

Currently we need to assume ground literals since we will be producing a program transformation over the residual program based on the set of confirmed abducibles, which will be sent directly to Smodels. Smodels needs all literals to be instantiated, or to have an associated domain for instantiation. This limitation is present in many other state-of-the-art logic programming systems and its solution is not the main point of this work. Nevertheless, it is worth considering that

XSB's unification algorithm can ground some of the variables during the top-down derivation.

The XASP package[16] allows the programmer to collect rules in an XSB clause store. When the programmer has determined that enough clauses have been added to the store to form a semantically complete sub-program, the program is then *committed.* This means that information in the clauses is copied to Smodels, codified using Smodels data structures so that stable models of those clauses can be computed and examined.

When both the relevant abducibles and preference rules are determined, a variation of transformation Γ is applied, with every encoded clause being sent to the XASP store, reset beforehand in preparation for the stable models computation. Once the residual program, transformed to enact preferences, is actually committed to Smodels, we obtain through XASP the set of abductive stable models, and identify each one by their abducible choices (i.e. those *consider/1* literals which were collected beforehand in the residual program).

10.7.4. *Inspection Points*

Given the availability of a top-down proof procedure for abduction, implementing inspection points becomes simply a matter of adapting the evaluation of derivation subtrees falling under *inspect/1* literals, by following Definition 10.6 at a meta-interpreter level.

Basically, considered abducibles evaluated under *inspect/1* subtrees are codified in the following manner:

$$\begin{aligned} consider(A) &\leftarrow abduced(A). \\ abduced(A) &\leftarrow not\ abduced\_not(A). \\ abduced\_not(A) &\leftarrow not\ abduced(A). \end{aligned}$$

All the *abduced/1* predicates are thus collected during computation of the residual program and are later checked against the abductive stable models themselves. Every *abduced(a)* predicate must have a corresponding abducible *a* for the model to be accepted, as defined by transformation Γ.

10.7.5. *A Posteriori Choice Mechanisms*

If only a single model emerges from computation of the abductive stable models, goal evaluation can successfully terminate. However, in situations where multiple models still remain, there is an opportunity to introduce a posteriori choice mechanisms, which can actually be domain-specific for any given program. We account for this specificity by providing an implementation hook which the user can adopt for introducing the specific code for this final selection process.

We have currently implemented a fully working set of a posteriori choice mechanisms, more specifically those detailed in Section 10.5, but additional preference mechanisms can be exploited in the future to choose between abductive stable models.

These are the bases of the ACORDA system,[17,18] a logic programming framework specifically designed to accomodate for abduction in evolving scenarios, using the perspectives previously outlined. In addition, we have also implemented our framework in Neg-Abdual[19] an implemented XSB-Prolog system combining constructive negation and the abduction over the well-founded semantics of Abdual.[20] All our examples have been tested with success in these systems.

10.8. Conclusions

We have shown that a priori preferences over abductive logic programs are an important tool for knowledge representation in modeling different kinds of choice situations where an agent needs to make a decision while considering its present and future context. Some limitations which had already been identified in[4] were addressed, namely how to express preferences between abducible sets. We have shown how previous results of extensions to the Stable Model semantics, using cardinality constraints and inspection points, can be used to govern and constrain abduction of abductive literals in the context of our framework. The incorporation of these results has also led to a simpler transformation of abductive logic programs into normal logic programs extended with cardinality constraints.

The broadening of our research direction to incorporate a posteriori preferences has also been presented here as a major topic of interest for research into prospective agents, which can not only consider their immediate context, but can also present a modicum of lookahead and meta-reasoning over available scenarios, using both qualitative and quantitative dimensions for decision making. We have also shown the importance of using selected computed consequences to constrain and condition preference handling itself, and how model computation can be complemented by performing experiments with the purpose of acquiring new information with which to enact more informed choices.

In,[5] the authors detect a problem with the IFF abductive proof procedure[21] of Kung and Kowalski, in what concerns the treatment of negated abducibles in integrity constraints (e.g. in their examples 2 and 3). They then specialize IFF to avoid such problems and prove correctness of the new procedure. The problems detected refer to the active use an IC of some not A, where A is an abducible, whereas the intended use should be a passive one, simply checking whether A is proved in the abductive solution found. To that effect they replace such occurrences of not A by not provable(A), in order to ensure that no new abductions are allowed during the checking.

Our work on Inspection Points generalizes the scope of the problem they solved and solves the problems involved in this wider scope. For one we allow for passive checking not just of negated abducibles but also of positive ones, as well as passive checking of any literal, whether or not abducible, and allow also to single out which occurrences are passive or active. Thus, we can cater for both passive and active

ICs, depending on the use desired. Our solution uses abduction itself to solve the problem, making it general for use in other abductive frameworks and procedures. The declarative semantics of our approach is supplied by a straight forward, meaning preserving, program transformation whose correctness is apparent.

Finally, we have shown the advantages of implementing our framework as a hybrid Prolog-Smodels system via the XASP package, in order to efficiently combine backwards- and forwards-chaining reasoning, respectively for constraining the abducibles to only those which are relevant for a given goal, and then to compute the consequences of assuming them in each possible scenario.

Although in this work we insisted on a partial order for preferences, in[4] we have already shown that this need not necessarily be so, namely by specifying different conditions for revising contradictory preferences.[22] The possible alternative revisions, required to satisfy the conditions, impart a non-monotonic or defeasible reading of the preferences given initially. Such a generalization permits us to go beyond a simply foundational view of preferences, and allows us to admit a coherent view as well, inasmuch several alternative consistent stable models may obtain for our preferences, as a result of each revision.

In,[23] arguments are given as to how epistemic entrenchment can be explicitly expressed as preferential reasoning. And, moreover, how preferences can be employed to determine believe revisions, or, conversely, how belief contractions can lead to the explicit expression of preferences.[24] provides a stimulating survey of opportunities and problems in the use of preferences, reliant on AI techniques.

References

1. J. J. Alferes and L. M. Pereira. Updates plus preferences. In eds. M. O. Aciego, I. P. de Guzmán, G. Brewka, and L. M. Pereira, *Procs. of JELIA'00*, LNAI 1919, pp. 345–360, Málaga, Spain, (2000). Springer.
2. G. Brewka. Logic programming with ordered disjunction. In ed. M. Kaufmann, *Proc. 18th National Conference on Artificial Intelligence, AAAI-02*, (2002). URL `citeseer.ist.psu.edu/brewka02logic.html`.
3. G. Brewka, I. Niemelä, and M. Truszczynski. Answer set optimization. In *Proc. IJCAI-2003*, pp. 867–872, (2003). URL `citeseer.ist.psu.edu/brewka03answer.html`.
4. P. Dell'Acqua and L. M. Pereira, Preferential theory revision, *Journal of Applied Logic*. **5**(4), 586–601, (2007).
5. F. Sadri and F. Toni. Abduction with Negation as Failure for Active and Reactive Rules. In eds. E. Lamma and P. Mello, *An Implementation for Abductive Logic Agents*, LNAI 1792, pp. 49–60. Springer-Verlag, (1999).
6. M. Gelfond and V. Lifschitz. The stable model semantics for logic programming. In eds. R. Kowalski and K. A. Bowen, *5th Intl. Logic Programming Conf.*, pp. 1070–1080. MIT Press, (1988).
7. L. M. Pereira, G. Lopes, and P. Dell'Acqua. Pre and post preferences over abductive models. In eds. J. Delgrande and W. Kieling, *Multidisciplinary Workshop on Advances in Preference Handling (M-Pref'07) at 33rd Intl. Conf. on Very Large Data Bases (VLDB'07)*. VLDB, (2007). URL `http://www.mycosima.com/vldb2007-preferences/`.

8. I. Niemelä and P. Simons. Extending the Smodels system with cardinality and weight constraints. In ed. J. Minker, *Logic-Based Artificial Intelligence*, pp. 491–521. Kluwer Academic Publishers, (2000). URL `citeseer.ist.psu.edu/niemel00extending.html`.
9. L. M. Pereira and A. M. Pinto. Inspection points and meta-abduction in logic programs. Submitted, (2008).
10. D. Poole, The independent choice logic for modelling multiple agents under uncertainty, *Artificial Intelligence.* **94**(1-2), 7–56, (1997). URL `citeseer.ist.psu.edu/poole97independent.html`.
11. A. Seki and A. Takeuchi. An algorithm for finding a query which discriminates competing hypotheses. Technical Report TR 143, Institute for New Generation Computer Technology, Japan, (1985).
12. W. Styron, *Sophie's Choice.* (Bantam Books, New York, 1980).
13. L. F. Castro and D. S. Warren. An environment for the exploration of non monotonic logic programs. In ed. A. Kusalik, *Proc. of the 11th Intl. Workshop on Logic Programming Environments (WLPE'01)*, (2001).
14. I. Niemelä and P. Simons. Smodels: An implementation of the stable model and well-founded semantics for normal logic programs. In eds. J. Dix, U. Furbach, and A. Nerode, *4th Intl. Conf. on Logic Programming and Nonmonotonic Reasoning*, LNAI 1265, pp. 420–429, Berlin, (1997). Springer.
15. T. Swift, Tabling for non-monotonic programming, *Annals of Mathematics and Artificial Intelligence.* **25**(3-4), 201–240, (1999).
16. L. Castro, T. Swift, and D. S. Warren. *XASP: Answer Set Programming with XSB and Smodels.* `http://xsb.sourceforge.net/packages/xasp.pdf`.
17. G. Lopes and L. M. Pereira. Prospective logic programming with ACORDA. In eds. G. Sutcliffe, R. Schmidt, and S. Schulz, *Procs. of the FLoC'06 Ws. on Empirically Successful Computerized Reasoning, 3rd Intl. J. Conf. on Automated Reasoning*, number 192 in CEUR Workshop Procs., (2006).
18. L. M. Pereira and G. Lopes. Prospective logic agents. In eds. J. M. Neves, M. F. Santos, and J. M. Machado, *Progress in Artificial Intelligence, Procs. 13th Portuguese Intl.Conf. on Artificial Intelligence (EPIA'07)*, LNAI 4874, pp. 73–86, Guimarães (December, 2007). Springer.
19. Neg-Abdual. Available at `http://centria.di.fct.unl.pt/ lmp/software.html`, (2008).
20. J. J. Alferes, L. M. Pereira, and T. Swift, Abduction in Well-Founded Semantics and Generalized Stable Models via Tabled Dual Programs, *Theory and Practice of Logic Programming.* **4**(4), 383–428, (2004).
21. T. H. Fung and R. Kowalski, The IFF Proof Procedure for Abductive Logic Programming, *The J. of Logic Programming.* **33**(2), 151–165, (1997).
22. P. Santana and L. M. Pereira. Emergence of cooperation through mutual preference revision. In eds. M. Ali and D. Richard, *Procs. of the 19th Intl. Conf. on Industrial & Engineering Applications of Artificial Intelligence & Expert Systems (IEA/AIE'06)*, LNAI, Annecy, France, (2006). Springer.
23. H. Rott, *Change, Choice and Inference.* Oxford University Press, Oxford, (2001).
24. J. Doyle, Prospects for preferences, *Computational Intelligence.* **20**(3), 111–136, (2004).

Chapter 11

SUPERVISED NEURAL NETWORK LEARNING: FROM VECTORS TO GRAPHS

Monica Bianchini, Marco Maggini, and Lorenzo Sarti
Dipartimento di Ingegneria dell'Informazione
Università di Siena
Via Roma 56 - I-53100 Siena (Italy)
monica@dii.unisi.it
maggini@dii.unisi.it
sarti@dii.unisi.it

Recursive neural networks are a powerful tool for processing structured data, thus filling the gap between connectionism, which is usually related to poorly organized data, and a great variety of real–world problems, where the information is naturally organized in entities and relationships among entities. According to the recursive paradigm, the input information consists of directed positional acyclic graphs (DPAGs), on which recursive networks are trained, following the partial order defined by the links of the graph. In this chapter, we propose a unified framework for learning in supervised neural networks, based on the BackPropagation algorithm, starting from static multi–layered architectures to recursive network. In fact, BackPropagation can be naturally specialized to train complex architectures able to process structured data, in the form of sequences and graphs. In this way, the problem of learning in connectionism, no matter what type of data is "natural" to represent a given application domain, is mantained in the elegant framework of continuous optimization. Finally, an application of the recursive model to image processing is presented, giving evidence to the fact that encoding images by trees or graphs gives rise to a more robust and informative representation, aimed at facilitating object detection and image classification.

11.1. Introduction

Nature seems to obey harmoniously to elegant optimization laws that like the Fermat's minimum time principle in optics, or the least action principle in mechanics, rely on the minimization of proper functions. One might wonder whether this ubiquitous formulation of continuous optimization is a man's trend to unify or is instead the inherent solution of most natural problems. No matter what the true answer is, there is no doubt that a similar elegance and generality can also be gained for the solution of an impressive number of different problems, not necessarily related to natural phenomena. In spite of this generality, however, it seems that not all the problems are well–suited for solutions based on continuous optimization.

In the last twenty years, the renewal of interest in neural networks, that rely hardly on continuous optimization, witnesses the attention that the scientific community has been paying on problem solving by continuous optimization. By and large, continuous optimization is regarded as the natural framework for neural computation, which has been challenging traditional symbolic approaches in many interresting real–world problems. One of the common belief that seems to drive the research in that field is that neural networks give rise to a general problem solving methodology that is likely to be applied successfully to many different cases. Optimization–based problem solving has also been applied successfully in different fields, like numerical analysis, computational geometry, and information retrieval. The solution of a given problem in the framework of continuous optimization takes place by constructing a function that, once optimized, makes it possible to determine the solution. Basically, determining such a function seems to be related to the creative process of designing algorithms in the classic discrete setting of computation.

Nevertheless, learning in connectionism is usually related to poorly organized data, like arrays or sequences, even if, in several application domains the information which is relevant for solving problems is encoded, often implicitly, into the relationships among the basic entities. The simplest dynamic data type is the sequence, which is a natural way of modeling temporal domains. In speech recognition, for instance, words flow naturally constituting a temporal sequence of acoustic features. Analogously, in molecular biology, the proteins are organized in symbol strings, where each symbol is a representation of one out of the 20 amino–acids. Most interesting intelligent information processing tasks are based on the presence of some forms of structure that, however, cannot easily be grasped at a symbolic level. The data associated with real–world problems are often inherently structured. For instance, in automated deduction systems, theorem proving can be represented by a tree of inference rules;[1] in chemical compounds analysis, the molecular structure is modeled by a graph;[2] in document analysis, a graphical representation can be associated to the whole structure of the document so as to particular parts of it (e.g., logos[3]). On the other hand, the truly subsymbolic nature of many of those problems makes it very hard to extract clean structured symbolic data. In fact, feature encoding of such data can produce compact vectors, even if the encodings are often problem–dependent, time–consuming and heuristic. Moreover, some information is usually lost when complex data structures such as sequences, trees, or graphs of arbitrary size are encoded in fixed dimensional vectors. The way this kind of information is usually processed can be regarded neither as strictly symbolic nor as subsymbolic, neither sequential nor parallel. A structured pattern can be tought of as an arrangement of elements, deeply dependent by the interactions among them, and by the intrinsic nature of each element. Hence, the causal, hierarchical, and topological relations among parts of a given pattern yield significant information. The human brain is a complex graph of elementary neurons, and also the data to be processed can often be regarded as complex structures of elementary units.

Recently, new connectionist models, which can deal with structured information in the supervised learning framework have been introduced.[4–7] These models process structured information following a recursive computational schema, that takes into account both the local information associated to each atomic entity and the information induced by the topological arrangement of the elements, inherently contained into the structure. A feedforward network, called *recursive neural network*, is fed, at each time instant, with a node of the graph and with some of its past internal states, following a strategy similar to the way in which recurrent networks process sequential inputs. However, whereas recurrent networks process the data following the natural sequential order, recursive networks naturally follow the partial order defined by the arcs of the graph. In this way, recursive networks take advantage of the structure which is defined by the graph topology.

In this chapter, we will introduce a "taxonomy" of supervised neural network architectures, starting from static feedforward networks, fed with arrays, and proceeding with recurrent and recursive networks, tailored to process sequences and graphs, respectively. In all these cases, at each time instant, each neuron in the network evolves an internal state, that depends, in different ways for different architectures, from the local information which is presently injected into the neuron and, for dynamic architectures, also from the "past history" of the neuron itself. For all the presented architecture, we will explicitly describe how the information flows into the network, modifying the state of each neuron, and then frame the learning problem in the continuous optimization setting, describing the training rules (Backpropagation, Backpropagation Through Time, Backpropagation Through Structure) used to adapt the network parameters.

The chapter is organized as follows. In the next section, different neural network models are described, based on the data they are expected to process, whereas in Section 11.3 a detailed description of how learning takes place, using the appropriate version of the Backpropagation algorithm, is reported. Finally, in Section 11.4 some possible applications of the recursive model are briefly sketched, related to various research field, while the problem of how representing images by graphs and how use such representation for automatic image classification and object detection is chosen as a case–study and, therefore, extensively deepened.

11.2. Neural Network Models

In this section we define the formalism adopted throughout the chapter. Basically, we are interested in experiments involving neural networks that can be represented concisely by $\mathcal{E} = \{\mathcal{N}, \mathcal{L}_e, E\}$, being $\mathcal{N}$ the neural network, $\mathcal{L}_e$ the learning environment exploited to train the neural network, and E the cost function that measures how well the provided examples are fitted.

11.2.1. *Input data types*

Throughtout this chapter we will examine neural network architectures able to process three different types of data, i.e. vectors, sequences, and Directed Positional Acyclic Graphs (DPAGs). We will refer to these architectures as *static networks*, *recurrent networks* and *recursive networks*, respectively. In order to capture the very nature of learning in artificial neural networks, a fundamental issue to be faced is how data are organized and how the information they collect is injected and processed by the network. When dealing with static architectures, the input data are represented by feature vectors, $X_p \in I\!R^m$. In the case of recurrent networks, each input example is a sequence $\mathcal{S}_p = \{S_p(1), \ldots, S_p(T_p)\}$, where $S_p(t) \in I\!R^m$. Finally, when considering recursive networks, the input data appear in the form of directed positional acyclic graphs that are defined as follows.

Let $\mathcal{G}_p = (V_p, A_p, L_p)$ be a directed graph, where V_p is the set of nodes (or vertexes), $A_p \subseteq V_p \times V_p$ represents the set of directed arcs as ordered pairs of nodes, and $L_p : V_p \rightarrow I\!R^m$ is a node labeling function that attaches an m–dimensional real valued feature vector to each node in V_p. Given any node $v \in V_p$, $\mathrm{pa}[v] = \{u \in V_p | (u, v) \in A_p\}$ is the set of the *parents* of v, while $\mathrm{ch}[v] = \{u \in V_p | (v, u) \in A_p\}$ represents the set of its *children*. The *outdegree* of v, $\mathrm{od}[v]$, is the cardinality of $\mathrm{ch}[v]$, and $q = \max_v \mathrm{od}[v]$ is the maximum outdegree. A directed acyclic graph $\mathcal{G}_p$, with the arc set A_p, is positional if, for each vertex v, an injective function $\tau_v : A_p(v) \rightarrow I\!N$ is defined on the arcs leaving from v. Bounded DPAGs are defined by restricting the range of τ_v to the finite set $Q = \{1, 2, \ldots, q\} \subset I\!N$, for some fixed q. The set $J_v = Q \setminus \tau_v(A_p(v))$ contains the indexes of the children of v that are missing. These empty positions will be associated to a special arc that is referred to as *nil pointer*, allowing us to consider the same outdegree q for all nodes in V_p. As it will be detailed in the following, the architecture of recursive neural networks requires to process graphs whose nodes have exactly the same outdegree q. This motivates the introduction of nil pointers that properly encode the fact that no outgoing arc is present in a given position. As a consequence, the adjacency list of a DPAG is actually a vector, whose entries are either a child node in V_p or the nil pointer.

Finally, in the graph classification and regression settings, we usually require the DPAG to possess a *supersource*, i.e., a vertex s such that every other vertex of $\mathcal{G}_p$ can be reached by a directed path starting from s. Such a requirement is strictly related to the processing scheme which will be subsequently described. Note that if a DPAG does not have a unique supersource node, it is still possible to define a convention for adding an extra vertex s with a minimal number of outgoing edges, such that s is a supersource for the expanded DPAG.[8]

11.2.2. *The neural network $\mathcal{N}$*

We consider neural networks whose neurons are grouped into sets called layers. With reference to index l, we distinguish between the input layer ($l = 0$), the output layer

$(l = L)$, and the hidden layers $(0 < l < L)$. Among the hidden layers, we consider a special layer S, called *state layer*, that is used to encode and propagate the partial processing of complex data structures. We assume that the network is fed at discrete time by an atomic data d, which can be the entire input data, for static networks, or a part of it, both for recurrent and recursive architectures. In particular, for recurrent networks, d coincides with $S_p(t)$, i.e. with a particular *frame* belonging to the input sequence, whereas for recursive architectures $d \in V_p$, i.e. it represents a particular node in the graph. For each d and l, we consider

$$A_l(d) = [a_{1(l)}(d), \ldots, a_{n(l)}(d)]',$$
$$X_l(d) = [x_{1(l)}(d), \ldots, x_{n(l)}(d)]',$$

being $A_l(d), X_l(d) \in I\!R^{n(l)}$ the *activation* and the *output* vector of the neurons in layer l, respectively. The following model is assumed for the activation

$$\begin{aligned} a_{i(1)}(d) &= \mathcal{F}(W_{i(1)}, X_0(d); \{X_S(\mathrm{ch}[d])\}), \\ a_{i(l)}(d) &= \mathcal{F}(W_{i(l)}, X_{l-1}(d)), \end{aligned} \qquad (11.1)$$

where $W_{i(l)}$ is the weight vector associated with neuron $i(l)$ and $\{X_S(\mathrm{ch}[d])\}$ is the ordered set of the network states produced while processing the *predecessors* of the atomic data d. In the case of static networks, $\{X_S(\mathrm{ch}[d])\}$ is just the empty set, whereas it coincides with $X_S(p, t-1)$ for sequences, and, in the case of DPAGs, with the vector $[X_S(p, \mathrm{ch}_1[d]), \ldots, X_S(p, \mathrm{ch}_q[d])]$ that collects the state vectors of the children of d. Function $\mathcal{F}(\cdot)$ depends on the particular model of each neuron and defines the way of combining the inputs received from all the other neurons or external inputs. The assigned initial state of the network is referred to as $X_0(0), \ldots, X_L(0)$. The output of neuron $i(l)$ is related to its activation as follows

$$x_{i(l)} = f(a_{i(l)}),$$

where $f(\cdot) : I\!R \to [\underline{r}, \overline{r}]$ is a $\mathcal{C}^2$ function and $f'(a_{i(l)}) \neq 0$ in $I\!R$. For example, a *squash–like* function[9] satisfies these hypotheses.

Thus, as shown by equation (11.1), we consider a multi–layered feedforward architecture that allows connections only between adjacent layers. Different topologies can be considered as, for example, networks with forward connections from one layer to all the following layers or with shortcut connections. Nevertheless, the layered architecture proposed here, widely used in applications, is enough general and simple at the same time to be representative of a wider class of networks, to which the results described in the following are easily extended.

Static networks

In the previous general framework, static networks, whose neuron activation does not involve the state propagation, commonly adopts the following choice for the function $\mathcal{F}(\cdot)$

$$a_{i(l)}(d) = \mathcal{F}(W_{i(l)}, X_{l-1}(d)) = w_{i(l)} + \sum_{j(l-1)=1}^{n(l-1)} w_{i(l),j(l-1)} x_{j(l-1)}(d)$$
$$= W'_{i(l)} \hat{X}_{l-1}(d), \qquad (11.2)$$

being $W_{i(l)} \in I\!R^{n(l-1)+1}$. The parameter $w_{i(l),j(l-1)}$ is the weight of the connection between the neurons $j(l-1)$ and $i(l)$, and $w_{i(l)}$ is the neuron bias. In this case, the function $\mathcal{F}$ realizes simply the dot product of $\hat{X}_{l-1} = [X'_{l-1}, 1]'$ and $W_{i(l)}$, with '1' accounting for the bias. The output of neuron $i(l)$ is often related to its activation by means of the sigmoidal function

$$x_{i(l)} = f(a_{i(l)}) = \frac{1}{1 + \exp(-a_{i(l)})},$$

whose output range is [0,1]. An interesting property of this function is that its derivative can be easily computed as $f'(a) = f(a) \cdot (1 - f(a))$. Another common choice is to use the hyperbolic tangent as squashing function

$$x_{i(l)} = f(a_{i(l)}) = \tanh(a_{i(l)}),$$

which yields values in [-1,1] and whose derivative is $f'(a) = 1 - f^2(a)$.

A different choice of function $\mathcal{F}(\cdot)$ leads to Radial Basis Function networks (RBFs). The activation of the locally–tuned units of these networks follows the equation

$$a_{i(l)}(d) = \mathcal{F}(W_{i(l)}, X_{l-1}(d)) = \frac{1}{\sigma^2_{i(l)}} \sum_{j(l-1)=1}^{n(l-1)} (x_{j(l-1)}(d) - w_{i(l),j(l-1)})^2$$
$$= \frac{\|X_{l-1}(d) - \overline{W}_{i(l)}\|^2}{\sigma^2_{i(l)}}, \qquad (11.3)$$

being $W_{i(l)} = [\overline{W}'_{i(l)}, \sigma_{i(l)}]' \in I\!R^{n(l-1)+1}$, $w_{i(l),j(l-1)}$ the weight of the connection between the neurons $j(l-1)$ and $i(l)$, and $\sigma_{i(l)}$ the 'width' of the Gaussian. In this case the output neuron function is

$$x_{i(l)} = f(a_{i(l)}) = \exp(-a_{i(l)}),$$

that, for $a_{i(l)} \geq 0$, ranges in [0,1]. In this case the derivative is simply $f'(a) = -f(a)$. Usually RBF networks are composed by a hidden layer of locally–tuned units and by an output layer of linear or sigmoidal neurons.

The general architecture of multi–layered static network is depicted in Figure 11.1. It can be noticed that, in this case, no state layer is considered.

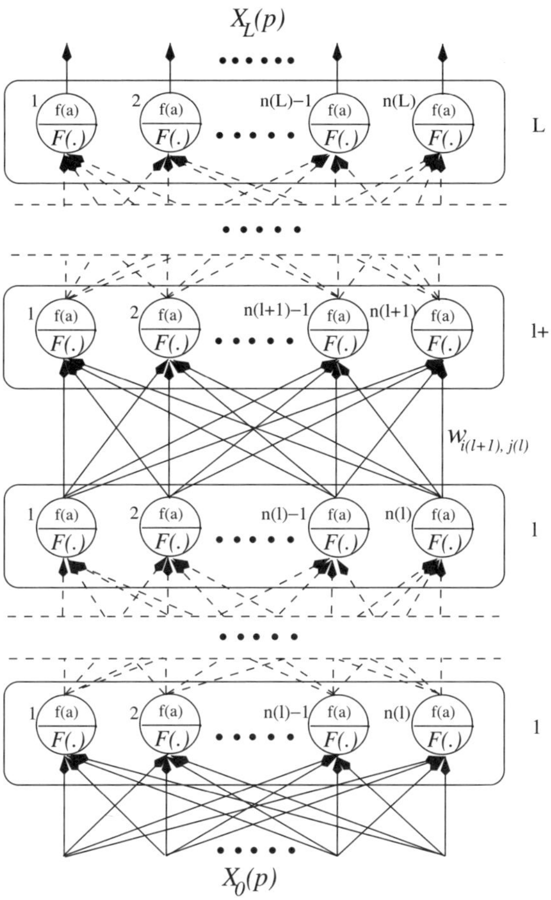

Fig. 11.1. Generic architecture of a multi–layered feedforward neural network without shortcut connections.

Recurrent networks

Recurrent neural networks process sequences exploiting an internal state representation to properly encode the history of past frames. Given a sequence of m–dimensional real vectors $S_p(1), \ldots, S_p(T)$, at each time step t, the neural network updates its internal state encoded in the hidden layer S, $X_S(p,t)$, and computes an output vector $X_L(p,t)$ which, due to the layered architecture, depends on the current state. The current state is updated by processing the current input frame $S_p(t)$ and the state at the previous time step $X_S(p,t-1)$, that are the inputs to the first layer of the neural network. Basically the layers $1, \ldots, S$ implement the *state transition function*, whereas the following layers, $S+1, \ldots, L$ realize the *output function*. The state should encode the relevant information in the past frames, needed to perform a given processing of the input sequence. The state must be set to a given initial configuration $X_S(0)$ before processing each sequence. The neural network can be trained to map the input sequence to another sequence with the same length, i.e. $X_L(1), \ldots, X_L(T)$, thus performing an *I/O–transduction* or time series filtering. Otherwise, it may be considered only the last output frame $X_L(T)$, mapping each input sequence to a vector in $\mathbb{R}^{n(L)}$. This last approach can be used to implement a sequence classifier that assigns an input sequence to one among a set of predefined classes.

In the case of a recurrent neural network, the neurons' activations evolve according to

$$\begin{aligned} a_{i(1)}(p,t) &= \mathcal{F}(W_{i(1)}, X_0(p,t), X_S(p,t-1)), \\ a_{i(l)}(p,t) &= \mathcal{F}(W_{i(l)}, X_{l-1}(p,t)), \end{aligned}$$

where $X_0(p,t) = S_p(t)$ is the input frame at time t. If $\mathcal{F}$ in the first layer is implemented by a linear function, we have

$$a_{i(1)}(p,t) = w_{i(1)} + (W^x_{i(1)})' X_0(p,t) + (W^b_{i(1)})' X_S(p,t-1),$$

where $W^b_{i(1)} \in \mathbb{R}^{n(S)}$ collects the *feedback connections*, $W^x_{i(1)} \in \mathbb{R}^m$ the *connections from external inputs*, and $w_{i(1)}$ is the bias. The output function can be chosen as in the case of static networks.

While processing an input sequence, the same neural network is exploited at each time step to compute both the state vector $X_S(p,t)$ and the output vector $X_L(p,t)$. This procedure yields a computational scheme that can be visualized as the *unfolding* of the given feed forward neural network along the time sequence. Each processing step adds a new replica of the recurrent neural network, that is attached to the previous one through the state layer. The external input frame is fed to the corresponding replica in the time unfolding. In other words, the complete processing of a sequence can be described as a static processing performed by a modular neural network obtained by the replication of the same module for each frame in the sequence. Clearly each module shares the same weights with all the other modules in the unfolding. The architecture of a recurrent neural network and the corresponding unfolding network obtained while processing an input sequence are shown in Figure 11.2.

Recursive networks

Recursive neural networks are a generalization of recurrent networks able to process graphs. In particular, we restrict our description to models able to deal with DPAGs, even if more general architectures have been proposed in the literature to process non-positional graphs with or without labeled edges (DAGs, DAGs-LE), cyclic and undirected graphs.[10–12] Recently a more general model, the Graph Neural Network (GNN), able to deal with general classes of graphs, has been proposed.[6,7]

Recursive processing takes place following the graph connectivity. In particular, the computation is performed starting from the graph *leaves*, i.e. the nodes having no descendants, up to the graphs *sources*, i.e. the nodes having no parents.

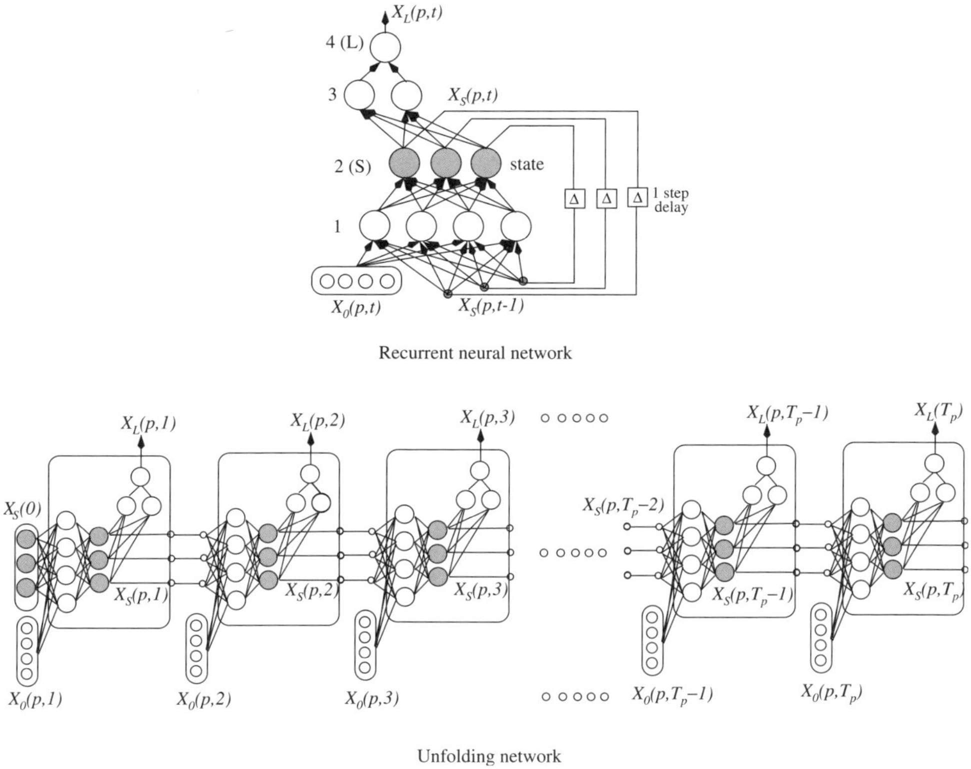

Fig. 11.2. Architecture of a recurrent neural network and the corresponding unfolding network when processing an input sequence.

This schema requires to process the graph nodes following their inverse topological order. In this order each node is preceded by its descendants and followed by its ancestors. As it happens for recurrent networks, the state layer plays a crucial role in the processing phase. In fact, the network state $X_S(p, v)$, for each node v, should encode the relevant information present in the subgraph containing node v and its descendants. Thus, the state for a node v is computed from the states of its children $X_S(\mathrm{ch}_1[v]), \ldots, X_S(\mathrm{ch}_q[v])$ and the feature vector stored in the node itself, $X_0(p, v)$. Unless different schemas are adopted, this type of computation requires the input graph to be directed, to define a unique direction for the state propagation (i.e. from the children to the parents), positional, to assign each child's state to a precise argument of the state transition function, and acyclic, to avoid loops in the state computation that would create self dependencies for the state values.[a] As introduced in Section 11.2.1, we assume that the outgoing arc positions corresponding to missing arcs are represented by nil pointers with an associated *initial*

[a]In a more general framework this requirement could be removed by considering a relaxation process that iterates the state transition function until the states converge to a stable configuration.

(or nil) state $X_S(0)$ that is predefined. In particular, the state of the leaf nodes is computed feeding the neural network with the local label and the nil state $X_S(0)$ for all the children.

At each node v of the graph, the neurons' activations are computed as

$$\begin{aligned} a_{i(1)}(p,v) &= \mathcal{F}(W_{i(1)}, X_0(p,v), X_S(p,\mathrm{ch}_1[v]),\ldots,X_S(p,\mathrm{ch}_q[v])), \\ a_{i(l)}(p,v) &= \mathcal{F}(W_{i(l)}, X_{l-1}(p,v)). \end{aligned}$$

As noticed before, unless the function $\mathcal{F}(\cdot)$ is commutative with respect to the arguments $X_S(p,\mathrm{ch}_k[v])$, $k=1,\ldots,q$, the child nodes must be associated to a given position, assuming $X_S(p,\mathrm{ch}_k[v]) = X_S(0)$ when the child in the k–th position is missing. In general, a different permutation of the node's children yields a different state. Moreover, it is assumed the same number of children q for any node in the input graphs. Thus, the value of q must be chosen a priori, by considering the maximum outdegree of the input graphs. In some applications, this may be obtained by applying a pruning technique that limits the nodes' outdegrees to q, by removing those links that are considered less informative. Clearly, the pruning techniques are problem dependent.

If we assume that $\mathcal{F}(\cdot)$ performs a linear combination of the inputs, the activations of the neurons in the first layer are computed as

$$a_{i(1)}(p,v) = w_{i(1)} + B_{i(1)}X_0(p,v) + \sum_{k=1}^{q} A^k_{i(1)} X_S(p,\mathrm{ch}_k[v]),$$

where $A^k_{i(1)} \in I\!R^{n(S)}$, $k=1,\ldots,q$, are the *pointer* vectors that are used to propagate the states of the node's children, $B_{i(1)} \in I\!R^m$ is used to process the node's label, and $w_{i(1)}$ is the neuron's bias. The weight vector of neuron $i(1)$ is $W_{i(1)} = [(A^1_{i(1)})',\ldots,(A^q_{i(1)})', B'_{i(1)}, w_{i(1)}]'$. Also in this case, the outputs of the neurons are generally obtained from their activations by applying a squashing function.

As in the case of recurrent neural networks, the recursive processing scheme yields an *unfolding network* that is built by interconnecting the same neural network module through the state layer using the input graph topology. For each node v, a replica of the neural network performs the local computation in order to compute the node state and the node output. Each replica of the neural network in the unfolding network shares exactly the same set of weights with the other replicas. Clearly, when processing different input graphs, the same recursive network generally yields unfolding networks with a different topology. In fact, the unfolding network can be considered only a visualization of the processing schema (see Figure 11.3).

The first S layers of the neural network are used to implement the *state transition function* that maps the state vectors of the node's children and the local feature vector into the node's state vector $X_S(p,v)$. This part of the neural network realizes an *enconding network* that maps each subgraph rooted at a given node into its state vector. The successive layers, $l = S+1,\ldots,L$, compute the *output function* given the node's state in layer S. The output network produces a vector $X_L(p,v) \in I\!R^{n(L)}$

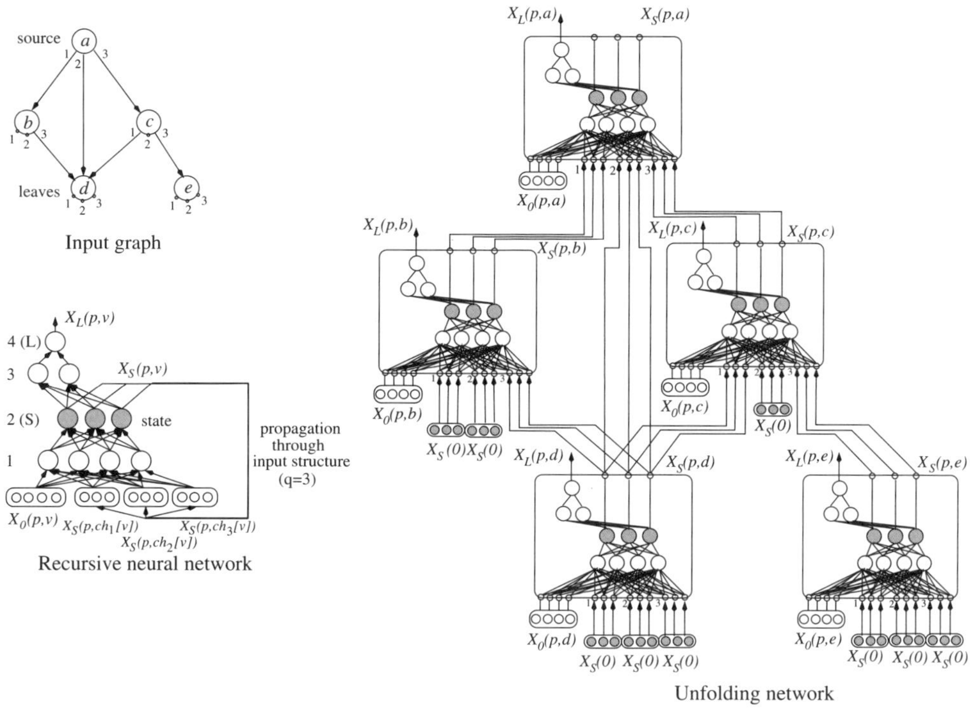

Fig. 11.3. Architecture of a recursive neural network and the corresponding unfolding network when processing an input graph. The recursive neural network has a three dimensional state and it is able to process DPAGs with maximum outdegree $q = 3$.

for each node, that represents the output of the recursive computation. Hence, the processing schema realizes a so called I/O–transduction, that maps an input graph into an isomorphic graph with different labels. Alternatively, the output vector can be evaluated only at the graph supersource providing a mapping of the input graphs into an Euclidean space of appropriate dimension.

11.3. Learning with Neural Networks

Independently of the structure of the processed data objects, learning for neural networks is defined in a supervised framework, in which it is assumed to have assigned a learning set $\mathcal{L}_e$ containing labeled data examples. Depending on the task at hand, the learning problem can be formulated as a *regression task* or a *classification task*. In the first case, the provided supervision is in the form of real valued vectors that represent the target for the neural network output, whereas in the latter case a class label is provided for each example. However, the classification task can be conveniently treated as a regression task once that the class labels are properly coded by a real valued vector. For instance, for two–class classification problems, the class membership can be encoded using only one output neuron by assigning two real values, usually corresponding to the limits of the codomain of the neuron output function, to the two class labels. When a multiclass problem is considered, if there are C classes, a common approach is to associate each class to one output neuron such that the target vector for an input pattern belonging to class c has its c–th component set to the higher output value and all the other components set to the lower limit of the ouput range.

Depending on the specific structure of the input data, supervision can be available in different ways. When dealing with data represented by feature vectors, the training examples are pairs of the form (X_p, D_p), $p = 1, \dots, P$, where $X_p \in \mathbb{R}^m$ is the input vector and $D_p \in \mathbb{R}^{n(L)}$ is the corresponding target for the neural network output. In the case of recurrent networks, each input example is a sequence and the targets can be provided at different time steps along the sequence. Thus if $\mathcal{S}_p$ is the input sequence $S_p(1), \dots, S_p(T_p)$, where $S_p(t) \in \mathbb{R}^m$, the supervision is represented by a subsequence $D_p = \{D_p(t_1), \dots, D_p(t_K)\}$, $t_1, \dots, t_K \in [1, T_p]$, where $D_p(t_i) \in \mathbb{R}^{n(L)}$ is the output target at time t_i. Basically, the supervised schema for sequences allows the specification of an output target for a subset of the time steps. Finally, when considering structured inputs in the form of DPAGs, each example in the learning set is a pair composed by a graph $\mathcal{G}_p$ and an associated set of supervisions on its nodes $D_p = \{D_p(v_1), \dots, D_p(v_K)\}$, $v_1, \dots, v_K \in V_p$, where V_p is the set of the nodes of $\mathcal{G}_p$ and $D_p(v_k) \in \mathbb{R}^{n(L)}$ is the target value for node v_k.

A general definition for the training set can be formulated by introducing the following concepts. Let I_p be an input pattern for the neural network and $\mathcal{I}$ the corresponding input space, i.e. $I_p \in \mathcal{I}$. In the general case, I_p is a structured object composed by atomic elements with links among them. The set of the atomic elements is denoted by $V_p = \{v_1^p, \dots, v_{N_p}^p\}$, where v_k^p represents the k–th element composing the structured input I_p. In the case of static networks, I_p is a vector in $\mathbb{R}^m$ and V_p contains a single element labeled with the input vector X_p; for recurrent neural networks, I_p is a sequence of T_p m–dimensional vectors and V_p is the set of these T_p elements; for recursive neural networks I_p is a DPAG whose nodes are labeled with feature vectors in $\mathbb{R}^m$ and V_p is the set of nodes in the graph with the associated vector labels. The corresponding supervision is assigned as target vectors for the neural network outputs on a subset of the elements in V_p. Thus the target set is $D_p = \{D_p(v_{s_1}^p), \dots, D_p(v_{s_K}^p)\}$, where $v_{s_1}^p, \dots, v_{s_K}^p \in V_p$ and $D_p(v_{s_k}^p) \in \mathbb{R}^{n(L)}$. Hence, the learning set can be formally defined as a set of input–target pairs

$$\mathcal{L}_e = \{(I_p, D_p), I_p \in \mathcal{I}, D_p \in \mathcal{D}(I_p)\ p = 1, \dots, P\},$$

where I_p is the input, D_p the corresponding target set, and $\mathcal{D}(I_p)$ is the target space, that represents the partial mappings of the elements in I_p to vectors in $\mathbb{R}^{n(L)}$. Each component of $D_p(v_{s_k}^p)$ belongs to $[\underline{r}, \overline{r}]$.

The output–target data fitting is estimated by means of the cost function

$$E(\mathcal{L}_e, \Theta) = \sum_{p=1}^{P} E((I_p, D_p), \Theta) = \sum_{p=1}^{P} \sum_{k=1}^{K} d(X_L(p, v_{s_k}^p, \Theta), D_p(v_{s_k}^p)),$$

where $X_L(p, v^p_{s_k}, \Theta)$ is the neural network output for the input pattern p, computed for the s_k–th input data element $v^p_{s_k}$ as specified by the associated supervision target $D_p(v^p_{s_k})$. The vector Θ collects all the neural network free parameters, i.e. it can be obtained as the concatenation of the connection weight matrices for all the neural network layers. The total error on the learning set is a linear function of the error contribution for each example (I_p, D_p), that is also a linear combination of the errors measured on each data element $v^p_{s_k}$ in the input I_p for which a supervision target D_p is provided. The choice of the distance measure $d(\cdot)$ in $\mathbb{R}^{n(L)}$ plays a very crucial role in practice and depends significantly on the problem at hand. A common choice is that of considering the distance induced by an L_q norm $(1 \leq q \leq \infty)$. In the case of $q = 2$, that is most frequently considered, the cost is given by

$$E_Q(\mathcal{L}_e, \Theta) = \frac{1}{2} \sum_{p=1}^{P} \sum_{k=1}^{K} \sum_{j(L)=1}^{n(L)} \left[x_{j(L)}(p, v^p_{s_k}, \Theta) - d_{j(L)}(v^p_{s_k}) \right]^2. \tag{11.4}$$

The use of different values of q has been widely evaluated in literature[13,14] in a number of different domain problems. It turns out that the noise in the target domain can be reduced by rising power values less than 2, whereas the sensitivity of the partition planes to the geometry of the problem may be increased with higher power values.

By the definition of the cost function, the learning problem is cast as a non–linear optimization task where the cost function $E(\mathcal{L}_e, \Theta)$ is to be minimized with respect to the parameters Θ. The difficulty of this problem is due to the high dimensionality of the parameter vector Θ and to the complexity of the error surface, that can present local minima, large plateaux and complex symmetries. In this scenario, the most successfull approach is to apply a local optimization technique based on gradient descent. Luckily the gradient of the cost function $E(\mathcal{L}_e, \Theta)$ can be computed efficiently taking into account the modular architecture of the neural network that is composed by many simple interconnected computing units. Once the cost function gradient $\nabla_\Theta E(\mathcal{L}_e, \Theta)$ is computed for the current parameter vector Θ^e, the new hypothesis in the learning process can be computed as

$$\Theta^{e+1} = \Theta^e - \eta^e \cdot \nabla_\Theta E(\mathcal{L}_e, \Theta)|_{\Theta^e}, \tag{11.5}$$

where e is usually referred to as *learning epoch* and η^e is a parameter called *learning rate*. Usually, the neural network weights are randomly initialized at epoch $e = 0$ using an uniform distribution in $[w_{min}, w_{max}]$. Many different schemas have been proposed to improve the basic BackPropagation gradient descent algorithm, for instance based on an automatic adaptation of the learning rate parameter[15] or exploiting higher order information.[16] Among the adaptive learning rate algorithms, the *Resilient Propagation* (RPROP) algorithm[17,18] constitutes an useful, and actually frequently employed, variation of the simple gradient descent approach of equation (11.5), aimed at performing a local adaptation of the weights based only on the temporal behaviour of the gradient sign. In fact, many algorithms have been proposed so far to deal with the problem of appropriately carrying out the weight update, exploiting some sort of parameter adaptation during learning, the most common of which presuppose to adjust a possibly weight specific learning rate, according to the observed behaviour of the error function. Instead, what is often disregarded is that the weight step Δ_θ^e, along the weight coordinate $\theta \in \Theta$, depends both on the learning rate and on the partial derivative $\frac{\partial E}{\partial \theta}$, so that the effect of the learning rate adaptation can be significantly affected by the unpredictable behaviour of the derivative modulus. In order to avoid the problem of the *blurred adaptivity*, the RPROP algorithm changes the weight step Δ_θ^e directly, i.e. based only on the sign persistency (or not) of $\frac{\partial E}{\partial \theta}$ (but not on its modulus). To accomplish this task, the individual update value Δ_θ^e is introduced for each weight as

$$\Delta_\theta^e = \begin{cases} \eta^+ \cdot \Delta_\theta^{e-1} & \text{if } \left.\frac{\partial E}{\partial \theta}\right|_{e-1} \cdot \left.\frac{\partial E}{\partial \theta}\right|_{e} > 0 \\ \eta^- \cdot \Delta_\theta^{e-1} & \text{if } \left.\frac{\partial E}{\partial \theta}\right|_{e-1} \cdot \left.\frac{\partial E}{\partial \theta}\right|_{e} < 0 \\ \Delta_\theta^{e-1} & \text{otherwise} \end{cases} \tag{11.6}$$

where $0 < \eta^- < 1 < \eta^+$. Based on equation (11.6), every time the partial derivative with respect to a certain weight θ changes its sign, which indicates that the last update was too big and the descending algorithm has jumped over a minimum, Δ_θ^e is decreased by a factor η^-, whereas if, for two consecutive steps, the derivative remains constant in sign, Δ_θ^e can be slightly increased, to accelerate convergence in locally monotonic regions. Following equation (11.6), the weight itself is modified as follows

$$\theta^{e+1} = \theta^e - \Delta_\theta^e \cdot \text{sign}\left(\left.\frac{\partial E}{\partial \theta}\right|_e \right),$$

i.e., at epoch e each weight is changed in the opposite direction of the corresponding gradient component and the amount of the update is given by the adapted local learning rate Δ_θ^e. Additionally, in case of a change in sign, there should be no adaptation in the succeeding learning step, that can be achieved, in practice, by setting $\left.\frac{\partial E}{\partial \theta}\right|_{e-1} = 0$ in the adaptation rule of equation (11.6). In order to reduce the number of freely adjustable parameters, fixed values for $\eta^+(= 1.2)$ and $\eta^-(= 0.5)$ were also suggested and experimentally proved to be effective and robust (slight vatiations of such values did neither improve nor deteriorate the effectiveness and the efficiency of RPROP).[18]

However, the main drawback of all gradient based approaches are due to their locality, that makes them hindered mainly by the presence of local minima in the error surface. Nevertheless, in most applications, these issues are rarely a problem and can be overcome by starting the learning process from different initial weight vectors, thus providing a more extensive exploration of the hypothesis space.

The gradient of the cost function can be computed effeciently by exploiting the neural network structure. Firstly, as observed before, the total cost function is obtained by summing the contributions of each example in the learning set. Thus the total gradient can be computed by accumulating the contributions of each single supervised input I_p

$$\nabla_\Theta E(\mathcal{L}_e, \Theta) = \sum_{p=1}^{P} \nabla_\Theta E((I_p, D_p), \Theta).$$

When the parameters are updated using the gradient computed on all the examples in the learning set for each epoch e, the learning process is referred to as *batch learning.* In some cases, especially when the number of the available examples is large and the learning set is supposed to be redundant, a *block* or *pattern* mode can be applied. In the first case, the parameter vector is updated using the gradient accumulated on a subset (i.e. a block) of examples, in the second one the update takes place using the gradient for a single pattern. These two approaches are clearly approximations and sometimes can lead to instabilities in the learning process.

To compute the gradient of the cost fuction $E((I_p, D_p), \Theta)$, let us consider a generic parameter $\theta \in \Theta$ corresponding to a connection weight or to a bias. When considering recurrent and recursive neural networks, the network computation is repeated for each data element v_k^p using the same set of parameters Θ, leading to the *unfolding* structure illustrated in the previous sections. In order to modularize the gradient computation, we can decouple the weights corresponding to the processing of each data element v_k^p of the input I_p by the neural network. Let us denote by $\Theta(v_k^p)$ the weights of the neural network replica used to process the element v_k^p. Thus the cost function for the input I_p can be espressed as a function of the parameter vectors $\Theta(v_k^p)$, $k = 1, \ldots, |V_p|$, i.e. we consider $E((I_p, D_p), \Theta(v_1^p), \ldots, \Theta(v_{|V_p|}^p))$ with the weight sharing constraints $\Theta(v_1^p) = \ldots = \Theta(v_{|V_p|}^p) = \Theta$. By applying the derivative rule for function composition we obtain

$$\nabla_\Theta E((I_p, D_p), \Theta) = \sum_{k=1}^{|V_p|} \nabla_{\Theta(v_k^p)} E((I_p, D_p), \Theta(v_1^p), \ldots, \Theta(v_{|V_p|}^p)), \tag{11.7}$$

that states that the gradient of the cost function with respect to the neural network weights Θ can be computed by accumulating the gradients with respect to each replica of the shared weights in the unfolded network. In the following, to simplify the notation, we will refer to the partial error function on the example (I_p, D_p) as E_p, omitting the list of variables on which it depends.

The derivative of the cost function with respect to a given weight replica $\theta(v_k^p)$ can be computed by applying the same approach as in the classical BackPropagation for feedforward neural networks. In fact, due to the neural network architecture, in which each weight is local to a given neuron, the weight $\theta(v_k^p)$ will affect directly only the activation $a_\theta(v_k^p)$ of the corresponding neuron when processing the input element v_k^p. Thus, we can rewrite the partial derivative of the partial cost function with respect to $\theta(v_k^p)$ as

$$\frac{\partial E_p}{\partial \theta(v_k^p)} = \frac{\partial E_p}{\partial a_\theta(v_k^p)} \cdot \frac{\partial a_\theta(v_k^p)}{\partial \theta(v_k^p)} = \delta_{a_\theta(v_k^p)} \cdot \frac{\partial a_\theta(v_k^p)}{\partial \theta(v_k^p)}, \tag{11.8}$$

where $\delta_{a_\theta(v_k^p)}$ is often referred to as the *generalized error* for the neuron whose activation is $a_\theta(v_k^p)$ and represents the influence of the same activation on the partial cost function. The term $\frac{\partial a_\theta(v_k^p)}{\partial \theta(v_k^p)}$ depends on the particular function realized by the neuron and on the nature of the parameter $\theta(v_k^p)$. For instance, when the activation function is a linear combination of the inputs, as in equation (11.2), and $\theta(v_k^p) = w_{i(l),j(l-1)}$ we obtain

$$\frac{\partial a_{i(l)}(v_k^p)}{\partial w_{i(l),j(l-1)}} = x_{j(l-1)}(v_k^p),$$

while, for the bias $w_{i(l)}$, it is simply

$$\frac{\partial a_{i(l)}(v_k^p)}{\partial w_{i(l)}} = 1.$$

As a different example, for a radial basis function neuron, as specified in equation (11.3), we have

$$\frac{\partial a_{i(l)}(v_k^p)}{\partial w_{i(l),j(l-1)}} = -2 \cdot \frac{(x_{j(l-1)}(v_k^p) - w_{i(l),j(l-1)})}{\sigma_{i(l)}^2},$$

and

$$\frac{\partial a_{i(l)}(v_k^p)}{\partial \sigma_{i(l)}} = -2 \cdot \frac{\|X_{j(l-1)}(v_k^p) - \overline{W}_{i(l)}\|^2}{\sigma_{i(l)}^3}.$$

To complete the computation of the derivative, a method to express the generalized error $\delta_{a_\theta(v_k^p)}$ for any neuron is required. The propagation of the generalized errors is the core of the *backward step* of the BackProgation algorithm. Even if the procedure is the same, depending on the nature of the input pattern, the propagation algorithm is called BackPropagation (BP) in the case of static inputs, BackPropagation Through Time (BPTT) in the case of time series, and BackPropagation Through Structure (BPTS) in the case of graphs. The generalized errors are defined as the derivative of the partial error with respect to the neurons' activations, thus it is convenient to consider E_p as dependent on the activations $a_{i(l)}(v_i^p)$ instead of the neural network weights $\theta(v_k^p)$, i.e. we consider E_p as $E((I_p, D_p), a_{1(1)}(v_1^p), \ldots, a_{L(n(L))}(v_{|V_p|}^p))$. Then, we can expand the derivative for the following three cases, applying the chain rule.

(1) $a_{i(l)}(v_k^p)$ is the activation of an output neuron (i.e. $l = L$).
In this case, if no supervision is specified for the data element v_k^p then $\delta_{i(L)}(v_k^p) = 0$, otherwise the generalized error can be computed directly taking into account

that the only term in the cost function depending on $a_{i(L)}(v_k^p)$ is the one corresponding to the supervision $D_p(v_k^p)$. In fact, if we consider the quadratic cost function of equation (11.4), we obtain

$$\delta_{i(L)}(v_k^p) = f'(a_{i(L)}(v_k^p))[x_{i(L)}(p, v_k^p) - d_{i(L)}(v_k^p)] \ , \tag{11.9}$$

where $f'(\cdot)$ is the derivative of the neuron output function. Thus, in this case the value $\delta_{i(L)}(v_k^p)$ is actually obtained by considering the approximation error for the corresponding output.

(2) $a_{i(l)}(v_k^p)$ is the activation of a neuron in a hidden layer.
In this case, the output of the neuron $i(l)$ is fed to the neurons of the $(l+1)$–th layer. Hence, by considering that the activation $a_{i(l)}(v_k^p)$ directly affects the activations of the neurons in the following layer, using the chain rule, we can expand the derivative as

$$\frac{\partial E_p}{\partial a_{i(l)}(v_k^p)} = \sum_{j(l+1)=1}^{n(l+1)} \frac{\partial E_p}{\partial a_{j(l+1)}(v_k^p)} \cdot \frac{\partial a_{j(l+1)}(v_k^p)}{\partial x_{i(l)}(v_k^p)} \cdot \frac{\partial x_{i(l)}(v_k^p)}{\partial a_{i(l)}(v_k^p)} \ .$$

If we use the definition of the generalized error, this equation can be rewritten as

$$\delta_{i(l)}(v_k^p) = f'(a_{i(l)}(v_k^p)) \cdot \sum_{j(l+1)=1}^{n(l+1)} \delta_{j(l+1)}(v_k^p) \cdot \frac{\partial a_{j(l+1)}(v_k^p)}{\partial x_{i(l)}(v_k^p)} \ , \tag{11.10}$$

that is the classical backward propagation rule for the generalized errors. The coefficients used to propagate backwards the generalized errors $\delta_{j(l+1)}(v_k^p)$ to the previous layer are the derivatives $\frac{\partial a_{j(l+1)}(v_k^p)}{\partial x_{i(l)}(v_k^p)}$, that depend on the function implemented by the neurons in the $(l+1)$–th layer to combine the outputs of the neurons in the layer l. When the activation is computed as a linear combination of the inputs as in equation (11.2), the derivative is

$$\frac{\partial a_{j(l+1)}(v_k^p)}{\partial x_{i(l)}(v_k^p)} = w_{j(l+1),i(l)} ,$$

yielding the well know backpropagation rule

$$\delta_{i(l)}(v_k^p) = f'(a_{i(l)}(v_k^p)) \cdot \sum_{j(l+1)=1}^{n(l+1)} \delta_{j(l+1)}(v_k^p) \cdot w_{j(l+1),i(l)} \tag{11.11}$$

inside the v_k^p replica of the neural network. In the case of RBF neurons of equation (11.3) the coefficients are

$$\frac{\partial a_{j(l+1)}(v_k^p)}{\partial x_{i(l)}(v_k^p)} = 2 \cdot \frac{x_{i(l)}(v_k^p) - w_{j(l+1),i(l)}}{\sigma_{j(l+1)}^2} ,$$

and thus the error propagation is modulated by the alignment between the neuron input and its centroid.

(3) $a_{i(l)}(v_k^p)$ is the activation of a neuron in the state layer.
In this case the generalized error propagation is determined by the input data structure. In fact, the output of neuron $i(l)$ is used to compute a state component which is propagated to the element's parents (i.e. the state at the next time step in the case of sequences and the states of the parent nodes in the case of DPAGs). Hence, in this case, the affected layer is the first layer in the neural network replicas corresponding to the current node's parents. Thus, in the chain rule expansion of the derivative we must consider all the variables affected by $a_{i(l)}(v_k^p)$, obtaining

$$\frac{\partial E_p}{\partial a_{i(l)}(v_k^p)} = \sum_{v \in pa[v_k^p]} \sum_{j(1)=1}^{n(1)} \frac{\partial E_p}{\partial a_{j(1)}(v)} \cdot \frac{\partial a_{j(1)}(v)}{\partial x_{i(l)}(v_k^p)} \cdot \frac{\partial x_{i(l)}(v_k^p)}{\partial a_{i(l)}(v_k^p)},$$

that yields the backpropagation rule through the data structure

$$\delta_{i(l)}(v_k^p) = f'(a_{i(l)}(v_k^p)) \cdot \sum_{v \in pa[v_k^p]} \sum_{j(1)=1}^{n(1)} \delta_{j(1)}(v) \cdot \frac{\partial a_{j(1)}(v)}{\partial x_{i(l)}(v_k^p)}, \tag{11.12}$$

where, as in the previous case, the term $\frac{\partial a_{j(1)}(v)}{\partial x_{i(l)}(v_k^p)}$ depends on the adopted neuron model. When the activation is computed as a linear combination of the neuron's inputs, the formula is a straightforward extension of the well known backpropagation rule of equation (11.11)

$$\delta_{i(l)}(v_k^p) = f'(a_{i(l)}(v_k^p)) \cdot \sum_{v \in pa[v_k^p]} \sum_{j(1)=1}^{n(1)} \delta_{j(1)}(v) \cdot w_{j(1),i(l)}^{\#(v_k^p)},$$

with the only difference that there is a double propagation that considers both the network topology (i.e. the inner summation) and the structure of the current input (i.e. the outer summation), with a set of weights $w_{j(1),i(l)}^{\#(v_k^p)}$ that depend on the position $\#(v_k^p)$ of the current node in its parent's children list.

If we consider the analytical form of the error propagation equations, it can be noticed that the algorithm can be implemented with a modular structure. In fact, if we assume that the neural network is a computational block that is applied recursively on the data elements, in the forward pass the network state vectors $X_S(v)$ are propagated from the child nodes to their parents, each through the associated input components in the first neural network layer, as shown in the unfolding procedure. In the backward step, the generalized errors $\delta_k(v)$ are computed inside the network block following the standard backpropagation scheme as detailed in the points (1) and (2) of the previous list. If we consider the state propagation inputs in the first neural network layer as dummy units, we can compute the associated generalized errors as

$$\bar{\delta}_{i,u}(v) = \sum_{j(1)=1}^{n(1)} \delta_{j(1)}(v) \cdot \frac{\partial a_{j(1)}(v)}{\partial x_i(u)},$$

where i is the index identifying the state component and u the position of the corresponding state vector in the node child list. This computation can be performed inside the neural network replica at node v since it uses only local variables. Finally, the generalized errors $\bar{\delta}_{i,u}(v)$ available in correspondence of the block state inputs can be propagated to the child nodes in the structure by simply rewriting the structure propagation equation (11.12) as

$$\delta_{i(l)}(v_k^p) = f'(a_{i(l)}(v_k^p)) \cdot \sum_{v \in pa[v_k^p]} \bar{\delta}_{i(l),\#(v_k^p)}(v),$$

that evidences the error propagation along the input data structure arcs. The structure of the error propagation is shown in Figure 11.4.

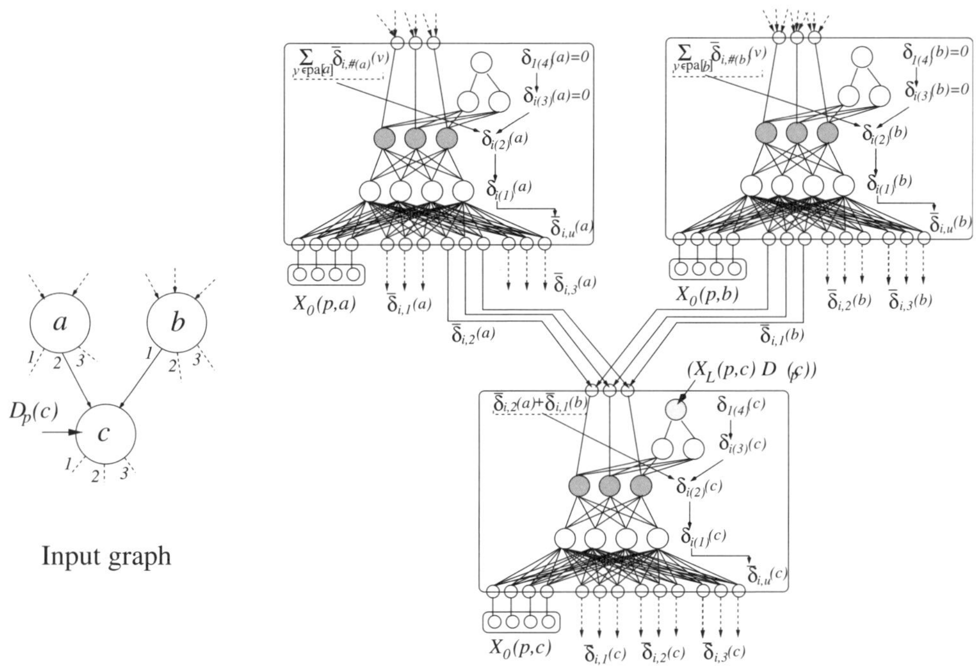

Fig. 11.4. Propagation of the generalized errors through the unfolding structure. The $\delta_{i(l)}(v)$ for the output layers are not null only for the node c, for which a supervision target $D_p(c)$ is provided. The generalized errors are propagated inside each neural network replica following the standard BackPropagation scheme, whereas the $\bar{\delta}_{i,u}(v)$ are propagated and accumulated following the input graph connections.

In general, a layer can be subject to any combination of the three propagation patterns. In this case the computation can be simply performed by summing up all the contributions.

Finally, in order to compute the gradient for a given input data pattern, the following two steps must be performed.

(1) *Forward step.* The input structure is processed from the leaves to the sources following the inverse topological order (in the case of sequences it corresponds to the natural time flow). The neurons' outputs are computed and stored in the network replicas at each input data element.
(2) *Backward step.* The generalized errors $\delta_{i(l)}(v)$ for each replica v of neuron $i(l)$ are set to zero and the backward propagation is performed following the topological order of the input data elements, i.e. from the source to the leaf nodes in the case of DPAGs, backwards in time for time sequences. When a supervision $D_p(v)$ is available for the data element v, the variables $\delta_{i(l)}(v)$ for the neurons in the corresponding replica of the neural network output layer are updated using the contribution of the error as in equation (11.9). For each neural network replica v, the propagation of the $\delta_{i(l)}(v)$ is performed backward through the network layers as described in equation (11.10). During the same phase, the gradient components for each connection are updated by equation (11.8). It should be noticed that the weight sharing among the different neural network replicas requires to accumulate the gradients of each weight computed for each data element as described by equation (11.7). Thus it is not needed to have a different gradient component for each replica v of the network, as it is also for the weights themselves, that are stored just once. Finally, when the first layer is reached, the $\delta_{i(l)}(v)$ are backpropagated to the children of node v as described in equation (11.12).

11.4. Processing Graphs: Application Domains

In the following, we shortly review and briefly sketch some real–world problems that will be useful in order to understand the characteristics of the application domains of recursive neural networks. Next, we will deeply examine the problem of dectecting object in images and of classifying images, using *ad hoc* recursive neural network architectures.

Logo Recognition

Pattern recognition is an important source of applications suited for an adaptive processing of data structures. This was recognized early with the introduction of syntactic and structural pattern recognition,[19,20] that are based on the premise that the structure of an entity is very important for both the classification and the description of the related information.

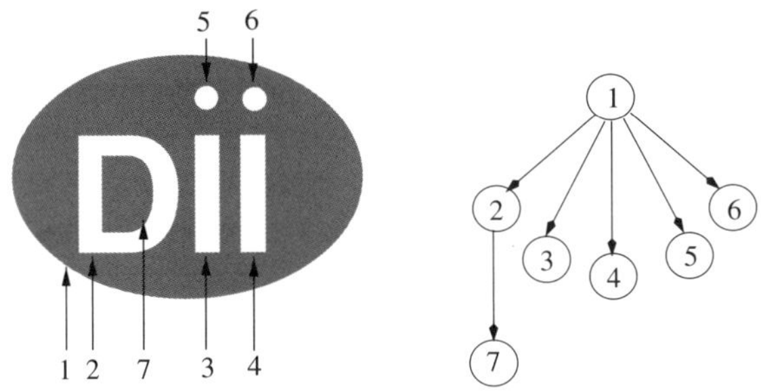

Fig. 11.5. Representation of a logo with a contour–tree structure.

Figure 11.5 shows a logo with a corresponding structural representation, called a contour–tree.[3] The contour–tree is constructed by associating a node with each exterior or interior contour extracted from the logo instance. Nodes in the tree are labeled by a feature vector of reals. In fact, the contour–tree representation contains both the topological structured information related to the logo appearance and the continuous values pertaining to each contour node.[b] Hence, symbolic and subsymbolic information coexist in the contour–tree representation. This representation is invariant with respect to roto–translations and scaling (that do not cause loss in information). Of course, the extraction of robust representations from patterns is not a minor problem. The presence of a significant amount of noise is likely to deeply affect representations that are strongly based on symbols. Hence, depending on the problem at hand, the structured representation that we derive should emphasize the symbolic or the subsymbolic information.

Problem Solving in Artificial Intelligence

Another interesting application is related to the classification of logical terms. Theorem provers and expert systems can exploit term classification in order to devise heuristics that speed up the search of a solution. In fact, when the search–tree is very large, some heuristics must be used in order to identify the most promising paths. In practice, the heuristics suggest the next path by imposing an ordering on the facts (i.e., the subgoals) that can be used at the next step of the computation. An interesting solution consists of learning the heuristics from data samples obtained by solutions which were already found (positive examples are states or inference steps on solution paths within the search–tree, while negative samples account for nodes on failure paths). Logical terms in the search–tree can be represented by directed acyclic graphs (see Figure 11.6) and, therefore, the implementation of the heuristics gives rise to a classification problem on structured objects. In

[b]The feature vector typically contains parameters like the perimeter, the surrounded area, a synthetic representation of the curvature plot, etc.

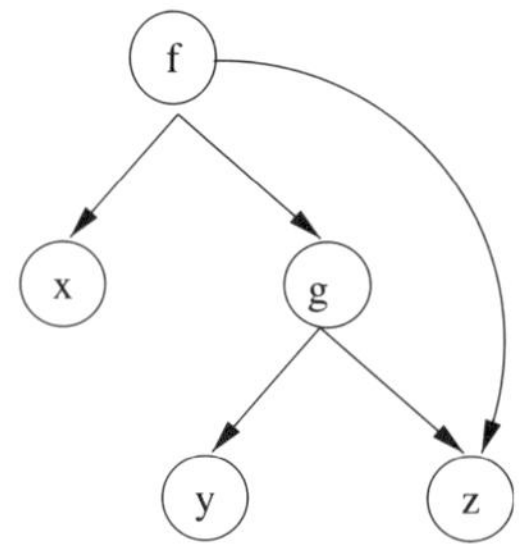

Fig. 11.6. The graphical representation of the term $f(x, g(y, z), z)$.

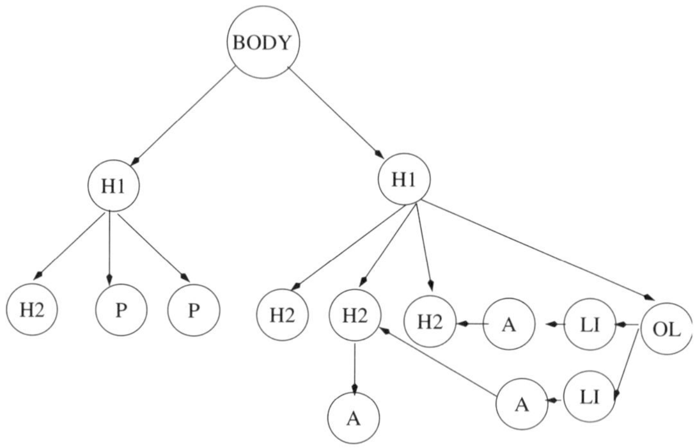

Fig. 11.7. The graphical structure of an HTML page as it can be derived from its HTML tags.

this framework, Goller[1] proposed a new solution that is based on recursive neural networks.

Internet Surfing With Recommender Systems

In the last few years, prototypal services for Internet surfing, called recommender systems, were developed.[21–24] They help the users to search information on the Internet. A recommender system must be capable of emulating the user behaviour while searching the Web. In fact, a similar system explores autonomously the Web and assigns a predicted interest level to each page using schemes based on learning from examples. Web pages can naturally be represented by graphs which can be deduced directly by the HTML tags (see Figure 11.7). In such a representation the nodes denote the logic contexts (paragraphs, sections, lists, etc.), while the arcs denote the inclusion relationships between the contexts and the connections established by the hyperlinks. The labels of the nodes contain the words that are enclosed in the corresponding contexts.

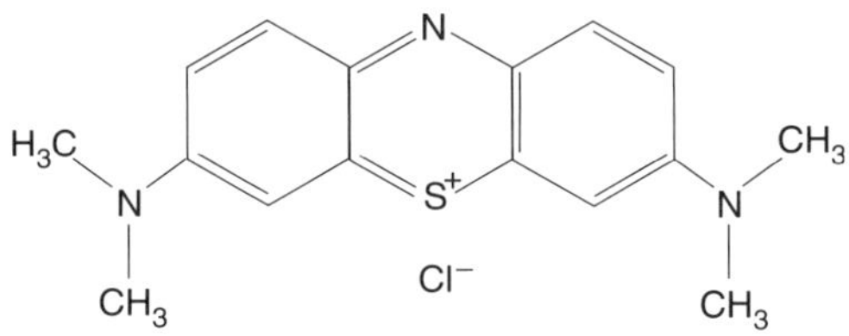

Fig. 11.8. The graphical structure of a chemical compound (methylene blue).

Chemistry and Bioinformatics

Chemical compounds are usually represented as undirected graphs. Each node of the graph is an atom or a group of atoms, while arcs represent bonds between atoms (see Figure 11.8).

One fundamental problem in chemistry is the prediction of the biological activity of chemical compounds. Quantitative Structure–Activity Relationship (QSAR) is an attempt to face the problem relying on compound structures. The biological activity of a drug is fully determined by the micromechanism of interaction of the active molecules with the bioreceptor. Unfortunately, discovering this micromechanism is very hard and expensive. Hence, because of the assumption that there is a direct correlation between the activity and the structure of the compound, the QSAR approach is a way of approaching the problem by comparing the structure of all known active compounds with inactive compounds, focusing on similarities and differences between them. The aim is to discover which substructure or which set of substructures characterize the biomechanism of activity, so as to generalize this knowledge to new compounds.[25,26] In the bioinformatic field, the contact map of each protein is represented by an undirected graph, that can be used to predict the protein secondary and tertiary structure[27,28] and, in particular, to classify those aminoacyds chains that represent alpha–helixes and beta sheets, and to predict the presence of disulfide bonds.[29]

Image analysis

In the last decade, graph–based representation of images received a growing interest, since such representation allows us to collect both symbolic and structural information, at the same time being invariant with respect to rotations, translations, and scalings of the original image. With the aim of obtaining a graphical representation of images, a preprocessing phase (image segmentation) to extract a set of homogenous regions must be carried out. At the end of this step, each region collects a set of pixels that are "similar" with respect to a certain visual property (color, texture, etc…). Clearly, also the kind of structure used to represent the regions plays a crucial role and it should be carefully chosen. In the following, both the segmentation methods and some graphical structures particularly suited to represent images will be described.

The segmentation process is particularly critical for image processing and very often it determines the quality of the final results. A theoretical definition of segmentation[30] is as follows. If $P()$ is a homogeneity predicate defined on groups of connected pixels, then a segmentation is a partition of the whole set of the pixels F into connected subsets or regions $(S_1, S_2, \ldots, S_n)$ such that

$$\cup_{i=1}^{n} S_i = F, \quad S_i \cap S_j = \emptyset, \ (i \neq j).$$

The predicate $P(S_i)$, that measures the homogeneity of the set S_i, is true for each region, and $P(S_i \cup S_j)$ is false if S_i and S_j are adjacent.

Unfortunately, according to Fu et al.,[31] "the image segmentation problem is basically one of psychophysical perception, and therefore not susceptible to a purely analytical solution". Thus, there is no universal theory on image segmentation yet. All of the existing methods are, by nature, ad hoc, and they are strongly application–dependent, i.e. there are no general algorithms that can be considered effective for all images. However, from the late seventies, a wide variety of methods were proposed; here we report a classification for color image segmentation, presented by Cheng et al.[32] in a recent survey. The main color image segmentation methods can be classified as follows.

- Histogram thresholding;
- Color space clustering;
- Region growing, splitting, or merging;
- Edge detection;
- Hybrid techniques based on fuzzy methods, physics approaches, and neural networks.

In the following, we describe an effective segmentation method, that can be useful in order to extract a graphical representation of images. The algorithm can be sketched as follows.

- An initial clustering of the pixels belonging to each image is performed, with the aim of obtaining a set of clusters whose pixels are homogeneous with respect to their colors. The clustering can be carried out using the well–known K–means algorithm.[33]
- Then, a region growing procedure is performed in order to reduce the number of regions.

Practically, the number of clusters extracted during the first step can be chosen approximately equal to the number of regions in which the image should be correctly partitioned. Nevertheless, this choice is not so crucial with respect to the whole segmentation process, since the successive region growing phase decreases the number of obtained regions. In fact, the number of regions computed via the clustering process is greater than the number of clusters, since each cluster is divided into a certain number of connected components (regions).

At the end of the segmentation process, a graphical structure that represents the relative positions of the regions can be extracted. Generally, each node of the structure is associated to a distinct region, while the presence of an edge between a pair of nodes represents the adjacency of the associated regions or some different relationship. Finally, a real valued vector (label) can be associated to each node, in order to describe visual and geometrical properties of the regions, and eventually to each edge, with the aim of representing some spatial relationships between the regions which correspond to the linked nodes. The extracted structure can be directed or undirected.

In the following, we describe two kinds of structures particularly useful for representing images, i.e. Region Adjacency Graphs (RAGs)[34] and MultiResolution Trees (MRTs).[35]

A RAG is an undirected structure, that is extracted from the segmented image as follows (see Figure 11.9).

- A labeled node is associated to each region. The label is constituted by a real vector that collects information about the visual and geometrical properties of the region.
- Pairs of nodes that correspond to adjacent regions are linked together with an undirected edge.

It is worth noting that two regions R_1, R_2 are considered adjacent if, for each pixel $a \in R_1$ and $b \in R_2$, there exists a path connecting a, and b, entirely lying into $R_1 \cup R_2$.

Both the topological arrangement of the regions and the symbolic visual information are represented by a RAG. Moreover, the RAG structure is invariant under translations and rotations (the labels are generally not invariant), that is an useful property for any kind of image representation technique.

An MRT is a hierarchical data structure that is generated during the segmentation process, so it describes the whole region growing procedure, differently from a RAG, that the represents just the final result of the segmentation. By this point of view, MRTs are related to quad–trees[36] or X–Y trees,[37] that are generally exploited to represent document images. Given an initial partitioning of the input image, for instance performing a clustering procedure, an MRT is built by performing the following steps (see Figure 11.10).

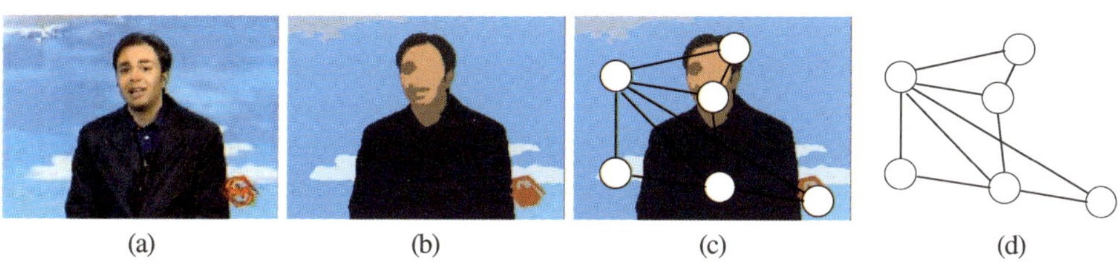

Fig. 11.9. (a) Input image — (b) Segmented image — (c,d) Corresponding RAG.

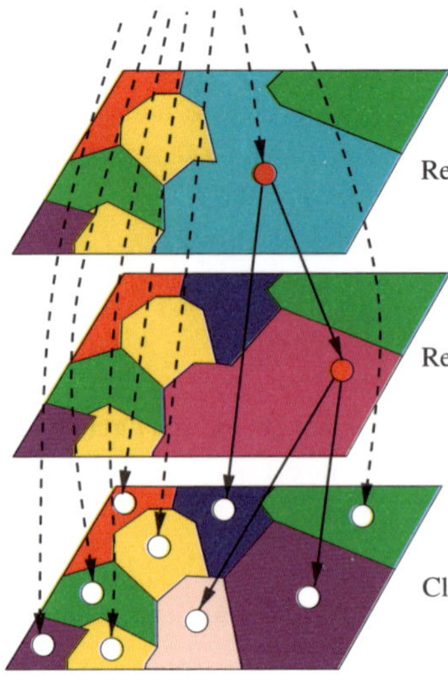

Fig. 11.10. MRT generation.

- Each initial region is associated to a leaf of the tree.
- During the region growing phase, any time two regions are merged together, a new node associated to the new region is added to the tree. The new node is added to the tree as the parent of the nodes that correspond to the merged regions.
- At the end of the region growing step, a virtual node is added as the root of the tree. Nodes corresponding to the set of regions obtained at the end of the segmentation process become the children of the root node.

Each node of an MRT, except for the root node, is labeled using a real–valued vector. Also in this case, the label describes some properties of the associated region. Instead, the root node has an artificial label (generally all the entries are set to 1.0), since it is not associated to a specific region. MRTs do not represent the spatial arrangement of the regions; however, such information can be inferred by a recursive neural network, considering both the geometrical features associated to the nodes and the MRT structure.

MRTs present two main advantages with respect to RAGs. First, the number of regions that describe a certain image does not represent a parameter for the MRT generation. In fact, the region growing process can be stopped when a very small number of regions is obtained, and considering that the "correct" segmentation is however stored at some level of the tree. Secondly, MRTs are directed structures, whereas RAGs are not, and recursive neural networks are able to deal only with directed graphs or with trees. Hence, if the images are represented by means of RAGs, a preprocessing phase that transforms each RAG into a directed structure is needed.[34]

Finally, we need to remark that the MRTs extracted using the previously described procedure are binary trees that have a high depth. Since recursive neural networks may suffer from the *long term dependency* problem,[38] the depth of the MRTs can be reduced modifying the region growing procedure so that at each step a group of regions, instead of a pair, are merged together following a predefined homogeneity criterion.[35] Using this method, each node v of the MRT possesses a certain number of children that depends on the number of regions that are merged together to obtain the new region associated to v. This procedure allows us both to speed–up the segmentation process and to reduce the MRT depth.

Both RAGs and MRTs can be used as inputs for a recursive neural network trained to classify images or to locate objects into images. The image classification problem consists in assigning to an input image a certain category, belonging to a predefined set of possible categories. In order to setup a learning environment, each image must be represented by a graphical structure. Moreover, for each graph, a target label that represents the image category must be assigned to the supersource node (see Section 11.2.1). In fact, in order to classify images, we can train a recursive neural network that performs a supersource transduction.

If the images are represented by MRTs, then the target labels are assigned to the root nodes of the MRTs. On the contrary, if RAGs have been chosen as the image representation, then a preprocessing step must be carried out in order to transform each RAG $\mathcal{G}$ into a DPAG (see Figure 11.11), before assigning the target labels to the supersource nodes.[39] The transformation can be performed as follows.

- A starting node (root) of the RAG is chosen and added as the supersource of the DPAG. The target label is associated to such node.
- An ordering rule among the adjacent regions is selected. For example the adjacent regions can be ordered by scanning the region boundary clock–wise, starting from the vertical axis.
- The DPAG is constructed recursively using a breadth–first visit of the RAG nodes starting from the root node. When a new node a is visited, this node is added to the DPAG, then the edges from a to the nodes b_k that have not been already visited are considered. The direction of the edges is chosen to be from a to b_k.

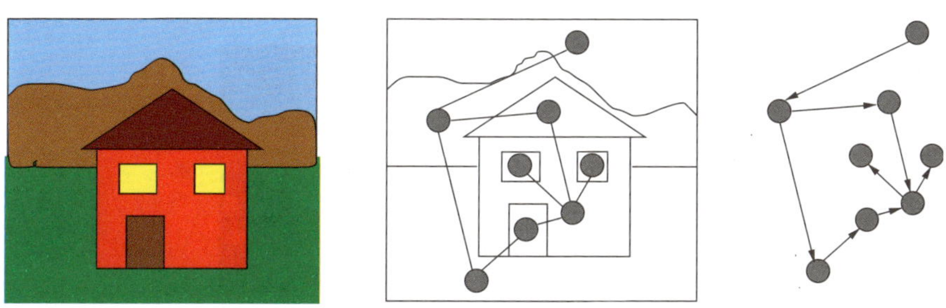

(a) (b) (c)

Fig. 11.11. (a) Input image — (b) Corresponding RAG — (c) The DPAG obtained performing a breadth–first visit of the RAG. The node that corresponds to the "sky region" is chosen as root.

Moreover the order of these arcs is defined using the ordering established by the previous rule.

The DPAGs obtained from the RAGs can be processed using a recursive neural network. However, since the state transition function requires to set a maximum outdegree q, during the transformation from RAGs to DPAGs, if the number of adjacent regions is greater than q, some links must be pruned. In the pruning phase, the boundary in common among the region i and its adjacent regions is considered. The links between nodes corresponding to adjacent regions are removed starting from those having the shortest common boundary until the outdegree becomes q. Finally, the starting region is chosen using a heuristic criterion, in order to select a node that is likely to be associated to an"important" region of the image. For instance, the node that corresponds to the region located in the center of the image can be selected.

When a set of directed structures, together with their associated target labels, are collected, a recursive neural network can be trained to predict the image category. In image classification problems, MRTs are generally preferable to RAGs, as shown in Bianchini et al.,[35] since the heuristic transformation from RAGs to DPAGs introduces a bias in the image representation, due to the arbitrary assignment of the edge direction.

RAGs and recursive neural networks can be also exploited to localize objects in images. Such kind of task consists in determining the bounding boxes or the contours of the target objects in an input image, assuming that at least one object is present inside the image. In order to train a recursive neural network to locate objects, a target label must be associated to each node of the RAGs that belong to the training set. The target label associated to a node specifies whether the corresponding region represents a part of the target object or a background area (see Figure 11.12). Instead, if a certain node corresponds to a region that contains both an object part and the background, the target label can be associated evaluating if the region mainly covers the object or not. Practically, the recursive neural network is trained to predict if a certain region corresponds to a part of the target object or

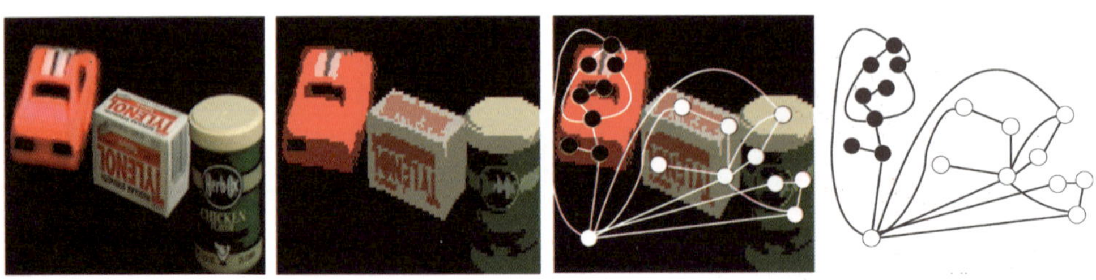

(a) (b) (c) (d)

Fig. 11.12. (a) Input image — (b) Segmented image — (c,d) Extracted RAG with associated targets, assuming to train a recursive neural network to locate the "red toy car". The nodes that correspond to the car and the background nodes have different labels.

not, as a binary classification problem. At the end of the node labeling, a technique that allows us to transform a given RAG into a forest of *recursive equivalent*[11] trees is exploited. Such transformation is needed in order to obtain directed structures; in particular a RAG that collects n nodes is transformed into n trees. Each root node of the trees (the supersources) has a target label that is equal to the one associated to the corresponding node in the original RAG. Given a trained recursive neural network and an image, an object can be localized as follows. First of all, the RAG that represents the image content is extracted and transformed into its equivalent forest of trees. Then, for each tree, the neural network is exploited to predict if the root of the tree corresponds to a part of the object or not, and finally the regions predicted as a part of the object are merged together, in order to determine the correct position of the whole object.

References

1. C. Goller. *A Connectionist Approach for Learning Search–Control Heuristics for Automated Deduction Systems.* PhD thesis, Technische Universität, München, Germany, (1997).
2. T. Schmitt and C. Goller. Relating chemical structure to activity: An application of the neural folding architecture. In *Proceedings of the Fifth International GI–Workshop on Fuzzy-Neuro Systems*, pp. 170–177, (1998).
3. E. Francesconi, P. Frasconi, M. Gori, S. Marinai, J. Sheng, G. Soda, and A. Sperduti. Logo recognition by recursive neural networks. In eds. K. Tombre and A. K. Chhabra, *Graphics Recognition Algorithms and Systems, LNCS 1389*, pp. 104–117. Springer, Heidelberg, (1998).
4. P. Frasconi, M. Gori, and A. Sperduti, A general framework for adaptive processing of data structures, *IEEE Transactions on Neural Networks.* **9**(5), 768–786, (1998).
5. A. Küchler and C. Goller. Inductive learning in symbolic domains using structure–driven recurrent neural networks. In eds. G. Görz and S. Hölldobler, *KI–96: Advances in Artificial Intelligence, LNCS 1137*, pp. 183–197. Springer, Berlin, (1996).
6. F. Scarselli, M. Gori, A. Tsoi, and M. Hagenbuchner, The graph neural network model, *IEEE Transactions on Neural Networks.* **20**(1), 61–80, (2009).
7. F. Scarselli, M. Gori, A. Tsoi, and M. Hagenbuchner, Computational capabilities of graph neural networks, *IEEE Transactions on Neural Networks.* **20**(1), 81–102, (2009).
8. A. Sperduti and A. Starita, Supervised neural networks for the classification of structures, *IEEE Transactions on Neural Networks.* **8**(3), 714–735, (1997).
9. D. E. Rumelhart, G. E. Hinton, and R. J. Williams. Learning internal representations by error propagation. In eds. D. Rumelhart and J. McClelland, *Parallel Distributed Processing*, vol. 1, chapter 8, pp. 318–362. MIT Press, Cambridge, (1986).
10. M. Bianchini, M. Gori, and F. Scarselli, Processing directed acyclic graphs with recursive neural networks, *IEEE Transactions on Neural Networks.* **12**(6), 1464–1470, (2001).
11. M. Bianchini, M. Gori, L. Sarti, and F. Scarselli, Recursive processing of cyclic graphs, *IEEE Transactions on Neural Networks.* **17**(1), 10–18, (2006).
12. M. Bianchini, M. Maggini, L. Sarti, and F. Scarselli, Recursive neural networks for processing graphs with labelled edges: Theory and applications, *Neural Networks.* **18**(8), 125–130, (2005).

13. S. Hanson and D. Burr. Minkowski–r back–propagation: Learning in connectionist models with non-Euclidean error signals. In ed. D. Anderson, *Neural Information Processing Systems*, pp. 348–357, (1987).
14. P. Burrascano, A norm selection criterion for the generalized delta rule, *IEEE Transactions on Neural Networks*. **2**(1), 1040–1050, (1991).
15. R. A. Jacobs, Increased rates of convergence through learning rate adaptation, *Neural Networks*. **1**(4), 295–307, (1988). ISSN 0893-6080. DOI: 10.1016/0893-6080(88)90003-2. URL `http://www.sciencedirect.com/science/article/B6T08-485RHDS-3/2/b423578abafad85dbf18003b11db5b98`.
16. M. Hagan and M. Menhaj, Training feedforward networks with the marquardt algorithm, *IEEE Transactions on Neural Networks*. **5**(6), 989–993, (1994).
17. M. Riedmiller and H. Braun. A direct algorithm method for faster backpropagation learning: the RPROP algorithm. In *Proceedings of the IEEE International Conference on Neural Networks*, vol. 1, pp. 586–591, (1993).
18. M. Riedmiller. RPROP — implementation details. Technical report, University of Karlsruhe, (1994).
19. K. Fu, *Syntactic Pattern Recognition and Applications*. (Prentice–Hall, Englewood Cliffs, N.J., 1982).
20. R. Schalhoff, *Pattern Recognition: Statistical, Structural and Neural Approaches*. (John Wiley & Sons, 1992).
21. M. Gori, M. Maggini, and E. Martinelli. Connectionist–based information retrieval. In *Proceedings of SPIE, Applications of Softcomputing*, vol. 3165, pp. 78–85, (1998).
22. E. Han, D. Boley, M. Gini, R. Gross, K. Hastings, B. Mobasher, J. Moore, G. Karypis, and V. Kumar, Document categorization and query generation on the World Wide Web using WebACE, *Journal of Artificial Intelligence Review*. **11**, 365–391, (1999).
23. T. Joachims, D. Freitag, and T. Mitchell. WebWatcher: A tour guide for the World Wide Web. In *Proceedings of the International Joint Conference on Artificial Intelligence*, pp. 770–777, (1997).
24. M. Pazzani and D. Billsus, Learning and revising user profiles: The identification of interesting websites, *Machine Learning*. **27**, 313–331, (1997).
25. A. M. Bianucci, A. Micheli, A. Sperduti, and A. Starita, Analysis of the internal representation development by neural networks for structures applied to QSAR studies of benzodiazepines, *Journal of Chemical Information and Computer Sciences*. **41**(1), 202–218, (2001).
26. A. Micheli, F. Portera, and A. Sperduti. A preliminary experimental comparison of recursive neural networks and a tree kernel method for QSAR/QSPR regression tasks. In ed. D-side, *Proceedings of the 12th European Symposium on Artificial Neural Networks*, pp. 293–298, Bruges (Belgium), (2004).
27. G. Pollastri, P. Baldi, A. Vullo, and P. Frasconi. Prediction of protein topologies using generalized IOHMMs and recursive neural networks. In *Neural Information Processing Systems*, pp. 1449–1456, (2002).
28. A. Vullo and P. Frasconi. A Bi–Recursive neural network architecture for the prediction of protein coarse contact maps. In *Proceedings of the 1st IEEE Computer Society Bioinformatics Conference*, pp. 187–196, (2002).
29. A. Vullo and P. Frasconi, Disulfide connectivity prediction using recursive neural networks and evolutionary information, *Bioinformatics*. **20**(5), 653–659, (2004).
30. N. R. Pal and S. K. Pal, A review on image segmentation techniques, *Pattern Recognition*. **26**(9), 1277–1294, (1993).
31. K. S. Fu and J. K. Mui, A survey on image segmentation, *Pattern Recognition*. **13**(1), 3–16, (1981).

32. H. Cheng, X. Yang, Y. Sun, and J. Wang, Color image segmentation: Advances and prospects, *Pattern Recognition.* **34**, 2259–2281, (2001).
33. R. Duda and P. Hart, *Pattern Classification and Scene Analysis.* (Wiley, New York, 1973).
34. C. de Mauro, M. Diligenti, M. Gori, and M. Maggini, Similarity learning for graph-based image representations, *Pattern Recognition Letters.* **24**(8), 1115–1122, (2003).
35. M. Bianchini, M. Maggini, and L. Sarti. Object recognition using multiresolution trees. In *Structural, Syntactic, and Statistical Pattern Recognition, LNCS 4109*, pp. 331–339, Heidelberg, (2006). Springer.
36. R. Finkel and J. Bentley, Quad trees a data structure for retrieval on composite keys, *Acta Informatica.* **4**(1), 1–9, (1974).
37. G. Nagy and S. Seth. Hierarchical representation of optically scanned documents. In *Proceedings of the Seventh International Conference on Pattern Recognition*, pp. 347–349, (1984).
38. Y. Bengio, P. Simard, and P. Frasconi, Learning long-term dependencies with gradient descent is difficult, *IEEE Transactions on Neural Networks.* **5**(2), 157–166, (1994).
39. M. Gori, M. Maggini, and L. Sarti, A recursive neural network model for processing directed acyclic graphs with labeled edges, *Proceedings of the International Joint Conference on Neural Networks.* **2**, 1351–1355, (2003).

Chapter 12

PARACONSISTENT ARTIFICIAL NEURAL NETWORKS AND APPLICATIONS

Jair Minoro Abe[1] and Kazumi Nakamatsu[2]
[1] *Graduate Program in Production Engineering, ICET — Paulista University*
R. Dr. Bacelar, 1212, CEP 04026-002 São Paulo — SP — Brazil
[1] *Institute for Advanced Studies — University of São Paulo — Brazil*
jairabe@uol.com.br
[2] *School of Human Science and Environment/H.S.E.*
University of Hyogo — Japan
nakamatu@shse.u-hyogo.ac.jp

In this expository work we present a new class of ANN — Paraconsistent Artificial Neural Networks based on Paraconsistent Annotated Evidential Logic $E\tau$ that is capable of manipulating imprecise, contradictory and paracomplete information. Basic concepts are discussed followed with some applications in Biomedicine.

12.1. Introduction

The Artificial Neural Network — ANN has been extensively studied in AI so it has found interesting applications in many unexpected directions. ANN can be described as a computational system consisting of a set of highly interconnected processing elements, called artificial neurons, which process information as a response to external stimuli. An artificial neuron is a simplistic representation that emulates the signal integration and threshold firing behavior of biological neurons by means of mathematical structures. Like their biological counterpart, artificial neurons are bound together by connections that determine the flow of information between peer neurons. Stimuli are transmitted from one processing element to another via synapses or interconnections, which can be excitatory or inhibitory [Erganian & Mahmoudi 05], [Fausett 94].

The advantage of neural networks over conventional programming lies on their ability to solve problems that do not have an algorithmic solution or the available solution is too complex to be found [Subasi, Alkan Kokluyakaya & Kiymik 05]. Neural networks are well suited to tackle problems that people are good at solving, like prediction and pattern recognition. Neural networks have been applied within the medical domain, engineering, automation, robotics, data compression, etc.

Several theories of artificial neural networks have been proposed with different characteristics. In this paper we discuss a new theory of ANN based on a paraconsistent annotated logic $E\tau$ with some applications.

12.2. Background

Paraconsistent Artificial Neural Networks — PANN is a new artificial neural network introduced in [Da Silva Filho & Abe 01]. Its basis leans on paraconsistent annotated logic $E\tau$ [Abe 92], also discussed in another chapter of this book. Roughly speaking, the logic $E\tau$ is a logic that allows contradictions and paracompleteness without trivialization. Also it is possible manipulate imprecise concepts in its interior.

The atomic formulas of the logic $E\tau$ are of the type $p_{(\mu,\lambda)}$, where $(\mu, \lambda) \in [0,1]^2$ which is a lattice with the order relation $(\mu_1, \lambda_1) \leq (\mu_2, \lambda_2) \Leftrightarrow \mu_1 \leq \mu_2$ and $\lambda_1 \leq \lambda_2$.

We introduce the Uncertainty Degree: $G_{un}(\mu, \lambda) = \mu + \lambda - 1$; Certainty Degree: $G_{ce}(\mu, \lambda) = \mu - \lambda$ and we consider the following 12 output states:

Table 12.1. Extreme and Non-extreme states.

Extreme states	Symbol	Non-extreme states	Symbol
True	V	Quasi-true tending to Inconsistent	$QV \to T$
False	F	Quasi-true tending to Paracomplete	$QV \to \bot$
Inconsistent	T	Quasi-false tending to Inconsistent	$QF \to T$
Paracomplete	$\bot$	Quasi-false tending to Paracomplete	$QF \to \bot$
		Quasi-inconsistent tending to True	$QT \to V$
		Quasi-inconsistent tending to False	$QT \to F$
		Quasi-paracomplete tending to True	$Q\bot \to V$
		Quasi-paracomplete tending to False	$Q\bot \to F$

The control values are:

- V_{cic} = maximum value of uncertainty control $= Ft_{ct}$
- V_{cve} = maximum value of certainty control $= Ft_{ce}$
- V_{cpa} = minimum value of uncertainty control $= -Ft_{ct}$
- V_{cfa} = minimum value of certainty control $= -Ft_{ce}$

For the discussion in the present paper we have used: $Ft_{ct} = Ft_{ce} = ½$.

All states are represented in the next figure.

For the discussion in this work we have used: $C_1 = C_3 = ½$ and $C_2 = C_4 = -½$.

12.3. The Paraconsistent Artificial Neural Cells

In the paraconsistent annotated structure, the main aim is to know how to measure or to determine the certainty degree concerning a proposition, if it is False or True. Therefore, for this, we take into account only the certainty degree G_{ce}. The

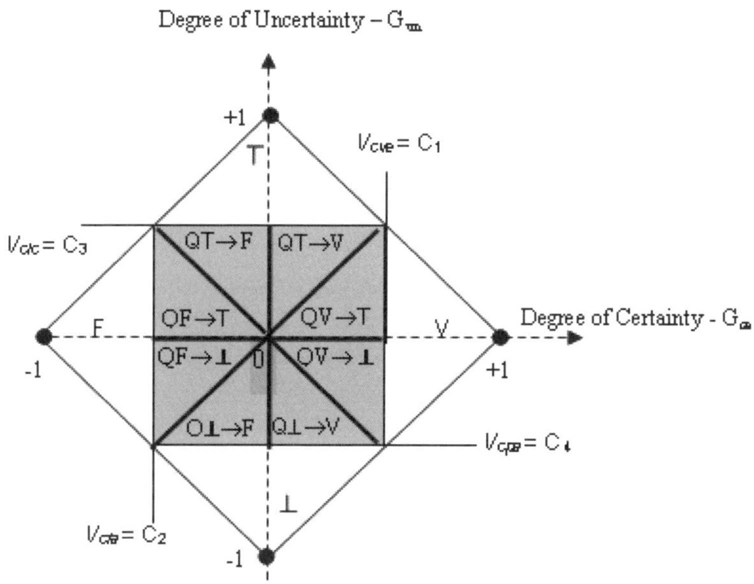

Fig. 12.1. Representation of the certainty degrees and of contradiction degrees.

uncertainty degree G_{un} indicates the measure of the inconsistency or paracompleteness. If the certainty degree is low or the uncertainty degree is high, it generates an indefinition.

The resulting certainty degree G_{ce} is obtained as follows:

```
If: Vcfa ≤ Gun ≤ Vcve or Vcic ≤ Gun ≤ Vcpa ⇒ Gce
     = Indefinition
For: Vcpa ≤ Gun ≤ Vcic
If: Gun ≤ Vcfa ⇒
Gce = False with degree Gun
Vcic ≤ Gun ⇒ Gce = True with degree Gun
```

The algorithm that expresses a basic Paraconsistent Artificial Neural Cell - PANC - is:

```
* /Definition of the adjustable values * /
  Vcve = C1 * maximum value of certainty control * /
  Vcfa = C2 * / minimum value of certainty control * /
  Vcic = C3 * maximum value of uncertainty control * /
  Vcpa = C4 * minimum value of uncertainty control* /
     * Input /Variables * /
   μ, λ
     * Output /Variables *
    Digital output = S1
    Analog output = S2a
```

```
    Analog output = S2b
* /Mathematical expressions * /
begin:
0 ≤ μ ≤ 1 e 0 ≤ λ ≤ 1
Gun = μ + λ − 1
Gce = μ − λ
              * / determination of the extreme states * /
if Gce ≥ C1 then S1 = V
if Gce ≥ C2 then S1 = F
if Gun ≥ C3 then S1 = T
if Gun ≤ C4 then S1 = ⊥
If not: S1 = I − Indetermination
        Gun = S2a
        Gce = S2b
```

A PANC is called basic PANC when given a pair (μ, λ) is used as input and resulting as output: G_{un} = resulting uncertainty degree, G_{ce} = resulting certainty degree, and X = constant of Indefinition, calculated by the equations $G_{un} = \mu+\lambda-1$ and $G_{ce} = \mu - \lambda$.

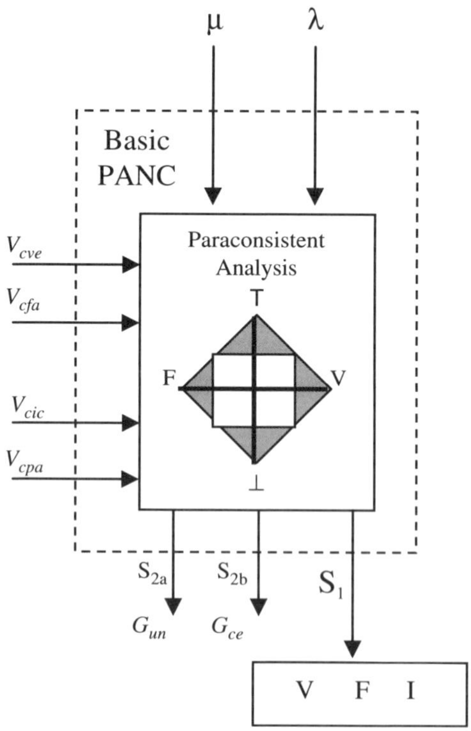

Fig. 12.2. The basic paraconsistent artificial neural cell.

12.4. The Paraconsistent Artificial Neural Cell of Learning

A Paraconsistent Artificial Neural Cell of Learning — PANC-l is obtained from a basic PANC.

In this learning cell, sometimes we need the action of the operator Not in the training process. Its function is to do the logical negation in the resulting output sign. For a training process, we consider initially a PANC of Analytic Connection the one not undergoing any learning process.

According to the paraconsistent analysis, a cell in these conditions has two inputs with an indefinite value ½.

So, the basic structural equation yields the same value ½ as output, having as result an indefinition.

12.5. The Learning of a PANC-l

The learning cells can be used in the PANN as memory units and pattern sensors in primary layers. For instance, a PANC-l can be trained to learn a pattern by using an algorithm. For the training of a cell we can use as pattern real values between 0 and 1. The cells can also be trained to recognize values between 0 and 1.

The learning of the cells with extreme values 0 or 1 composes the primary sensorial cells. Thus, the primary sensorial cells consider as pattern a binary digit where the value 1 is equivalent to the logical state True and the value 0 is equivalent to the logical state False. The repeated appearance of the input 0 means that the resulting favorable evidence degree is going to increase gradually in the output reaching the value 1. In these conditions we say that the cell has learned the falsehood pattern.

The same procedure is adopted when the value 1 is applied to the input repeated times. When the resulting favorable evidence degree in the output reaches the value 1 we say that the cell has learned the truth pattern. Therefore a PANC can learn two types of patterns: the truth pattern or the falsity pattern. In the learning process of a PANC, a learning factor — LF can be introduced that is externally adjusted. Depending on the value of LF, it gives the cell a faster or slower learning. In the learning process, given an initial favorable evidence degree $\mu_r(k)$, we use the following equation to reach $\mu_r(k) = 1$, for some k.

So, for truth pattern we have

$$\mu_r(k+1) = \frac{(\mu_1 - \mu_r(k)_c)LF + 1}{2}$$

where $\mu_r(k)_c = 1 - \mu_r(k)$, and $0 \leq LF \leq 1$. For falsity pattern, we have

$$\mu_r(k+1) = \frac{(\mu_{1c} - \mu_r(k)_c)LF + 1}{2}$$

where $\mu_r(k)_c = 1 - \mu_r(k)$, $\mu_{1c} = 1 - \mu_1$, and $0 \leq LF \leq 1$

So we can say that the cell is completely learned when $\mu_r(k+1) = 1$.

If $LF = 1$, we say that the cell has a natural capacity of learning. Such capacity decreases as LF approaches 0. When $LF = 0$, the cell loses the learning capacity and the resulting favorable evidence degree will always have the indefinition value ½.

12.6. Unlearning of a PANC-l

Even after having a cell trained to recognize a certain pattern, if insistently the input receives a value totally different, the high uncertainty makes the cell unlearn the pattern gradually. The repetition of the new values implies in a decreasing of the resulting favorable evidence degree. Then, the analysis has reached an indefinition. By repeating this value, the resulting favorable evidence degree reaches 0 meaning that the cell is giving the null favorable evidence degree to the former proposition to be learned. This is equivalent to saying that the cell is giving the maximum value to the negation of the proposition, so the new pattern must be confirmed. Algorithmically, this is showed when the certainty degree G_{ce} reaches the value -1. In this condition the negation of the proposition is confirmed. This is obtained by applying the operator Not to the cell. It inverts the resulting favorable evidence degree in the output. From this moment on the PANC considers as a new pattern the new value that appeared repeatedly and unlearning the pattern learned previously. By considering two factors, LF — learning factor and UF — unlearning factor, the cell can learn or unlearn faster or slower according to the application. These factors are important in giving the PANN a more dynamic process.

Using the concepts of basic PANC, we can obtain the family of PANC considered in this work, as described in Table 12.2 below:

Table 12.2. Paraconsistent artificial neural cells.

PANC	Inputs	Calculations	Output
Analytic connection — PANNac	$\mu, \lambda, Ft_{ct}, Ft_{ce}$	$\lambda_c = 1 - \lambda$, G_{un}, G_{ce}, $\mu_r = (G_{ce} + 1)/2$	If $\|G_{ce}\| > Ft_{ce}$ then $S_1 = \mu_r$ and $S_2 = 0$; If $\|G_{un}\| > Ft_{ct}$ and $\|G_{un}\| > \|G_{ce}\|$ then $S_1 = \mu_r$ and $S_2 = \|G_{un}\|$, if not $S_1 = ½$ and $S_2 = 0$
Maximization — PANNmax	μ, λ	none	If $\mu > \lambda$, then $S_1 = \mu$, if not $S_1 = \lambda$
Minimization — PANNmin	μ, λ	none	If $\mu < \lambda$, then $S_1 = \mu$, if not $S_1 = \lambda$

12.7. Operating PANN

Now we show some basic aspects of how PANN operates. Let us take three vectors: $V_1 = (8, 5, 4, 6, 1)$; $V_2 = (8, 6, 4, 6, 5)$; $V_3 = (8, 2, 4, 6, 9)$. The favorable evidence is calculated as follows: given a pair of vectors, we take '1' for equal elements and '0' for the different elements, and we figure out its percentage.

Comparing V_2 with $V_1 : 1 + 0 + 1 + 1 + 0 = 3$; as a percentage: $(3/5)^*100 = 60\%$
Comparing V_3 with $V_1 : 1 + 0 + 1 + 1 + 0 = 3$; as a percentage: $(3/5)^*100 = 60\%$

The contrary evidence is the weighted addition of the differences between the different elements, in module:

Comparing V_2 with $V_1 = 0 + 1/10 + 0 + 0 + 4/10 = (5/10)/5 = 10\%$
Comparing V_3 with $V_1 = 0 + 3/10 + 0 + 0 + 8/10 = (11/10)/5 = 22\%$

Therefore, visually we see that vector V_2 is 'more similar' to V_1 than V_3. We use a PANN to recognize this similarity mathematically.

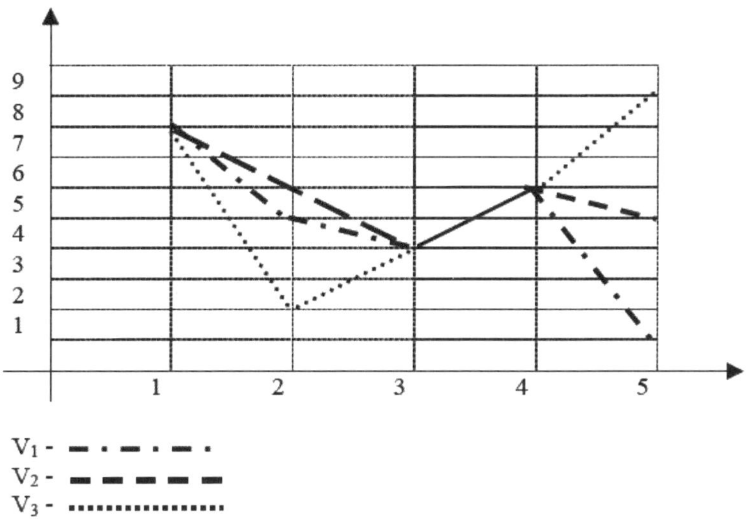

Fig. 12.3. Vectors V_1, V_2, V_3.

Representing the learning of the vectors $V_1 = (8, 5, 4, 6, 1)$; $V_2 = (8, 6, 4, 6, 5)$; $V_3 = (8, 2, 4, 6, 9)$ in layers.

Let's consider the following three waves.

The analyzed wave is the wave that will be submitted to PANN for recognition. The learned wave 1 and learned wave 2 were previously stored in the control database (normality pattern). Observe that 'visually', Learned wave 1 is more 'similar' than Learned wave 2. We introduce the concepts in order to deal mathematically with this similarity.

The favorable evidence is obtained by the identical positions.

The contrary evidence is obtained by difference sum (in module) of the correspondent different positions.

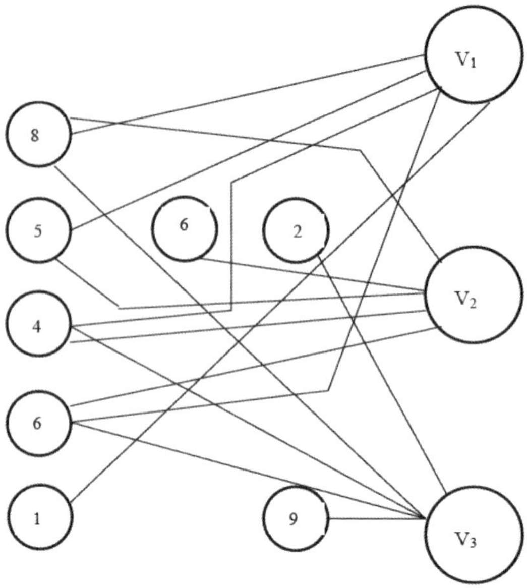

Fig. 12.4. PANN and layers.

Table 12.3. Table of the analyzed waves.

Wave's name	Position 1	Position 2	Position 3	Position 4	Position 5
Analyzed wave	8	5	4	6	1
Learned wave 1	8	6	4	6	5
Learned wave 2	8	2	4	6	9

Comparing the Analyzed wave and Learned wave 1, we have:

Table 12.4. Analyzed wave × Learned wave 1.

Wave's name	Position 1	Position 2	Position 3	Position 4	Position 5	Total
Analyzed wave	8	5	4	6	1	—
Learned wave 1	8	6	4	6	5	—
Favorable evidence	1	0	1	1	0	3
Contrary evidence	0	1	0	0	4	5

Comparing the Analyzed wave and Learned wave 2, we have:

Table 12.5. Analyzed wave x Learned wave 2.

Wave's name	Position 1	Position 2	Position 3	Position 4	Position 5	Total
Analyzed wave	8	5	4	6	1	—
Learned wave 2	8	2	4	6	9	—
Favorable evidence	1	0	1	1	0	3
Contrary evidence	0	3	0	0	8	11

Normalizing the values by the division of the favorable evidence (μ) and of the contrary evidence (λ) for the number of elements of the wave, we have:

Table 12.6. Normalized values: Learned wave 1 and Learned wave 2.

Case	μ	λ	Normalization FE	Normalization CE
Analyzed wave × Learned wave 1	3	5	0.6	1
Analyzed wave × Learned wave 2	3	11	0.6	2.2

Therefore, we notice that the wave with the maximum favorable evidence and the minimum contrary evidence is the learned Wave 1, in other words, this is the most similar wave to the analyzed Wave.

12.8. Why PANN Can be Useful

A problem in EEG analysis, as well as any other measurement devices are limited and subject to the inherent imprecision of the several sources involved: equipment, movement of the patient, electric registers and individual variability of physician visual analysis. Such imprecision can often include conflicting information or paracomplete data. The majority of theories and techniques available are based on classical logic and so they cannot handle adequately such sets of information, at least directly. So, as PANN is based on logic $E\tau$, it can deal with uncertainty, inconsistent and paracomplete data.

We show some tests made with recognizing PANN. In this analysis, all stored waves are perfect sinusoidal waves with 1Hz of variance comprising Delta, Theta, Alpha, and Beta waves. Thus we can study their approximate frequency as well as which group the wave belongs to.

The lattice τ was adapted to the following displayed in the table below.

Table 12.7. Extreme and Non-extreme states.

Extreme states	Symbol	Non-extreme states	Symbol
True	V	Quasi-true	QV
False	F	Quasi-false	QF
Inconsistent	T	Quasi-false tending to Paracomplete	QF$\perp$
Paracomplete	$\perp$	Quasi-inconsistent tending to True	QTV
		Quasi-inconsistent tending to False	QTF
		Quasi-paracomplete tending to True	Q$\perp$V
		Quasi-paracomplete tending to False	Q$\perp$F

For instance, we show the study of 4 waves.

The sequence of figures shows an EEG exam (light grey; not perfect sinusoidal wave) being compared with the most similar wave of the data group (slightly dark grey; perfect sinusoidal waves).

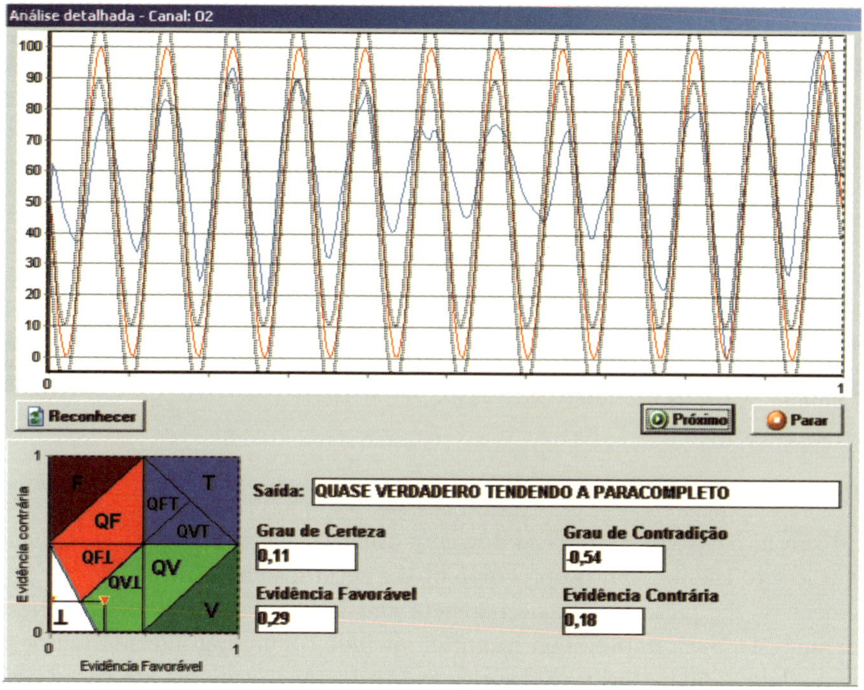

Fig. 12.5. Wave 1.

Table 12.8. Analysis.

	Wave 1
Output	Almost-true tending to Paracomplete
Certainty degree	0.11
Uncertainty degree	−0.54
Favorable evidence	0.29
Contrary evidence	0.18

12.9. Methodology

For the purpose of EEG analysis we've made some previous treatment of signals, observing that the process does not allow the loss of any essential wave characteristics for our analysis.

The capturing of the data is obtained from magnetic archives (suitable software for physical capture of the signals) or manually (TXT archives — American National Standard Code for Information Interchange).

As the actual EEG examination values can vary highly, in module, in the range $10\,\mu$V to $1500\,\mu$V, we normalize the values between $100\,\mu$V and $-100\,\mu$V using a simple linear conversion, to facilitate the manipulation and to visualize in the screen:

$$x = \frac{100.a}{m}$$

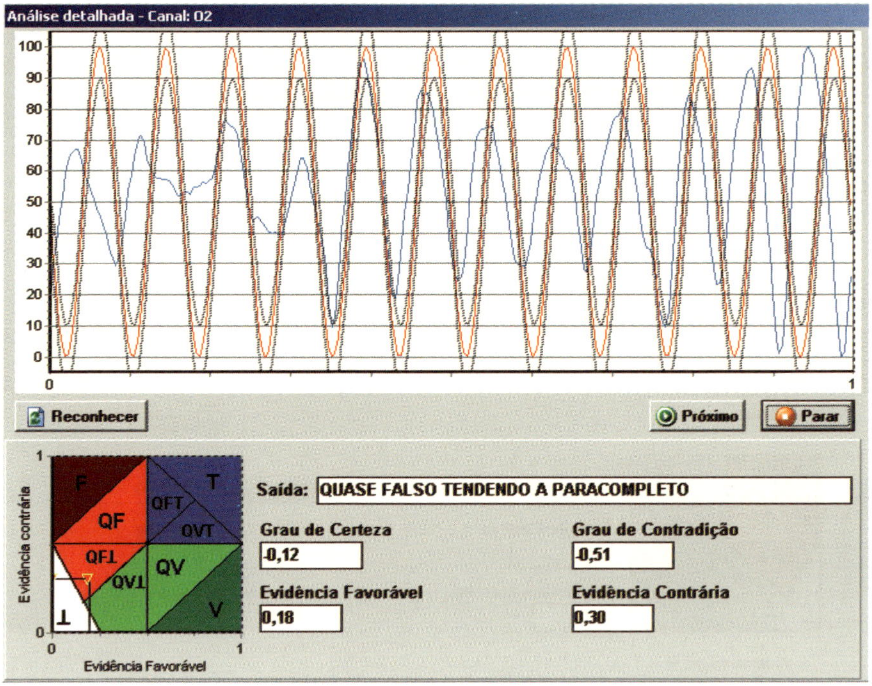

Fig. 12.6. Wave 2.

Table 12.9. Analysis.

Wave 2	
Output	Almost-false tending to Paracomplete
Certainty degree	−0.12
Uncertainty degree	−0.51
Favorable evidence	0.18
Contrary evidence	0.30

where m is the maximum value of the exam; a is the current value of the exam. So, x is the current normalized value. The minimum value of the exam is taken as zero value and the remaining values are translated proportionally in order to eliminate the negative cycle.

12.10. Data analysis, Expert System, and Wave Morphology

In analyzing EEG signals, one important aspect to take into account is their morphology. To do this task, it is interesting to build one very simple Expert System, which allows us to verify "abnormalities", such as spikes and artifacts. Also, it analyses the signal behavior, verifying in which band it belongs (delta, theta, alpha and beta).

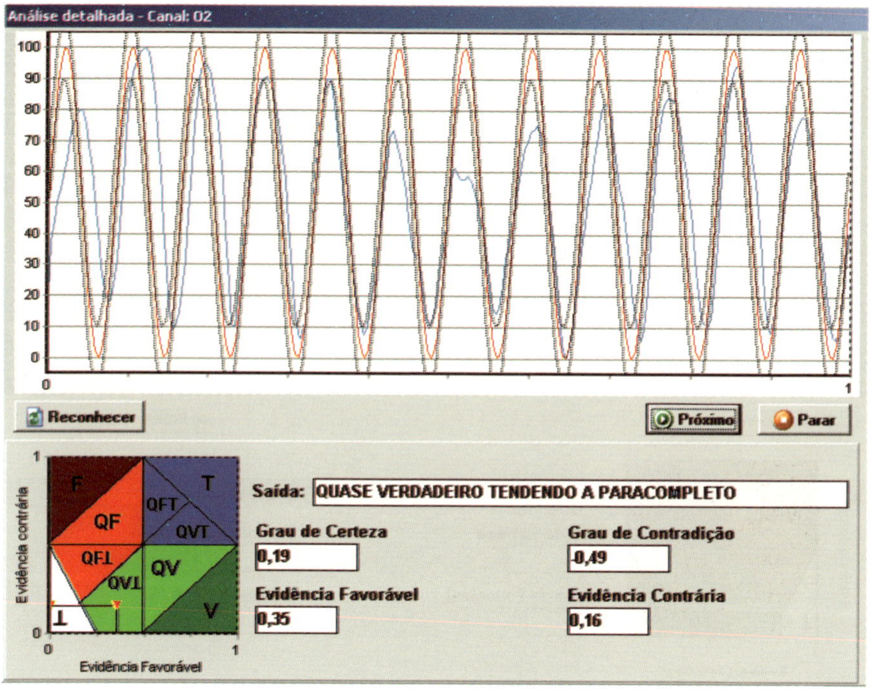

Fig. 12.7. Wave 3.

Table 12.10. Analysis.

Wave 3	
Output	Almost-true tending to Paracomplete
Certainty degree	0.19
Uncertainty degree	−0.49
Favorable evidence	0.35
Contrary evidence	0.16

The process of the morphological analysis of a wave is made by comparing with a certain pattern set of waves (considered normal).

For this comparison we take now as favorable evidence the number of wave peaks, i.e., the closer to the peaks quantity, the higher is the favorable evidence degree:

$$\mathrm{EF} = 1 - ((|\mathrm{bd} - \mathrm{vt}|)/(\mathrm{bd} + \mathrm{vt})),$$

where:

vt = number of wave peaks of the exam
bd = number of the wave peaks being compared (pattern stored in the database)

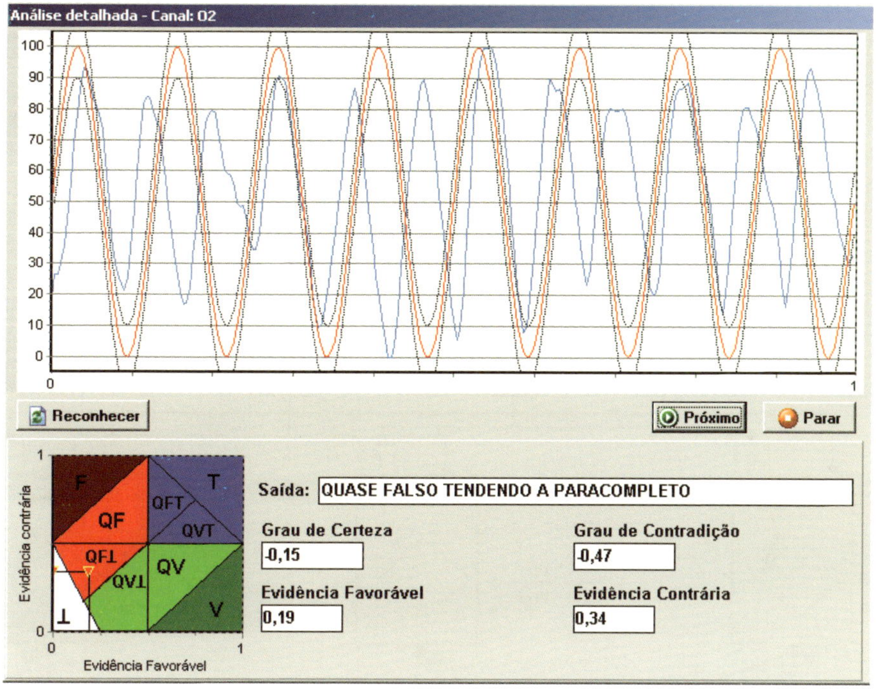

Fig. 12.8. Wave 4.

Table 12.11. Analysis.

	Wave 4
Output	Almost-false tending to Paracomplete
Certainty degree	−0.15
Uncertainty degree	−0.47
Favorable evidence	0.19
Contrary evidence	0.34

We also use an expert system for detecting the diminishing average frequency level. An expert system verifies the average frequency level of Alpha waves and compares them with a fixed external one (external parameter wave).

Expert system for Alpha band concentration

This expert system will be utilized for Alpha band concentration in the exam. For this, we consider the quotient of the sum of fast Alpha and Beta waves by slow Delta and Theta waves, i.e., (Alpha + Beta)/(Delta + Theta).

Expert System for Theta band concentration

This expert system will be utilized for Delta band concentration in the exam. We consider (slow waves Delta + slow waves Theta)/(fast waves Alpha + fast waves Beta).

Data Analysis

When analyzing information from sources, we can be faced with contradictory, fuzzy or paracomplete data. However we have to take a decision. For this, suppose that we have three pieces of information PA, PB, and PC, and that we are analyzing PA and PB. If we cannot decide between this expert information, we take into account the third PC in the following way.

The first layer is composed of three analytical connection PANC: C1, C2, and C3 whose signals are analyzed by means of the Basic Structural Equation:

$$\text{BSE} : S = (\mu - (1 - \lambda) + 1)/2$$

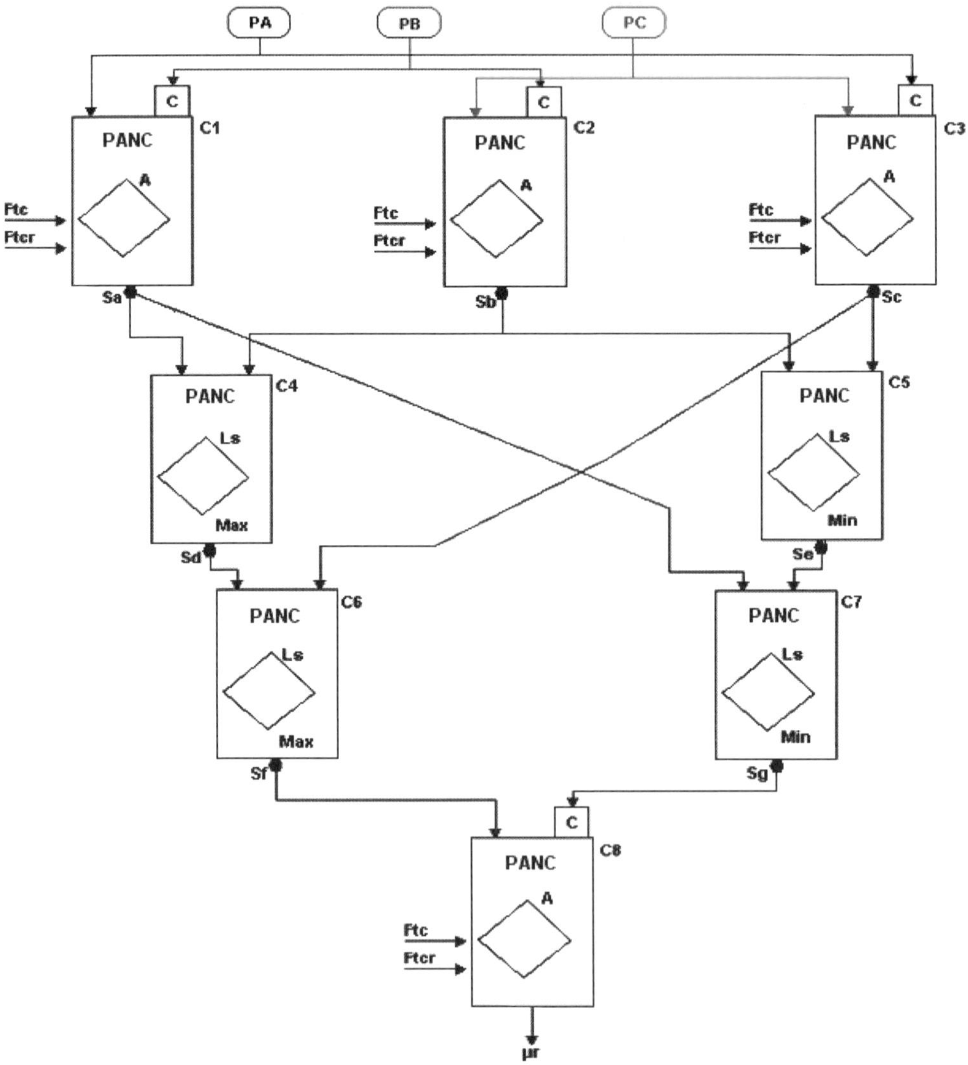

Fig. 12.9. PANN and decision-making.

Resulting in the output signals SA, SB, and SC.

In the internal layers, the cells C4 and C6 constitute the Maximization Neural Unit (it takes the maximum value SG among output values SA, SB, and SC) and the cells C5 and C7, the Minimization Neural Unit (it takes the minimum value SE among output values AS, SB, and SC).

Another form to define its interpretation of the analysis is to use the resultant value (μ_r) and complement it, then generate a complemented resultant value (λ_r). In this manner we will have a resultant favorable evidence (μ_r) and a resultant contrary evidence (λr), as exemplified below.

We applied this methodology in two subjects (blind study), one of them normal and the other one with AD, and the method was able to distinguish both, including the normal subject in normal standard of waves and the AD ones in the abnormal model.

As an illustrative example, with suitable adaptations, an earlier version of the method was tested on real EEG exams in the aid of Alzheimer Disease diagnosis. Detailed exposition and discussion is to be found in [Anghinah 03]. Here we mention only the performance results. 10 EEGs were tested and the system classified correctly as normal at a rate of 80% and 20% as false-positive. More 10 EEGs was tested and the system classified correctly as AD at a rate of 80 % and 20% as false-negative, as in the Table 12 below [Abe, Lopes & Anghinah 07].

Table 12.12. Diagnosis — Normal × Probable AD patients [Abe, Lopes & Anghinah 07].

	Gold standard		
	AD patient	Non AD patient	Total
Negative	2	2	4
Positive	8	8	16
Total	10	10	20
Negative	10%	10%	20%
Positive	40%	40%	80%
Total	50%	50%	100%
Sensitivity:	0,8		
Specificity:	0,8		

As preliminary test, normal patient patterns — waves from 8 Hz to 12 Hz, allowing for a variance of 0.5 Hz — were considered as representing the Alpha band. We store 9 positive cycle wave patterns plus 9 negative cycle wave patterns. In this test, 10 second periods were selected, containing Alpha band characteristics. The PANN was capable of recognizing waves belonging to the Alpha band with favorable evidence of 0.30 and contrary evidence of 0.19, while for waves not exhibiting Alpha patterns, we obtained an average favorable evidence of 0.19 and contrary evidence of 0.32, thus demonstrating that PANN was efficient for recognizing Alpha waves.

12.11. PANN and Speech Recognition

The following is based on [Prado 07], [Abe, Prado & Nakamatsu 06]. For the present application, we've developed a computational system with the capability of capturing and converting a speech signal as a vector. After this, we analyze the percentage recognition results shown below. With these studies we point out some of the most important features of PANN: firstly, the PANN recognition becomes 'better' in every new recognition step, which is a consequence of discarding contradictory signals and recognizing them by proximity, without trivializing the results. Finally, the performance and efficiency of the PANN is enough to recognize in real time, any speech signal.

Now we show how the PANN was efficient in formants recognition. The tests were made in Portuguese and 3 pairs of syllables were chosen 'FA-VA', 'PA-BA', 'CA-GA' presenting one articulation and differences in sonority (see Table 12.13). The speaker is an adult, male, 42 years old, Brazilian, from São Paulo city.

Table 12.13 shows the recognizing capability. The recognition in the first column is 100% because the PANN is empty and the syllables are just being learned. The process of recognition is made in the computer system as follows: in step 2 the speaker says, for instance, the syllable 'FA'. Then the PANN gives an output with the calculations (favorable/contrary evidences, G_{ce}, G_{un}) and asks the speaker (operator) if the data is acceptable or not. If the answer is 'Yes', the PANN keeps the parameters for the next recognition. If the answer is 'Not", the PANN recalculates the parameters in order to criticize the next recognition, till such data belongs to the False state (fig. 12.1), preparing for the next step to repeat the process (in this way, improving the recognition). This is performed by the neural cell PNACde (see Table 12.13):

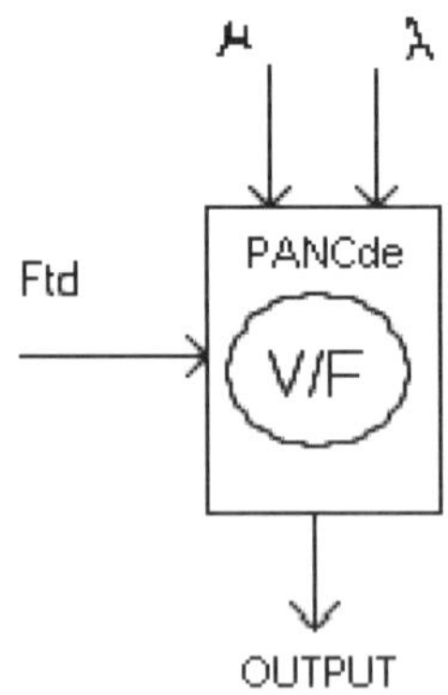

Fig. 12.10. PANCde.

These adjustments are made automatically by the PANN except for the learning steps, which relies on interaction with the operator to provide the Yes/No data. Thus, from the second column on, PANN is recognizing and learning occurs simultaneously, as well as adaptation; this is the reason that the recognition factor increases.

Table 12.13. Syllable recognition.

	Steps										
Syllable	1	2	3	4	5	6	7	8	9	10	Average
FA	100%	87%	88%	91%	90%	92%	95%	94%	94%	95%	91,78%
VA	100%	82%	85%	87%	88%	90%	94%	92%	96%	95%	89,89%
PA	100%	83%	86%	90%	88%	95%	89%	94%	95%	95%	90,56%
BA	100%	85%	82%	87%	90%	89%	92%	95%	95%	97%	90,22%
CA	100%	82%	87%	89%	88%	92%	90%	94%	92%	95%	89,89%
GA	100%	84%	88%	92%	90%	89%	95%	95%	95%	92%	91,11%

Table 12.14. Recognition of pairs of syllables with one different speech articulation.

	Steps										
Pairs	1	2	3	4	5	6	7	8	9	10	Average
FA-VA	70%	67%	72%	59%	65%	71%	64%	69%	66%	63%	66,60%
PA-BA	51%	59%	49%	53%	48%	52%	46%	47%	52%	48%	50,50%
CA-GA	62%	59%	61%	62%	58%	59%	60%	49%	63%	57%	59,00%

In the example, we can see that after the sixth speech step, the PANN is able to recognize efficiently every signal with recognition factor higher than 88%. Every signal lower than this factor can be considered as unrecognized.

Table 12.14 shows the recognition factor percentage when PANN analyzes a syllable with one different speech articulation. As we can see, when the PANN has learned the 'FA' syllable (10 times) and it is asked to recognize the 'VA' syllable, the recognition factor is never higher than 72%. For the remaining pairs of syllables, this factor is even lower. The results below are taken from [Abe Prado & Nakamatsu 06].

The variation of the analyzed values are interpreted by the PANN and adjusted automatically by the system. Due to the structure of the PANN, the network is able to identify small variations between the pairs of syllables chosen. One central reason is its capability of proximity recognition and discarding contradictory data without trivialization. In the examples above, we can define as recognized if the factor is higher than 88%, and non-recognized, if the factor is lower than 72%. The difference of 16% (between recognition and non-recognition) is enough to avoid mistakes in the interpretation of the results. Thus, PANN shows itself as a superior system in being capable to manipulate the factors described showing high accuracy in data analysis.

12.12. PANN and Craniofacial Variables

In this section we apply PANN in cephalometric analysis. The basis is [Mario 07], [Abe, Ortega, Mário, Del Santo 05], [Mário, Abe, Ortega, Del Santo 08]. Craniofacial discrepancies, either skeletal or dental, are usually assessed in lateral cephalograms by cephalometric analysis. In Orthodontics, this quantitative analysis compares individual measurements to a pattern, i.e., a norm for that variable, assessed in a sample of patients who have the same age, gender, and race. However, cephalometric measurements hold significant degrees of imprecision, vagueness, and even inconsistency, making its clinical application less effective than expected by clinicians.

A better scenario would be to know how much the value of a variable of a certain patient deviates from its norm. It would be better if the "noise" carried by a cephalometric value could be quantified, in its own context. Besides inherent imprecision of measurements, we are also faced with individual variability of physicians' visual analysis and subjective interpretation.

So, developing a mathematical structure able to provide qualitative analysis, as well as quantitative information and modelling the imprecision, inconsistencies, and abnormality evidences of these variables could be relevant and useful.

In this section we employ PANN to analyse cefalometrical variables as an auxiliary tool for orthodontic diagnosis, suggesting a contextual cephalometric diagnosis.

In order to analyze skeletal and dental changes we selected a set of cephalometric variables based on expertise knowledge (Figures 12.11 and 12.12). These cephalometric variables are usually collected by experts through characteristic points in an X-Ray image of the patient's head [Ricketts 61].

The selected cephalometric variables feed the PANN in the following three units: Unit I, considering the antero-posterior discrepancy, Unit II, considering vertical discrepancy, and Unit III, taking into account dental discrepancy (see Figure 12.5).

Unit I comprises 2 levels. The first one involves the ANB and Wits variables. At the second level, there is a combination of the result of level 1 and the variables

1. Basion
2. Sella
3. Nasion
4. Posterior Nasal Spine
5. Anterior Nasal Spine
6. Inter-Molars
7. Inter-Incisors
8. Gonion
9. Menton
10. Gnathion
11. A Point
12. B Point
13. Pogonion
14. Incisal Edge - Upper Incisor
15. Apex - Upper Incisor
16. Incisal Edge - Lower Incisor
17. Apex - Lower Incisor

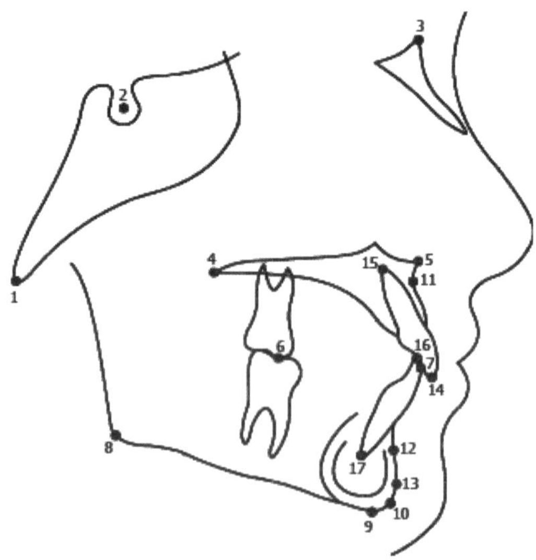

Fig. 12.11. Cephalometric Variables.

1. Anterior Cranial Base
2. Palatal Plane (PP)
3. Oclusal Plane (OP)
4. Mandibular Plane (MP)
5. Cranial Base
6. Y Axis
7. Posterior Facial Height
8. Anterior Facial Height - Median Third
9. Anterior Facial Height - Lower Third
10. Anterior Facial Height
11. SNA
12. SNB
13. Long Axis - Upper Incisor
14. Long Axis - Lower Incisor
15. A Point - Pogonion Line

Wits: distance between the projections of the A and B Points on the occlusal plane.

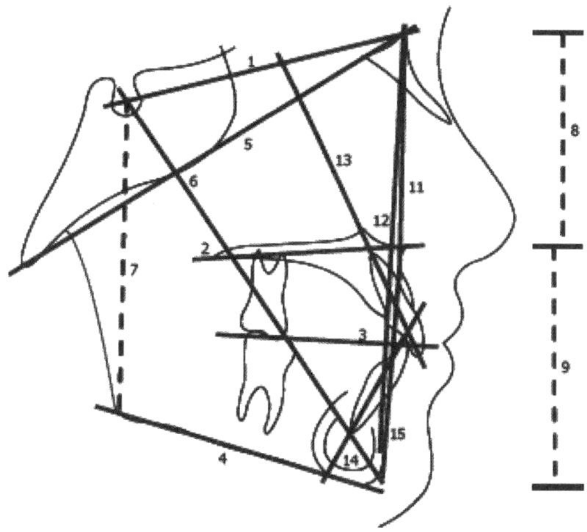

Fig. 12.12. Cephalometric analysis.

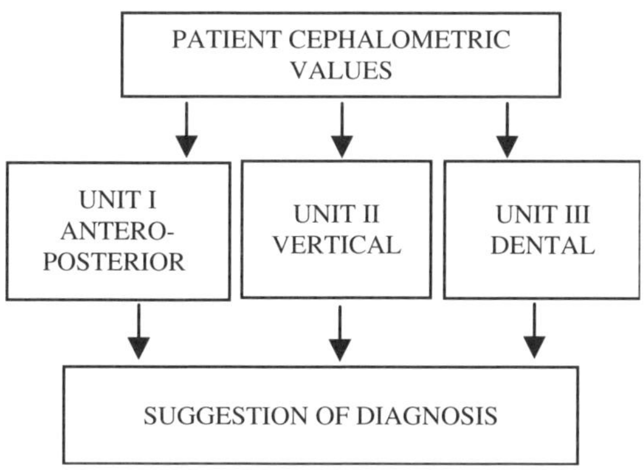

Fig. 12.13. Functional macro view of the PANN architecture.

SNA and SNB. The output of the second level relates to the position of the maxilla and the mandible, classifying it as: *well positioned*, *protruded*, *retruded*, *tending to protruded* and *tending to retruded*. The classes *protruded* and *retruded* is accompanied by their respective degrees of evidence of abnormality. Moreover, the classes "tending to" suggest the assessment of the outputs of the Units II and III.

The variables considered in Unit II are divided in three different groups: I) Se-Go/Na-Me proportion. The value of this proportion may result in a normal, vertical or horizontal face; II) Y Axis. The value of this angle may also result in a normal, vertical or horizontal face; and III) angles SeNa/PP, SeNa/OP and SeNa/MP. Each one of the three angles may also result in normal, vertical or horizontal face. The combination of the output from Groups I, II and III will also result in a normal, vertical or horizontal face.

The variables considered in Unit III are also divided into a three different groups: I) U1.PP and U1.SN angles and linear measurement U1-NA, taking into account the SNA angle (Unit I). The upper incisors may be in a normal position, proclined or retroclined. II) L1.APg, L1. NB and L1.GoMe angles and linear measurements L1-APg, L1-NB, taking into account the SNB angle.

The lower incisors may be in a normal position, proclined or retroclined. III) angle U1.L1. This value results in three possible positions: normal, proclined and retroclined. The combination of the outputs of the Groups I, II and III will result in normal, proclined, retroclined, tending to proclined and tending to retroclined.

Each unit has the following components, as shown in Figure 12.6:

Standardization: the difference between the data from the patient radiographs and its relative norm, by age and gender; and

Data modeling: matrices with possible degrees of evidence of abnormality.

Modeling Tool: matrices with all possible values of evidence of abnormality related to the patient variables, using the standard deviation values and providing the evidence degree of abnormality.

Learning Cell: "learn" the functions of the matrices which contain the degrees of evidence of abnormality of the variables.

Neural Network of Evidence: file the learned information and return with the degree of evidence according to the standard deviation of each variable in a contextualized form, i.e., considering the degrees of evidence of abnormality of the other variables.

Evaluation System: provides the diagnosis, inferred from the neural degrees of evidence of abnormality of each unit.

We are considering two variables, age and gender. So, given any age in the interval 6 to 18 years old and gender information, the developed system works with 4 craniofacial measurements in Unit I, resulting in 46 boolean inference rules, providing 33 possible outputs of types of malocclusion (diagnosis). In Unit II, 5 craniofacial measurements allow 90 inference rules and 4 possible outputs of diagnosis. In the Unit III, 9 cephalometric measurements were assessed, giving 87 inference rules and 12 possible classifications. In total, the decision-making support system is based upon 18 craniofacial cephalometric measurements, 223 inference rules and 49 possible diagnostic outputs.

In order to evaluate the system performance 120 orthodontic protocols with the cephalometric measurement information containing all variable values were analyzed by the model and three orthodontic experts. Kappa agreement concordance was evaluated comparing the diagnostic proposal of the experts and the system [Siegel & Castellan 88], [Fleiss 81].

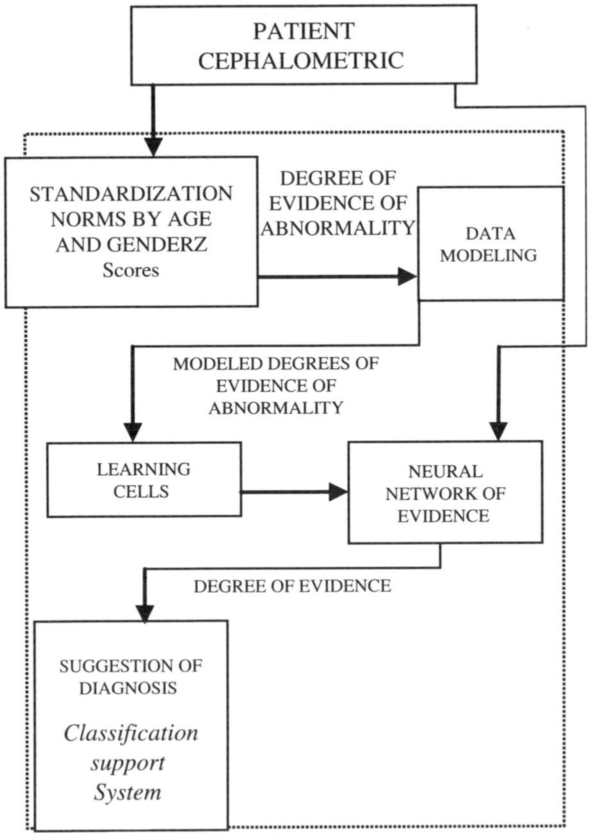

Fig. 12.14. Functional microview.

Table 12.15. Kappa values of agreement where E_1, E_2, and E_3 indicates the experts.

	E1 & model	E2 & model	E3 & model	E1&E2&E3
Anteroposterior region	0.343	0.289	0.369	0.487
Mandible position	0.296	0.245	0.306	0.404
Maxilla position	0.343	0.289	0.369	0.421
Vertical discrepancy	0.75	0.372	0.67	0.534
Superior incisive	0.443	0.216	0.454	0.468
Inferior incisive	0.451	0.084	0.457	0.418
Lip position	0.924	0.849	0.885	0.838

The analyzed data is composed of 44,17% males and 55,83% females, with age ranging between 6 to 53 (all Brazilians); 18.33% are more than 18 years old (corresponding to 22 in 120), which were treated by the model as 18 years old for simplicity. The sample is composed for the most part of a white group.

Cephalometrics is a most useful tool for orthodontic diagnosis, since it assesses craniofacial skeletal and dental discrepancies. However, conventional cephalometrics suffers important limitations, mostly due to the fact that the cephalometric variables are not assessed under a contextualized scope and exhibit important variation when compared to sample norms. So, its clinical application is relative, subjective, and routinely less effective than expected. Consequently, disagreement among orthodontists about diagnosis is very common.

From our point of view, this is a perfect scenario to evaluate the paraconsistent neural network capacity to perform with uncertainties, inconsistencies, and paracompleteness in a practical problem.

Kappa values of agreement where E_1, E_2, and E_3 indicate the experts shows that PANN performance is good and although the system needs more accurate validation, the preliminary results shown encourages that Paraconsistent Artificial Neural Network is a promising theory in the Cephalometric analysis, opening a new promising tool of research in health area.

12.13. Conclusions

Paraconsistent Artificial Neural Networks are a new kind of ANN based on Paraconsistent Annotated Evidential Logic $E\tau$. The logic $E\tau$ is capable of manipulating impreciseness, inconsistency and paracompleteness internally without trivilialization. Its composition as a 2-sorted language makes this property become critical when complexity of hardware implementation of paraconsistent logical circuits and computational implementation are the main concern. In the three example applications discussed, PANN has performed with suitable results, thus demonstrating it to be a promising theory, opening new areas for research.

References

1. J. M. Abe, "*Fundamentos da Lógica Anotada*" (Foundations of Annotated Logics), (in Portuguese) Ph. D. Thesis, University of São Paulo, São Paulo, 1992.
2. J. M. Abe, Paraconsistent Artificial Neural Networks: an Introduction, *Lecture Notes In Artificial Intelligence* 3214, Springer, Eds. J.G. Carbonell & J. Siekmann, ISSN 0302-9743, ISBN 3-540-23206-0, pp. 942–948, 2004.
3. J. I. Da Silva Filho & J. M. Abe, *Fundamentos das Redes Neurais Paraconsistentes* — Destacando Aplicações em Neurocomputação, (in Portuguese) Editôra Arte & Ciência, ISBN 85-7473-045-9, 247 pp., 2001.
4. R. Sylvan & J. M. Abe, On general annotated logics, with an introduction to full accounting logics, *Bulletin of Symbolic Logic*, 2, 118–119, 1996.
5. C. Stephens & N. Mackin, The validation of an orthodontic expert system rule-base for fixed appliance treatment planning. *Eur J Orthod*, 20:569–578, 1998
6. M. Del Santo Jr., Influence of the Occlusal Plane on ANB and Wits Assessments of Anteroposterior Relationship of the Jaws. *American Journal of Orthodontics and Dentofacial Orthopedics*, in press, 2005.
7. D. R. Martins, G. R. P. Janson, R. R. Almeida, A. Pinzan, J. F. C. Henriques, M.R. Freitas, *Atlas de Crescimento Craniofacial.* (in Portuguese) Ed. Santos, São Paulo, SP, 1998.

8. M. C. Mario, Modelo de análise de variáveis craniométricas através das Redes Neurais Artificiais Paraconsistentes, Ph. D. Thesis, University of São Paulo, 2006.
9. Y.Sorihashi Y, C. D. Stephens, K. Takada, An inference modeling of human visual judgment of sagittal jaw-base relationships based on cephalometry: Part II. *Am. J. Orthod. Dentofac. Orthop.* 117:303–11, 2000.
10. C. Steiner, Cephalometrics for you and me, *Am. J. Orthod.* 39:729–55, 1953.
11. Jacobson, "Wits" appraisal of jaw disharmony, *Am. J. Orthod.* 67:125–38, 1975.
12. Jacobson , The application of the "Wits" appraisal. *Am. J. Orthod.* 70:179–89, 1971.
13. R. M. Ricketts, Cephalometric analysis and synthesis, *Angle Orthod.* 31:141–56, 1961.
14. Abe, J.M., N. Ortega, M.C. Mario & M. Del Santo Jr., Paraconsistent Artificial Neural Network: An Application in Cephalometric Analysis, Lecture Notes In Artificial Intelligence 3684, Sub-serie da Lecture Notes In Computer Science, Springer, ISBN: 0302-9743, 716–723, 2005.
15. Weinstein J, Kohn K, Grever M. Neural Computing in Cancer Drug Development: Predicting Mechanism of Action. *Science* 1992, 258, 447–451.
16. Baxt WJ., Application of Artificial Neural Network to Clinical Medicine, *Lancet.* 1995, 346, 1135–1138.
17. Duffy FH, Albert MS, Mcnulty G, Garvey AJ. Age differences in brain electrical activity of healthy subjects. *Ann Neural* 1984; 16:430–438.
18. Nuwer MR, Comi G, Emerson R, Fuglsang-Frederiksen J, GuériT M, Hinrichs H, Ikeda A, Luccas FJC, Rappelsberger P. IFCN standards for digital recording of clinical EEG. *Electroencephalogr. Clin. Neurophysiol.*, 1998; 106:259–261.
19. Nitrini R, Caramelli P, Bottino CM, Damasceno BP, Brucki SM, Anghinah R; Academia Brasileira de Neurologia.Diagnosis of Alzheimer's disease in Brazil: Diagnostic criteria and auxiliary tests. Recommendations of the Scientific Department of Cognitive Neurology and Aging of the Brazilian Academy of Neurology. *Arq Neuropsiquiatr.* 2005 Sep; 63(3A):713–9.
20. Montenegro MA, Cendes F, Guerreiro MM, Guerreiro CAM. *EEG na Prática Clínica*, LEMOS Editorial 2001, 303pp.
21. Anghinah R. Estudo da densidade espectral e da coerência do eletrencefalograma em indivíduos adultos normais e com doença de Alzheimer provável (in Portuguese), Ph.D. Thesis, Faculdade de Medicina da Universidade de São Paulo, São Paulo, Brazil, 2003.
22. Da Costa N.C.A., J.M. Abe, J.I. Da Silva Filho, A.C. Murolo, C.F.S. Leite, *Lógica Paraconsistente Aplicada*, ISBN 85-224-2218-4, Editora Atlas 1999, 214 pp.
23. Erganian A, Mahmoudi B. Real-time ocular artifact suppression using recurrent neural network for electro-encephalogram based brain-computer interface. *Med. Biol. Eng. Comput.* 2005, 43(2):296–305.
24. Fausett L. Fundamentals of Neural Network Architectures, Algorithms, and Applications, USA, s.ed. Prentice Hall, 1994.
25. Subasi A, Alkan A, Koklukaya E, Kiymik MK. Wavelet neural network classification of EEG signals by using AR model with MLE preprocessing, *Neural Networks* 2005, Vol. 18(7), 985–997.
26. Ventouras EM, Monoyou EA, Ktonas PY, Paparrigopoulos T, Dikeos DG, Uzunoglu NK, Soldatos CR. 25 Sleep spindle detection using artificial neural networks trained with filtered time-domain EEG: A feasibility study, *Comput Methods Programs Biomed.* 2005, 78(3), 191–207.
27. Abe, J.M., H.F.S. Lopes & R. Anghinah, Paraconsistent Artificial Neural Network and Alzheimer Disease: A Preliminary Study, Dementia & Neuropsychologia, Vol 3., 241–247, 2007.

28. Abe J.M., J.C.A. Prado & K. Nakamatsu, Paraconsistent Artificial Neural Network: Applicability in Computer Analysis of Speech Productions, Lecture Notes in Computer Science 4252, 844–850, Springer, 2006.
29. Prado, J.C.A., Constelação Fônica e Redes Neurais Artificiais: Aplicabilidade na Análise Computacional da Produção da Fala, Ph.D. Thesis, University od São Paulo, São Paulo, Brazil, 2007.
30. Mario, M.C., J.M. Abe, N. Ortega, & M. Del Santo Jr., Paraconsistent Artificial Neural Network as Auxiliary in Cephalometric Diagnosis, 2008, to appear.
31. Fleiss, J.L., *Statistical methods for rates and proportions.* John Wiley, pp. 212–236. New York; 1981.
32. Siegel, S., Castellan N. *Nonparametric Statistics for the Behavioral Sciences.* 2.ed.: McGraw-Hill, pp. 284–285, New York; 1988.
33. Abe, J.M., J.C.A. Prado & K. Nakamatsu, Paraconsistent Artificial Neural Network: Applicability in Computer Analysis of Speech Productions, Lecture Notes in Computer Science 4252, pp. 844–850, Springer, 2006.

Chapter 13

PARACONSISTENT ANNOTATED EVIDENTIAL LOGIC $E\tau$ AND APPLICATIONS IN AUTOMATION AND ROBOTICS

Jair Minoro Abe[1] and Kazumi Nakamatsu[2]
[1]*Graduate Program in Production Engineering, ICET — Paulista University*
R. Dr. Bacelar, 1212, CEP 04026-002 São Paulo — SP — Brazil
[1]*Institute For Advanced Studies — University of São Paulo — Brazil*
jairabe@uol.com.br
[2]*School of Human Science and Environment/H.S.E.*
University of Hyogo — Japan
nakamatu@shse.u-hyogo.ac.jp

Automation and Robotics are fields that are growing very fast, both theoretically and in applications. Very imaginative theories have been proposed where classical logic is not suitable; perhaps the most famous is the so-called Fuzzy set theory, as proposed by Zadeh. This chapter presents a new non-classical logic, namely the Paraconsistent Annotated Evidential Logic $E\tau$, which is capable of manipulating imprecise, contradictory and paracomplete information without trivilialization. We illustrate with some applications in Automation and Robotics with some automated mobile robots showing their specific capabilities. We think that paraconsistent logical systems are very useful for applications in general opening applied techniques.

13.1. Introduction

For thousands of years humans have had a fascinating desire in constructing machines as close as possible to human beings. In particular, they have researched to find logical structures to 'imitate' human thinking and actions, in order for implementation in robots and automated processing. As the majority of automated processes and robotics are based on classical logic, in a certain sense their actions are predictable; all commands are of the type yes/no. So, robots built with classical logic are too limited, they cannot handle conflicting information, have tremendous difficulty in dealing with impreciseness, and are not able to manage directly with lack of information. So, we need more flexible logical systems for robots for them to perform more complex tasks in a smooth way without becoming involved with conflicts, imprecision and paracompleteness.

In this chapter we present some applications, mainly in robotics, of a new class of non-classical logic, namely the paraconsistent annotated logics. It was incorporated into some interesting autonomous mobile robots, which can manipulate imprecise, inconsistent and paracomplete data. One of the robot series dubbed Emmy,[a] based on a particular annotated system, namely, the paraconsistent annotated evidential logic $E\tau$, began with the 1st prototype studied in [Da Silva Filho 99], [Da Silva Filho & Abe 01]. Subsequently, some improvements were made in its 2nd prototype Emmy II [Torres 04] and in this paper we sketch the 3rd prototype in discussing a navigation system. Many of the topics, mainly the basic ones, were discussed in [Abe & Da Silva Filho 03]. We include a recent study of an application of the paraconsistent logical controller in a robot with temperature sensors [Berto 07]. Such a controller also was implemented in an electronic device for locomotion help for blind/deaf people.

13.2. Paraconsistent Annotated Logics

Annotated logics are a family of non-classical logics, being in general, paraconsistent, paracomplete, and non-alethic logics. The latter systems are among the most original and imaginative systems of non-classical logic developed in the past century.

An extensive study of annotated logics was made in [Abe 92]. Some applications are summarized in [Abe 97]. In view of the applicability of annotated logics to differing formalisms in computer science, it has become essential to study these logics more carefully, mainly from the foundational point of view.

Annotated systems are surprisingly useful logics. Technically, they constitute 2-sorted logics, having a very clear syntax and semantics; thus, it is possible to investigate their metatheory. They admit an elegant Hilbertian axiomatics, which are sound and complete (when the lattice is finite) regarding their semantics [Abe 92]. They also have a natural deduction system, which make possible more natural inferences, as well as better computability [Akama, Nakamatsu & Abe 07]. In fact, it has been implemented in a logic programming dubbed Paralog. Annotated high order logics can be established. Also annotated set theory was studied; these encompass *in totum* the Fuzzy set theory in different ways [Akama, Abe & Murai 03]. An axiomatic version of certain class of Fuzzy systems was obtained via annotated logics [Akama & Abe 00]. Annotated logics also allow defeseable, deontic, non-monotonic, linear reasoning, among others in a different approach than usual [Nakamatsu, Seno, Abe & Suzuki 03], [Nakamatsu, Mita, Shibata & Abe 03].

It has been used to build electronic logical gates Not, And Or which permit 'inconsistent' signals in a non-trivial manner; logical analyzer — Para-analyzer; logical controller — Para-control, logical simulator — Para-sim, all allowing the manipulation of inconsistent, imprecise and paracomplete signals [Da Silva Filho 99].

[a]The name Emmy is in homage to the mathematician Amalie Emmy Nöether (1882–1935).

Other logical controllers and integrated circuits were obtained. All these ideas were experimented in real-life robots or in simulators. Underlying these studies, an automata theory based on annotated systems was developed, close to weighted automata. Finally, Artificial Neural Networks were introduced based on annotated logics, with promising results [Da Silva Filho & Abe 01]. A concise presentation of the foundations of annotated systems is to be found in [Abe 92, 01].

13.3. Paraconsistent Annotated Evidential Logic Eτ

One of the most useful annotated systems is when we consider the lattice τ below and the corresponding logic is called paraconsistent annotated evidential logic Eτ. The atomic formulas of the logic $E\tau$ are of the type $p_{(\mu,\lambda)}$, where $(\mu, \lambda) \in [0,1]^2$ and $[0,1]$ is the real unitary interval (p denotes a propositional variable). $p_{(\mu,\lambda)}$ can be intuitively as read: "It is assumed that p's favorable evidence is μ and contrary evidence is λ."

Thus,

- $p_{(1.0,\ 0.0)}$ can be read as a true proposition,
- $p_{(0.0,\ 1.0)}$ as false,
- $p_{(1.0,\ 1.0)}$ as inconsistent,
- $p_{(0.0,\ 0.0)}$ as paracomplete, and
- $p_{(0.5,\ 0.5)}$ as an indefinite proposition.

Also we introduce the following concepts:

- Uncertainty Degree: $G_{un}(\mu, \lambda) = \mu + \lambda - 1\, (0 \leq \mu, \lambda \leq 1)$;
- Certainty Degree: $G_{ce}(\mu, \lambda) = \mu - \lambda\, (0 \leq \mu, \lambda \leq 1)$;

An order relation is defined on $[0,1]^2 : (\mu_1, \lambda_1) \leq (\mu_2, \lambda_2) \Leftrightarrow \mu_1 \leq \mu_2$ and $\lambda_1 \leq \lambda_2$, constituting a lattice that will be symbolized by τ.

With the uncertainty and certainty degrees we can obtain the following 12 output states: *extreme state* and *non-extreme states*, showed in the Table 13.1.

Table 13.1. Extreme and Non-extreme states.

Extreme States	Symbol	Non-extreme states	Symbol
True	V	Quasi-true tending to Inconsistent	QV $\rightarrow$ T
False	F	Quasi-true tending to Paracomplete	QV $\rightarrow \perp$
Inconsistent	T	Quasi-false tending to Inconsistent	QF $\rightarrow$ T
Paracomplete	$\perp$	Quasi-false tending to Paracomplete	QF $\rightarrow \perp$
		Quasi-inconsistent tending to True	QT $\rightarrow$ V
		Quasi-inconsistent tending to False	QT $\rightarrow$ F
		Quasi-paracomplete tending to True	Q$\perp \rightarrow$ V
		Quasi-paracomplete tending to False	Q$\perp \rightarrow$ F

All states are represented in the next figure.

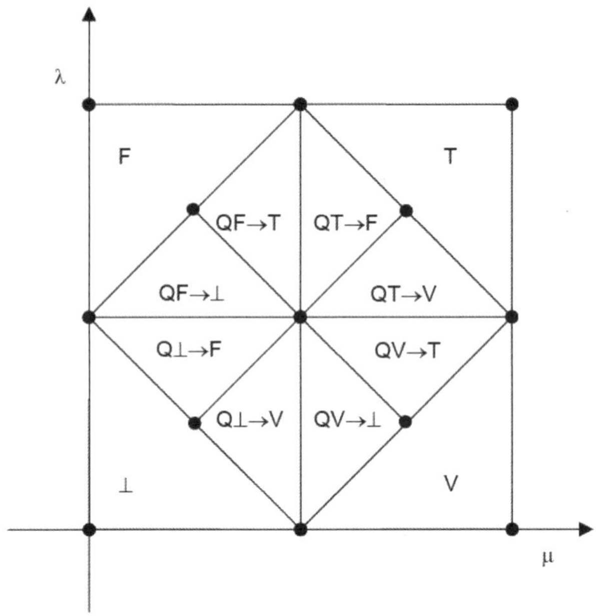

Fig. 13.1. Lattice τ.

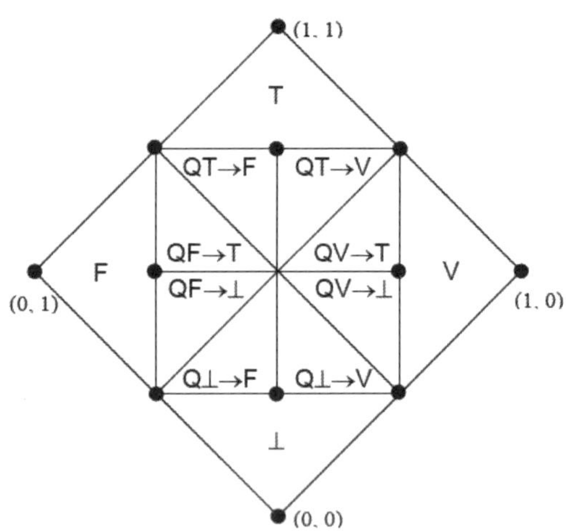

Fig. 13.2. Lattice τ.

Some additional control values are:

V_{cic} = maximum value of uncertainty control = C_3
V_{cve} = maximum value of certainty control = C_1
V_{cpa} = minimum value of uncertainty control = C_4
V_{cfa} = minimum value of certainty control = C_2

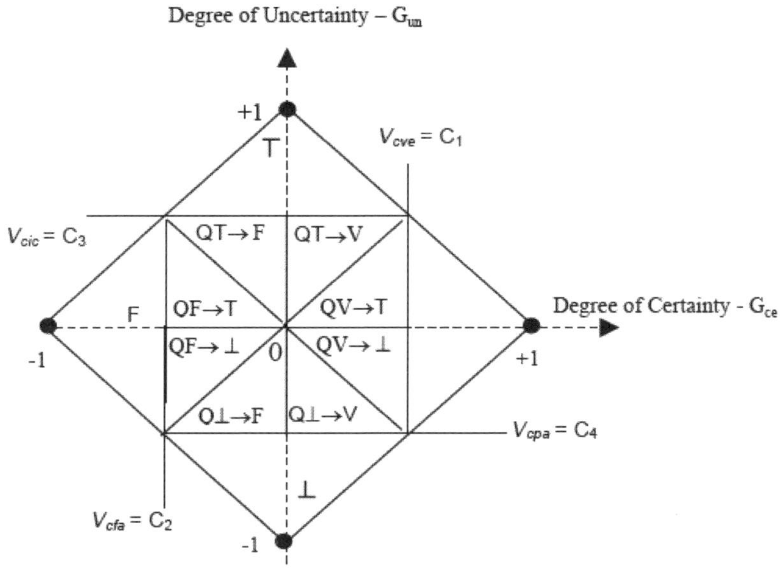

Fig. 13.3. Certainty and Uncertainty degrees.

13.4. The Paraconsistent Logical Controller — Paracontrol

The Paracontrol is an eletric-eletronic materialization of the Para-analyzer algorithm [Da Silva Filho 99], which is basically an electronic circuit, which treats logical signals in the context of logic Eτ. Such a circuit compares logical values and determines domains of a state lattice corresponding to output value. Favorable evidence and contrary evidence degrees are represented by voltage. Certainty and Uncertainty degrees are determined by analogues of operational amplifiers. The Paracontrol comprises both analog and digital systems and it can be externally adjusted by applying positive and negative voltages. The Paracontrol was tested in real-life experiments with an autonomous mobile robot Emmy, whose favorable/contrary evidences coincide with the values of ultrasonic sensors and distances are represented by continuous values of voltage.

13.5. The Autonomous Mobile Robot Emmy

The Paracontrol controller was applied in a series of autonomous mobile robots. The Emmy autonomous mobile robot has been presented previously [Da Silva Filho 99] and it consists of a circular mobile platform of aluminum 30 cm in diameter and 60 cm high.

While moving in a non-structured environment the robot Emmy gets information about the presence/absence of obstacles using the sonar system called Parasonic [Abe & Da Silva Filho 03]. Parasonic is able to detect obstacles in an autonomous mobile robot's path by transforming the distances to the obstacle into electric signals of the continuous voltage ranging from 0 to 5 volts. The Parasonic is basically

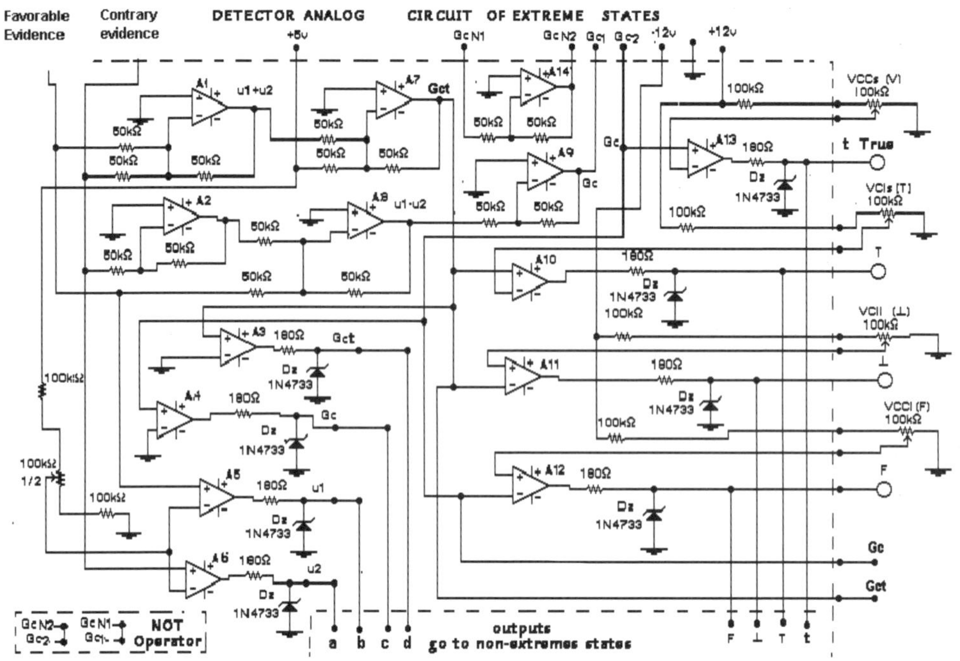

Fig. 13.4. Paracontrol circuit.

composed of two ultrasonic sensors of type POLAROID 6500 controlled by an 8051 microcontroller. The 8051 is programmed to carry out synchronization between the measurements of the two sensors and the transformation of the distance into electric voltage. Emmy has undergone improvements and the 2nd prototype is described in what follows.

The robot Emmy uses the para-control system to negotiate traffic in non-structured environments avoiding collisions with human beings, objects, walls, tables, etc. The form of reception of information on the obstacles is non-contact in nature is the method to obtain and to treat signals from ultra-sonic or optical sensors in order to avoid collisions.

The system of the robot's control is composed by the Para-sonic, Para-control and supporting circuits. See Figure 13.5.

- Ultra-Sonic Sensors — the two sensors detect the distance between the robot and the object through the emission of a pulse train of ultra-sonic sound waves and the return reception of the signal (echo).
- Signals Treatment — The treatment of the captured signals is made through the Para-sonic. The microprocessor is programmed to transform the time elapsed between the emission of the signal and the reception of the echo in an electric signal of the 0 to 5 volts for belief degree, and from 5 to 0 volts for disbelief

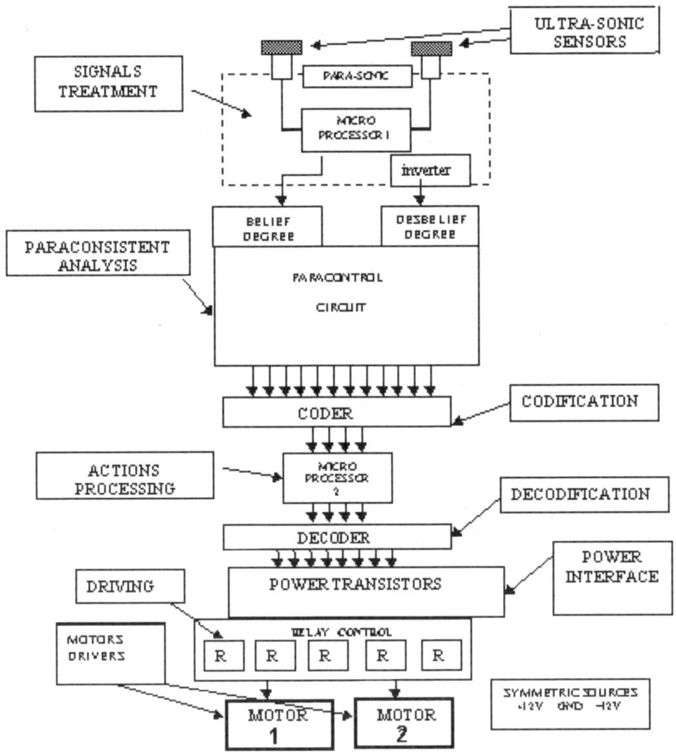

Fig. 13.5. Emmy robot control system schematic.

degree. The width of each voltage is proportional to the time that has elapsed between the emission of a pulse train and its reception by the sensors.

- Paraconsistent Analysis — The paraconsistent control logic makes a logical analysis of the signals according to the logic $E\tau$.
- Codification — The coder circuit changes the binary word of 12 digits to a code of 4 digits to be processed by the personal computer.
- Action Processing — The microprocessor is programmed conveniently to work the relay in sequences to establish actions for the robot.
- Decodification — The circuit decoder change the binary word of 4 digits in signals to charge the relay in the programmed paths.
- Power Interface — The power interface circuit is composed of transistors that amplify the signals making possible the relay's control by digital signals.
- Driving — The relays ON and OFF the motors M_1 and M_2 according to the decoded binary word.
- Motor Drives — The motors M_1 and M_2 move the robot according to the sequence of the relays control.
- Sources — The robot Emmy is supply by two batteries forming a symmetrical source of tension of activate ± 12 Volts DC.

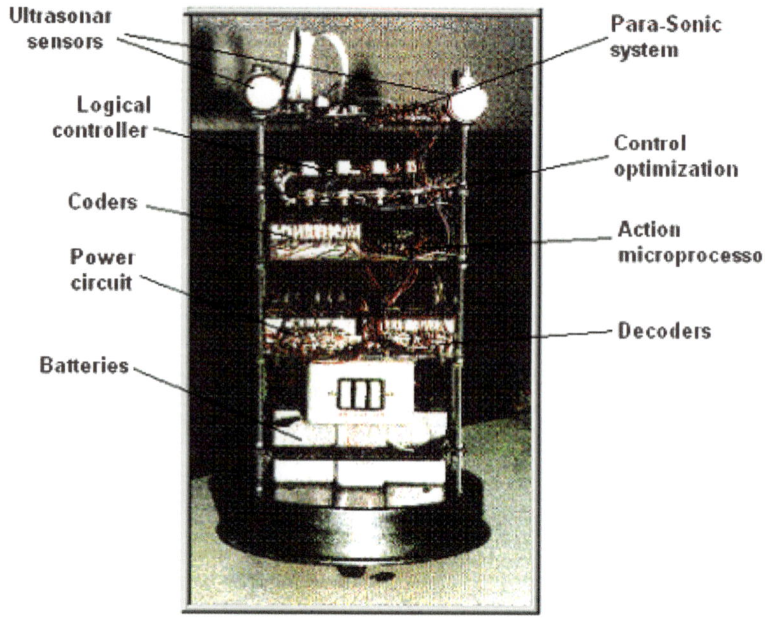

Fig. 13.6. Robot Emmy.

As the project is built in hardware besides the paracontrol it was necessary to install components for supporting circuits allowing the resulting signals of the paraconsistent analysis to be addressed and indirectly transformed in action.

In this first prototype of the robot Emmy it was necessary to incorporate a coder and a decoder such that the referring signals to the logical states resulting from the paraconsistent analysis were processed by a microprocessor of 4 inputs and 4 outputs.

13.6. Robot Emmy II

The platform used to assemble the Emmy II robot measures approximately 23 cm high and 25 cm in diameter (circular format). The main components of Emmy II are a 8051 microcontroller, two ultrasonic sensors, and two DC motors. Figure 1 shows the Emmy II basic structure. These signals from the 2 sensors are used to determine the favorable evidence degree μ and the contrary evidence degree λ regarding the proposition "There is no obstacle in front of the robot".

Then the Paracontrol, recorded in the internal memory of the microcontroller uses in order to determine the robot movements. Also, the microcontroller is responsible for applying power to the DC motors.

The Figure 13.7 shows the simplified block diagram of Emmy II.

The decision-making of the Emmy II regarding the movements to undertake is according to the decision state lattice shown in Figure 13.10.

Each robot movement lasts approximately 0,4 seconds.

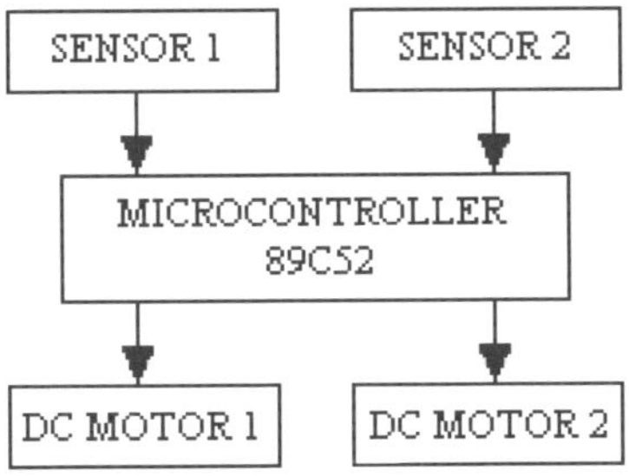

Fig. 13.7. Block diagram.

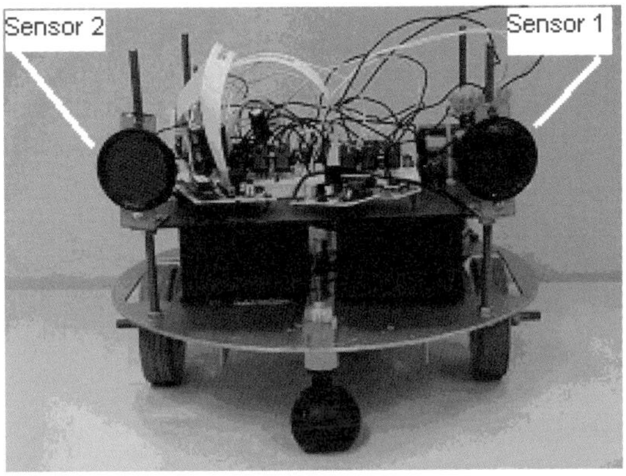

Fig. 13.8. Frontal vision of Emmy II.

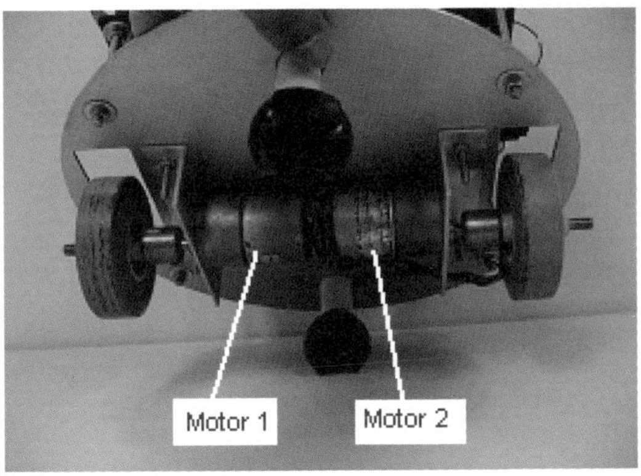

Fig. 13.9. Inferior vision of Emmy II.

The source circuitry supplies 5, 12 and 16,8 Volts DC to the other robot Emmy II circuits.

The "Polaroid 6500 Series Sonar Ranging Module" [POL 96] was used in the Emmy II robot. This device has three inputs (Vcc, Gnd and Init) and one output

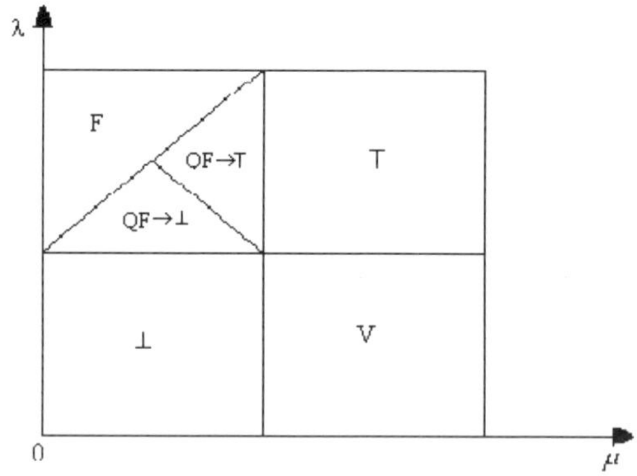

Fig. 13.10. Logical output lattice.

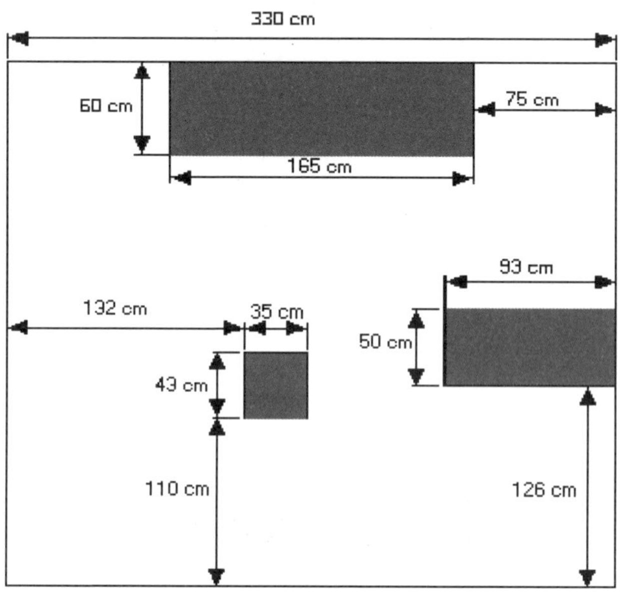

Fig. 13.11. Test environment.

(Echo). When INIT is taken high by the microcontroller, the sonar ranging module transmits 16 pulses at 49,4 kilohertz. After receiving the echo pulses, which cause the ECHO output to go high, the sonar ranging module sends an ECHO signal to the microcontroller. Then, with the time interval between INIT sending and ECHO receiving, the microcontroller is able to determine the distance between the robot and the obstacle.

The microcontroller 89C52 from 8051 family is responsible to control the Emmy II robot. Its input/output port 1 is used to send and receive signals to and from the Sensor Circuitry and Power Circuitry. Buffers are used in the interface between the microcontroller and the other circuits.

Table 13.2. Logical states and action.

Symbol	State	Action
V	True	Robot goes ahead
F	False	Robot goes back
$\perp$	Paracomplete	Robot turns right
T	Inconsistent	Robot turns left
$QF \to \perp$	Quasi-false tending to paracomplete	Robot turns right
$QF \to T$	Quasi-true tending to inconsistent	Robot turns left

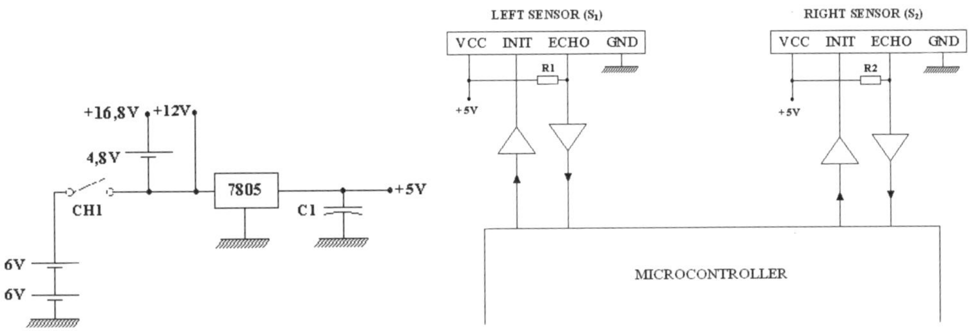

Fig. 13.12. Emmy II source circuitry electric scheme and sensor circuit.

The microcontroller I/O Port 1 has 8 pins. The function of each pin is as follows:

- Pin 0: INIT of sonar ranging module 1 (S_1).
- Pin 1: ECHO of sonar ranging module 1 (S_1).
- Pin 2: INIT of sonar ranging module 2 (S_2).
- Pin 3: ECHO of sonar ranging module 2 (S_2).
- Pin 4: When it is taken high (+5 Volts), DC motor 1 has power supplied.
- Pin 5: When it is taken high (+5 Volts), DC motor 2 has power supplied.
- Pin 6: When it is taken low (0 Volts), DC motor 1 spins around forward while pin 4 is taken high (+5 Volts). When it is taken high (+5 Volts) DC motor 1 spins around backward while pin 4 is taken high (+5 Volts).
- Pin 7: When it is taken low (0 Volts), DC motor 2 spins around forward while pin 5 is taken high (+5 Volts). When it is taken high (+5 Volts) DC motor 2 spins around backward while pin 5 is taken high (+5 Volts).

Two DC motors supplied by 12 Volts DC are responsible for Emmy II robot movements. The Paracontrol, through the microcontroller, determines which DC motor must be supplied and in which direction it must spin.

Basically, the power interface circuitry is comprised of power field effect transistors — MOSFETs.

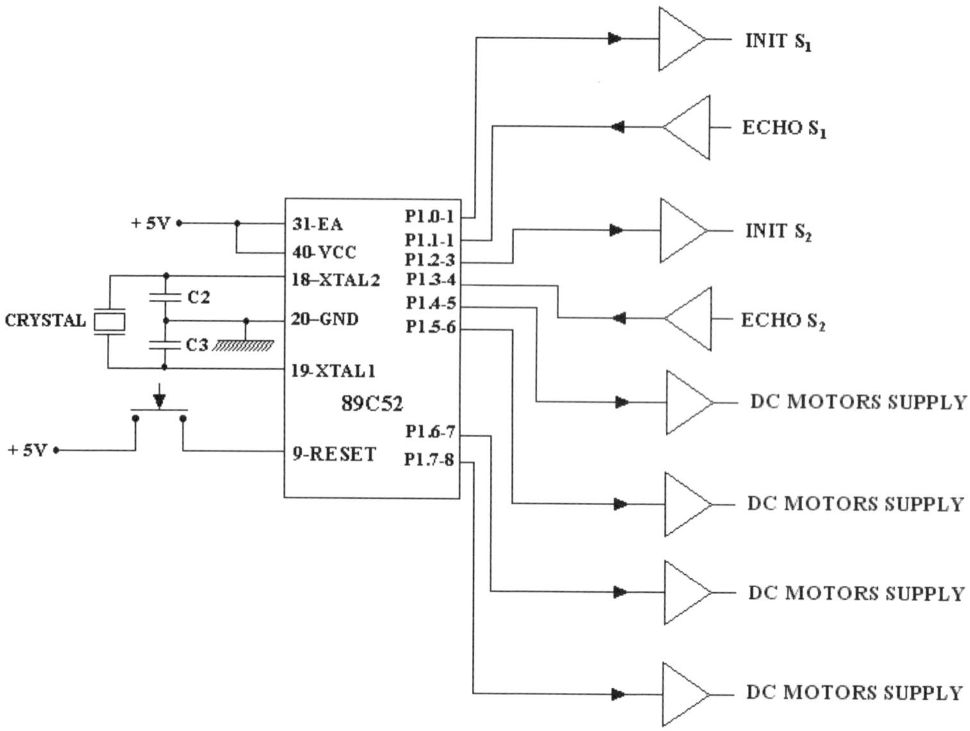

Fig. 13.13. Robot Emmy II main microcontroller connections.

The microcontroller 89C52 internal memory can store numbers of 8 bits. If we represent these numbers in hexadecimal, it means that we can store numbers from 00h to FFh. Internal memory location 30h of the 89C52 microcontroller was chosen to store the favorable evidence grade value.

When the microcontroller 89C52 I/O port 1 pin 0 is taken high, meaning that sonar ranging module 1 transmits sonar pulses, 30h memory position starts to increase by in 1 unit per 118 μsec.

The sonar ranging module 1 ECHO is connected to microcontroller 89C52 I/O port 1 pin 1. So, when this pin is taken high, meaning that sonar echo pulses have just returned, incrementing of the 30h internal memory position ceases.

The time for the sound to travel 4 centimeters is 118 μsec. Hence, the value stored in the microcontroller 89C52 30h internal memory position multiplied by 2 represents the distance in centimeters between sonar ranging module 1 and the obstacle in front of it.

The maximum value that can be stored in the microcontroller 89C52 30h internal memory position is 3Fh (to allow arithmetic calculations). So, the sonar ranging module 1 can measure distances between 00h and 3Fh, or, 0 cm and 126 cm.

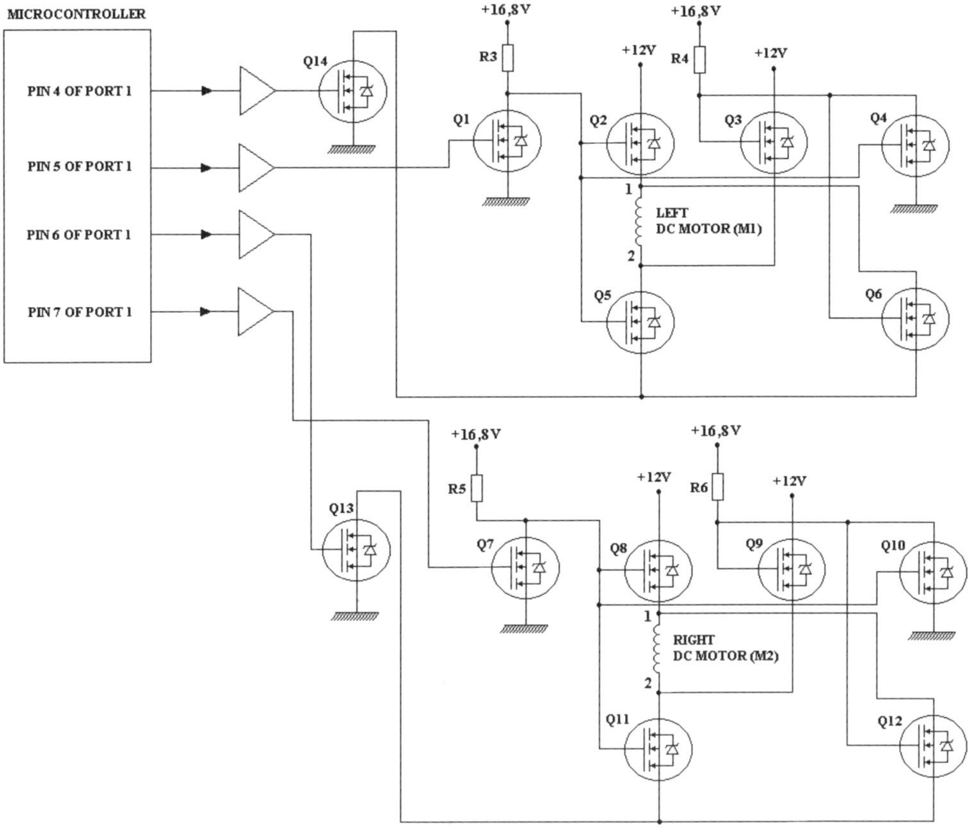

Fig. 13.14. Emmy II robot power interface circuitry.

Therefore, 00h stored in the microcontroller 89C52 30h internal memory position means that the favorable evidence grade value (μ) on the proposition "There is no obstacle in front of the robot" is 0. And 3Fh stored in the microcontroller 89C52 30h internal memory position means that the favorable evidence grade value (μ) on the proposition "There is no obstacle in front of the robot" is 1. In the same way, intermediate values between 00h and 3Fh stored in the microcontroller 89C52 30h internal memory position means intermediate favorable evidence grade values (μ) on the proposition "There is no obstacle in front of the robot"

When the microcontroller 89C52 I/O port 1 pin 2 is taken high, meaning that sonar ranging module 2 transmits sonar pulses, 31h memory position starts to be increased by 1 unit per 118 μseg.

The sonar ranging module 2 ECHO is connected to microcontroller 89C52 I/O port 1 pin 3. So, when this pin is taken high, meaning that sonar echo pulses have just returned, incrementing of the microcontroller 89C52 internal position 31h ceases.

The time for the sound to travel 4 centimeters is 118 μsec. Hence, the value stored in the microcontroller 89C52 31h internal memory position multiplied by 2 represents the distance in centimeter between sonar ranging module 2 and the obstacle in front of it.

The maximum value that can be stored in the microcontroller 89C52 31h internal memory position is 3Fh. Then, the sonar ranging module 2 can measure distances between 00h and 3Fh, or, 0 cm and 126 cm.

The microcontroller 89C52 31h internal memory position is complemented. Thus, it represents the contrary evidence grade value (λ) on the proposition "There is no obstacle in front of the robot". Consequently, 00h stored in the microcontroller 89C52 31h internal memory position means that the contrary evidence grade value (λ) on the proposition. "There is no obstacle in front of the robot" is 0. And 3Fh stored in the microcontroller 89C52 31h internal memory position means that the contrary evidence grade value (λ) on the proposition "There is no obstacle in front of the robot" is 1. In the same way, intermediate values between 00h and 3Fh stored in the microcontroller 89C52 31h internal memory position means intermediate contrary evidence grade values (λ) on the proposition "There is no obstacle in front of the robot".

13.7. Autonomous Mobile Robot Emmy III

The aim of the Emmy III robot version is to be able to move from an origin point to an end point, both predetermined, in a non-structured environment. The first prototype of the autonomous mobile robot Emmy III is composed of a planning system and a mechanical construction.

The planning system considers an environment to be divided in cells. This first version considers all cells to be free. Firstly, the robot must able to move from a point to another without encountering any obstacle. This environment is divided into cells as in [Elfes 89] and a planning system gives the sequence of cells the robot must travel to reach the end cell. This idea was applied in [Desiderato & Oliveira 06], [Maran *et al.* 06].

Then it asks for the initial point and the target point. After that a sequence of movements is given. Also a sequence of pulses is sent to the step motors that are responsible for moving the physical platform of the robot. So, the robot moves from the initial point to the target point.

The physical construction of the first prototype of the Emmy III robot is basically composed of a circular platform of approximately 286 mm diameter and two step motors. The figure below shows the Emmy III first prototype.

The planning system is recorded in a notebook. And the communication between the notebook and the physical construction is made through the parallel port. A potency driver is responsible for gettting the pulses from the notebook and sending them to the step motors.

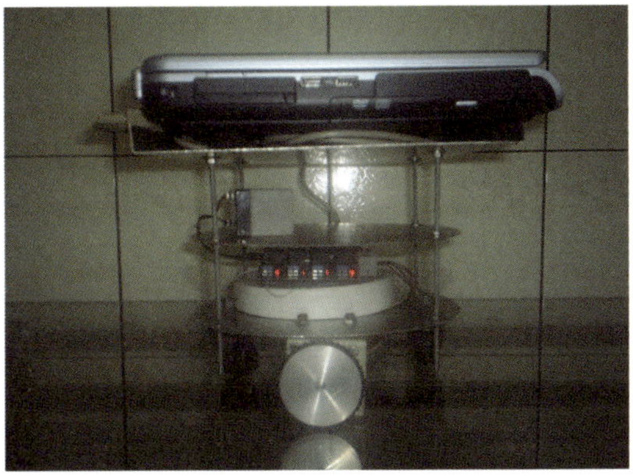

Fig. 13.15. Robot Emmy III.

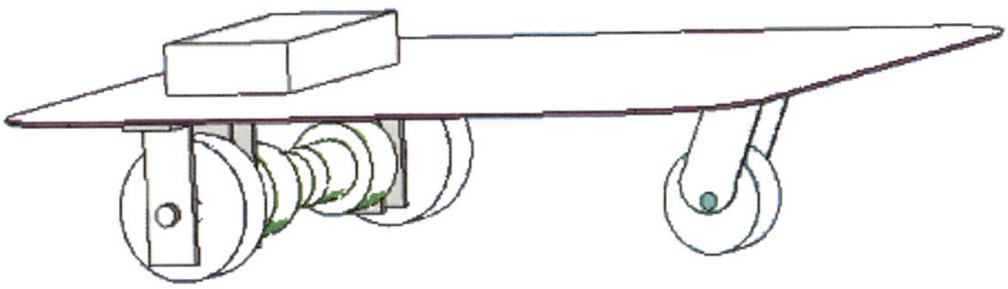

Fig. 13.16. Emmy III — basic structure.

Similarly to the first prototype, the second prototype of the autonomous mobile robot Emmy III is basically composed of a planning system and a mechanical structure. The planning system can be recorded in any personal computer and the communication between the personal computer and the mechanical construction is done through a USB port.

The planning system considers the environment around the robot to be divided into cells. So, it is necessary to inform the planning system which the cell the robot is in and likewise for the target cell. Then, the planning system gives a sequence of cells that the robot must follow to go from the origin cell to the target cell. The planning system considers all cells to be free.

The mechanical construction is basically composed of a steel structure, two DC motors and three wheels. Each motor has a wheel fixed in its axis and there is free wheel. The Figure 13.16 shows the mechanical structure.

There is electronic circuitry on the steel structure. The main device of the electronic circuit is the microcontroller PIC18F4550 that is responsible for receiving the schedule from the planning system and activating the DC motors. Also there is a potency driver between the microcontroller and the DC motors.

The final goal is to obtain a robot that is able to move between two predetermined arbitrary points, but this time with some obstacles.

13.8. Paraconsistent Autonomous Mobile Robot Hephaestus[b]

The Paracontrol was applied in the study of a robot with temperature sensors. The robot, dubbed Hephaestus, consists of a 20 cm × 30 cm rectangular platform in which there are a microcontroller PIC 16f877A, two pairs of dual temperature sensors (to get the environment temperature) LM35 and two continuous eletric current engines with speed reduction. The sensor LM35 can measure the temperature from 0°C to 100°C, with variation of the tension sign on the output of 10mV/°C. The thermometer scale are composed of unitary numbers according to the characteristics of LM35, and for each 1°C of temperature variation there will be a corresponding 10 mV variation on the tension, which makes it possible to obtain, therefore, temperatures between 2°C and 60°C.

The robot is described in the following four stages, these being:

- Feed circuit
- Sensor circuit
- Control circuit
- Performance circuit

The aim of the feed circuit is to provide the necessary electrical supplies to guarantee that the circuits of the robot can perform. The Hephaestus circuits need potentials of 5 and 12 Volts.

Information on the robot's environment is sent to the microcontroller, which processes according to the Paracontrol, resulting in a decision of which movement the robot has to do. Next, there is a description of the operation of the program located in the memory of the microcontroller.

This is the main step of Hephaestus' circuit as it is responsible for calculating the temperature of where the robot is. It is also responsible for transforming this temperature into values of favorable and contrary evidence degree, execution of the Para-analyzer algorithm and generating the signals for starting the engines.

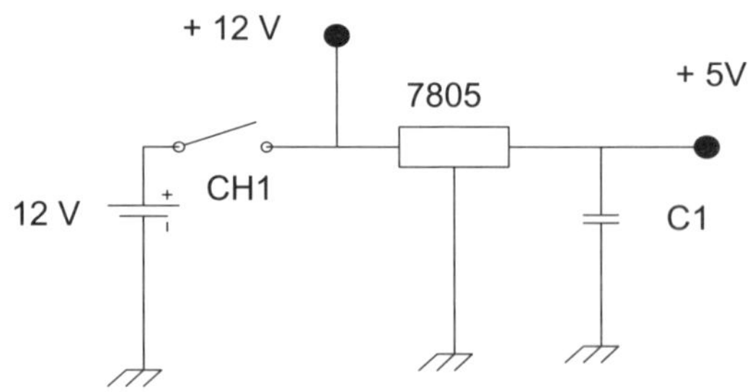

Fig. 13.17. Electric diagram of the feed circuit used for the Hephaestus robot.

[b]The ancient Greek god of fire.

Two engines that are fed by a continuous electric supplies ranging from 0 to 12 Volts are responsible for the movements of Hephaestus robot. The engines can make the robot go back and forward and it can also have its speed changed. Paracontrol is responsible, through the microcontroller, for determining which engine should operate, which way it should turn and which speed should be applied. A circuit D/A converter was built using an adder chain to process the incoming digital signals convert them first into analog signals, a PWM (Pulse Width Module) circuit to help controlling the speed and a bridged circuit H to control where the engines should turn to. Each engine has their independent circuits operated by the microcontroller outputs.

The logical controller Paracontrol will quantify the values of favorable evidence (μ) and of contrary evidence (λ) degrees, corresponding to the signs of the physical signal values originating from the temperature sensors, resulting in action. This analysis is made according to the output states lattice τ. Information about temperature is captured by two pairs of dual sensors. Each pair gives the value of favorable evidence degree and contrary evidence degree: (μ_1, λ_1) and (μ_2, λ_2).

The circuit of measuring and controlling the temperature assumed a base of 50°C, that is to say, the robot acts an action if the temperature is equal to or greater than 50°C. The possible actions are go forward, backward or turn aside.

The temperature sensors were placed on the extremities at the base of the robot, in such a manner that can be identify different temperatures in the front or in the back, on the left or right sides of the robot, which will determine its decision. In this case, the different degrees of favorable or contrary evidences of the process can be submitted to a logic operation NOT, OR and AND in order to obtain resulting values (μR e λR). The kind of logic operation applied will depend on the process that is under control. In the case of controlling the temperature developed in this work, the logic operation OR was defined due to the fact that if one of the sensors fails, the measuring of the temperature will be performed by the second sensor without loss of efficiency. For that reason it is considered the logical operator OR (maximization of the terms).

Table 13.3 presents the converted values of the continuous potential of the analog input signs.

During this stage of the process, the converted input converted signals are analyzed by applying the Para-analyzer algorithm. From the values of converted

Table 13.3. Values of the tension of the converted input analog signs.

Evidence Logical Signal		0	0.1	0.2	0.3	0.4	0.5	0.6	0.7	0.8	0.9	1
Voltage Signal (V)		0	0.3	0.6	0.9	1.2	1.5	1.8	2.1	2.4	2.7	3
Binary signal	LSB	0000	1001	0011	1101	0111	0001	1011	0101	1110	1000	1111
	MSB	0000	0001	0011	0100	0110	1000	1001	1011	1100	1101	1111
Hexadecimal signal		00	19	33	4D	67	81	9B	B5	CE	D8	FF

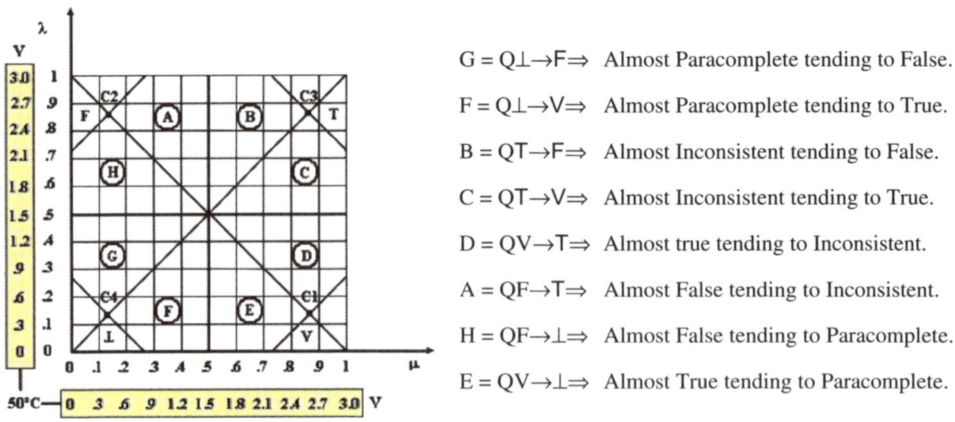

Fig. 13.18. Lattice τ with control values of 0.75.

Table 13.4. Parameters of the limit values of control.

Parameters of the control limit values			
C1	C2	C3	C4
11000001	00111111	11000001	00111111

reference tension and reference temperature potential, we determine the maximum value of μ and λ. In this work it was considered that the values for limiting the control would be of 0.75 of the maximum and minimum values of the degrees of certainty and contradiction, aiming to provide excellent tolerance to the process of temperature control.

With the definition of the limit values, in this work, the unitary square of the Cartesian plan has the form presented on Figure 13.3.

The True, False, Inconsistent and Paracompleteness areas were reduced because of the values imposed on the parameters C1, C2, C3 and C4. The goal was to diminish the margin of error between the measured and the reference temperatures. This tolerance is defined according to the process under control and is done in the software, so it is not necessary to adjust the hardware of the controller, which is an advantage.

The programming of the processed measuring and analysis system is based on the Para-analyzer algorithm. The Para-analyzer algorithm develops the paraconsistent expressions to define in which region of the reticulate in the unitary square of the Cartesian plan is the point of the parameters μ and λ, representing the Cartesian coordinates of the axes "x" and "y", respectively.

The analysis of the input parameters will be started after the definition of the limit values of control (C1, C2, C3 e C4). The parameters of the limit values of control are defined as follows:

C1 = Vscc (upper value of certainty control);
C2 = Vicc (lower value of certainty control);
C3 = Vscct (upper value contradiction control);
C4 = Vicct (lower value of contradiction control).

In the microcontroller program, the resulting values of the favorable and contrary evidence degrees are determied, as well as the degrees of certainty and of contradiction.

Aiming to verify how operational Hephaestus robot is, tests were done in a $1 \times 0{,}6$ m. tunnel especially constructed for the robot. An electric heater was placed in one of the extremities, serving as a source of heat, and the temperature presented was about 70°C. Figure 13.8 shows the environment built for the tests. The results are presented in Table 13.5. These tests were done basically to verify how long, in seconds, the robot takes to move backward from the heat while moving in an environment.

The results and tests that were obtained confirmed the importance of using the Paraconsistent Logic in projects of autonomous mobile robots and process controllers, efficiently processing data that has problems such as uncertainty, inconsistencies and paracompleteness. Thus, the establishment of an autonomous mobile

Fig. 13.19. Picture of Hephaestus robot moving into the tunnel.

Table 13.5. Table with the results of the tests.

Test	Time	Result	Description
1	13	succeeded	Succeeded on going back and forth.
2	18	succeeded	Succeeded on going back and forth.
3	60	Failed	It did not return because the printed circuit board was heated up.
4	15	succeeded	Succeeded on going back and forth.
5	30	succeeded	Succeeded on going back and forth, but moves took longer.
6	+60	Failed	It did not return and stood still inside the tunnel.

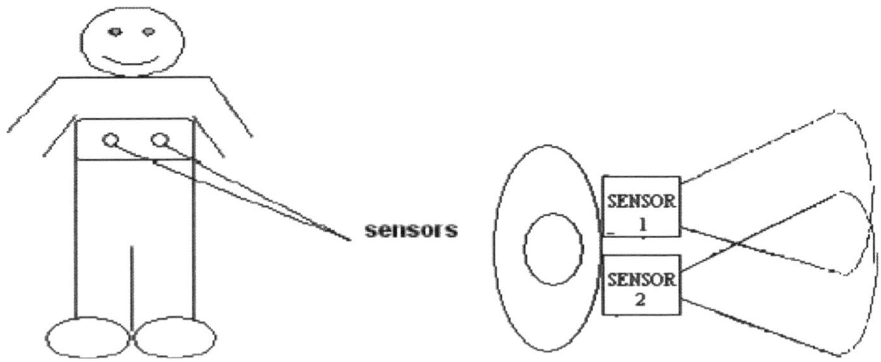

Fig. 13.20. Keller: position of the sensors.

robot was achieved without external supervision and with a possibility of operating in a non-structured environment, at relatively low financial cost and with good efficiency.

13.9. Keller — Electronic Device for Blind and/or Deaf People Locomotion

The Paracontrol was also used for the construction of an electronic device which we dubbed Keller,[c] for helping blind and/or dumb people in their locomotion. The main components of Keller are a microcontroller of the 8051 family, two ultrasonic sensors, and two vibracalls. Figure 13.20 shows the Keller basic structure.

The ultrasonic sensors are responsible for verifying whether there is any obstacle in front of the person for the area the sonars are acting. The signals generated by the sensors are sent to the microcontroller. These signals are used to determine the favorable evidence degree μ and the contrary evidence degree λ regarding the proposition "There is no obstacle in front of the person". Then the Paracontrol, recorded in the internal memory of the microcontroller, is used in order to determine people's movements, through a signal generated by the vibracalls, also provided by the microcontroller. The vibracalls can be placed in a confortable place for the user. In the prototype Keller, the two ultrasonic sensors were placed on the person's chest (Fig. 13.21). Thus, the main motivation in studying this prototype is as a very simple implementation of the Paracontrol. Keller is a promising device in aiding blind and/or deaf people in their locomotion. Some tests were made in real situations with promising results. However, we are aware we need improve with more accurate sensors, as well as how to detect obstacles on the ground and numerous other questions.

[c]Keller is in homage to Helen Adams Keller (1880–1968).

13.10. Conclusions

The applications considered show the power and even the beauty of Paraconsistent systems. They have provided other ways than usual electronic theory, opening applications of non-classical logics in order to overcome situations not covered by other systems or techniques. All the topics treated in this chapter are being improved as well as many other projects in a variety of themes. We hope to say more in forthcoming papers.

References

1. J. M. Abe, "Fundamentos da Lógica Anotada" (Foundations of Annotated Logics), in Portuguese, Ph. D. Thesis, University of São Paulo, São Paulo, 1992.
2. J. M. Abe & J. I. da Silva Filho, Manipulating Conflicts and Uncertainties in Robotics, *Multiple-Valued Logic and Soft Computing*, V.9, ISSN 1542–3980, 147–169, 2003.
3. J. M. G. Desiderato & E. N. De Oliveira, Primeiro Protótipo do Robô Móvel Autônomo Emmy III. Trabalho de Conclusão de Curso. Universidade Metodista de São Paulo, São Bernardo do Campo, 2006.
4. A. Elfes, Using occupancy grids for mobile robot perception and navigation, Comp. Mag., vol. 22, No. 6, pp. 46–57, June 1989.
5. L. H. C. Maran, P. A. Riba, R. G. Collett & R. R. De Souza, Mapeamento de um Ambiente Não-Estruturado para Orientação de um Robô Móvel Autônomo Utilizando Redes Neurais Paraconsistente. Trabalho de Conclusão de Curso. Universidade Metodista de São Paulo, São Bernardo do Campo, 2006.
6. Manual do sensor de ultra-som com sistema POLAROID 500, 1996.
7. C. R. Torres, *Sistema Inteligente Paraconsistente para Controle de Robôs Móveis Autônomos*, MSc. Dissertation, Universidade Federal de Itajubá — UNIFEI, Itajubá, 2004.
8. J. M. Abe, Some Aspects of Paraconsistent Systems and Applications, *Logique et Analyse*, 157(1997), 83–96.
9. J. M. Abe & J. I. da Silva Filho, Manipulating Conflicts and Uncertainties in Robotics, *Multiple-Valued Logic and Soft Computing*, V.9, ISSN 1542-3980, 147–169, 2003.
10. J. I. da Silva Filho — Métodos de Aplicações da Lógica Paraconsistente Anotada de Anotação com Dois Valores LPA2v com Construção de Algoritmo e Implementação de Circuitos Eletrônicos, in Portuguese, Ph. D. Thesis, Universidade de São Paulo, São Paulo, 1999.
11. J.I. da Silva Filho & J. M. Abe — "Emmy: a paraconsistent autonomous mobile robot", in Logic, Artificial Intelligence, and Robotics, *Proc. 2nd Congress of Logic Applied to Technology — LAPTEC'2001*, Edts. J.M. Abe & J.I. Da Silva Filho, Frontiers in Artificial Intelligence and Its Applications, IOS Press, Amsterdan, Ohmsha, Tokyo, Vol. 71, ISBN 1 58603 206 2 (IOS Press), 4 274 90476 8 C3000 (Ohmsha), ISSN 0922-6389, 53–61, 287p., 2001.
12. J. M. Abe, C. R. Torres, G. L. Torres, K. Nakamatsu & M. Kondo, Intelligent Paraconsistent Logic Controller and Autonomous Mobile Robot Emmy II, Lecture Notes in Computer Science 4252 ISSN 0302-9743, pp.851–857, Springer, 2006.
13. J.M. Abe, & M.F. Berto, Paraconsistent Autonomous Móbile Robot and Temperature Sensors, presented in 6th Congress of Logic Applied to Technology — LAPTEC2007, Santos, Brazil.

14. M. F. Berto, Aplicação da Lógica Paraconsitente Anotada Evidencial Eτ no Controle de Sensores de Temperatura na Atuação de Robôs Móveis, MSc. Dissertation, in Portuguese, Paulista University, São Paulo (2007).
15. S. Akama, J. M. Abe & T. Murai, On the relation of fuzzy and annotated logics, Proc. of the 7th International Conference Artificial Intelligence and Soft Computing, 46–51 2003.
16. S. Akama & J. M. Abe, Fuzzy annotated logics, Proc. Of 8th International Conference on Information Processing and Management of Uncertainty in Knowledge Based Systems, IPMU'2000, Organized by: Universidad Politécnica de Madrid (Spain), July 3–7, 2000, Madri, Espanha, Vol. 1, 504–508, 2000.
17. K. Nakamatsu, T. Seno, J. M. Abe & A. Suzuki, Defesiable deontic traffic signal control based on a paraconsistent logic program EVALPSN, IPMM'03, The Fourth International Conference on Intelligent Processing and Manufacturing of Materials, Sendai, Japan, 18–23, 2003.
18. K. Nakamatsu, H. Suito, J. M. Abe & A. Suzuki, Intelligent Real-time Traffic Signal Control Based on a Paraconsistent Logic Program EVALPSN, Lecture Notes in Computer Science, Springer-Verlag Heidelberg, Volume 2639, 719–723, 2003.
19. K. Nakamatsu, Y. Mita, T. Shibata & J. M. Abe, Defesiable deontic action control based on paraconsistent logic program and its implementation, 2003 International Conference on Computational Intelligence for Modelling, Contrl & Automation — CIMCA 2003, Proceedings, M. Mohammadia (Ed), ISBN 1740880684, 233–246, 2003.
20. J. M. Abe & J. I. Da Silva Filho, Simulating Inconsistencies in a Paraconsistent Logic Controller, International Journal of Computing Anticipatory Systems, vol. 12, ISSN 13735411, ISBN 2-9600262-1-7, 315–323, 2002.
21. S. Akama, K. Nakamatsu & J. M. Abe, A Natural Deduction System for Annotated Predicate Logic, Lecture Notes in Computer Science, ISSN 0302-9743, Volume 4693, Springer-Verlag, 861–868, 2007.
22. S. Akama, J. M. Abe & T. Murai, On the relation of fuzzy and annotated logics, Proc. of the 7th IASTED International Conference Artificial Intelligence and Soft Computing — ASC 2003, 46–51, 2003.

Chapter 14

ADAPTIVE INTELLIGENT LEARNING SYSTEM FOR ONLINE LEARNING ENVIRONMENTS

Fatma Cemile Serçe
Department of Information Systems Engineering, Atilim University
P.O. Box 06836, İncek Gölbaşı, Ankara, Turkey
fcserce@atilim.edu.tr

Ferda Nur Alpaslan
Department of Computer Engineering, Middle East Technical University
06800, Ankara, Turkey
alpaslan@ceng.metu.edu.tr

Lakhmi C. Jain
University of South Australia
Adelaide, SA, Australia
Lakhmi.Jain@unisa.edu.au

The present chapter deals with an Adaptive Intelligent Learning System (AILS) which is designed to be used with any Learning Management System (LMS). The adaptiveness provides unique method of identifying and monitoring the learner's learning processes according to their respective learning ability. AILS is a multi-agent system. This was developed in the form of JADE agents. The chapter presents the learning model, the system components, agent behavior in learner scenarios, the ontologies used in agent communications, and adaptive strategies. A sample application of AILS to a dummy LMS is also presented.

14.1. Introduction

Distance learning using the emerging computer technologies offers many promising possibilities in education. A crucial issue in educational technology is the ability to provide instructional environments and conditions that can provide for individually both different educational goals and learning ability to cope with different abilities. The learning management systems provide educational services for a wide range of students. They can also help students achieve their goals by providing knowledge in an adaptive and individualized way [1].

Adaptiveness is a crucial requirement in online learning environments (OLE). In [1], it is claimed that virtual learning environments (VLE) are at their best teaching effectiveness when they can adapt to the needs of the individual learner.

OLEs should be able to identify these learning needs and to customize the solutions which foster successful learning and performance. This property is necessary with or without the aid of an instructor to supplement instruction [2]. It is explained in [3] that the ultimate goals of an online learning environment are to achieve adaptive learning and help learners to create understanding of the material.

Adaptive intelligent learning systems have the advantage of both artificial intelligence and learning management systems. AI is seen as a way to extend and improve LMS. The intelligence of the LMS is in part attributed to its ability to adapt to the learner's needs during the learning process. Flexible OLE can be established according to any prior knowledge of the subject domain, the individual student or the groups of students, and the teaching methodology in use [4].

Pedagogical agents are autonomous agents which aid human learning by interacting with the students in the context of an intelligent learning environment. They extend and improve previous work as an intelligent tutoring system in a number of ways. They can adapt their behavior to the dynamic state of a learning environment. They take advantage of those learning opportunities as they arise. They can support collaborative learning and individualized learning. This is because multiple students and agents can interact in a shared environment.

Some adaptive systems have already been developed. Examples are iWeaver[5], INSPIRE[6], MANIC[7], ARTHUR[8], CS388[9], AEC-ES[10]. The adaptation strategy of these systems is based on a learning style. They differ in which learning style model they choose to use and how they adapt. For example, ARTHUR, iWeaver, MANIC, CS388 are based on sensory preferences; AEC-ES categorize the learners as either field-dependent (FD) or independent (FI), for example.

Almost all of the presented systems except for iWeaver and MANIC, assess the learning styles using psychometric questionnaires. The disadvantage of this approach is that the learners are classified into stereotypical groups. Also the assumptions about their learning styles are not updated following their interaction with the system [11].

These adaptive systems have important adaptive features. It is however not possible to integrate any of them with an existing LMS. These are designed as stand-alone systems. However, there are many LMSs used in practice and it could be effective to plug adaptive features into these existing and widely used learning management systems.

These existing adaptive systems are not compatible with any learning standard. This is true for learner information, or the standard storing of learning material. For multiple use interoperability of the learning content and learner information, learning standards are essential for the researchers.

In this chapter, a multi-agent intelligent learning system, named AILS, is proposed. Unlike the other systems, AILS has been designed to work interchangeably with an existing LMS. To obtain this modularity, an interface for an adaptive system has been developed. The agent has been given the task of inter AILS-LMS communication. There is no standard interface defined for interaction between the adaptive systems and the LMS, which makes it necessary to add extensions for the specific LMS to establish AILS-LMS communication.

The aim of AILS is to provide e-learning services for online learners. The main operations to be adapted can be summarized as modeling the learner, presenting the adapted instructional material, searching for keywords, and communicating with other systems. In AILS environment, all AILS agents interact with each other to accomplish their individual roles. They use the available resources to provide the learners' needs.

AILS was analyzed and designed using an agent-oriented software engineering method, namely Gaia [12, 13]. The AILS system has been developed to satisfy systems requirements and design issues. AILS's agents have been implemented as JADE agents. JADE [14] is a middleware that facilitates the development of multi-agent systems. The agent behaviors are learners' actions using the LMS. These are explained in the form of "scenarios". AILS is still under development. In this study, scenarios such as the "user login", "user action to view the lecture note" and "user search request entering a keyword". These are explained and their implementation results are also given. In order to better explain the system and provide an easy test, a dummy LMS has been developed. During the development phase, we have used this dummy LMS to test the system. A sample course on "Java Programming provided the basis of test. The sample results obtained from this utilization of AILS are explained.

14.2. Agents

There are many definitions of agents. Some researchers define agents as the software systems for performing the tasks delegated to them. Researchers have implemented agents which are flexible and autonomous and can work in dynamic and unpredictable environments. A number of multiagent systems are in use today in virtually every area such as healthcare, business, games, engineering, aviation, healthcare, education, and so on.

Agents have grown exponentially in the last few years due to their properties such as autonomy, mobility, reasoning, learning and so on. Mobility is a very useful property as it allows agent to move around to perform a given task. The learning ability permits agent to learn from the interactions and by experience in order to improve its performance. Reasoning is very helpful in dynamic decision making in unknown environments. It is a common practice to group agents to take collective decision in a multiagent environment. Each agent is usually responsible to interact with other agents and environment through cooperation, coordination and learning. It is our belief that the agent-based theory and system design techniques are not ready for larger projects. For example, there is a need to evolve reasoning based techniques. Thus, it becomes difficult to design autonomous systems. An overview of the future for agent technology is presented in a paper, published in Hindustan Journal, in Volume 2, 2008. It is also stated in the paper that the future for intelligent agent technology will be limited to niche areas and that a team programming paradigm offers a better alternative for tackling general complex system applications such as sensor network management and enterprise management. The readers are urged to refer Section 9 on resources for further reading on agents.

14.3. AILS: Adaptive Intelligent Learning System

In this section, the description of the AILS architecture is given. Its functional components, the agents and their roles, the communication and interactions of the agents are provided. Since AILS was developed as a multi-agent system, we used an agent-oriented software engineering modeling technique as the model of the system. The existing software development techniques such as, object-oriented analysis and design, were not suitable for multi-agents systems' analysis and design. There is a fundamental mismatch between the concepts used by object-oriented developers and that of the agent-oriented view. For this reason, Gaia was used for the agent-oriented software engineering technique. This was used in an attempt to define a general methodology which was specially tailored for the analysis and design of multi-agent systems (MASs). In Gaia, the analysis and design may be thought of as a process of developing increasingly detailed models of the system to be constructed [12]. The models of the Gaia methodology is given in Figure 14.1. The architecture of AILS was produced using these Gaia models [15]. In this section, these models are given in order to understand the AILS process and its environment.

14.3.1. *Components of AILS*

The aim of AILS is to provide e-learning services for online learners. The main components of the system are as follows:

- **Student Modeling:** AILS provides online learners with a personalized learning experience. AILS maintains a learner profile. It allows the learner to specify his/her preferences, but it also exhibits a self-learning ability to determine the learner's characteristics by monitoring the behavior.
- **Adapted Instruction:** AILS adapts the learning services according to the learners' needs. These are defined according to student profile which is modeled by the system. Adapting services involves the content adaptation arising from the use

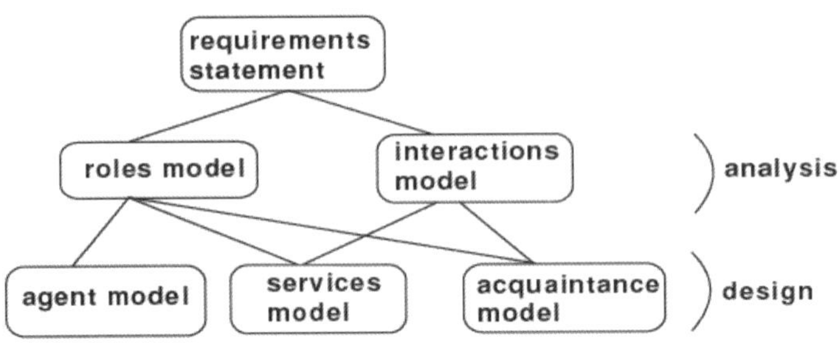

Fig. 14.1. Gaia Models.

of SCORM compliant lecture notes. AILS chooses the most appropriate lecture note by using metadata of the content and the learner's profile information.
- **Communication with other AILSs:** The system can communicate with other AILSs. It maintains keywords of the learning information stored in these systems. This is done in order to provide an inter-AILS search functionality.
- **Integration with LMSs:** The system may be integrated using LMS with the the following features, such as SCORM (Sharable Content Object Reference Model) compliant, "upcoming events" tools that can be used to display any message and to provide keyword search capability.
- **Learning Object Search:** A learner can search the learning objects available by entering keywords of the search tool in the LMS. The system responds to the user search request using the resources existing in the repository together with the resources existing in other AIL which the system is able to communicate.

14.3.2. *The Architecture*

An Adaptive Intelligent Learning System generally consists of two modules. These are the agents module and the environment module shown in Figure 14.2.

The agent module has the necessary software agents required to support adaptive learning in LMS. The communication and coordination between AILS agents occur in the agent module. The second module in the system is the environment module. The environment module consists of the following entities and is shown in Figure 14.3:

- **LMS:** AILS is a system built on the top of an LMS. It is where the learning content is delivered and managed. A combination of software tools have a variety of functions related to online learning.

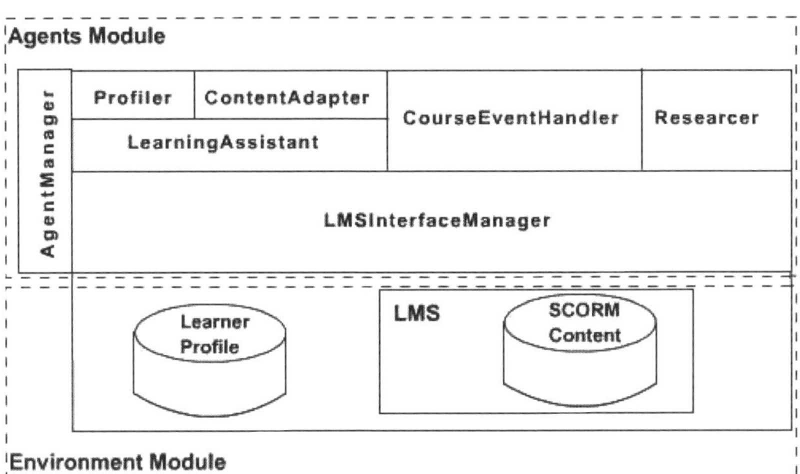

Fig. 14.2. The AILS System Overview.

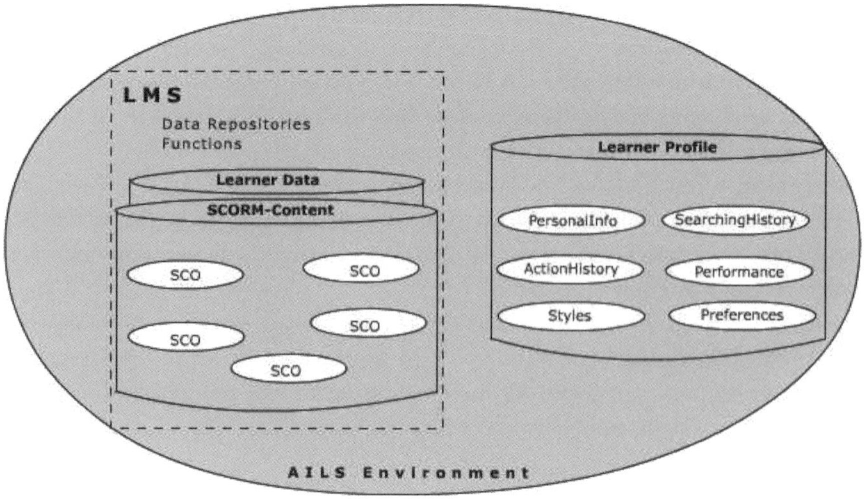

Fig. 14.3. The Environment Model of AILS.

- **Learner Profile:** The Learner profile collects data about a student's learning experience, learning preferences, learning style, and learning habits. The AILS system creates, stores and manages the learner profile repository. The system obtains information about the learner from the LMS' database. It, then, updates the learner profile repository.
- **SCORM Content:** SCORM is the Shareable Content Object Reference Model. It is a collection of specifications that enable inter operability, accessibility, and reusability of web-based learning content. The individual learning units of the SCORM-compliant content are called SCOs and in this study it is assumed that the SCOs of each course are kept in the LMS section. AILS does not store actual content objects. AILS system only stores the general information on the SCOs and the keywords. The location is arranged so that it can conduct the search operation.

14.3.3. *The Agents*

AILS contains six agents which are: LMSInterfaceManager, LearnerAssistant Agent, ContentAdapterAgent, ProfileAgent, ResearcherAgent, and AgentManager. The descriptions and roles of each agent are as follows:

- **LMSInterfaceManager** is the communication agent in the LMS. The other AILS agents communicate with the LMSInterface Manager to manage the related functions in LMS section. In addition, it informs the other agents about LMS events.
- **LearnerAssistantAgent** records the learner's progress. It coordinates the learners modeling and content adaptation processes. Because of the

complexity, it contains two worker agents which are ContentAdapterAgent and ProfileAgent.

- **ContentAdapterAgent** prepares an adapted content template for the learner which uses the learner profile information received from ProfileAgent. It applies adaptive strategies which are defined for AILS. This agent also provides a content template for ResearcherAgent which deals with adaptive searching.
- **ProfileAgent** initializes and updates the learner profile. It provides learner profile information which is requested by the other agents.
- **ResearcherAgent** keeps track of the learners searching criteria and it provides an ongoing search using the learner's search concepts, even when the learner is not online.
- **AgentManager** launches new agents, registers new services, and manages other agents.

Figure 14.4 shows agent types which are used in AILS. The agent realizes these agent types during run-time. For each online learner, each type of agent is instantiated once. There are as many agents as the number of online learners in the LMS at runtime.

14.3.3.1. *The Agent's Roles*

The role model identifies the basic skills required by the system. An agent can play a number of roles. To accomplish this, agents typically need to interact with each other in order to exchange knowledge and coordinate activities within the model.

The AILS agents have four major roles. The role models of each of the agent roles are given in the Figures 14.5–14.8. Note that the ContentAdapterAgent and the ProfileAgent's roles are summarized in the role of learning assistant agent.

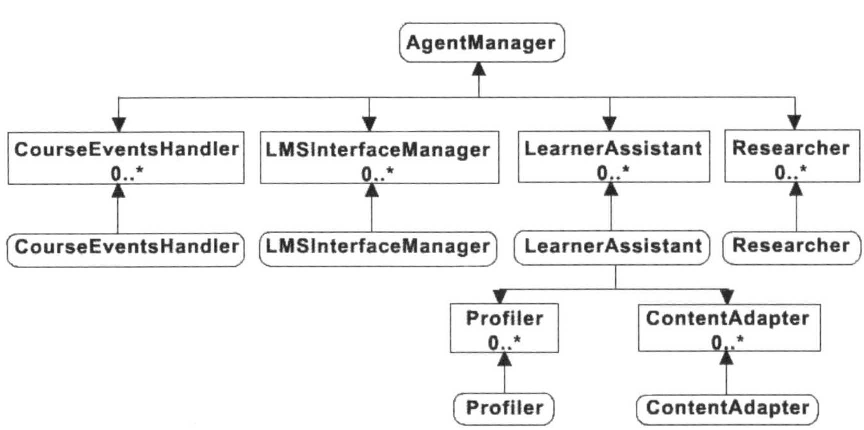

Fig. 14.4. The Agent Model.

Role: LearnerAssistant(LA)

Description: This is the most complex role serving to online learners in AILS, LA initializes and manages learner's profiles and learner's sequence of actions. LA retrieves learners data from LMS module and puts these data into profile informations. LA gets learner's learning request and provides learners adapted learnin services usin the profile, and action history data.
LA has a self-learning ability on learner's learning process. It learns the learner's learning prefrences through learner selections and learning experiences

Protocols & Actvities: InitializeLearnerProfile, ReceiveNewCourseEvents, GetLearnerRequest, GetLearnerData, AdaptContent, GetLearnerProfile, GetContentt, RespondToLearner, UpdateLearnerProfile, LearnByML

Permissions: reads *profileData* //profile data of the learner
creates *profileData*
updates *profileData*
reads *content* //lecture notes of the course which are in LOs
updates *content*
reads *acquaiantances* *// acquaiantances data structure*

Responsibilities:

Liveness:

LEARNERASSISSTANT = GetLearnerData.InitializeLearnerProfile.
(ServeLearner)^{w} || (ReceiveNewCourseEvents)^{w}
|| (LearnByLearnerProfile)^{w}

SERVELEARNER = GetLearnerRequest || GetLearnerProfile.AdaptInstruction.
(RespondToLearner || UpdateLearnerProfile)

ADAPTINSTRUCTION = GetContent.AdaptContent.GetLearnerProfile

LEARNBYLEARNERPROFILE = LearnByML.UpdateLearnerProfile

Safety:
A successful connection with the database and LMS is established

Fig. 14.5. The Role Model of LEARNERASSISTANT.

Role: LMSInterfaceManager(LIM)

Description: This role provides establishing a communication platform between AILS nad LMS. It receives actions from LMS module and then activates the corresponding function in AILS module communicating wiht other roles(agents) in AILS. It also provides AILS to activate some functions in LMS.

Protocols & Actvities: SetLMSEvent, SetAILSEvent, ActivateAILSFunctions, ActivateLMSFunction

Permissions: reads *acquaiantances* *// acquaiantances data structure*

Responsibilities:

Liveness:

LMSINTERFACEMANAGER = (SetLMSEvent.ActivateAILSFunction)^{w} |
(SetAILSEvent.ActivateLMSFunction)^{w}

Safety:
A successful connection with LMS is established

Fig. 14.6. The Role Model of LMSINTERFACEMANAGER.

Role: Researcher(RS)

Description: This role provides online learners and facilitator to search for some learning objects both in internal repository and also in the other AILS's repositories. RS also keeps track of each individual's searching keywords in the profile information and provdes continous search on these subjects. As the user configures, whenever a new learning object is inserted into the repositories, it informs the individual about this new resource.

Protocols & Actvities: ReceiveSearchRequest, SearchInternalRepository, SearchExternalRepository, RetrieveRemoteObjects,RespondToLearner,UpdateProfile,SearchRecentResources, GetSearchResults

Permissions:
reads *learningObjects* // *learning objects in the repository*
reads *acquaiantances* // *acquaiantances data structure*
reads *profileData* // *profile data of the user*
updates *profileData* // *update data about keywords searched*

Responsibilities:

Liveness:

RESEARCHER = ReceiveSearchRequest.(Search)$^{\omega}$.
(UpdateProfile || RespondToLearner)

SEARCH = (SearchInternalRepository || SearchExternalRepository).
RetrieveRemoteObjects.(SearchRecentResources)$^{\omega}$.
GetSearchResults

Safety:

A successful connection with the database and other AILS is established

Fig. 14.7. The Role Model of RESEARCHERAGENT.

Role: AgentManager(AGM)

Description: This role serves as a personal assistant of the multi agent system, initializing the system parameters, launching new agents when necessary and monitoring the agents' operation. Agents may register their services with the Agent and query the Agent to find out what services are offered by other agents. It also gets acquainted with specifc agents.

Protocols & Actvities: RegisterAgent, QueryAgent, SaveNewAcquaintance, IntroduceNewAgent, Monitor

Permissions:
creates *acquaiantances* // *acquaiantances data structure*
reads *acquaiantances*
updates *acquaiantances*

Responsibilities:

Liveness:

AGENTMANAGER = (GetAcquaintance. (MeetSomeone)$^{\omega}$) || (Monitor)$^{\omega}$

GETACQUAINTANCE = RegisterAgent.QueryAgent.[IntroduceNewAgent]

MEETSOMEONE = IntroduceNewAgent.SaveNewAcquaintance

Safety:

true

Fig. 14.8. The Role Model of AGENTMANAGER.

14.3.3.2. *Interactions among Agents*

Inevitably there are dependencies and relationships between the various roles in a multi-agent system. Indeed, such interchange is central to the way in which the system functions. Interactions obviously need to be captured and represented in the

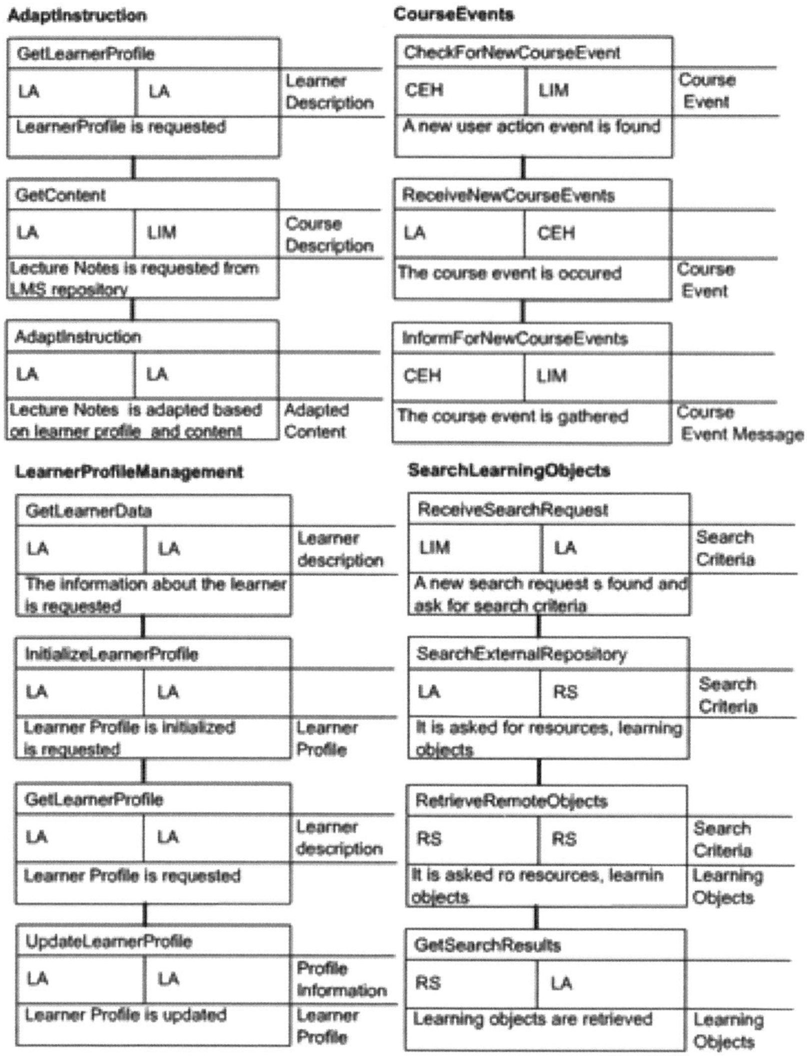

Fig. 14.9. The Interaction Model.

analysis phase. In Gaia [13], such links between the various roles are represented in the interaction model. The interaction model of AILS is shown in Figure 14.9.

14.3.3.3. *Services associated with Agent Roles*

The aim of the Gaia services model is to identify the services associated with each agent's role. It also specifies the main properties of these services. A service is a function of the agent.

The services provided by the AILS agents are given in Table 14.1. Each of the service in AILS identifies the inputs, outputs, pre-conditions, and post-conditions of

Table 14.1. The Service Model

Service	Get Course Content Event
Inputs	—
Outputs	The description of the event
Pre-condition	A learner assistant agent, LMS interface agent, course event handler agent are instantiated and associated with the AILS module. A course event relevant to the learner's action has happened
Post-condition	—
Service	Search Learning Object
Inputs	The searching criteria
Outputs	A list of learning objects
Pre-condition	A Researcher agent is instantiated and associated with the learner.
Post-condition	User view a learning object
Service	Adapt Instruction
Inputs	The learner profile and instructional materials
Outputs	The adapted instruction
Pre-condition	A Learner Assistant agent is instantiated and associated with the online learner. The online learner requests to view the lecture noted in LMS.
Post-condition	The learner view the adapted lecture notes.
Service	LMS Integration
Inputs	The course events
Outputs	LMS events
Pre-condition	The LMS Interface Agent is instantiated and associated with the AILS module.
Post-condition	The functions in LMS are activated.

each service. These services are derived from the protocols, activities, responsibilities and the liveness properties of role models described in the Section 3.3.1.

14.3.3.4. *The Acquaintances of Agents*

The acquaintance models define the communication links that exist between the agent types. This model does not define either the messages sent or received. It simply indicates that the communication pathways exist. Accordingly, the communication among the agents of AILS is identified in the acquaintance model (Figure 14.10).

14.4. Implementation

AILS's agents are developed as JADE agents. JADE (Java Agent Development Framework) is a software development framework for use in developing multi-agent systems and applications conforming to FIPA (Foundations of Intelligent and Physical Agents) standards for intelligent agents. In this study, the version JADE 3.5 has been used. JADE has been one of the most commonly used agent middleware. It has a well-structured agent management mechanism which provides a runtime environment, a library of classes that programmers can use, and a suite of graphical

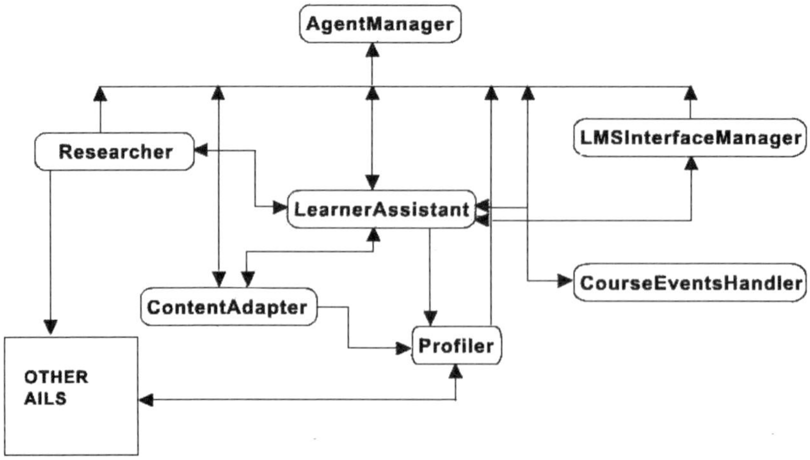

Fig. 14.10. The Acquaintance Model.

tools that allow the administration and monitoring of the activity of the running agents.

The implementation of AILS involves learner modeling, agent behavior in terms of the scenarios, and the ontologies used in agent communication, an LMS-AILS connector interface, and content adaptation strategies. Each of the implementation issues are explained in detail in the following sections.

In order to test the system, we have developed a simple dummy LMS. We have done a demo application of AILS using this LMS. At the end of this section, we provide sample screens obtained from this demo application.

14.4.1. *Learner Modeling*

Learner modeling is crucial in order to provide adaptive instruction. Each student needs an individualized student model. The learner model consists of information about the learner's style, domain knowledge, progress, preferences, goals, interests and other information about the learner. This is crucial when adapting the learning process of the learner for an LMS.

We modeled the learner using three factors derived from the conceptual model proposed in [16]. These factors are: behavioral factors, knowledge factors and personality factors.

14.4.1.1. *Behavioral Factors*

The behavioral factors involve the actions performed by the learner in the LMS module. The action history of the learner is kept using the following attributes:

- Username of learner
- LMS resource name (Search Tool, Login, Lecture Notes)

- Action name (Search, View Lecture Notes, Login)
- Time the action begins
- Time the action ends
- Concept (if this is applicable)
- Action identification

14.4.1.2. *Knowledge Factors*

AILS system uses a SCORM-compliant LMS. In the design of the AILS, it has been decided to keep the learning objects repository in the LMS part. The AILS keeps the domain knowledge of five attributes. It is possible to search for these in the resource repository. The attributes of each of the knowledge units are:

- The concept
- The content
- The type (image, text, exercise, activity, question, fact, definition, exam, advance organizer, concept map)
- The sign ("UNDERSTOOD", "NOT UNDERSTOOD", "MISUNDERSTOOD")
- The location, path where the learning resource is returned in the LMS part.

14.4.1.3. *Personality Factors*

The personality factors are composed of the learning standards and the learning styles.

The Learning Standard: These are standards defined in order to model the learner which may be IMS LIP, IEEE PAPI, for example. Here, we have chosen IMS LIP (Instructional Management Systems Learner Information Package) to be the learning standard package. It is possible to convert IMS LIP to IEEE PAPI if required.

The IMS LIP specification addresses the interoperability of the Internet-based learner information systems with other systems in the Internet learning environment. The IMS LIP has information on data such as biographic and demographic relevant to learning; other objectives and aspirations; qualifications, certifications and licenses which have been awarded by recognized authorities; learning-related activity. This may be informal education, training, work experience, and military or civic service in any state of completion; transcript; information describing hobbies and recreational activities; skills, knowledge, and abilities acquired in the cognitive, affective, psychomotor domains; membership of professional organizations; on a set of passwords and security keys [16].

Although the standards cover a great deal of information, they are incapable of providing a personalized education [17]. For example, the learning style is very important for adaptation. The IMS LIP does not provide this required information. We decided to use the personality factors which included the learner's IMS LIP and the learning style information.

The Learning Style: Learners have many different ways of perception, construction and retention of knowledge. These differences which occur during the learning process are unique to each individual and are based on many factors such as previous experience, mental ability, and personal characteristics. In order to provide adapted instruction, the learning styles must be suitable and the instructors must ask "how can this learner achieve better results?". This is better than "why is this learner not high-achiever?".

There are several different learning style models used in the literature. These are

- Myers-Briggs Type Indicator [19]
- Kolb's Learning Style Model [20]
- Honey and Mumford's Typology of Learners [21]
- Felder-Silverman Model [22]
- Dunn, Dunn and Price Model [17]

We have adopted the Felder-Silverman model as the core learning style. This is a widely used model. A well-known questionnaire is used to categorize individual learning style based on this idea. It is called the Felder-Solomon Index of Learning Style (ILS)[23].[a] This questionnaire is validated in reference [24].

According to [22], the Felder-Silverman model classifies students according to where they fit into a number of scales which pertain to the ways in which they receive and process information. In [22], Felder and Silverman provide definitions of learning styles. These are described as follows:

- **Active:** Active learners try things out to see how they work. They also like to work with others.
- **Reflective:** Reflective learners like to think things through first.
- **Sensing:** Sensors like to learn the facts, and to use well established methods. They are practical and careful workers.
- **Intuitive:** Intuitors tend to work fast and be innovative. They can often handle abstract and mathematical concepts successfully.
- **Visual:** Visual learners prefer diagrams, pictures, graphs and films.
- **Verbal:** Verbal learners learn better using words both verbal and written.
- **Sequential:** Sequential learners like to work in linear steps which follow logically.
- **Global:** Global learners like to begin and absorb material nearly at random. They then assemble the complete solution.

The Felder-Silverman learning style model categorizes the student's learning style on a sliding scale of four dimensions. Examples of this are the Active-Reflective, Sensing-Intuitive, Visual-Verbal and Sequential-Global.

[a]Copyright\copyright North Carolina State University (Authored by Richard M. Felder and Barbara A. Soloman). Reprinted by permission of North Carolina State University.

14.4.2. *Agent Behaviors: SCENARIOS*

AILS agents work cooperatively to perform operations such as login, search keywords and view lecture notes. For the sake of simplicity, each operation is explained as scenarios such as "login", "search" and "view lecture note".

14.4.2.1. *Login Operation*

1. User enters username and password
2. LMS invokes LMSINTERFACEMANAGER about the login and it provides the user with information by means of the AILSLMSConnector
3. LMSINTERFACEMANAGER sends the login action to LEARNER ASSISTANT-AGENT
4. LEARNERASSISTANTAGENT receives the new action requirement
5. LMSINTERFACEMANAGER sends the user instance to LEARNERASSISTANT-AGENT
6. LEARNERASSISTANTAGENT receives user concept instance
7. LEARNERASSISTANTAGENT sends the user concept to PROFILEAGENT
8. LEARNERASSISTANTAGENT sends the new action instance to PROFILE-AGENT
9. PROFILEAGENT receives user concept
10. PROFILEAGENT receives login action instance
11. PROFILEAGENT checks whether user exists. If the user exists, it retrieves learner profile.
12. PROFILEAGENT updates the learner's action history
13. PROFILEAGENT sends profile to LEARNERASSISTANTAGENT
14. LEARNERASSISTANTAGENT receives the learner profile (LEARNERASSISTANTAGENT will provide profile information to the others)

Figure 14.11 depicts the agent communication for the login scenario.

14.4.2.2. *View Lecture Notes*

1. Learner selects "View Lecture Notes"
2. LMS informs LMSINTERFACEMANAGER through AILSLMSConnector and provide view lecture note action instance
3. LMSINTERFACEMANAGER sends action information to LEARNER ASSISTANTAGENT
4. LEARNERASSISTANTAGENT receives action information
5. LEARNERASSISTANTAGENT sends action history instance to profiler to update the learner's profile information
6. LEARNERASSISTANTAGENT sends action history to CONTENTADAPTER-AGENT to display the appropriate content
7. PROFILEAGENT receives view content action instance
8. PROFILEAGENT receives the profile request

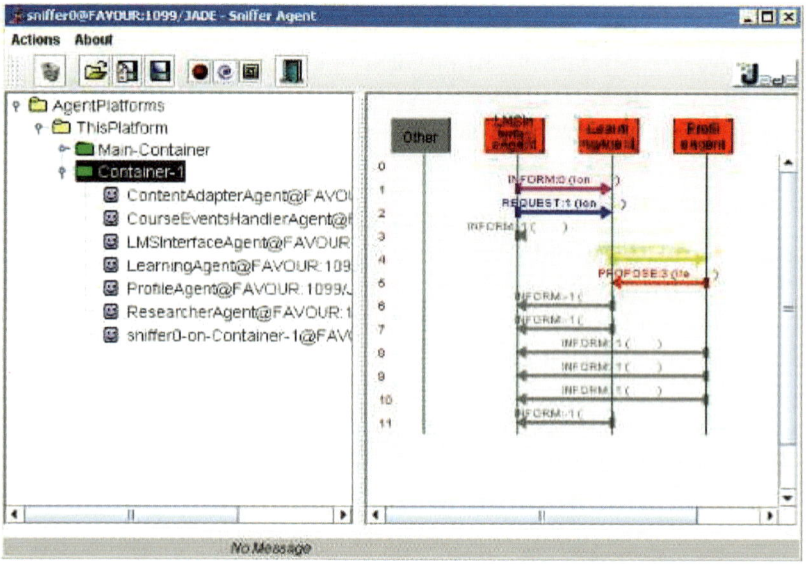

Fig. 14.11. The Sniffing Login Behavior.

9. ProfileAgent updates the profile
10. ProfileAgent sends the updated profile information
11. ContentAdapterAgent receives the action history
12. ContentAdapterAgent asks for the profile information to ProfileAgent
13. ContentAdapterAgent receives the profile information
14. ContentAdapterAgent finds the appropriate content resource, ie. adapt content according to the profile information
15. ContentAdapterAgent sends resource name to Learner AssistantAgent
16. LearnerAssistantAgent receives content resource
17. LearnerAssistantAgent sends resource to LMSInterfaceManager
18. LMSInterfaceManager receives the resource to display
19. LMS displays the learning content resource invoked by LMSInterfaceManager

Figure 14.12 shows the sequence diagram for agent communication while performing the viewing lecture notes scenario.

14.4.2.3. *Search Keywords*

The agent communication for this scenario is given in Figure 14.13.

1. LMSInterfaceManager receives action history instance (through constructor)

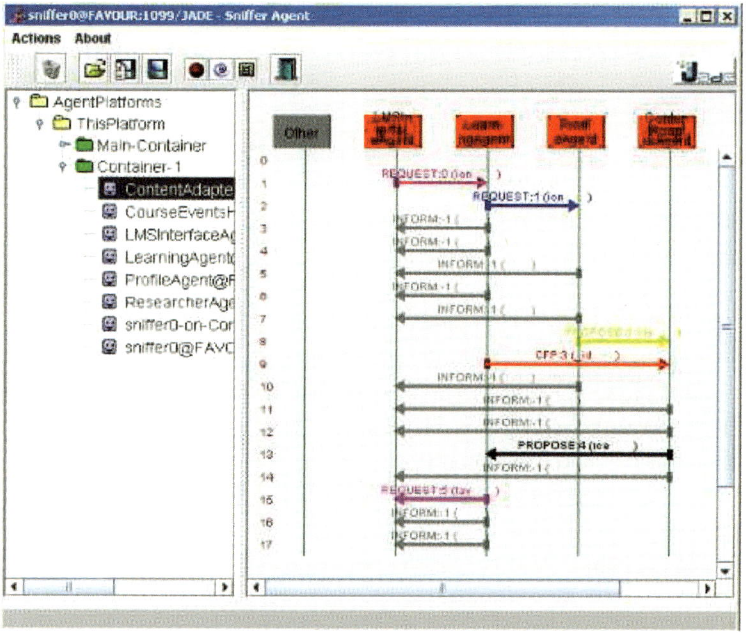

Fig. 14.12. The Sniffing View Lecture Notes Behavior.

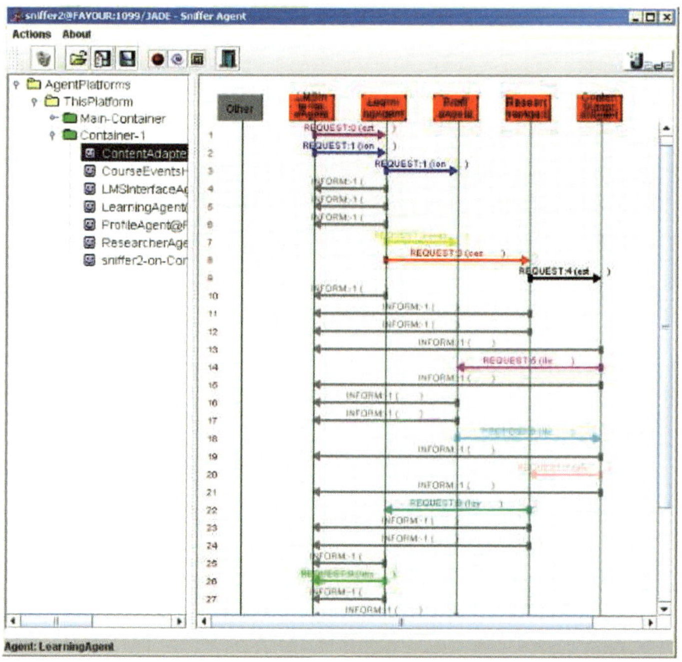

Fig. 14.13. The Sniffing Search Keyword Behavior.

2. LMSInterfaceManager sends action information e (involving lms ID) to LearnerAssistantAgent
3. LMSInterfaceManager sends search case information to LearnerAssistantAgent
4. LearnerAssistantAgent receives action information
5. LearnerAssistantAgent sends the action history instance to profiler to update the profile information
6. ProfileAgent receives the actions
7. ProfileAgent updates the profile action history
8. LearnerAssistantAgent receives search case instance
9. LearnerAssistantAgent sends search case to ResearcherAgent
10. ResearcherAgent receives the search case
11. ResearcherAgent sends search case to ContentAdapter Agent agent
12. ContentAdapterAgent receives the profile-content template
13. ContentAdapterAgent request the learner's profile
14. ProfileAgent receives the profile request
15. ProfileAgent finds and sends the profile to ContentAdapter Agent
16. ContentAdapterAgent receives the learner profile
17. ContentAdapterAgent prepares a content template
18. ContentAdapterAgent sends the resulting content template to Researcher
19. ResearcherAgent receives the content template and adds keyword information to the template
20. ResearcherAgent sends the resulting template to Learner Assistant-Agent
21. LearnerAssistantAgent receives the template
22. LearnerAssistantAgent sends the content template to LMSInterfaceManager to find LMS resources
23. LMSInterfaceManager receives the template
24. LMSInterfaceManager asks LMS to display the resources which are suitable to that template
25. LMS displays the appropriate content

14.4.3. *The AILS Ontologies*

Messages exchanged by JADE agents have a format specified by the agent communication language (ACL). This is defined by the FIPA (http://www.fipa.org) international standard for agent interoperability. This format is comprised as a number of fields such as the sender, the list of receivers, the content, content language, and the ontology for example. A JADE ontology describes the vocabulary of the symbols used in the content and their meaning. Both the sender and the receiver must attribute the same meaning to the symbols for the communication to be effective.

In AILS, six JADE ontologies are developed to meet the requirements of the system. The ontology maps are given in the following figures. The ontologies are the LearnerProfileOntology (Figure 14.14), the LearningContentOntology (Figure 14.15), the LMSUserOntology (Figure 14.16), the SeachCaseOntology (Figure 14.17), and the ActionHistoryOntology (Figure 14.16).

14.4.4. *The AILS Adaptation Strategies*

The adaptive strategies of AILS is divided into four main categories:

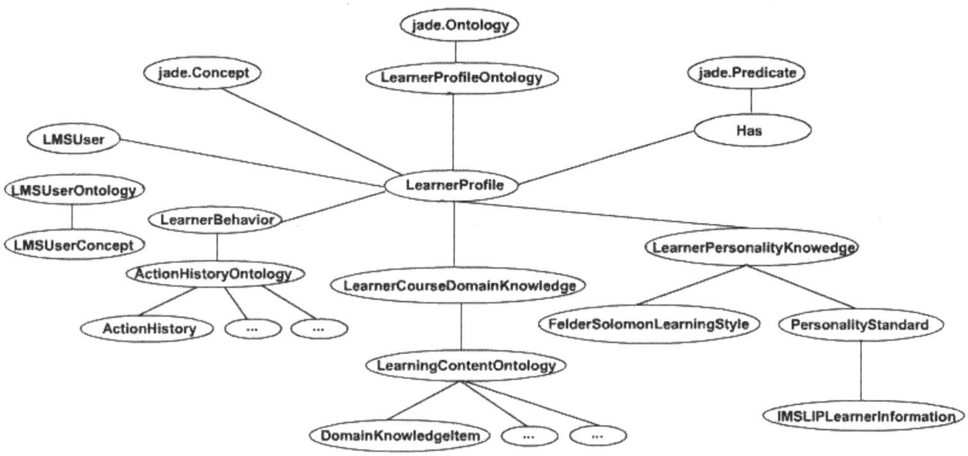

Fig. 14.14. The Ontology Map for the Learner Profile.

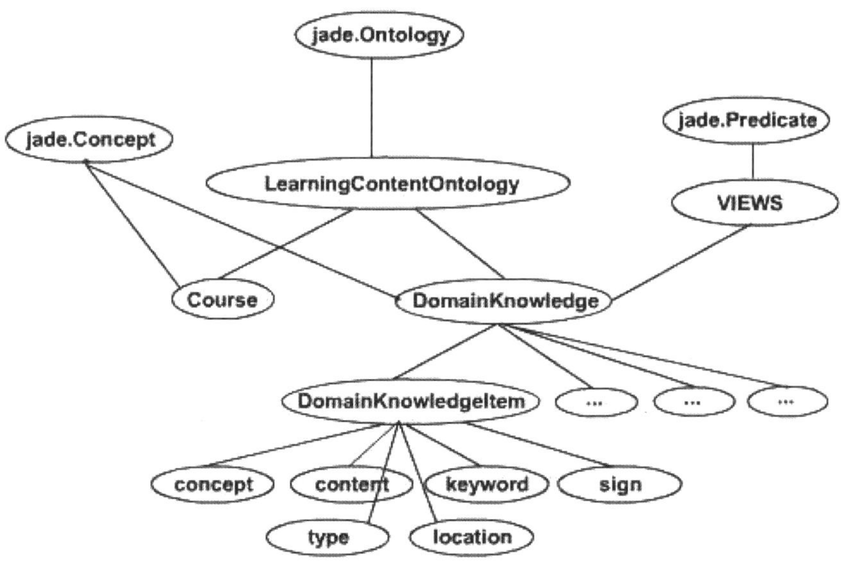

Fig. 14.15. The Jade Ontology Map for the Learner Profile.

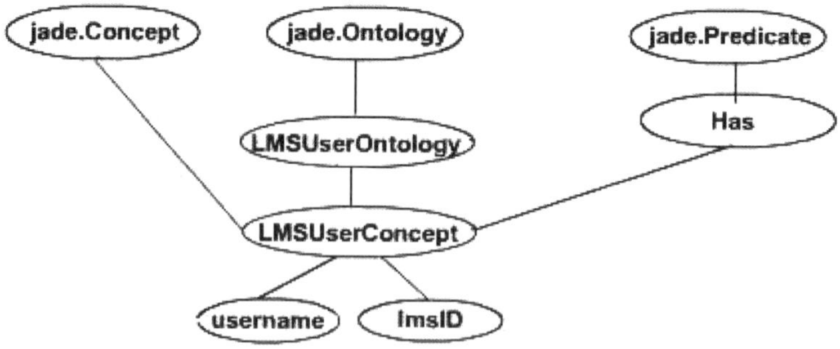

Fig. 14.16. The Jade Ontology Map for the LMS User (Learner).

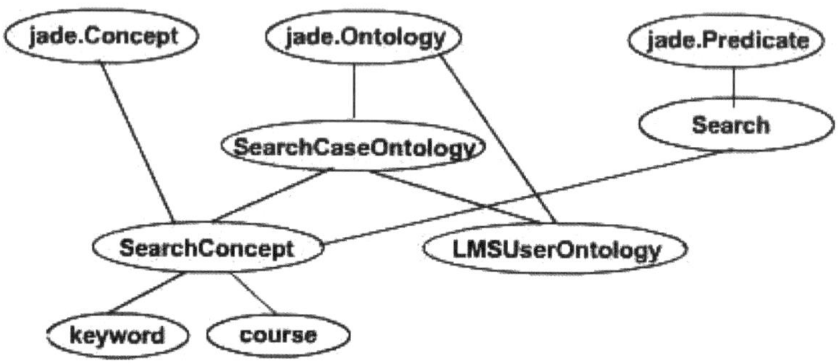

Fig. 14.17. The Jade Ontology Map for the Learner Search Action.

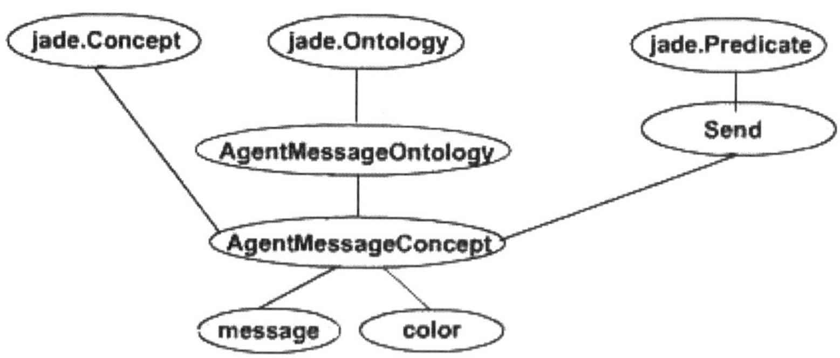

Fig. 14.18. The Jade Ontology Map for the Agent Messages in the LMS.

- Content Adaptation
- Presentation Adaptation
- Participation Adaptation
- Perspective Adaptation

14.4.4.1. *Content Adaptation*

The content adaptation of AILS relies on sensing and intuitive dimension of the learner's style. Everyone uses both faculties, but most people tend to favor one over the other. According to Felder [25],

- *Sensors* like facts, data, and experimentation; *intuitors* prefer principles and theories. *Sensors* like solving problems by standard methods and dislike "surprises", *intuitors* like innovation and dislike repetition.
- *Sensors* are patient with detail but do not like complications; intuitors are bored by detail and welcome complications
- *Sensors* are good at memorizing facts; *intuitors* are good at grasping new concepts.
- *Intuitors* are more comfortable with symbols than are *sensors.*

We figured out the applicable learning content resources for the sensing and intuitive type of learners as in Table 14.2. According to the learner's style on this sensing-intuitive dimension, the list of appropriate learning resources is listed to the learner. Apart from the resources listed, the content adaptation of AILS involves the complexity level of the content. As explained above, unlike sensing learners, intuitive learners are not comfortable with the complex content. This is also taken into consideration while presenting the adapted content to the learner.

In content adaptation, the learner's prior knowledge is also important. The content is adapted according to the learner's sign on the learning resource (which may be "UNDERSTOOD", "NOT UNDERSTOOD" and "MISUNDERSTOOD"). For example, if the learner wants to view a lecture note on a subject and if the learner's understanding level on this subject is marked as "MISUNDERSTOOD", then AILS displays a learning resource about the same subject accordingly. If it is marked as "UNDERSTOOD", the system displays the more advanced topics on this subject.

14.4.4.2. *Presentation Adaptation*

The presentation of the content covers the visual and audio properties of the content resource. In other words, a visual content or an audio resource is displayed to the user.

The learning style constitutes one of the aspects considered in the presentation adaptation. The visual and verbal dimension is the corresponding dimension in the learning style.

Table 14.2. The Learning Resources for Sensing and Intuitive Learners

Learning Resource Type	**Dimension**
DATA, EXPERIMENT, PROCEDURE, CONCEPT	Active
THEORY, PRINCIPLE, INNOVATION, CONCEPT(new), MATHEMATICAL MODELS	Reflective

According to the Felder-Silverman learning style, *visual* learners remember best what they see: such as, pictures, diagrams, flow charts, time lines, films, demonstrations. If something is simply said to them they will probably forget it. *Auditory* learners remember much of what they hear and more of what they hear and then say. They get a lot out of discussion, prefer verbal explanation to visual demonstration, and learn effectively by explaining things to others.

Table 14.3 lists the appropriate learning resource types for visual and verbal learners.

The learner preferences such as favoring simple or complex visual designs, color choices (if applicable), language are also considered when personalizing the content.

14.4.4.3. *Participation Adaptation*

The complex mental processes by which perceived information is converted into knowledge can be conveniently grouped into two categories: active experimentation and reflective observation [20].Active experimentation involves doing something in the external world with the information discussing it or explaining it or testing it in some way, and reflective observation involves examining and manipulating the information introspectively.

Felder [25] explains the active learners as experimentalists and the reflective learners as theoreticians. He stated that active learners need to be active in order to learn, and reflective learners need to have an opportunity to think about the information being presented. It is obvious that when the students in a class are passive, then neither active learners nor reflective learners can learn effectively. As explained in the learning style model, the active learners work well in groups. However, the reflective learners work better by themselves or with at most one other person.

Table 14.4 summarizes the type of learning resources with corresponding learner type (active and reflective).

Felder [25] suggests to present material that emphasizes both practical problem solving (active) and fundamental understanding (reflective). It might be good to

Table 14.3. The Learning Resources for Visual and Verbal Learners

Learning Resource Type	**Dimension**
IMAGE, DIAGRAM, CHART	Visual
AUDIO, TEXT	Verbal

Table 14.4. The Learning Resources for Active and Reflective Learners

Learning Resource Type	**Dimension**
EXAMPLE, EXERCISE, ACTIVITY, DISCUSSION, PROBLEM SOLVING	Active
QUESTION, EXAMPLE, LINKS, CRITIQUE	Reflective

present lectures with occasional pauses for thought (reflective) and brief discussion or problem-solving activities (active).

An exceptionally effective technique for reaching active learners is to have students organize discussion groups in the forum and let them come up with collective answers to questions posed by the instructor. To support this, the AILS agents ask the learner to work with the other learners who have common goals which are hold in the IMS LIP standard information.

14.4.4.4. *Perspective Adaptation*

Most formal education involves the presentation of material in a logically ordered progression, with the pace of learning dictated by the clock and the calendar. When a body of material has been covered the students are tested on their mastery and then move to the next stage. Some students are comfortable with this system; they learn sequentially, mastering the material more or less as it is presented [25].Others, however, cannot learn in this manner. They learn in fits and starts: they may be lost for days or weeks, unable to solve even the simplest problems or show the most rudimentary understanding. They may then understand the material well enough to they apply it to problems that leave most of the sequential learners baffled. These are the global learners [25].

Sequential learners may be strong in convergent thinking and analysis; global learners may be better at divergent thinking and synthesis. Sequential learners learn best when material is presented in a steady progression of complexity and difficulty; global learners sometimes do better by jumping directly to more complex and difficult material. The curricula, course syllabi, textbooks, lecture notes prepared for online education are generally designed as sequential, which is already suitable for sequential learners.

In order to reach the global learners, it might be good to provide overall picture of the course or goal of a lesson before presenting the steps, doing as much as possible to establish the context and relevance of the subject matter and to relate it to the students' experience.

Another way to support global learners is to explain their learning process to them. Felder [25] says "while global learners are painfully aware of the drawbacks of their learning style, it is usually a revelation to them that they also enjoy advantages that their creativity and breadth of vision can be exceptionally valuable to future employers and to society".

Table 14.5 summarizes the type of learning resources with corresponding learner type (sequential and global).

14.4.5. *AILS-LMS Interface*

AILS-LMS Connector is an interface developed for integration of AILS to the LMS.

The fields in the information about the learner, course, content, and searching operation are summarized in Table 14.6. AILS receives these fields, performs

Table 14.5. The Learning Resources for Sequential and Global Learners

Learning Resource Type	Dimension
SEQUENTIAL CONTENT, TOC	Sequential
CONCEPT MAP, SYLLABUS, ADVANCE ORGANIZER, LEARNING STYLE	Global

Table 14.6. The Information in AILS-LMS Communication

Information	Fields
Learner	username, id number, first name, last name, email, icq, skype, msn, phone1, phone2, institution, department, address, city, country, lang, time zone, description, time modified, course id, grade, time access, start time, lesson time
Learner Action	learner ID, first access, last access, last login, current login, course id, action
Course	id, full name, short name, description, format, instructors, students, start date, number of sections, language, time created, time modified, SCORM content
Content	SCORM name, reference, version, max grade, grade method, maximum attempt, summary, time modified, SCO manifest, maximum time allowed, time limitation, next SCO, previous SCO, visible sections, hidden sections
Search	course id, learner id, time, keywords
Message	Message learner id, AILS message, time created

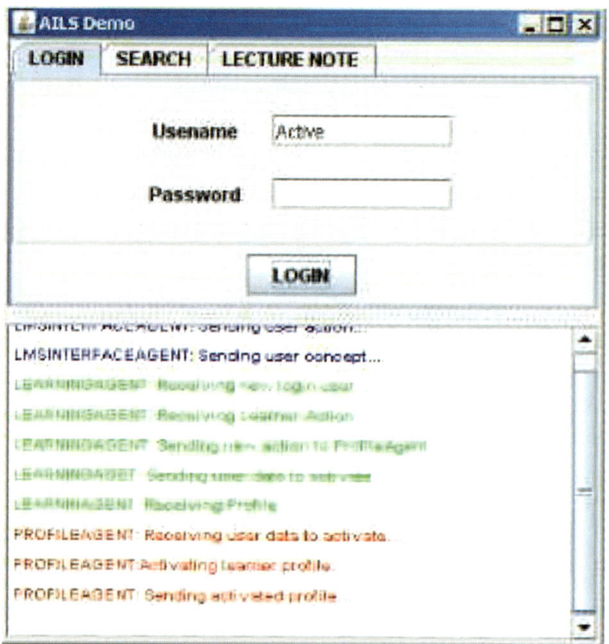

Fig. 14.19. The Login Screen.

adaptation tasks and provides results by updating the necessary fields. LMSINTERFACEMANAGER works as AILS-LMS Connector.

Since there is no standard interface defined for interaction among the adaptive systems and the LMS, it is necessary to add extensions to the LMS in order to establish AILS-LMS communication. In this extension, the LMS is supposed to provide necessary methods required in AILSLMSConnector.

In order to set up the communication link between AILS and LMS, AILSLMSConnector should be implemented by the agent, LMSINTERFACEMANAGER on AILS part, and also the interface specifically implemented on the LMS part.

14.5. A Sample Session of AILS

To design, develop, and test the system easily, a dummy LMS was developed. The dummy LMS keeps the user information and the lecture notes. AILS system communicates with the dummy LMS through the connector. The LMS has login, search, and view lecture note functionalities. Additional functionalities, such as assignment, announcement, forum will also be added to the system as a future work.

The dummy LMS keeps the learning resources and user information, whereas AILS keeps the learner profile and the domain knowledge. The learner profile is the

Fig. 14.20. The Search Tool.

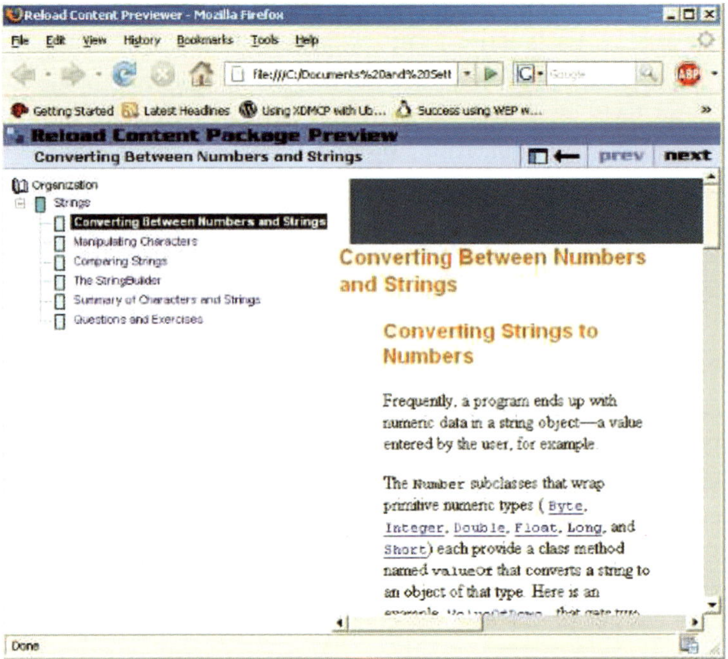

Fig. 14.21. The Searching for the learner whose participation style is Active.

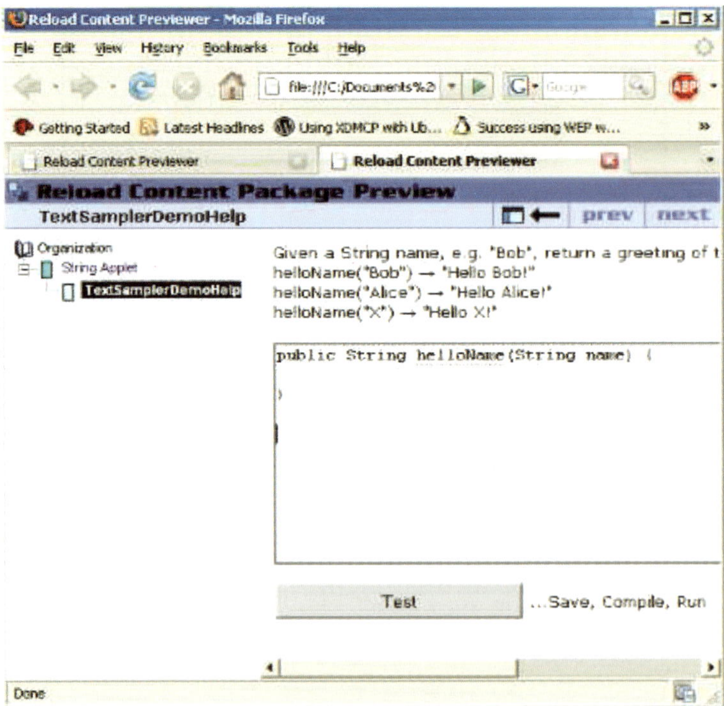

Fig. 14.22. The Lecture Notes for the learner whose participation style is "Active".

most important part of AILS. In that section, the samples screens for the learner actions and the resulting the AILS agent's actions are given.

14.5.1. *Login Tool*

Figure 14.19 shows the login screen in the dummy LMS. The learner enters the username and password information, AILS retrieves the username information and activate the learner profile accordingly. The activated profile is used in the other functionalities such as searching and viewing content.

14.5.2. *Search Tool*

Figure 14.20 shows the search screen in the dummy LMS. The activated learner enters the keyword to search. The AILS agents create a content resource template which best fit to the learner profile. The dummy LMS keeping the resource repository uses that template to display the appropriate resource. To be adaptive, AILS uses the learner profile to construct the personalized content template.

Figure 14.21 displays an example for the search results found from the search of a learner who has participation type as "Active".

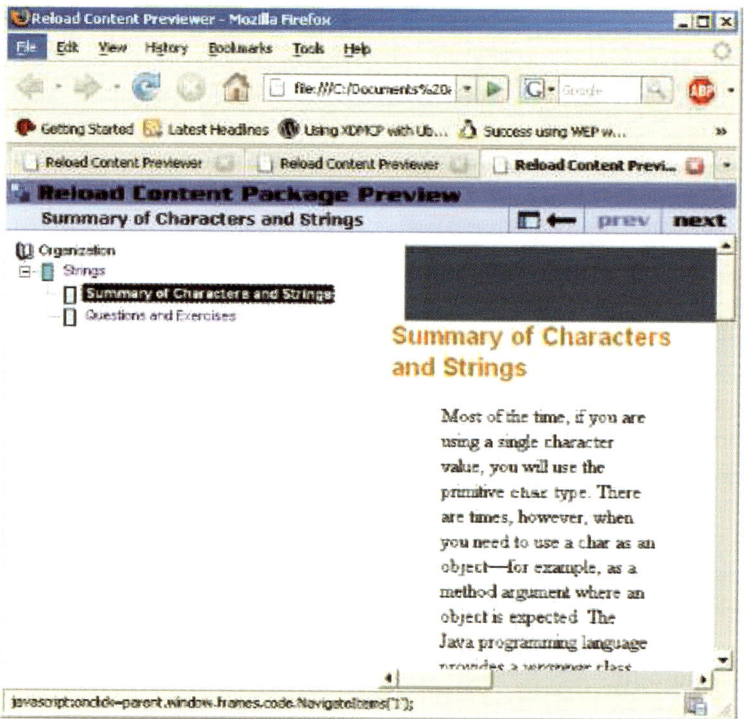

Fig. 14.23. The Lecture Notes for the learner whose participation style is "Reflective".

14.5.3. *Lecture Notes Tool*

When a learner wants to view the lecture notes, AILS uses the learner profile information to find and display the personalized content. Figure 14.22 shows the personalized content for the learner who has participation type, "Active".

Another example for personalized content is given for the user having participation type as, "Reflective" (See Figure 14.23).

14.6. Conclusion

There is a need for providing education to students in different places around the world. To support teaching and facilitating the learners, learning management systems must provide adaptive learning features. This yields the necessity for smart learning environments offering personal services with capabilities to learn, reason, have autonomy, and be totally dynamic. The adaptive intelligent technologies might diminish the barriers of learning in distance education.

The purpose of this study is to develop an adaptive learning system, named AILS, that enrich online learning management systems with personalization. AILS is aimed to fulfill the needs of adaptiveness of OLE. It is based on the learning objects and the learning profiles to achieve the goal of adaptive learning in virtual learning environments. The main requirement is to provide learners to be uniquely modeled and to be monitored according to their respective learner profile.

AILS has six agents: LMSInterfaceManager, Learner AssistantAgent, ContentAdapterAgent, ProfileAgent, ResearcherAgent and AgentManager. These agents have specific roles such as, registering agent's services, activating learner profile on learner's login action, adaptive searching keyword locally and globally (inter-AILS communication), and viewing lecture notes behaviors. In this study, we have implemented some of the behaviors such as registering agent's services, LMS-AILS communication, activating existing learner's learner profile, conducting adaptive search locally in the current AILS, and viewing adapted lecture notes which are adapted according the learner's profile and adaptive strategies. We have also developed and used a dummy LMS in order to test the system. To establish agent-communication, we have developed JADE ontologies.

As a future work we plan to use data mining techniques to learn association rules in the learner profile. The adapting strategies will be updated accordingly when new associations are obtained. We also plan to use AILS in a course offered through an LMS, such as Moodle [25] or OLAT [26]. This yields to gather more realistic information about the effectiveness and usability of the system.

Acknowledgement

The initial results of this research are reported in the International Journal of Hybrid Intelligent Systems as well as International Conference on eLearning, eLearn07, 2007.

References

1. I. Park and M.J. Hannafin. Empirically based guidelines for the design of interactive multimedia. *Educational technology Research and Development*, 41(3):6385, 1993.
2. D. Xu, H. Wang and M. Wang. A conceptual model of personalized virtual learning environments. *Expert Systems with Applications*, 29(Issue 3):525–534, October 2005.
3. C.S. Lin and M.S. Kuo. Adaptive networked learning environments using learning objects, learner profiles and inhabited virtual learning worlds. *Fifth IEEE International Conference on Advanced Learning Technologies* (ICALT'05), pages 116–118, 2005.
4. B. Han. Student modeling and adaptivity in web-based learning systems. *Master thesis*, Massey University, New Zealand, 2001.
5. C. Wolf. iweaver: towards 'learning style'-based e-learning in computer science education. In ACE '03: *Proceedings of the fifth Australasian Conference on Computing Education*, pages 273–279, Darlinghurst, Australia, Australia, 2003. Australian Computer Society, Inc.
6. M. Grigoriadou, K. Papanikolaou, H. Kornilakis, and G. Magoulas. Inspire: an intelligent system for personalized instruction in a remote environment, 2001.
7. M. Stern, J. Steinberg, H. Lee, J. Padhye, and J. Kurose. Manic: Multimedia asynchronous networked individualized courseware, 1997.
8. C. Y. Gilbert, Juan E. Han. Arthur: Adapting instruction to accommodate learning style. *In WebNet 99 World Conference on the WWW and Internet Proceedings*, page 7, Honolulu, Hawaii, October 24-30 1999.
9. R.A Howard, C.A. Carver, and E. Lavelle. Enhancing student learning by incorporating learning styles into adaptive hypermedia. *In Proceedings of 1996 ED-MEDIA World Conf. on Educational Multimedia and Hypermedia*, pp. 118–123, Boston, USA, 1996.
10. E. Triantafillou, A. Pomportsis, and E. Georgiadpu. Aes-cs: Adaptive educational system base on cognitive styles. In AH2002 Workshop, Adaptive Systems for Web education, *2nd International Conference, Adaptive Hypermedia and Adaptive Web Based Systems*, volume 20, 2002.
11. S.M. Parvez and G.D. Blank. A pedagogical framework to integrate learning style into intelligent tutoring systems. *J. Comput. Small Coll.*, 22(3):183–189, 2007.
12. N.R. Jennings. Agent-Oriented Software Engineering. In Francisco J. Garijo and Magnus Boman, editors, *Proceedings of the 9th European Workshop on Modelling Autonomous Agents in a Multi-Agent World: Multi-Agent System Engineering (MAAMAW-99)*, volume 1647, pages 1–7. Springer-Verlag: Heidelberg, Germany, 30–2 1999.
13. M. Wooldridge, N.R. Jennings, and D. Kinny. The gaia methodology for agent-oriented analysis and design. *Autonomous Agents and Multi-Agent Systems*, 3(3):285–312, 2000.
14. C.G. Trucco, T. Rimassa and G. Bellifemine, F. Jade programmer's guide, 2002.
15. F.C. Serçe and F.N. Alpaslan. An Adaptive Intelligent Learning System Model for Online Learning Environments Using Gaia, Proceedings of International Conference on eLearning, eLearn07, 2007.
16. T.Q. Gu; Sumner. Support personalization in distributed e-learning systems through learner modeling. *Information and Communication Technologies*, 1:610–615, April 2006.
17. R. Dunn. Learning styles: Theory, research, and practice. *National Forum of Applied Educational Research Journal*, 13(1):3–22, 2000.
18. IMS Global Learning Consortium. IMS learner information package specification. http://www.imsglobal.org/profiles/, accessed December 2006.

19. B. Myers. *Gifts Differing: Understanding Personality Type.* Davies-Black Publishing, reprint edition (may 1, 1995) edition, 1980.
20. D.A. Kolb. Experiential *Learning:Experience as the Source of Learning and Development.* Prentice-Hall, Englewood Cliffs, N.J., 1984.
21. P. Honey and A. Mumford. *Manual of Learning Styles.* P. Honey, London, 1982.
22. R.M. Felder and L.K. Silverman. Learning and teaching styles in engineering education. *Journal of Engineering Education*, 78(7):674–681, 1988.
23. R.M. Felder and B.A. Soloman. Index of learning styles questionnaire. http://www.ncsu.edu/felder-public/ILSpage.html, accessed December 2006.
24. M. Zywno. A contribution to validation of score meaning for felder-soloman's index of learning styles. *In Proceedings, 2003 ASEE Annual Conference, American Society for Engineering Education*, June 2003.
25. R.M. Felder and L.K. Silverman. Learning and teaching styles in engineering education. *Engineering Education*, 78(7):674–681, 1998.
26. Moodle. A Free Open Source Learning Management System, http://moodle.org.
27. OLAT, Online learning and training, an open source learning management system, University of Zrich OLAT, 2000.

14.7. Resources

Following is a sample of resources on agents and educational research.

Journals

- IEEE Transactions on Education, IEEE Press, USA
- Education for Information, IOS Press, The Netherlands
- International Journal of Artificial Intelligence in Education, IOS Press, The Netherlands.
- Journal of Interactive Technology and Smart Education, Troubador Publishing Ltd, UK.

Special Issue of Journals

- Aroyo, L., Graesser, A. and Johnson, L., Intelligent Educational Systems of the Present and Future, IEEE Intelligent systems, Volume 22, number 4, July/August 2007.
- Palade, V., Ghaoui, C. and Jain, L.C., Intelligent Instructional Environments, Journal of Interactive Technology and Smart Education, Troubador Publishing Ltd, UK, Volume 1, Issue 3, August 2004.

Conferences

- KES International Conference Series www.kesinternational.org/

Conference Proceedings

- Jedrzejowicz, P., Nguyen, N.T., Howlett, R.J. and Jain, L.C. (Editors), Agent and Multi-Agent Systems: Technologies and Applications, Volume 1, Lecture Notes in Artificial Intelligence, **LNAI 6070**, Springer-Verlag, 2010.

- Jedrzejowicz, P., Nguyen, N.T., Howlett, R.J. and Jain, L.C. (Editors),, Agent and Multi-Agent Systems: Technologies and Applications, Volume 2, Lecture Notes in Artificial Intelligence, **LNAI 6071**, Springer-Verlag, 2010.
- Håkansson, A., Nguyen, N.T., Hartung, R., Howlett, R.J. and Jain, L.C. (Editors, Agents and Multi-Agents Systems: Technologies and Applications, Lecture Notes in Artificial Intelligence, Springer-Verlag, Germany, 2009.
- Nguyen, N.T., Jo, G.S., Howlett, R.J. and Jain, L.C. (Editors), Agents and Multi-Agents Systems: Technologies and Applications, Lecture Notes in Artificial Intelligence, Springer-Verlag, Germany, 2008, 1000 pp.
- Jain, L.C., Gini, M., Faltings, B.B., Terano, T., Zhang, C., Cercone, N. and Cao, L. (Editors), Proceedings of the 2008 IEEE/WIC/ACM International Conference on Intelligent Agent Technology, IAT 2008, IEEE Computer Society, USA, 2008.
- Nguyen, N.T., Grzech, A., Howlett, R.J. and Jain, L.C., Agents and Multi-Agents Systems: Technologies and Applications, Lecture Notes in Artificial Intelligence, Springer-Verlag, Germany, 2007, 2007, 1046 pp., ISBN: 978-3-540-72829-0.

Books

- Srinivasan, D. and Jain, L.C. (Editors), Advances in Multi-agent Systems and Applications, Springer-Verlag, 2010.
- Nguyen, N.T. and Jain, L.C. (Editors), Intelligent Agents in the Evolution of Web and Applications, Springer-Verlag, Germany, 2009.
- Jain, L.C, Tedman, R. and Tedman, D. (Editors), Evolution of Teaching and Learning in Intelligent Environment, Springer-Verlag, 2007.
- Khosla, R., Ichalkaranje, N. and Jain, L.C. (Editors), Design of Intelligent Multi-Agent Systems, Springer-Verlag, Germany, 2005.
- Ghaoui, C., Jain, M., Bannore, V. and Jain, L.C., (Editors), Knowledge-Based Virtual Education, Springer-Verlag, Germany, 2005.
- Resconi, G. and Jain, L.C., Intelligent Agents: Theory and Applications, Springer-Verlag, Germany, 2004.
- Ko, C.C., Creating Web-based Laboratories, Springer-Verlag, London, 2004.
- Hasebrook, J.P. and Maurer, H.A., Learning Support for Organisational Learning, Series on Innovative Intelligence, World Scientific Publishing Company Singapore, Volume 8, 2003.
- Jain, L.C. and Kacprzyk, J. (Editors), New Learning Paradigms in Soft Computing, Springer-Verlag, Germany, 2002.
- Jain, L.C., Chen. Z. and Ichalkaranje, N. (Editors), Intelligent Agents and Their Applications, Springer-Verlag, Germany, 2002.
- Jain, L.C., Howlett, R.J., Ichalkaranje, N., and Tonfoni, G. (Editors), Virtual Environments for Teaching and Learning, World Scientific Publishing Company Singapore, 2002.

- Jain, L.C., Innovative Teaching and Learning in Intelligent Environment, Springer-Verlag, 2000.

Book Chapters

- Tolk, A., Madhavan, P., Tweedale, J. And Jain, L.C., Agents and Decision Support Systems, Wiley-VCH Verlag, Chapter 14, 2009, pp. 399–431. ISBN: 978-3-527-40781-1.
- Khazab, M., Tweedale, J. and Jain, L.C., Dynamic Applications using Multi-agent Systems Personalization, Chapter 4, Springer-Verlag, 2009, pp. 67–79.
- Tweedale, J., Ichalkaranje, N., Sioutis, C., Urlings, P. and Jain, L.C., Future Directions: Building a Decision Making Framework using Agent Teams, Springer-Verlag, 2008, pp. 381–402, Chapter 14.
- Jarvis, B., Jarvis, D. and Jain, L., Teams in Multi-Agent Systems, in IFIP International Federation for Information Processing, Vol. **228**, Intelligent Information Processing III, Springer, 2006, pp. 1–10, Chapter 1.
- Tedman, D. and Jain, L.C., An Introduction to Innovative Teaching and Learning, in Teaching and Learning, Springer, 2000, pp. 1–30, Chapter 1.

Journal Papers

- **Jarvis, J., Jarvis, D.,Ronnquist, R. and Jain, L.C., The Future for Intelligent Agents, Hindustan Journal, Volumr 2, July, 2008, pp. 1–6.**
- Jain, L.C., "The Knowledge-Based Intelligent Engineering: New Engineering", The Journal of the Institution of Electronics and Telecommunication Engineers: Technical review, Vol 15, No.5, 1998, pp. 345–348.
- Jain, L.C., "Hybrid Intelligent Techniques in Teaching and Research," IEEE AES, USA, Vol. 10, No. 3, March 1995, pp. 14–18.
- Jain, L.C. and Garud, G.N., "CAD and CAL Tools for Electrical Engineering Students," Journal of the Institution of Engineers, Vol. 75, November 1994, pp. 52–54.
- Jain, L.C. and Bowden, B.S., "Development of Expert System Course for Electronic Students," International Journal of Electrical Engineering Education, UK, Vol. 31, No. 1, January 1994, pp. 34–45.
- Rowland, J.G. and Jain, L.C., "Artificial Intelligence Languages in Engineering Education," International Journal of Electrical Engineering Education, UK, Vol. 31, No. 2, April 1994, pp. 138–143.
- Jain, L.C. and Bowden, B.S., "A Framework for Knowledge-Based Approach for Designing Electronic Circuits and Systems," Measurement, The Journal of the International Measurement Confederation, Elsevier Science Publisher, Vol. 12, 1993, pp. 9–24.
- Rowland, J.G. and Jain, L.C., "Knowledge-Based Systems for Instrumentation Diagnosis, System Configuration and Circuit and System Design,"

International Journal of Engineering Applications of Artificial Intelligence, No. 5, Vol. 6, 1993, pp. 437–446.

- Jain, L.C. and Harpas, P., "Personal Computer Software for Electronic Circuit Design in Printed Circuit Form," Computer-Aided Engineering Journal, IEE, UK, August 1988, pp. 148–152.
- Sydenham, P.H. and Jain, L.C., "CAENIC - User Characterised, Electronic Systems Development Software," Computer-Aided Engineering Journal, IEE, UK, October 1988, pp. 200–205.

Conference Papers

- Khazab, M., Tweedale, J. and Jain, L.C., Interoperable Intelligent Agents in a Dynamic Environment, New Advances in Decision Technologies, Proceedings of the First International Symposium on Intelligent Decision Technologies, Himeji Japan, April 2009, Springer-Verlag, 2009, pp. 183–191.
- Tweedale, J. and Jain, L.C., Multilingual Agents in a Dynamic Environment, New Advances in Decision Technologies, Proceedings of the First International Symposium on Intelligent Decision Technologies, Himeji Japan, April 2009, Springer-Verlag, 2009, pp. 193–202.
- Quteishat, A., Lim, C.P., Tweedale, J. and Jain, L.C., A Multi-agent Classifier System Based on the Trust-Negotiation-Communication Model, Applications of Soft Computing: Updating the state of the art, Springer-Verlag, Germany, pp. 97–106, 2009.
- Leng, J., Li, J. and Jain, L.C., A Role-Based Framework for Multi-agent Teaming, KES 2008, Springer-Verlag, Part III, LNAI 5179, 2008, pp. 642–649.
- Tweedale, J. and Jain, L.C., Dynamic Agent Architectures for Complex Systems, Proceedings of the Eight Joint Conference on Knowledge-Based Software Engineering, Athens, August 2008, IOS Press, The Netherlands, 2008, pp. 173–178.
- Consoli, A., Tweedale, J. and Jain, L., The Link Between Agent Coordination and Cooperation, in IFIP International Federation for Information Processing, Vol. 228, Intelligent Information Processing III, Springer-Verlag, 2006, pp. 11–19, Chapter 2.
- Urlings, P., Sioutis, C., Tweedale, J., Ichalkaranje, N. and Jain, L., A Future Framework for Interfacing BDI Agents in a Real-Time Teaming Environment, Journal of Network and Computer Applications, Elsevier, UK, Volume 29, Issue 2, 2006, pp. 105–123.
- Fyfe, C. and Jain, L.C., Teams of Intelligent Agents which Learn using Artificial Immune Systems, Journal of Network and Computer Applications, Elsevier, UK, Volume 29, Issue 2, 2006, pp. 147–159.
- Jarvis, J., Rönnquist, R., McFarlane, D., Jain, L.C., A Team-Based Holonic Approach to Robotic Assembly Cell Control, Journal of Network and Computer Applications, Elsevier, UK, Volume 29, Issue 2, 2006, pp. 160–176.

- Jarvis, B. and Jain, L., Trust in LORA: Towards a Formal Definition of Trust in BDI Agents, Lecture Notes in Artificial Intelligence, KES 2006, Part II, LNAI 4252, Springer-Verlag, Germany, 2006, pp. 458–463.
- Ichalkaranje, N., Sioutis, C., Tweedale, J., Urlings, P. and Jain, L., The Equilibrium of Agent Mind: The Balance Between Agent Theories and Practice, Lecture Notes in Artificial Intelligence, KES 2006, Part II, LNAI 4252, Springer-Verlag, Germany, 2006, pp. 450–457.
- Consoli, A., Tweedale, J. and Jain, L., An Overview of Agent Coordination and Cooperation, Lecture Notes in Artificial Intelligence, KES 2006, Part III, LNAI 4253, Springer-Verlag, Germany, 2006, pp. 497–503.
- Phillips-Wren, G. and Jain, L., Innovations in Intelligent Agents and Web, Lecture Notes in Artificial Intelligence, KES 2006, Part III, LNAI 4253, Springer-Verlag, Germany, 2006, pp. 504–506.
- Jarvis, B., Corbett, D. and Jain, L., "Beyond Trust: A Belief-Desire-Intention Model of Confidence in an Agent's Intentions", Lecture Notes in Artificial Intelligence, KES 2005, Springer-Verlag, Germany, Vol. 3682, pp. 844–850, 2005.
- Farooque, S., Jain, L.C. and Abraham, A., Collaborative Agent Learning Using Hybrid Neurocomputing, Proceedings of the IADIS International Conference on Applied Computing, February 2005, pp. 377–384
- Farooque, S., Abraham, A. and Jain, L.C., "Collaborative Agent Learning using Neurocomputing", Neural Information Processing, Proceeding of the 11th International Conference, ICONIP, India, Springer-Verlag, LNCS 3316, Germany, 2004, pp. 619–624.
- Alpaslan, F. and Jain, L.C., Agent Based Virtual AI Classroom, Proceedings of the International Workshop on Hybrid Systems, Adelaide, Australia, December 2001, Springer-Verlag, 2002, pp. 485–496.
- Nedic, Z., Machotka, J. and Jain, L.C., Modelling Learning Processes, Proceedings of the Fourth UICEE Annual Conference on Engineering Education", Bangkok, Thailand, February 2001, pp. 139–143.
- Nedic, Z., Raza, S.R., Jain, L.C., and Machotka, J, "Computer-Aided Tutorials with Randomised Parameters," Proceedings of the Third UICEE Annual Conference on Engineering Education, UNESCO International Centre for Engineering, Australia, February 2000, pp. 229–232.
- Jain, L.C. and Garud, G.N., CAD and CAL Tools for Electrical Engineering Students, Journal of the Institution of Engineers, India, Vol. 75, November 1994, pp. 52–54.
- Jain, L.C. and Bowden, B.S., Development of Expert System Course for Electronic Students, International Journal of Electrical Engineering Education, UK, Vol. 31, No. 1, January 1994, pp. 34–45.
- Rowland, J.G. and Jain, L.C., Artificial Intelligence Languages in Engineering Education, International Journal of Electrical Engineering Education, UK, Vol. 31, No. 2, April 1994, pp. 1

- Rowland, J.G. and Jain, L.C., Artificial Intelligence Languages Education, Proceedings of PRCEE, Adelaide, 1992, pp. 201–206.
- Jain, L.C. and Bowden, B.S., Introductory Course on Expert Systems in the University of South Australia, Proceedings of PRCEEE, Adelaide, 1992, pp. 194–200.

Chapter 15

AUTOMATIC TEST PROGRAM GENERATION: HOW ARTIFICIAL EVOLUTION MAY OUTPERFORM EXPERIENCE

Danilo Ravotto, Ernesto Sanchez and Giovanni Squillero
Politecnico di Torino — DAUIN
Corso Duca degli Abruzzi, 24
10129 Torino, ITALY
ravotto@gmail.com; ernesto.sanchez@polito.it
giovanni.squillero@polito.it

Within the design arena of modern devices based on cutting-edge processor cores, the availability of effective verification, validation and test methodologies able to automatically generate test programs represents an interesting advantage, since it can dramatically reduce the overall time for design and manufacturing, while improving yield and quality.

This chapter proposes several approaches able to automatically generate test programs for critical components of today's microprocessors. The methodologies proposed here are based on the use of an artificial tool that exploiting the simulation of biological evolution, is proven able to outperform the choices of skilled test engineers.

The experimental results presented for real complex microprocessors, show the effectiveness and the suitability of the artificial evolution to the automatic test program generation.

15.1. Introduction

Testing microprocessors and microprocessor cores is a challenging task for both producers, that want to be sure about the quality of the product they deliver, and users, that may be interested in checking the possible presence of faults in devices they exploit for their applications. Moreover, the continuous technological improvements allow to embed many module components (cores) into a single chip (System-on-Chip, SoC) leading to new issues in the testing due to the limited accessibility of each component.

The design of electronic devices is commonly carried out through an incremental process delivering a more complex description of the device at every step. The methodology allows designers to progress towards the final product in a gradual manner, decreasing at every step the abstraction level of the description, up to obtain the final model. The first design step produces the most abstract description, which describes the general behavior of the device leaving internal details out, whereas the last steps provide low-level descriptions, with details close to the actual implementation of the circuit. Clearly, the lower the abstraction level, the higher the complexity of the resulting model.

However, current Verification, Validation and Test (VV&T) techniques are rarely able to take advantage of the higher-level models in order to reduce the efforts required to validate the lower-level ones, and hence of the final product. Therefore, current VV&T techniques are hardly able to adequately support the generation of valid set of stimuli for cutting-edge processor cores since they require a vast quantity of resources in terms of both computational time, as well as, human intervention. Moreover, it is a well know fact that one of the major bottlenecks in the processor design cycle is related to the generation of valid stimuli sets for fulfilling VV&T requirements. For these reasons, the availability of effective verification, validation and test methodologies able to take advantage of high level descriptions of processor cores represents an interesting advantage, since they can dramatically reduce the overall time for design and manufacturing, while improving yield and quality.

This chapter proposes possible solutions to some validation and test issues of critical components of today's microprocessors. The methodologies proposed here are designed to be flexible, fast and to reduce the computational efforts required for the generation of the complete test set. The applicability of the proposed approaches has been widely tested and the experimental results gathered on real complex designs show the effectiveness of the proposed methodologies.

Section 15.3 presents a simulation-based method able to efficiently exploit high-level descriptions of multithreaded microprocessor cores for generating effective test programs of critical modules. The methodology is mostly automatic and makes use of the feedback from the simulation. The feedback-based generation process is driven by an evolutionary algorithm, and exploits a high-level simulator. The method is focused on the generation of multithreaded assembly test programs targeting the maximization of a set of selected verification metrics. The evolutionary algorithm is able to generate, represent and effectively manipulate multithreaded programs. Moreover, in order to minimize the required computational effort, the approach is organized in a series of runs where, at each step, the best test program generated is added to the final test set. To demonstrate the suitability and the scalability of the approach, two rather complex microprocessors, the OpenSPARC T1 and OpenSPARC T2 developed by SUN Microsystems, have been considered as case study. The high-level models of the two microprocessors are publicly available from SUN Microsystems [1]. The test sets generated tackle two different typicals internal modules. The experimental results clearly show that the method is effective both in reaching high figures when a given coverage metric is considered, and in generating test programs able to cover very specific corner cases, that could hardly be managed through a random approach. The methodology requires a very limited effort in terms of manual intervention, while the required computational power can be traded-off by running it separately for different components of the microprocessor.

Section 15.4 firstly presents the applicability of two different software-based testing approaches in a typical SoC architecture built around a RISC processor core with three popular communication interfaces: UART, HDLC and Ethernet controllers. The first approach [8] proposes a fully automated test program generation flow built around an evolutionary algorithm that iteratively generates test programs for the embedded cores of the SoC targeting at the maximization of a set of coverage metrics. The second approach [3] proposes a systematic test generation flow guided by a set of pre-computed test routines for some individual subcomponents of the cores. This case study not only brought out the limitations of these two approaches when applied in more complex communication cores within a SoC, but also raised the perspective of a possible combination of the two approaches. Then, a new effective test program generation methodology that takes the advantages of both previous approaches is proposed. It exploits the deterministically generated test routines of the second approach to guide the automatic test program generation of the first approach to achieve higher fault coverage with limited human effort. The methodology exploits an evolutionary algorithm as test program generator and the whole approach is fully automated requiring a very low human effort. The advantages of the proposed methodology against the previous approaches are demonstrated on the same communication cores in terms of fault coverage, test code size, and test application time.

15.2. Background

15.2.1. *Verification, Validation and Test methodologies*

Considering VV&T in general, it is possible to state that almost all techniques can be classified either as formal or simulation-based.

Formal methodologies exploit mathematical techniques to prove specific properties, such as the absence of deadlocks or the equivalence between two descriptions. These methodologies usually need: a model of a design, a description of the environment where the design is supposed to operate in, and some properties that the design is intended to fulfill. Based on these information, these techniques search some input patterns able to violate the properties. Such a proof implicitly considers all possible inputs and all accessible states of the circuit. Classical approaches for demonstrating that such input stimuli exist are random simulation, or directed test. Formal verification is an alternative approach that can be used both to search for inputs sequences that violate the properties, and prove that the properties always hold in the case when no such stimuli exist. Formal verification can be applied to designs described at many different levels of abstraction, ranging from the gate level, to RTL implementations, and in some cases even to transaction level models described in standardized programming languages [7].

Simulation-based approaches [10], on the other hand, rely on the simulation of a set of stimuli to uncover misbehaviors. An application called *simulator* runs on a workstation and allows the user to simulate a device as it would actually work in a real system. In these approaches, the user:

- can describe dynamic stimulus that is to be applied to the device and then monitor the response of the device as the stimulus is applied over time.

- can specify signals that are to be driven to produce the stimulus.
- has access to the values driven on each signal within the device at any given time during the simulation; using these values, the user can check for proper operation of the device under verification.

A simulation-based approach may therefore be able to demonstrate the presence of a bug, but will never be able to prove its absence. Indeed, the user may assume that a bug does not exist with level of confidence related to the quality of the simulated test set. The generation of a qualifying test set is the key problem with simulation-based techniques. Different methodologies may be used to add contents to such test sets, ranging from deterministic to pseudo-random.

Feedback-based techniques are used to iteratively improve a test set in order to maximize a given target measure. The improvement procedure is fully automatic and exploits the result of the simulation to drive the creation of new stimuli. However, simulation of low-level descriptions could require enormous efforts in terms of computational time, memory occupation and hardware.

Theoretically, formal techniques are able to guarantee their results, while approaches based on simulation can never reach complete confidence. However, the former require an enormous computational power, and therefore may not be able to provide result for a complex design. Moreover, to enable their use, formal methodologies are routinely used on simplified models or with simplified boundaries conditions. Thus, their result is absolutely exact, but the model used to calculate it was not. This introduces a certain amount of uncertainty in the process [13].

Nowadays, simulation-based techniques dominate the VV&T arena for microprocessors, with formal methods bounded to very specific components in the earliest stages of the design. When tackling microprocessors, a set of stimuli has most likely the form of an assembly program. Such test programs, whose introduction dates back to the 80's [32], are executed by the device under test, but their purpose is not to calculate a certain value or to solve a problem, but rather to excite the different supported functions, and to expose the internal information. Test programs have been exploited in the so-called Software-based Self-test (SBST) [27], where the goal is to make evident a difference between a faulty device and a working one. In a SBST approach, a proper test program is developed, which is then uploaded in the code memory (or cache) of the processor, and then executed by the processor under test. Results stemming from the program execution are checked, allowing the detection of possible faults. If the program is properly generated, this approach can be very effective, combining high detection capabilities and short test time with flexibility and independence on Design for Testability (DfT) solutions. Therefore, the real issue is often how to generate effective test programs.

SBST techniques may benefit from cross-fertilization with validation techniques, since both require generating effective test stimuli able to adequately excite the different functions and parts of a module. Simulation-based methodologies aim at uncovering design errors by thoroughly exercising the current model of the circuit.

Briefly speaking, a design, test and validation process based on circuit simulation requires three basic elements: input information (also called *set of stimuli*), the model of the device under evaluation (also called design or *device under test/validation*), and finally the *response checker*, which generates the pass/fail information regarding the inspection process based on the comparison between the observed and expected behavior.

Simulation-based methodologies are strongly dependent on the quality of the set of stimuli used to excite the design. These methodologies are seldom exhaustive and only consider a limited sub-set of possible circuit behaviors. Since the quality of the results first depends on the percentage of applied stimuli with respect to an exhaustive set, they almost never achieve 100% confidence of correctness.

Coverage metrics were firstly defined in software testing as the measure of how thorough exercised a given piece of code is by a test; this goal can be achieved by quantifying the capacity of a given input stimuli to activate specific properties of the program code [15]. Similarly, borrowing the idea from software testing, a coverage metric for hardware test and validation must be defined to assure the adequacy of the set of stimuli, and the collected information about coverage can be exploited as a useful test criterion. The selection of the coverage metric exploited during the generation process depends on the possibilities of the described model.

Code coverage metrics are directly derived from metrics used in software testing. These metrics identify which code structures belonging to the circuit description are exercised by the set of stimuli, and whether the control flow graph corresponding to the code description has been thoroughly traversed. The structures exploited by code coverage metrics range from a single line of code to if-then-else constructs.

Functional coverage metrics, as the name recalls, target design functionalities during the test and validation process. These metrics are composed of a set of very precise test cases that exercise the design in well defined situations guaranteeing that the design under evaluation complies with some design functionalities. Often, the functions to be covered by these test cases are summarized in a check table.

15.3. Test Program Generation for Microprocessor Validation

15.3.1. *Motivation*

Currently, market pressures are coercing original equipment manufacturers (OEM) to shrink as much as possible the production cycle of electronic commodities, since customers continuously require new and more advanced features, and fulfilling their requirements would greatly benefit on the usage of off-the-shelf devices. Consequently, OEMs are constrained to design new devices based on more performing processor cores. However, inclusion of cutting-edge processor cores in modern devices may derive in longer production times since verification, validation and testing (VV&T) methodologies are not able to keep the same pace of production methodologies. In fact, production methodologies for electronic devices are backed by emerging methodologies that carry out the whole design process through a step-by-step process delivering a more complex description of the device at every step. The whole production methodology allows designers to progress towards the final product in a gradual manner, decreasing at every step the abstraction levels of the delivered description, up to obtain the final device. Unfortunately, VV&T methods are not yet able to fully take advantage of the descriptions delivered at different abstraction levels in order to diminish the efforts required to guarantee the correct functioning of the final product.

On the design of modern devices based on cutting-edge processor cores, the lack of VV&T methodologies able to take advantage of high level descriptions of the processor core represents a really critical issue. In fact, modern processors are designed exploiting complex architectural paradigms that mostly prevent the use of not only formal methods for VV&T, but even random-based ones. Moreover, VV&T methodologies relying on the simulation of low-level descriptions of the device under test could require enormous efforts in terms of time, memory occupation and hardware in order to reach the expected results. On the other side, relying on high level descriptions and exploiting the coverage metrics [31] may provide several advantages, such as reducing the required computational effort and allowing an early identification of possible problems. For these reasons, focusing on VV&T methods able to exploit high-level models of the device under consideration could become a key factor in the design flow. Previous considerations are particularly valid for multithreaded processors, where the search space to be explored is even larger than for other processors. In a traditional processor without pipeline the status of the device mainly depends on the single instruction which is under execution at any given time: thus, the search space to be explored is composed of all instructions with all addressing modes with all operand values. In a pipelined processor the status depends on the set of instructions within the pipeline, and the search space increases, comprising all sequences of instructions, and for each one all addressing modes with all operands values; the maximum sequence length depends on the pipeline length. In a multithreaded pipelined processor the search space explodes even more, including all the possible combinations of sequences of instructions (one for each thread).

In this section, a simulation-based method for automatically generate test programs for VV&T of the critical modules present in modern computer architectures is presented. The methodology, exploiting high-level descriptions of processor cores, is able to generate test programs for the targeted modules. The generation process is driven by a feedback value computed resorting to a high level simulator. The method is focused to the generation of multithreaded assembly test programs targeting the maximization of a set of coverage metrics, and exploits a suitable technique to generate, represent and manipulate multithreaded programs. In order to minimize the required computational effort, the proposed approach is organized in a series of runs where, at each step, the best test program generated is retained in the final test set.

The approach has been evaluated on two microprocessors developed by SUN Microsystems, the OpenSPARC T1 and the OpenSPARC T2. The high-level models of the two microprocessors are publicly available from SUN Microsystems [1]. The generated test sets target two different characteristics internal modules.

The experimental results clearly show that the method is effective both in reaching high figures when a given coverage metric is considered, and in generating test programs able to cover very specific corner cases, that could hardly be managed through a random approach. The method requires a very limited effort in terms of manual intervention, while the required computational power can be traded-off by running it separately for different components of the microprocessor.

15.3.2. *Background*

15.3.2.1. *Design Validation*

Design validation is one of the most time consuming and expensive tasks in the development of new electronic systems, and in particular, in the case of multi-threaded processor cores.

Formal verification techniques [9, 17, 33] are able to guarantee high levels of confidence in the correctness of the result. Despite that, the applicability of such techniques on multithreaded designs can be impractical or may need considerable human effort. Thus, industry adopts formal methods when facing only single modules or when boundary conditions allow to significantly constraint the validation task.

On the other hand, the common practice is to tackle corner cases resorting to hand-written tests, since random generation can hardly cover them. Usually these tests are written by skilled designers that have a deep knowledge of the device under validation, thus requiring considerable human efforts during the development phase. Moreover, test cases developed by hand are not always portable, making difficult to reuse them.

Constrained-random test generation [2, 5, 14, 34] develops random tests following constrained generation processes. These techniques usually exploit rigid templates in which the structure of each program fragment of be generated, as well as the parameters to be randomized, are previously defined. Such techniques are very challenging when targeting real complex designs.

An evolution of the constrained-random methods is based on the exploitation of feedback metrics able to drive the generation process [16]. These techniques dynamically and automatically adapt the generation process resorting to coverage feedback results, in order to improve the generated programs. For example, code coverage and transition coverage have been used, among others, during the generation of tests cases for processor cores [6]. These techniques generate and simulate a huge quantity of programs but, the generation process leaves users with only a small set of test programs.

15.3.2.2. *Basics on OpenSPARC processor cores*

The OpenSPARC processor core family is designed exploiting the chip multithreaded (CMT) processor architecture paradigm. The processor family fully implements Sun's Throughput Computing initiative for the horizontal system space. Throughput Computing is a technique that takes advantage of the thread-level parallelism.

The basic principle behind Throughput Computing is that exploiting ILP (Instruction Level Parallelism) and deep pipelining has reached the point of diminishing returns, and as a result current microprocessors do not utilize their underlying hardware very efficiently. For many applications, the processor is usually idle most of the time waiting on memory, and even when it is executing, the processor is often able to only utilize a small fraction of its wide execution units. So rather than building a large and complex ILP processor that sits idle most of the time, a number of small, single-issue processors that employ multithreading are built in the same chip area. Combining multiple processors on a single chip with multiple strands or threads per processor, allows very high performance for highly threaded applications. This approach is usually called thread-level parallelism (TLP).

The one disadvantage to employing TLP over ILP is that execution of a single thread will be slower on the TLP processor than an ILP processor. However, current processors running at high frequencies, a strand capable of executing only a single instruction per cycle is fully capable of completing tasks in the time required by the application, making this disadvantage a nonissue for nearly all applications [1].

15.3.3. *Proposed Approach*

In this section an automatic approach able to generate test programs aiming at verifying noteworthy modules belonging to complex computer architectures is presented. In particular, the presented approach deals with processor cores implementing the multithread paradigm. The real challenge in a multithreaded environment is to find a series of test programs to be scheduled on the processor threads such that it is able to fully excite a specifically targeted module. Moreover, with the intention of generating appropriate stimuli for every thread, it is important to correctly manipulate programs running contemporarily in all threads, in order to thoroughly excite appropriate execution conditions for every thread. This is not easily performed neither resorting to a random-based approach nor by hand, even though a deep knowledge of the processor core is available.

Figure 15.1 reports the conceptual view of the proposed methodology. The whole approach is divided in two main parts: the first one is devised by hand, and is related to the general framework configuration; the second one is devoted to automatically generate test programs and it is based on a multi-run process where an evolutionary algorithm is exploited for the generation of assembly programs.

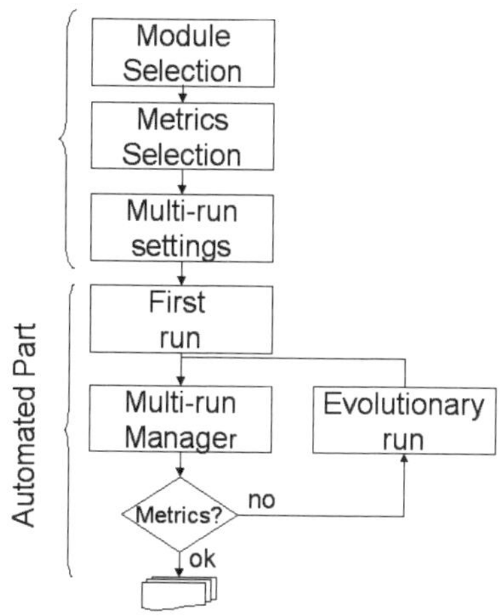

Fig. 15.1. Conceptual view of the proposed methodology.

The main sub-phases of the first part are described in the following:

- **Module selection**: The test engineer has to select the target module.
- **Metrics selection**: After selecting the module to be targeted, the test engineering has to select the metric(s) to be maximized. The choice is highly dependent on the description style of the high level model and in some cases it may be more effective to exploit a set of coverage metrics, instead of only one [31].
- **Multi-run settings**: Test engineer must fulfill a series of requirements in order to setup the generation environment:
  - Develop the constraint library, describing the processor assembly syntax by means of a set of macros that describe processor instructions
  - Determine a set of privileged instructions among the described ones to use from the second iteration of the loop on
  - Characterize every generation run, setting up information related to the length of the programs as well as to the programs size, etc.
  - Define termination conditions for both every single run, and the overall multi-run experiment

As mentioned before, the proposed approach limits human intervention to only this section of the general framework.

The second part of the generation process is divided into a series of steps (i.e., evolutionary runs): in every one, a series of test programs are generated, evaluated and improved; at the end, only the best program is preserved in the final test set. It should be noted that since herein the proposed approach targets a multithread processor core, the programs generated are composed by different pieces of assembly code that conform independent threads and, the real big challenge faced in every evolutionary run is to evolve all threads in a combined manner. In the automated part the generation process is handled by the *multi-run manager* that, using the configuration information previously defined by the test engineer, launches a series of evolutionary runs exploiting the evolutionary algorithm until satisfactory figures are obtained in the chosen metrics. Thus, the main goal of the multi-run manager is to activate every evolutionary run with the right parameters, and to collect the final set of test programs.

The rationale for this step-by-step methodology is that in general it is not possible to completely solve the problems resorting to only one test program, especially when complex modules are targeted. Moreover, this multi-run execution allows the evolutionary algorithm to better focus on the targeted module, thus reducing the generation time, since long test programs must be only evaluated during the first run.

Once the test engineer provides with configuration information to the multi-run manager, the automatic process is ready to start. Progressively, the multi-run manager, diminishes the dimension of the required test program, trying to excite specific and not already covered parts of the design. It is important to notice that at the end of each run the covered items are actually marked as covered, so that the generation process focuses only on those parts that are not yet excited, speeding up simulations. Each run is terminated when the evolutionary algorithm reaches a condition in which it is considered that it is not possible to further improve the targeted metrics or when a maximum number of programs have been simulated.

Figure 15.2 presents in more details the environment defined to implement the proposed approach, highlighting the elements that intervene in the generation loop.

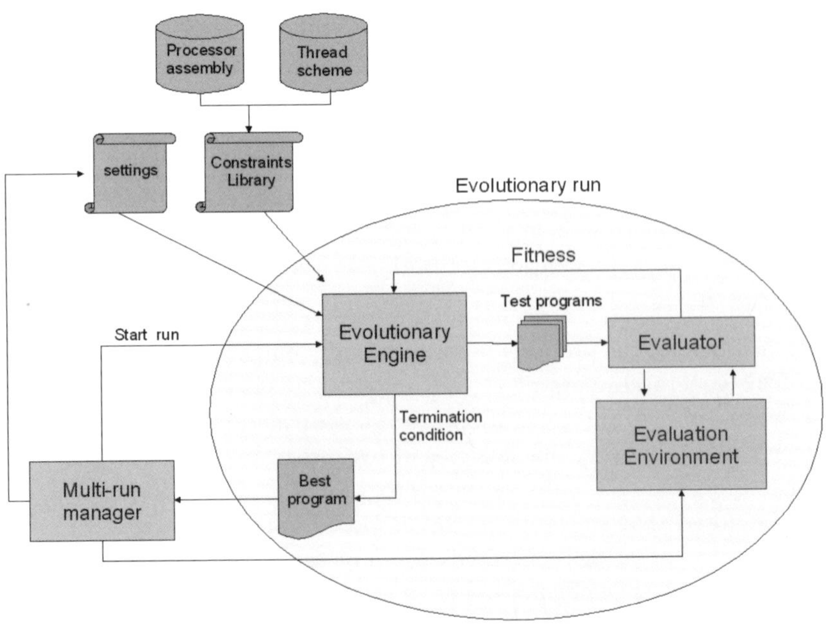

Fig. 15.2. Multi-run architecture.

Every evolutionary run launches the evolutionary engine as automatic test program generator. This engine has been previously introduced in [11] and is able to automatically generate test programs starting from a set of constraints that describe the syntactic appearance of the test programs. This information is organized in the user-defined constraints library. The constraints library lists all syntactically correct instructions and all the valid operands for each one. Running in a multithreaded environment, also the thread scheme and the number of threads to be generated must be encoded in the constraints library. Based on both the constraints library and the settings provided by the multi-run manager, the evolutionary engine generates syntactically correct test programs; each of them is evaluated by the *Evaluator* using an evaluation environment, and a measure of how well it solves the given problem is fed back to the evolutionary engine. The evolutionary engine allows to simultaneously maximize all the selected metrics with a defined priority. The evolutionary engine keeps a set of best test programs (ranked on their fitness) that are selected, manipulated, and combined at every iteration in order to generate new and eventually improved test programs. At the end a single test program is provided as output, which represents the local optimum found by the tool. This program is picked by the multi-run manager and retained in the final test set. In addition, the multi-run manager reconfigure the evaluation environment in order to remove the covered items, eventually changes the settings, and starts a new evolutionary run (if the maximum number of runs are not reached or the targeted values of the metrics are not yet saturated).

15.3.3.1. *The feedback-based generation algorithm*

The algorithm is able to generate, manipulate and improve multithreaded test programs; every single one is internally represented as the union of N different programs, being N the number of processor threads. Every one of these programs must be enriched and improved contemporarily in order to reach a useful final test program that exploits the actual interdependency among the different instructions on every single thread.

For multithreaded test program generation purposes, the start point has been an evolutionary optimizer presented in [11]. This evolutionary algorithm exploits a suitable internal representation for programs, based on graphs, and basically works as follows: an initial set of random test programs (called individuals) is created and evaluated with respect to an evaluation function; then, the best fitted programs are chosen to be improved by the application of one of the so called genetic operators mutation and crossover. This process iterates until a stopping condition is reached. The approach has been able to successfully deal with the generation of test programs tackling different goals: processor testing [28], verification [11] and post-silicon verification [25]; however, in order to deal with multithreaded processors, it is necessary to introduce special features that allow the evolutionary optimizer to effectively manipulate and evolve test programs for multithreaded processor cores.

15.3.3.2. *New multithread-oriented features*

Test programs for SMT processor cores mainly differ from predecessor ones in the composition of every test program: in the SMT case, every single test program is composed of a set of N threads, whereas the effectiveness of the test program is evaluated considering the parallel execution of all the threads. Conceptually, every single thread executes an independent program whereas sharing hardware resources with other threads. However, from the generation point of view, it is important to improve the single thread, as well as to maximize the interaction among the parallel executions of the different threads.

In the version of the tool oriented to SMT processors which is proposed here the evolutionary engine is able to act separately on both the whole test program and the single thread by exploiting different genetic operators. The evolutionary tool is able first to create random (but correct) multithreaded test programs and then to independently evolve them by applying suitable thread-oriented evolutionary operators. In other words, new genetic operators work specifically on independent threads by inserting, deleting and mutating instructions. In addition, the evolutionary engine also exploits an inter-thread crossover operator, allowing code recombination between threads, thus permitting the evolutionary optimizer to exchange valid pieces of code between the different threads of a single program, as well as between threads in different test programs. In this way, the method leverages easy exchange of outstanding pieces of code between different threads.

Figure 15.3 graphically represents the application of the crossover operator between two threads of the same program. Let us suppose that TP1 (in the left part of the figure) represents the selected program that undergoes the thread-oriented crossover. TP1 is composed of two threads (th 1 and th 2), each one of them containing three sections represented with some geometric figures. Let us also suppose that in both threads of TP1 the dashed rectangles demark the pieces of code selected to be exchanged during the application of the mentioned crossover inter-thread operator. The resulting individual (TP1$'$) is depicted in the right part of the figure.

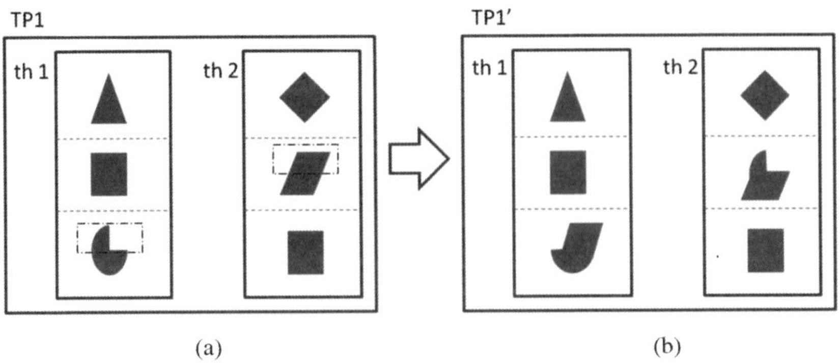

Fig. 15.3. Thread-oriented crossover.

It is interesting to highlight that the evolutionary optimizer is able to safely exchange pieces of code between threads since every test program and in particular every thread is internally represented as an acyclic graph. The graph-based representation of test programs allows the evolutionary tool to copy, transfer, and even eliminate code segments without losing important details regarding the program structure for pieces of code such as subroutines and jumps. On the other hand, the whole test program is also eventually modified by the application of genetic operators that, as in the previous versions of the evolutionary optimizer, elaborate on the complete program structure. In order to measure the interaction between threads, the evaluation of every program is performed on the whole multithreaded test program, and the feedback information gathered are used by the evolutionary tool to adapt the generation process.

15.3.4. *Case Study 1 — The OpenSPARC T2 processor*

15.3.4.1. *OpenSPARC T2*

The OpenSPARC T2 is a 64 bit open-source microprocessor developed by SUN Microsystems and freely available at [1].

OpenSPARC T2 implements the chip multithreading (CMT) paradigm and counts on eight SPARC physical processor cores. Each SPARC core is able to support eight strands (i.e., a virtual processor able to run a software thread). Internally, every processor core divides its eight strands in two groups of four strands each. In order to increase the throughput, each SPARC core contains two integer execution pipelines (one per strand group), one floating-point execution pipeline, and one memory pipeline (shared by the eight threads).

The OpenSPARC T2 can execute up to 64 simultaneous threads, but in every SPARC core at most two strands can be active at the same clock cycle. The strands are switched at each clock cycle choosing the least recently issued available strand within each group. A strand is marked as unavailable if it encounters a long latency event, such as a cache miss, a floating point operation, etc.; no instructions will be picked from an inactive strand until the long-latency event is terminated. The execution of the others strands continues normally while the long latency operation is executed.

Each SPARC processor core has a 16 KB, 8-way associative instruction cache, a 8 KBytes, 4-way associative data cache, a 64-entry fully associative instruction TLB, and a 128-entry fully associative data TLB, that are shared by the eight strands. The eight processor cores are connected through a crossbar to an on-chip unified 4 MBytes, 16-way associative L2 cache (64-byte lines) banked eight ways. The L2 cache connects to four on-chip DRAM controllers.

The main components of the OpenSPARC T2 core are:

- **The Instruction Fetch Unit (IFU)**, which distributes instructions to the rest of the core, while updating both the program counters (PCs) for each thread, and the instruction caches. The IFU is composed of three subunits: the fetch unit, the pick unit, and the decode unit.

- **The Execution Unit (EXU)**, which executes almost all the integer operations, with the exception of the multiply and division instructions.
- **The Load and Store Unit (LSU)**, which manages memory accesses between the processor core and the data caches (L1 and L2).
- **The Cache Crossbar (CCX)**, which connects the 8 SPARC cores to the 8 banks of the L2 cache. A maximum of 8 operations on the memory can be processed simultaneously.
- **The Floating-Point and Graphics Unit (FGU)**, which implements the SPARC V9 floating-point instruction set, integer multiply, division, and population count (POPC) instructions.
- **The Trap Logic Unit (TLU)**, which handles exceptions, trap requests, and traps for the SPARC core. A trap is a vectored transfer of control to supervisor software through a trap table. The TLU maintains the processor state related to traps as well as the Program Counter (PC) and the Next Program Counter (NPC).
- **The Memory Management Unit (MMU)**, which handles all memory accesses validating the permission to access to the requested address

15.3.4.2. *Module Selection*

The module within the OpenSPARC T2 processor chosen for the evaluation of the proposed methodology has been the pick unit, which is the module mainly responsible for the management of the parallel execution of the different threads. In addition, this module is distinctive of a multithreaded processor and is not present in other types of processors. Targeting the pick unit is particularly critical, since the test program requires properly combining the execution of several threads, making this unit the most suitable part of the OpenSPARC T2 where to evaluate the approach. The module is located within the instruction fetch unit and at each cycle is in charge of selecting two threads out of eight for the execution. Its correct behavior is essential in order to guarantee both proper functioning and maximum performance of the processor. Figure 15.4 reports the block diagram of the IFU.

The threads are organized in two thread groups of four threads. The pick unit is fed by eight Instruction Buffers (IBs), one for each threads, each of them has eight entries (the IB entry 0 contains the oldest instruction fetched). The pick unit contains a state machine for each thread with two possible states: READY or WAIT. A thread can be picked if it is in a READY state and the IB entry 0 is valid. A thread is in a WAIT state if any wait conditions exists for the thread and it remains in a WAIT state until the condition or the conditions that caused the transition to WAIT are resolved or the thread is flushed. Among the two thread groups, at each clock cycle the Least Recently Picked (LRP) READY thread is picked (non speculative threads

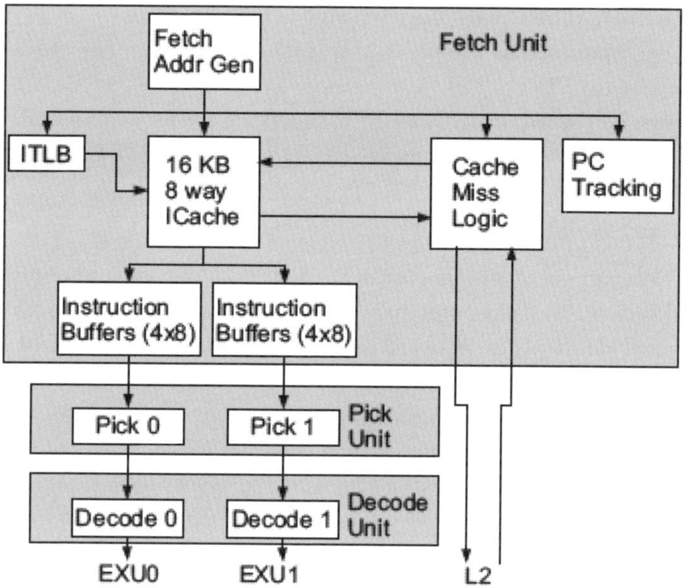

Fig. 15.4. Block diagram of the instruction fetch unit.

have higher priority than speculative threads). The pick process is independent between the two groups, thus some hazards could arise but they are resolved by the decode unit. A thread in the READY state can be picked no matter the type of instructions in the instruction buffers. Afterwards, the instruction is decoded and if some dependencies and resource limitations are detected the thread may move into the WAIT state iff the condition or the conditions giving rise to the hazard or hazards are not resolved during the current cycle. There are several conditions that may cause a thread to transition to the WAIT state. These conditions are grouped as below:

- **Post synchronization wait conditions**: certain instructions transition the thread in a WAIT state after they are picked
- **Speculation not enabled wait conditions**: if speculation is not enabled certain types of instructions (integer loads, floating-point operations, and conditional branches) may transition a thread in a WAIT state
- **Dependency wait conditions**: a thread enters the WAIT state as soon as a dependency between the new instruction and any of the instructions in the pipeline is detected
- **Divide wait conditions**: a thread transitions to the WAIT state as soon as a divide instruction is picked
- **Delayed Control Transfer Instruction (CTI) wait conditions**: when a delayed CTI instruction without a delay slot is picked the thread transitions to the WAIT state

- **LSU synchronization wait conditions**: when the latency of a LSU operation is greater then the normal latency of an integer operation the thread transitions to the WAIT state
- **Store buffer full wait conditions**: if the store buffer of the LSU is full and a new valid store is detected the thread transitions to the WAIT state.

15.3.4.3. *Metric Selection*

The RT level description of the pick unit is written in Verilog and counts 3,688 lines. As mentioned before, the pick unit needs to check a number of thread conditions.

The RT-level description is organized in a significant amount of signal conditional assignments that check all the possible WAIT/READY conditions. For this reason the obvious choice for the main metric to drive the generation process is the Condition Coverage, which measures which possible values each conditional expression takes during simulation, tracking the sensitized vectors (those that change the result of the overall expression) for each conditional expressions. This metric is very meaningful for the pick unit, since, as mentioned before, a thread could transition to the wait state due to several conditions and it is necessary to ensure that all of them are covered by the test programs in order to fully excite the unit. Moreover, in the T2 pick unit descriptions other coverage metrics like Statement Coverage, Branch Coverage, FSM coverage have a number of items to be monitored negligible, or null. On the other hand, the number of conditions is 29,394.

15.3.5. *Evolutionary tool*

The proposed approach was developed using the version 3 of a tool named μGP [12], that implements the evolutionary engine; the tool counts about 42,000 lines of C++ code, and about 90 classes. μGP^3 bases its evolutionary process on the concept of constrained tagged graph, that is a directed graph every element of which may own one or more tags, and that in addition has to respect a set of constraints. A tag is a name-value pair whose purpose is to convey additional information about the element to which it belongs, such as its name. Tags are used to add semantic information to graphs, augmenting the nodes with a number of parameters, and also to uniquely identify each element during the evolution. The constraints may affect both the information contained in the graph elements and its structure. Graphs are initially generated in a random fashion; subsequently, they may be modified by genetic operators, such as the classical mutation and recombination, but also by different operators, as required by the specific application. The tool architecture has been specially thought for easy addition of new genetic operators as needed by the application. The activation probability and strength for every operator is an endogenous parameter. The genotype of every individual is described by one or more constrained tagged graphs, each of which is composed by one or more sections. Sections allow to define a global structure for the individuals that closely follows the structure of any candidate solution for the problem.

15.3.5.1. *Constraints*

The purpose of the constraints is to limit the possible productions of the evolutionary tool, and also provide them with semantic value. The constraints are provided through a user-defined library that provides the genotype-phenotype mapping for the generated individuals, describes their possible structure and to define which values the existing parameters (if any) can take. Constraint definition is left to the user to increase the generality of the tool. The constraints are divided in sections, every section of the constraints matching a corresponding section in the individuals. Every section may also be composed of subsections and, finally, the subsections are composed of macros. Constraint definition is flexible enough to allow the definition of complex entities, such as the test blocks described above, as individuals. Different sections in the constraints, and correspondingly in the individual, can map to different entities. In this specific case the constraints define three sections: a program configuration part, a program execution part and a data part or stimuli set. The first two are composed of assembly code, the third is written as part of a VHDL testbench. Though syntactically different, the three parts are interdependent in order to obtain good solutions.

15.3.5.2. *Fitness*

Individual fitnesses are computed by means of an external evaluator: this is usually a script that runs a simulation using the individual as input and collects the results, but may be any program able to provide the evolutionary core with proper feedback. This complete decoupling between the evolutionary engine and the fitness evaluator makes it possible to use μGP^3 with most existing simulation tools. The fitness of an individual is represented by a sequence of floating point numbers optionally followed by a comment string. This is currently used in a prioritized fashion: one fitness A is considered greater than another fitness B if the n-th component of A is greater than the n-th component of B and all previous components (if any) are equal; if all components are equal then the two fitnesses are considered equal. The two value sequences may then be considered symbols in a string, and the fitnesses may be thought of as such strings, which are then compared lexicographically. For uniform comparison all those strings have to be equal length, meaning that the number of values in the individual fitness has to be specified before every run. It may, however, vary between one run and the next.

15.3.5.3. *Evolutionary Scheme*

The evolutionary tool is currently configured to cultivate all individuals in a single panmictic population, although it can be configured to use an island model.

The population is ordered by fitness. Choice of the individuals for reproduction is performed by means of a tournament selection; the tournament size τ is also endogenous. The population size μ is set at the beginning of a run, and the tool employs a variation on the plus $(\mu + \lambda)$ strategy: a configurable number λ of genetic operators are applied on the population. Since different operators may produce different number of offspring the number of individuals added to the population is variable. All new unique individuals are then evaluated, and the population resulting from the union of old and new individuals is sorted by decreasing fitness. Finally, only the first μ individuals are kept. To ensure that only different individuals are evaluated a hash is computed for each individual, only comparing individuals for equality if their hashes are the same. To promote diversity, individuals that are genetically equal to already existing ones, called clones, may have their fitness scaled by a fixed value in the range [0.0,1.0]. The possible termination conditions for the evolutionary run are: a target fitness value is achieved by the best individual; no fitness increase is registered for a predefined number of generations; the evolution process is executed for the maximum number of generations. At the end of every generation the internal state of the algorithm is saved in XML format for subsequent analysis. This also provides a minimal measure of tolerance to system crashes.

15.3.5.4. *Test program generation environment*

The constraints library written for the OpenSPARC T2 core counts 4,213 lines and implements the SPARC V9 instruction set architecture and the OpenSPARC T2 specific instructions. The constraints are divided in sections, and for each one, the number of different instances in the test program generated must be set. In each section there are defined some macros, each of them representing a valid instruction or set of instructions. To correctly generate the test programs two sections have been defined: the first represents a thread (the tool has been set up to generate eight different segments of this type) and the second represents subroutines (the tool has been set up to generate ten different segments of this type). Writing the constraint library required less than two days to a T2 programmer: remarkably, this operation is needed once for a given Instruction Set Architecture.

The evolutionary tool produces test programs containing eight different threads and, during the evolutionary process, the tool modifies programs by mutating instructions or parameters on different threads as well as by applying the crossover operator between two threads. Thus, it is able to evolve each thread independently, combining the execution of every thread while executing the entire program. At the end of each evolutionary run, the best test program is retained in the final test set.

15.3.5.5. *Experimental results*

All the reported experiments have been performed on a HP xw 8600 Workstation with 8 GB of RAM running Linux. The evaluation environment used is the

Table 15.1. Results obtained by the proposed approach on the pick unit.

Run [#]	Condition Coverage (%)	Code Size [KB]	Test Program duration [CC]
1	90.40	53.376	25,165
2	93.79	11.168	13,680
3	95.90	10.680	9,568
4	96.37	6.976	4,712

OpenSPARC T2 verification suites provided by Sun Microsystems and specifically the cmp1 testbench environment, where only one core out of eight available in the T2 is simulated [1].

Table 15.1 shows the results obtained on the pick unit when applying the proposed methodology. The table reports the condition coverage in percentage, the code size and the duration (in terms of clock cycles) gathered at the end of every evolutionary run. The condition coverage reported is the cumulative value obtained by the set of programs collected until the referred run and not to the single program; the code size and the test program duration, on the other hand, are related to the best test program produced in each run.

The methodology is able to achieve 96.37% of condition coverage by executing the complete test set that lasts 53,125 clock cycles. The amount of computational time required for generating the complete test set is about 5.5 days: 2.5 days for the first run and about 1 day for each following run. This discrepancy between the first run and the following is due to the greater length of the program in the first run. The required time for performing the simulation of the programs corresponds to about 6 to 8 minutes for the first run and about 2 to 3 minutes for the following runs, since these are shorter programs. The generation of the complete test set requires the simulation of about 2,000 programs; it is interesting to note that thanks to the exploited workstation, during the generation more than one program at a time is evaluated. The additional computational effort required by the evolutionary engine is negligible.

In order to assess the proposed methodology a comparison against a random test has been performed. The test programs have the same length of the test programs generated in the first run of the proposed approach and in the final test set are retained those programs able to improve the condition coverage. As in the case of the proposed approach 2,000 test programs have been simulated. Table 15.2 reports a comparison of the results obtained by the two approaches, highlighting the condition coverage obtained and the number of clock cycles.

It can be seen from Table 15.2 that the approach proposed in this section is able to achieve a significantly higher condition coverage figure than the random one, while the test length of the generated test program is significantly lower. This difference can partly be attributed to the ability of the evolutionary engine of guiding the search towards the most promising areas, and partly to the multi-run mechanism,

Table 15.2. Results obtained by the random approach and by the proposed approach.

Strategy	**Condition Coverage (%)**	**Test Program duration [CC]**
Random	89.9	640,000
Proposed	96.37	53,125

which allows boosting up the coverage at every run while reducing the program length (with respect to the first run).

15.3.5.6. *Covered Corner case*

An analysis performed on the conditions left uncovered by the random approach (while covered by the evolutionary-based approach) shows that the random approach is not able to cover the store buffer full wait condition. The load and store unit has, for each thread, a buffer with eight entries containing all outstanding store instructions. The pick logic, on its side, maintains a four bit speculative store counter per thread, which is incremented each time a store is picked. When this counter reaches the value eight and a new valid store is detected, the pick unit transitions the thread to the WAIT state. This condition is not easily reachable using a random approach: in fact, it needs a very specific sequence of memory operations. On the other side, the proposed approach is perfectly able to excite this condition, mainly due to the generation process, where the best test programs are combined and/or manipulated.

15.3.6. *Case Study 2 — The OpenSPARC T1 processor*

15.3.6.1. *OpenSPARC T1*

OpenSPARC T1 is a 64 bit open-source microprocessor core developed by SUN Microsystems and freely available at [29]. The OpenSPARC T1 implements the 64-bit SPARC V9 architecture. If instantiated in the full version, the processor contains eight SPARC processor cores, able to support four threads. Each SPARC core has an instruction cache, a data cache, and a fully associative instruction and data translation lookaside buffers (TLB). The eight SPARC cores are connected through a crossbar to an on-chip unified level 2 cache (L2-cache).

Each SPARC core has single issue and six stages pipeline. The six stages are: Fetch, Thread Selection, Decode, Execute, Memory and Write Back. Additionally, every SPARC has the following units:

- Instruction fetch unit (IFU) includes the following pipeline stages: fetch, thread selection, and decode. The IFU also includes an instruction cache complex.
- Execution unit (EXU) includes the execute stage of the pipeline.

- Load/store unit (LSU) includes memory and writeback stages, and a data cache complex.
- Trap logic unit (TLU) includes trap logic and trap program counters.
- Stream processing unit (SPU) is used for modular arithmetic functions for crypto.
- Memory management unit (MMU).
- Floating-point front-end unit (FFU) interfaces to the FPU.

Every single thread, on the other hand, is supported by a register file (with eight register windows), with most of the general purpose and special purpose registers replicated per thread. The four threads share the instruction, the data caches, and the TLBs. Each instruction cache is 16 KBytes with a 32-byte line size. The data caches are write through, 8 KBytes, and have a 16-byte line size.

15.3.6.2. *Module and metric selection*

The OpenSPARC T1 processor core is described at RT-level in Verilog; in this case, the chosen module is the switch logic unit that belongs to the instruction fetch unit (IFU) and is mainly responsible for managing the four threads that are executed inside the processor core by performing the processor thread selection policy. The thread selection policy is as follows: the available threads are selected at every clock cycle giving priority to the least recently executed thread. The threads become unavailable due to the long latency operations like loads, branch, MUL, and DIV, as well as to the pipeline stalls like cache misses, traps, and resource conflicts. The loads are speculated as cache hits, and the thread is switched-in with lower priority.

The switch logic unit is described at RT-level in four different files and counts 1,428 lines, and as in the first experiment, this is a distinctive module in a multi-threaded processor core, making this module crucial to the correct behavior of the processor core. The selected coverage metric regards with toggling the 7,387 store elements presented in the switch logic module. Once again, considering the processor description, this is the most important coverage metric available in the RT-level model, other verification metrics are negligible or not available for the considered module.

15.3.6.3. *Experimental results*

All the reported experiments have been performed on a HP xw 8600 Workstation with 8 GB of RAM running Linux. The evaluation environment used is the OpenSPARC T1 verification suites provided by Sun Microsystems and specifically the cmp1 testbench environment, where only one core out of eight available in the T1 is simulated [1].

Table 15.3 shows the results gathered on the switch logic unit of the OpenSPARC T1 processor core. Every line of Table 15.3 reports the results obtained by the best programs obtained in every experiment step (Run #). In particular, the toggle coverage reported is the accumulated value of the test set up to the considered

Table 15.3. Results obtained on the OpenSPARC T1 processor.

Run [#]	Toggle Coverage (%)	Code Size [KB]	Test Program duration [CC]
1	75.12	9.03	3,672
2	80.92	25.61	4,410
3	82.37	27.07	4,067
4	84.25	26.18	3,810
5	85.67	8.89	3,656

run. Additionally, for every program the code size, and the duration (in terms of clock cycles) is also reported.

The complete test set lasts 19,615 clock cycles to be executed, requiring about 96KB. The complete test set achieves about 86% on the targeted metric. The complete experiment required about 4 days on the described workstation, requiring the simulation of about 4,000 programs. Compared with the pure random strategy, the proposed approach improves the coverage results in about 8%.

15.3.7. *Case Study 3 — The OpenSPARC T1 processor with hardware acceleration*

15.3.7.1. *Evaluation Environment*

This section reports the characteristics of a prototypical framework specifically set up to reduce the generation time required for the generation of the final test set implementing the concepts introduced in the previous section. The framework consists of a hardware platform where the targeted processor is emulated, and the programs generated by the proposed evolutionary tool are evaluated, gathering relevant information able to guide the evolutionary optimizer. Thus, differently from the previous case study, in this case the programs generated are not simulated but are directly emulated on the hardware platform.

The emulation board must be able to provide an external device with a feedback useful for the validation purposes. Clearly, the feedback information provided to the outside of the board strictly depends on the faced validation problem. Performance Counters and similarly accessible internal registers may be exploited by the validation engineer as processor internal information to efficiently provide the generator tool with appropriate feedback information during the program generation.

As mentioned before, the generation process needs to be guided by selecting appropriate metric(s) that is (are) able to measure the goodness of a given program for the targeted validation task. Being based on a hardware platform, the easiest and most general choice is to make this metric accessible from the user software. For these reasons the framework has been set up to experimentally evaluate the following two possibilities for measuring the quality of the programs:

- **Exploiting the already available Performance Counters**. Depending on the targeted validation task a combination of the accessible Performance Counters or the use of only one is possible. Performance Counters are today a common feature in practically every complex processor. Normally, they are used to identify the causes of performance degradation, but since they count particular events like cache misses, pipeline stalls, floating-point operations performed, etc., they can also be used for validation purposes;

- **Instrumenting the target processor in order to expose an internal register to the user software.** Clearly, this choice is much more expensive than the previous, since it requires a deeper understanding of the processor architecture and description, and a change in its model; therefore, it should be followed only when the available PCs are not meaningful for the targeted validation problem. Clearly, it is essential that the validation engineer identifies a noteworthy register whose activity is directly related to the targeted validation task. Moreover, the kind of monitor to be implemented needs to be accurately selected trading-off among the hardware overhead introduced, human effort needed for the implementation and the complexity of the monitor. It is important to remark that the monitoring system is only used for validation purposes and not included in the final design.

It is important to highlight that the presented framework is flexible enough to easily allow the validation engineer to target different validation problems by only changing the exploited performance counters or the monitored register.

15.3.7.2. *FPGA board*

The OpenSPARC T1 processor has been synthesized on a Xilinx ML505 board using the OpenSPARC T1 hardware package 1.7 provided by Sun Microsystems and freely available at [30]. The FPGA is a Xilinx Virtex-5 XC5VLX110T. The system emulated consists of one OpenSPARC T1 core with 4 threads. In order to fit the FPGA board the L2 cache is emulated with a Microblaze core that processes the memory transactions. The block diagram of the system provided in the package is reported in Figure 15.5. In the figure the block denoted as ccx2mb is an adapter between the cache crossbar interface of the OpenSPARC T1 core and the Microblaze FSL bus.

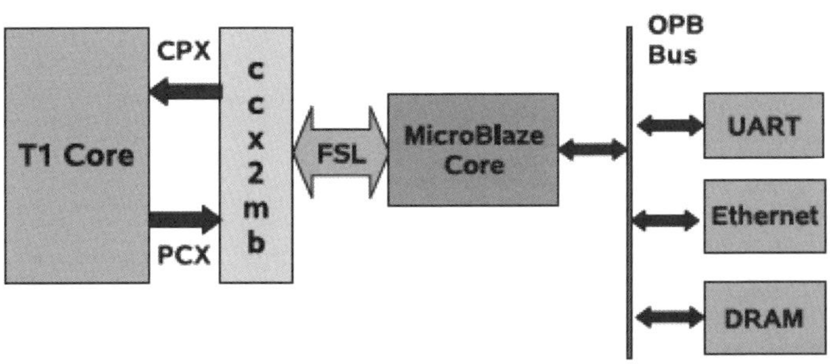

Fig. 15.5. System block diagram for OpenSPARC T1.

The firmware of the Microblaze core performs the following operations: interpret and support the memory transactions of the OpenSPARC T1 core maintaining the caches coherency; send the initial wake-up signal to the OpenSPARC T1 after reset; communicate device interrupt to the OpenSPARC T1 core; download and run standalone assembly programs. In addition to the previous operations, the firmware code of the Microblaze core has been modified in order to read the adopted metric at the end of the program execution. The value read is then communicated to the host PC through the serial connection.

15.3.7.3. *Evolutionary tool*

The evolutionary tool reads a library that describes the instruction set architecture of the target processor (in this case the SPARC V9). The library encodes also the thread scheme and the mechanism adopted to collect information about the program execution. The skeleton of each generated program is reported below:

```
# General processor configuration
# Fork four threads
# General Thread:
# Initialize metric monitored
# Evolutionary tool's generated code
# Write metric into specific memory location
```

For the code generated by the evolutionary tool a maximum number of possible instructions for each thread has been set up. The programs generated by the evolutionary tool are downloaded into the FPGA board through a JTAG cable, while the feed-back information are provided to the evolutionary tool by the firmware of the Microblaze core on a serial line. All these communications are handled by a series of Perl scripts.

15.3.7.4. *Internal information gathering scheme*

In order to gather information about the execution of a particular program, both PCs and an internal register monitor are used. The first option is the easiest one and it does not require any modification of the OpenSPARC T1 core. For the second choice, on the other hand, the methodology exploits the Address Space Identifiers (ASIs) present in the SPARC architecture that allow, among others, to expose an internal register to user-level and privileged software. Because ASI accesses take the form of 64-bit load and store instructions, they offer a convenient and flexible way to read and write both small and large internal structures. The source code of the processor has been modified in order to implement a module in charge of counting the number of transitions executed by a particular register. The register that contains the number of transitions has been made accessible by the user software through the ASI mechanism. The register selected has been a 64 bits register located in the instruction fetch unit that contains global information about the threads. The register holds the current thread configuration (speculation enabled/disabled, instruction cache miss behavior, etc.) and the current state of the thread The monitor is written in Verilog and the overhead introduced in the synthesized version of the processor is negligible.

15.3.7.5. *Experimental results*

In order to assess the proposed methodology three different experiments have been performed targeting the maximization of a set of particular metrics. All the experiments performed are particularly focused on the multithreaded architecture, with the purpose of evaluating the ability of the proposed methodology in the generation of assembly programs that properly combine the execution of several threads. In the first experiment, the generation goal is the maximization of a tricky performance counter available in the OpenSPARC T1 architecture that counts the number of clock cycles that a thread spend in the wait state during the program execution. In particular, the performance counter specifically relates to a particular stall condition related to the load and store unit. In SMT processors the load store unit must be able to handle a series of memory accesses avoiding to become the bottleneck stage during the program execution. In the case of the T1 processor, the load/store unit implements an eight entry buffer mainly containing store instructions; every time the buffer is full, and a new valid store instruction is detected, the thread enters the wait state. By maximizing the mentioned performance counter, the proposed framework generates a final program that strongly stresses the load/store unit. The second experiment targets an interesting equation that combines two available performance counters. Ntcc counts the number of clock cycles. NIex is the number of completed instructions. By maximizing the ratio between the two PCs, it is possible to assess the ability of the evolutionary tool to create assembly programs in which the threads frequently enter the wait state, since this situation reduces the number of executed instructions while increasing the clock cycles required by the program execution. A thread may be moved to a wait state due to some of the following conditions: execution of a long latency operation, cache misses, traps, resource conflicts between threads; therefore, the generated program likely excites the mentioned situations. In the third experiment, the instruction fetch unit has been tackled, one of the most relevant units of SMT processors. This unit is mainly responsible for the management of the different threads to be executed in parallel. In addition, this module is distinctive of a multithreaded processor and is not present in other types of processors. Unfortunately, there isn't any PC related to this unit. For this reason, it has been implemented a register monitor able to snoop an internal processor register located in the fetch unit that stores the current thread configuration and the current state of every thread. The implemented monitor counts the number of transitions of this register, trying to measure the global threads activity. It is important to highlight that in the first and second experiment no additional implementations are required with the purpose of gathering information from the

Table 15.4. Results obtained by the proposed approach.

	1st generation	Best individual	Percent increase
1st metric	62,017	247,172	398
2st metric	688,004	1,631,240	237
3st metric	21,785	38,627	177

Table 15.5. Results of the comparison between the random and the proposed approach.

	Random	Proposed method	Percent difference
1st metric	224,601	247,172	10.04
2st metric	1,409,296	1,631,240	15.74
3st metric	34,398	38,627	12.29

program emulation. On the contrary, the final experiment required the implementation of a special monitor resorting to an Address Space Identifier available in the T1 processor. Table 15.4 shows the results obtained by the proposed methodology when the three aforementioned problems are faced. The table reports for every metrics the maximum value obtained in the first generation, mainly composed of random programs, and those obtained after the evolution process, highlighting also the percentage increase between the beginning and the end of each experiment. The generated programs have a maximum size of 500 instructions and their evaluation on the board requires, in average, 25 seconds to be completed. The tool generated about 1,000 programs and the whole generation process took about 7 hours.

The proposed methodology is able to significantly improve the quality of the generated programs. The reader should note that for the first and the second metric the numbers achieved are affected by the fact that the L2 cache in the FPGA is just emulated and not implemented in the processor model, due to space requirements of the selected hardware platform. In order to further assess the proposed methodology, a comparison against a purely random method has been performed. With the aim of executing a fair comparison, the random approach has been setup to generate 1,000 programs, each of them with a maximum size of 500 instructions. In this way the time needed to evaluate a single program as well as the total time required for the whole generation are the same for both approaches. It is important to mention that two approaches required the same human effort to be executed. Table 15.5 reports the results obtained by the two methodologies highlighting the obtained results of the best individual generated by each of them for the three problems considered.

It can be seen from Table 15.5 that the proposed approach is able to achieve higher figure of merit with respect to the best random one, when the same number of programs is generated. This difference can be partly attributed to the feedback mechanism that is able to guide the evolutionary tool to search towards the most promising areas, and partly to new thread oriented features of the tool that allow the exchange of outstanding pieces of code between threads. It is important to remark that, the human effort as well as the amount of time required by the two approaches are practically the same, while the obtained results are significantly lower for the random approach.

15.4. Test of peripheral cores in SoCs

15.4.1. *Motivation*

Production cost and time-to-market are the driving factors in the development of the new complex integrated circuits. The System-on-Chip (SoC) design paradigm meets these demands since it allows the integration and reuse of digital cores such as embedded processors, memories, analog modules as well as controllers for different communication interfaces. While simplifying the design and verification phases, the SoC design paradigm makes manufacturing testing a really big challenge, mainly because of the limited accessibility of the embedded cores from the SoC boundary.

Hardware-based testing of embedded cores in SoCs is commonly based on the adoption of the IEEE 1500 Standard [21] and the Built-In-Self-Test (BIST) [18] technique. Although the IEEE 1500 standard along with hardware BIST architectures offer an automated, well-establish test solution, the associated circuit modifications may introduce non-trivial area overheads, performance degradation, and excessive power consumption during test operation mode. In addition, at-speed testing is not always economically feasible, since it requires the usage of high-speed, high-cost functional testers, while companies are forced to renew regularly the fleet of their testers in order to follow up the increasing speed of IO interfaces. Thus, many research groups are currently studying new test approaches that might substitute or supplement the hardware-based structural solutions adding test quality without increasing test cost.

One of such new test solutions is commonly known as Software-Based Self-Test (SBST), and has been proven effective for the manufacturing testing of the most common embedded cores in SoCs, the processor cores [4]–[23]. According to SBST solution, the test programs are loaded in the processors instruction cache, then the processor executes them, and eventually, the test program responses are stored in the processor's data cache. The transfer of test program and test data between the tester and the on-chip caches are performed through low-rate tester interfaces; thus it does not require high-cost functional testers. The primary advantages of SBST issue from its non-intrusive functional nature and its capability to use the processor resources and instruction set to test itself in the normal operating mode. Moreover, test programs developed for manufacturing testing maybe reused for on-line periodic testing.

Some recent approaches [3, 8, 19, 20, 22, 24] have proposed the extension of SBST at the SoC level for the testing of other non-programmable cores of the SoC. The main objective of these approaches is to use the embedded processor and the available SoC interconnections in order to generate and distribute test vectors to the peripheral cores and collect their test responses. The basic idea behind these approaches significantly differs from previous ones, where special test access mechanisms are implemented in the cores, and an external tester plays the role of test source and sink.

This chapter, firstly presents the applicability of two different software-based testing approaches in a typical SoC architecture built around a RISC processor core with three popular communication interfaces: UART, HDLC and Ethernet controllers. The first approach [8] proposes a fully automated test program generation flow built around an evolutionary algorithm that iteratively generates test programs for the embedded cores of the SoC targeting at the maximization of a set of coverage metrics. The second approach [3] proposes a systematic test generation flow guided by a set of pre-computed test routines for some individual subcomponents of the cores. This case study not only brought out the limitations of these two approaches when applied in more complex communication cores within a SoC, but also raised the perspective of a possible combination of the two approaches. Then, a new effective test program generation methodology that takes the advantages of both previous approaches is proposed. It exploits the deterministically generated test routines of the second approach to guide the automatic test program generation of the first approach to achieve higher fault coverage with limited human effort. The methodology exploits an evolutionary algorithm as test program generator and the whole approach is fully automated requiring a very low human effort. The advantages of the proposed methodology against the previous approaches are demonstrated on the same communication cores in terms of fault coverage, test code size, and test application time.

15.4.2. *System-on-Chip Architecture*

Figure 15.6 shows the block diagram of a typical SoC architecture. The increasing need for transferring data between electronic devices makes embedded communication peripheral cores a mandatory feature of most SoCs. Depending on the application requirements, SoC design may incorporate communication peripheral cores for low-rate interfaces hanging on the low-speed bus (e.g., UART, I2C, Microwire) and/or communication cores for high-rate interfaces hanging on the high-speed bus (e.g., Ethernet, USB, HDLC). All these peripheral cores communicate with the embedded processor cores of the SoC and occupy a considerable area of the entire SoC that needs to be extensively tested.

For the purpose of this chapter, a sample SoC benchmark is selected and it is built around a fully-pipelined RISC processor, an OpenRISC 1200. It includes a 64KB on-chip RAM, and three illustrative communication peripherals: a UART controller, an HDLC controller and an Ethernet controller. The interconnection backbone of the SoC is based on a 32-bit Wishbone bus. The processor core, the communication cores and the Wishbone model are all publicly available at [26]. The scaling complexity level of these three communication cores, from the simple UART to the more sophisticated Ethernet controller, enables us to identify the

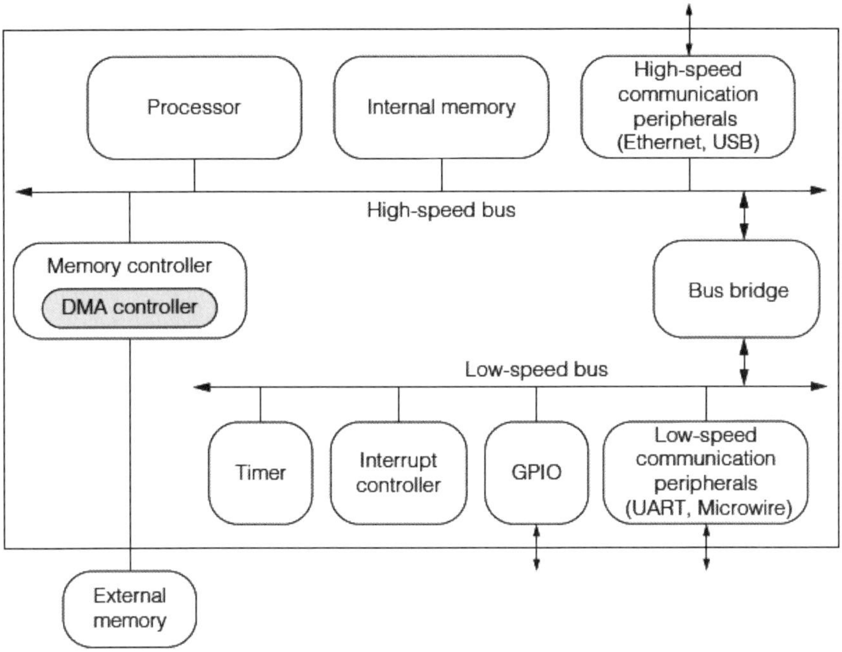

Fig. 15.6. Block diagram of a typical SoC architecture.

Table 15.6. Synthesis statistics for the SoC benchmark.

Core	Sizes(gates)	Stuck-at faults
UART	7,771	20,558
HDLC	22,010	57,868
Ethernet	35,617	77,574
OpenRISC	35,657	85,224
Total	101,055	241,224

limitations of the different methodologies. The core complexity mainly depends on the corresponding communication protocol, while its size is also affected by the parameters of the embedded FIFO structures (most communication cores include FIFOs for rate compensation between the embedded processor and the connecting devices). The representative parameters for the FIFOs of the three cores are: 8-bit width, 16-entries FIFO for the UART core, and 32-bit width, 32-entries FIFOs for the HDLC and Ethernet cores.

Table 15.6 shows the sizes (in gate count and stuck-at fault count) of the processor and communication cores of the SoC benchmarks. The synthesis of the SoC has been performed using a 0.35-micron technology library; the operating frequency of the SoC is 102 MHz.

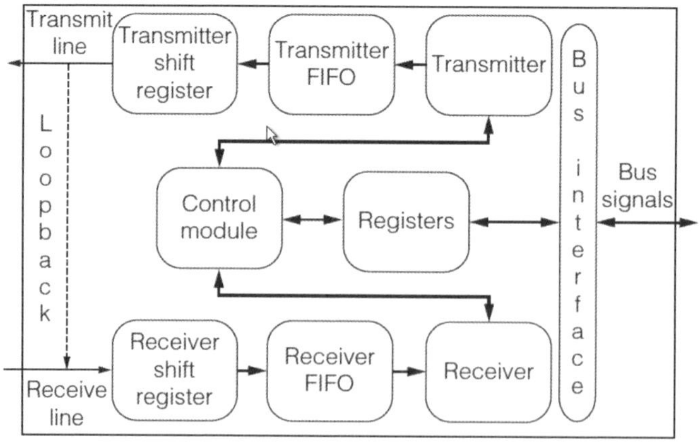

Fig. 15.7. Typical communication peripheral architecture.

A typical communication peripheral core (Figure 15.7) incorporates

- a bus interface logic for the access of the internal core's registers
- FIFO structures
- transmit and receive controllers for the transmit and receive related operations of the core
- a control module that allows the setting of the transmission and the reception according to the value specified by the user in the core's register.

Finally, a loopback line that directly connects the transmit and receive communication line is available for debug and test purpose is assumed available.

15.4.3. *Previous Work and Communication Peripherals Test Challenges*

Processor-based testing of peripheral cores has not received the same attention as self-testing of the processor cores themselves. Recent approaches in the literature [3, 8, 19, 20, 22, 24] have proposed the extension of SBST targeting peripheral cores. Jayaraman *et al.* [22] proposed a functional software-based test approach and applied it to a simple SoC design which includes a GL85 processor (a model of Intel 8085) and a 8251-compatible USART. The experimental results showed that functional tests achieve unacceptably low stuck-at fault coverage (67.89%) for the peripheral core. Hwang and Abraham [20] developed a bus-based test architecture which uses the embedded processor as the test source and sink and the existing bus as the TAM. The approach was demonstrated on an ARM-based SoC with two ISCAS89 cores, a USB Controller and a Viterbi encoder. Huang *et al.* [19] proposed an SBST approach for a bus-based SoC where an embedded MIPS processor delivers the test patterns which are encapsulated in test packets. The core test wrappers decode the test packets and execute the tests. Experimental results for various

ISCAS89 cores with respect to stuck-at and path delay fault models have been provided. Lai and Cheng [24] used the mechanism proposed in 4.4 in a DLX-based SoC to apply scan vectors in four ISCAS89 cores. The methodology proposed in this chapter is based on two previous approaches presented in [8] and [3]. In [8] a fully automated methodology has been presented for the generation of test programs for peripheral cores (PIA, UART) embedded in a SoC. The methodology is based on the exploitation of the correlation between high-level metrics (Toggle, Expression, Condition, Branch and Statement Coverages) and the gate-level fault coverage. It uses an evolutionary tool that generates test programs maximizing selected coverage metrics on a high-level description of the SoC. The evolutionary optimizer generates test programs starting from information about the peripheral core and the processor assembly syntax. The generation process is based on feedback coming from a simulator, and finishes when either the attained values of the targeted high-level metrics are satisfactory, or after a predefined time limit. When the correlation between high-level metrics and fault coverage exists, the methodology is able to achieve acceptable fault coverage with a very low human effort; the code size of the test programs is not the best achievable, since the approach is hardly able to generate loop-based (thus compact) self-test programs. In [3] a generic deterministic flow has been presented for the application of processor-based testing to communication peripheral cores. In this approach, the test sets for the individual subcomponents of the communication peripheral core are pre-computed and pre-generated. Test data are loaded into the on-chip data memory or generated during the execution of the test program. Consequently, test data can be applied in two different ways, as a compact loop-based test program which fetches the test data from the data memory, or as an unrolled test program using immediate instruction format. This methodology achieves high fault coverage, generally with a small and compact test program. Unfortunately, the deterministic nature of this approach requires non-trivial high human effort for writing the final test program. Firstly a quantitative analysis using the tools and methodologies proposed in [3, 8] is reported.

Table 15.7 summarizes the characteristics of the two approaches. In this chapter a comparative results on the previous approaches [22]–[24] are not included since they have been evaluated in a different set of benchmarks. Furthermore, most of them [19, 20, 24] propose the insertion of test wrappers to provide test access mechanism to the peripheral cores I/Os, in contrast with approaches [8] and [3]

Table 15.7. Comparison of automatic and deterministic approaches.

Feature	Automatic	Deterministic
Fault Coverage	Medium	High
Self-test program	Large	Small
Development time	Days	Months
Human effort	Low	High

that present a pure functional test method without using special test hardware or putting the cores in special test mode. The only previous approach [22] that proposes a pure functional software-based method and could be directly compared with approaches [8] and [3] has been applied to a very low complexity core, a USART core (about 3k gates), achieving very low fault coverage (less than 68%).

To study the scalability and identify the drawbacks of the two approaches each of them has been carefully evaluated on the three communication cores of the selected SoC benchmark. The following Tables summarize the comparative results of the study for the two approaches.

Table 15.8 reports the values of each high-level metric for the considered core achieved by the automatic approach of [8]. The high-level metrics can effectively guide the generation process, but this achieves relatively low fault coverage (see second column of Table 15.10) for all the peripheral cores.

Table 15.9 compares the test program development time of the two approaches. The automatic approach is able to generate the final test program in just a few days with very low human effort, while the deterministic approach requires several days of test generation with considerable human effort. Generally, the test development time for the deterministic approach is about 30 times greater than the automatic approach, basically due to the time required by the manual test program generation. The figures in Table 15.9 provide a rough indication of the time consumed for the particular benchmark SoC used with the three communication peripherals. These numbers may vary depending on the test engineer's expertise, but the difference in test development time between the two approaches is clear. Table 15.10 compares the fault simulation results for the automatic and the deterministic approaches. The deterministic approach achieves more than 93 % of stuck-at fault coverage in all communication cores, while the fault coverage results of the automatic approach are

Table 15.8. High-level metrics results for the peripheral cores (automatic approach).

Coverage Metric	UART	HDLC	Ethernet
Toggle (%)	90.0	96.2	87.9
Expression (%)	92.0	96.4	99.4
Condition (%)	95.5	94.4	94.5
Branch (%)	97.2	97.5	97.8
Statement (%)	97.3	95.8	97.7

Table 15.9. Comparison of test program development time, in approximate days.

Core	Automatic [8]	Deterministic [3]
UART	1	30
HDLC	0.5	15
Ethernet	2	45

Table 15.10. Comparison of fault coverage.

Core	Automatic [8]	Deterministic [3]
UART	86.35	93.92
HDLC	68.65	97.34
Ethernet	86.57	93.39

considerably lower, especially for the HDLC core, in which the correlation between the high-level metrics and the fault coverage is relatively weak. This poor correlation is confirmed by the fact that on this core, the automatic approach obtains very high values for each high-level metric fault covehat on this core the automatic approach achieve a very high value for each high-level metrics (Table 15.8) but the respective fault coverage is only 68.65%.

In summary, the automatic approach can achieve relatively low fault coverage but with a low engineering effort in short time. The deterministic approach guarantees high fault coverage, with smaller test program and data, but requires a long and expensive test development time. Merging the advantages of the two approaches while avoiding their drawbacks is the basic idea behind the methodology proposed in this chapter. From the case study it is possible to conclude that the main difficulties in the application of a processor-based testing approach to the communication peripherals of a SoC are:

- Limited controllability and observability of the internal registers of the cores from the SoC boundary.
- Low transmit and receive frequency compared to the frequency of the embedded processor. This fact poses an extra challenge in the application of the method, since the embedded processor has to wait for long intervals before the next pattern can be applied to the peripheral core.
- Testing the receiver can only be done if it is possible to drive the receiver line. Similarly, testing the transmitter can only be done if it is possible to observe the transmitter line.
- Peripherals that need to exchange a large amount of data between the system memory and the core's internal memory may use a DMA mechanism, thus introducing new issues in the testing phase.
- The communication peripheral cores can be set, according to the protocol, in many different operating modes; a good test program needs to exploits all of them.

A processor-based test approach can take advantage of the loopback mode of the cores in order to overcome the problem of driving the receiver line or monitoring the transmitter line and test them in a closed-loop without external test equipment support. The loopback mode is a typical debugging mechanism available in the majority of communication cores and its utilization for processor-based testing can be a generic solution.

15.4.4. *Proposed Test Program Generation Methodology for Communication Peripherals*

In this section a hybrid test program generation methodology which effectively combines the advantages of the previous approaches is proposed. Figure 15.8 presents the test program generation flow of the hybrid methodology. The generation flow is similar to that used in [8] where the evolutionary core generates test programs starting from the processor assembly syntax and using information about the peripheral core. In the hybrid methodology the evolutionary tool is guided using information from the pre-generated test programs of the deterministic methodology. For this purpose a user-defined constraints library is introduced which defines the skeleton of the test programs and the parameters that the tool must optimize. The constraints library reduces significantly the search space size of the evolutionary tool which can now generate more effective test programs in a smaller number of iterations.

The constraints library consists of a number of test blocks. Each test block unit is composed of two parts, as shown in Figure 4: a configuration part and a

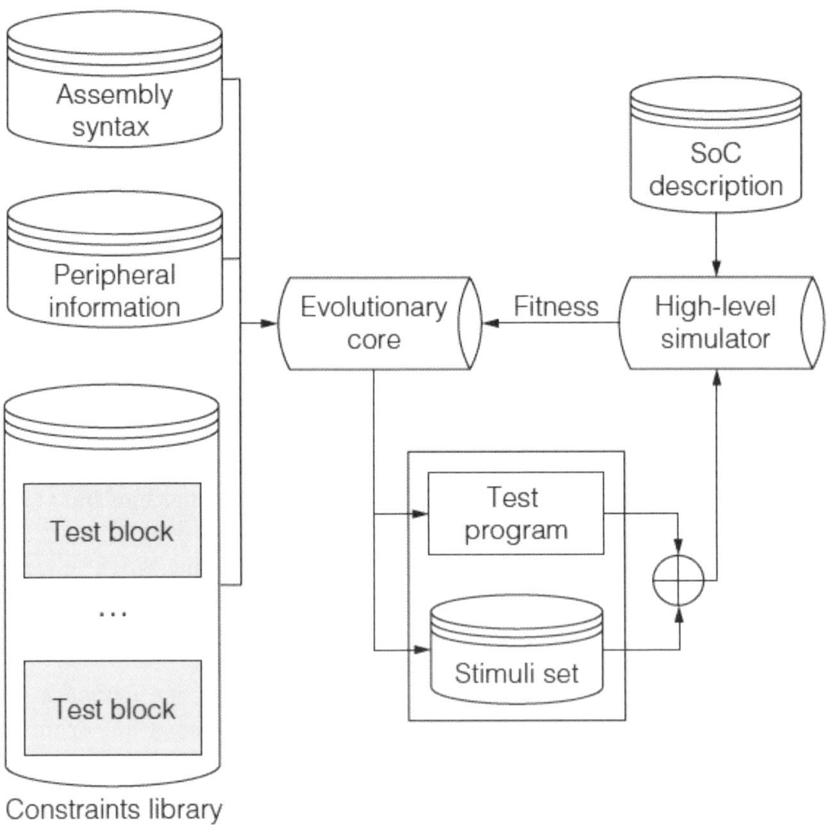

Fig. 15.8. Hybrid test program generation flow.

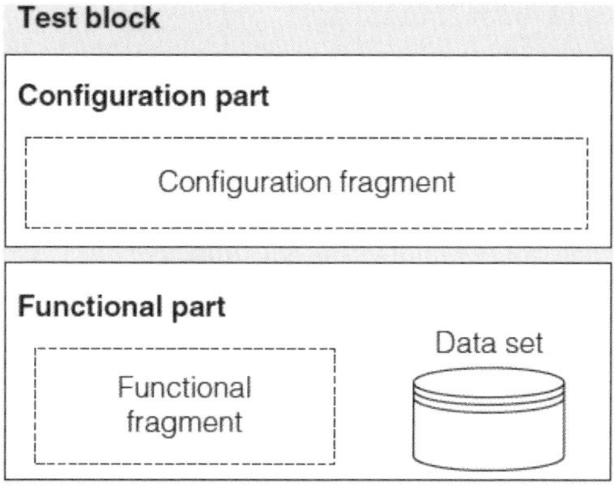

Fig. 15.9. Conceptual scheme of a test block.

functional part. The configuration part includes a program fragment that defines the configuration modes used by the peripheral; the functional part contains one or more program fragments that exercise the peripheral's functionalities as well as test vectors that must be applied to the core.

Each section of a test block is matching a corresponding section in the test programs. Every new test program generated is evaluated using a high-level simulator. The evaluation stage assigns a fitness to every individual (an individual corresponds to a test program), i.e., a measure of how well it solves the given problem. The fitness is based on the following high-level metrics: Toggle Coverage, Expression Coverage, Condition Coverage, Branch Coverage and Statement Coverage. Metrics are considered in order of importance. The Toggle Coverage is the first metric maximized since — as proved in the case study — it is the most difficult to improve and is most correlated with gate-level fault coverage; then the Branch, Condition, Expression and Statement coverage are maximized. The proposed algorithm starts from randomly generated programs, then evolves them according to the classical evolutionary scheme through generations. At each generation individuals (i.e., test programs) are selected according to the fitness and manipulated using proper operators, thus leading to the generation of new (and possibly improved) individuals. The procedure ends when a given time limit is elapsed or when a predefined number of test programs are generated without any improvement of the high-level metrics (steady state). At the end a single test program is provided as an output. The pre-computed test routine templates of the deterministic methodology are transformed into test blocks of the constraints library. Hence, the constraints library includes four test blocks, one test block for each individual step of the deterministic methodology as described in [3]:

- Test block for configuration modes
- Test block for FIFOs testing

- Test block for protocol error handling
- Test block for bus interface logic.

15.4.4.1. *Test block for configuration modes*

The first test block aims at testing the peripheral core in all possible configurations, targeting the logic which implements the communication protocol. It describes the registers of the peripheral cores, its memory address and its read/write access protocol. It also describes the memory space where to read and write the corresponding transmit and receive data, in case the peripheral core incorporates a DMA controller. The DMA controller is often implemented in communication cores that need to exchange a large amount of data between core's FIFOs and the memory of the SoC, without heavily loading the processor. The evolutionary optimizer writes and reads the memory mapped registers of the peripheral core trying to maximize the high-level metrics. Clearly, this step requires a very low human effort since the task of learning the operating modes of the peripheral has been left to the tool. Moreover, test engineers have not to identify the specific operating modes for the core under test and write the corresponding test routines.

15.4.4.2. *Test block for FIFOs testing*

The second test block aims at testing the FIFO structures embedded in the peripheral core. In fact, in a communication peripheral core there are normally two FIFO structures, one at each side of the communication peripheral, that occupy a considerable portion of the design. The automatic approach relies only on high-level metrics and for this part of the design the correlation with the gate-level fault coverage is normally very poor, especially with the increased depth of the FIFO, with a significant impact on the final fault coverage. For this reason it has been adopted an enriched deterministic test set to test FIFO arrays (proposed in [3]), that is able to reach more than 99% of stuck-at fault coverage for any FIFO size and depth. The corresponding second test block of the constraints library describes this deterministic test pattern. A pseudo-code template of the test block is reported below:

```
parameter=fifo_depth
tableA=[0x0000,0xFFFF,0xFFFF,0x0000,...]
tableB=[0xFFFF,0x0000,0x0000,0xFFFF,...]
Initialize general registers
Activate loopback mode
# Apply test set #1
  for i =0 to fifo_depth
        tx_fifo_reg = tableA[i];
# Capture the responses
  for i =0 to fifo_depth
        read rx_fifo_reg
```

```
# Apply test set #2
  for i =0 to fifo_depth
        tx_fifo_reg = tableB[i];
# Capture the responses
  for i =0 to fifo_depth
        read rx_fifo_reg
```

15.4.4.3. *Error Handling Activation*

The third test block targets the faults in the error handling logic of the core. Errors are due to the physical communication channel that can alter the packets. One way to stimulate the error handling logic is based on changing the value of the register that determines the parameters of transmission and reception (mode register). The basic idea is to modify the value of this register after a transmission has been started, waiting for the end of the reception and then reading the register that contains the status of the line. Since the transmission is started with some parameters, and the reception is done according to other parameters, one or more errors may be activated. The corresponding third test block of the constraints library follows the pseudo-code reported below.

```
  valueA = uGP_parameterA
  valueB = uGP_parameterB
# Initialize general registers
  Activate loopback mode
  Mode Register = valueA
  Write tx_fifo_reg
# Change the Mode Register
  Mode Register = valueB
  Read status_regs
```

The evolutionary tool optimizes the two parameters ($\mu GP\_parameterA$ and $\mu GP\_parameterB$) in order to provoke an error. The previous routine is repeated xerr times, which is equal to the number of all possible errors detected by the peripheral.

15.4.4.4. *Bus Interface Logic Testing*

The fourth test block targets the testability problems related to the bus interface logic. These problems are mainly due to the address decoding logic, since each core has a pre-assigned memory space. Moreover, if the core incorporates a DMA controller, different blocks of memory can be set as the DMA region. In order to excite the faults in the address decoding logic all address bits have to be set to both 0 and 1 values. Thus, a solution derived from [3] is adopted; different regions of the embedded data memory have been assigned as DMA area and for each one a transfer (packet transmission) is executed. The corresponding fourth test block of the constraints library follow the pseudo-code reported below:

```
  RegionA = uGP_parameterA
  RegionB = uGP_parameterB
# Define virtual DMA region
  Configure the DMA region [RegionA] of TX FIFO
  Configure the DMA region [RegionB] of RX FIFO
# Apply Test Data
  Stores the transmit frame to memory
  Transmit frame
# Capture Test Responses
  Load the receive frame from memory
```

The evolutionary tool optimizes the two DMA regions in order to set all address bits to both 0 and 1 values.

15.4.5. *Evolutionary Tool*

For the automatic generation of the test blocks an evolutionary tool named μGP^3 presented in the section 15.3.5 has been employed.

15.4.6. *Experimental Evaluation*

The proposed methodology was evaluated on the same SoC benchmark introduced in Section 15.4.2, and the experimental results summarized in the following Tables demonstrate its effectiveness with respect to the previous approaches [8] and [3], in terms of fault coverage, test development time, test code size, and test application time. A prototypical tool was developed, which implements the described method; the tool has been written in C++ and amounts to about 50,000 lines of code. All the experiments have been performed on a PC equipped with an Athlon XP 3000 processor and 1GB of RAM, running Linux. Fault simulation results have been gathered using the TetraMax tool by Synopsys, while the high-level metrics have been computed resorting to the ModelSim tool by Mentor. Table 15.11 shows detailed fault simulation results with respect to the stuck-at fault model for the UART core. The hybrid methodology achieves similar fault coverage with respect to the deterministic approach, but in much less test development time (see Table 15.14). Most subcomponents of the UART core achieve more than 90% stuck-at fault coverage, while the total fault coverage is 93.13%. Especially for the FIFOs, which occupy a significant portion of the core, the test programs achieve very high fault coverage (>99.5%). Table 15.12 presents detailed fault simulation results for the HDLC core. The hybrid methodology achieves 97.43% stuck-at fault coverage (the much larger FIFO of the core has positive impact on the total fault coverage). The proposed approach achieves a fault coverage figure which is almost 30% above the fault

Table 15.11. Fault simulation results for the UART core.

Component	Faults	Fault Coverage (%)
Wishbone I/F	2,058	90.18
Registers	4,490	87.32
Tx Engine	1,560	96.59
Tx FIFO cntr	566	98.94
Tx FIFO	2,850	99.72
Rx Engine	3,066	88.13
Rx FIFO cntr	2,546	95.68
Rx FIFO	2,850	100.00
Debug	536	86.57
Total	20,558	93.13

Table 15.12. Fault simulation results for the HDLC core.

Component	Faults	Fault Coverage (%)
Tx Synch	40	90.00
Tx Core	1,590	88.61
Rx Channel	1,216	80.17
Rx Synch	40	90.00
Wishbone I/F	1,812	85.26
Tx FIFO	25,118	99.24
Tx CRC	1,606	84.68
Rx FIFO	25,308	98.90
Rx CRC	1,118	93.29
Total	57,868	97.43

coverage of the automatic approach. The latter cannot achieve much better results because the correlation between the high level metrics of the FIFO and the respective fault coverage is weak. Once more, the test development time compared to the deterministic approach is relatively lower (Table 15.14).

Table 15.13 reports detailed fault simulation results for the Ethernet core. For this core, it is also necessary to describe in the constraints library the DMA controller access mechanism. The fault coverage in this case is slightly lower (91.79%) than the deterministic approach, since the existence of the DMA adds complexity (or implicit access), which constitutes an obstacle for the hybrid methodology. The test development time is increased due to the description of the DMA access mechanism in the constraints library, but in any case is much less than for the deterministic approach (see Table 15.14).

Table 15.15 summarizes the fault coverage of the three different approaches. The last row reports the total fault coverage considering all the communications peripheral cores. The proposed methodology achieves a fault coverage which is nearly as high as in the deterministic approach with limited engineering effort, as in the automatic approach. Thus, the experimental results demonstrate the effectiveness of the proposed hybrid approach for this part of the SoC design.

Table 15.13. Fault simulation results for the Ethernet core.

Component	Faults	Fault Coverage (%)
Wishbone I/F	11,858	88.26
Registers	8,120	95.10
Tx Engine	5,002	75.31
Tx FIFO	20,465	98.81
Rx Engine	3,564	72.00
Rx FIFO	20,465	99.12
Management	1,868	97.22
Control	3,884	74.12
Status	1,306	38.31
Glue Logic	1,042	94.13
Total	77,574	91.79

Table 15.14. Test program development time of the different approaches.

	Test Development Time (approx. days)		
Core	Automatic	Deterministic	Proposed
UART	1	30	1.5
HDLC	0.5	15	0.8
Ethernet	2	45	6

Table 15.15. Fault coverage of the different approaches.

Core	Fault Coverage (%)	Automatic	Deterministic	Proposed
UART	86.35	93.92	93.13	
HDLC	68.65	97.34	97.43	
Ethernet	86.57	93.39	91.79	
Total	79.89	94.92	94.05	

Table 15.16. Test program statistics of the different approaches.

	Automatic		Deterministic		Proposed	
Core	Code (words)	Time (msec)	Code (words)	Time (msec)	Code (words)	Time (msec)
UART	5,969	2,91	3,429	22,55	6,330	6.00
HDLC	1,816	1,95	623	3,21	1,823	2,04
Ethernet	15,190	9.38	2,358	59.30	12,129	29.21

Table 15.16 details the statistics of the test programs generated by each studied approach. The automatic approach compared to the deterministic one generates longer test programs for each core, but the test application time is much lower. As a consequence, the hybrid approach also generates longer test programs, but with

a shorter test application time. This point is crucial, since the test application time is normally very important in manufacturing testing. In the case of the Ethernet core the difference between the code size of the automatic/hybrid approach and the code size of the deterministic approach is larger, since the presence of the DMA controller adds more constraints to the generation process and the tool is not able to produce well optimized code.

The reader should note that when considering the total time required by the Automatic and Proposed approaches (as reported in Table 15.14), the time required to setup the environment (mainly to create the constraints library) is negligible with respect to the computational time to run the tool and gather the feedback from the simulator.

15.5. Conclusions

This chapter proposed several possible solutions to some validation and test issues caused by the new design trend. The chapter has focused on test program generation methodologies for critical part of modern processors, with the objective to define effective methodologies able to reduce as much as possible their cost and their application time while still maintaing their ability to validate or test. Moreover, the methodologies proposed have been extensively evaluated on real complex designs in order to asses their effectiveness and their scalability. The results gathered on state of the art microprocessors demonstrate the wide applicability and scalability of the proposed methodologies.

Section 15.3 presented an automatic methodology able to generate effective test programs for the validation of complex modules in multithreaded processors. The methodology requires a limited and low-skilled manual intervention (basically to describe the assembly syntax) and can easily focus on specific critical targets (thus reducing the required computational effort), while still maintaining the ability of generating test programs for the whole processor. Despite the complexity of generating effective multithreaded test programs, the results gathered on a real complex designs, show the effectiveness and the scalability of the proposed method, in terms of both achieved coverage figures, and ability to cover very specific corner cases. The method is also shown to significantly outperform a random approach in terms of achieved results when a comparable number of test programs is evaluated.

Section 15.4 has described an effective methodology for the generation of test programs for different types of communication peripheral cores of a SoC. The proposed methodology successfully combines two different approaches, one fully automated approach and one semi-automated deterministic approach. The experimental results on different types of communication peripheral cores (UART, HDLC and Ethernet controllers) show the effectiveness of the proposed methodology.

References

1. (2010). http://www.opensparc.net/, .
2. Adir, A., Almog, E., Fournier, L., Marcus, E., Rimon, M., Vinov, M. and Ziv, A. (2004). Genesys-pro: innovations in test program generation for functional processor verification, *Design & Test of Computers* **21**, 2, pp. 84–93.
3. Apostolakis, A., Psarakis, M., Gizopoulos, D. and Paschalis, A. (2007). Functional processor-based testing of communication peripherals in systems-on-chip, *IEEE Trans. on Very Large Scale Integration (VLSI) Systems* **15**, 8, pp. 971–975.
4. Bayraktaroglu, I., Hunt, J. and Watkins, D. (2006). Cache resident functional microprocessor testing: Avoiding high speed io issues, in *Proc. IEEE Intl. Test. Conf.*, pp. 112–121.
5. Behm, M., Ludden, J., Y.Lichtenstein, Rimon, M. and Vinov, M. (2004). Industrial experience with test generation languages gar processor verification, in *Design Automation Conference*, pp. 36–40.
6. Benjamin, M., Geist, D., A.Hartman, Wolfsthal, Y., Mas, G. and Smeets, R. (1999). A study in coverage-driven test generation, in *Proceedings of 36th Design Automation Conference, 1999*, pp. 970–975.
7. Bjesse, P. (2005). What is formal verification, in *ACM SIGDA Newsletter*, p. Issue 24.
8. Bolzani, L., Sanchez, E., Schillaci, M., Reorda, M. S. and Squillero, G. (2007). An automated methodology for cogeneration of test blocks for peripheral cores, in *Proc. IEEE Intl. On-Line Testing Symposium*, pp. 265–270.
9. Campenhout, D. V., Mudge, T. and Hayes, J. (1999). High-level test generation for design verification of pipelined microprocessors, in *Proc. 36th Design Automation Conf. (DAC 99)* (ACM Press), pp. 185–188.
10. Carter, B. H. and Hemmady, S. (2005). *Metric-Driven Design Verification* (Springer US).
11. Corno, F., Sanchez, E., Reorda, M. S. and Squillero, G. (2004). Automatic test program generation: A case study, *Design & Test of Computers, IEEE* **21**, 2, pp. 102–109.
12. Corno, F., Sanchez, E. and Squillero, G. (2005). Evolving assembly programs: how games help microprocessor validation, *Evolutionary Computation, IEEE Transactions on* **9**, pp. 695–706.
13. Dhiraj, K. P. and Ian, G. H. (2009). *Practical Design Verification* (Cambridge University Press).
14. Emek, R., Jaeger, I., Naveh, Y., Bergman, G., Aloni, G., Katz, Y., Farkash, M. and ans A. Goldin, I. D. (2002). X-gen: A random test-case generator for systems and socs, in *High-Level Design Validation and Test Workshop*, pp. 145–150.
15. Goodenough, J. B. and Gerhart, S. L. (1977). Toward a theory of testing: Data selection criteria, in *Current trends in programming methodology*, Vol. 2, pp. 44–79.
16. Guzey, O. and Wang, L. (2007). Coverage-directed test generation through automatic constraint extraction, in *High Level Design Validation and Test Workshop*, pp. 151–158.
17. Harmani, N. (2001). Verifying a simple pipelined microprocessor using maude, in *Lecture Notes in Computer Science 2267* (Springer-Verlag), pp. 128–142.
18. Hetherigton, G., Fryars, T., Tamarapalli, M., Kassab, M., Hassan, A. and Rajski, J. (1999). Logic bist for large industrial designs: Real issues and case studies, in *Proc. IEEE Intl. Test Conf.*, pp. 358–367.
19. Huang, J., Iyer, M. and Cheng, K. (2001). A self-test methodology for ip cores in bus-based programmable socs, in *Proc. IEEE VLSI Test Symp.*, pp. 198–203.
20. Hwang, S. and Abraham, J. (2001). Reuse of addressable system bus for soc testing, in *Proc. IEEE Intl. ASIC/SoC Conf.*, pp. 215–219.

21. IEEE (2005). *IEEE 1500 Standard for Embedded Core Test*, URL `http://www.ieee.org`.
22. Jayaraman, K., Vedula, V. and Abraham, J. (2002). Native mode functional self-test generation for systems-on-chip, in *Proc. Intl. Symp. on Quality Electronic Design*, pp. 280–285.
23. Kranitis, N., Paschalis, A., Gizopoulos, D. and Xenoulis, G. (2005). Software-based self-testing of embedded processors, *IEEE Trans. on Computers* **54**, 4, pp. 461–475.
24. Lai, W. and Cheng, K. (2001). Instruction-level dft for testing processor and ip cores in system-on-a-chip, in *Proc. IEEE/ACM Design Automation Conf.*, pp. 59–64.
25. Lindsay, W., Sanchez, E., Reorda, M. S. and Squillero, G. (2004). Automatic test programs generation driven by internal performance counters, in *MTV'04: 5th International Workshop on Microprocessor Test and Verification*, pp. 8–13.
26. Opencores (2010). Projects openrisc, uart, hdlc, ethernet and wishbone models, URL `http://www.opencores.org/projects/`.
27. Parvathala, P., Maneparambil, K. and W., L. (2002). Frits a microprocessor functional bist method, in *Proc. IEEE Intl. Test Conf.*, pp. 590–598.
28. Sanchez, E., Reorda, M. S. and Squillero, G. (2005). *Test Program Generation From High-level Microprocessor Descriptions*, chap. 179 (Springer), ISBN 1-85233-899-7, pp. 83–106.
29. Sun (2010a). Opensparc t1 microarchitecture specification, URL `http://opensparc-t1.sunsource.net/specs/OpenSPARCT1_Micro_Arch.pdf/`.
30. Sun (2010b). Opensparc t1 project, URL `http://www.opensparc.net/opensparc-t1/`.
31. Tasiran, S. and Keutzer, K. (2001). Coverage metrics for functional validation of hardware designs, *Design & Test of Computers, IEEE* **18**, 4, pp. 36–45.
32. Thatte, S. and Abraham, J. (1980). Test generation for microprocessors, *IEEE Transactions on Computers* **C-29**, pp. 429–441.
33. Velev, M. and Bryant, R. (2000). Formal verification of superscalar microprocessors with multicycle functional units, exception, and branch prediction, in *Proc. 37th Design Automation Conf. (DAC 00)* (ACM Press), pp. 112–117.
34. Yuan, J., Pixley, C., Aziz, A. and Albin, K. (2003). A framework for constrained functional verification, in *International Conference on Computer Aided Design*, pp. 142–145.

Chapter 16

DISCOVERY OF COMMUNICATIONS PATTERNS BY THE USE OF INTELLIGENT REASONING

J. Fulcher, M. Zhang, Q. Bai and F. Ren
Intelligent Systems Research Centre,
University of Wollongong NSW 2522 Australia
{john,minjie,quan,fr510}@uow.edu.au

An agent-based model of communications traffic generated within a social network is described, which caters for various knowledge discovery techniques to be used in order to extract 'interesting' temporal patterns contained within anomalous data records. Attendant Java-based software — `NetShow` — is presented which enables analysis and display of static network configuration, links between network nodes, and correlations between the traffic passing through nominated nodes. In addition, display of network dynamics is facilitated via the incorporation of swarm techniques. Related issues of 'similarity', 'familiarity' and 'contact lists' are discussed within the context of Social Network Analysis and Link Mining. Finally, several suggestions are made for extending the work reported earlier in the chapter.

16.1. Data Ming and Knowledge Discovery in Databases

Data is accumulating at TerraByte (PetaByte even) rates on a weekly, daily, or even hourly basis, depending on the application domain of interest — in any case, far beyond our ability to keep pace with extracting meaningful information from the relevant data bases and/or data warehouses in question. Given this phenomenon (which, by the way, shows no signs of abating, but rather will continue to increase in accordance with Moore's Law), the huge up swell of interest in Data Mining (DM) during the past two decades should come as no surprise. DM combines techniques developed separately within the fields of DataBases, Machine Learning (ML) and Statistics, and especially those capable of handling large data sets.[1,2]

The fundamental aim with DM is to discover 'patterns of interest' hidden within the data. Just what constitutes 'interesting' is nevertheles open to interpretation. More specifically, various stakeholders can hold quite different views in this regard — for instance a Data Miner may not have the necessary background/experience/insight to interpret such patterns. Furthermore, the 'customer' will often not themselves know what specific patterns to look for — in which case the DM exercise is essentially one of exploration.

There are a number of critical considerations in DM, most fundamentally the nature of 'information' itself. First and foremost, data and information are *not* synonymous. According to Fulcher, information can be defined as 'data + meaning' (and by extension, knowledge as 'information + understanding'; wisdom as 'knowledge + experience + insight', and so forth).[3] In other words, data of itself is of little value; the 1s and 0s representing the data stored within the data bases/warehouses of interest need to be interpreted *in context* before they will be of value to stakeholder(s). An analogy can be drawn here with internet search engines. The common fallacy underlying the latter can be paraphrased as follows: "there is so much information out there on the net, all one has to do is go searching". Not so — in fact the World Wide Web (WWW) can be characterized as exhibiting a very small signal-to-noise ratio — SNR (to use signal processing terminology). The adage "not to believe everything one reads" in the print media needs to be applied even more stringently when viewing outputs produced by internet search engines. Indeed, one needs to exercise considerable caution in assigning credibility and authority as a matter of course (i.e. by default), devoid of any requisite critical evaluation.

A second fundamental issue with DM is data pre-processing, for indeed DM is only one step (albeit it the most central one) of the broader Knowledge Discovery in Databases (KDD) process, as outlined in Table 16.1.

Prior to DM, we need to undertake data pre-processing, else we run the risk of encountering 'garbage in, garbage out'. Ideally we would like to be working with a 'knowledge rich' data set, but more typically this will not be the case; there will be missing, redundant, inaccurate, inconsistent and/or noisy data in practice. Error bounds checking, noise filtering, data normalization, and format conversion are some typical data 'cleaning' steps that need to be undertaken prior to DM proper.

Another consideration is the use of simple data visualization tools prior to invoking DM techniques in earnest. For example, data clustering can often become immediately apparent by way of 2D(3D) visualization; likewise scatter plots can reveal dependencies within the data set.

Turning now to the central step of the KDD process, we have at our disposal several DM techniques from which we can choose — Association Rules (AR), Decision Trees (DT) and Artificial Neural Networks (ANN) being three of the more popular. The former originated from the field of Databases, the next from the Machine Learning (ML) and Statistics fields, and the latter from the connectionist branch

Table 16.1. KDD Process.

1	Data selection (extraction)
2	Data pre-processing (cleaning, enrichment, coding, etc.)
3	Data Mining proper
4	Interpretation of discovered patterns
5	Presentation/reporting of results

of Artificial Intelligence — AI (and which in more recent times, constitutes one of the three central pillars of Computational Intelligence — CI[4]).

Up to now we have assumed *static* patterns within the data set of interest. Quite a lot of data captured in the real world is however inherently temporal in nature. Not surprisingly, there have been quite a number of techniques developed that facilitate the discovery of *dynamic* (that is, time-varying) patterns. In statistics, the Auto-Regressive Moving Average (ARMA) and Box-Jenkins models have proved popular, while common CI approaches include software agents,[5–7] Time-Delay Neural Networks (TDNNs),[8] Higher-Order Neural Networks (HONNs),[9] Support Vector Machines (SVMs),[10,11] Evolutionary Algorithms (EAs),[12–18] and Genetic Programming (GP).[19–22]

16.1.1. *Communications Data*

The first data set of interest in the present study is communications traffic — a total of 50,000 records — over a 2-year period. The data format is as shown in Table 16.2.

Data always exists for the first two columns in the record, but not necessarily with the third. Uni-directional data is the norm, although some bi-directional data is also contained within this data set. The anonymous data reflects communications activity over a single (tightly interconnected) network during the specified time period. Note that this does not necessarily correspond to people *per se*, but communications services. This network comprises a tightly-connected 'inner circle', with branches to a peripheral group, although some communications do emanate from outside this inner core. In other words, the data describes a social network of sorts.[23] Accordingly, in the next section we turn our attention to Social Network Analysis (SNA).

A second, more extensive (258,052 record), 2-year communications traffic data set was also made available during the course of this study (Table 16.3); it differs from the first data set in that the type of communication is also recorded.

16.2. Social Network Analysis

At its most fundamental level, Social Network Analysis (SNA) models 'entities' (such as people), and the relationships between them, as nodes and links

Table 16.2. Data Format #1.

ID	Date	Time & duration

Table 16.3. Data Format #2.

(sanitized) date & time	'from' service	'to' service

(respectively) on a graphical model.[24–26a] Such graphs can be likened to the entity-relationship (ER) diagrams commonly encountered in databases.[27–29]

While SNA emphasizes network structure, it also encompasses concepts such as 'between-ness', 'closeness', 'centrality' (placement of the most important/significant entities in the middle of graphs), 'clustering coefficient', 'cohesion', 'path length', 'radiality', 'reach', and the like.

Commonly encountered SNA software tools[30b] include `UCINet`,[c] `Pajek`,[31d] and the *network* software[e] developed for the `R` statistical analysis package.

SNA has been successfully applied to criminal investigations.[32,33]

16.3. Intelligent Reasoning Methods

Numerous approaches have been used for temporal data mining, including graph-based methods,[34,35] fractals,[36] Hidden Markov Models (HMMs),[37,38] Bayesian (Belief) networks,[37,39–41] transactional networks,[42] matrix decompositions,[43] and even Semantic Web-related techniques (e.g. RDF, RDFS and OWL).[44] In the present study we focus on the following techniques: (i) link mining, (ii) software agents, (iii) swarms, and (iv) Artificial Neural Networks.

16.3.1. *Link Mining*

Link (or relational) Mining is a specialized form of DM better suited to the mining of highly structured, heterogeneous data sets in which the objects (entities) are linked in some manner.[45] In graphical terms, objects (entities) correspond to nodes, whereas links (relationships) are represented by edges. 'Similarity' depends not only on comparing attributes but also on links (relationships) between objects (entities) — issues of familiarity and similarity are discussed further in Sects. 16.5.4 and 16.9, respectively. Link Mining has been driven by developments in fields as diverse as link analysis, graph mining, relational learning, inductive logic programming, but most especially hypertext and web mining. Link analysis is used in law enforcement, counter-intelligence, fraud detection and the like.[46] Fawcett and Provost investigated cellular (mobile) phone fraud.[47,48] It has the capability of *automatically* constructing link diagrams from large databases. A typical link analysis tool is `NetMap`.[49]

Link Mining has been applied to web page rankings produced by internet search engines, epidemiology (the spread of communicable diseases), and bibliography searching.

[a] http://www.orgnet.com/sna.html
[b] http://www.insna.org/INSNA/soft_inf.html
[c] http://www.analytictech.com/ucinet/ucinet.htm
[d] http://vlado.fmf.uni-lj.si/sub/networks/pajek/
[e] http://cran.r-project.org/src/contrtib/Descriptions/network.html

Typical Link Mining tasks include data classification (of entities over time), cluster analysis, identification of the number of links and their type, determination of link strengths, and the tracking of activities 'of interest'. However as Senator points out, no comprehensive framework yet exists for combining the link mining tasks commonly required for real-world applications; it remains an *ad hoc* activity.[50]

16.3.2. *Software Agents*

Software agents, as their name suggests, perform actions on behalf of their human 'owners'. They owe their origins to work in distributed AI during the 1980s.[51,52] Agents are autonomous, possess knowledge about their environment such that they are able to interact 'intelligently' with it, and are able to learn and adapt their behaviour over time.[5–7] Many different types of software agent have been developed over the years, a popular one being the so-called BDI (beliefs-desires-intentions) agent. A critical consideration with agents is the granting of permission to execute on remote hosts; oftentimes firewalls and other system security measures will prevent this by default, unless *explicit* permission is sought (and granted).

More specifically, software agents are characterized by the following:

(1) *persistence* — they run continuously and decide *for themselves* when they should perform various activities;
(2) *autonomy* — they are capable of task selection, prioritization, goal-directed behaviour, and decision making *without* human intervention;
(3) *social ability* — they are capable of *interacting* with other agents and/or humans in order to satisfy their design objectives;
(4) *reactivity* — they are able to *perceive* their environment, and *respond* in a timely fashion to changes that occur therein, in order to satisfy their design objectives.

Since the environment of interest — communications — is inherently complex and dynamic in nature, then agents are particularly well suited to the present study.

16.3.3. *Swarms*

Swarms were inspired initially by the flocking/herding/collective behaviour of social insects. The resulting 'collective intelligence' of the swarm (flock, herd) is able to achieve performance way beyond that of its constituent members, by virtue of the interactions between the (*un*-intelligent) individuals, for example through the laying down 'pheromone' trails.[53,54] Unlike evolutionary approaches,[12–14] it is the behaviour of *present*-generation members that determines swarm behaviour, not on the inherited characteristics ('genetic' information) passed on to successive population generations. Nevertheless, the two approaches are inherently iterative, data-driven and bottom-up in nature. Several (open source) packages are in widespread use, one of the more popular being the (open source, java-based) Recursive Porous Agent Simulation Toolkit — `RePast`.[f]

Of particular interest to the present project is the fact that swarms have been widely applied in agent-based simulations (ABSs) in the past,[55,56] and hold potential for tracking both individual and group life cycles. In other words, they are essentially dynamic, and hence well suited to the modelling/display of how social networks change over time (in contrast, say, to the *static* network displays produced by `JUNG`); this capability is explored further in Sect. 16.7.2.

16.3.4. *Artificial Neural Networks*

Like swarms, Artificial Neural Networks (ANNs) take their inspiration from Nature. They are simplified models of biological neural networks ('brains'), comprising 'neurons' (nodes) along with interconnecting weights, which are trained using representative input-output exemplars.[57–59] Rather than being a model-driven, algorithmic approach, they constitute a bottom-up, data-driven one.[3] As with Data Mining (Sect. 16.1), pre-processing is critical for good performance — even more so, since ANNs typically require *many* training iterations to arrive at an acceptable solution (corresponding to a global minimum in the energy/solution landscape). The input patterns are associated with their corresponding output pairs within the network weights (but in a holistic, rather than in a one-to-one sense). Despite their long training times, ANNs are able to respond to new input data (patterns) immediately, indeed they are well known to excel at pattern recognition, and this is the reason for their utilization in the *Knowledge Discovery* Module of our MAS-based System model.

16.4. Multi-Agent System (MAS) Network Model

Figure 16.1 shows our overall system block diagram. Although we chose a Multi-Agent framework, our design is not limited to MASs — as indicated in the *Knowledge Discovery* Module — rather a range of techniques can be accommodated under this umbrella, including (but not restricted to):

- Link mining,
- Correlation matrix,
- Markov Chains (for prediction purposes),
- Swarms,
- Artificial Neural Networks (as outlined in Sect. 16.3.4),
- Evolutionary Algorithms,
- Machine Learning techniques.

[f] http://repast.sourceforge.net/

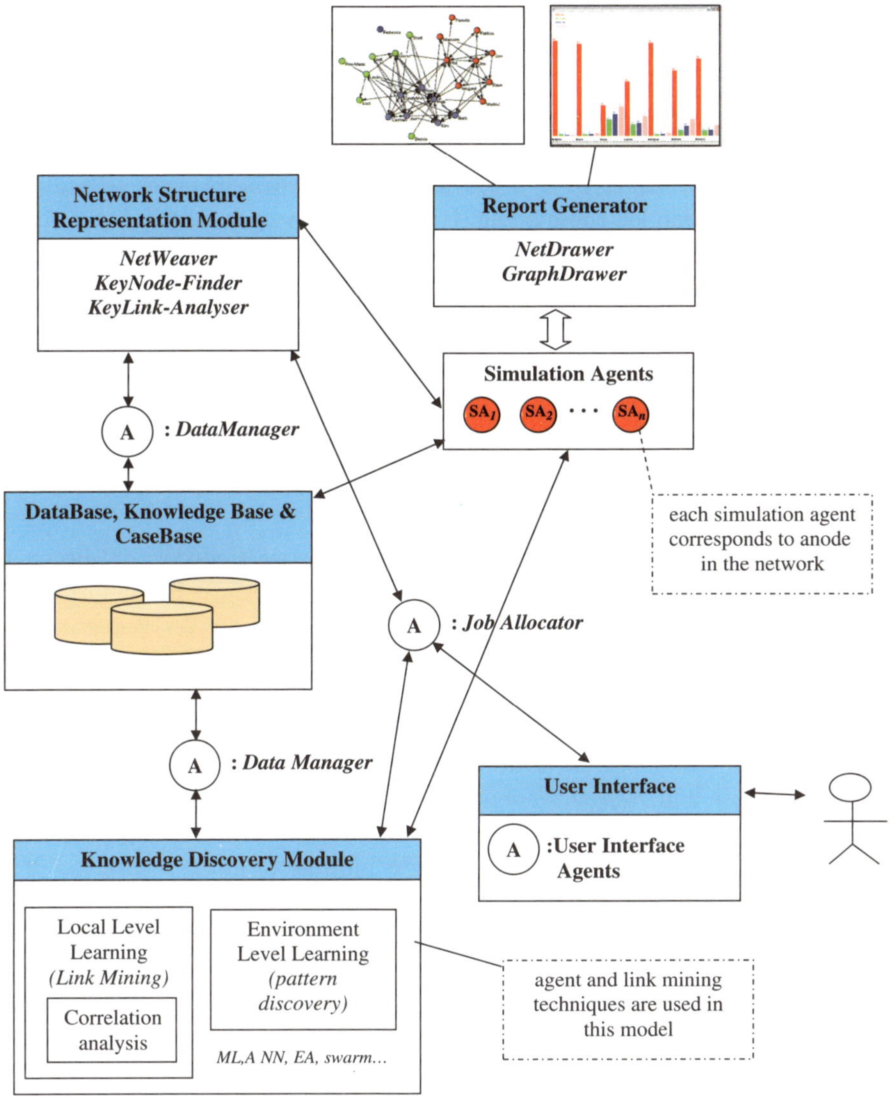

Fig. 16.1. System Block Diagram.

For instance, there may well be benefit in combining agent-based learning (a local technique) with Data Mining (a global technique) — more specifically, in order to discover more about relationships between nodes (accounts, people and phone owners), changing trends, and other previously undiscovered information/knowledge.

Agent learning could proceed on two levels:

(1) system (*based on time*), and
(2) local (*based on traffic*).

The first would focus on how both traffic and links between nodes change over time, while the second would concentrate on correlations between nodes. A combination of agents and link mining (or other data mining method) could prove useful with the former, while a combination of agents and pattern matching (or other Machine Learning method) may prove useful with the latter.

We invoked three different types of agent facilitators in order to realize our social network MAS model, these being (i) UserInterface, (ii) JobAllocator, and (iii) DataManager. The *UserInterface* agent receives and pre-processes user requests, prior to forwarding these on to the *JobAllocator*. *JobAlloactor* is a middle-level agent that allocates user requests to the most appropriate model. The *DataManager* agent manages the data contained within the data/knowledge/case bases (*Network Structure Representation, Correlation Analysis* and *Report Generator* modules), and supplies data to the upper-level module(s) appropriate for specific user requests.

The Network Structure Representation module incorporates *NetWeaver, KeyNodeFinder* and *KeyLinkAnalyser* agents:

- *NetWeaver* reads raw data (.txt files) and performs any necessary pre-processing. More specifically, it converts data files into appropriate SNA format (e.g. .net(work) files in Pajek format, as used as inputs in `Jung`). *NetWeaver* is able to receive user requests from the *JobAllocator* and to provide the required .net files to *NetDrawer* (in the *Report Generator* module). *NetWeaver* also provides data to the *DataBase* module for historical analysis.
- *KeyNodeFinder* identifies 'important' network nodes and sends the KeyNode list to the *DataBase* and *Report Generation* modules; currently 'important' is characterized by (i) heavy call traffic, (ii) the node in question being a 'key' node ('hub'), (iii) user-selected, or (iv) the largest number of links.
- *KeyLinkAnalyser* identifies and analyzes 'important' links between nodes, and provides results to the *DataBase* and *Report Generation* modules.

The *Correlation Analysis* module comprises both *HighCorrelationFinder* and *PairCorrelationAnalyzer* agents. The former identifies pairs with high correlation coefficients and reports results to the *DataBase* and *Report Generator* modules, whereas the latter identifies correlation coefficients between user-specified node pairs.

In the *Report Generator* module, *NetDrawer* receives data from *NetWeaver* and displays the corresponding network graph; *GraphWeaver* produces correlation graphs based on data received from the *Correlation Analysis* module(s).

Initially, `AgentBuilder`[g] was used in the development of the above modules (and agents), but subsequently Java proved less restrictive. A general-purpose agent was used at first, but later specialized agents were employed for each respective module. Initially we created 'agent' classes as 'beliefs' within `AgentBuilder`, in which

---

[g]http://www.agentbuilder.com/

the user selects 'ontology' — i.e. via `NetShow` — which implies classes/methods (in other words, instances of objects); subsequently we utilized Java classes.

16.5. `NetShow` Software

During the course of this study we developed software for the automatic extraction of network descriptors, followed by display of the social network under study, using the system model of Figure 16.1. If we are able to model the communications network with sufficient accuracy, then this should facilitate the 'reverse-engineering' of data leading up to some external event, in order to predict 'similar' occurrences in the future, but note that this pre-supposes the ready availability of representative case studies (we will return to consider this issue in Sec. 16.8). Our longer-term goal is therefore to develop an agent-based framework for first modeling then ultimately predicting the behaviour of the social network of interest. To this end, call frequency is deemed to be more significant than call duration. Accordingly, we display call frequencies (for caller-callee pairs) as the interconnection/link strengths (magnitudes) between nodes in the SNA graph.

`NetShow` automatically extracts the following dynamic descriptors: network architecture, links, traffic, key node analysis and correlation coefficients. Figure 16.2 shows a typical display produced by `NetShow`.

It becomes immediately apparent that there are two 'central' nodes in this particular network, these being ON19074N6A and OA2N05N855; likewise the highest call traffic (link strengths) are 76.0 (0746A287N7 → 0746A9619N), 72.0 (0A2557458N → 0A2N796N19), 56.0 (0A2N197052 → 0N19074N6A), and so on. Network pruning (say by removing all links less than 5.0 in strength) renders this even more apparent.

Moreover, since `NetShow` was developed using `UCINet`, it is capable of performing classical 'cut point' SNA. For example, the following 12 'k-cores'[h] were automatically extracted from Figure 16.3, where the link strengths have been removed for improved readability (but are still available within the program): 0A2N05N855, 0746505N78, 07465N5N19, 0A2N8NN702, 0N19074587, 0A2N877N7N, 0A2N858-N6A, 0A2N197052, 0A2N19A052, 0A2N5357N7, 0N19074N6A, 0A2N796N19, 0A2N750687, and 07465011A6.

The Java Universal Network Graphical package (`JUNG`)[i] was also used to facilitate the development of `NetShow`, however the latter is capable of displaying not only static 'snapshots' of the network, but also dynamic displays of network architecture and link strengths, with screen updates taking place every 30–60 seconds.

[h] Given a graph $G = \{V, E\}$ with vertices set V and edges set E, the k-core is computed by pruning all vertices (along with their respective edges) with degree (that is, number of connections to other nodes) less than k.
[i] http://jung.sourceforge.net/

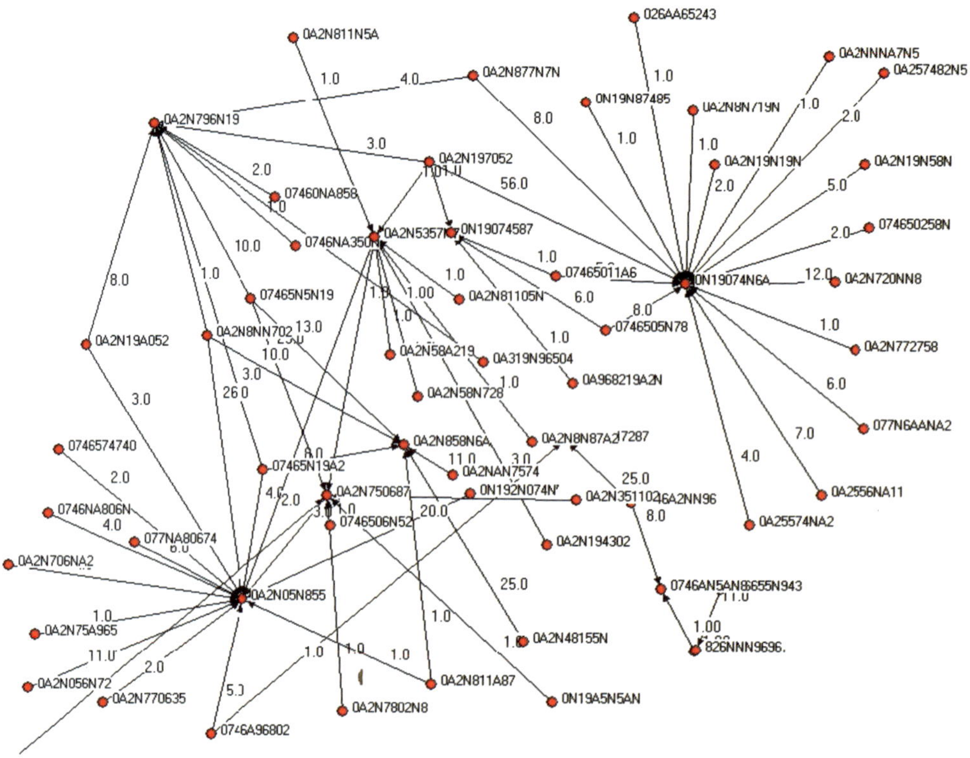

Fig. 16.2. Network Node Diagram (over a 7-month period).

16.5.1. `NetShow 2.0` *Software*

The original `NetShow` program was expanded and refined, with `NetShow 2.0` being completely re-written using more recent releases of both Java (V1.5) and `JUNG` (V2.0), leading to the following improvements:

Figure 16.4 shows how users are able to create images from the current graph within the `NetShow 2.0` GUI, while Figure 16.5 shows the automatic report generator in action (i) through (iv).

(i) it is now linked with a *bona fide* database (`MySQL`) rather than reading data records from a .txt file;
(ii) it can now automatically generate report files, which can include network structure graphs, related network analysis results, as well as a log of all operations performed by the user (refer to the leftmost panel in the ensuing screen dumps);
(iii) it allows users to select nodes and include related information about them in reports;
(iv) visualization functions have been greatly improved, courtesy of new functions incorporated into `JUNG V2.0`.

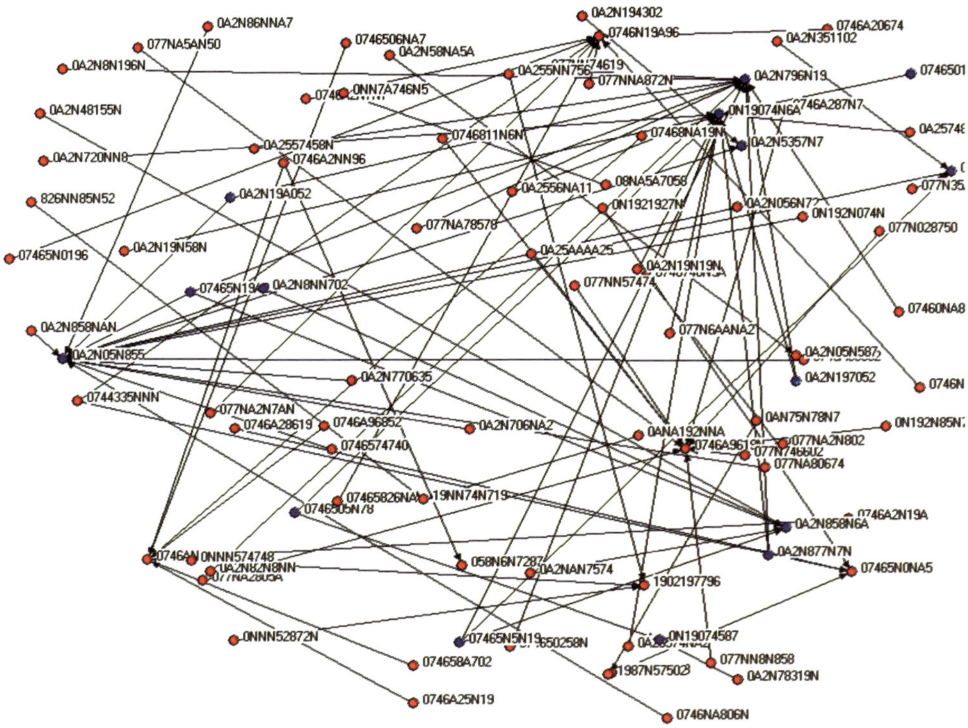

Fig. 16.3. Network Node Diagram (2002; link weights >2.0).

Since the database contains two different types of data (telecommunications traffic), we need to utilize two different kinds of agents. However in order to cater for future revision and/or expansion, both need to be based on an abstract class in which all common attributes and functions are defined. Once generated, an agent needs to possess the following capability:

(1) read data from the database and fill the agent's attributes accordingly,
(2) analyze the data, summarize the current situation and predict possible future scenarios,
(3) communicate with other agents, sharing data and analyses where appropriate (and if permitted), and
(4) visualize and represent analysis results (in various formats/styles, as appropriate).

16.5.2. *Network Visualization*

`JUNG V2.0` supports different presentation/display options, including tree, radial, balloon, label, minimum spanning, FR2 and (preferred) isometric (Figure 16.6).

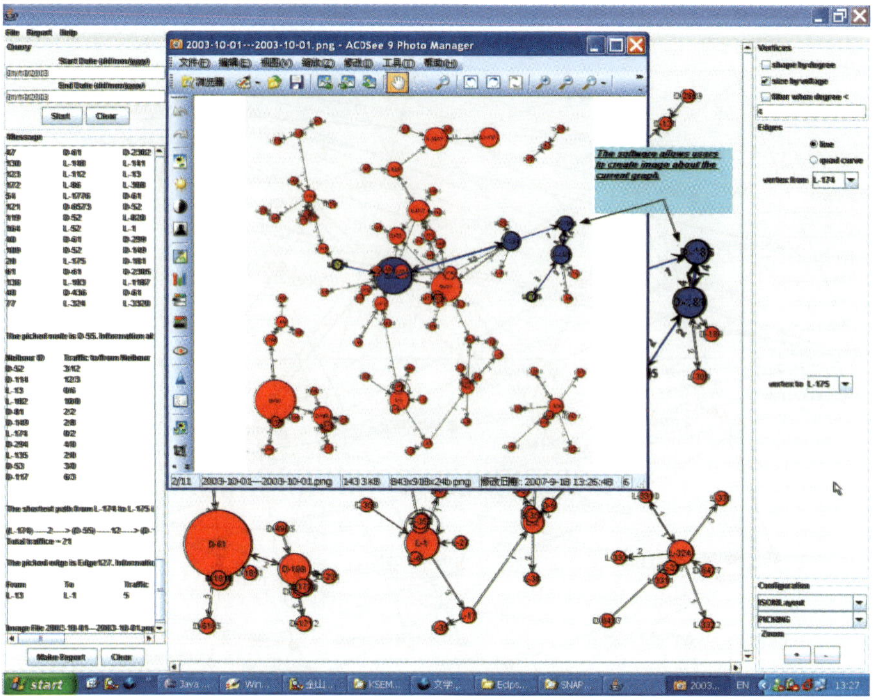

Fig. 16.4. Image Creation.

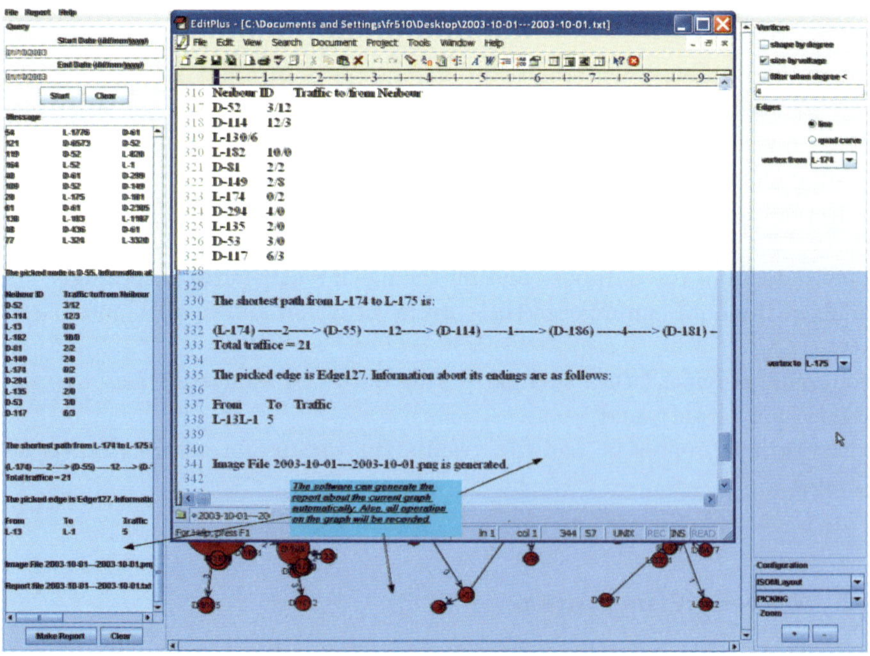

Fig. 16.5. Automatic Report Generation.

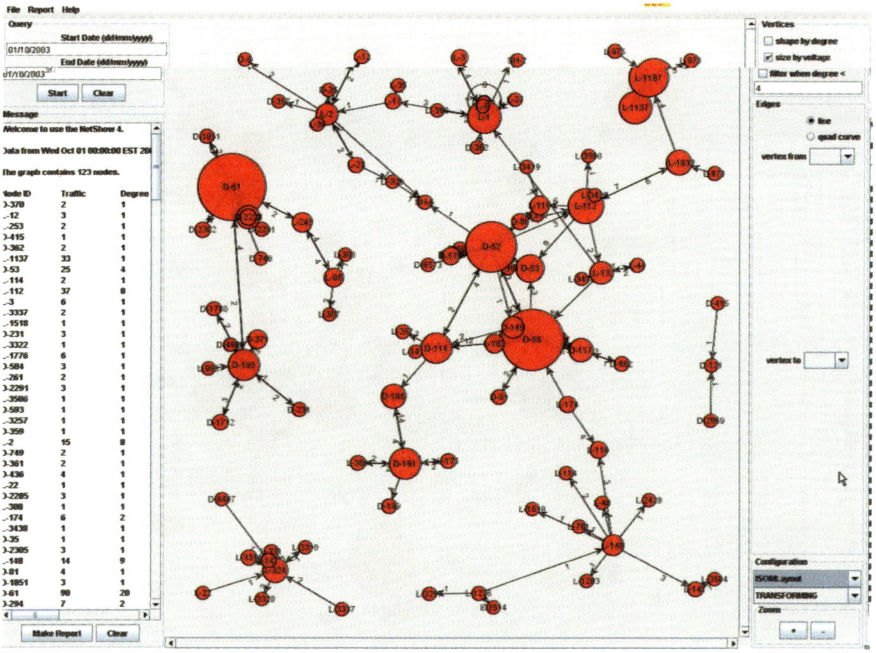

Fig. 16.6. ISOMLayout.

Visualization functions within `NetShow 2.0` have been improved as follows:

- it uses node size and shape to demonstrate both the degree of the node and the traffic passing through it (Figure 16.7):
  (a) nodes with larger node sizes have higher traffic;
  (b) nodes with more shape edges have higher degrees.
- it uses line thickness to represent link weights (Figure 16.8);
- it allows users to select a node, and to highlight that particular node's neighbours (Figure 16.9);
- it can show the 'shortest path' between two user-selected nodes. Furthermore, the shortest path will be highlighted and related path information included in the report (Figure 16.10).

16.5.3. *Pattern Discovery from Contact Lists*

We now present an example of how `NetShow 2.0` can be used for pattern discovery. Contact Lists for individual network nodes can be generated, then pattern analysis performed. It is also possible to use Association Rules (ARs) to analyze contact associations of individual nodes, as follows:

(1) first convert the daily contacts of the node in question into a transaction,
(2) then discover ARs from the node's transactions.

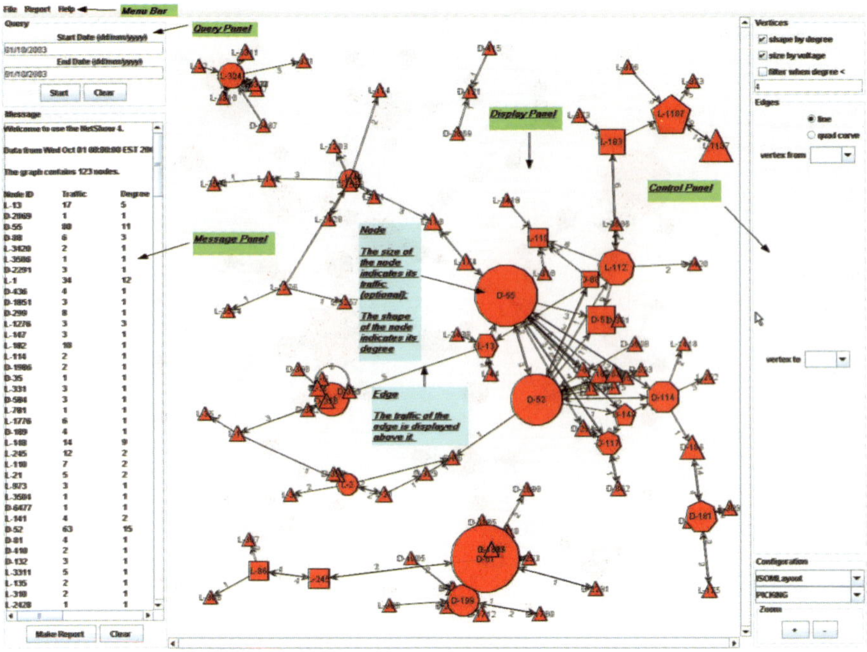

Fig. 16.7. Node Size and Shape.

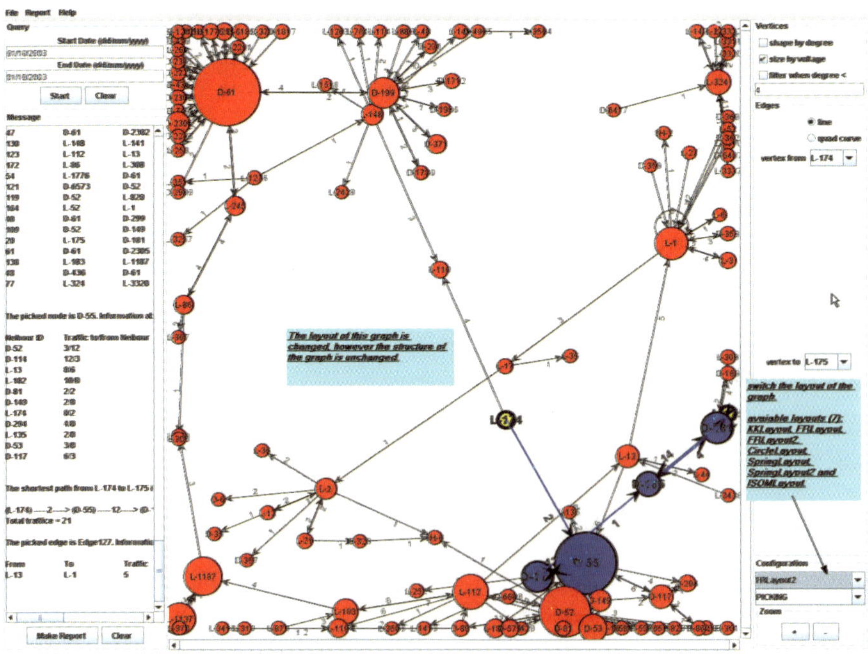

Fig. 16.8. Link Weights.

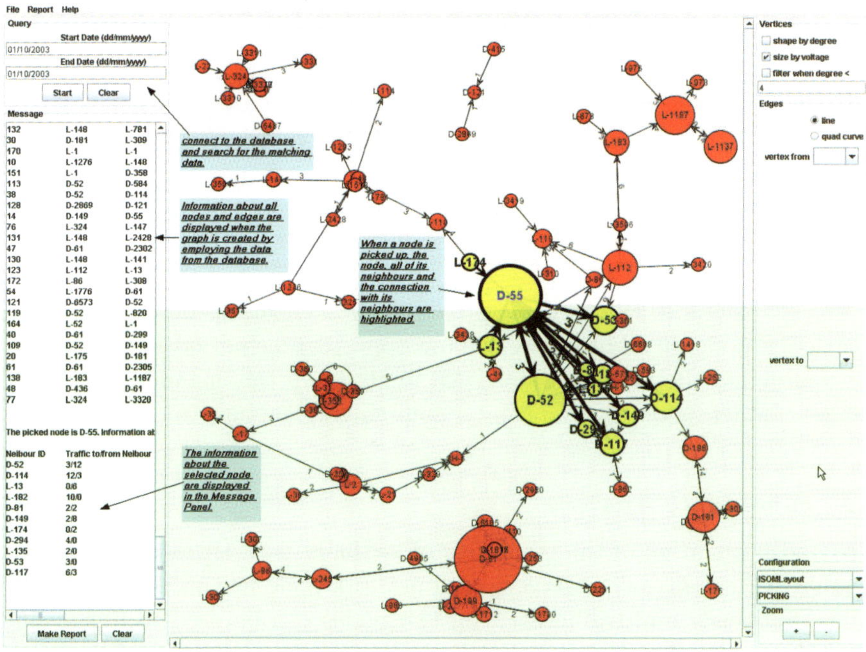

Fig. 16.9. Neighbours.

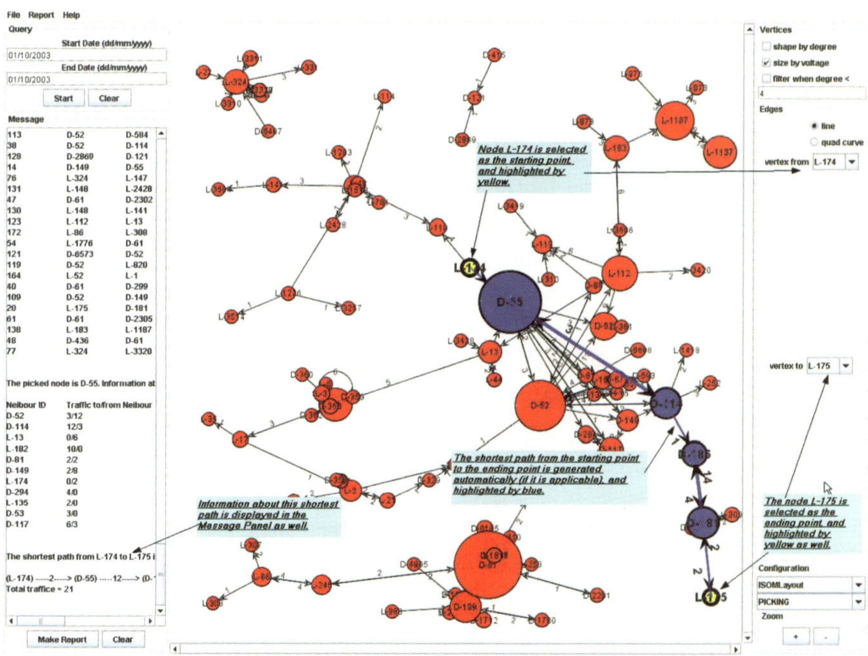

Fig. 16.10. Shortest Path.

Table 16.4. Tracking Familiarity Over Time.

January 2000		February 2000	
Contact	Familiarity	Contact	Familiarity
A	0.57	A	0.64
B	0.23	B	0.20
C	0.15	C	0.16
D	0.05	D	0.10

Using this method, we can expect to discover associations between the contacts of a particular node. In addition, we can incorporate some period selection processes before conducting the aforementioned AR analysis. In this manner, we are able to determine contact associations during a nominated time interval.

16.5.4. *Familiarity Analysis*

As a second illustration of the usefulness of `NetShow 2.0`, let us consider the issue of 'familiarity'. As already observed, each network node has an associated *Contact List*. The frequency and traffic between two nodes can be used to first define the degree-of-familiarity between two nodes, then this information added to the node's *Contact List*. The *Contact List* can then be sorted according to familiarity degree. In this manner, we can track changes in the relationships between nodes and their contacts. Table 16.4 shows an example of tracking familiarity over time.

16.6. Communications Analysis

16.6.1. *First Communications Data Set*

In order to characterize the social network of interest, ideally we would like data to be made available for a reference population, in order to compare and contrast the behaviour(s) of our group with 'standard' communications patterns; to couch this in immunity-based computing (IBC) terminology: *self* versus *non-self*.[60] For instance, a typical person will tend to make regular calls to friends, family, and businesses (and most likely with this order of frequency). They will also typically opt for economical telephone services (be that via a plan or using a pre-paid service). Lastly, they will both expect and deliver quick responses (witness the rapid turnaround times with which teenagers text each other).

Given the above, any investigation of the data will necessarily be looking for anomalous patterns over different time frames — in other words, daily/weekly/seasonal (and similar) variations.

A preliminary static analysis of this first telephone call traffic data set revealed a high proportion of short-duration (<30 second) calls. Next, the ten most frequent callers during a 6-month period were identified. The top four of these are shown in Figure 16.11, with traffic split into bi-directional links where known.

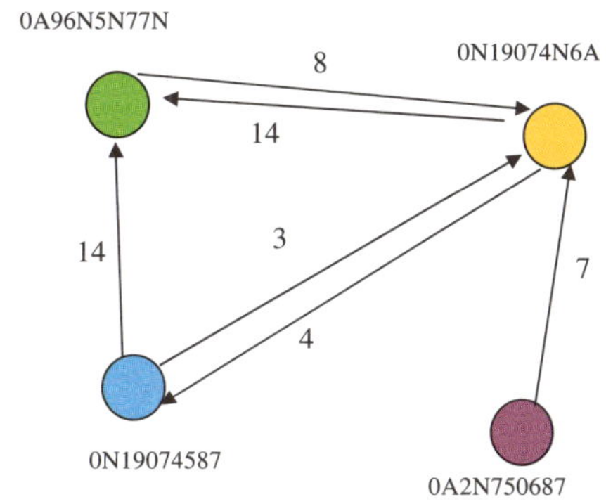

Fig. 16.11. Link Diagram (January–September 2004).

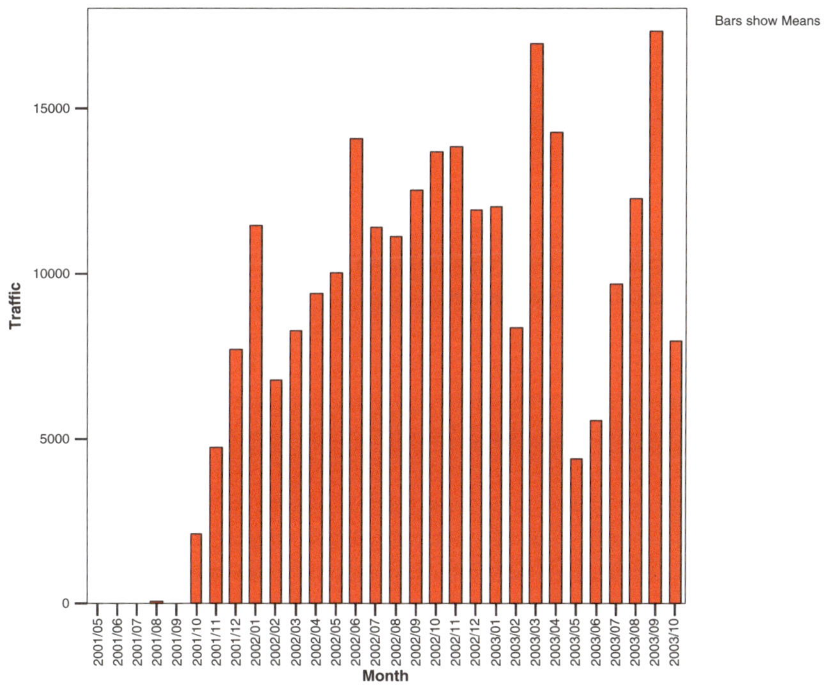

Fig. 16.12. Communications Traffic (24-month period).

Dynamic data analysis is of more interest than static however. In keeping with our own admonition *viz-a-viz* to visualize the data prior to invoking DM techniques in earnest, we simply plot the total traffic over the 2-year period, looking for general patterns of activity/inactivity. Figure 16.12 shows a gradual/steady increase during the first 12–14 months, with peaks at January and June of the first year, as well as March and September of the second year, together with a trough at May of

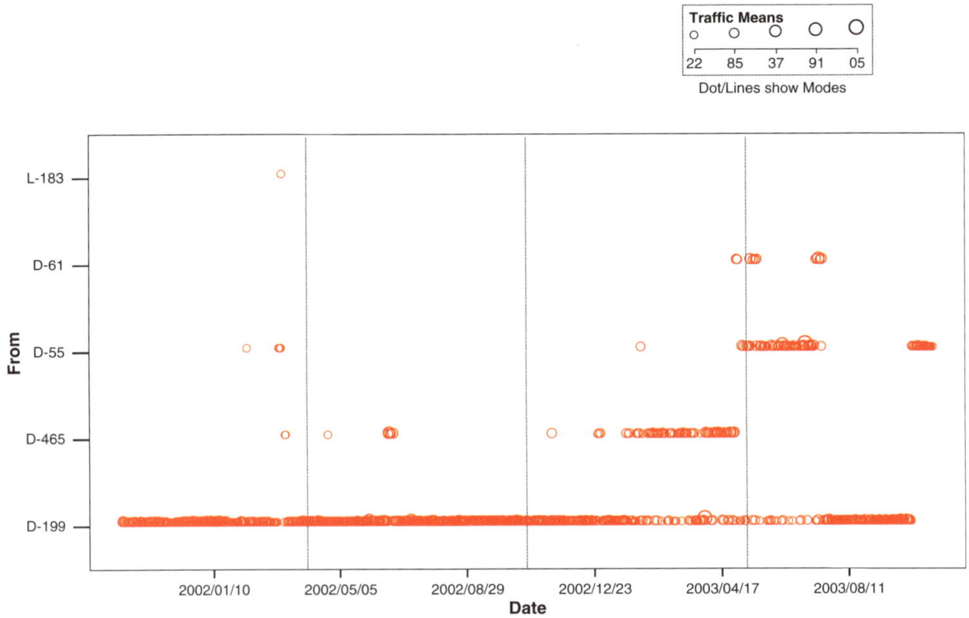

Fig. 16.13. Communications Traffic (13-month period).

the second year. Figure 16.13 shows traffic call patterns for five nominated nodes (services) during a shorter time frame.

Changes in network architecture reflect changes in the composition of the social network over time; likewise changes in link strengths reflect variations in inter-group communications. Secondly, correlations between callers may reveal useful information. For instance, do certain external events trigger call activity in other network nodes?

Four of the ten most frequent callers from March through August 2004 were plotted over the same time period, using an SPSS smoothing function[j] for ease of readability (Figure 16.14). It is obvious that not only are these data correlated, but that a call (trigger) sequence occurred, from node 19NN74N719 → 0586058N19 → 0586058N74 → 0NN7NA11NN.

One particular correlation was examined further, using the in-built SPSS cross-correlation tool. Figure 16.15 shows the correlation between nodes 0586058N19 and 0586058N74 during the period January–September 2004. A correlation coefficient of 40% is evident (Figure 16.16), however beyond ±2–3 days this value becomes somewhat meaningless. This is an unexpectedly strong correlation, and reveals the fact that the signal-to-noise ratio of the data is quite high.

An even more interesting result is that on several occasions strong correlations are observed for nodes with no *direct* connection between them. In other words,

[j]SPSS uses the T4253H smoothing function to create new series values based on a compound data smoother.[61]

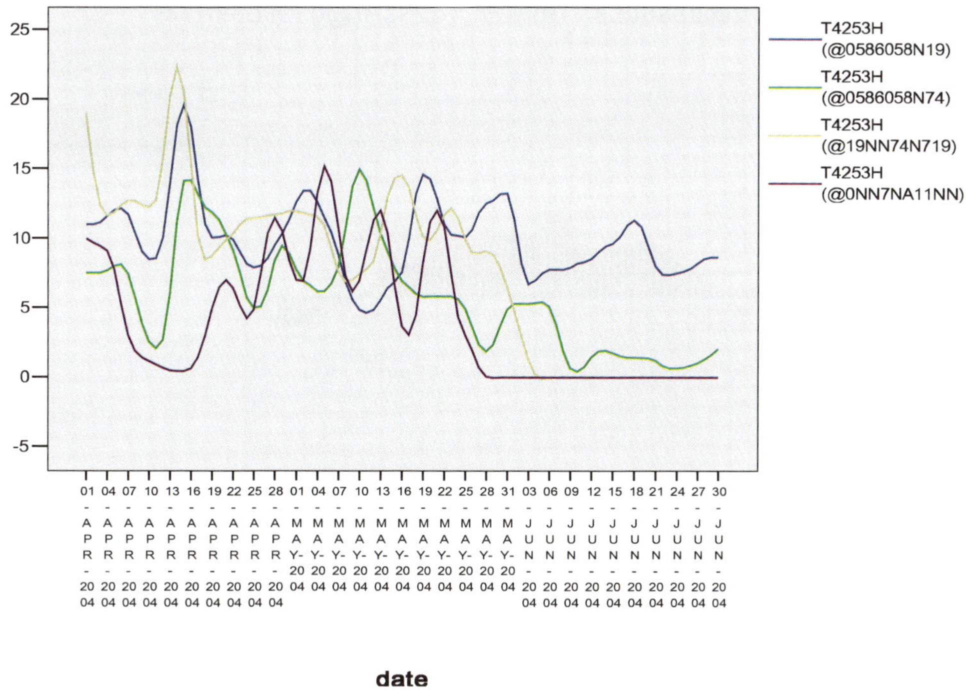

Fig. 16.14. The Top 4 Callers (April–June 2004).

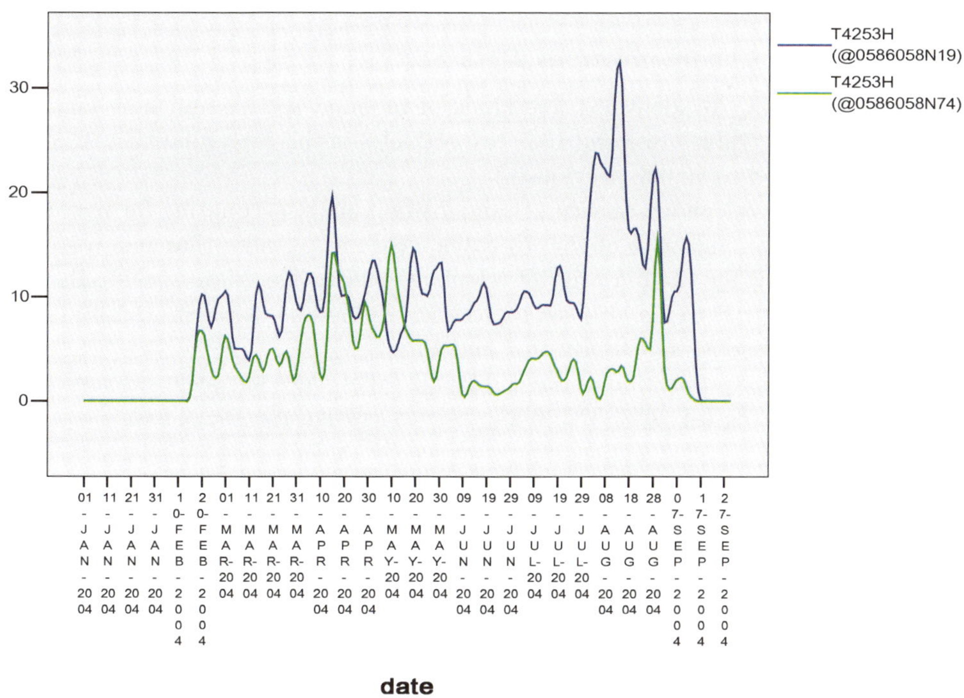

Fig. 16.15. 0586058N19 and 0586058N74 (January–September 2004).

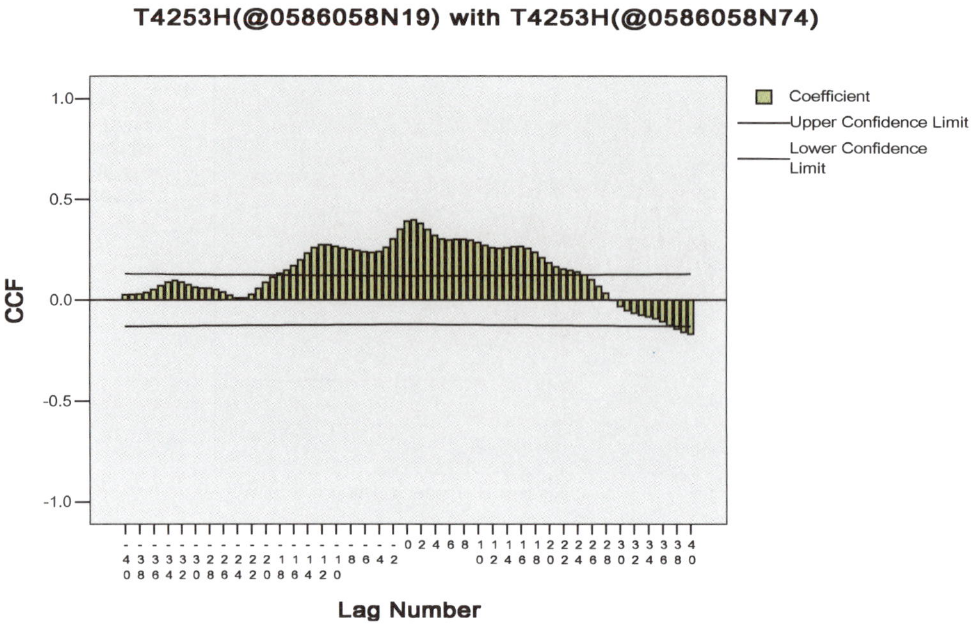

Fig. 16.16. 0586058N19 and 0586058N74 (January–September 2004).

indirect links (via a third-party common to both nodes) exist within this particular social network.

16.6.2. *Second Communications Data Set*

First and foremost, we identified five kinds of inbound service and nine kinds of outbound service (Tables 16.5 and 16.6, respectively) in this larger data set (258,000 records as opposed to 50,000 previously).

Two types of telephone service (*D* and *L*) dominate, accounting for over 99% of the *From* traffic, and 93% of the *To* traffic. We further note that just a couple of services account for the majority of telephone call traffic. More specifically, the number one Caller (D-61) is also the top callee; the number 3 caller (D-199) is the number 2 callee, and so on (Tables 16.7 and 16.8).

Thirdly, a large proportion of the call traffic takes place between the same type of service (D-D 53%; L-L 32%; D-L 6.3%; L-D 4%; D-I 1.4%; D-SN 0.62%, etc.).

Table 16.5. Inbound services.

Service Type	Number of IDs
D	4,273
L	2,046
OP	19
Y	14
I	3

Table 16.6. Outbound services.

Service Type	Number of IDs
D	5,026
L	2,845
L2	152
SN	152
OP	114
L1	84
SH	33
I	9
N	8

Table 16.7. The Top 10 Callers.

Rank	ID	Traffic	Rank in Caller List
1	D-61	23,704	1
2	D-465	14,283	7
3	D-199	10,686	2
4	D-55	7,325	4
5	L-183	6,785	3
6	L-13	6,285	11
7	D-181	5,967	10
8	L-973	5,702	28
9	D-1799	4,849	9
10	D-114	4,307	19

Table 16.8. The Top 10 Callers.

Rank	ID	Traffic	Rank in Caller List
1	D-61	19,511	1
2	D-199	10,378	3
3	L-183	7,916	5
4	D-55	7,836	4
5	L-1187	6,793	20
6	L-185	5,874	15
7	D-465	4,484	2
8	L-313	5,251	8
9	D-1799	4,230	9
10	D-181	4,160	7

A cross-correlation analysis of high-frequency callers did not yield any significant results, even invoking a 2-3 day time lag either side of zero. Nevertheless, as Figure 16.17 shows, reasonably high correlations are observed during certain time periods.

We experimented with manual determination of an 'optimum' time resolution (12, 6, 3, 1 month, 2 weeks, 1 week), and found that 1 month was preferable to 3 (since the latter tended to average out to a 'false' correlation level). We automated this process by way of a scripting (pre-processor) program which (i) automatically

Table 16.9. The Top 20 *From* Accounts.

Rank order	From Account	#Transactions	Rank order in *To* list
1	E-15	9	—
2	E-6	8	—
3	E-69*	7	A
4	E-12	6	—
5	E-24	6	—
6	E-14*	5	C
7	E-2	4	—
8	E-57	4	—
9	E-61	4	—
10	E-68	3	—
11	E-77	3	—
12	E-19*	2	S
13	E-34	2	—
14	E-39	2	—
15	E-43	2	—
16	E-50	2	—
17	E-71	2	—
18	E-87	2	—
19	E-92	2	—
20	E-10	1	—

Table 16.10. The Top 20 *To* Accounts.

Rank order	To Account	#Transactions	Rank order in *From* list
A	E-69*	13	3
B	E-13	4	—
C	E-14*	4	6
D	E-58	4	—
E	E-40	3	—
F	E-78	3	C
G	E-1	2	—
H	E-29	2	—
I	E-35	2	—
J	E-5	2	—
K	E-53	2	—
L	E-59	2	—
M	E-72	2	—
N	E-93	2	—
O	E-96	2	—
P	E-11	1	—
Q	E-16	1	—
R	E-18	1	—
S	E-19*	1	12
T	E-21	1	—

detects the highest correlations, and (ii) identifies the period(s) during which these occurs. It should be noted that this will involve variable starting times (i.e. they will not necessarily coincide with the start of a month). Again, no results of significance were forthcoming.

Table 16.11. The Top 20 *From* — *To* Pairs.

From Account	*To* Account	#transactions
E-69	E-69	6
E-12	E-13	4
E-14	E-14	4
E-57	E-58	4
E-77	E-78	3
E-2	E-1	2
E-6	E-29	2
E-34	E-35	2
E-12	E-53	2
E-15	E-59	2
E-2	E-69	2
E-68	E69	2
E-87	E-69	2
E-71	E-72	2
E-92	E-93	2
E-10	E-11	1
E-15	E-16	1
E-17	E-18	1
E-19	E-19	1
E-20	E-21	1

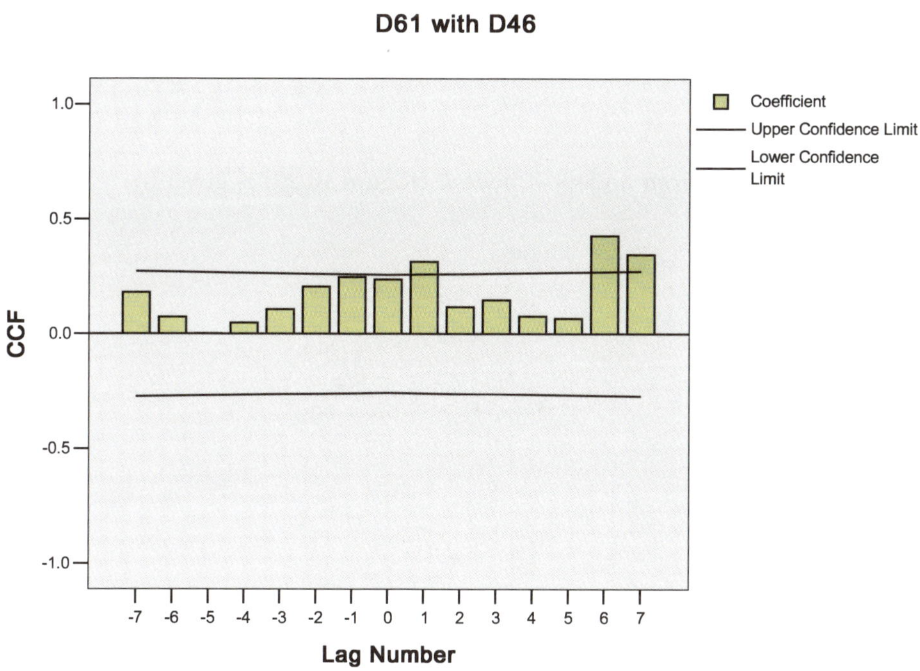

Fig. 16.17. D61-D46 Cross Correlation (1.3.03–30.4.03).

16.7. Network Dynamics

16.7.1. *Adding Meaningful Link and Path Weights to a Transaction Network*

As the current data-set is 'poor' in meaningful annotations, we first defined then added some meaningful weights to both network links and paths. The formula for calculating link/path weights is as follows:

Direct links (simplest case)

$$w_{ij} = \frac{t_{ij}\beta}{T_i} \quad \beta \leq 1; \; w_{ij} \leq 1 \tag{16.1}$$

where t_{ij} is the traffic from node-i to node-j, w_{ij} is the link weight from node-i to node-j, and T_i is the total (outgoing) traffic from node-i; β is a transaction type-related coefficient, and has different meanings for different kinds of transactions, and can be user-defined (Figure 16.18). The reason for adding this parameter is to denote the importance of a certain transaction. For example, in communication transactions, longer phone calls have higher β values.

Path weights

A *path* is defined as a route between two nodes. (Note that multiple paths can exist.)

Simple path weight

pw_{ik}, the link-weight from node-i to node-k (Figure 16.20) is given by:

$$\begin{aligned} pw_{14} &= w_{12}w_{24} \\ &= \left(\frac{t_{13}}{T_1}\right)\left(\frac{t_{24}}{T_2}\right) \\ &= \left(\frac{t_{13}}{t_{12}+t_{13}}\right)\left(\frac{t_{24}}{t_{24}+t_{25}}\right). \end{aligned} \tag{16.2}$$

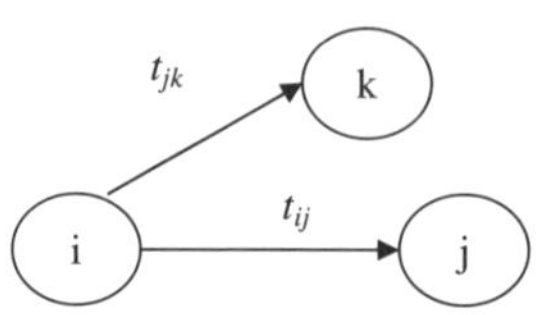

Fig. 16.18. Link/Path Weights.

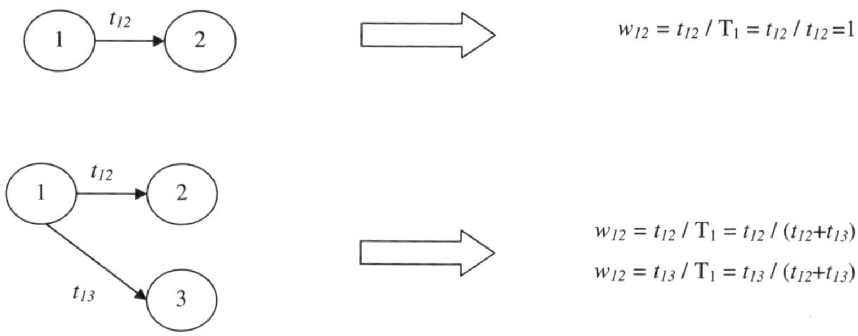

Fig. 16.19. Example Link/Path Weights.

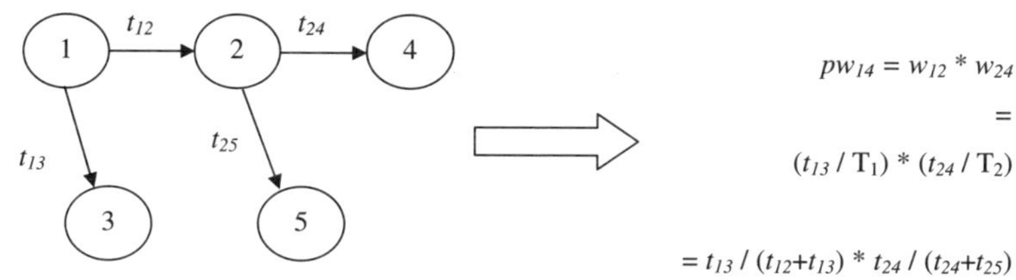

Fig. 16.20. Single Path.

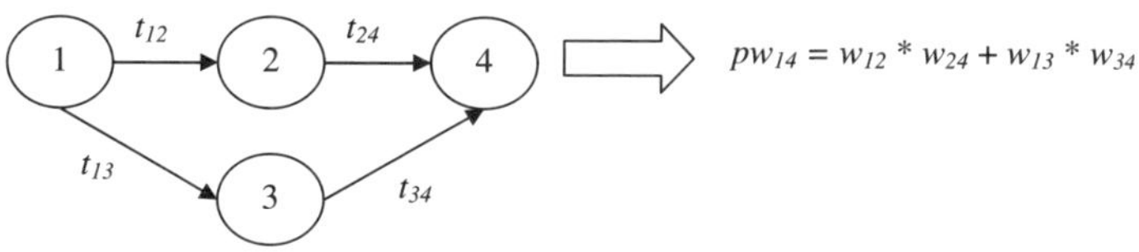

Fig. 16.21. Multiple Paths.

Multiple paths between two nodes

If there is more than one path between two nodes (Figure 16.21), the path weight between these two nodes is the sum of all individual path weights:

$$pw_{14} = w_{12}w_{24} + w_{13}w_{34} \tag{16.3}$$

Path weights for paths with overlapping links

If paths between two nodes have overlapping link(s), the link weight of the overlapped link is divided by the overlapped time when calculating the path weight. In the example of Figure 16.22, there are three paths between node-1 and node-4, with all three paths sharing a common link — l_{12} (the link between node-1 and node-2). Hence, in calculating pw_{14}, w_{12} needs to be divided by 3.

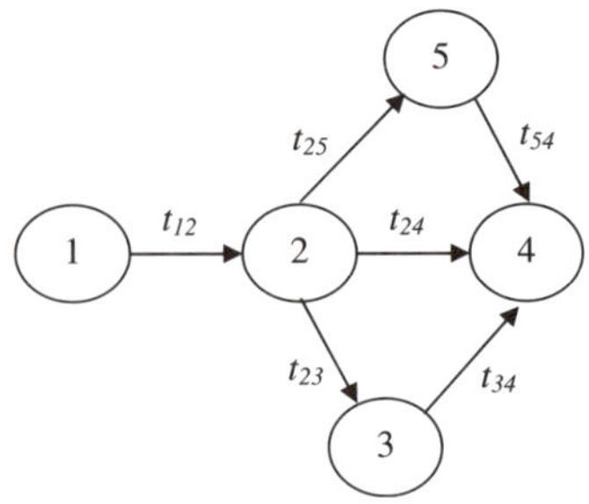

Fig. 16.22. Overlapping Paths.

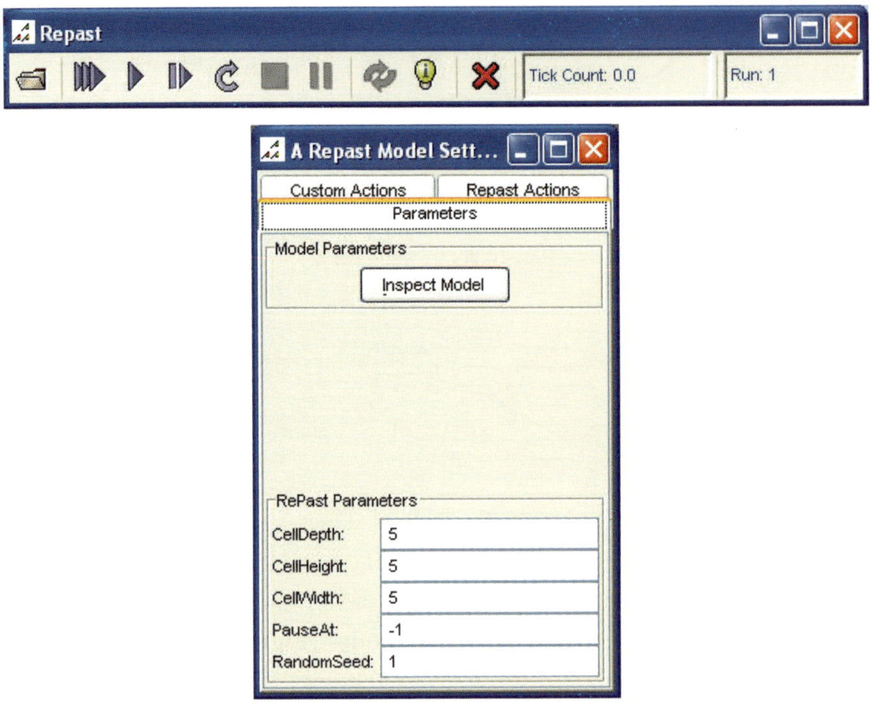

Fig. 16.23. Main SWARM User Interface.

16.7.2. *Building SWARM Simulations to Display Network Dynamics*

The main purpose of incorporating SWARM techniques in this project is twofold: (i) to demonstrate (model) dynamic network changes, and (ii) to (potentially) enable network prediction. The tool currently being used in this endeavour is `Repast`. We have performed some preliminary work to embed a SWARM simulation into our system. Figure 16.23 shows the SWARM Graphical User Interface, whereas

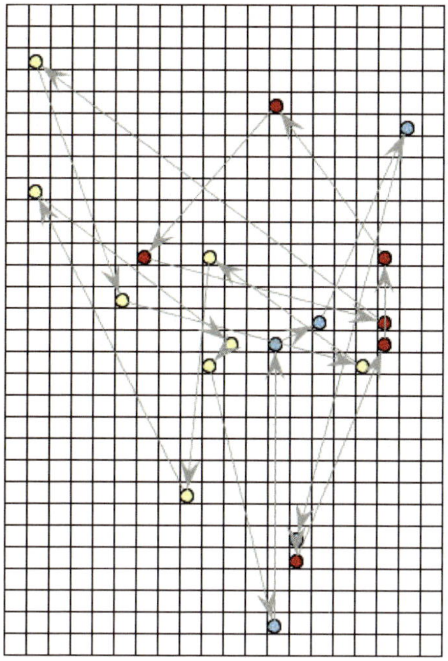

Fig. 16.24. 2D Network View.

Figures 16.24 through 16.26 are snapshots of typical (dynamic) 2-Dimensional and 3-Dimensional network views.

Each agent (node) in the output window represents a person in the network. Node positions show the relationships between people in the network; the more contacts that people make, the closer the relationship between them. The time tick of the output update can be daily based. The input of the simulation is transaction data sets. Accordingly, the simulation can show not only the network life-cycle, but also how relationships among members of the network change over time.

16.8. Public Domain Data

In Sect. 16.5, we stated that a long-term aim with the present project was to model social networks with sufficient accuracy to enable the 'reverse-engineering' of time series data leading up to known real-world events (given the availability of well documented case studies, that is). The data sets at our disposal however can be best characterized as 'data rich, knowledge poor', as previously observed. The question arises then as to whether more attribute-rich data sets are available, and more especially available in the public domain (which would overcome the need for data anonymity). The short answer is that such data sets *do* exist, one example being `Tweed`.[62, 63k] However we should point out that this latter data set differs markedly

[k] http://www.uib.no/people/sspje/tweed.htm

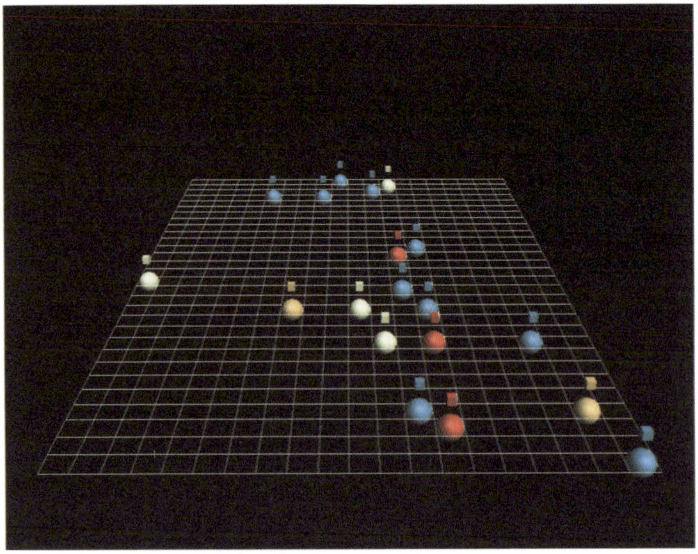

Fig. 16.25. 3D Network View (nodes only).

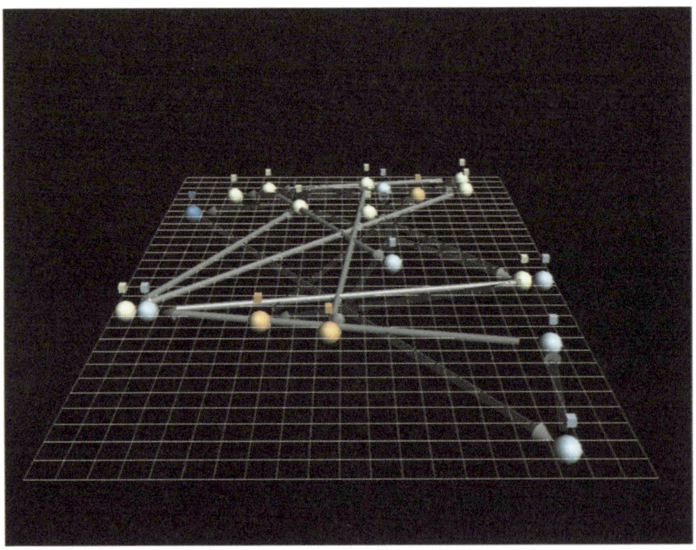

Fig. 16.26. 3D Network View (nodes and links).

from the communications traffic data sets considered up to now. First and foremost, `Tweed` records many *different* social networks, not just the one (as with the present study). Secondly, the records contain case summaries ('macro' data, if you will), rather than 'micro' data.

16.9. Conclusion and Suggestions for Further Work

We have devised a model to enable the development of agent-based systems for modelling (and by extension, prediction), and which can incorporate not only agents, but also neural network, evolutionary algorithm, swarm, machine learning (e.g. pattern recognition), data mining/link mining, and statistical techniques.

There are several promising avenues to explore in future studies, in order to build upon the foundations we have laid with `NetShow 2.0`, these being:

(1) validation of our agent-based model (ABM) and system, beyond just the (limited) data sets available during the course of the present study,
(2) adaptation of `NetShow 2.0` in order to handle other data sets (e.g. `Tweed`),
(3) correlation of data with 'events', in order to analyze patterns leading up to such events, with a view to predicting 'similar' such occurrences in the future,
(4) combining agent-based learning (= a local technique) with Data/Link Mining (= a global technique), in order to discover more about relationships between nodes (accounts, people and phone owners), changing trends, and other previously undiscovered information/knowledge,
(5) comparing system- *versus* local-level agent learning, combining agents with link mining in the former, and agents and pattern matching for the latter,
(6) combining agent-based learning with Computational Intelligence (e.g. ANNs, evolutionary algorithms, swarms) and/or Machine Learning techniques, to enhance knowledge discovery from network data,
(7) integration of telephony and geographical location data (although we appreciate the inherent problems in adequately de-referencing/aliasing the latter),
(8) further research into how different data types can be combined, namely nominal, ordinal, interval and ratio (the data sets we have been working with up to now have been almost exclusively nominal),
(9) development of alternative data visualization/presentation formats/views (e.g. nodes being denoted by different colours and/or shapes),
(10) further development of swarm-based visualization of network dynamics (in contrast to `NetShow 2.0`, which is essentially restricted to static displays),
(11) further investigation into pattern discovery from *Contact Lists*,
(12) further development of *degree-of-familiarity* and the tracking of same over time, as well as other metrics (e.g. 'similarity').

Let us consider one of these recommendations in a little more detail (8 above). Both the telephone call traffic and financial transaction data sets involve nominal data. The question arises as to whether the techniques developed above apply equally well to data sets which incorporate other data types, such as ordinal, interval, and ratio data (again, this could be relevant to data sets such as `Tweed`). In other words, can objects (entities) described in terms of different data types be combined together? Fundamentally, this is not feasible unless there is some underlying *relationship* between the data in question (which could, for example be time).

One potential approach is to use fuzzy reasoning combined with agent-based leaning.[64,65] This would involve the following four steps:

(1) Define four *fuzzy variables* to describe each data group, say N, O, I, and R,
(2) Build *fuzzy membership functions* for each fuzzy variable, based on the features of each individual data group.
(3) Derive *fuzzy rules* based on the problem domain, in which relationships exist between the four data groups.
(4) Build *reasoning mechanisms* to evaluate these data by use of *agent learning* and *reasoning*, assuming the measurements can be related to different purposes. Supposing we choose a BDI agent to implement the data evaluation, then

    - Fuzzy rules correspond to *belief*,
    - Measurement guidelines, such as 'trend of changes', correspond to *intention*, and
    - Different views or actions relating to multiple measurement purposes can be modelled as *plan*.

For subsequent data analysis, we might use four variables to describe an object in 4-dimensional space, then analyze the projection of the object onto lower-dimension space, in order to classify or cluster these objects by applying Data Mining techniques based on the requirements for data analysis. (Note that the above proposed method is predicated upon *relationships* actually existing between these four data types.)

Acknowledgment

The authors would like to acknowledge the support of the Intelligent Systems Research Centre at the University of Wollongong.

References

1. C. Westphal and T. Blaxton, *Data Mining Solutions: Methods and Tools for Solving Real-World Problems.* (Wiley, New York, NY, 1998).
2. I. Whitten and E. Frank, *Practical Data Mining: Machine Learning Tools and Techniques.* (Morgan Kaufmann, San Francisco, CA, 2005).
3. J. Fulcher. Computational Intelligence: An Introduction. In eds. J. Fulcher and L. Jain, *Computational Intelligence: a Compendium.* Springer-Verlag, Berlin, Germany, (2008).
4. J. Fulcher and L. Jain, Eds., *Computational Intelligence: A Compendium.* (Springer-Verlag, Berlin, Germany, 2008).
5. L. Padgham and M. Winikoff, *Developing Intelligent Agent Systems: A Practical Guide.* (Wiley, New York, NY, 2004).
6. M. Wooldridge, *An Introduction to Multiagent Systems.* (Wiley, Chichester, UK, 2002).
7. M. Zhang, B. Bai, F. Ren and J. Fulcher. Collaborative Agents for Complex Problem Solving. In eds. C. Mumford and L. Jain, *Collaborative Computational Intelligence.* Springer-Verlag, Berlin, Germany, (2008).

8. D. Clouse, C. Giles, B. Horne and G. Cottrell, Time delay neural networks: representation and induction of finite-state machines, *IEEE Trans. Neural Networks.* **8**, 1065–1070, (1997).
9. M. Zhang, Ed., *Artificial Higher-Order Neural Networks for Economics and Business.* (Idea Group International, Hershey, PA, 2008).
10. S. Abe, Ed., *Support Vector Machines for Pattern Classification.* (Springer-Verlag, New York, NY, 2005).
11. J. Shawe-Taylor and N. Christianni, Eds., *Support Vector Machines and Other Kernel Based Learning Methods.* (Cambridge University Press, Cambridge, UK, 2000).
12. T. Back, *Evolutionary Algorithms in Theory and Practice.* (Oxford University Press, New York, NY, 1996).
13. L. Davis, Ed., *Handbook of Genetic Algorithms.* (Van Nostrand Reinhold, New York, NY, 1991).
14. D. Fogel, *Evolutionary Computation: Toward a New Philosophy of Machine Intelligence.* (Wiley, New York, NY, 1999).
15. D. Goldberg and K. Deb, *Foundations of Genetic Algorithms: A Comparative Analysis of Selection Schemes Used in Genetic Algorithms.* (Morgan Kaufmann, San Mateo, CA, 1991).
16. J. Holland, *Adaptation in Natural and Artificial Systems (2nd ed.).* (MIT Press, Cambridge, MA, 1992).
17. Z. Michalewicz, *Genetic Algorithms + Data Structures = Evolution Programs.* (Springer-Verlag, Berlin, Germany, 1996).
18. M. Mitchell, *Introduction to Genetic Algorithms.* (MIT Press, Cambridge, MA, 1995).
19. W. Banzhaf, *Genetic Programming: An Introduction.* (Morgan Kaufmann, San Francisco, CA, 1998).
20. J. Koza, *Genetic Programming: On the Programming of Computers by Means of Natural Selection.* (MIT Press, Cambridge, MA, 1992).
21. W. Langdon, *Data Structures and Genetic Programming: GP+ Data Structures= Automatic Programming.* (Kluwer Academic Publishers, Boston, MA, 1998).
22. W. Langdon, R. Poli, N. McPhee and J. Koza. Genetic Programming: An Introduction and Tutorial, with a Survey of Techniques and Applications. In eds. J. Fulcher and L. Jain, *Computational Intelligence: a Compendium.* Springer-Verlag, Berlin, Germany, (2008).
23. V. Buskens, *Social Networks and Trust.* (Kluwer, Boston, MA, 2002).
24. L. Freeman, *The Development of Social Network Analysis: A Study in the Sociology of Science.* (Empirical Press, Vancouver, Canada, 2004).
25. J. Scott, *Social Network Analysis: A Handbook (2nd ed.).* (Sage, Newberry Park, CA, 2000).
26. S. Wasserman and K. Faust, Eds., *Social Networks Analysis: Methods and Applications.* (Cambridge University Press, Cambridge, UK, 1994).
27. C. Date, *Introduction to Data Base Systems.* (Addison Wesley, Reading, MA, 2004).
28. D. Kroenke, *Data Base Processing: Fundamentals Design and Implementation.* (Prentice Hall, Englewood Cliffs, NJ, 2004).
29. P. Rob, *Data Base Systems: Design, Implementation, and Management.* (Course Technology, Cambridge, MA, 2004).
30. M. Huisman and M. Van Duijn. Software for Social Network Analysis. In eds. P. J. Carrington, J. Scott and S. Wasserman, *Models and Methods in Social Network Analysis,* pp. 270–316. Cambridge University Press, New York, NY, (2005).
31. W. Nooy, A. Mrvar and V. Batageli, Eds., *Exploratory Social Network Analysis with Pajek.* (Cambridge University Press, Cambridge, UK, 2005).

32. J. Mena, *Investigative Data Mining for Security and Criminal Detection.* (Butterworth-Heinemann, Boston, MA, 2003).
33. M. Sageman, *Understanding Terror Networks.* (University of Philadelphia Press, Philadelphia, PA, 2004).
34. D. Farley, Breaking al qaeda cells: a mathematical analysis of counterterrorism, *Operations: Studies in Conflict Terrorism.* **26**, 399–411, (2003).
35. N. Memon and H. Larsen. Practical approaches for analysis, visualization and destabilizing terrorist networks. In *Proc. 1st Intl. Conf. Availability, Reliability and Security (ARES'06)*, pp. 906–913, Los Alamitos, CA, (2006). IEEE Computer Society Press.
36. V. Latora and M. Marchiori, How the science of complex networks can help develop strategies against terrorism, *Chaos, Solitons and Fractals.* **20**, 69–75, (2004).
37. K. Pattipati, P. Willet, J. Allanach, H. Tu and S. Singh. Hidden Markov Models and Bayesian Networks for counter-terrorism. In eds. R. Popp and J. Yen, *Emergent Information Technologies and Enabling Policies for Counter-Terrorism*, pp. 25–50. IEEE Press, Piscataway, NJ, (2006).
38. L. Rabiner and B. Juang, An introduction to hidden markov models, *IEEE ASSP Magazine.* **January**, 4–16, (1986).
39. F. Jensen, *Bayesian Networks and Decision Graphs.* (Springer-Verlag, Berlin, Germany, 2001).
40. K. Korb, *Bayesian Artificial Intelligence.* (CRC Press, Bocas Raton, FL, 2004).
41. R. Neapolitan, *Learning Bayesian Networks.* (Prentice Hall, Englewood Cliffs, NJ, 2003).
42. T. Mifflin, C. Boner, G. Godfrey and M. Greenblatt. Detecting terrorist activities in the twenty-first century: a theory of detection for transactional networks. In eds. R. Popp and J. Yen, *Emergent Information Technologies and Enabling Policies for Counter-Terrorism*, pp. 349–365. IEEE Press, Piscataway, NJ, (2006).
43. D. Skillicorn. Social network analysis via matrix decomposition. In eds. R. Popp and J. Yen, *Emergent Information Technologies and Enabling Policies for Counter-Terrorism*, pp. 367–391. IEEE Press, Piscataway, NJ, (2006).
44. J. Golbeck, A. Mannes and J. Hendler. Semantic web technologies for terrorist network analysis. In eds. R. Popp and J. Yen, *Emergent Information Technologies and Enabling Policies for Counter-Terrorism*, pp. 125–137. IEEE Press, Piscataway, NJ, (2006).
45. L. Getoor, Link mining: a new data mining challenge, *SIGKDD Explorations.* **4**, 1–6, (2003).
46. M. Sparrow, The application of network analysis to criminal intelligence: an assessment of the prospects, *Social Networks.* **13**, 251–274, (1991).
47. T. Fawcett and F. Provost. Combining data mining and machine learning for effective user profiling. In *Proc. 2nd Intl. Conf. on Knowledge Discovery and Data Mining (KDD'96)*, Portland, OR, 2.
48. T. Fawcett and F. Provost. Activity monitoring: Noticing interesting changes in behavior. In *Proc. 5th Intl. Conf. on Knowledge Discovery and Data Mining (KDD'99)*, San Diego, CA, 15.
49. T. Senator, H. Goldberg, *et al.* The financial crimes enforcement network ai system (fais), *AI Magazine.* **16**, 21–39, (1995).
50. T. Senator, Link mining applications: Progress and challenges, *SIGKDD Explorations.* **7**, 76–83, (2006).
51. J. Ferber, *Multi-Agent Systems: An Introduction to Distributed Artificial Intelligence.* (Addison Wesley, Reading, MA, 1999).
52. M. Wooldridge and N. Jennings, Intelligent agents: theory and practice, *The Knowledge Engineering Review.* **10**, 115–152, (1995).

53. A. Abraham, C. Grosan and V. Ramos, Eds., *Swarm Intelligence in Data Mining.* (Springer-Verlag, Berlin, Germany, 2006).
54. A. Engelbrecht, *Fundamentals of Computational Swarm Intelligence (2nd ed.).* (Wiley, Hoboken, NJ, 2005).
55. F. Luna and B. Stefansson, Eds., *Economic Simulations in Swarm: Agent-Based Modelling and Object Oriented Programming.* (Kluwer Academic Publishers, Dordrecht, The Netherlands, 2000).
56. P. Terna, Simulation tools for social scientists: building agent-based models with swarm, *J. Artificial Societies and Social Simulation.* **1**, (1998).
57. S. Haykin, *Neural Networks: A Comprehensive Foundation (2nd ed.).* (Prentice Hall, Englewood Cliffs, NJ, 1999).
58. G. Orr and K.-R. Mueller, Eds., *Neural Networks: Tricks of the Trade.* (Springer-Verlag, Berlin, Germany, 1998).
59. J. Principe. N. Euliano and W. Lefebre, *Neural and Adaptive Systems: Fundamentals Through Simulations.* (Wiley, New York, NY, 2000).
60. Y. Ishida, *Immunity-Based Computing: A Design Perspective.* (Springer-Verlag, Berlin, Germany, 2004).
61. P. Velleman and D. Hoaglin, *Applications, Basics, and Computing of Exploratory Data Analysis.* (Duxbury Press, Boston, MA, 1981).
62. V. Krebs, Mapping terrorist networks, *Connections.* **24**(3), 43–52, (2002).
63. J. Engene, Five decades of terrorism in europe: the tweed data set, *J. Peace Research.* **40**, 109–121, (2007).
64. M. Zhang and W. Li, Identifying potential synthesis cases in distributed excpert systems: a fuzzy logic approach, *Knowledge Based Systems.* **14**, 359–365, (2001).
65. F. Ren, M. Zhang and Q. Bai. A fuzzy-based approach for partner selection in multi-agent systems. In *Proc. 6th IEEE/ACIS Intl. Conf. on Computer and Information Science (ACIS'07)*, pp. 457–464, Los Alamitos, CA, (2007). IEEE Computer Society Press.

Chapter 17

ADAPTIVE APPROACH TO QUALITY ENHANCEMENT AND STORAGE OF SIGNATURES AND FINGERPRINT IMAGES

Roumen Kountchev
Department of Radio Communications, Technical University of Sofia
Bul. Kl. Ohridsky 8, Sofia 1000, Bulgaria
rkountch@tu-sofia.bg

In this work is presented one new approach for processing the images of fingerprints, signatures, and handwritten texts, based on histogram modification and on adaptive filtration followed by image segmentation. The visual quality of the images obtained in result of the processing is significantly enhanced. The archiving is based on new method for adaptive lossless data compression, which suits the statistic of the processed images very well and ensures high compression ratio. The comparison with JPEG 2000 LS proved the method efficiency.

17.1. Introduction

Significant part of the biometric information used nowadays is visual. It comprises mainly still images of fingerprints (fingers, palms.), signatures, faces, etc. The efficient processing, compression and storage of biometric information of any kind is of high importance for different application areas — matching, secure authorization and identification, archiving, etc. In most cases when such documents are scanned, the retaining of their visual quality after compression becomes one of the greatest challenges.

Various approaches are used when biometric information is prepared for storage: the signatures or latent fingerprints are usually preprocessed and compressed with software based on some of the compression standards (JPEG, JPEG 2000, WSQ, etc.); the raw signature data (taken by table-pad or from scanned paper document) are usually preprocessed by translation, rotation and/or scaling (Fuentes *et al.*) in order to afford some degree of invariance to changes with each realization. Some algorithms also require segmentation, for instance into components that are perceptually significant according to a motor model. Features can be extracted from the preprocessed signature, which can then be represented according to two broad paradigms (Plamondon and Lorette; Lau *et al.*): in the function-oriented paradigm the signals, extracted from signature data are considered as functions of time, whose values directly constitute the feature vectors; in the parametric paradigm, local or global parameters are computed from the measured signals and used as features. The extracted local parameters represent the properties of sampling points, pen strokes, or components and can be extracted from the signature local segments. The successful extraction of the significant parameters depends mainly on the image quality.

The quality of fingerprints, obtained by using contemporary sensors is good, but many problems exist when photos of latent fingerprints have to be processed, matched, etc. Specific quality of latent fingerprint images is the insufficient contrast and (correspondingly) the narrow range of brightness levels. The SC37 working draft on the fingerprint image data standard (ISO/IEC standard) specifies the main requirements to fingerprint image quality as follows: the signal-to-noise ratio should be equal to, or higher than 125; the dynamic grayscale range of the image data should be 200 gray levels for at least 80% of the captured image and 128 gray levels for at least 99% of the captured image. The typical latent fingerprint images do not meet these requirements. Various techniques have already been developed for quality improvement of grayscale images, based on contrast enhancement, histogram equalization, local histogram equalization (J. Yang *et al.*), etc. Another traditional approach is to apply a kind of image filtration. The processing is rather efficient when the ridges are not clear and minutiae extraction is difficult because of the poor image quality. For the processing are usually used Gabor filters. In result the visualization becomes much better and matching is easier to be performed. The main disadvantage is that some details might be lost forever. For example, because of the poor quality some small fingerprint details, such as the sweat pores (admissible points of identification in court) are not always visible. Quality improvement of low-contrast images is usually achieved by transforming the brightness of every pixel in accordance with certain functions (Bovik; Bow; Pratt; Gonzalez and Woods; Sharma). These functions could be classified as follows:

— Linear, piecewise linear and non-linear transforms: logarithmic, hyperbolic, exponential, power-low transformations, etc;
— Statistical (adaptive) transforms: histogram equalization and modification, bi-histogram transform (Kim) and transforms, based on local statistics information (mean value, standard deviation, etc.);
— Membership functions for contrast adaptation by: minimization of image fuzziness, fuzzy histogram hyperbolization and a rule-based approach (Rowberg and Malcolm).

Transforms of the types, pointed above, are based on the global image histogram data or on the local histogram statistics (Young *et al.*). All these methods for image contrast enhancement offer very good results for image quality improvement where natural grayscale images are concerned, without being a universal tool where biometric visual information is processed.

Two new approaches are presented in this work for processing images of fingerprints, signatures and handwritten samples: the first — based on histogram modification, and the second — on adaptive filtration followed by image segmentation. The two methods provide software tools for processing of fingerprints or signatures and facilitate other operations, such as matching, local parameters extraction, etc. The visual quality of the images obtained in result of the processing is significantly enhanced. The archiving of the processed images is based on a new method for adaptive lossless data compression, which corresponds to the processed images statistic and ensures high compression ratio.

17.2. Image Histogram Modification

The techniques, aimed at the adaptive histogram modification, presented below, are based on the specific features of the processed images (mainly fingerprints and signatures). The algorithm is based on histogram analysis and modification.

The typical histogram of the latent fingerprint or signature usually has two maximums, indicating the main objects in the processed image: the fingerprint (signature, handwriting) and the background. The histogram modification starts with the histogram separation into corresponding parts followed by brightness change in accordance with fixed predefined relations. The basic advantage of this approach is that there is no loss of brightness information — the brightness levels in the original image are only transformed in higher or lower ones, retaining the image minutiae.

The calculation of the image histogram permits to define the minimum and maximum brightness values in the processed image and to evaluate its quality in accordance to the ISO/IEC SC37 fingerprint image data standard. The processing starts with the calculation of the image histogram $h(b)$ for $b = 0, 1, \ldots, b_{\max}$, where $(b_{\max} + 1)$ is the number of gray levels.

The histogram modification, offered in this work, is performed in 2 steps. The **first step** is used to define which part of the histogram represents the image background and to reduce the number of the corresponding brightness levels, retaining the most significant information (the object), and the **second step** is aimed at the contrast enhancement of this most significant part of the image. In many cases (when the size of the part of the image, corresponding to the background is negligible) the histogram of the fingerprint image has one maximum only, and the detection of the brightness values corresponding to the main object is not difficult. The case where the image histogram has two or more maximums is more complicated and requires additional analysis. An example test fingerprint image (300×300 pixels, 8 bpp) and its histogram are shown accordingly in Figs. 17.1 and 17.2.

The positions of the maximums in the image histogram are used to set the points defining the main objects in the processed image (the fingerprint/signature and the background) and therefore special adaptive segmentation is performed. The boundaries of the segments are defined by the points, which separate the parts of the histogram that correspond to the main objects in the processed image (in this

Fig. 17.1. Test fingerprint image

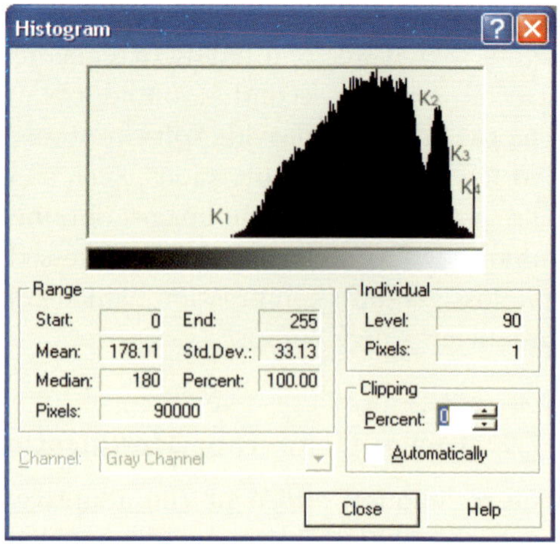

Fig. 17.2. The brightness histogram image with the boundary points

case — the fingerprint and the background). As it is seen in Figure 17.2, there are two maximums in the histogram (corresponding to the two objects) and many brightness values, which are not used in the processed image (in the left part of the histogram). The image segmentation is performed following the algorithm, presented below.

Three important assumptions are first made:

— The image represents a monochrome background with a darker object (fingerprint, signature, manuscript, etc.) on it;
— The minimum difference between the mean brightness values for the object and the background is δ;
— Sequences of brightness values, represented in the image histogram by small number of pixels only (less than 0,05% of the total number of pixels in the image for a single brightness value), are merged as corresponding to the separate objects.

The ***consecutive steps*** of the algorithm are:

Step 1. Calculation of the brightness level $b = \alpha_1$, for which $h(\alpha_1) = \max$ in the range $0 \leq b \leq b_{\max}$. Here α_1 is the mean brightness value for the background.

Step 2. Calculation of the brightness level $b = \alpha_2$, for which $h(\alpha_2) = \max$ in the range $0,0005(M \times N) \leq b \leq (\alpha_1 - \delta)$. Here α_2 is the mean brightness value for the object and the term $(M \times N)$ represents the number of pixels in the image.

Step 3. Detection of the brightness levels $b = \alpha_3$ for which $h(\alpha_3) \leq 0,0005(M \times N)$ in the range $0 \leq b \leq \alpha_2$.

Step 4. Detection of the brightness levels $b = \alpha_4$, for which $h(\alpha_4) \leq 0,0005(M \times N)$ in the range $\alpha_2 \leq b \leq b_{\max}$. (This step is not obligatory, but is very useful if such areas exist in the image histogram).

Step 5. Calculation of the histogram modes for $\alpha_2 \leq b \leq (\alpha_1 - \delta)$, after which for every mode (sub-range of $h(b)$, which has a local maximum) the corresponding area is calculated:

$$S_i = \sum_{b=\beta_i}^{\beta_{i+1}} h(b) \quad \text{for } i = 1, 2, \ldots, P. \tag{17.1}$$

Here β_i and β_{i+1} are the brightness levels, which define the boundaries of the sub-range "i". These boundaries coincide with the closest minimums to the left and to the right of the local maximum; P is the number of sub-ranges (respectively, the modes); S_1 corresponds to the mode of the object, placed between the range boundaries, and S_P — to the mode of the background placed between its' boundaries: from β_P up to $\beta_{P+1} = b_{\max}$ (in case another range had been defined in step 4, the upper boundary is a brightness value, lower than $b_{\max}$).

Step 6. Calculation of the histogram modes for $\alpha_1 \leq b \leq \alpha_2$, after which for every mode (sub-range of $h(b)$, which has a local maximum) the corresponding area (as in Eq. 17.1) is calculated. Similar calculations are performed for the range $\alpha_1 \leq b \leq b_{\max}$, if such a range had been defined.

Step 7. The value of P (obtained in Steps 5 and 6) is analyzed as follows:

<u>For $P > 4$</u>: rcursive mode merging is performed in the range $\beta_2 \leq b \leq \beta_P$ according to the relation:

$$S_i^*(k) = \begin{cases} S_i^*(k-1) \cup S_{i-1}^*(k-1), & \text{if } S_{i-1}^*(k-1) > S_{i+1}^*(k-1); \\ S_i^*(k-1) \cup S_{i+1}^*(k-1), & \text{if } S_{i-1}^*(k-1) \leq S_{i+1}^*(k-1), \end{cases} \tag{17.2}$$

$$\text{for } i = 1, 2, \ldots, P(\kappa),$$

where $S_i^*(k)$ is the area of the merged mode "i" in the iteration "k", and $P(k)$ is the number of the modes for the iteration "k" if the condition $P(0) = P$ is satisfied.

The modes merging is ended in the iteration "k", where $P(k) = 4$ and the analysis is thus finished. Values higher than 4 mean that too many objects in the image had been detected (not just an object and a background).

<u>For $P = 4$</u>: the values of β_2, β_3 and β_P are defined and the analysis is thus finished.

<u>For $P < 4$</u>: new boundaries should be calculated

$$\beta_2^* = (\alpha_2 + \delta) < \beta_P \quad \text{and} \quad \beta_3^* = (\alpha_2 - \delta) < \beta_P \tag{17.3}$$

and the analysis ends.

In result are obtained the areas, corresponding to the 2 main objects of the brightness histogram, as follows:

— For the **object** (fingerprint or signature) — from β_3 up to β_2 (or β_2^*, i.e. the upper boundary of the mode $S_1^*(k)$);
— For the **background** — from β_P (or β_P^*, i.e. the lower boundary of the mode $S_P^*(k)$), up to $b_{\max}$.

In the range $(\beta_P - \beta_2)$ are placed the gray levels, corresponding to the noises or representing some irregularities in the background.

In result, the histogram of the example test image (Fig. 17.2) is divided into 5 main ranges:

Range 1 — the brightness levels, practically not used in this case (for this test image these are the brightness levels 0–99). The brightness values in this group are represented by less than 1% of the image pixels. The range boundaries are at points 0 and k_1;

Range 2 — the brightness levels, corresponding to the fingerprint (brightness levels 100–199). The range boundaries are k_1 and k_2. The point k_2 is placed at the minimum, which indicates the start of the part in the histogram, corresponding to the second object in the image (the background);

Range 3 — the brightness levels, corresponding to the object background (brightness levels 200–234). The range boundaries are at points k_2 and k_3;

Range 4 — the brightness levels, corresponding to the background, but represented with relatively small number of pixels (brightness levels 235–249). The range boundaries are placed at points k_3, k_4;

Range 5 — the brightness levels, whose brightness information could be neglected, i.e. too small number of pixels with such brightness in the processed image (brightness levels 250–255). The range boundaries are k_4 and 255.

The histogram is then modified as follows:

Range 1 — represented by 1 brightness level only (brightness = 0). In the example test image there are no brightness values in this range;

Range 2 — all brightness values are retained;

Range 3 — represented by 5 brightness values, placed at equal distances;

Range 4 — represented by 2 brightness values, placed at equal distances;

Range 5 — represented by 1 brightness value only (in this case, the brightness $b_{\max} = 255$).

The image, obtained in result of the 5-part histogram modification is shown in Fig. 17.3, and the new histogram — in Fig. 17.4. As a result, ranges 3–5 are

Fig. 17.3. The test image after histogram modification (visually same as Fig. 17.1)

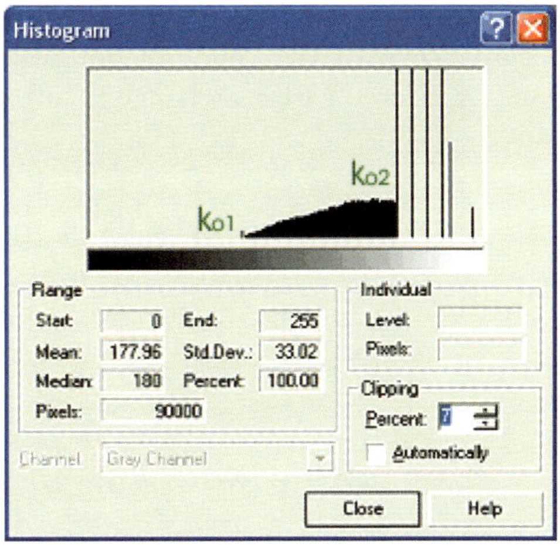

Fig. 17.4. The histogram of the processed image

represented by a small number of brightness values, but the image visual quality is retained.

The processing continues with the second step — the contrast enhancement for the most significant part of the image. For this, the part of the histogram, which represents the main object (the fingerprint or the signature), placed in Range 2, is stretched, and the remaining ranges are accordingly skewed. To perform this, it is assumed that the image histogram comprises 3 main parts, i.e. (for the example) part A, containing brightness values in the range of 0–99 (Range 1), part B, containing brightness values in the range of 100–199 (Range 2), and part C containing the remaining brightness values in the range of 200–255 (i.e., the merged Ranges 3, 4 and 5). The separating points for part B (the object) are k_{01} and k_{02} (they correspond to points k_1 and k_2 from the initial histogram). The contrast enhancement is performed by changing the brightness level, b for every pixel in these three parts according to a transformation table, as follows:

$$g(b) = \begin{cases} g_A(b) & \text{if } 0 \leq b < k_{01}; \\ g_B(b) & \text{if } k_{01} \leq b \leq k_{02}; \\ g_C(b) & \text{if } k_{02} < b \leq b_{\max}. \end{cases} \tag{17.4}$$

Here $g_A(b), g_B(b)$ and $g_C(b)$ represent the relations which define the brightness transformation in parts A, B and C. In order to improve the object image quality, the end points k_{o1}, k_{o2} of the part B are moved so as to make this part wider — up to $(k_{01} - \delta_1) \geq 0$ and $(k_{02} + \delta_2) \leq 255$, skewing respectively parts A and C. In this case δ_1 and δ_2 are parameters, which influence the contrast enhancement for the part B and hence, change the contrast range for both remaining parts A and C. Each brightness transformation table is defined in accordance with the condition for histogram equalization in the corresponding part $(A, B$ or $C)$ with a changed (correspondingly stretched or skewed) dynamic range:

$$g_A(b) = (k_{01} - \delta_1)\sum_{l=0}^{b} h_A(l), \tag{17.5}$$

$$g_B(b) = (k_{02} - k_{01} + \delta_1 + \delta_2) \sum_{l=k_{01}-\delta_1}^{b} h_B(l) + (k_{01} - \delta_1) \tag{17.6}$$

$$g_C(b) = (b_{\max} - k_{02} - \delta_2) \sum_{l=k_{02}+\delta_2}^{b} h_C(l) + (k_{02} + \delta_2) \tag{17.7}$$

In particular, in case that the histogram of the corresponding part is constant, i.e. for:

$$h_A(b) = \frac{1}{k_{01}} \quad \text{for } b = 0, 1, \ldots, k_{01} - 1; \tag{17.8}$$

$$h_B(b) = \frac{1}{k_{02} - k_{01}} \quad \text{for } b = k_{01},\ \ k_{01} + 1, \ldots, k_{02}; \tag{17.9}$$

$$h_C(b) = \frac{1}{b_{\max} - k_{02}} \quad \text{for } b = k_{02} + 1,\ \ k_{02} + 2, \ldots, b_{\max}, \tag{17.10}$$

the brightness transformation table for each part is linear and the relations are defined as follows:

$$g_A(b) = \left(\frac{k_{01} - \delta_1}{k_{01}}\right) b; \tag{17.11}$$

$$g_B(b) = \left(\frac{k_{02} - k_{01} + \delta_1 + \delta_2}{k_{02} - k_{01}}\right) b - \left(\frac{\delta_2 k_{01} + \delta_1 k_{02}}{k_{02} - k_{01}}\right); \tag{17.12}$$

$$g_C(b) = \left(\frac{b_{\max} - k_{02} - \delta_2}{b_{\max} - k_{02}}\right) b + \left(\frac{\delta_2 k_{02} + \delta_2^2}{b_{\max} - k_{02}}\right). \tag{17.13}$$

In result, the brightness levels range for part B is stretched and corresponding inverse operations are carried out for the two remaining parts of the histogram: $(0, k_{o1} - 1)$ and $(k_{o2} + 1, b_{\max})$, i.e. they are skewed. The image obtained in result of the processing, is shown in Fig. 17.5, and its' histogram — in Fig. 17.6.

The main advantage of the presented method is that the visual quality of the processed image is significantly improved without loss of visual information. The existing minutiae (ridges, sweat pores, etc.) are retained.

In case that the image histogram has only one maximum, it shall be initially separated into three parts $(A, B$ and $C)$ and the processing then continues in the already described way (Eq. 17.4 and the next ones).

Fig. 17.5. The processed image

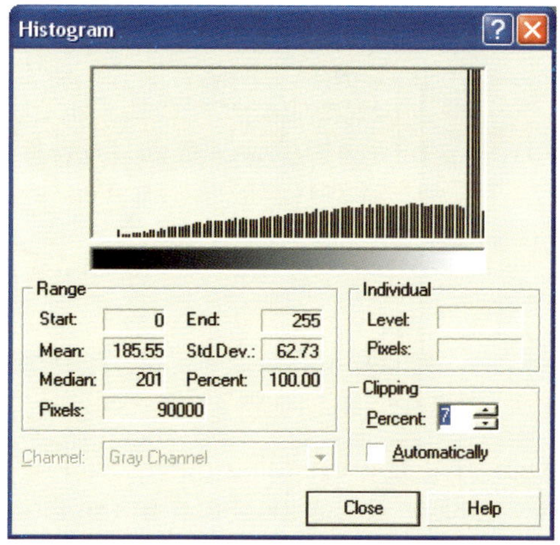

Fig. 17.6. The histogram of the image in Fig. 17.5

17.3. Image Filtration and Segmentation

The methods for image filtration, presented below, usually precede other very important operations (image segmentation, matching, etc.). Such approach is very useful when the image histogram is complicated (for example, when it has two or more maximums) and in result the correct image segmentation is not reliable. The presented approach is based on the use of two consecutive filters: the first, aimed at the adaptive noise filtration, and the second — at the equalization of the image background illumination. In result the high-frequency noise is suppressed first, and the background equalization follows.

17.3.1. *Adaptive noise filtration*

The first part of the processing is aimed at the additive Gaussian noise removal. In general, the methods for local-adaptive filtration (Young *et al.*, Lim) suit this aim very well: the main objects (signatures or fingerprints) are not affected because their basic characteristics, such as dimensions, area, shape, contrast, etc., are quite different from those of the noise.

In this work below is presented a method for the suppression of the additive Gaussian noise, using two-dimensional adaptive fuzzy filter (2DAFF) (Kountchev *et al.*), with a sliding window of size $(2N_1 + 1) \times (2N_2 + 1)$ pixels. The aims are to retain the sharp brightness transitions, and to smooth the small changes corresponding to noise features. The algorithm, presenting the filter performance, is given below:

$$y(i,j)=\begin{cases}\left\lfloor\dfrac{\sum_{r=-N_1}^{N_1}\sum_{s=-N_2}^{N_2} f(i+r,j+s)x(i+r,j+s)}{\sum_{r=-N_1}^{N_1}\sum_{s=-N_2}^{N_2} b(i+r,j+s)}\right\rfloor \\ \qquad\qquad \text{if } \sum\limits_{r=-R}^{R}\sum\limits_{s=-S}^{S} f(i+r,j+s)\geq T, \\ \left\lfloor (1/MN)\sum\limits_{r=-N_1}^{N_1}\sum\limits_{s=-N_2}^{N_2} x(i+r,j+s)\right\rfloor \qquad \text{- in all other cases,}\end{cases} \tag{17.14}$$

where: $\lfloor \circ \rfloor$ is a rounding operator; T — a threshold; $x(i,j)$ and $y(i,j)$ are pixels belonging to the original and to the filtered image respectively;

$$f[\Delta(i+r,j+s)]=\begin{cases}1, & \text{if } \Delta(i+r,\ j+s)\leq\alpha; \\ \dfrac{\Delta(i+r,j+s)-\beta}{\alpha-\beta}, & \text{if } \alpha\leq\Delta(i+r,j+s)\leq\beta; \\ 0, & \text{if } \Delta(i+r,j+s)\geq\beta,\end{cases} \tag{17.15}$$

is the membership function with parameters α and $\beta(\beta>\alpha)$, which define the filter fuzziness; the argument Δ is the module of the difference between the central pixel $x(i,j)$ in the filter window and the pixel $x(i+r,j+s)$, which is at a distance (r,s):

$$\Delta(i+r,j+s)=|x(i,j)-x(i+r,j+s)| \quad \text{for } r=\overline{-N_1,N_1} \quad \text{and} \quad s=\overline{-N_2,N_2} \tag{17.16}$$

The values of parameters α and β are defined by the relations:

$$\alpha=\chi-\varepsilon, \quad \beta=\chi+\varepsilon,$$

where χ represents the center of the filter fuzziness, for which the function $\mu(\chi)=0,5$, and ε defines the boundaries of the deviation from χ. The value of the parameter χ is defined by the relation:

$$\chi=\frac{1}{2}|D_{\max}|.$$

Here $D_{\max}$ represents the maximum brightness difference between the background and the "meaning" part (the objects) in the image. The parameters ε and T of the 2DAFF are selected depending on the perception noise level and the filter window width. The perception noise level (represented by ε) for 8 bpp grayscale images is 2–3 brightness levels; the threshold is set to be $T=9$ for filter window of size 3×3 pixels and smaller, for larger filter windows (for example, for window of size 15×15 pixels it is about 2).

17.3.2. *Equalization of the image background illumination*

The uneven background illumination normally exists in the images of scanned documents (signatures, manuscripts, etc.) and in the photos of latent fingerprints. The correction of the uneven illumination (background equalization) is necessary because it simplifies the next operations — the matching of the searched objects or their efficient compression and archiving. For the equalization of the image background illumination in this work is proposed a 2D linear digital filter of a non-recursive kind. The filter performance is presented by the relation:

$$\begin{aligned} z(i,j) &= d\left[x(i,j) - \frac{1}{A}\sum_{m=-N_1}^{N_1}\sum_{n=-N_2}^{N_2} x(i+m,j+n)\right] + \mu_x \\ &= d[x(i,j) - \mu_x(i,j)] + \mu_x \end{aligned} \tag{17.17}$$

Here:

$x(i,j)$ is the brightness of the element (i,j) from the original image (before the background illumination correction);
$z(i,j)$ — the brightness of the element (i,j) from the corrected image;
μ_x — the mean brightness of the original image. The value of μ_x is defined as:

$$\mu_x = \frac{1}{2}(x_{\max} - x_{\min}) \tag{17.18}$$

where:

$x_{\max}$ and $x_{\min}$ are respectively the pixels with maximum and minimum brightness value in the processed image;
$\mu_x(i,j)$ — the mean local brightness in the window around the pixel $x(i,j)$;
$A = (2N_1+1)\times(2N_2+1)$ — the number of pixels in a rectangular window of size $(2N_1+1)\times(2N_2+1)$;
d — a coefficient, representing the contrast enhancement applied to the small details in the image $(d \geq 1)$.

The filtration, presented with Eq. (17.17) could be accelerated, transforming the relation in a recursive form:

$$\begin{aligned} \mu_x(i,j) &= \mu_x(i-1,j) + \mu_x(i,j-1) - \mu_x(i-1,j-1) + \frac{1}{A}[x(i+N_1,j+N_2) \\ &\quad - x(i+N_1,j-N_2-1) - x(i-N_1-1,j+N_2) \\ &\quad + x(i-N_1-1,j-N_2-1)]. \end{aligned} \tag{17.19}$$

Then from Eqs. (17.17) and (17.19) follows:

$$\begin{aligned} z(i,j) &= z(i-1,j) + z(i,j-1) - z(i-1,j-1) + d\Bigg\{x(i,j) - x(i-1,j) \\ &\quad - x(i,j-1) + x(i-1,j-1) - \frac{1}{A}[x(i+N_1,j+N_2) \end{aligned}$$

$$-x(i-N_1-1,j+N_2)-x(i+N_1,j-N_2-1)$$
$$\left.+x(i-N_1-1,j-N_2-1)]\right\}. \tag{17.20}$$

The number of the components in the last equation is always 11 and it does not depend on the filter window size, $(2N_1+1)\times(2N_2+1)$. In case that $N_1 \gg 1$ and $N_2 \gg 1$, the number of operations becomes:

$$\eta = \frac{1}{11}[(2N_1+1)(2N_2+1)+2] \approx 0,36N_1N_2 \tag{17.21}$$

For example, for $N_1 = N_2 = 63$ and in accordance with Eq. (17.21), is calculated $\eta \approx 1429$, i.e. the number of additions necessary for the recursive filter performance is more than 1400 times smaller.

For further simplification it is possible instead of using one, two-dimensional filter to use two, one-dimensional filters — the first one in horizontal direction and the second — in vertical direction, represented correspondingly by the relations:

$$\begin{aligned} y(i,j) &= d_1\left[x(i,j)-\frac{1}{(2N_1+1)}\sum_{m=-N_1}^{N_1} x(i+m,j)\right]+\mu_x;\\ z(i,j) &= d_2\left[y(i,j)-\frac{1}{(2N_2+1)}\sum_{n=-N_2}^{N_2} y(i,j+n)\right]+\mu_y, \end{aligned} \tag{17.22}$$

where, $d = d_1 \times d_2$, and $y(i,j)$ is the pixel (i,j) of the image obtained in result of the first filtration (in horizontal direction), with mean brightness, μ_y. By analogy with Eq. (17.20), the relations in Eq. (17.22) are presented in recursive form:

$$\begin{aligned} y(i,j) &= y(i-1,j)-\frac{d_1}{A_1}[x(i-N_1-1,j)-x(i+N_1,j)];\\ z(i,j) &= z(i,j-1)-\frac{d_2}{A_2}[y(i,j-N_2-1)-y(i,j+N_2)], \end{aligned} \tag{17.23}$$

where:

$$A = A_1 \times A_2; \quad A_1 = 2N_1+1, \quad A_2 = 2N_2+1.$$

In result, the filtration is significantly accelerated because of the reduced number of operations performed:

$$\eta = \frac{2N_1+3}{3}+\frac{2N_2+3}{3} = \frac{2(N_1+N_2)+6}{3} = 0,66(N_1+N_2)+2. \tag{17.24}$$

For example, if $N_1 = N_2 = 63$, then $\eta \approx 85$, i.e. in result of the recursive approach, the number of additions is more than 80 times smaller. Algorithm efficiency is enhanced because of the doubled use of the same filter.

The described filtration causes some distortion at the image edges. In order to avoid this, the matrix of the processed image should be artificially made larger (Gonzalez and Woods) adding pixels in both directions (horizontal and vertical). In result, the original matrix of size $M_1 \times M_2$ becomes of size $(M_1+2N_1)\times(M_2+2N_2)$. The easiest way is to add zeros in both directions (Young *et al.*), but in the image obtained after filtration, usually appear distortions, called zero-padding artifacts. The better way is to use image replication of size equal to that of the filter window side instead of zeros.

The filter parameters N_1, N_2 and d_1, d_2 are defined in accordance with the background uneven illumination, which should be corrected.

17.3.3. *Image segmentation*

The histogram of the processed image after the background equalization has usually only one maximum, which corresponds to the image background and in result, the detection of a second maximum pointing at the searched object(s) is not possible. For this reason, the segmentation is based on the Triangle algorithm (Young *et al.*), modified for this application. The modification concerns faster detection of the values of the segmentation thresholds.

An assumption is made, that the image represents a monochrome background with dark object on it.

The segmentation is based on the image histogram $h(b)$ analysis, performing the operations, described below. The analysis starts with the calculation of the image histogram. Then follow the operations:

Two points are defined in the image histogram $h(b)$:

<u>First point</u> (h_0, b_0) — the maximum of the histogram, which is usually the mean value of the equalized background;

<u>Second point</u> (h_1, b_1), defined by the relation:

$$h_1(b_1) = h_0(b_0); \quad h_1(b_1) < h_0(b_0) \quad \text{for } b_1 < b_0. \tag{17.25}$$

A straight line is defined, which joins the two points: (h_1, b_1) and (h_0, b_0).The equation of this line is presented by the well-known relation:

$$Ab + Bh + C = 0, \tag{17.26}$$

where:

$$b_1 \leq b \leq b_0, \quad A = h_1 - h_0, \quad B = b_0 - b_1, \quad C = h_0 b_1 - h_1 b_0. \tag{17.27}$$

The distance to the so-defined line is calculated for each point of the histogram $h(b)$, in accordance with the relation:

$$D(b) = \frac{Ab + Bh + C}{\sqrt{A^2 + B^2}}. \tag{17.28}$$

The value θ of the variable b is defined, for which the distance $D(\theta) = \max$. This value is the segmentation threshold, for which from Eq. 17.28 is obtained:

$$h(\theta + 1) - h(\theta) \approx \frac{h_0 - h_1}{b_0 - b_1} \quad \text{for the range } b_1 \leq b \leq b_0 \tag{17.29}$$

The image segmentation threshold is used to separate respectively the dark object(s) and the background.

The image is then binarized in accordance with the relation:

$$p(i,j) = \begin{cases} 1, & \text{if } b(i,j) \leq \theta; \\ 0, & \text{if } b(i,j) > \theta, \end{cases} \tag{17.30}$$

The binary image $p(i,j)$ contains the detected and separated dark objects.

17.4. Lossless Compression of Biometric Images

The method for lossless compression of biometric image data is based on new content-based run-length encoding (CBRLE). It is aimed at the compression of data, which is N-dimensional sequence of n-bits binary words (numbers) whose values are placed in the range $(-2^{n-1}, 2^{n-1} - 1)$. The method comprises two consecutive stages. In the first one, the input data is transformed so as to increase the compression performed in the second one. The processed data are so transformed, that the data volume is not increased but sequences of equal numbers of maximum length are obtained. The transform is reversible and is performed by applying the following operations:

— The not used (free) values and the sequences of free values (if there are any) in the image histogram are detected and for every such sequence the start value and length are defined;
— The start value of the longest sequence of free values in the histogram is defined. In case that there is more than one sequence of maximum length, it is accepted that the one with the smallest start value will be used for the next operations;
— The input data is transformed in accordance with a new method, called Size-saving prediction coding. For this, every number from the input data sequence is replaced by its difference from the preceding number. In case, the result is outside the range of the input numbers, the sum of the current value and "1" is calculated with negative sign.
— The histogram of the transformed data sequence is calculated;
— The free values and the sequences of such values (if there are any) in this histogram are detected. The start value and length are defined for each sequence;
— The start value of the longest sequence of free values in the histogram is defined. In case, that there is more than one sequence of maximum length, it is accepted that the one with the smallest start value will be used for the next operations;
— In case no sequences or single free values in both histograms have been detected, the data are not suitable for compression in accordance with the CBRLE

method, the processing is stopped and the output sequence remains the same as the input one. Else, the coding continues in accordance with the following operations:

— The more suitable data sequence is selected for further processing (the input one or the one obtained after performing the Size-Saving Prediction coding). The choice is made depending on the histogram analysis result — the availability of a single free value or of a sequence of such values, and in case there are such sequences in both histograms — the one with a longer sequence is preferred.
— The most frequent value in the chosen sequence is detected;
— The difference between every number of the selected sequence and the most frequent value is calculated without carry;
— A third histogram is calculated (for the new data sequence, obtained after the calculation of the above-mentioned differences) and single free values or sequences of free values there are detected together with the start value of the longest sequence.

With this the first stage of the processing is finished.

In the second stage the header is composed and the transformed data are coded. Therefore the following operations are carried out.

Composition of the header

The header comprises a control word containing the information about the detected sequences of same values, their lengths and start values.

Data coding. The coding is performed as follows:

- Every sequence of zeros in the transformed data, whose length is smaller or equal than that of the longest sequence of free values in the histogram, is replaced by a number, equal to the sum of the start value of the sequence of free values and the length of the compressed sequence, reduced by one;
- Every sequence of zeros in the transformed data, which is longer than the sequence of free values in the histogram, but is shorter than 2mn, is represented by 2m words, n-bits each (here m is a number bigger or equal to 1): the first word contains the start value of the longest sequence of free values, the next $(m-1)$ words contain zeros, and the remaining m words — the length of the coded sequence, decreased by one;
- Every sequence of same values, different from zero and longer than 4, but shorter than $2mn$, is replaced by $(2m+2)$ words, n-bits each. The first word contains the start value of the longest sequence of free values, in the second word is stored the number "1", in the next $(m-1)$ words is stored "0", in the next m words — the length of the coded sequence, and in the last word — the coded value.
- In case, that the sequence of same values, different from zero, is shorter or equal to 4, it shall not be coded and remains unchanged.

The decoding is performed in reverse order.

The block diagram of the method for adaptive lossless data compression is shown in Fig. 17.7. In the block diagram are used the following abbreviations:

SSP encoder — Size-saving prediction encoder; **Control software** — software implementation of the algorithm for data and histograms analyses; **SW1** and **SW2** — logical switches, which change the data flow direction in accordance with the commands obtained from the control software; **DARL encoder** — data adaptive run-length encoder.

A specific feature of the described method for data coding and decoding is that it comprises only operations like "sorting" and "addition", while operations like "multiplication" and "division" are not used. In result, the software implementation of the presented method is fast, and decoding is much simpler than coding.

The main advantage of the method is that it permits to obtain high compression ratio in the case, when the processed data contains long sequences of same numbers. These basic characteristics distinguish the presented method from the famous data compression methods, based on the run-length encoding.

One more advantage is that the coded data sequence is never longer than the original one. The algorithm is universal, because it is suitable for processing of any kind of data: numbers, images, etc.

17.5. Experimental Results

The software implementation of the presented methods proved their efficiency. Some of the obtained results are presented below.

17.5.1. *Histogram modification and segmentation*

The experiments with the algorithm for histogram modification were carried out with the image database of the Laboratory for Image and Sound Processing at Technical University of Sofia (more than 400 fingerprint images and 200 signatures). All images are grayscale (8 bpp); the brightness histograms were obtained with Corel Photo Paint 10.

In result of the histogram analysis of the image in Fig. 17.8.a, the places of points $k_1 - k_4$, i.e. $k_1 = 96$; $k_2 = 228$; $k_3 = 244$; $k_4 = 250$ were defined.

After the detection of the three-part histogram boundaries, the main part (brightness values in the range 100–228) was defined and stretched. The resulting image is shown in Fig. 17.9.a. The same test image was processed in accordance with the algorithm for filtration and segmentation. The parameters of the 2DAFF used for the noise suppression were: brightness threshold value $T = 16$; filter window width: $(2N_1 + 1) \times (2N_2 + 1) = 9$. The window width of the filter for background equalization was $(2N_1 + 1) \times (2N_2 + 1) = 99$ pixels. The visual quality of the image obtained in result (Fig. 17.9.b) is quite different from that of the original, but the local parameters extraction is easier.

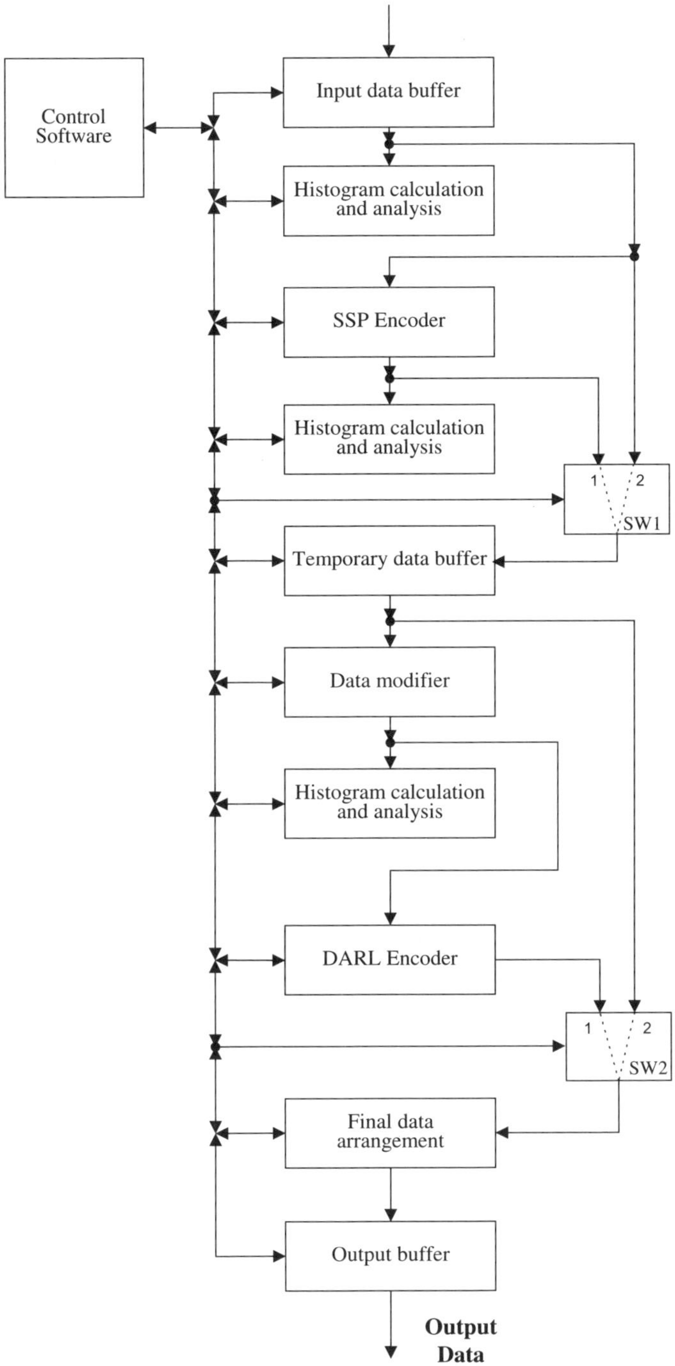

Fig. 17.7. Block diagram of the algorithm for lossless data compression

Fig. 17.8.a Test image of size: 308 × 308 pixels, 8 bpp

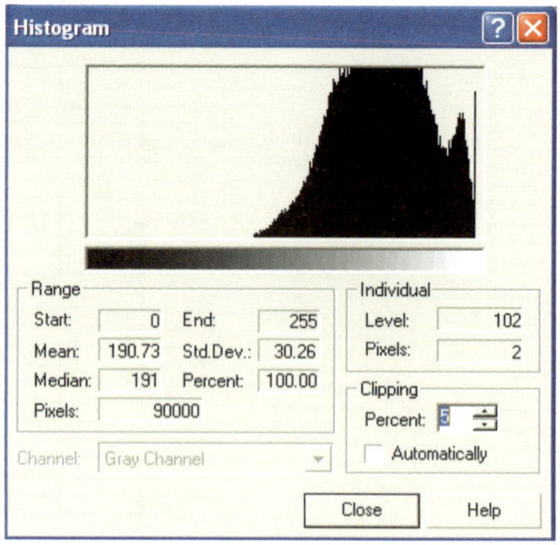

Fig. 17.8.b The image histogram

Fig. 17.9.a Image obtained after histogram modification

Fig. 17.9.b The same test image after filtration and segmentation

Similar results were obtained for the test image shown in Fig. 17.10. The values of the 4 points, corresponding to the objects detected, were: $k_1 = 132$; $k_2 = 216$; $k_3 = 234$; $k_4 = 245$. After merging the parts of the histogram, which correspond to the background, the main object was defined in the brightness range 145–210, which was stretched. The result is shown in Fig. 17.11 and the corresponding histograms — in Fig. 17.12.a, b.

Fig. 17.10. Test image of size: 280 × 280 pixels

Fig. 17.11. Image obtained after histogram modification

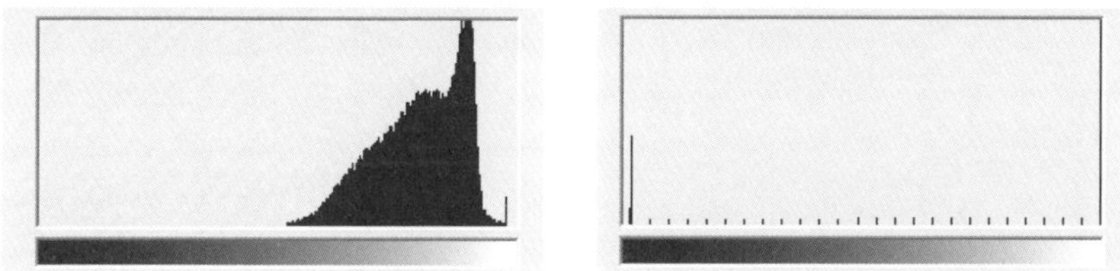

(a) (b)

Fig. 17.12. Image histograms: (a) of the image on Fig. 17.10; (b) of the image on Fig. 17.11

Special attention deserves the test image from Fig. 17.13: after histogram modification the image, shown in Fig. 17.14 was obtained. The brightness range corresponding to the main object was found to be between values 10 and 90.

Similar results were obtained using the method comprising filtration and segmentation: the original image (300 × 320 pixels, 8 bpp), shown in Fig. 17.13, after processing with filter of size 511 pixels (in both directions) and image segmentation, gave the result, shown in Fig. 17.15. After the processing the image quality was enhanced and some specific features (in this case-small scars) were easily noticeable. Additional interest lies in the fact, that the sweat pores were retained in both cases.

The results obtained for the next test image (Fig. 17.16a, b) are shown below.

One more example is shown in Fig. 17.17 (the original) and the image obtained after histogram modification (Fig. 17.18). The original image has two latent fingerprints in it (one of them extremely pale), but after processing the main fingerprint is quite clear and its' further processing is easier. The time necessary for the processing of the test images is less than 1 s.

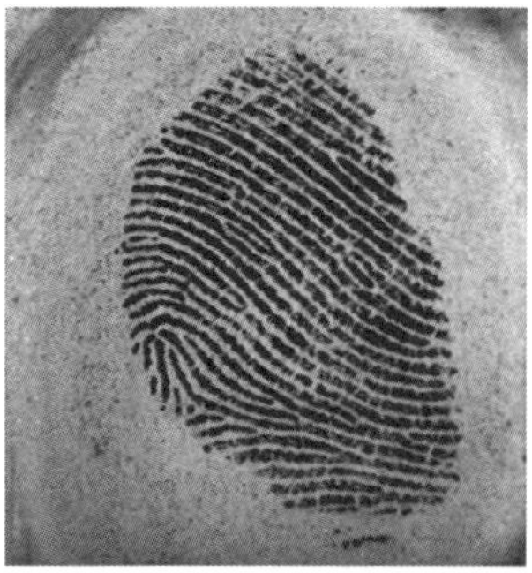

Fig. 17.13. Original image

Fig. 17.14. Image obtained after histogram modification

Fig. 17.15. The test image after filtration and segmentation

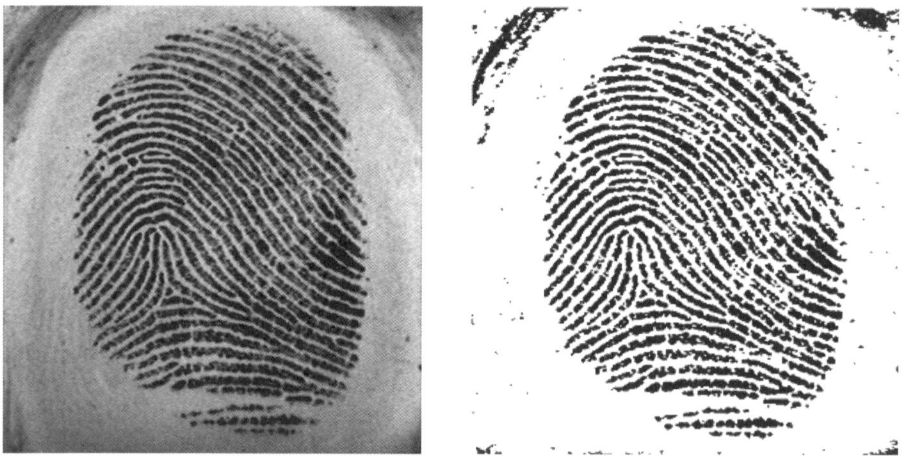

(a) (b)

Fig. 17.16. Test image of size 280 × 300 pixels, 8 bpp. (a) Original test image, (b) After filtration and segmentation

The same approach is suitable for processing of signature images. Some of the test images and the corresponding results obtained are shown in the figures below.

The contrast of the original images is slightly enhanced so that to make the background "noise" easily visible. The images obtained in result of the histogram modification retain the visual quality of the main object, but the background is cleared. This improves the image quality and influences the compression efficiency.

The image shown in Fig. 17.20 was processed in similar way, with same parameter values. After the processing it is easier to evaluate the pen pressure in the signature and it seems more natural. The time necessary for the processing of the test images is less than 1 s.

Fig. 17.17. Original test image (256 × 364 pixels, 8 bpp)

Fig. 17.18. After histogram modification

(a)

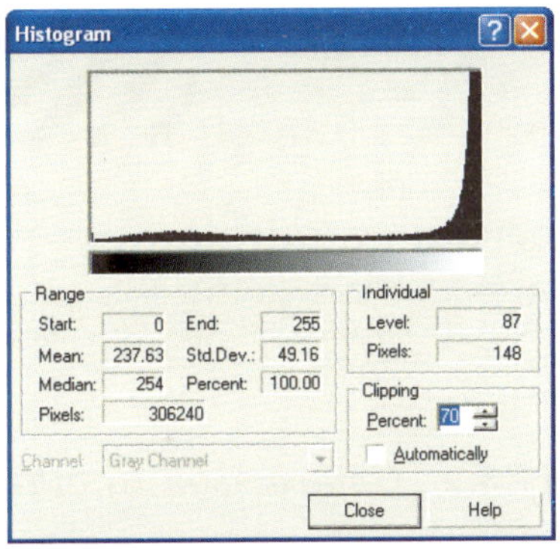

(b)

(c)

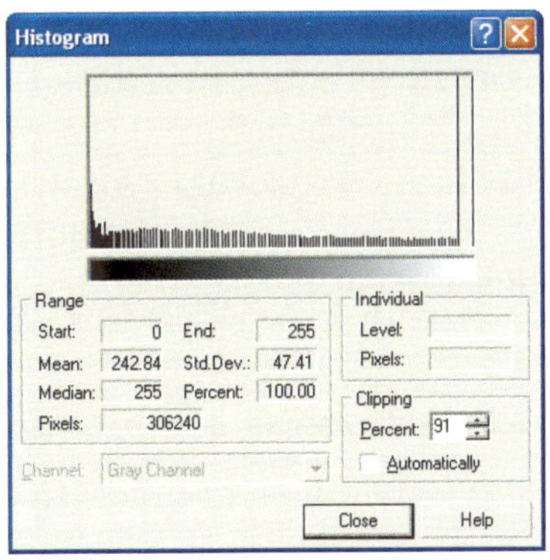

(d)

Fig. 17.19. (a) The original test image, (b) The histogram of the test image, (c) The image obtained after histogram modification (the main object was detected in the brightness range 12–150); (d) The histogram of the processed image.

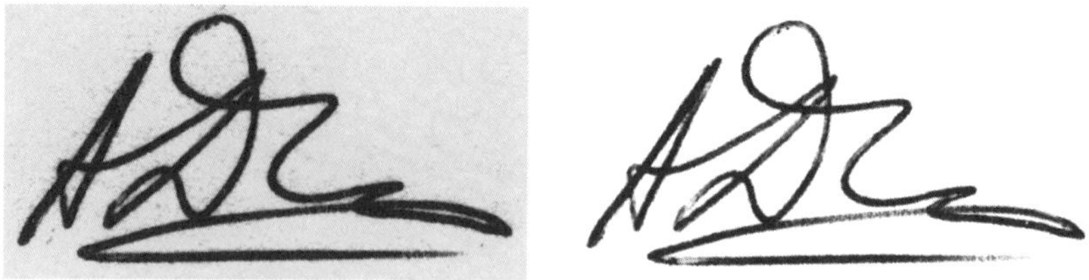

(a) (b)

Fig. 17.20. (a) The original test image; (b) the image obtained in result of the histogram modification

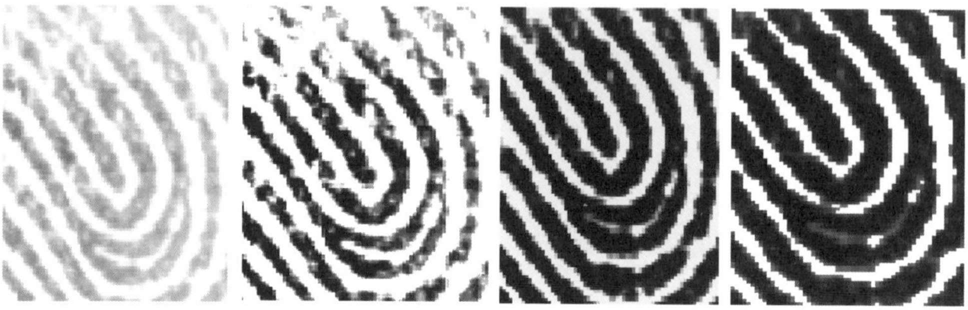

(a) (b) (c) (d)

Fig. 17.21. Results obtained for an enlarged part of fingerprint test image, 140 × 180 pixels: (a) Original image FP11 (enlarged part); (b) After treatment with the new method; (c) After contrast enhancement; (d) After local contrast enhancement.

17.5.2. *Comparison to other similar techniques*

The comparison of the new method performance was done with the methods for global and local contrast enhancement. The original test fingerprint image, chosen for the comparison, had low contrast and contained complicated ridges and many sweat pores. In result of the histogram modification the visual quality of the image was enhanced and the sweat pores retained. The visual quality of the images obtained after global or local contrast enhancement was worse.

The program Corel Photo Paint 10 was used for global and local contrast enhancement.

17.5.3. *Lossless compression*

The Content-based run-length encoding (CBRLE) method for lossless compression is very efficient when images of signatures and fingerprints have to be stored. Its efficiency was compared to that of the JPEG 2000 standard (lossless). Comparison with the WSQ standard has not been included in this work, because the WSQ compression is lossy. The comparison was done with the lossless version of the

Table 17.1. Result obtained after lossless compression of fingerprint images

No.	Image	Image size (pixels)	Original file size [B]	CBRLE compression		Compression JPEG2000	
				CR	bpp	CR	bpp
1	FP101_1	300 × 300	90 000	1,30	6,15	1,94	4,12
2	FP101_7	300 × 300	90 000	1,16	6,89	1,45	5,51
3	FP301_1	448 × 478	214 144	1,11	7,20	1,44	5,55
4	FP301_7	448 × 478	214 144	1,03	7,76	1,41	5,67
5	FP302_3	448 × 478	214 144	1,14	7,01	1,42	5,63
6	FP101_1P	300 × 300	90 000	15,49	0,51	2,48	3,22
7	FP101_7P	300 × 300	90 000	16,06	0,49	2,20	3,63
8	FP301_1P	448 × 478	214 144	22,61	0,35	3,91	2,58
9	FP301_7P	448 × 478	214 144	22,02	0,36	3,85	2,07
10	FP302_3P	448 × 478	214 144	17,22	0,46	2,80	2,85

JPEG 2000 standard only. For this were used the software implementation of the new method, and LuraTech Algovision — for the JPEG 2000 standard.

The investigation was carried out on more than 200 fingerprint images and on about 100 signatures. The results obtained for the CBRLE method and for the JPEG 2000 standard within the investigated image classes were very consistent. In Table 17.1 below are given the results for 5 fingerprint test images only. The first 5 rows in the table show the results obtained for the original images. In the lower 5 rows are given the results obtained for the same 5 images after their processing in accordance with the methods for histogram modification and lossless compression, presented above. It is easy to see that for the original images the JPEG 2000LS offers a little higher compression, but the compression ratios obtained after processing are much higher for the new method. Considering that the processing performed aims to remove the information redundancy, the compression ratio obtained is very good.

As already pointed, the histogram modification retains the brightness values corresponding to the object and changes only their positions in the histogram, i.e. makes the image clearer. The real changes in the processed image concern the background only — in result, it is represented by reduced number of brightness values. The compression ratio obtained with JPEG 2000LS is higher for the processed images as well, but the results are not as impressive as these, with the CBRLE. Just for comparison, the mean compression ratio obtained for fingerprint images with the lossy WSQ standard, is 16 (Hopper *et al.*).

Graphic presentation of the results obtained for the processed images is given in Fig. 17.22.

Similar results were obtained for images of signatures. In Table 17.2 below are given the results for 8 test images. In the first 8 rows are given the results obtained for the test images before, and in the next 8 rows — after their processing. The mean JPEG 2000LS compression ratio (CR) and correspondingly — the bitrate is a little better for the original images, but after processing the results are significantly changed. The visual quality of the processed images is retained and along with

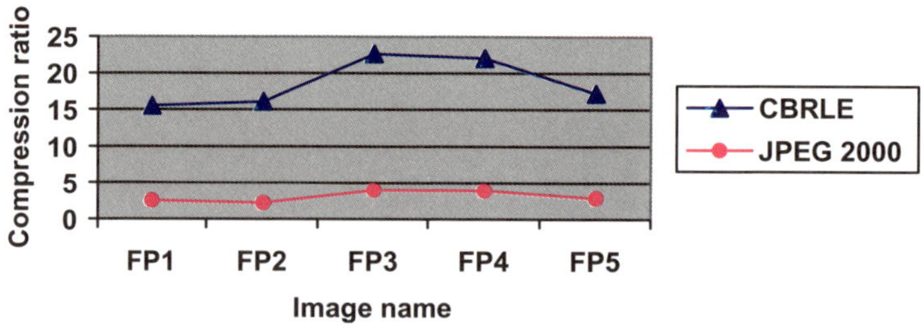

Fig. 17.22. Graphic presentation of the comparison results obtained for fingerprints

Table 17.2. Result obtained after lossless compression of signature images

No.	Image (Grayscale, 8 bpp)	Image size (pixels)	Original file size [KB]	compression CBRLE		Compression JPEG2000LS	
				CR	bpp	CR	bpp
1	St140	567 × 319	941	13,29	0,60	18,30	0,44
2	St163	849 × 197	739	7,68	1,04	11,80	0,68
3	St171	1017 × 401	1160	10,03	0,79	14,60	0,54
4	St18	567 × 319	177	1,26	6,34	1,70	4,70
5	St23	679 × 355	707	7,72	1,03	14,60	0,55
6	St26	514 × 353	532	6,72	1,19	10,90	0,73
7	St63	516 × 380	574	5,78	1,38	8,88	0,90
8	St155	660 × 343	663	6,59	1,21	10,70	0,74
9	St 140P	567 × 319	941	43,66	0,18	29,00	0,27
10	St 163P	849 × 197	739	19,85	0,40	15,60	0,51
11	St171P	1017 × 401	1160	35,10	0,22	25,00	0,32
12	St18P	567 × 319	177	10,72	0,74	7,12	1,12
13	St23P	679 × 355	707	34,44	0,23	23,70	0,33
14	St26P	514 × 353	532	23,72	0,33	15,20	0,52
15	St63P	516 × 380	574	21,08	0,37	16,20	0,49
16	St155P	660 × 343	663	17,64	0,45	14,80	0,54

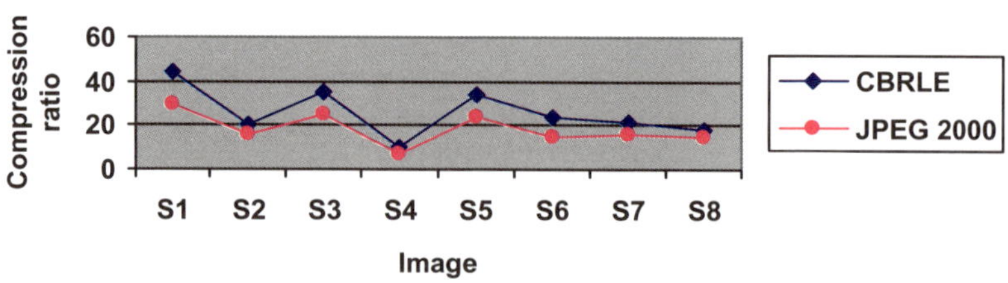

Fig. 17.23. Graphic presentation of the comparison results obtained for signature images

this, some of their most significant qualities are visualized better and make other operations like matching, etc., easier.

The graphic presentation of the results obtained for the compression ratio after the image processing is shown in Fig. 17.23.

Fig. 17.24.a Original test image St18

Fig. 17.24.b The processed test image St18

Fig. 17.25.a Original test image St63

Fig. 17.25.b The processed test image St63

The method efficiency is more important for signatures than for fingerprints. This is because the signature takes relatively small part of the image, while the fingerprint is usually placed over the whole picture and the background is smaller. In result of the processing, the background of the signature, which is usually rough and seems gray, is cleared and the signature appears much better. Besides, some parts of the signature become clearer and pen movement and pressure could be easily traced. This is quite clear in Fig. 17.24, where in the original image the curve seems equally dark, while after processing the parts, where the pen pressure has been lower, are lighter and thinner. Similar result is shown in Fig. 17.20, above. The results obtained for the test image St63 are shown in Fig. 17.25a, b and again confirm the conclusions. For better visibility the contrast of the original images was slightly enhanced.

The examples show that in result of the processing the meaning information in the processed image is retained unchanged. In case that the image background contains some special visual information, for example a picture, etc., the presented method permits after signature extraction, to process the remaining part of the image with some kind of lossy compression and then to recover the processed image. Some investigations, already carried out, prove that the visual quality and the compression ratio in such cases are enhanced as well.

17.6. Conclusion

The experimental results proved the efficiency of the proposed methods. Compared to other methods for contrast enhancement (Yang *et al.*; Pratt), the new method which is based on **histogram modification** has specific advantages:

— *Lower computational complexity* than that of the methods for contrast enhancement which are based on image filtration;
— *Adaptation* — the image processing is performed using the results from the histogram analysis, which permits the values of the participating parameters to be calculated and set automatically in correspondence to the image data statistics;
— *Retaining of the object features* — unlike the famous methods for contrast enhancement, which are based on image filtration, there are no changes in the frequency domain of the processed image and in result all the available information about the main object is preserved. Along with this, the visual quality of the images processed in accordance with the histogram modification method is better than the one, obtained with methods, based on global or local contrast enhancement.

The second method for processing of fingerprints and signatures is based on **image filtration and segmentation** and has the following specific advantages:

— It offers reliable capacity for features extraction: the image, obtained after the segmentation, can be used for detection and evaluation of some of the basic fingerprint minutiae.
— Another advantage is that the binarized image could be used as a mask, in order to facilitate other operations, concerning the object (fingerprint or signature) extraction and matching.

The most important features of the presented algorithm for **lossless data compression** are:

— The basic advantage is that in result of the presented lossless coding the processed data volume could never be increased with more than one word (in case that the processed data are not suitable for compression) while in all other cases the data volume is significantly decreased.
— One more advantage is that the new method is suitable for various applications: processing of data sequences, images (color or grayscale), texts, graphic information, scanned documents, etc.
— The method efficiency is very high when graphics, texts or biometric images are compressed, i.e. when the processed images contain only limited number of brightness values.
— The computational complexity of the algorithm is low, because it requires no interpolations, decimations, multiplications or divisions and in result the time necessary for the processing of one image of size up to 512×512 is less than 1s.

— The compression is adaptive to the processed data contents.
— The coding side does not create a coding table and therefore such table is not necessary for decoding and should not be transferred to the receiving side, which enhances the method efficiency.

The presented method has already been tested for various applications: still images of texts and graphics, biometric information (fingerprints, signatures, and voice samples), compound images (comprising texts, graphics and pictures) (Kountchev *et al.*, 2006, 2008). The software implementation of the method has proved its efficiency for all those applications and is suitable for real-time use.

The future development of the method will be directed to the definition of global constant parameters for the two image classes presented here (fingerprints and signatures). This will permit the use of smaller headers in the compressed data files and will ensure in result higher compression ratios and faster processing. Along with this, such approach will allow the development of a global format for efficient lossless compression of biometric data.

Acknowledgement

This work was supported by the National Fund for Scientific Research of the Bulgarian Ministry of Education and Science (Contract BY-TH 202).

References

Bovik Al. (Ed.). (2000) *Handbook of Image and Video Processing.* Academic Press.

Bow S. (2002) *Pattern Recognition and Image Preprocessing*, Marcel Dekker.

Fuentes D., Mostefa D., Kharroubi J., Garcia-Salicetti S., Dorizzi B., Chollet G. (2002) Identity verification by fusion of biometric data: on-line signatures and speech, *Proc. of COST Workshop on Advent of Biometrics on the Internet*, pp. 83–86.

Gonzalez R. and Woods R. (2002) *Digital Image Processing*, Prentice-Hall.

Hopper T., Brislawn C., Bradley J. (1993) WSQ Gray-scale fingerprint image compression Specification. *FBI, Doc. No. IAFIS-IC-0110.*

International Standards Organization, ISO/IEC JRC1/SC37. (2004) *Working draft, Biometric data interchange formats-Part 4: Finger image data, Switzerland.*

Kim Y. (1997) Contrast Enhancement Using Brightness Preserving Bi-Histogram Equalization. *IEEE Trans. on Consumer Electronics*, Vol. 43, No. 1, pp. 1–8.

Kountchev R., Milanova M., Todorov Vl. and Kountcheva R. (2006) Adaptive Fuzzy Filter for the Reduction of Blocking Artifacts in Images Compressed with IDP Decomposition and JPEG. *WSEAS Trans. on SP.* Issue 7, Vol. 2, pp. 941–948.

Kountchev R., Todorov Vl., and Kountcheva R. (2008) Adaptive preprocessing of scanned documents. *Proc. of WSEAS Intern. Conf. on Mathematical Methods and Computational Techniques in Electrical Engineering*, pp. 34–39.

Lau K. K., Yuen P. C., and Tang Y. Y. (1990) Advances in Handwriting Recognition. An Efficient Function-based On-line Signature Recognition System, *World Scientific*, pp. 559–568.

Lim J. *Two-dimensional Signal and Image Processing.* Prentice Hall.

Plamondon R., and Lorette G. (1989) On-line signature verification. *Proc. of the IEEE International Carnahan Conference on Security Technology*, Zurich, Switzerland, pp. 183–191.

Pratt W. (2007) *Digital Image Processing*, J. Wiley & Sons.

Rowberg A. and Malcolm B. (2004) Distortion-free Image Contrast Enhancement, *EuroPACS-MIR in the Enlarged Europe*, P. Inchingolo and R. Pozzi-Mucelli (Eds), pp. 357–360.

Sharma G. (Ed.). (2003) *Digital Imaging Color Handbook*, CRC Press, New York.

Yang J., Lin L., Jiang T. and Fan Y. (2003) A Modified Gabor Filter Design Method for Fingerprint Image Enhancement, *Pattern Recognition Letters*, Vol. 24, pp. 1805–1817.

Young I., Gerbrands J., and van Vliet L. (1996) *Image Processing Fundamentals.* Delft, Netherland.

Chapter 18

KNOWLEDGE REPRESENTATION FOR ELECTRONIC CIRCUITS IN LOGIC PROGRAMMING

Takushi Tanaka
Department of Computer Science, Fukuoka Institute of Technology
3-30-1 Wajiro-Higashi, Higashi-ku, Fukuoka Japan, 811-0295,
tanaka@fit.ac.jp

This chapter presents a brief introduction to logic programming and its application to knowledge representation for electronic circuits. We first show how circuits can be represented in terms of predicate logic. Basic concepts in logic programming such as facts, rules, and goals are introduced with circuits as examples. Next, we develop a grammatical method for circuit representation. Circuits are viewed as sentences, and their elements as words. Circuit structures are defined by a logic grammar called DCSG for word-order free languages. A set of grammar rules, when converted into Prolog clauses, forms a logic program which performs top-down parsing. Furthermore, we extend DCSG by introducing semantic terms into the grammar rules. These semantic terms define relationships between syntactic structures and their meanings. By assuming circuit functions to be the meanings of the circuit structures, knowledge of circuit structures and functions is coded into grammar rules. When a given circuit is parsed, not only its syntactic structure is determined, but also its functions are derived as meanings of the sentence.

18.1. Introduction

Prolog is a programming language based on predicate logic. It is well suited for symbolic computations that handle problems between objects and their relations. Programs written in ordinary programming language define procedures to solve problems, while Prolog programs define problems themselves and relationships between objects. Prolog programs are executed by a mechanism of theorem proving. Instead of defining the procedure explicitly, solving problems with the mechanism of theorem proving is called logic programming.

Using electronic circuits as examples, we introduce the basic concepts of logic programming. We first show how circuits are represented in terms of predicate logic. A given circuit is defined by a set of Prolog facts. Circuit structures are defined by Prolog rules. Finding structures in the given circuit is realized by Prolog goals. The examples presented here also show the problems and limitations of this method.

Next, we develop a new method for circuit representation to overcome these problems. In this new method, circuits are viewed as sentences in a word-order free language, and their elements as words. Circuit structures are defined by grammar rules. In order to implement this method, we introduce a logic grammar called DCSG (Definite Clause Set Grammar) for word-order free languages.[4] A set of grammar rules, when converted into Prolog clauses, forms a logic program which performs top-down parsing.

Electronic circuits, both analogue and digital, are designed as a hierarchical structure of functional blocks. Each functional block consists of sub-functional blocks, and each sub-functional block is also decomposed further into sub-sub-functional blocks until the individual parts are reached. Since these hierarchical structures are analogous to syntactic structures of language, DCSG is useful for defining syntactic structures of electronic circuits. However, DCSG is not able to represent knowledge about circuit functions.

Since we need a method to represent knowledge of the relationship between circuit structures and their functions, we develop a new circuit grammar[8] by extending DCSG. The new circuit grammar has semantic fields to define relationships between syntactic structures and their meanings. By assuming circuit functions to be the meanings of the circuit structures, knowledge concerning circuit structures and functions is coded into the grammar rules. When a given circuit is parsed, not only its syntactic structure but also its functions are derived as meanings of the sentence.

18.2. Circuit Representation in Prolog

18.2.1. *Facts*

Using electronic circuits as examples, we will introduce basic concepts of logic programming in Prolog. All objects in circuits are represented by logical nouns called terms, and all relationships between those objects are represented by logical predicates. The terms and the predicates are combined to form atomic formulas, which are logical sentences. The following six atomic formulas represent the circuit *ca*39 shown in Figure 18.1.

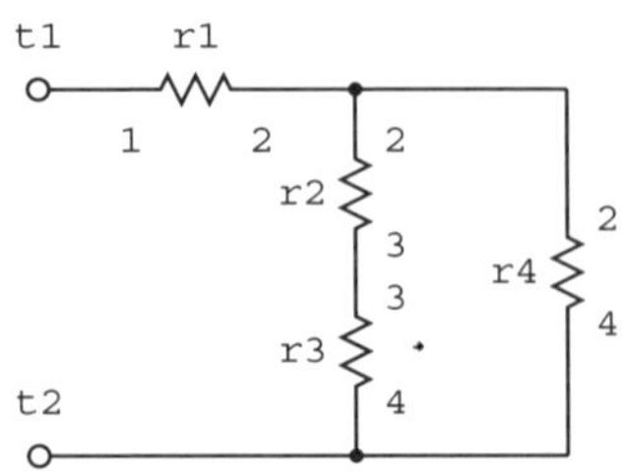

Fig. 18.1. Circuit ca39

$resistor(r1, 1, 2).$
$resistor(r2, 2, 3).$
$resistor(r3, 3, 4).$
$resistor(r4, 2, 4).$
$terminal(t1, 1).$
$terminal(t2, 4).$

The atomic formula "$resistor(r1, 1, 2)$" consists of a predicate symbol "$resistor(...)$" and three constant terms "$r1$", "1" and "2". The atomic formula states that the resistor named $r1$ is connected to node 1 and node 2. The atomic formula "$terminal(t1, 1)$" states that the external terminal named $t1$ is connected to node 1. These six logical sentences describe whole circuit topology of $ca39$. When we consider the circuit in Figure 18.1, these sentences are called "facts", and placed in the Prolog database, which holds true sentences.

18.2.2. *Goal*

Since we are considering the circuit $ca39$ in Figure 18.1, all the formulas that represent the circuit are already placed in the Prolog database. The following atomic formula prefixed by "?−" is called a goal clause, which asks the Prolog system whether the resistor $r1$ is connected to the nodes 1 and 2. Here, "?−" is a prompt for goal clause generated by the Prolog system. When the goal clause is given, the Prolog system looks for the goal in the Prolog database. Since $resistor(r1, 1, 2)$ is in the database, the goal becomes *true*, and the Prolog system answers *yes*.

$? - \ resistor(r1, 1, 2).$
yes

We can find the nodes to which the resistor "$r2$" is connected by the following goal clause with variables A and B. These variables are also terms which construct atomic formulas. Here, the variables are represented by a string beginning with an upper case letter.

$? - \ resistor(r2, A, B).$

The variables A and B in the goal clause are assumed to be bound by existential quantifiers in predicate logic. Namely, the Prolog system is asked to prove the sentence

$$(\exists A)(\exists B)resistor(r2, A, B)$$

which states that "there exist nodes A and B connected to the resistor $r2$". The Prolog system looks in the database and finds the fact $resistor(r2, 2, 3)$ as an instance for which $resistor(r2, A, B)$ becomes *true*, and the system outputs the variable-value bindings as:

$$A = 2$$
$$B = 3$$
yes

These variable-value bindings are made by a mechanism called unification. Unification discovers a variable substitution for the two formulas $resistor(r2, 2, 3)$ and $resistor(r2, A, B)$ that makes them equal.

The following goal clause finds a resistor "X" connected to the nodes 3 and 4 as:

$$?-\ resistor(X, 3, 4).$$
$$X = r3$$
yes

Note, however, that if the node order is reversed, the goal fails to find the resistor connected to nodes 4 and 3:

$$?-\ resistor(X, 4, 3).$$
no

Since resistors are non-polar elements, we want to refer to resistors regardless of their node order. This can be done by defining a new predicate using rules.

18.2.3. *Rules*

A rule in Prolog is a conditional sentence called a definite clause, which is also placed in the Prolog database. A rule consists of a left-hand side called a head, a special symbol " : –", and a right-hand side called a body. The head consists of an atomic formula which is the result of the conditional sentence. The symbol ": –" is the logical connective of implication "←", although the order of the condition and result are reversed. The body, which is the conditional part, consists of atomic formulas. Facts in Section 18.2.1 can be viewed as a special case of rules which do not have conditions.

The following rule defines the new predicate "$res(R, A, B)$" which can refer either $resistor(R, A, B)$ or $resistor(R, B, A)$. Here, the symbol ";" works as a logical connective "*or*". This rule can be read if $resistor(R, A, B)$ or $resistor(R, B, A)$ is *true*, $res(R, A, B)$ becomes *true*. Procedurally, in order to show that $res(R, A, B)$ is *true*, the Prolog system tries to show that either $resistor(R, A, B)$ or $resistor(R, B, A)$ is *true*. The variables R, A, and B in facts and rules are assumed to be bound by universal quantifiers in predicate logic.

$$\begin{aligned} res(R, A, B)\ :-\ & resistor(R, A, B); \\ & resistor(R, B, A). \end{aligned}$$

When the following goal is given, the goal is unified with the head of this rule, and the body of the rule becomes a new goal.

$$?-\ res(X, 4, 3).$$

The body generates the following disjunction as the new goal.

$$?-\ resistor(X, 4, 3);\ resistor(X, 3, 4).$$

The first sub-goal of disjunction fails because there is no fact unified with $resistor(X, 4, 3)$ in the database, but the second sub-goal $resistor(X, 3, 4)$ succeeds, and outputs as:

$X = r3$
yes

Similar rules can be defined for other non-polar elements such as capacitors and inductors:

$$\begin{aligned} cap(C, A, B)\ :-\ & capacitor(C, A, B); \\ & capacitor(C, B, A). \end{aligned}$$

$$\begin{aligned} ind(L, A, B)\ :-\ & inductor(L, A, B); \\ & inductor(L, B, A). \end{aligned}$$

The following rule defines the predicate "$anyElm(X, A)$" which refers to any element X connected to node A. Here, we assume the external terminal "$terminal(X, A)$" to be a kind of any element. The underscores "_" in atomic formulas are anonymous variables which are not of concern in the rule.

$$\begin{aligned} anyElm(X, A)\ :-\ & terminal(X, A); \\ & res(X, A, \_\ ); \\ & cap(X, A, \_\ ); \\ & ind(X, A, \_\ ). \end{aligned}$$

18.2.4. *Predicates for circuit structures*

In order to find resistors connected in series (Figure 18.2) in the circuit *ca*39, we attempt to satisfy the following conjunctive goal. Here, "," between two atomic formulas works as the logical connective "*and*".

$$?-\ res(X, A, B),\ res(Y, B, C).$$

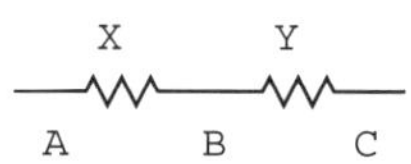

Fig. 18.2. Resistors connected in series

This conjunctive goal successfully finds resistors connected in series. The first sub-goal $res(X, A, B)$ finds the resistor $r2$ and its connecting nodes 2 and 3, then the second sub-goal $res(Y, B, C)$ finds the resistor $r3$ connecting node 3 as:

$$\begin{aligned}&A = 2\\&B = 3\\&C = 4\\&X = r2\\&Y = r3;\end{aligned}$$

The ";" after $Y = r3$ is typed by the user and directs the Prolog system to find another answer. The Prolog system discard the first answer, and tries to find another answer using the Prolog backtracking mechanism. The conjunctive goal also outputs unexpected answers such as:

$$\begin{aligned}&A = 1\\&B = 2\\&C = 1\\&X = r1\\&Y = r1\end{aligned}$$

In the series circuit shown in Figure 18.2, neither elements X and Y nor nodes A and C may be the same. Since the condition "$not\ A = C$" topologically includes the condition "$not\ X = Y$", we next try proving the following goal:

$$?-\ \ res(X, A, B),\ res(Y, B, C),\ not\ A = C.$$

The Prolog system ceases to output undesired answers of the above type, but it still outputs another type of unexpected answers as follows.

$$\begin{aligned}&A = 1\\&B = 2\\&C = 3\\&X = r1\\&Y = r2\end{aligned}$$

No element other than X and Y may be connected to the central node B of the series circuit. This can be expressed by first defining an element Z other than X and Y connected to B as:

$$\begin{aligned}otherElm(Z, B, X, Y) :-\ \ &anyElm(Z, B),\\&not\ Z = X,\\&not\ Z = Y.\end{aligned}$$

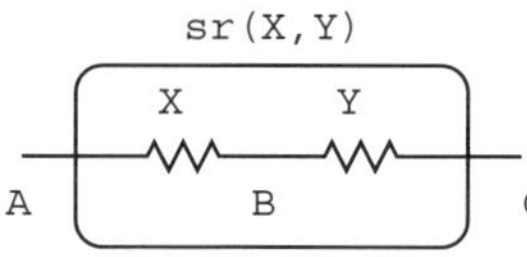

Fig. 18.3. Series connection of resistors

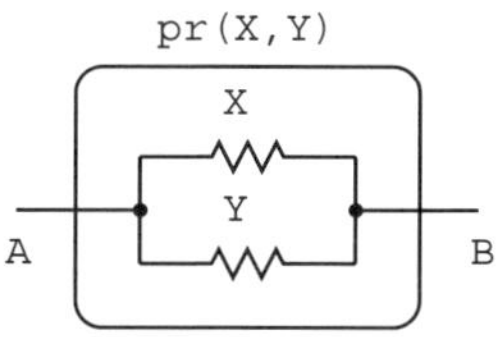

Fig. 18.4. Parallel connection of resistors

Now, we can define the predicate $rSeries(sr(X,Y),A,C)$ for series circuit of resistors by adding the conditions "$not\ A = C$" and "$not\ otherElm(\_\ ,B,X,Y)$" to reject undesired answers.

$$\begin{aligned} rSeries(sr(X,Y),A,C) :- \ & res(X,A,B),\\ & res(Y,B,C),\\ & not\ A = C,\\ & not\ otherElm(\_\ ,B,X,Y). \end{aligned}$$

The first argument "$sr(X,Y)$" of the predicate $rSeries(\ldots)$ is the name given to the series circuit of resistors X and Y. The name is a form of compound term made of the function symbol $sr(\ldots)$ and variables X and Y. Namely, the name is given by depending on the value of X and Y. This is a technique to give an unique name to a new object, and the function is called a Skolem function.

The predicate for parallel connections is also defined as the same manner:

$$\begin{aligned} rParallel(pr(X,Y),A,C) :- \ & res(X,A,B),\\ & res(Y,A,B),\\ & not\ X = Y. \end{aligned}$$

18.2.5. *Difficulties in circuit representation using predicates*

First we defined terms and predicates for circuit elements, and then we defined predicates for abstract elements and circuit structures. If we could define predicates for all concepts in circuits in the same manner, we would have an axiomatic system of circuits which can automatically derive true sentences on circuits.

However, our circuit representation faces difficulties when we try to define predicates for relationships between two circuits such as equivalent circuit. In order to refer to relationships between circuits, we need a mechanism to identify a set of facts for a specific circuit.

One method is to introduce a new argument for circuit identification into each circuit predicate as:

$$
\begin{aligned}
&resistor(ca39, r1, 1, 2).\\
&resistor(ca39, r2, 2, 3).\\
&resistor(ca39, r3, 3, 4).\\
&resistor(ca39, r4, 2, 4).\\
&terminal(ca39, t1, 1).\\
&terminal(ca39, t2, 4).
\end{aligned}
$$

The first argument $ca39$ of each predicate indicates that each element belongs to the circuit $ca39$.

Although we can refer to relationships between two circuits using identification, we have another problem, namely rewriting circuits. In circuit analysis, we often rewrite circuits into equivalent circuits. Rewriting facts in the Prolog database can be done using the built-in predicates "*assert*" and "*retract*". However using *retract* is problematical from a logical point of view, because using *retract* means erasing facts which were given as true sentence, and the erased facts can never be used again. In the following sections, we resolve these problems by assuming circuits to be a kind of formal language.

18.2.6. *Changing circuit representation*

We have already introduced two kinds of terms, constants and variables, to represent objects in circuits. These terms are viewed as nouns to make logical sentences. Predicate logic has another kind of term, called a compound term, which consists of a function symbol and other terms. In the previous sections, the atomic formula $resistor(r1, 1, 2)$ is a logical sentence which states that "$r1$" is a resistor connected to node 1 and node 2. Here, the $resistor(\ldots)$ worked as a predicate symbol. As predicates and functions are the same in style, we will change $resistor(\ldots)$ from a predicate symbol to a function symbol. As a result, $resistor(r1, 1, 2)$ becomes a compound term instead of an atomic formula. Unlike mathematics, in logic, functions are not computed. The compound term works as a noun phrase, while the atomic formula works as a simple sentence. An atomic formula states a relationship between objects while a compound term represents a new object using other objects. Therefore, the compound term $resistor(r1, 1, 2)$ can be read "resistor $r1$ connected to node 1 and 2" as though it were a noun phrase.

Prolog supports a special data structure called "list". A list is a sequence of any number of terms surrounded by "[" and "]". We can use a list as a new circuit representation for ca39 as follows:

$$
\begin{aligned}
&[resistor(r1, 1, 2),\ resistor(r2, 2, 3),\ resistor(r3, 3, 4),\\
&\ resistor(r4, 2, 4),\ terminal(t1, 1),\ terminal(t2, 4)].
\end{aligned}
$$

Here, the list is used to represent the set of all the elements that make up the circuit *ca*39. In the next section, we develop a mechanism called DCSG to treat these lists as a word-order free sentence.

18.2.7. *Lists*

Since we will use lists to represent circuits, we discuss lists further here. A list can be viewed as a compound term made by iteratively applying the special function ".". For example, the list $[a,\ b,\ c]$ consists of the function "$.(a,\ .(b,\ .(c, [\ ])))$". Here, a list without elements is called the empty list and is simply written as "[ ]". Using the empty list, the special function "$.(c, [\ ])$" makes the list "$[c]$". The function "$.(b,\ [c])$" makes the list "$[b,\ c]$". And the function "$.(a,\ [b,\ c])$ makes the list "$[a,\ b,\ c]$".

The first element of a list is called the head of the list. The remaining part is another list and is called the tail. That is, a list consists of "$.(Head,\ Tail)$". The special symbol "|" also separates a list into a head and a tail and composes a list as shown in the following goal clauses:

$?-\ [a] = [Head \mid Tail].$
$Head = a$
$Tail = [\ ].$

$?-\ [a,\ b,\ c] = [Head \mid Tail].$
$Head = a$
$Tail = [b,\ c]$

$?-\ X = [a,\ b \mid [c,\ d]].$
$X = [a,\ b,\ c,\ d]$

$?-\ X = [a \mid [b \mid [c \mid [\ ]]]].$
$X = [a,\ b,\ c]$

18.3. Logic Grammar DCSG

18.3.1. *Word-order free language*

Most Prolog systems provide a mechanism for parsing context-free languages called DCG(Definite Clause Grammar).[3] A set of the grammar rules, when converted into Prolog clauses, forms a logic program which executes top-down parsing. Here, we develop a logic grammar DCSG(Definite Clause Set Grammar)[4] for word-order free language similar to the method of DCG.

First, we will introduce a concept of word-order free language. A word-order free language **L(G')** is defined by modifying the definition of a formal grammar. We define a context-free word-order free grammar **G'** to be a quadruple $< V_N, V_T, P, S >$ where: V_N is a finite set of non-terminal symbols, V_T is a finite set of terminal symbols, P is a finite set of grammar rules of the form:

$$A \longrightarrow B_1, B_2, ..., B_n. \quad (n \geq 1)$$
$$A \in V_N, \quad B_i \in V_N \cup V_T \quad (i = 1, ..., n)$$

and $S(\in V_N)$ is the starting symbol. The above grammar rule means that the symbol A is rewritten not with the string of symbols "$B_1, B_2, \ldots, B_n$", but with the set of symbols $\{B_1, B_2, \ldots, B_n\}$. A sentence in the language **L(G')** is a set of terminal symbols which is derived from S by successive application of grammar rules. Here the sentence is a multi-set which admits multiple occurrences of elements taken from V_T. Each non-terminal symbol used to derive a sentence can be viewed as a name given to a subset of the multi-set.

18.3.2. *DCSG conversion*

When a set of grammar rules is given to a Prolog system, the DCG mechanism is used to convert the grammar rules into Prolog clauses. We now develop a conversion for word-order free languages that is analogous to DCG conversion. The general form of the conversion procedure from a grammar rule

$$A \longrightarrow B_1, B_2, ..., B_n. \tag{1}$$

to a Prolog clause is:

$$\begin{aligned} subset(A, S_0, S_n) \; :- \; & subset(B_1, S_0, S_1), \\ & subset(B_2, S_1, S_2), \\ & \ldots \\ & subset(B_n, S_{n-1}, S_n). \end{aligned} \qquad (1)'$$

Here, all symbols in the grammar rule are assumed to be non-terminal symbols. If "$[B_i]$"$(1 \leq i \leq n)$ is found in the right hand side of grammar rules, where "B_i" is assumed to be a terminal symbol, then "$member(B_i, S_{i-1}, S_i)$" is used instead of "$subset(B_i, S_{i-1}, S_i)$" in the conversion.

The arguments $S_0, S_1, \ldots, S_n$ in $(1)'$ are multisets of V_T, represented as lists of elements. The predicate "*subset*" is used to refer to a subset of an object set which is given as the second argument, while the first argument is the name of its subset. The third argument is a complementary set which is the remainder of the second argument less the first; e.g. "$subset(A, S_0, S_n)$" states that "A" is a subset of S_0 and that S_n is the remainder.

The predicate "*member*" is defined by the Prolog clauses (2) and (3) below. It has three arguments. The first is an element of a set. The second is the whole set. The third is the complementary set of the first argument.

$$member(M, [M|X], X). \tag{2}$$
$$member(M, [A|X], [A|Y]) \; :- \; member(M, X, Y). \tag{3}$$

When the clause (1)' is used in parsing, an object sentence (multiset of terminal symbols) is given to the argument S_0. In order to find the subset A in S_0, the first sub-goal finds the subset B_1 in S_0 then put the remainder into S_1, the next sub-goal finds B_2 in S_1 then put the remainder into $S_2, \ldots,$ and the last sub-goal finds B_n in S_{n-1} then put the remainder into S_n. That is, when a grammar rule is used in parsing, each non-terminal symbol in the grammar rule makes a new set from the given set by removing itself as its subset. While, each terminal symbol used in the grammar rule also makes a new set from the given set by removing itself as its member.

DCSG uses the predicates *subset* and *member* to convert grammar rules into Prolog clauses, but the differences between DCG and DCSG are minimal. If we replace the predicate *subset* with *substring* and remove the clause (3) from the definition of *member*, the conversion will be equivalent to DCG conversion, although ordinary DCG does not use the predicate of *substring* for simplification.

18.3.3. *Backward chaining and top down parsing*

Usually, a Prolog program consists of two kinds of definite clauses, called facts $\{F_1, F_2, \ldots, F_n\}$ and rules $\{R_1, R_2, \ldots, R_m\}$ both viewed as axioms. The execution of a Prolog program can be viewed as a process of deriving a theorem by backward chaining from the axioms. The top-down parsing of a word-order free sentence somewhat resembles the process of backward chaining. The object sentence is given as a set of terminal symbols $\{W_1, W_2, \ldots, W_n\}$. The starting symbol "$S$" is decomposed into terminal symbols using grammar rules $\{G_1, G_2, \ldots, G_m\}$ until they coincide with the given sentence. That is, the set of facts corresponds to the sentence, and the set of backward chaining rules corresponds to the set of grammar rules. Deriving theorems in backward chaining corresponds to identifying non-terminal symbols in the top-down method.

There is an important difference between backward chaining and top-down parsing. Backward chaining allows multiple use of the same fact to derive a theorem, while in a context-free language, each terminal symbol in a sentence contributes only once to the reduction of non-terminal symbols. This characteristic is very useful for avoiding a common looping problem in backward chaining, a problem which is caused by multiple use of the same fact.

18.3.4. *The looping problem*

When problems to be solved are formalized and expressed in Prolog, we often encounter a certain kind of looping problem. We will clarify a cause of this looping problem using voltage derivation as an example.

Assume that the voltages in a circuit are those given in Figure 18.5. We might consider representing this voltage data by the following facts:

$voltage(1, 2, 20).$
$voltage(3, 2, 15).$
$voltage(3, 4, 8).$

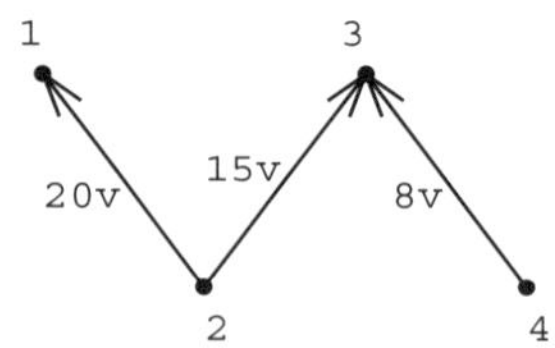

Fig. 18.5. Voltages on a circuit

"$voltage(1, 2, 20)$" states that the voltage between node 1 and node 2 is 20 volts. In order to derive the voltage data independently of the node order, we could consider defining "$volt$" by the following rules:

$$\begin{aligned} &volt(A, B, V) \ :- \ voltage(A, B, V). \\ &volt(A, B, -V) \ :- \ voltage(B, A, V). \end{aligned}$$

Furthermore, we will define the predicate "v" that derives voltages between two arbitrary nodes A and C as

$$\begin{aligned} &v(A, C, V) \ :- \ volt(A, C, V). \\ &v(A, C, V + W) \ :- \ volt(A, B, V), \ v(B, C, W). \end{aligned}$$

It turns out, however, that these definitions will not work as intended. In order to derive the voltage between 1 and 4, we will attempt to execute the following goal clause:

$$?- \ v(1, 4, X).$$

As the voltage between 1 and 4 is not given, the goal is decomposed into sub-goals by the second definition of "$v(\ldots)$". The first sub-goal succeeds by "$volt(1, 2, 20)$" binding node B with 2. The second sub-goal "$v(2, 4, W)$" is also decomposed into sub-goals by the second definition of "$v(\ldots)$" again. The first sub-goal succeeds as "$volt(2, 1, -20)$" using the same voltage data, and the second sub-goal becomes the same as the initial goal. Thus, the system loops.

One method to avoid this problem is to erase voltage data as they are used, so that the same datum is not used twice. This can be done by replacing the definition of "$volt(\ldots)$" with

$$\begin{aligned} volt(A, B, V) \ :- \ & voltage(A, B, V). \\ & retract(voltage(A, B, V)). \\ volt(A, B, -V) \ :- \ & voltage(B, A, V). \\ & retract(voltage(B, A, V)). \end{aligned}$$

But the erased data cannot be recovered in backtracking.

Another common method keeps track of the data used, so that the same datum is not used twice. In this method, the definition of "$v(\ldots)$" is replaced with the following clauses, and we acquire the voltage between nodes 1 and 4 by the goal clause "$?- \ v(1, 4, X, [\,])$". The fourth argument of the goal is a list of nodes which have already been used to calculate the voltages and must not be used twice. This method has the disadvantage of requiring the overhead of explicitly keeping track of the data used:

$$\begin{aligned} v(A, C, V, \_) &:- \ volt(A, C, V). \\ v(A, C, V + W, T) &:- \ volt(A, B, V), \\ &\qquad not\ member(B, T, \_), \\ &\qquad v(B, C, W, [A|T]). \end{aligned}$$

In the next section, we show how to avoid this problem by viewing problem solving as a generalized parsing problem.

18.3.5. *Solution of the looping problem*

To solve the above looping problem, we introduce a change of representation which involves viewing the voltage derivation problem not as a backward search problem, but as a parsing problem. The node-voltage data are not representated by a set of facts. Each expression "$voltage(A, B, V)$" forms a compound term. The voltages in Figure 18.5 are represented by a set of those terms using a list:

$$vData([voltage(1, 2, 20),\ voltage(3, 2, 15),\ voltage(3, 4, 8)]).$$

Accordingly, we represent the voltage derivation not as clauses for backward chaining but as grammar rules for parsing. The following grammar rules correspond to the clauses of "$volt(\ldots)$":

$$\begin{aligned} volt(A, B, V) &\longrightarrow [voltage(A, B, V)]. \\ volt(A, B, -V) &\longrightarrow [voltage(B, A, V)]. \end{aligned}$$

"$voltage(A, B, V)$" surrounded by "[" and "]" is a terminal symbol, while "$volt(A, B, V)$" is a nonterminal symbol. Here, we have introduced universally quantified variables (A, B, and V) into the grammar rules. These variables are instantiated when they are applied to object sentences. According to the DCSG conversion procedure, the grammar rules are converted into the following clauses:

$$\begin{aligned} subset(volt(A, B, V), S0, S1) &:- \ member(voltage(A, B, V), S0, S1) \\ subset(volt(A, B, -V), S0, S1) &:- \ member(voltage(B, A, V), S0, S1) \end{aligned}$$

In order to derive the voltage between two arbitrary nodes, we define the following grammar rules corresponding to the clauses of "$v(\ldots)$":

$$\begin{aligned} v(A, C, V) &\longrightarrow volt(A, C, V). \\ v(A, C, V + W) &\longrightarrow volt(A, B, V), v(B, C, W). \end{aligned}$$

These grammar rules are converted into the following clauses:

$$\begin{aligned} subset(v(A, C, V), S0, S1) &:- \ subset(volt(A, C, V), S0, S1). \\ subset(v(A, C, V + W), S0, S2) &:- \ subset(volt(A, B, V), S0, S1), \\ &\qquad subset(v(B, C, W), S1, S2). \end{aligned}$$

Deriving the voltage X between nodes 1 and 4 is accomplished by identifying the nonterminal symbol "$v(1,4,X)$" in the word-order free sentence of voltage data as follows:

$$?-\ vData(VD),\ subset(v(1,4,X),VD,\_\,).$$

$$X\ =\ 20\ +\ (-15\ +\ 8)$$

Terminal symbols associated with the nonterminal symbol "$v(1,4,X)$" are removed sequentially from the object sentence. Therefore, the looping problem due to using the same data repeatedly does not occur. This method is similar to removing data using the predicate "*retract*(...)" described in the previous section. But the removed data can be recovered by backtracking.

We have overcome a common looping problem in backward chaining by simply rewriting backward chaining rules as grammar rules and by viewing a set of facts as a word-order free sentence. The looping problem caused by multiple use of the same fact is avoided by viewing DCSG as a generalized parsing problem.

18.4. Finding Structures in Circuits

18.4.1. *Circuits represented as sentences*

Since we have a mechanism for parsing word-order free languages, we will treat a circuit represented as a list as a word-order free sentence. The circuit *ca*49 in Figure 18.6 is represented by the fact (4) shown below. The predicate *ca*49([...]) states that the word-order free sentence "[...]" represents circuit *ca*49. The compound term $battery(b1,1,2)$ represents the battery $b1$ connected its positive terminal to the node 1 and its negative terminal to the node 2.

$$\begin{aligned}&ca49([resistor(r1,1,3),\ resistor(r2,2,3),\ resistor(r3,1,4),\\&\qquad resistor(r4,2,4),\ resistor(r5,3,4),\ battery(b1,1,2)]).\end{aligned}\qquad(4)$$

18.4.2. *Grammar rules without recursion*

Since a resistor is a non-polar element, we need to refer a resistor regardless of its node order. The non-terminal symbol $res(R,A,B)$ can refer either $resistor(R,A,B)$

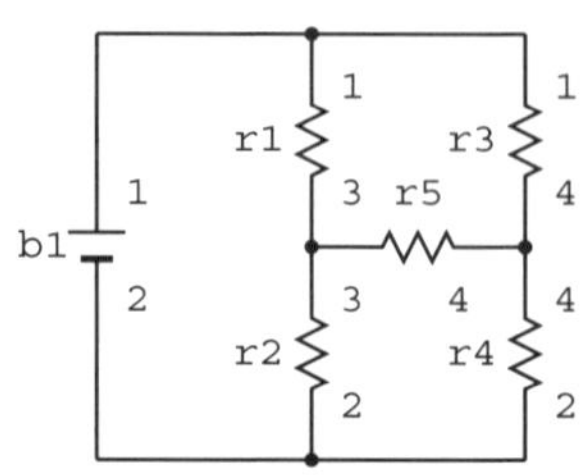

Fig. 18.6. Circuit ca49

or $resistor(R, B, A)$. Here, DCSG allows the symbol ";" as abbreviation of two grammar rules with the same left hand side. The non-terminal symbol $batt(...)$ enables us to refer the same battery with different names. If a battery is referred to with reverse node-order, the name of battery is prefixed by a minus symbol. The non-terminal symbol $anyElm(X, A)$ can refer to any element X connected to the node A.

$$\begin{aligned} res(R, A, B) \longrightarrow\ & [resistor(R, A, B)]; \\ & [resistor(R, B, A)]. \end{aligned}$$

$$batt(E, A, B) \longrightarrow [battery(E, A, B)].$$
$$batt(-E, A, B) \longrightarrow [battery(E, B, A)].$$

$$\begin{aligned} anyElm(X, A) \longrightarrow\ & res(X, A, \_); \\ & batt(X, A, \_). \end{aligned}$$

18.4.3. *All elements connected to a node*

In circuit analysis, Kirchhoff's current law (KCL) requires that the sum of all branch currents into a node be zero. In order to apply KCL at a specific node, we must find all elements connected to the node. The non-terminal symbol $allElm(X, A)$ successfully finds these elements connected to the node A. Here, X is replaced by a list of the all elements. If the same rules were written in backward chaining by replacing the symbol "$\longrightarrow$" with ":-", a looping problem due to using the same data repeatedly would occur.

$$allElm([\,], A) \longrightarrow not\ anyElm(\_, A).$$
$$\begin{aligned} allElm([X|Y], A) \longrightarrow\ & anyElm(X, A), \\ & allElm(Y, A). \end{aligned}$$

The following goal finds all elements connected to node 1 in the circuit $ca49$ as:

$$?-\ ca49(CT), subset(allElm(X, 1), CT, \_).$$

$$X = [r1, r3, b1]$$

18.4.4. *Paths and loops*

According to Kirchhoff's voltage law (KVL), the sum of branch voltages along to a loop must be zero. In order to find loops in a circuit, we first define the non-terminal symbol $path(X, A, B)$ which finds routes between two nodes A and B.

$$path([X], A, B) \longrightarrow anyElm(X, A, B).$$
$$\begin{aligned} path([X, B|Y], A, C) \longrightarrow\ & anyElm(X, A, B), \\ & path(Y, B, C). \end{aligned}$$

The variable X in $path(X, A, B)$ is substituted for by elements and nodes from the starting node A to the ending node B. The following goal attempts to find all routes from the starting node 1 to the ending node 2 in the circuit $ca49$.

$$?-\ ca49(CT), subset(path(X, 1, 2), CT, \_).$$

$$X = [b1];$$
$$X = [r1, 3, r2];$$
$$X = [r1, 3, r5, 4, r4];$$
$$X = [r1, 3, r5, 4, r3, 1, b1]$$

The first answer $[b1]$ shows the path from the node 1 via element $b1$ to the node 2. The second answer shows the path from the node 1 via $r1$, node 3, and $r2$ to the node 2. The first three answers show paths from the node 1 to the node 2, but the fourth answer is not desired. It goes back to the starting node 1 and then goes to the ending node 2 via $b1$. Since parsing sentence does not use the same word twice, no element in the circuit appeared twice in the answer. The problem is that our definition does not inhibit the use of the same node twice.

Since we want to get correct answers which do not include loops in the paths, we modify the grammar rules as follows:

$$\begin{aligned} &path([X], A, B, \_) \longrightarrow anyElm(X, A, B). \\ &path([X, B|Y], A, C, T) \longrightarrow anyElm(X, A, B), \\ &\qquad quote\ \ not\ member(B, T, \_), \\ &\qquad path(Y, B, C, [A|T]). \end{aligned}$$

The "*quote*" in the grammar rule is a command to the DCSG converter. It directs the converter to insert the following strings as is. Namely, "$not\ member(B, T, \_)$," is inserted as a Prolog clause in the DCSG conversion. The last argument of *path* is substituted for by a list of nodes which are already in the path and not to be used twice as a relay node in the process of finding path.

The following goal successfully finds all paths from the node 1 to the node 2 in the circuit $ca49$. The last argument of *path* is substituted for by "[2]" which shows the ending node 2 must not be used as a relay node.

$$?-\ ca49(CT), subset(path(X, 1, 2, [2]), CT, \_).$$

$$X = [b1];$$
$$X = [r1, 3, r2];$$
$$X = [r1, 3, r5, 4, r4];$$
$$X = [r3, 4, r4];$$
$$X = [r3, 4, r5, 3, r2];$$
$$No$$

If the ending node is equal to the starting node, the goal enumerates all loops through the node as follows:

$$?-\ ca49(CT), subset(path(X,1,1,[\,]),CT,\_).$$

$X = [r1, 3, r5, 4, r3];$
$X = [r1, 3, r5, 4, r4, 2, -b1];$
$X = [r1, 3, r2, 2, -b1];$
$X = [r1, 3, r2, 2, r4, 4, r3];$
$X = [r3, 4, r4, 2, -b1];$
...

18.5. Circuit Grammar for Knowledge Representation

This section extends DCSG to create a new formalism called circuit grammar. This circuit grammar has fields for semantic terms, and defines not only syntactic structures but also the relationships between those structures and their meaning. Unlike DCSG, circuit grammar does not distinguish terminal symbols from non-terminal symbols in the grammar rules. This makes it possible to handle the meanings of words in the same way we handle the meanings of structures.

18.5.1. *Semantic field in left-hand side*

The semantic terms are placed in curly brackets in a grammar rule as follows.

$$A, \{F_1, F_2, ..., F_m\} \longrightarrow B_1, B_2, ..., B_n. \tag{5}$$

This grammar rule can be read as stating that the symbol A with meaning $\{F_1, F_2, \ldots, F_m\}$ consists of the syntactic structure $B_1, B_2, \ldots, B_n$. This rule is converted into a Prolog clause as follows:

$$\begin{aligned} ss(A, S_0, S_n, E_0, [F_1, F_2, ..., F_m|E_n]) :-\ & ss(B_1, S_0, S_1, E_0, E_1), \\ & ss(B_2, S_1, S_2, E_1, E_2), \\ & \ldots, \\ & ss(B_n, S_{n-1}, S_n, E_{n-1}, E_n). \end{aligned} \tag{5'}$$

Since this conversion differs from that used in DCSG, we use the predicate "*ss*" instead of "*subset*". When a rule is used in parsing, the goal $ss(A, S_0, S_n, E_0, E)$ is executed, where the variable S_0 is replaced by an object set (object circuit) and the variable E_0 is replaced by an empty set. The subsets "$B_1, B_2, \ldots, B_n$" are successively identified in the object set S_0. After all of these subsets are identified, the remainder of these subsets (the complementary set) is put into S_n. While, the semantic information from B_1 is added with E_0 and put into E_1, the semantic information of B_2 is added with E_1 and put into $E_2, \ldots$, and the semantic information of B_n is added with E_{n-1} and put into E_n. Finally, the semantic information

$\{F_1, F_2, \ldots, F_m\}$, which is the meaning associated with symbol A, is added and all of the semantic information is put into E.

18.5.2. *Semantic field in right-hand side*

Semantic terms on the right-hand side define the semantic conditions for the grammar rule. For example, the following rule (6) is converted into the Prolog clause (6)' as follows.

$$A \longrightarrow B_1, \{C_1, C_2\}, B_2. \tag{6}$$

$$\begin{aligned} ss(A, S_0, S_n, E_0, E_n) \ :- \ & ss(B_1, S_0, S_1, E_0, E_1), \\ & member(C_1, E_1, \_\ ), \\ & member(C_2, E_1, \_\ ), \\ & ss(B_2, S_1, S_2, E_1, E_2). \end{aligned} \qquad (6)'$$

When the clause (6)' is used in parsing, the conditions C_1 and C_2 are tested to see if the semantic information E_1 meets these conditions after identifying the symbol B_1. If it succeeds, the parsing process goes on to identify the symbol B_2.

18.5.3. *Terminal symbols with semantic fields*

In the new circuit grammar, terminal symbols are defined as grammar rules without a right-hand side to rewrite. This means that terminal symbols are not distinguished from non-terminal symbols. Both terminal and non-terminal symbols are converted with the predicate "*ss*". The terminal symbol A with meaning $\{F_1, F_2, ..., F_m\}$ is written as (7).

$$A, \{F_1, F_2, ..., F_m\}. \tag{7}$$

This rule is converted into the following clause (7)' in circuit grammar.

$$ss(A, S_0, S_1, E_0, [F_1, F_2, ..., F_m|E_0]) \ :- \ member(A, S_0, S_1). \qquad (7)'$$

That is, when the rule (7) is used in parsing, the terminal symbol A is searched for in the object set S_0. If it is found, the complementary set is returned from S_1, and the semantic term $\{F_1, F_2, ..., F_m\}$ associated with A is added with the current semantic information E_0 to make the fifth argument of "*ss*". Thus, we can handle the meaning of words in the same way we handle the meanings of structures.

18.5.4. *English interface for semantic term*

Circuit functions are electrical behaviors that are useful to users of the circuit. We first define semantic terms for voltages and currents, the basic components of electrical behaviors of circuits. The compound term "$voltage(A, B)$" represents the voltage at node A to node B. The compound term "$current(A, X)$" represents the current flowing from node A into element X.

The most important knowledge for understanding circuit behavior consists of understanding the causal relationships between these voltages and currents. Some causal relationships are used for their function and others for their side effects. These causal relationships are often represented by numerical formulae and characteristic curves. Since we do not intend to develop a precise circuit simulator but rather aim to create a model for understanding circuits, we formalize only the dependencies between these voltages and currents using the predicate "*cause*". We also use the predicates "*control*" and "*enable*" to represent dependencies between circuits and functions.

Although only a small number of terminal symbols are required to represent circuit topologies, unlike the circuit representation problem, handling semantics, that is, circuit functions, requires that many more functional blocks be defined and that many more terms and predicates for functions be provided. Since the number of these terms and predicates increases as we integrate more knowledge, we define English interfaces whenever we introduce new terms and predicates. The special predicate "$t2e(U, V)$" (Term to English) shows how to read the semantic term U as the noun phrase V in English, and the special predicate "$p2e(U, V)$" (Predicate to English) shows how to read the semantic predicate U as the sentence V in English.

$$t2e(voltage(A, B), ['voltage\ at', A, to, B]).$$
$$t2e(current(A, X), ['current\ from', A, to, X]).$$
$$p2e(cause(A, B), [A, causes, B]).$$
$$p2e(control(A, B), [A, controls, B]).$$
$$p2e(enable(A, B), [A, enables, B]).$$

18.6. Grammar Rules

18.6.1. *Circuits as Functional Blocks*

Electronic circuits are designed as a hierarchical structure of functional blocks. We now develop grammar rules using the functional blocks appearing in the circuit *cd*42, which is a simple operational amplifier (Figure 18.7). The circuit *cd*42 is represented as the following word-order free sentence. Here, the compound term $npnTr(q1, 3, 5, 6)$ represents the NPN transistor named $q1$ with the base connected to node 3, the emitter to node 5, and the collector to node 6 respectively.

$$\begin{aligned}\text{cd42}([\ & resistor(r1, 2, 10), resistor(r2, 9, 1), npnTr(q1, 3, 5, 6),\\ & npnTr(q2, 4, 5, 7), npnTr(q3, 10, 1, 5), npnTr(q4, 10, 1, 10),\\ & npnTr(q5, 10, 1, 8), npnTr(q6, 8, 9, 2), pnpTr(q7, 7, 2, 6),\\ & pnpTr(q8, 7, 2, 7), pnpTr(q9, 6, 2, 8), terminal(t1, 3),\\ & terminal(t2, 4), terminal(t3, 2), terminal(t4, 9),\\ & terminal(t5, 1)]).\end{aligned}$$

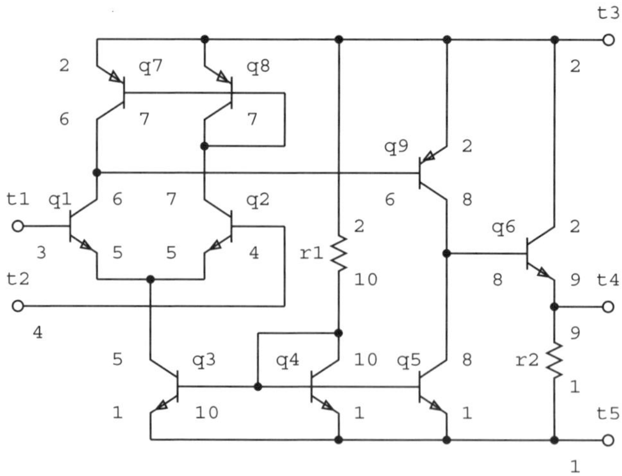

Fig. 18.7. Circuit cd42

18.6.2. *Terminal Symbols*

Unlike the DCSG formalism described in Section 18.3, this new circuit grammar does not distinguish terminal symbols from non-terminal symbols in grammar rules. Terminal symbols are defined by rules that do not have a right-hand side for rewriting. For example, the following grammar rule without a right-hand side defines a resistor as a terminal symbol.

$$resistor(R, A, B).$$

When a $resistor(R, A, B)$ is identified in a circuit, we can assume the existence of a voltage across the resistor and a current through the resistor. The voltage and the current obey the constraints of Ohm's law as represented by the equation $V = IR$. Since we intend to construct a model for understanding circuits, we only focus on the causality of the voltage and the current. Although it is difficult to decide which is the cause and which is the effect between the voltage and the current, we often say: *"Since a voltage is applied to a resistor, a current flows through the resistor"*, or *"Since a current flows through a resistor, a voltage appears across the resistor"*. These views form a causal chain of voltages and currents, which becomes important to understanding the electrical behavior of circuits. Fault diagnosis for an electronic circuit often consists of finding a defect in this causal chain.

These causalities can be added to the definition of *resistor* as semantic terms as follows.

$$\begin{aligned}&resistor(R, A, B),\\&\{\ cause(voltage(A, B), current(A, R)),\\&\quad cause(current(A, R), voltage(A, B))\}.\end{aligned}$$

Using the English interface described in Section 18.5, the compound term "$cause(voltage(A, B), current(A, R))$" can be read: *The voltage at A to B causes the current from A to R.* While "$cause(current(A, R), voltage(A, B))$" can be read: *The current from A to R causes the voltage at A to B.* Actually, these causalities are added to the rules for non-polar element defining $res(R, A, B)$ rather than the terminal symbol so that resistors can be referred to in reverse node-order.

When an NPN transistor is found in a circuit, we can assume the existence of its voltages and currents. Usually, relationships between these voltages and currents are represented by characteristic curves of the transistor. Here, we formalize only the qualitative aspects of these voltages and currents in the following grammar rules.

$$
\begin{aligned}
&npnTr(Q, E, B, C),\\
&\{\ state(Q, active),\\
&\quad gt(voltage(C, E), vst),\\
&\quad equ(voltage(B, E), vbe),\\
&\quad gt(current(B, Q), 0),\\
&\quad gt(current(C, Q), 0),\\
&\quad cause(voltage(B, E), current(B, Q)),\\
&\quad cause(current(B, Q), current(C, Q)),\\
&\quad cause(signal(voltage(B, E)), signal(current(B, Q))),\\
&\quad cause(signal(current(B, Q)), signal(current(C, Q)))\}.
\end{aligned}
$$

This grammar rule defines an NPN transistor in the active state. The compound terms "$gt(voltage(C, E), vst)$" represents that $voltage(C, E)$ is greater than the collector saturation voltage vst. The compound term "$equ(voltage(B, E), vbe)$" represents that $voltage(B, E)$ is equal to the base-emitter forward voltage vbe. The compound term "$cause(current(B, Q), current(C, Q))$" represents that $current(B, Q)$ causes $current(C, Q)$. Here, $current(B, Q)$ is intended to represent DC-current, while $signal(current(B, Q))$ represents a small signal current.

Since we have introduced new terms and predicates in the grammar rules, we also define the followings English interfaces.

$$
\begin{aligned}
&t2e(vst, ['collector\ saturation\ voltage']).\\
&t2e(vbe, ['base\ forward\ voltage']).\\
&t2e(signal(X), ['signal\ of', X]).\\
&p2e(state(Q, S), [Q, 'is\ in', S, state]).\\
&p2e(gt(A, B), [A, 'is\ greater\ than', B]).\\
&p2e(equ(A, B), [A, 'is\ equal\ to', B]).
\end{aligned}
$$

Grammar rules for NPN-transistors in the saturated state and the cutoff state are also defined in the same manner. Similar rules are also defined for PNP-transistors.

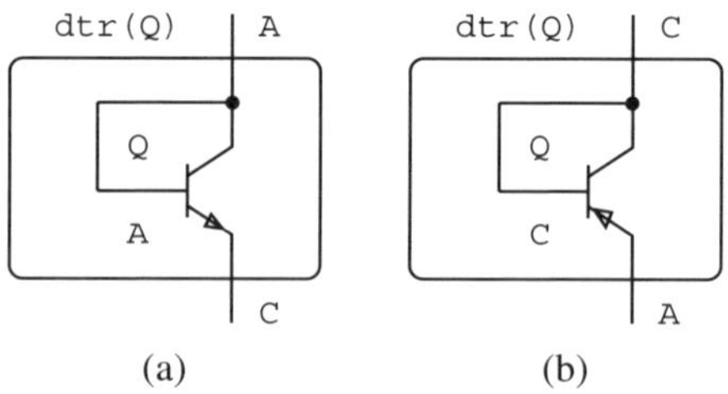

Fig. 18.8. Diode-connected transistor

If a terminal symbol $npnTr(Q, E, B, C)$ is identified in a given sentence, one of its semantic terms (active state, saturation state, and cutoff state) is added as a meaning of the terminal symbol non-deterministically. Usually, if the functional block containing a transistor requires a specific electrical state, the required state will be selected.

18.6.3. *Non-Terminal Symbols*

A transistor in which the base and collector are connected together works as a diode (Figure 18.8). The following grammar rule defines a diode-connected transistor "$dtr(dtr(Q), A, C)$" in the *conductive* state. The syntactic part of the right-hand side defines either an NPN-transistor Q or a PNP-transistor Q whose base and collector are connected to the same node. Here, Q is a name of the original transistor, and $dtr(Q)$ is a name given to the diode (Skolem function). The semantic terms in the left-hand side show that the current flows from A to $dtr(Q)$ in the conductive state, and that the current causes the $voltage(A, C)$. The semantic term $state(Q, active)$ in the right-hand side is an electrical condition which requires that the transistor Q must be in the *active* state. The grammar rule for a diode-connected transistor in the cutoff state is also defined in the same manner.

$$
\begin{aligned}
& dtr(dtr(Q), A, C), \\
& \{state(dtr(Q), conductive), \\
& \quad gt(current(A, dtr(Q)), 0), \\
& \quad cause(current(A, dtr(Q)), voltage(A, C))\} \longrightarrow \\
& \qquad\qquad\qquad (\, npnTr(Q, A, C, A);\ pnpTr(Q, C, A, C)\,), \\
& \qquad\qquad\qquad \{state(Q, active)\}.
\end{aligned}
$$

The following grammar rule defines a simple voltage regulator with a Vbe (0.6 volt) output. The syntactic part of the right-hand side defines the structure shown in Figure 18.9. The semantic condition $state(D, conductive)$ in the right-hand side requires that the diode D must be in conductive state. The semantic term $control(vreg(D, R), voltage(Out, Com))$ in the left-hand side defines a function of

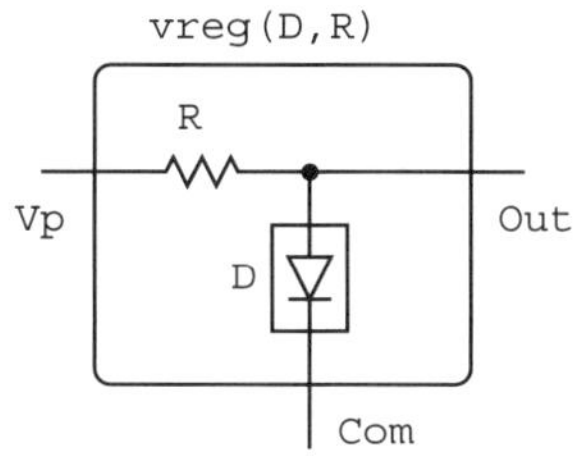

Fig. 18.9. Vbe-voltage regulator

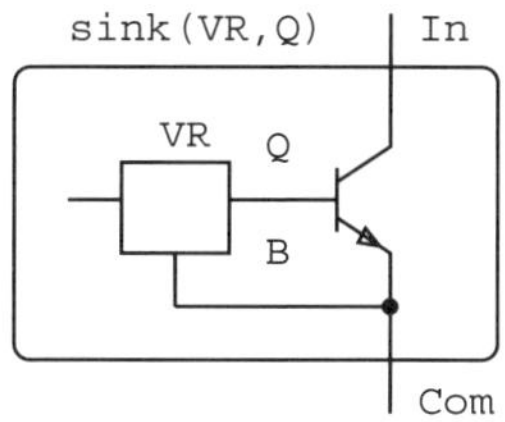

Fig. 18.10. Current source (sink type)

voltage regulator $vreg(D, R)$ that controls the voltage between Out and Com.

$$\begin{aligned}
&vbeReg(vreg(D,R), Vp, Com, Out),\\
&\{control(vreg(D,R), voltage(Out, Com))\} \longrightarrow\\
&\qquad dtr(D, Out, Com),\\
&\qquad \{state(D, conductive)\},\\
&\qquad res(R, Vp, Out).
\end{aligned}$$

The following grammar rule defines the current source (sink type) shown in Figure 18.10. The right-hand side of the grammar rule has a disjunction of syntactic and semantic conditions, which requires either an existence of a voltage regulator VR or the semantic information that the voltage between B and Com is controlled by a voltage regulator VR. This type of grammar rule becomes useful to parse context-dependent circuits (Section 18.7).

$$\begin{aligned}
&cSink(sink(VR,Q), In, Com),\\
&\{control(sink(VR,Q), current(In,Q))\} \longrightarrow\\
&\qquad (\ vbeReg(VR, _\,, Com, B);\\
&\qquad\ \ \{control(VR, voltage(B, Com))\}),\\
&\qquad npnTr(Q, B, Com, In),\\
&\qquad \{state(Q, active)\}.
\end{aligned}$$

The following grammar rule defines the active load (current mirror) shown in Figure 18.11. The semantic terms in the right-hand side are the electrical conditions that operate the circuit. The semantic term in the left-hand side shows that the

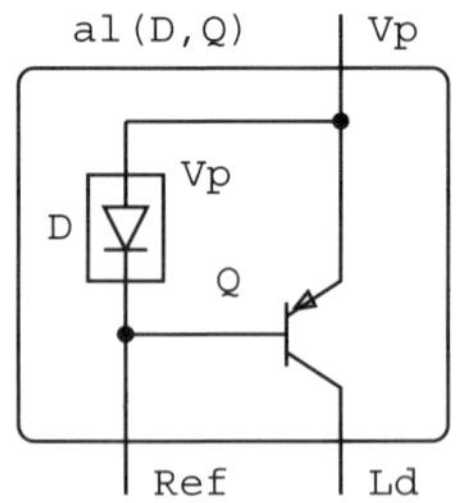

Fig. 18.11. Active load

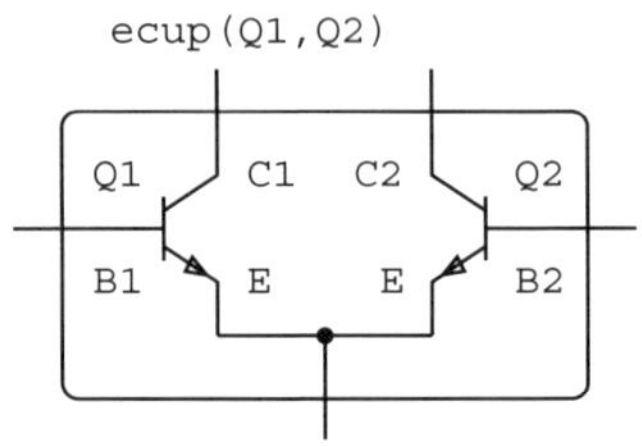

Fig. 18.12. Emitter coupled pair

current from Q to Ld is controlled by the active load $al(D,Q)$, and the current is caused by the current through Ref, and that these two currents become equal.

$$\begin{aligned}
&activeLoad(al(D,Q), Ref, Vp, Ld),\\
&\{\ control(al(D,Q), current(Q,Ld)),\\
&\quad cause(current(al(D,Q), Ref), current(Q,Ld)),\\
&\quad equ(current(Q,Ld), current(al(D,Q), Ref))\} \longrightarrow\\
&\qquad\qquad dtr(D, Vp, Ref),\\
&\qquad\qquad \{state(D, conductive)\},\\
&\qquad\qquad pnpTr(Q, Ref, Vp, Ld),\\
&\qquad\qquad \{state(Q, active)\}.
\end{aligned}$$

An emitter-coupled pair is defined as follows (Figure 18.12). Although it has no semantic terms, it is useful to define the syntactic structure.

$$\begin{aligned}
&eCoupledPair(ecup(Q1,Q2), B1, B2, E, C1, C2) \longrightarrow\\
&\qquad\qquad npnTr(Q1, B1, E, C1),\\
&\qquad\qquad npnTr(Q2, B2, E, C2).
\end{aligned}$$

A single-ended differential amplifier consists of an active load, an emitter-coupled pair, and a current source as shown in Figure 18.13.

$$\begin{aligned}
&sDiffAmp(sdAmp(EC, AL, CS), B1, B2, C1, Vp, Vm),\\
&\{input(sdAmp(EC, AL, CS),\\
&\qquad\qquad difference(voltage(B1, Vm), voltage(B2, Vm))),\\
&\quad output(sdAmp(EC, AL, CS), current(C1, sdAmp(EC, AL, CS))),
\end{aligned}$$

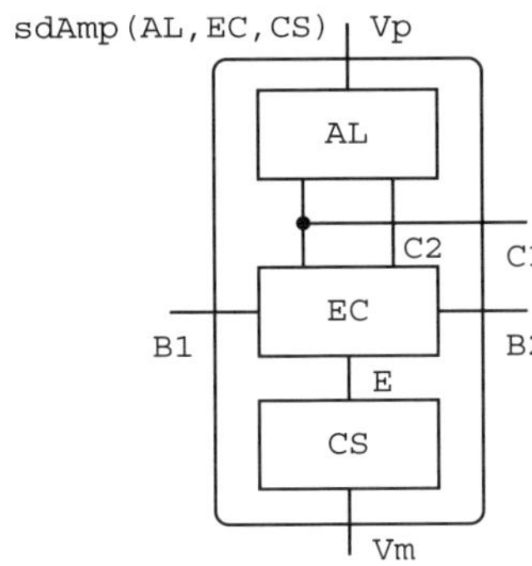

Fig. 18.13. Single-ended differential amplifier

$$\begin{aligned}
&cause(voltage(B1, B2), current(C1, sdAmp(EC, AL, CS))),\\
&suppress(CS, common_mode_gain(sdAmp(EC, AL, CS))),\\
&double(AL, current_gain(EC))\} \longrightarrow\\
&\qquad\qquad\qquad eCoupledPair(EC, B1, B2, E, C1, C2),\\
&\qquad\qquad\qquad activeLoad(AL, C2, Vp, C1),\\
&\qquad\qquad\qquad cSink(CS, E, Vm).
\end{aligned}$$

In order to parse the circuit $cd42$, the common-emitter, the emitter-follower, and the operational amplifier are defined similarly.

Since we have introduced new semantic terms and predicates for circuit functions, we now define their English interfaces as follows:

$$\begin{aligned}
&t2e(common_mode_gain(X), ['common_mode\ gain\ of', X]).\\
&t2e(current_gain(X), ['current_gain\ of', X]).\\
&t2e(voltage_gain(X), ['voltage_gain\ of', X]).\\
&t2e(input_impedance(X), ['input_impedance\ of', X]).\\
&t2e(output_impedance(X), ['output_impedance\ of', X]).\\
&t2e(load_impedance(X), ['load_impedance\ of', X]).\\
&t2e(difference(X, Y), [difference, X, from, Y]).\\
&p2e(high(X), [X, is, high]).\\
&p2e(low(X), [X, is, low]).\\
&p2e(amplify(X, Y), [X, amplifies, Y]).\\
&p2e(control(X, Y), [X, controls, Y]).\\
&p2e(input(X, Y), [X, is, input, to, Y]).\\
&p2e(output(X, Y), [X, outputs, Y]).\\
&p2e(suppress(X, Y), [X, suppresses, Y]).\\
&p2e(double(X, Y), [X, doubles, Y]).\\
&p2e(cause(C, E), [C, causes, E]).\\
&p2e(enable(X, Y), [X, enables, Y]).
\end{aligned}$$

18.7. Parsing Circuits

All of the grammar rules defined in the previous section are converted into Prolog clauses according to the circuit grammar conversion method described in Section 18.5. The clauses form a logic program that performs top-down parsing. The following goal parses the circuit *cd*42 and derives the circuit structure and functions. The first sub-goal substitutes the circuit *cd*42 into the variable *Circuit*. The first three arguments of the predicate *ss* are the same arguments as the predicate *subset* in DCSG. That is, the first argument is the name of a subset, the second argument is the whole set, and the third argument is its remainder. Since the third argument of *ss* is empty, the second sub-goal tries to identify the whole *Circuit* as a functional block X. Since the forth argument of *ss* is empty, no semantic information is given to parse the circuit. When the goal successfully parses the circuit, the circuit functions are substituted into Y as semantic information.

$$\begin{aligned} ?- \quad & cd42(Circuit), \\ & ss(X, Circuit, [\,], [\,], Y). \end{aligned}$$

$$\begin{aligned} X = \; & operationalAmp(\,opAmp(sdAmp(ecup(q1, q2), \\ & \qquad\qquad al(pdtr(q8), q7), \\ & \qquad\qquad sink(vreg(dtr(q4), r1), q3)), \\ & \qquad pnpCE(q9, sink(vreg(dtr(q4), r1), q5)), \\ & \qquad npnEF(q6, r2)), \\ & \quad 3,4,9,2,1) \end{aligned}$$

The value of X shows that the circuit *ca*42 is identified as an operational amplifier "*operationalAmp*(*opAmp*(. . .), 3, 4, 9, 2, 1)". The first argument *opAmp*(...) is a name given to the identified circuit, and the rests are the connecting nodes in the circuit. The given name keeps track of identified functional blocks, and is viewed as a parse tree which shows the syntactic structure of the circuit (Figure 18.14). Each node represents a functional block identified in the circuit *cd*42.

If a circuit is generated with only context-free grammar rules, its parse tree becomes a simple tree. Note, however, that the parse tree for *cd*42 has a structure with shared nodes. This means that two current sinks share a single voltage regulator. That is, the circuit designer used a context-dependent grammar rule which collapsed two voltage regulators into one voltage regulator. The semantic condition of the current sink (Figure 18.10) enables us to parse this context dependent circuit.[5]

Each time a functional block is identified in parsing, semantic information about the functional block is added. After parsing, the value of Y has a large amount of semantic information about the circuit as follows.

$$\begin{aligned} Y = \; & [input(opAmp(...), voltage(3, 4)), \\ & output(opAmp(...), voltage(9, 1)), \\ & cause(voltage(3, 4), voltage(9, 1)), \end{aligned}$$

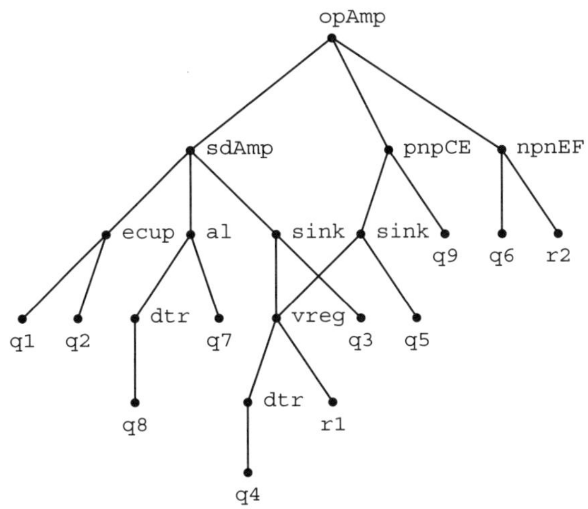

Fig. 18.14. Parse Tree for *cd*42

$enable(sdAmp(...), amplify(opAmp(...), differential\_inputs)),$
$enable(pnpCE(...), high(voltage\_gain(opAmp(...)))),$
$enable(npnEF(...), low(output\_impedance(opAmp(...)))),$
$input(npnEF(...), voltage(8, 1)),$
$output(npnEF(...), voltage(9, 1)),$
$cause(voltage(8, 1), voltage(9, 1)),$
$high(input\_impedance(npnEF(...))),$
$low(output\_impedance(npnEF(...))),$
$equ(voltage\_gain(npnEF(...)), 1),$
$cause(voltage(9, 1), current(9, r2)),$
$cause(current(9, r2), voltage(9, 1)),$
$state(q6, active),$
$gt(voltage(2, 9), vst),$
... 120 lines omitted ...)]

18.8. Functional Explanations in English

After parsing a circuit, we will have acquired a large amount of semantic information about the circuit *cd*42. This information consists of causal relationships between voltages and currents, electrical states of devices, and electrical behaviors and functions of the circuit. Using the English interface and the data defined by "*p*2*e*" and "*t*2*e*", the following sentences are generated from the semantic information.

$? - cd42function(X), write\_function(X).$

voltage at 3 *to* 4 *is input to opAmp*(...).
opAmp(...) *outputs voltage at* 9 *to* 1.

voltage at 3 *to* 4 *causes voltage at* 9 *to* 1.
sdAmp(...) *enables that opAmp*(...) *amplifies differential_inputs.*
pnpCE(...) *enables that voltage_gain of opAmp*(...) *is high.*
npnEF(...) *enables that output_impedance of opAmp*(...) *is low.*
voltage at 8 *to* 1 *is input to npnEF*(...).
npnEF(...) *outputs voltage at* 9 *to* 1.
voltage at 8 *to* 1 *causes voltage at* 9 *to* 1.
input_impedance of npnEF(...) *is high.*
output_impedance of npnEF(...) *is low.*
voltage_gain of npnEF(...) *is equal to* 1.
voltage at 9 *to* 1 *causes current from* 9 *to r*2.
current from 9 *to r*2 *causes voltage at* 9 *to* 1.
*q*6 *is in active state.*
voltage at 2 *to* 8 *is greater than vst.*
... 120 lines omitted ...

18.9. Conclusions

We have developed a grammatical method for circuit representation. Circuits were viewed as sentences, and their elements as words. Circuit structures were defined by a logic grammar called DCSG for word-order free languages. A set of grammar rules, when converted into Prolog clauses, forms a logic program which performs top-down parsing. Although DCSG can recognize the syntactic structures of circuits, it could not handle the relationships between circuit functions.

Since electronic circuits are designed to achieve specific functions, we need a method to represent knowledge of circuit structures and functions. The newly developed circuit grammar described here defines not only syntactic structures, but also the relationships between structures and meanings. That is, we take a circuit's function to be the meaning of that circuit. Knowledge of circuit structures and functions was then coded into grammar rules as semantic information. Using these grammar rules, sample circuits were parsed and their structures and functions were derived successfully.

Our English interface generated more than 130 sentences for the circuit functions of even the simple operational amplifier *cd*42. Although these sentences are not ordered in contents, each sentence is useful in understanding circuit behavior. We can also derive causal chains of voltages and currents from the parse tree and semantic information, and these chains will be useful in troubleshooting.

Since grammar rules are additive, knowledge of new circuits can be added as new grammar rules. As more grammar rules are defined, more circuits can be parsed. The knowledge and mechanisms developed here achieve automatic circuit understanding which can help engineers design and troubleshoot circuits.

References

1. S.A. Greibach, *A New Normal Form Theorem for Context-Free Phrase Structure Grammars*, JACM, vol. 12, pp. 42–52, 1965.
2. J. deKleer, *Causal and Teleological Reasoning in Circuit Recognition*, MIT, AI-TR-529, 1979.
3. F.C.N. Pereira and D.H.D. Warren, *Definite Clause Grammars for Language Analysis*, Artificial Intell., Vol. 13, pp. 231–278, 1980.
4. Takushi Tanaka, *Definite Clause Set Grammars: A Formalism for Problem Solving*, J. Logic Programming, Vol. 10, pp. 1–17, 1991.
5. Takushi Tanaka, *Parsing Circuit Topology in A Logic Grammar*, IEEE-Trans. Knowledge and Data Eng., Vol. 5, No. 2, pp. 225–239, 1993.
6. T. Tanaka and O. Bartenstein, *DCSG-Converters in Yacc/Lex and Prolog*, Proc. 12th International Conference on Applications of Prolog, pp. 44–49, 1999.
7. Takushi Tanaka, *A Logic Grammar for Circuit Analysis — Problems of Recursive Definition*, Lecture Notes in Artificial Intelligence Vol. 4693, Springer, pp. 852–860, 2007.
8. Takushi Tanaka, *Circuit grammar: knowledge representation for structure and function of electronic circuits*, Int. J. Reasoning-based Intelligent Systems, Vol. 1, Nos. 1/2, pp. 56–67, 2009.
9. Takushi Tanaka, *Deriving Electrical Dependencies from Circuit Topologies Using Logic Grammar*, Lecture Notes in Artificial Intelligence Vol. 5712, Springer, pp. 325–332, 2009.
10. Takushi Tanaka, *A Mechanism for Converting Circuit Grammars to Definite Clauses*, Lecture Notes in Computer Science, Vol. 6278, Springer, pp. 190–199, 2010.
11. Takushi Tanaka, *Deriving electrical dependencies from circuit topologies using logic grammar*, Int. Journal of Reasoning-based Intelligent Systems, Vol. 3, No. 1, pp. 28–33, 2011.
12. Takushi Tanaka, *Analyzing Circuit Structures as Language*, Circuit Analysis (Virginia E. Wright ed.), Nova Science Publishers Inc. NY, pp. 143–179, 2011.
13. *101 Analog IC Designs*, Sunnyvale, CA: Interdesign Inc., 1976.

Chapter 19

AN INTELLIGENT CBR MODEL FOR PREDICTING CHANGES IN TROPICAL CYCLONES INTENSITIES

James N.K. Liu, Simon C.K. Shiu, Jane You and Leon S.K. Law
Department of Computing, Hong Kong Polytechnic University
Hung Hom, Kowloon, Hong Kong
csnkliu@comp.polyu.edu.hk
csckshiu@comp.polyu.edu.hk
csyjia@comp.polyu.edu.hk

The intensity of a tropical cyclone is an important factor in determining its direction of movement. This chapter presents an intelligent case-based reasoning system to help tropical cyclone forecasters predict changes in tropical cyclone intensities. Using best track data from the Joint Typhoon Warning Centre, we extract data for the coordinates, speed, direction, season and atmospheric pressure of tropical cyclones, treating each six-hour observation as a case and classifying them in three groups, "South China Sea", "Philippines and Taiwan" and "Japan". Similar cases are constructed by using Apriori association rule (AR) mining model, which is built-in in data mining software — SPSS Clementine 10.1, to reveal the hidden relationships between the meteorological environment and TC intensity change. Rules generated from the model are used to predict changes of intensities of TCs. Experimental results indicate that the proposed intelligent model is able to provide reasonable forecast in severe weather. In particularly, it is found that the accuracy of forecast is strongly related to the track pattern and the number of cases that are not located in the three location groups.

19.1. Introduction

Changes in tropical cyclones (TCs) are difficult to predict because they are highly chaotic and their relationships of cause and effect are non-linear and hard to identify. Yet TCs are nonetheless recognizable natural phenomena and as such must be bound and defined by some kinds of principles of formation and operation. The information relevant to these principles must be, if suitable techniques can be developed and applied, available and amenable of classification and further use, for example in severe weather prediction

TC and severe weather prediction and assessment commonly makes use of a number of parameters, including that of intensity, which is defined as the maximum sustained 1-minute surface winds. One way to estimate intensity is by associating certain cloud features with conventional estimates of a TC's strength. Another is to use satellite imagery [4], relying on the interpretive skill of the meteorologist, who matches the image against certain pattern types, curved band, shear, eye pattern, etc [8]. This is in turn considered in terms of the TC life cycle, which is classified into eight categories (T-numbers) using satellite images and a decision tree. Ultimately, however, the combination of subjective analysis and the rapid development of TCs in severe weather conditions mean that the resulting predictions are always inaccurate.[16,17]

In this chapter we describe our Intelligent CBR Intensity Prediction Model, which involves four major tasks, pre-processing data, building a case base and data mining, checking accuracy and making adjustment based on meteorological principles. Our model applies data mining techniques to meteorological data and uncovers interesting inter-transactional association rules for weather prediction [5]. The data mining tools and algorithms can extract process-related features from raw observational data such as the TC best track data, Tropical Rainfall Measurement Mission (TRMM) data for 3D rainfall and 2D surface information. We then apply an approach similar to that presented in literature[16] in which the data mining tools or algorithms use visualization or logical inference [6] to search for hidden relationships among the features. Our model forecasts the intensity of TCs by using Case-Based Reasoning (CBR) [4]. This involves extracting cases based on historical meteorological information then using that library of cases in training to generate forecasting rules. Usually, the accuracy of a system should increase with the number of cases in the library. We will describe all of the main features of our model: the data, how we define features from the data, our data exploration tools, the design of case-based reasoning model, and how we construct cases using soft computing techniques including fuzzy logic and data mining. These all combine to allow us to uncover hidden and uncertain relationships between physical processes and tropical cyclone intensity with the goal of better prediction of changes in TCs.

The remainder of this chapter is organized as follows: In Section 19.2, we describe the categories of Tropical Cyclones that we consider in this work. Section 19.3 describes our case selection and experimental data sets. Section 19.4 describes in detail the design of our Intelligent CBR Intensity Prediction Model including the CBR cycle, data pre-processing, case base construction, data mining and forecast adjustments. Section 19.5 discusses our experiments and comparison results. Section 19.6 gives the conclusion and future work.

19.2. Categories of Tropical Cyclones

A tropical cyclone (TC) is a storm system characterized by a low pressure center and numerous thunderstorms that produce strong winds and flooding rain. It feeds on the heat released when moist air rises and the water vapor it contains condenses. TCs can be categorized[1] according to what are called their moving track patterns. In this chapter we refer to the tracking data of two leading meteorological organizations, the Joint Typhoon Warning Centre (JTWC) and the Hong Kong

Table 19.2a. TC categories (HKO)

Name	Wind Speed (km/h)	Pressure (hPa)
Tropical Depression (TD)	38–62	>=1000
Tropical Storm (TS)	63–87	999–991
Severe Tropical Storm (STS)	88–117	990–984
Typhoon (TY)	>117	>984

Table 19.2b. TC categories (JTWC)

Name	Wind Speed (km/h)	Pressure (hPa)
Tropical Depression (TD)	38–55	>=1000
Tropical Storm (TS)	56–100	999–984
Typhoon 1 (TY)	101–140	983–966
Typhoon 2 (TY)	141–165	965–941
Typhoon 3 (TY)	166–185	940–928
Typhoon 4 (TY)	186–235	927–912
Super Typhoon (STY)	>235	<912

Observatory (HKO). Each organization has its own model of categorization which includes the best tracking data denoting the locations of tropical cyclone center and the intensities of cyclones (i.e., the maximum 1-minute mean sustained 10-meter wind speed) at six-hour intervals. Table 19.2a lays out HKO's categories and criteria and Table 19.2b does the same for the JTWC. The main differences to note are that the JTWC has not categorized Severe Tropical Storm (STS) and that the two categorizations apply different average wind speeds.

19.2.1. *Classic Moving Track Patterns in the North-Western Pacific Ocean*

TCs in North-Western Pacific Ocean exhibit four classic types of moving track, which are differentiated initially based on their direction or style of travel: the northwest type, the changing direction type, the moving west type, and the monsoon type. The following describes each of these but our work will have only three classifications, "North west", "Moving west" and "Changing direction" and will combine the "Monsoon" and changing direction types because these differ only in their driving force and cannot be classified using only the information in tracking diagrams.

Figure 19.2a shows the track of a classic northwest type TC. It forms on the ocean east of the sea of Philippines and its direction is set by a sub-tropical ridge (shown in yellow in Figure 19.2b) which steers it NNW. When TC reaches the SW of the sub-tropical ridge, it moves NW across the South China Sea until it makes landfall. The TC will have an average speed of 15–20 km/h and most typically appears in July and August.

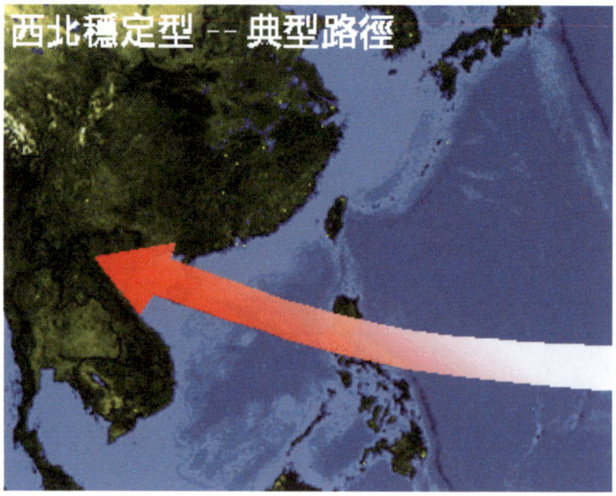

Fig. 19.2a. Direction of movement of a "North West" type[2] TC

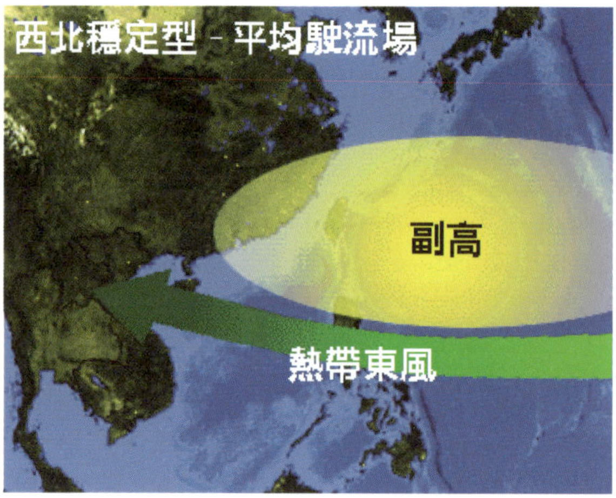

Fig. 19.2b. The direction of movement of the northwest type[2] is controlled by a sub-tropical ridge (in yellow)

Figure 19.2c shows the track of a classic "Changing direction" type. Again it forms east of the Philippines and directed by the sub-tropical ridge and moves NW at 10–20 km/h. The change of direction comes about when the sub-tropical ridge is itself pushed east by westerly winds (compare Figures 2d and 2e). The TC then slowly moves NE. When this type of TC is classified as a typhoon, it may be weakened by the high-altitude westerly winds which disrupt thick upper-layer clouds. As the result, the top and the bottom of the TC may separate and this will decrease the intensity of the TC. When the TC starts moving NW, it will gather speed and turn into a temperate cyclone.

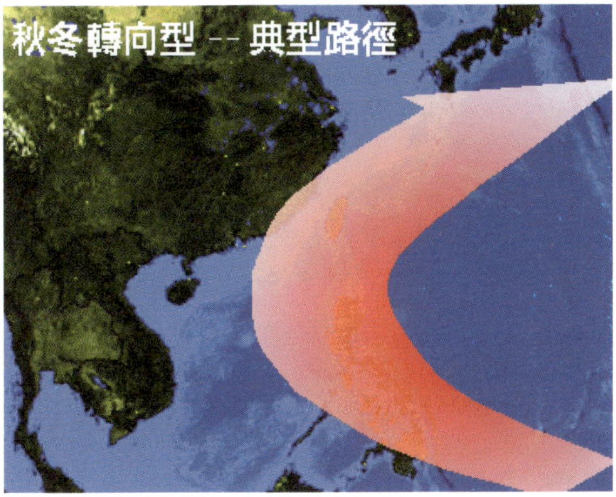

Fig. 19.2c. The direction of movement of the "Changing direction" type[2] of TC

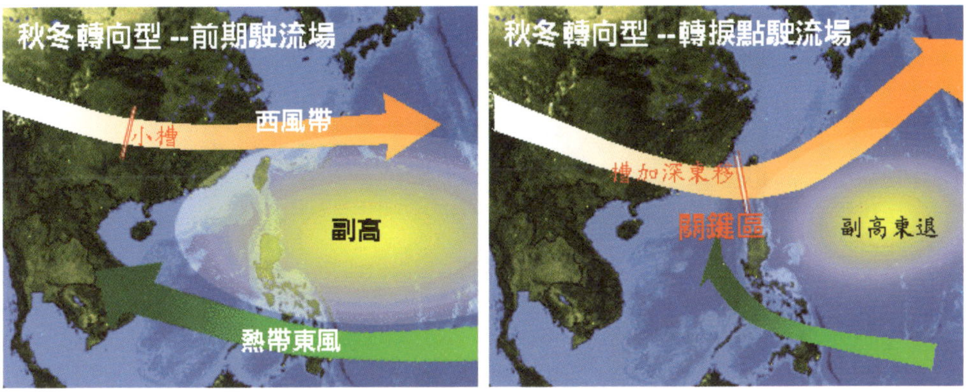

Fig. 19.2d, 2e. Direction of TCs before and after the sub-tropical ridge moves to the east[3]

Figure 19.2f shows the track of a classic "Moving west" type of TC. The TC formed east of the Philippines, moves either W or SWW and finally lands around Vietnam. The determining factors in whether this TC moves like a "Changing direction" type or a "Moving west" type is the strength of TC as passes through the north of Luzon, the main island of the Philippines, and the straits just north (Figure 19.2g). This type of TC usually appears in the late windy season. (late August to late November)

Figure 19.2h shows the track of a classic "Monsoon" type TC. It forms in the South China Sea or east of the Philippines. Directed by the southwest monsoon wind, it moves N at 10–15 km/h. Northwest of the sub-tropical ridge, it changes direction, moving quickly NE, and will turn into a temperate cyclone. Figure 19.2i shows the southwest monsoon wind (the green arrow) the sub-tropical ridge (in yellow), and the high level westerlies (in orange).

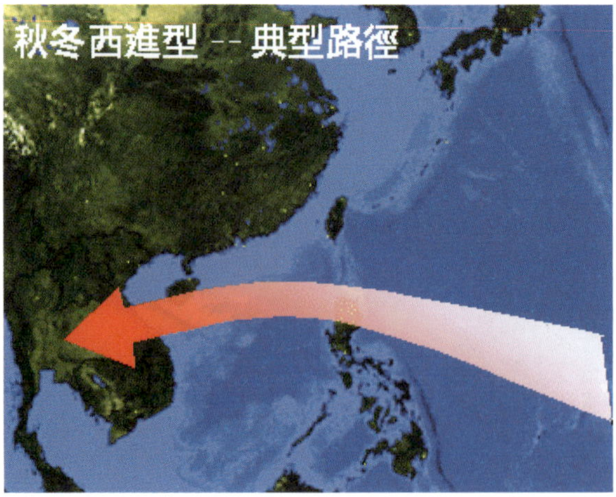

Fig. 19.2f. Movement trend of "Moving west" type[4]

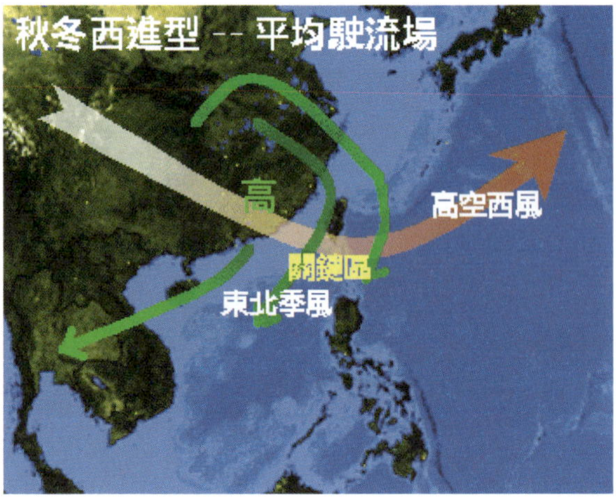

Fig. 19.2g. Two different driving forces of TC movements[4]

19.2.2. *TC Best Tracks*

The Joint Typhoon Warning Center[19] maintains an archive of tropical cyclone track data, commonly referred to as "best-tracks". Each best-track file contains TC center locations and intensities (i.e., the maximum 1-minute mean sustained 10-meter wind speed) at six-hour intervals. The geographical domain of the archive is the western North Pacific (WP), North Indian Ocean (IO) and Southern Hemisphere (SH). Most of the best-track data are published in JTWC's Annual Tropical Cyclone Reports (ATCRs) (US Pacific Fleet, Air Force, Meteorological Squadron One, 1947; North Pacific Typhoon Warning Service, 1948–49 and 50–51; JTWC, 1952–58; JTWC, 1959–2000).

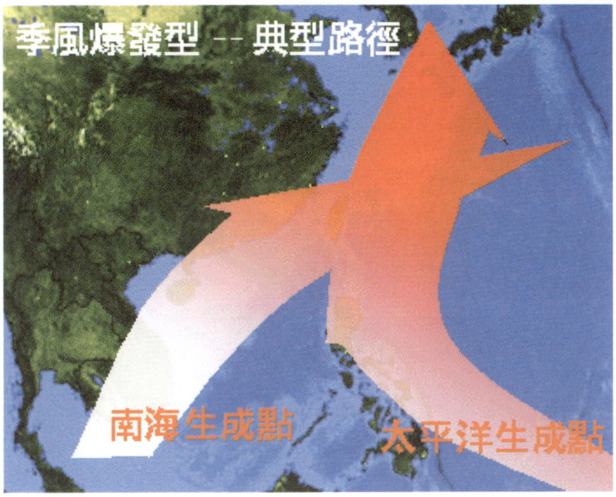

Fig. 19.2h. Movement trend of "Monsoon" type[4]

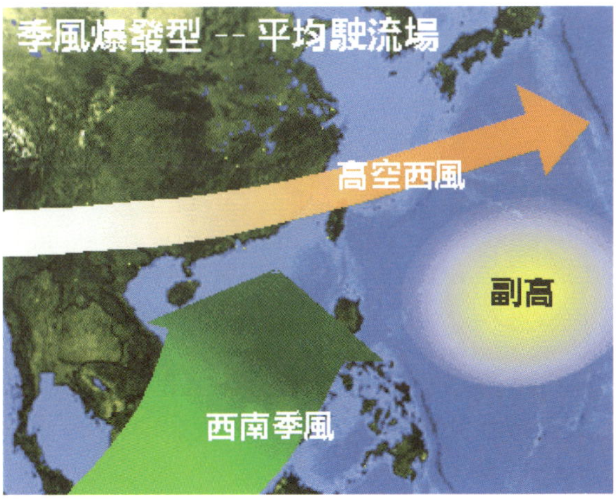

Fig. 19.2i. Sub-tropical ridge (yellow), southwest monsoon (green) and high level westerly winds (orange) that direct "Monsoon" type TC movements[4]

19.3. Case Selection and Experimental Data Sets

The cases used in this work incorporate synoptic information that is regarded as helpful in satellite interpretation, such as the direction of prevailing winds [10]. This kind of information is especially useful when satellite data indicates a loosely organized developing system. The selection of cases also involves both the subjective judgments of forecasters as to the similarity and usefulness of past cases to a current case and records of past decisions, actions or consequences associated with the best analogs. These can be reused or adapted to justify decisions or actions in current situations.

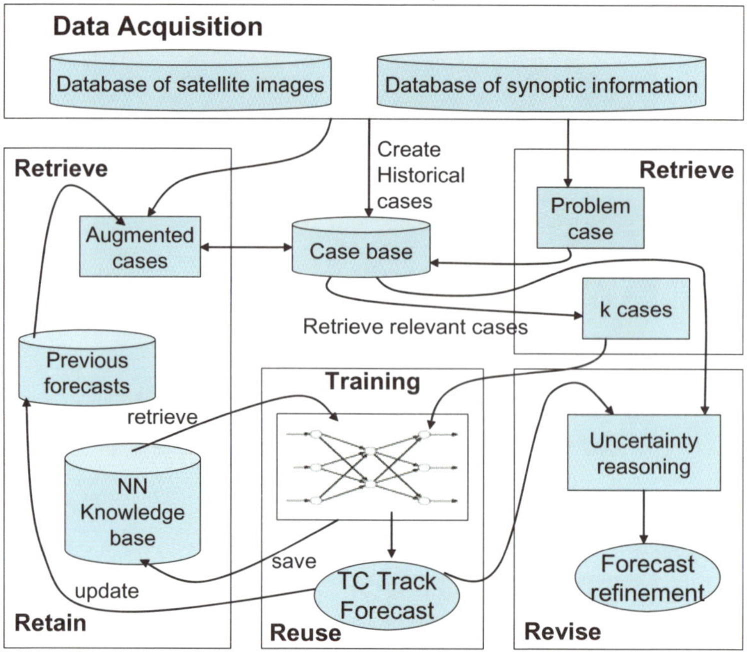

Fig. 19.3. An integrated forecasting framework

The case retrieval process aims to select the best analogs to the current best expert knowledge and operates in a series of cycles, as shown in Fig. 19.3 [9]. Cases or data are ordered in relations of rank in which one case or dataset may outrank another, thereby allowing the selection of the best analogs. Past cases are selected based on how well they match with and are useful to the current situation. This is done in a series of partial matchings of past cases with the current case, and by ranking across case dimensions until the system has retrieved a smaller set of matching and useful cases (e.g. [3, 11]).

Our tests use best track data from the Joint Typhoon Warning Centre for the period from 2002–2006. As discussed in Section 19.2.2, best track data is the observation data of each TC for every six hours. It is a comprehensive TC track analysis after considering all available observations and expert interpretation. Each observation is regarded as a case. Some data processing is required before the data can be used for data mining. In our work, we have created some new fields including IntensityChange, LandDistance and AverageSST, which are created in four steps that transform the raw data in the two datasets.

(1) Filling the intensity change of tropical cyclone
(2) Calculating the distance between the TC center and the nearest land mass
(3) Calculating the averaged SST within 1° radius of the TC center
(4) Discretization on numeric fields in order to uncover the relationship among different fields in the data.

The intensity change of TC is calculated by computing the difference between two consecutive values of tropical cyclone intensity [6]. The new value is stored in the new field called "IntensityChange" in the database:

$$\Delta I_t = I_{t+1} - I_t$$

where ΔI_t = TC intensity change at time t; I_t = TC intensity at time t

19.4. Design of the Intelligent CBR Intensity Prediction Model

The Intelligent CBR Intensity Prediction Model makes use of case-based reasoning, fuzzy logic and association mining. In the following we first describe the four processes through which CBR cycles of case data. In the subsequent subsections we describe the three procedures that the system is designed to simulate the CBR cycle:

1. Data pre-processing: Converting raw data and land data into cases
2. Case-store building and data mining: The cases are used to build the case base. The data mining software then access the case base to carry out data mining. Three association rules files are exported as text file.
3. Accuracy checking of generated rules from data mining and adjustments: The system retrieves a year of cases from case base as test cases and they will be used to test with the exported rules files and hence the accuracy is calculated.

19.4.1. *The Case-Based Reasoning (CBR) Cycle*

Figure 19.4a shows the four steps of a basic CBR cycle[5]: retrieval, reuse, revision and retention. A case consists of a problem and solution. Retrieval involves retrieving from the case base any cases that are similar in terms of the definition of the problem. Reuse refers to applying all the retrieved cases to problem. This may involve adapting the solution as needed to fit the problem. Revise refers to determine the solution of the problem by the retrieved cases. This may alter the properties of cases inside the case base. Retention, the final step in the process, involves retaining useful cases and store them in the case base and discard the non-useful cases, so that when there is a new problem, more useful cases can be used and thus increase the accuracy.

19.4.2. *Data pre-processing*

The model (Fig. 19.4b) requires two types of data: lands data and best track data. Lands data can be used directly after extraction from its text file. Best track data cannot be used for data mining directly. It is essential to process the data first. Best track data with 5990 records contains many parameters from which only useful parameters (listed in Table 19.4a) are extracted. After extraction of the data,

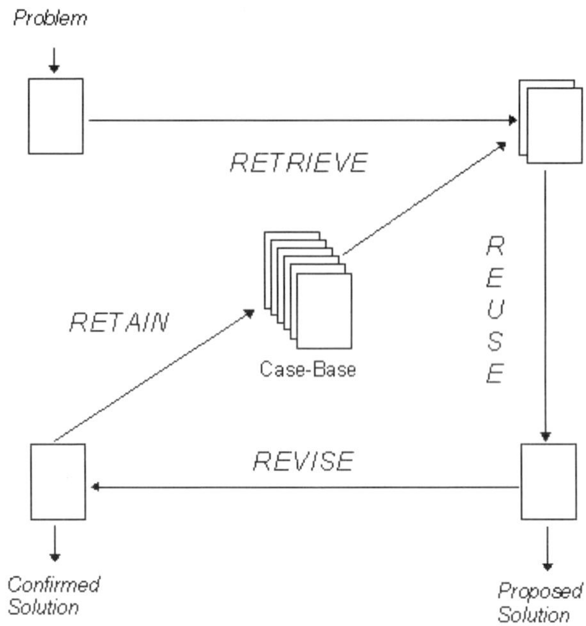

Fig. 19.4a. CBR Cycle [adapted from Aamodt & Plaza, 1994]

some data are processed to derive new parameters and there are a total of 11 parameters:

1. Latitude: Coordinate of TC, copied from *latitude* in extracted data
2. Longitude: Coordinate of TC, copied from *longitude* in extracted data
3. Direction of movement: the direction of movement of the TC is found by calculating the difference of *latitude* and *longitude* between two consecutive observations. There are a total of 9 values: N, NE, E, SE, S, SW, W, NW and NIL.
4. Speed: The speed of the TC is found by calculating the difference of *latitude* and *longitude* between two consecutive observations. There are four values: NIL, Low, Med, High.
5. Pressure: This involves linguistic variable X = $\{P\}$. The membership function is defined in the following and as shown in Figure 19.4c:

$$\text{High} = P \geq m + \frac{s}{2} \tag{19.1}$$

$$\text{Med} = m - \frac{s}{2} < P < m + \frac{s}{2} \tag{19.2}$$

$$\text{Low} = P \leq m - \frac{s}{2} \tag{19.3}$$

where P is broken down into three ranges by its own mean (m) and half of its standard deviation (s) over the entire observation sets [10].

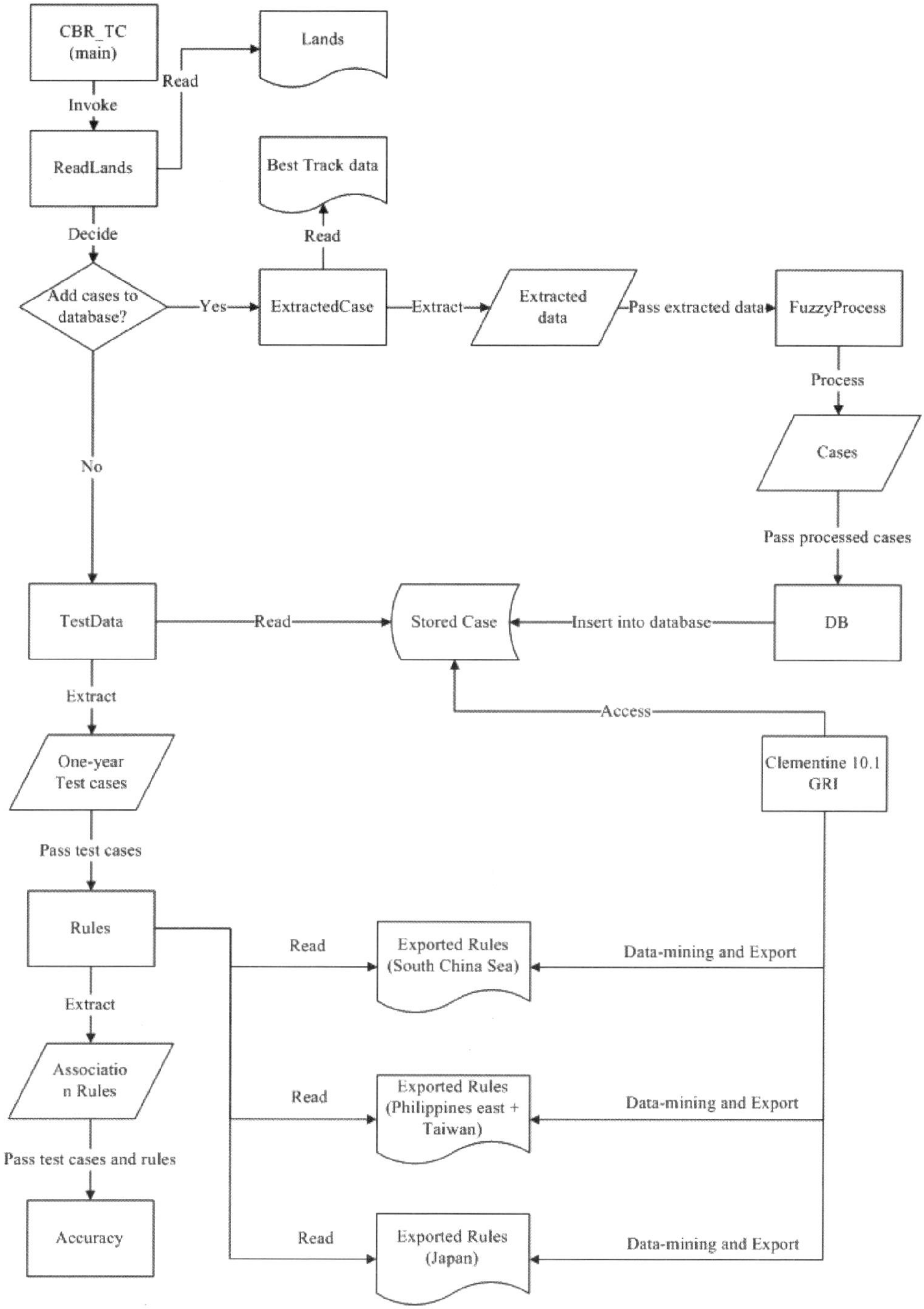

Fig. 19.4b. Model Design

Table 19.4a. Useful parameters extracted from best track data

Name	Description
Latitude and Longitude	Coordinates of the TC
Wind speed	Average wind speed, used for finding out the intensity change
Pressure	Minimum pressure of the TC centre. It involves linguistic variable X = {P} and the fuzzy set X with descriptions, P = {Low, Mid, High}
Time	Timestamp of the observation. Month and year of the TC observation are derived from it. For example, 2002010806 = 06:00 8 Jan 2002.
TC number	A number that assigned by JTWC to every TC, used for identifying the storm.

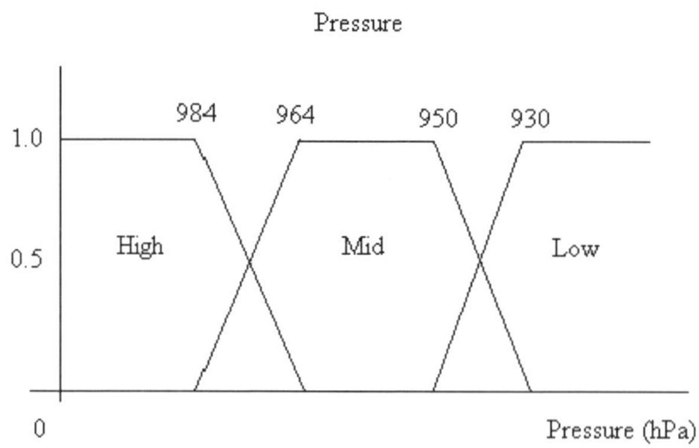

Fig. 19.4c. Membership function of pressure parameter

6. Intensity Change: This is found by calculating the difference of *wind speed* between two consecutive observations. There are three values: decrease, no change, and increase.
7. Time: Timestamp of the observation, copied from *time* in the extracted data
8. Month and Year: Extract the month and year value from the time data.
9. Season: Extract the month value from the *time* data. There are 4 values: Winter, Spring, Summer, Autumn.
10. Wind speed: Average wind speed of the TC copied from *wind speed* in the extracted data
11. TC number: Unique number of TC is copied from *TC number* in the extracted data

19.4.3. *Case base building and data mining*

After the best track data have been pre-processed, the output is then introduced to the case store. Accordingly, the case base has totally 5990 cases. Data mining tool (SPSS Clementine 10.1) is used to access the case base through ODBC and for mining association rules. Antecedents of the rules are Speed, Season, Direction of movement, Pressure High, Pressure Mid and Pressure Low while the Consequent is Intensity Change. The minimum antecedent support is set to 1%, minimum rule confidence is set to 40%, maximum number of antecedents is set to 5 and maximum number of rules is set to 100. The rules will be exported as text file afterwards. A typical sample of these rules is:

For TCs in South China Sea

Rule 1: if Pressure_Mid > 0.3 and Speed = Nil, then the Intersity_Change = NoChange with support count = 16%, Confidence = 95%

For TCs in Philippines east and Taiwan

Rule 2: if Pressure_High < 0.275 and Speed = Nil, then the Intersity_Change= NoChange with support count = 40%, Confidence = 95%

For TCs in Japan

Rule 3: if Direction = NE, then the Intersity_Change = Decrease with support count = 35%, Confidence = 75%

Cases are clustered into three groups according to their coordinates before mining. These groups are "South China Sea", "Philippines east and Taiwan" and "Japan", as shown in Table 19.4b and Figure 19.4d. For the cases that do not lie in these three regions, they will not be included for training rules.

The purpose of separating the cases into three groups is that TCs in these three regions have different moving track patterns (see Section 19.2.1). Cases with similar moving track patterns are grouped together so that it will be easier to mine the association rules. Each location group also has a different environment like different sea surface temperature, different distribution of lands, different vertical wind shear, different Coriolis force, etc. All of them are greatly affecting the track and the intensity of TC. Therefore, by separating the cases into location groups, the variation of the environment of the cases can be reduced. The association rules of three different groups will be exported respectively after the data mining process.

Table 19.4b. Corresponding regions of "South China Sea", "Philippines east and Taiwan" and "Japan"

Group name	**Region**
South China Sea	5N-25N, 100E-120E
Philippines east and Taiwan	10N-25N, 120E-140E
Japan	30N-45N, 120E-150E

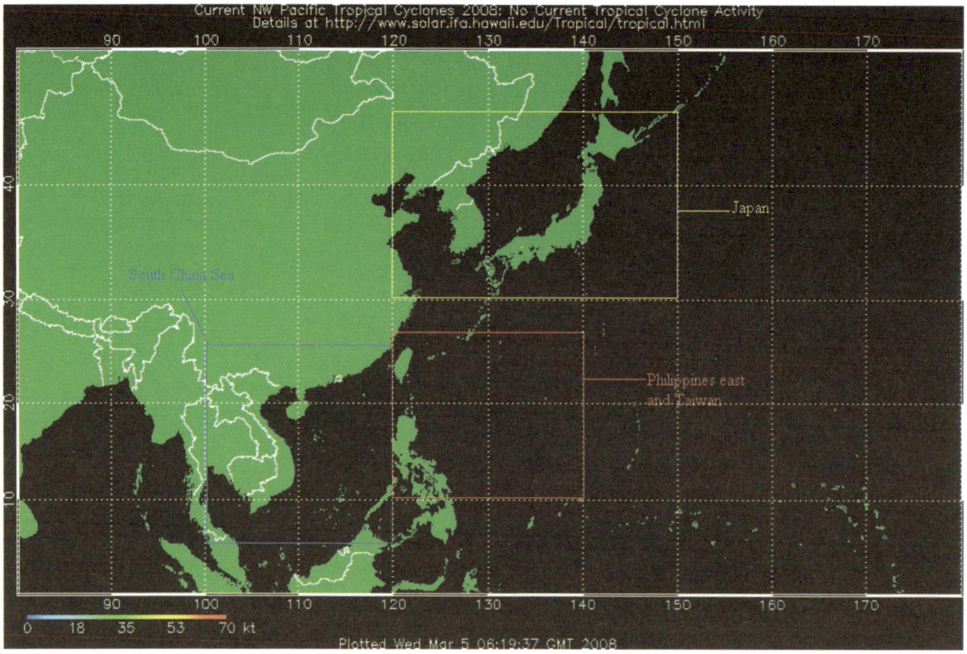

Fig. 19.4d. Corresponding Regions of "South China Sea", "Philippines east and Taiwan" and "Japan"[7]

Table 19.4c. Forecast adjustments when the TC is near the land

Speed	Adjustment
NIL	If the previous intensity change is "Decrease", the forecast will be "Decrease", else the forecast will be "NoChange"
Low	If the previous intensity change is "Increase", the forecast will be "NoChange", else the forecast will be "Decrease"
Med	Forecast will be "Decrease"
High	Forecast will be "Decrease"

19.4.4. *Checking the accuracy of exported rules from data mining and adjustments*

Before matching a test case with the rules, the "accuracy" process checks the test case coordinates to determine whether the TC is near the land or not. We made an assumption that the TC will be very likely to move onto the land when it is near the land. When the TC is on the land, it loses the energy supply from the sea and its intensity will decrease gradually. If the TC is near the land, it will check the speed and makes adjustments to the forecast. The adjustment details are shown in Table 19.4c. If the test case is not near the land, "Accuracy" will check which location group the case does belong to. The test case will be assigned into "Philippines east and Taiwan" group if it does not lie into any three groups. There

Table 19.4d. Adjustments for three location groups according to direction of movement

Location Group	**Direction**	**Adjustment**
South China Sea	N, NE, NW	Forecast count of "Decrease" + 2
Philippines east + Taiwan	All	Forecast count of "Increase" + 1
	W, NW	Forecast count of "Decrease" + 2
Japan	NE	Forecast count of "Decrease" + 2

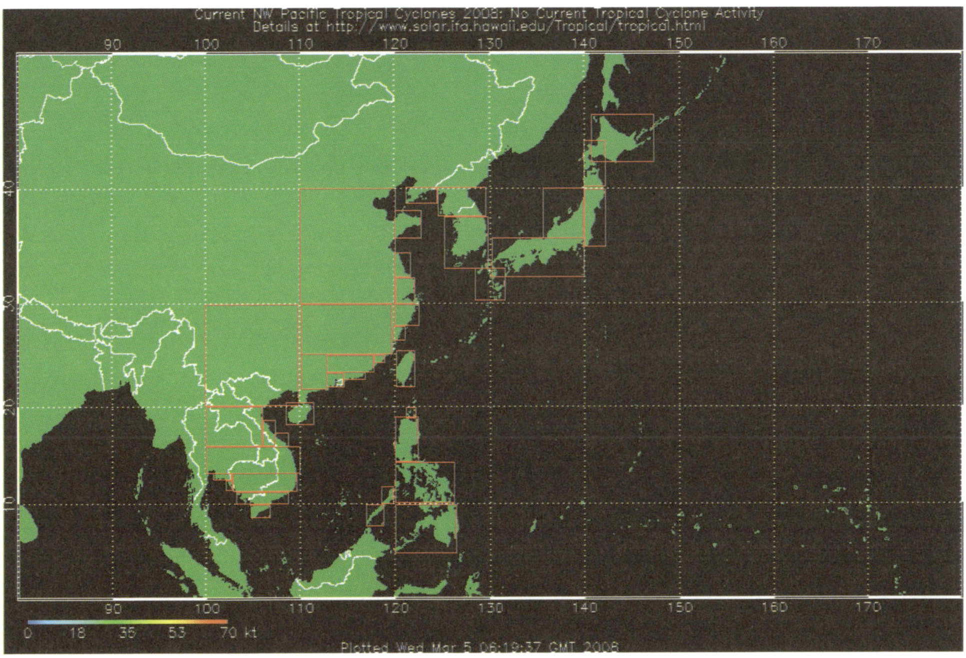

Fig. 19.4e. Land areas included in the land file[7]

are some adjustments according to the direction of movement after identifying the location groups. The adjustment details are shown in Table 19.4d. The reasons for the adjustments will be explained in Section 19.5.2.

When the TC is moving over the land, its intensity is expected to decrease. A file about the land area is made and the system will check whether the test cases are around these areas. If so, the forecast will be "decrease". In Figure 19.4e, all the areas that are bounded by squares are included in the file. For all the coordinates of the area, please refer to Appendix for detail. After the adjustments, the model reads the three exported files for the association rules and retrieves cases over the period to check for the accuracy. Each test case will be classified to any group before matching with the rules according to the geographical position of tropical cyclones. If the test case does not lie in the area of three groups, it will be regarded as the case in "Philippines and Taiwan" group. After classifying the group of the test cases, the model will use the rule file of that group to test the accuracy. It will match the test case with all the rules in the file one by one. If the conditions of the test case exactly match the antecedents of a rule, the consequent of the matched rule will be recorded (i.e. the intensity decrease, no change or increase). When the test case finished matching with all the rules, the model will count the frequency of three consequents. The highest frequency consequent will be the forecast. If the frequency of "no change" is the highest and same with other consequents, the forecast will be "no change" because most of the test cases have no intensity change. If the frequencies of "decrease" and "increase" are the same, the forecast will be "decrease".

19.5. Experimental Results

In this section, we will describe the performance of the model and the data that are used for data mining and accuracy tracking (i.e. 2002–2006 best track data from JTWC). The accuracy of this model is affected by the total number of cases in the year. Fewer cases increase the accuracy. Therefore, the total number of the cases in the year has to be taken into consideration in assessing the effectiveness of the model. We also consider whether location group adjustment improves the effectiveness of the model.

19.5.1. *Accuracy of the three location groups*

The number of cases per different location groups is shown in Table 19.5a and the results are shown in Table 19.5b. "Japan" has the fewest cases and this may account for its accuracy. "South China Sea" has fewer than 200 cases except in 2006. "South China Sea" is least accurate in 2003 and 2005. "Philippines east + Taiwan" has the most cases, especially in 2004. "Philippines east + Taiwan" is less accurate in 2002 and 2003 and more accurate than in 2004–2006. The distribution is similar to the overall accuracy.

a. South China Sea

The total number of cases in South China Sea group is shown in Table 19.5c and the results are shown in Table 19.5d. For the cases in a location group, they are categorized according to the intensity change: "Decrease", "NoChange" and "Increase",

Table 19.5a. Number of cases of three location groups in 2002–2006

	South	**Phil**	**Japan**	**Total**
2002	130	653	67	850
2003	136	683	45	864
2004	133	1515	114	1762
2005	161	997	94	1252
2006	371	821	70	1262

Table 19.5b. Accuracies of three location groups

	South	**Phil**	**Japan**	**Total**
2002	60.00%	50.38%	73.13%	53.65%
2003	55.15%	50.07%	62.22%	51.50%
2004	71.43%	70.43%	71.05%	70.54%
2005	54.66%	70.71%	71.28%	68.69%
2006	63.88%	66.75%	74.29%	66.32%

Table 19.5c. Total number of cases in South China Sea group

	Decrease	**NoChange**	**Increase**	**Total**
2002	17	85	28	130
2003	32	80	24	136
2004	26	84	23	133
2005	39	86	36	161
2006	79	255	37	371

Table 19.5d. Accuracies of South China Sea group

	Decrease	**NoChange**	**Increase**	**Total**
2002	52.94%	81.18%	0.00%	60.00%
2003	75.00%	56.25%	25.00%	55.15%
2004	61.54%	85.71%	30.43%	71.43%
2005	71.79%	67.44%	5.56%	54.66%
2006	36.71%	76.86%	32.43%	63.88%

which are based on the intensity difference between two consecutive data in best track data. The total accuracy for "Increase" is very low. The total accuracy for "NoChange" is above 65%, except 2003 which only has 56%. The accuracy for "Decrease" in 2002 and 2006 are lower. For 2002, there are only 17 cases for "Decrease" in this group so that its accuracy is rather low. For 2006, there are 79 cases for "Decrease" and the number of hit cases is 29 only, which means some anomalies may exist.

b. Philippines east + Taiwan

The total number of cases in Philippines east + Taiwan group is shown in Table 19.5e and the results are shown in Table 19.5f: The accuracies of "Increase" are similar for all years. The accuracies of "NoChange" in 2002 and 2003 are much lower than 2004–2006. The accuracy of "Decrease" in 2005 is much lower than average, some anomalies may exist.

c. Japan

The total number of cases in Japan group is shown in Table 19.5g and the results are shown in Table 19.5h: There are fewer cases in Japan group and only 2 cases for "Increase". The accuracy of "Increase" is not significant. The accuracy of "Decrease"

Table 19.5e. Total number of cases in Philippines east + Taiwan group

	Decrease	NoChange	Increase	Total
2002	116	309	228	653
2003	104	359	220	683
2004	140	1072	303	1515
2005	87	694	216	997
2006	67	565	189	821

Table 19.5f. Accuracies of Philippines east + Taiwan group

	Decrease	NoChange	Increase	Total
2002	44.83%	62.14%	37.28%	50.38%
2003	44.23%	57.94%	40.00%	50.07%
2004	32.86%	83.68%	40.92%	70.43%
2005	24.14%	86.17%	39.81%	70.71%
2006	46.27%	80.53%	32.80%	66.75%

Table 19.5g. Total number of cases in Japan group

	Decrease	NoChange	Increase	Total
2002	47	18	2	67
2003	28	15	2	45
2004	32	80	2	114
2005	27	65	2	94
2006	25	43	2	70

Table 19.5h. Accuracies of Japan group

	Decrease	NoChange	Increase	Total
2002	85.11%	50.00%	0.00%	73.13%
2003	100.00%	0.00%	0.00%	62.22%
2004	84.38%	66.25%	50.00%	71.05%
2005	92.59%	64.62%	0.00%	71.28%
2006	88.00%	69.77%	0.00%	74.29%

in 2003 is 100% while the accuracy of "NoChange" and "Increase" are 0%. Some anomalies appeared in 2003. The overall accuracy of "Decrease" is very high, averaging 90.01%. 2003 contains some anomalies and so could be unreliable for training. The accuracy of the "South" group is very low. The accuracy of the "Phil" group is good in later years (2004–2006), averaging 69.30%. The accuracy of the "Japan" group is good for all years (except 2003), averaging 72.44%. This result suggests that the "divide and conquer" approach (i.e. the three location groups approach) is a useful way to improve the accuracy of the model.

Table 19.5i. Adjustments for three location groups according to direction of movement

Group	Direction	Adjustment
South China Sea	N, NE, NW	Forecast count of "Decrease" + 2
Philippines east + Taiwan	All	Forecast count of "Increase" + 1
	W, NW	Forecast count of "Decrease" + 2
Japan	NE	Forecast count of "Decrease" + 2

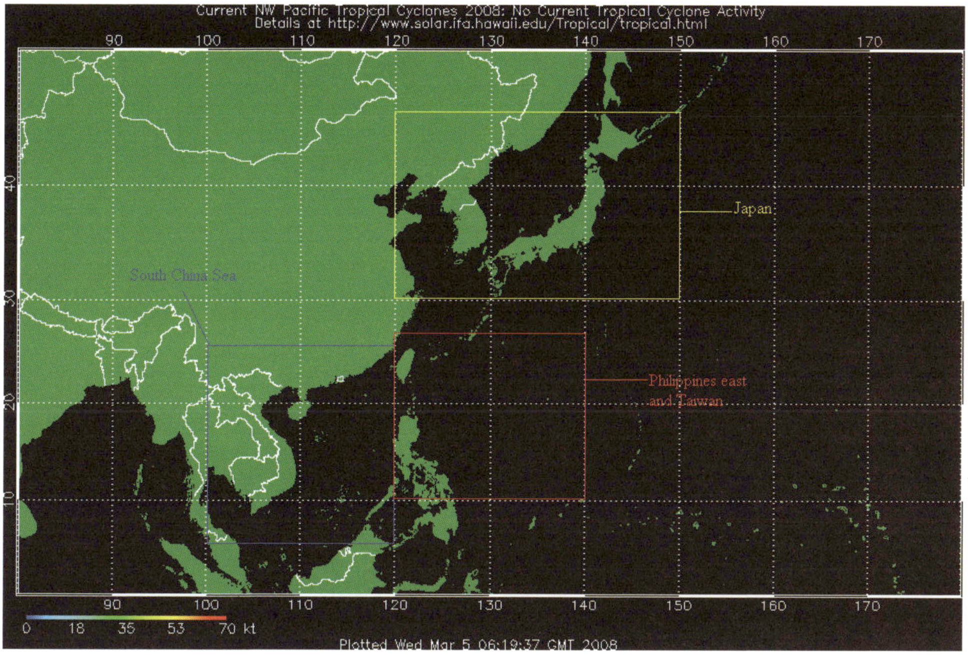

Fig. 19.5a. Map of North-Western Pacific Ocean[7]

19.5.2. *Effectiveness of the location groups adjustment*

The location group adjustment is based on the distribution of land. The adjustment is shown in Table 19.5i and the land distribution is shown in Figure 19.5a. If the TC moves with the direction stated in the adjustment, it will probably land. For the Philippines east + Taiwan group, "Increase" is increased by one because TCs usually form to the east of the Philippines. The sea surface temperature is usually high so TC intensity usually increases there. In this section, we will compare the number of hit cases with and without location adjustment as shown in Table 19.5j. With the location adjustment, the number of hit cases decreased slightly for all years. The location adjustment cannot improve the accuracy of the model. The

Table 19.5j. Comparison of hit cases between having location adjustment and no location adjustment

	O. De.	De.	O. No.	No.	O. In.	In.	O. Total	Total
2002	101	105	270	273	85	91	456	469
2003	98	97	253	256	94	97	445	450
2004	89	91	1022	1032	132	134	1243	1257
2005	74	76	698	704	88	82	860	862
2006	82	84	681	683	74	75	837	842

(Legend: O. = with location adjustment, De. = Decrease, No. = NoChange, In. = Increase)

Table 19.5k. Line colors of different intensity of TC

Type	Category	Line Color
Tropical Depression	TD	Green
Tropical Storm	TS	Yellow
Typhoon 1	TY1	Red
Typhoon 2	TY2	Light Red
Typhoon 3	TY3	Magenta
Typhoon 4	TY4	Light Magenta
Typhoon 5	TY5	White

result may be better if the direction of movement adjustments is included in the land adjustment, rather than location group adjustment:

19.5.3. *Data Analysis and Discussion*

In this section we will study the best track data by analyzing the variation of track pattern mentioned in Section 19.2.2 and the intensity of TC in the three location groups. The variation of intensity is satisfactory when the cases contain TCs in tropical depression, tropical storm and typhoon intensities. This is because typhoons have a wind eye while tropical depressions and tropical storms do not. The variation of track pattern should be as low as possible. If the variation is too high, it will be more difficult to mine the rules according to direction of movement. In Table 19.5k, TC intensities are shown in different colors:

19.5.3.1. *2002 Best track data*

Figure 19.5b shows the tracks of TCs on North-Western Pacific Ocean in 2002 [8] Best track data in 2002 is used to train the association rules. For "South China Sea" group, the variation of intensity is low: maximum intensity is TS only. There are only four TCs located in this group. This is too few to consider the variation of tracks. As there are so few cases in this group, it is difficult to train useful rules, which is the reason for the low accuracy in the "South China Sea" group. The variation of intensity in the "Philippines east + Taiwan" group is high which is satisfactory: from TD to TY5. The variation of track is too high: no obvious patterns can be seen. For the "Japan" group, the variation of intensity is satisfactory: from TD to TY2. The variation of track is low and satisfactory. Most of the TCs are of the

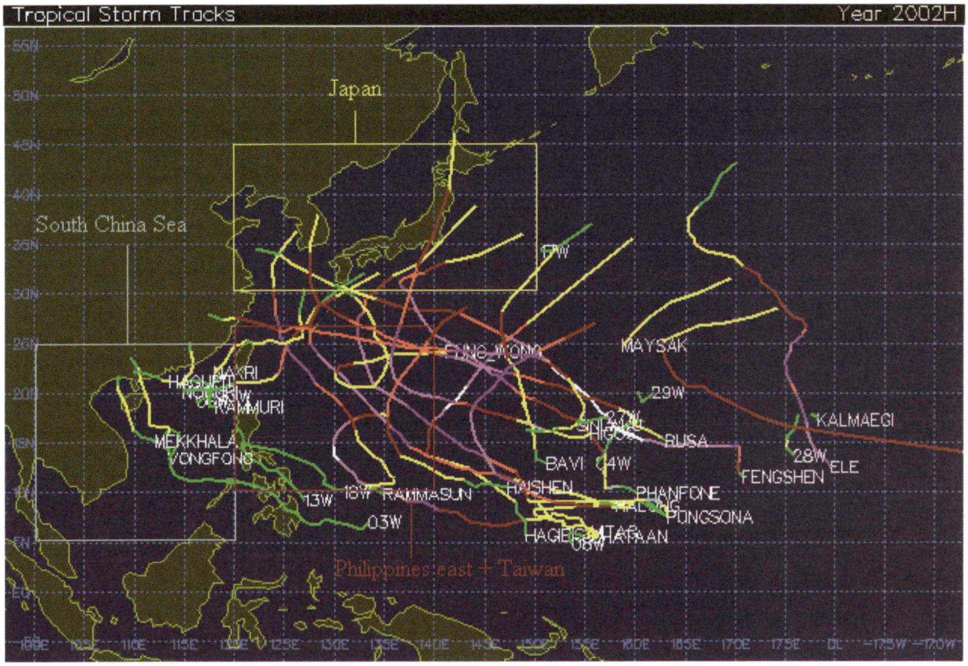

Fig. 19.5b. TC tracks on North-Western Pacific Ocean in 2002

"Changing direction" type. The intensity decreases when they move towards the NE. The accuracy for the 2002 forecast is 53.65% and it is lower than that of 2004–2006. Nearly half of the cases are not included in the three location groups and they are not used for training rules. These cases are difficult to predict accurately using the trained rules in the three location groups. This is probably why the accuracy of the 2002 forecast is lower.

19.5.3.2. *2003 Best track data*

Figure 19.5c shows the tracks of TCs on North-Western Pacific Ocean in 2003.[9] Best track data in 2003 is not used for training association rules. For the "South China Sea" group, the variation of intensity is satisfactory, from TD to TY4. The variation of track is low and satisfactory. Most of the TCs are of the "North west" type. For "Philippines east + Taiwan" group, the variation of intensity is high and satisfactory, from TD to TY5. The variation of track is too high. No obvious patterns can be seen. There are too many track patterns so the data of this location group in 2003 are not suitable for training rules. For the "Japan" group, the variation of intensity is satisfactory, from TD to TY3. The variation of track is low and satisfactory. Most of the TCs are "Changing direction" type. The intensity decreases when they move towards NE. 2003 is considered as having anomalies. The abnormal TC is PARMA. Its moving track is shown in Figure 19.5d. It has a long life span (11 days, resulting in 44 observations). At first, it moved towards the NE and kept

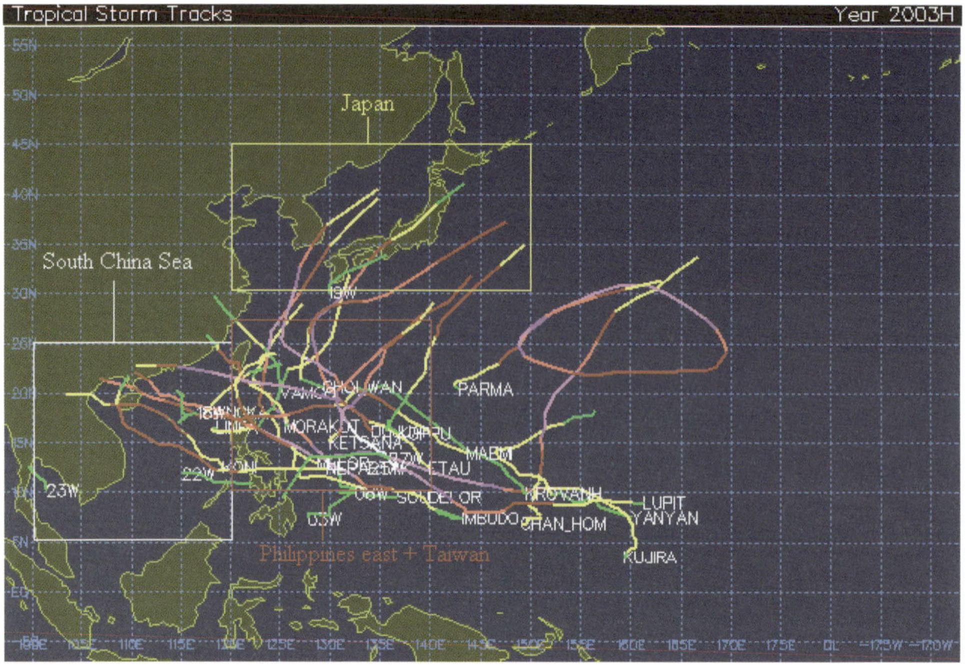

Fig. 19.5c. TCs tracks on North-Western Pacific Ocean in 2003

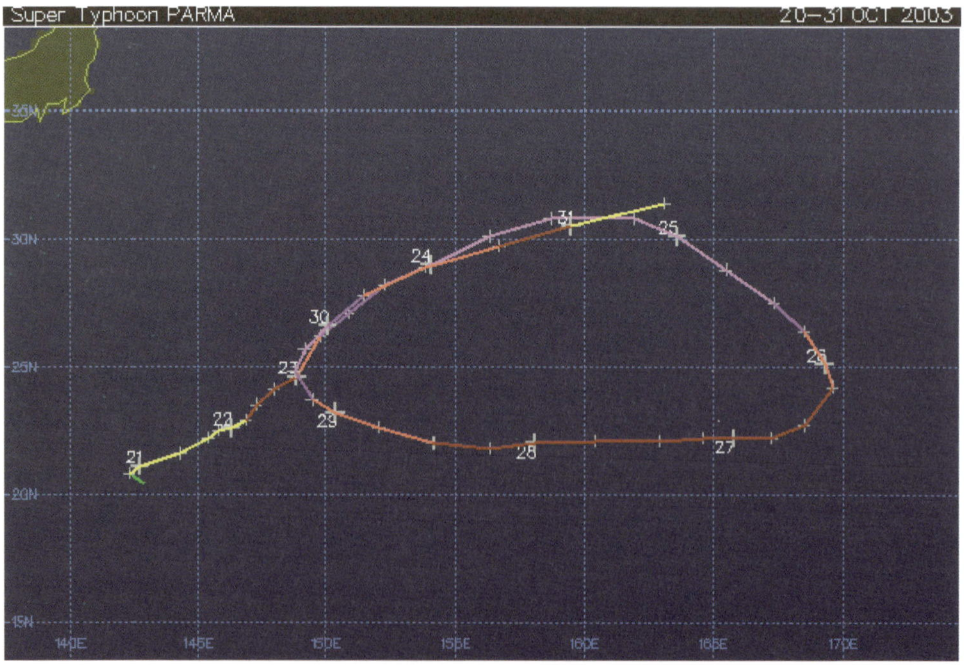

Fig. 19.5d. Moving track of STY PARMA[10]

increasing in intensity until it reached TY4 on 24 Oct. It started to turn clockwise and the intensity started decreasing on 25 Oct until it reached TY1 intensity. It kept moving towards the west on 27 October. The intensity increased again on 29 Oct and it continued to turn clockwise. It moved steady towards NE and dissipated on 31 Oct.[11] The development of PARMA was abnormal because it did not follow the tropical cyclone life cycle: it had two decay stages, two intensification stages and two mature stages. The track of PARMA is abnormal also because of its moving circular moving track which is very rare. As the result, it is difficult to forecast the intensity change of PARMA. As it had a long life span, it had more cases making the accuracy of 2003 lower.

19.5.3.3. *2004 Best track data*

Figure 19.5e shows the tracks of TCs on North-Western Pacific Ocean in 2004:[12] Best track data in 2004 is used for training association rules. For "South China Sea" group, the variation of intensity is satisfactory: from TD to TY2. The variation of track is satisfactory: it can be seen that most of the TCs are "Moving west" type. For "Philippines east + Taiwan" group, the variation of intensity is high which is satisfactory: from TD to TY5. The variation of track is satisfactory: though there are many TCs in this group, it can be seen that most TCs move towards NE at the later stage and they can be regarded as "Changing direction"

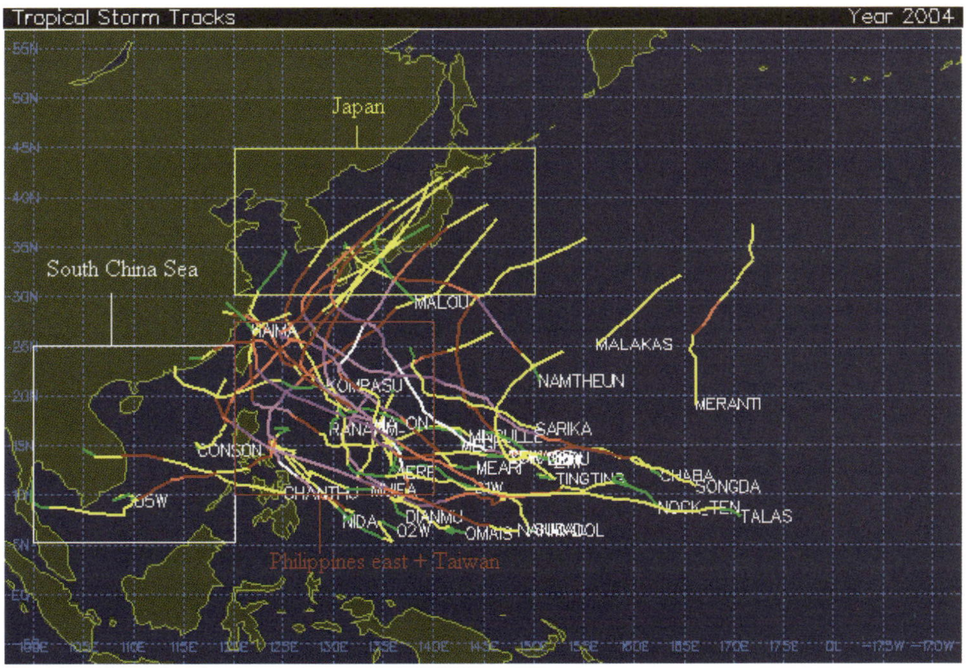

Fig. 19.5e. TCs tracks on North-Western Pacific Ocean in 2004

type. For "Japan" group, the variation of intensity is satisfactory: from TD to TY3. The variation of track is low and satisfactory: it can be seen that most of the TCs are "Changing direction" type. The intensity decreases when they move towards NE.

The accuracy of forecast in 2004 is 70.54% which is quite high. Though several TCs are not located in the three location groups, their moving track patterns are "Changing direction" type which is similar to the TCs in "Philippines east + Taiwan" group. Therefore, the rules of "Philippines east + Taiwan" group can forecast these cases with higher accuracy.

19.5.3.4. *2005 Best track data*

Figure 19.5f shows the tracks of TCs on North-Western Pacific Ocean in 2005:[13] Best track data in 2005 is not used for training association rules. For "South China Sea" group, the variation of intensity is satisfactory: from TD to TY2. The variation of track is satisfactory: it can be seen that most of the TCs are "Moving west" type. For "Philippines east + Taiwan" group, the variation of intensity is high which is satisfactory: from TD to TY5. The variation of track is satisfactory: most of them are "North west" type. For "Japan" group, the variation of intensity is satisfactory: from TD to TY2. The variation of track is low and satisfactory: it can be seen that

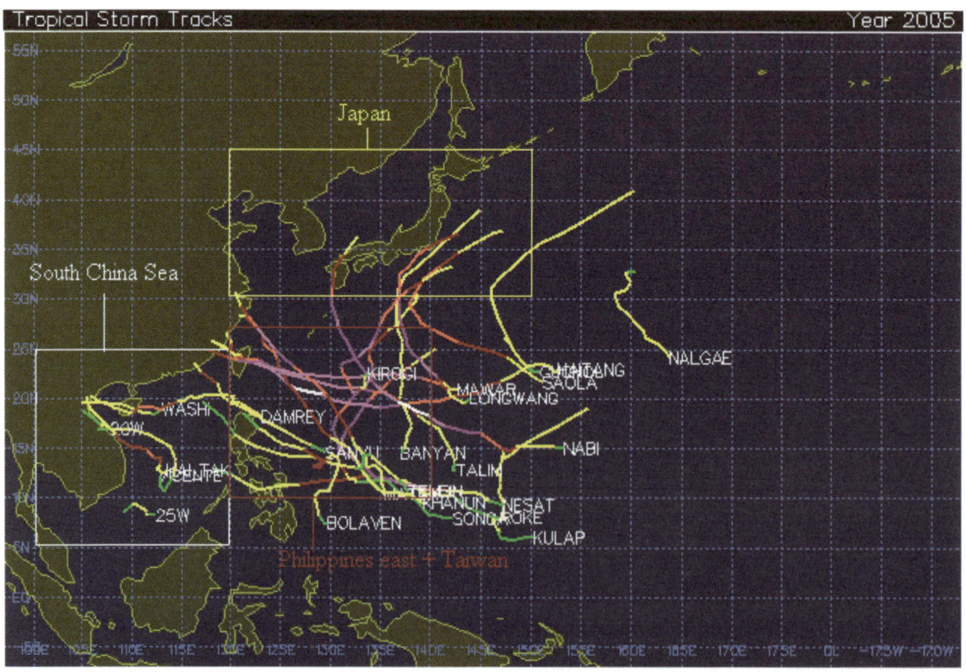

Fig. 19.5f. TCs tracks on North-Western Pacific Ocean in 2005

most of the TCs are "Changing direction" type. The intensity decrease when they move towards NE. The accuracy of forecast in 2005 is 68.69% which is quite high also. It can be seen that fewer cases are located outside the three location groups when comparing with those of previous years. This is probably the reason why the accuracy of forecast is high in 2005.

19.5.3.5. *2006 Best track data*

Figure 19.5g shows the tracks of TCs on North-Western Pacific Ocean in 2006:[14] Best track data in 2006 is not used to train association rules. For "South China Sea" group, the variation of intensity is satisfactory: from TD to TY4. The variation of track is high and not satisfactory: some TCs are "Moving west" type and some TCs are "North west" type. For "Philippines east + Taiwan" group, the variation of intensity is high which is satisfactory: from TD to TY5. The variation of track is high and not satisfactory: a half of them are "Moving west" type and some of them are "North west" type. For "Japan" group, the number of cases is too few to give significant variations. The accuracy of forecast in 2006 is 66.32% which is higher than that of 2002 and 2003 probably because of fewer cases located outside the three location groups. On the other hand, such result is lower than that of 2004 and 2005 due to relatively high variation of tracks.

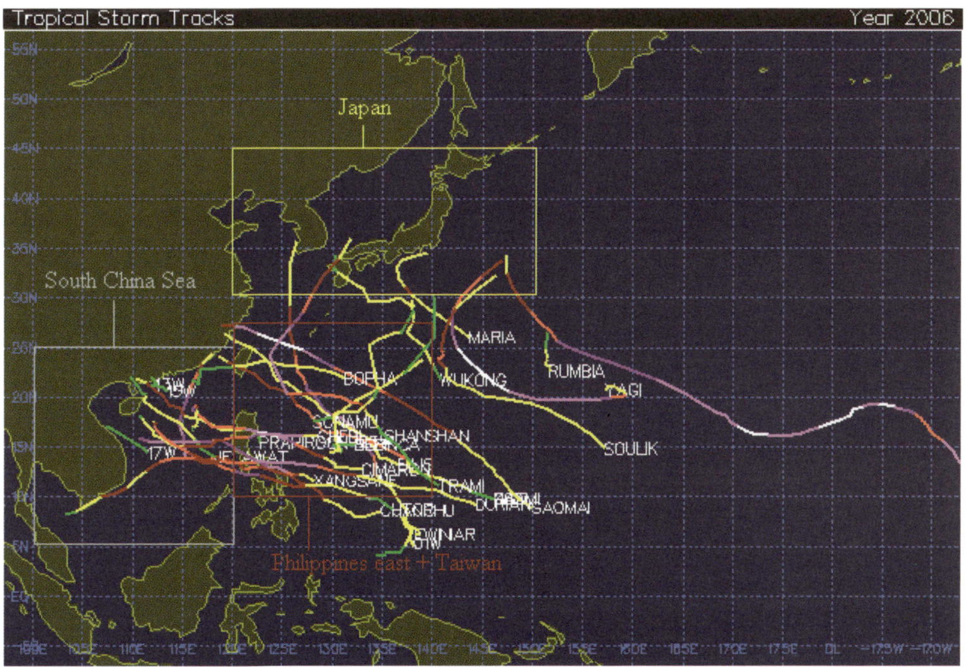

Fig. 19.5g. TCs tracks on North-Western Pacific Ocean in 2006

19.5.4. *Comparisons with other models*

We compare our approach with the STIPS (Statistical Typhoon Intensity Prediction Scheme) model from JTWC and a regression model developed by US Naval Research Laboratory. STIPS is a statistical-dynamical model for real time forecasting [7] developed by NOAA and enhanced by JTWC.

Input parameters

For STIPS, input parameters are atmospheric parameters like vertical wind shear, high level divergence, humidity, etc. Some calculated parameters like maximum potential intensity and relative eddy flux convergence.[15] The parameters are external to TC which means that they cannot be retrieved from TC data (i.e. best track data). The effectiveness of using these parameters is proved by meteorology. These data can only be retrieved from some professional organization as they require a lot of equipments and satellites to collect. Moreover, STIPS contains the moving track forecasting also which significantly helps the intensity forecasting. In the present study (Model Test), the data is extracted from the best track data of JTWC. Input parameters are the coordinates of TC, wind speed, pressure, direction of movement, etc. They are relatively simple and can be retrieved from TC data directly. These data can be easily downloaded from JTWC website. Some adjustments are added also to improve the accuracy.

Time frame

For STIPS, the time frame of forecasting is from 12-hour to 120-hour, with incremental value of 12 hours. The accuracies of short term forecasting (12-hour, 24-hour and 36-hour) are highest while the accuracies of long term forecasting are lowest (96-hour, 108-hour and 120-hour). In the present study, the time frame of forecasting has been kept at 6-hour with the expectation for better result. Accordingly, it is difficult to compare the results between two studies.

Regression model

Regression models for TC intensity forecasts are popular at operation centers because they provide quick and reliable guidance to the duty forecasters. The skills of such models rely on the selection of predictors. The selection is usually guided by experience, conceptual models, availability of data type, methods of data stratification and data quantity and quality. This can be subjective and empirical.[15] The time frame of forecasting is from 12-hour to 72-hour, with incremental values 12 hours. This model will predict the wind speed of TC.

Accuracy

For comparison, the testing data is the best track data of JTWC in 1990. The accuracies of STIFOR and JTWC are also included. The results are shown in Table 19.5l.

Table 19.51. The accuracies of the Proposed Model, STIFOR, JTWC and Regression Model for 6-hour, 12-hour and 24-hour forecasts [2]

	6-hour	12-hour	24-hour
Proposed Model	44%	42%	45%
STIFOR	—	55%	30%
JTWC	—	42%	30%
Regression model	—	49%	30%

In the present study, the accuracy of 6-hour forecast is lower than the accuracies from 2002-2006. This is because the training data are within 2002 and 2004, which have at least 12 years of difference. As this study is employing data mining approach, the accuracy depends largely on the quality of training data. The accuracy of 24-hour is the highest among other models. This is because the study simply predicts the change of intensity (i.e. "Decrease", "NoChange" and "Increase") whereas other models predict the exact wind speed of TC (e.g. 65 knots). Therefore, these results may not be that comparable.

19.6. Conclusion and Future Work

In this chapter, CBR is applied on TC intensity change forecasting. The best track data are processed into cases and then separated into three location groups: "South China Sea", "Philippines east + Taiwan" and "Japan". Association rules are generated for each different group and some adjustments are applied in order to find the ways to improve the accuracy. The accuracies are finally tested with each year of test cases. As demonstrated in experiments, the proposed intelligent model is able to provide reasonable forecast in severe weather. In particularly, it is found that the accuracy of forecast is strongly related to the track pattern and the number of cases that are not located in the three location groups. The accuracy is higher if the moving track pattern of the test case and the rules are the same. The model could be improved in two ways. In 2002 and 2004, the cases in "South China Sea" group have a very high moving track variation. The rules that are trained by these cases are not accurate enough. If the rules of "South China Sea" group are trained by data with a lower track variation (i.e. 2003 and 2006), the accuracy of "South China Sea" group will be improved. It is also possible to separate the cases into more location groups so that more rule files can be used for forecasting. For example, it is possible to separate the "Philippines east + Taiwan" into two groups to differentiate TCs of "Changing direction" type with "Moving west" and "North west". If possible, cases with abnormal moving tracks or abnormal life cycle should be removed from the training data.

Acknowledgement

The authors would like to acknowledge the partial support of CRG grant G-U186 of The Hong Kong Polytechnic University.

References

1. Agnar Aamodt and Enric Plaza, "Case-Based Reasoning: Foundational Issues, Methodological Variations, and System Approaches", *AICom — Artificial Intelligence Communications*, IOS Press, Vol. 7: 1, pp. 39–59, 1994. http://www.iiia.csic.es/People/enric/AICom.html
2. Jan-Hwa Chu, A Regression Model for the Western North Pacific Tropical Cyclone Intensity Forecast, Forecast Support Branch, Marine Meteorology Division, US Naval Research Laboratory, November 1994.
3. Corchado, J. M and Lees, B., "Adaptation of cases for case-based forecasting with neural network support", *in Soft Computing in Case Based Reasoning (S.K. Pal, S.D. Tharam, and D.S. Yeung, eds.),* Springer-Verlag, London, pp. 293–320, 2001.
4. Dvorak, V. F, "Tropical cyclone intensity analysis and forecasting from satellite imagery", *Mon. Wea. Rev.*, Vol. 103, pp. 420–430, 1975.
5. Feng, L, Dillon, T. and Liu, J., "Inter-transactional association rules for prediction and their application to studying meteorological data", *Data and Knowledge Engineering*, Elsevier, Vol. 37, pp. 85–115, 2001.
6. Jiang Tang, Ruixin Yang and Menas Kafatos (2005), "Data Mining for Tropical Cyclone Intensity Prediction", in *Proceedings of 6th Conference on Coastal Atmospheric and Oceanic Prediction and Processes*, Session 7.5, Jan 9–13, 2005.
7. Knaff, J., DeMaria, M. and Sampson, C., "An introduction to the statistical typhoon intensity prediction scheme (STIPS)", in *Proceedings of 26th conference on Hurricanes and Tropical Meteorology*, May 3–7, 2004, Miami, FL, USA.
8. Lee, R.S.T. and Liu, J.N.K., "Tropical Cyclone Identification and Tracking System using integrated Neural Oscillatory Elastic Graph Matching and hybrid RBF Network Track mining techniques", *IEEE Transaction on Neural Network*, Vol. 11, No. 3, pp. 680–689, 2000.
9. Liu, J.N.K., Shiu, S.C.K. and You, J., "Tropical Cyclone Forecaster Integrated with Case-Based Reasoning", in Mastorakis, N., MLadenov, V., Kontargyri, G.T., Eds. Springer-Verlag, Lecture Notes in Computer Science 5054, Vol. 28, Part III, pp. 235–244, 2009.
10. McGregor, J. L., Walsh, K. J. and Katzfey, J. J., "Climate Simulations for Tasmania", in *Proceedings of the 4th International Conference on Southern Hemisphere Meteorological and Oceanography*, American Meteorological Society, pp. 514–515, 2000.
11. San Pedro, J. and Burstein, F., "A Framework for Case-Based Fuzzy Multicriteria Decision Support for Tropical Cyclone Forecasting", in *Proceedings of the 36th Hawaii Int. Conf., System Sciences*, 2003.

[1] John Yeung, 熱帶氣旋之分級, Hong Kong Cyclone Observatory Centre [Retrieved on 20 Mar 2008] http://www.hkcoc.com/cyclonesort.htm

[2] John Yeung, 第二部份: 西北太平洋熱帶氣旋一般路徑及原因分析, Hong Kong Cyclone Observatory Centre [Retrieved on 19 Mar 2008] http://hkcoc.weather.com.hk/observatoryhome.htm

[3]夏季西進及秋冬轉向型, 西北太平洋熱帶氣旋一般路徑及原因分析, Hong Kong Cyclone Observatory Centre [Retrieved on 20 Mar 2008] http://hkcoc.weather.com.hk/tcmv2.1.htm

[4]秋冬西進及季風爆發型, 西北太平洋熱帶氣旋一般路徑及原因分析, Hong Kong Cyclone Observatory Centre [Retrieved on 20 Mar 2008] http:// hkcoc.weather.com.hk/tcmv2.2.htm

[5]The CBR Cycle, Case-based Reasoning Review, AI-CBR [Retrieved on 21 Mar 2008] http://www.ai-cbr.org/classroom/cbr-review.html

[6]Case-based Reasoning, Wikipedia [Retrieved on 21 Mar 2008] http://en.wikipedia.org/wiki/Case-based_reasoning

[7]NW Pacific Ocean, GIF storm track map, Institute of Astronomy, University of Hawaii [Retrieved on 21 Mar 2008] http://www.solar.ifa.hawaii.edu/Tropical/Gif/nwp.latest.gif

[8]2002 Hurricane/Tropical data for Western Pacific, Unisys [Retrieved on 31 Mar 2008] http://www.weather.unisys.com/hurricane/w_pacific/2002H/index.html

[9]2003Hurricane/Tropical data for Western Pacific, Unisys [Retrieved on 31 Mar 2008] http://www.weather.unisys.com/hurricane/w_pacific/2003H/index.html

[10]Super Typhoon PARMA Tracking information, Unisys [Retrieved on 6 Apr 2008] http://www.weather.unisys.com/hurricane/w_pacific/2003H/PARMA/track.gif

[11]Super Typhoon PARMA Tracking information, Unisys [Retrieved on 6 Apr 2008] http://www.weather.unisys.com/hurricane/w_pacific/2003H/PARMA/track.dat

[12]2004 Hurricane/Tropical data for Western Pacific, Unisys [Retrieved on 31 Mar 2008] http://www.weather.unisys.com/hurricane/w_pacific/2004H/index.html

[13]2005 Hurricane/Tropical data for Western Pacific, Unisys [Retrieved on 31 Mar 2008] http://www.weather.unisys.com/hurricane/w_pacific/2005H/index.html

[14]2006 Hurricane/Tropical data for Western Pacific, Unisys [Retrieved on 31 Mar 2008] http://www.weather.unisys.com/hurricane/w_pacific/2003H/index.html

[15]John Knaff from CIRA/Colorado State University, Mark DeMaria from NOAA/NESDIS, STIPS, JTWC [Retrieved on 9 Apr 2008] http://metocph.nmci.navy.mil/jtwc/pubref/References/stips.ppt

[16]Global Guide to Tropical Cyclone Forecasting, Bureau of Meteorology Research Centre, Australia http://www.bom.gov.au/bmrc/pubs/tcguide/globa_guide_ intro.htm

[17]Tropical Cyclone Formation Regions, NOAA http://www.srh.noaa.gov/jetstream/tropics/tc_basins.htm

[18]Alternative Data Mining Techniques for Predicting Tropical Cyclone Intensification. http://www.cs.colostate.edu/~kyriakos/projects/cs540.html [Retrieved on 15 July, 2007]

[19]JTWC Western-North Pacific Best tract data [Retrieved on 25 Mar 2008] https://metocph.nmci.navy.mil/jtwc/best_tracks/wpindex.html

Appendix

The field "LocName" and "Serial" are for identification purpose only. They do not involve in any forecasting process. "LatitudeS" and "LongtitudeS" are coordinate of the area's left top corner point. "LatitudeE" and "LongtitudeE" are the coordinate of the area's right bottom corner point.

LatitudeS	LongtitudeS	LatitudeE	LongtitudeE	LocName	Serial
200N	1000E	155N	1056E	Thai	1
183N	1056E	155N	1072E	Thai	2
172N	1072E	155N	1083E	Thai	3
155N	1000E	133N	1094E	Thai	4
133N	1011E	127N	1033E	Thai	5
127N	1022E	116N	1033E	Thai	6
133N	1033E	116N	1094E	Thai	7
116N	1033E	100N	1083E	Thai	8
100N	1055E	83N	1061E	Thai	9
205N	1083E	177N	1116E	HoiLam	1
300N	1000E	200N	1100E	China	1
300N	1100E	250N	1200E	China	2
250N	1100E	222N	1127E	China	3
250N	1127E	233N	1177E	China	4
233N	1127E	222N	1150E	China	5
233N	1150E	227N	1172E	China	6
250N	1177E	244N	1188E	China	7
300N	1200E	272N	1227E	China	8
272N	1200E	261N	1211E	China	9
261N	1200E	222N	1222E	China	10
400N	1211E	383N	1244E	China	11
377N	1200E	350N	1227E	China	12
350N	1200E	327N	1216E	China	13
327N	1200E	300N	1222E	China	14
400N	1244E	366N	1288E	Korea	1
366N	1250E	333N	1294E	Korea	2
333N	1283E	305N	1316E	Japan	1
350N	1300E	327N	1400E	Japan	2
400N	1355E	344N	1400E	Japan	3
400N	1400E	361N	1422E	Japan	4
455N	1400E	400N	1422E	Japan	5
477N	1411E	427N	1472E	Japan	6
122N	1183E	100N	1200E	Phil	1
100N	1200E	44N	1266E	Phil	2
138N	1200E	100N	1266E	Phil	3
200N	1200E	138N	1227E	Phil	4

Chapter 20

ANALYSIS OF SEQUENTIAL DATA IN TOOL MANUFACTURING OF VOLKSWAGEN AG

Kemal Ince[1], Thomas Schneider[2] and Frank Klawonn[3]
[1]*Volkswagen AG location Braunschweig, Components-Toolshop*
P.O. Box 4749, 38037 Braunschweig, Germany
kemal.ince@volkswagen.de
[2]*Volkswagen AG location Braunschweig, Components-Toolshop*
P.O. Box 4749, 38037 Braunschweig, Germany
thomas.schneider2@volkswagen.de
[3]*Department of Computer Science, Ostfalia University of Applied Sciences,*
Salzdahlumer Str. 46/48
38302 Wolfenbüttel, Germany
f.klawonn@ostfalia.de

Standardization is one of the important preconditions to assure efficient manufacturing in industry such as in toolmaking, because the creation of standardized and hence transparent manufacturing provides the opportunity to detect and implement optimizations. In this work we describe our aim, which was to find standard processing sequences for manufacturing in toolmaking by using appropriate data mining techniques. The data which has to be analysed was collected by a special extraction, transformation and loading process from archived text files that include CAM-information.

The realized results of this analysis are that up to 75% of the procedures in manufacturing moulds can be standardized.

20.1. Introduction

20.1.1. *Knowledge discovery and data mining*

Nowadays all industrial areas are affected by a data flood which often contains very useful information and knowledge. This is due to the technological progress in the information processing sector and the gain of networking in companies. For example Europe's VLBI has 16 telescopes, each of which produces 1 Gigabit/second of astronomical data over a 25-day observation session.[1,2] The acquirement of this concealed knowledge is leading to competitive advantage, but extracting it from such large databases needs a high effort, because the interesting information is hidden in large amounts of data.[5] KDD and especially data mining are very central techniques to find such information and knowledge in large databases. Manufacturing is also an application area which can gain significant competitive advantages when data mining techniques are applied in a suitable manner.[5] This work deals with the special area of mining sequential data in large databases. Sequence mining is a widely used technique which comes into operation in different areas like bioinformatics to analyse gene sequences[3] or the telecommunication branch to predict customer behaviour.[4] Also the data in the Components-Toolshop which is received from the CAM-System during the manufacturing process has the property to be analysed with sequence mining techniques. The sequences in this application area are composed of different manufacturing steps which are described in chapter two and especially in the Section 20.2.2 and 20.2.3 in detail.

20.1.2. *The application area*

Our area of application for sequence mining is the Components-Toolshop of Volkswagen AG in Braunschweig. This business area is responsible for producing tools to use them in other divisions of the Volkswagen company like the serial production for example. The Components-Toolshop has approximately 700 members of staff and includes a 30.000 m^2 production area so that it can be considered as one of the biggest tool shops in the world. To its product range belong among dies (like gearbox cases and engine boxes e.g.), injection moulds, permanent moulds and other formed components or tools. To exemplify such a component/tool a gearbox case, which was manufactured in the Components-Toolshop in Braunschweig, is shown in Figure 20.1.

Such tools could not be compared with components which were manufactured in serial production. In serial production or mass production it is easier to find standardized and automated approaches. Due to the uniqueness of the parts to be manufactured no mass production is possible. In parts production as it is in tool shop the question about finding standard approaches could not be responded in such facility, because each tool can differ from the other ones because of their individual properties like the surface structure for example. Nevertheless, in part production

Fig. 20.1. Gearbox case component/tool

there are also capabilities to standardize and hence to automatise the approaches of our application area. The data, which contains the description of the production steps, were extracted from text files and after transformation and cleaning stored in a database.

The next chapters deal with the description of our methods to analyse the data. In chapter two we give an introduction to the work sequence in Components-Toolshop of Volkswagen AG in Braunschweig. Chapter three discusses the data preprocessing steps and especially how the sequences, which have to be analysed, are generated. Chapter four describes the basic idea, an implemented algorithm, to find standardized approaches and shows the validated conclusions. Closing in chapter five we discuss the conclusions and give a brief outlook for concepts in future works.

20.2. The Work Sequence in the Components-Toolshop

The process of manufacturing a tool is a very complex one. In the following section we give an introduction to this process.

First of all there is the gross planning of the tool. In this gross planning step a qualified employee decides whether the requirements of the tool which has to be manufactured can be fulfilled or not. After this important decision, the tool has to be designed by a construction process using a CAD-system like CATIA[6] or Pro/E[7]. The result of the construction is a 3D data set of the tool which has to be manufactured. After these steps it is possible to plan the next steps of the production of the tool in detail, because now is it possible to specify which area of the tool has to be manufactured by using which milling strategy, milling cutter, blade diameter etc. With a conformable CAM-system it is possible to generate the NC-code which is responsible to control the milling machine and contains the different milling strategies. In this step the employee must select manually different milling strategies by defining various properties which are described in Section 20.2.1. Afterwards the CAM-system converts these strategies into the NC-program. The disadvantage of this approach is that the know-how of taking the right and especially the most efficient strategy depends on the experience of the employee which is not available for other employees. This know-how could be used to automatise standard cases. To centralize and making this know-how usable is the central aim of our approach. Figure 20.2 shows the whole work sequence in Components-Toolshop which is described above. Our connecting factor is between the programming and

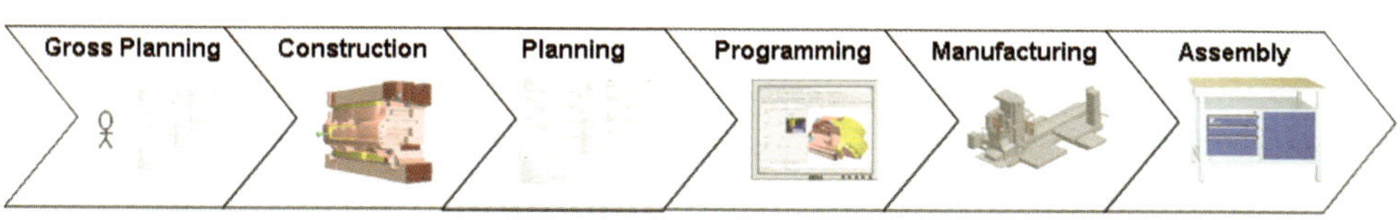

Fig. 20.2. The work sequence in Components-Toolshop

manufacturing step. In the next paragraph we will give a short introduction to the CAM-technology and how it works. With the generated NC-code it is possible to manufacture the tool on a milling machine. As a last step, if necessary, the assembly works should be conducted to deliver it the customer.

20.2.1. *NC, CNC, DNC and how it works*

An NC-controller is an electronic component which is able to read a data set of control commands, convert them into correct work flows and discharge them on a milling machine for example. This means that for each machining process on a CNC-machine, the NC-program has a command with the relevant information.[8]

There are three basic types of NC-machines in use. The standard NC-machine, which is controlled directly by NC-commands, the DNC-machine which is controlled via a computer and the CNC-machine which is controlled by a manufacturing computer that acts as a server.[8] In our case the NC-codes were sent via a network to the manufacturing computer to store them. From there they were sent to the machines which have to be controlled.[8] Nowadays mostly CNC-machines are used, which are also able to execute single commands. An advantage of this approach is that also complex instructions can be handled.

The CNC-unit of the milling machine is responsible for the interpretation of the NC-code. Due to this interpretation a mechanical treatment of the tempered steel by the machine drive is possible. An NC-program contains a collection of sequences which comply with processing steps. These processing assignments are coded by special keywords which consist of alphabetic and numeral characters. The meaning of these keywords is defined in the NC-control. Table 20.1 shows the most important keyword of the NC-language.

The ISO-Norm 6983 is responsible to define the NC-commands and the structure of the NC-programs in use.[9] Table 20.2 illustrates the standard structure of an NC-program consisting of elements which are specified in Table 20.1. The actual NC-program begins after a special character which informs the NC-controller that an NC-program will follow. The complete section in front of this character is defined as a comment and will be ignored by the NC-controller. After the beginning of the NC-program a series of NC-blocks follows, which include the relevant geometric and technical information that the CNC-controller requires for each machining step.[9] It is out of the scope of this contribution to describe the structure of these NC-blocks in detail so that we refer for information to.[9] The NC-program ends with a special character to indicate the end of the program.

20.2.2. *Components in manufacturing*

In this section we describe the central steps in manufacturing moulds to understand how such an eradication with the tools does work. First we describe in Figure 20.3 the most important components of standard milling machines.

Table 20.1. Important NC-commands

identification		function and meaning
%		beginning of an program
:	1 to 9999	main command
N		number
	...	...
G	0	quick motion
	1	interpolation of straight line
	...	...
	17	choosing the layer in X-Y
	18	choosing the layer in X-Z
	19	choosing the layer in Y-Z
	...	...
	80	abborting the operating cycle
	81 to 89	operating cycles
	...	...
F	0 to Fmax	feed motion in mm/min (in milling)
S	0 to Smax	spindle RPM in min-l
T	1 to 9999	Tool number
H	1 to 999	auxiliary function
L	01 to 99	number of the subprogram
M	[0-9]	holding point
(		begin comment
)		end comment
LF		line feed

Table 20.2. Standard structure of an NC-program[9]

% TP0147	NC-program beginning,
N10 G54 X80 Y100...	a series of NC-blocks
...	with the information for machining and
N75 G01 Z-10 F0.3 S1800 T03 M08	
...	
N435 M30	a command for ending the program.

There are two central types of milling machines the vertically and the horizontally working ones. In Figure 20.3 these two types of milling machines and their basic components are shown. Nowadays, the milling machines which work with NC-technology have e.g. three or five axes, independent of their different arbor types, but they basically are composed of the same components as shown in Figure 20.3.

The first step in manufacturing is that after planning and constructing the tool, the employee arranges the material on the work table of the milling machine. Afterwards he can start to mill when the NC-codes have been provided by the programmer. These NC-codes contain the information which traverse paths the spindle of the machine has to follow.

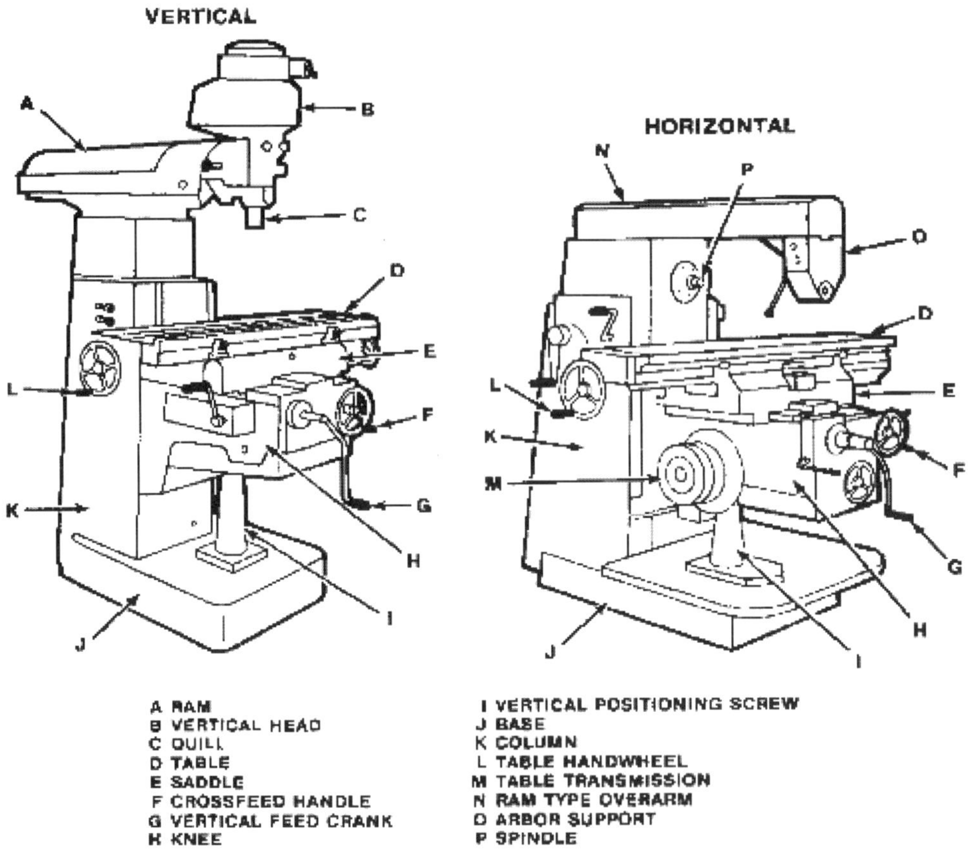

Fig. 20.3. Types of Milling Machines

In Section 20.2.1 we have already discussed the basic instructions of the NC-code and their meanings. Every job in which a tool has to be manufactured contains information about how the specific manufacturing process of the tool was designed.

20.2.3. *Sequences of operations*

In this section, we outline how an operation is built and which components are important for our analysis. Each tool is divided into different areas which are milled using suitable operations. These operation series for an area form a sequence which has to be analysed. Figure 20.4 serves as an example for such an operation sequence containing in (a) the milling strategy and in (b) the blade diameter of a manufactured area. Both of these parameters will be used in our analysis.

It is important to know that the sequences in the data sets can have variable lengths. The length of a sequence is defined as the number of objects which are used in this sequence. In our example shown in Figure 20.4 the length of both displayed sequences is three.

The manufacturing of a tool can be split into two main machining steps. The first one is concerned with milling as much material as possible in the shortest possible time and is named roughing. In milling strategies of this category the structure of the surface is not much important. The areas where the structure of the surface becomes importance are milled by other milling strategies like "finishing z constant", "residual material z constant" or others which belong to the category of smoothing. Figure 20.5 provides an example how the sequences by milling a part come together. In this example we have both machining steps, roughing and smoothing. The roughing step is not separated in multiple milling areas. This means that the

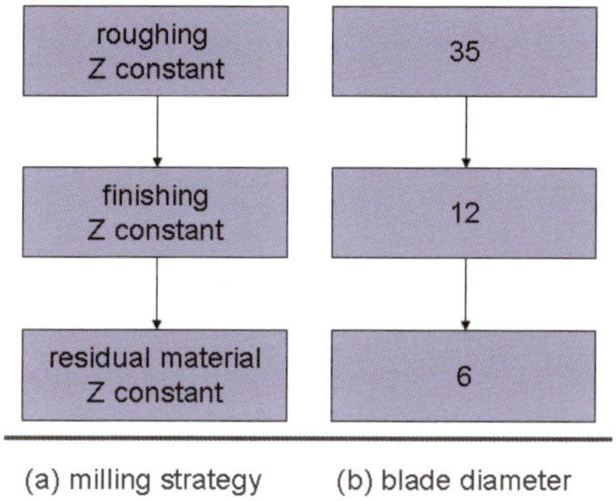

Fig. 20.4. A milling sequence

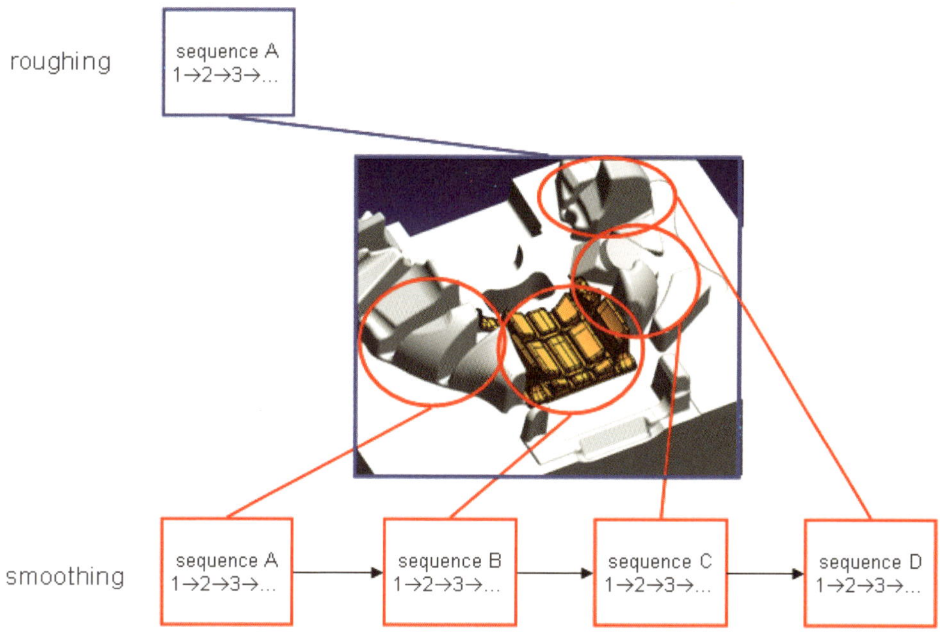

Fig. 20.5. Definition of sequences

Table 20.3. Data set with corresponding sequences

ID	machining step	operation	sequence
1	roughing	1	sequence A
2	roughing	2	sequence A
3	roughing	3	sequence A
4	smoothing	1	sequence A
5	smoothing	2	sequence A
6	smoothing	3	sequence A
7	smoothing	1	sequence B
8	smoothing	2	sequence B
9	smoothing	3	sequence B
10	smoothing	1	sequence C
11	smoothing	2	sequence C
12	smoothing	3	sequence C
13	smoothing	1	sequence D
14	smoothing	2	sequence D
15	smoothing	3	sequence D

whole part is milled with one roughing step composed of different operations. These abstract operations named from 1 to n build a sequence, here named sequence A. The smoothing step however is separated in multiple milling areas. The part is split into four milling areas which are manufactured with different operations, so that there exist four sequences, named sequence x with $x \in \{A, B, C, D\}$. These sequences are also composed of different operations, which are numbered from 1 to m.

The data set with the corresponding sequences from this example is shown in Table 20.3.

20.3. Data Preprocessing

The central aim of data preprocessing is to get a higher data quality by integrating the data which have to be analysed, avoiding inconsistencies and treating missing values.[10] For the special analysis of the sequences, which were described in Section 20.2.3, two attributes were considered, the "milling strategy" and the "blade diameter" The data preprocessing step can be split into three steps. In the first step the set of the possible attributes were processed, afterwards, in step two, the data basis for the analysis will be selected and at last in step three the data basis is transformed manually and by dint of an application into a special format, to carry out the analysis of the sequences.

20.3.1. *Step 1: Standardization of the domain*

The attribute "milling strategy" is nominal. This means that the values which this attribute can take are alphabetical characters and some additional characters like "–" for example. The domain of the attribute "blade diameter" however is numerical. The values of this attribute are integers. Preprocessing is a necessary step here, since the values for the milling strategy were not standardized. To illustrate this problem, Figure 20.6 shows different names for the same milling strategy "roughing z constant" which were used in different projects and have been found in the data sets. These different synonyms for the same milling strategy were caused by the manual entries provided by the employees when setting up the milling strategy for the corresponding manufacturing process.

The preprocessing step in detecting standardized values for the attributes works automatically. For this purpose, a special application has been implemented converting the different synonyms to standardized values based on regular expressions. The detailed operations behind these standardized values are not of importance in the context of the sequence analysis that we carry out here. After this preprocessing step both attributes have standardized values shown in Figure 20.7.

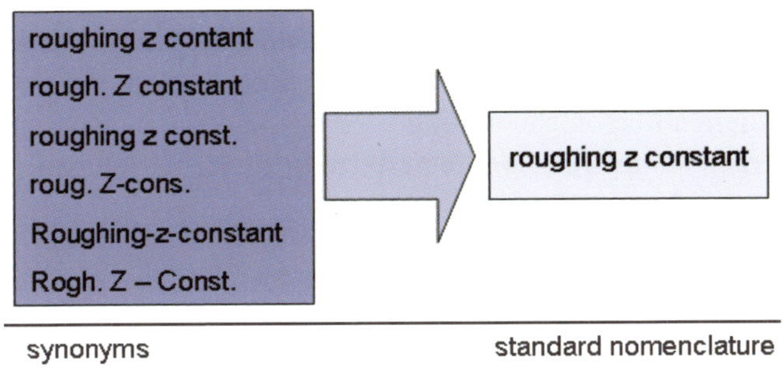

Fig. 20.6. Example of different synonyms

milling strategy
roughing z constant
roughing raw part
roughing outlines
roughing planar areas
roughing parallel
roughing bore jet
smooth finishing z constant
smooth finishing along curv
smooth finishing downwards
optimise z constant
optimise smooth finishing
residual material z constant
residual material contour

blade diameter
80
66
63
52
50
35
32
25
20
16
12
10
08
06
04
02

Fig. 20.7. Standard values of both attributes

20.3.2. *Step 2: Selection of the data set*

Based on expert knowledge and a previous analysis of the data it is known that the structure of the sequences depends on the "milling category" and the "machine type" on which the part has to be manufactured. The milling categories can be divided into different milling strategies. This means each of the n, with n ϵ {"roughing z constant", "smooth finishing z constant",...}, milling strategies belongs to one of the m, with m ϵ {"roughing", "smooth finishing", "optimize", "residual material"}, milling categories. Milling strategies belonging to the milling category "roughing" and milling strategies belonging to the milling categories "smooth finishing", "optimize" and "residual material" must be analysed separately. Another important attribute having a strong influence on the sequences to be analysed is the "machine type". Different machines are in use in Components-Toolshop and each machine type has its own special properties. Therefore, data from different machine types are analysed separately. This leads to a data set which contains sequences depending on the milling category and the machine type. Table 20.4 shows a small artificial data set before and Table 20.5 after splitting it according to the machine type and the milling category.

Table 20.4. Artificial data set before splitting

job_ID	blade diameter	milling category	milling strategy	machine type
1	80	roughing	roughing z constant	machine1
1	35	roughing	roughing z constant	machine1
1	10	roughing	roughing raw part	machine1
1	12	smooth finishing	smooth finishing z constant	machine1
1	8	optimse	optmise z constant	machine1
1	6	residual material	resudial material z const	machine1
2	80	roughing	roughing z constant	machine2
2	35	roughing	roughing z constant	machine2
2	10	roughing	roughing raw part	machine2
2	12	smooth finishing	smooth finishing z constant	machine2
2	6	optimse	optmise z constant	machine2
3	35	roughing	roughing z constant	machine1
3	10	roughing	roughing raw part	machine1
3	16	smooth finishing	smooth finishing z constant	machine1
3	10	optimse	optmise z constant	machine1
3	6	residual material	resudial material z const	machine1
4	35	roughing	roughing z constant	machine2
4	35	roughing	roughing z constant	machine2
4	12	roughing	roughing raw part	machine2
4	8	roughing	roughing raw part	machine2
4	20	smooth finishing	smooth finishing z constant	machine2
4	12	optimse	optmise z constant	machine2
4	6	residual material	resudial material z const	machine2

20.3.3. *Step 3: Data structure of the data set to be analysed*

After the data set has been split as described above, the sequences of interest are extracted. An artificial data set as it is shown in Table 20.6 encodes three sequences illustrated in Table 20.7.

Table 20.5. Artificial data sets after splitting

job_ID	blade diameter	milling category	milling strategy	machine type
1	80	roughing	roughing z constant	machine1
1	35	roughing	roughing z constant	machine1
1	10	roughing	roughing raw part	machine1
3	35	roughing	roughing z constant	machine1
3	10	roughing	roughing raw part	machine1
job_ID	**blade diameter**	**milling category**	**milling strategy**	**machine type**
2	80	roughing	roughing z constant	machine2
2	35	roughing	roughing z constant	machine2
2	10	roughing	roughing raw part	machine2
4	35	roughing	roughing z constant	machine2
4	35	roughing	roughing z constant	machine2
4	12	roughing	roughing raw part	machine2
4	8	roughing	roughing raw part	machine2
job_ID	**blade diameter**	**milling category**	**milling strategy**	**machine type**
1	12	smooth finishing	smooth finishing z constant	machine1
1	8	optimse	optmise z constant	machine1
1	6	residual material	resudial material z const	machine1
3	16	smooth finishing	smooth finishing z constant	machine1
3	10	optimse	optmise z constant	machine1
3	6	residual material	resudial material z const	machine1
job_ID	**blade diameter**	**milling category**	**milling strategy**	**machine type**
2	12	smooth finishing	smooth finishing z constant	machine2
2	6	optimse	optmise z constant	machine2
4	20	smooth finishing	smooth finishing z constant	machine2
4	12	optimse	optmise z constant	machine2
4	6	residual material	resudial material z const	machine2

Table 20.6. Transformed data set

ID	ID2	sequence state
1	100	A
1	101	A
1	102	B
2	200	A
2	201	B
2	202	B
2	203	C
3	301	B
3	303	C
3	302	B

Table 20.7. Correlation table

sequence object	blade diameter
A	35
B	20
C	12

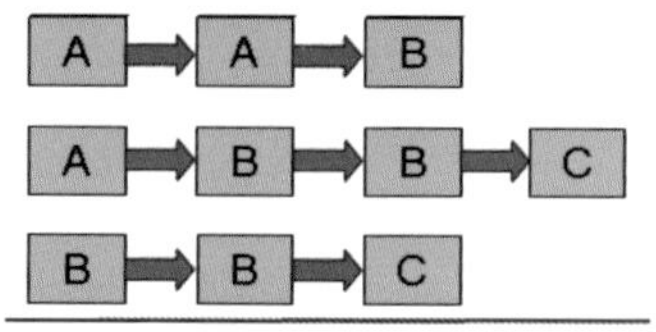

Fig. 20.8. Resulting sequences

"*ID*" refers to the job, i.e. all entries with the same "*ID*" form a sequence. "*ID2*" indicates the order in which the sequence is carried out and "*sequence state*" contains the attribute values that build the sequences to be analysed. This can for example be the "blade diameter", the "milling strategy", an attribute composed of aggregated values of multiple attributes and others.

If the "*sequence state*" is, for example, composed of the blade diameter, and the characters have correlations like in Table 20.7, the sequences shown in Figure 20.8 are obtained.

20.4. Analysis of Sequences

There are various approaches to find sequences in data, depending very much on the type of the data. In biology, alignment of sequences is a crucial problem as well as finding common subsequences in DNA.[3]

Process mining analyses sequences of steps in business production steps. The aim is often to extract process models from data or to check whether the data deviate from a given process model.[11–13]

There are also more complex models for sequence mining based on extensions of frequent itemset mining to investigate time relations like "happened before" or "happened during" between events with a duration.[14]

Hidden Markov models[15] try to identify sequences of state using additional intermediate hidden states that do not occur as states in the data set. In our application, we do not need such hidden states and apply simple Markov chains to analyse our sequences.

Our approach to analyse the sequences contains different steps which are described in this section. In Section 20.3 it was already explained how we obtain the data to be analysed. After these filtering and preprocessing steps the resulting data set is the input for our sequence finder algorithm. This algorithm delivers a probabilistic state machine in the form of a Markov chain, which represents a model of the real data set.

20.4.1. *The probabilistic state machine*

As described in Section 20.3 and illustrated in Figures 20.8 and 20.9 a sequence is composed of "*sequence states*" which can contain the blade diameter, the milling strategy, another attribute or composed attributes as values. Figure 20.10 shows such sequences containing in this case only the blade diameters as *sequence states*.

These sequences have different lengths. The sequence length is defined by the numbers of the included *sequence states*. This means that in our example in Figure 20.10 we have two sequences of length one, six sequences of length two, one sequence of length three and one sequence of length four.

The probabilistic state machine is a model which is generated by the sequences of the real data set. For our example this means that a state machine as shown in Figure 20.11 would be generated. States encode the actual position in a sequence. In the simplest case as in Figure 20.11 a state contains only the diameter at the actual position of the sequence. In this case, the state machine does not take previous diameters in the sequence into account when changing from one state to another. We also consider state machines where a state corresponds to a "history" of more than one diameter. The transitions of the state machine (lines) are labelled with the probabilities for the corresponding state changes. The meaning of these probabilities and how they are calculated is explained in Section 20.4.2.

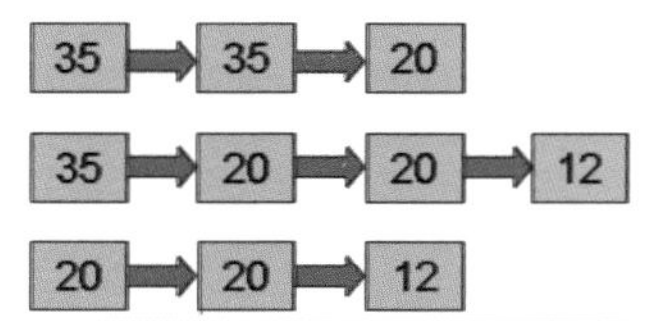

Fig. 20.9. Sequences with blade diameter

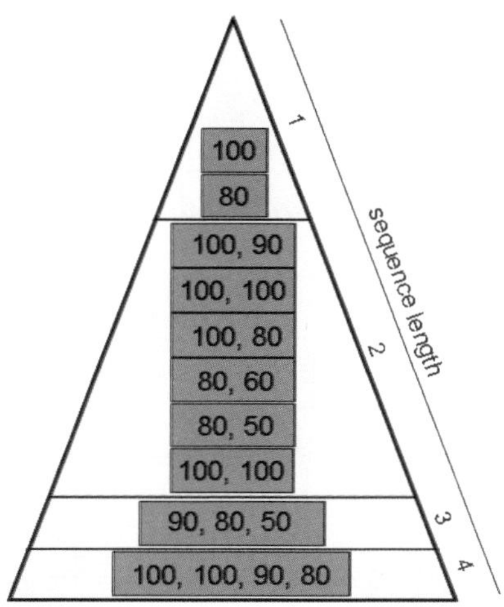

Fig. 20.10. Example of sequences

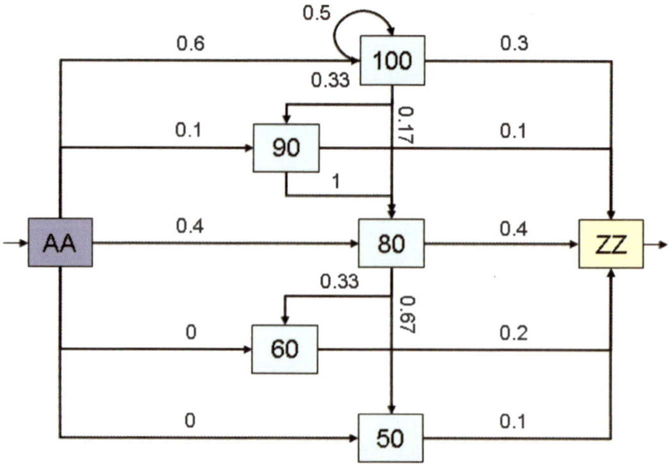

Fig. 20.11. Probabilistic state machine

The state "*AA*" represents the initial state and the state "*ZZ*" represents the final state. Both are not used for describing an actual *sequence state*, they are only needed for marking the beginning and the end of a sequence.

20.4.2. *Building the model*

20.4.2.1. *The predecessor and the prepredecessor sequence state*

For the construction of the probabilistic state machine we either consider only the direct predecessors or also the prepredecessors. As an example we consider the sequence of length four shown in Table 20.8.

In Table 20.8 the blade diameter "*100*" with the index "*0*" is considered to be the predecessor of the blade diameter "*100*" with the index "*1*". In the same way, the blade diameter "*100*" with index "*1*" is considered to be the predecessor of the blade diameter "*90*" with index "*2*".

In this example only the sequence *states* "*90*" and "*80*" have prepredecessors. The blade diameter "*100*" with index "*0*" is the prepredecessor of the *sequence state* "*90*" and the blade diameter "*100*" with index "*1*" is the prepredecessor of the *sequence state* "*80*".

20.4.2.2. *The probability matrix*

How the state machine is constructed depends on the model assumption whether it is sufficient to consider the predecessor only or whether the prepredecessor must also be taken into account to predict the next diameter.

Table 20.8. Example sequence

index	0	1	2	3
sequence	100	100	90	80

Table 20.9. Example sequences

100, 90, 80
100, 90, 60
100, 100, 90

Table 20.10. Probability matrix predecessor

	AA	100	90	80	60	ZZ
AA	XX	1.0	0	0	0	XX
100	XX	0.25	0.75	0	0	0
90	XX	XX	0	0.5	0.5	0.33
80	XX	XX	XX	0	0	0.33
60	XX	XX	XX	XX	0	0.33
ZZ	XX	XX	XX	XX	XX	XX

Starting probability
Ending probability
Transition probability

When only the predecessor is taken into account, the extremely small artificial data set in Table 20.9 leads to the state transition probabilities given in Table 20.10.

The entry "*XX*" in the probability matrix indicates zero probabilities, i.e. transitions which have not occured in the data set. The states AA and ZZ refer to the initial and final state, respectively. This means that each sequence is assumed to start in state AA and to end in state ZZ.

Based on this probability matrix the resulting probabilistic state machine is shown in Figure 20.12. Zero probabilities "*0*" are not shown in this figure.

When the prepredecessor is also considered, then a state corresponds to two values in a sequence. The simple example in Table 9 leads to the probability matrix shown in Table 20.11.

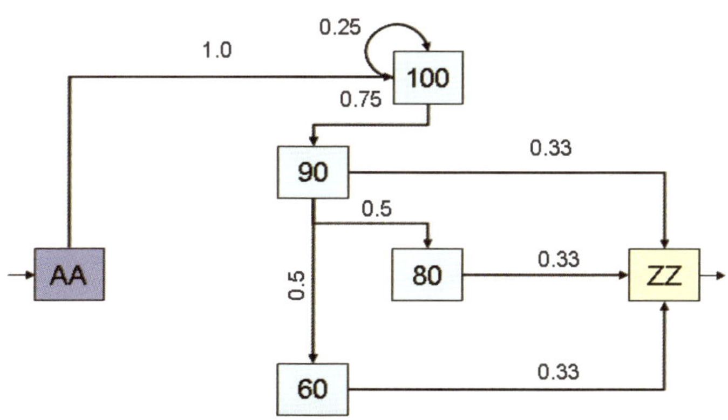

Fig. 20.12. Probabilistic state machine predecessor

Table 20.11. Probability matrix prepredecessor

	AA	100	90	80	60	ZZ
AA, AA	XX	XX	XX	XX	XX	XX
AA, 100	XX	0.33	0.67	0	0	0
AA, 90	XX	XX	0	0	0	0
AA, 80	XX	XX	XX	0	0	0
AA, 60	XX	XX	XX	XX	0	0
AA, ZZ	XX	XX	XX	XX	XX	XX
100, 100	XX	0	1.0	0	0	0
100, 90	XX	XX	0	0.33	0.33	0.33
100, 80	XX	XX	XX	0	0	0
100, 60	XX	XX	XX	XX	0	0
100, ZZ	XX	XX	XX	XX	XX	XX
90, 90	XX	XX	0	0	0	0
90, 80	XX	XX	XX	0	0	1.0
90, 60	XX	XX	XX	XX	0	1.0
90, ZZ	XX	XX	XX	XX	XX	XX
80, 80	XX	XX	XX	0	0	0
80, 60	XX	XX	XX	XX	0	0
80, ZZ	XX	XX	XX	XX	XX	XX
60, 60	XX	XX	XX	XX	0	0
60, ZZ	XX	XX	XX	XX	XX	XX

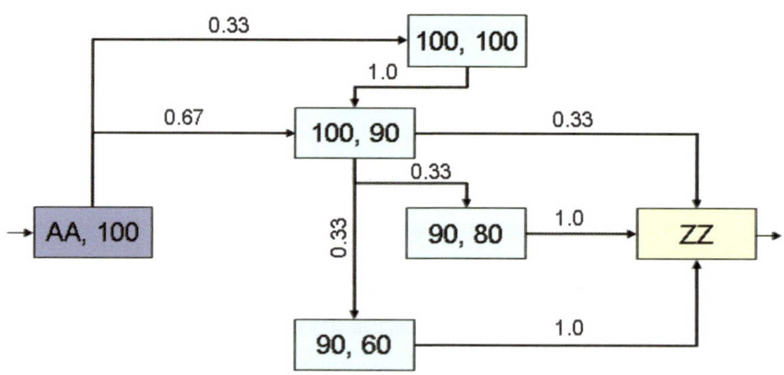

Fig. 20.13. Probabilistic state machine prepredecessor

The graph of the probabilistic state machine resulting from this probability matrix is shown in Figure 20.13.

Depending on the analysis method, whether the predecessor or the prepredecessor is considered, the model is builded. The pseudo code of building the model by using the prepredecessor method is described in Figure 20.14. The *lengthVector* contains the probabilities for the different lengths of the sequences which occur in

```
probVector <- (1,0,...,0)
lengthVector <- (0,...,0)
FOR i=0 TO maxLength-1 DO
  probVectorNew <- (0,0,...,0)
  FOR j=0 TO countDiameter-1 DO
    FOR k=0 TO countDiameter-1 DO
      FOR l=0 TO countDiameter-1 DO
        probVectorNew[j*countDiameter+k]=probVectorNew[j*countDiameter+ k]
             + probMatrix[l,j,k] * probVector[l*countDiameter+j]
      END-FOR
    END-FOR
  END-FOR
  probVector <- probVectorNew
  FOR j=1 TO countDiameter DO
    lengthVector[i] = lengthVector [i] + probVector[k*countDiameter-1]
  END-FOR
END-FOR
lengthVector[maxLength] = 1
FOR i=0 TO maxLength -1 DO
  lengthVector[maxLength] = lengthVector[maxLength] - lengthVector[i]
END-FOR
```

Fig. 20.14. Pseudo code for prepredecessor

Table 20.12. Results of chi-square test

sequence length	observed frequency data set	observed probability	probabilites in lengthVector	chi-square Value
0	0	0,000	0,000	0
1	420	0,376	0,376	0,000
2	274	0,246	0,219	3,488
3	154	0,138	0,127	1,037
4	106	0,095	0,098	0,077
5	55	0,049	0,070	6,747
≥6	107	0,096	0,110	2,055
sum	1116	1	1	**13,404**

the data set. These values were used to execute the chi-square test which is described in Section 20.4.3.

20.4.3. *Verifying the model*

The chi-square test can be used to compare two probability distributions. We apply this test in order to check whether our simple Markov model reflects the probabilistic structure in the data appropriately. In order to compare the data with the derived Markov model, we generate simulated data from the Markov model and compare the distribution of the lengths of the simulated sequences with the distribution of the sequence lengths in the data set. Table 20.12 shows such a test and its results.

The test has six degrees of freedom and the chi-square value is 13.404 leading to a p-value of less than 4%. This low p-value indicates that the simple trained Markov model does not seem to yield the same distribution. Because of this result we decided to try another approach to find standards in the data set, which is described in following section.

20.4.4. *Generating Rules*

In Section 20.4.2 we have discussed the probability matrix and how it was calculated. The probability matrix stores the transition probabilities depending on the transitions in the sequences of the observed data set. Our idea was to generate rules, which depend on these probabilities in the probability matrix and having a minimum confidence $C_{\min}$ and a minimum support $S_{\min}$. The $C_{\min}$-value makes sure that the successor is definitely predictable. The $S_{\min}$- value however makes sure that rules are only derived from transitions which occur frequently. If a transition e.g. occurs a few times in a data set with 10,000 transitions, it is suggested that this transition is a special case and that it does not make sense to standardize this transition in a rule. If we have transitions of the form $A \to B$ the confidence and the support will be calculated like in Figure 20.15 described with N_A the frequency of A, $N_{A,B}$ the frequency of A or B and $P(\{A,B\})$ the probability of A or B. N is the number of samples.

The generation of the rules can be handled in two different ways. In the first alternative we use 10-fold cross validation, i.e. the data set is randomly split into 10 disjoint subsets. One subset is removed and the remaining 90% of the data are used as training set to construct the rules. The 10% removed data are used as test data to calculate the actual confidence and support of the rules that were found. This is carried out 10 times, where each of the 10 subset is used as a test set once. The second alternative is generating rules without using 10-cross validation. Both versions generate many of rules. Depending on the approach whether the predecessor or the prepredecessor is considered, the generated rules have the form as shown in Figure 20.16.

The characters $C \in \{A, B, \ldots, Z\}$ in the figure represent sequence states. Presently we are able to cover up to 75% of the transitions in the data set by dint

$\frac{N_{A,B}}{N_A} = P(B \mid A)$	$\frac{N_{A,B}}{N} = P(\{A,B\})$
(a) confidence	(b) support

Fig. 20.15. Confidence and support of the rules

X → Y	W **AND** X → Y
W → A **OR** B	U **AND** V → A **OR** B
(a) predecessor	(b) prepredecessor

Fig. 20.16. Forms of generated rules

of these "few" generated rules. The remaining 25% of the uncovered transitions in the data set were considered as outliers or special cases in mould manufacturing.

20.5. Conclusions and Outlook

Our aim in this work was to find standard approaches in manufacturing moulds. Our first step was to collect the data which includes the information how a part is manufactured. Especially information about the used milling cutter, the milling strategy and other technological and strategically values were collected. In step two we split the data in different data sets. This split was carried out by regarding the milling category and the milling machine, on which the part was manufactured. Afterwards we analysed the collected data using different methods which were described detailed in the corresponding chapters. The result is that by dint of the probabilistic state machine we found a few rules which represent the manufacturing steps in the analysed data. Via these rules it is possible to cover up to 75% of the analysed data. This means we are able to generate rules and communicate these rules to the field workers to support their decisions during programming a part in the CAM-system. For example the data set of machine X delivers the rules R with $R \in \{r_1, r_2, \ldots, r_n\}$ and $n \in \{1, 2, \ldots\}$, which is representing the numbers of founded rules. These rules could be presented the field worker to support him in manufacturing new parts. This means the worker does not have to develop for every part, which has to be produced, an individual approach. Vantages are that the manufacturing steps are much more transparent and optimizations could be detected easier.

An important aspect which has to be cleared in future work is which of the delivered rules were actually used. This means it must be checked whether there are existing rules which were never used, or only used by specific workers for example. This could be realized by analyzing data again, which will be collected after introducing the "rule system". It is also important to know whether the delivered standards were as well used to manufacture the 25% which are not really covered up.

Taking everything in account we can say that standardization of mould manufacturing is a difficult problem which could not solved in a simply way. A very important aspect to get best possible results is to use consistent data with high quality for the analysis. But even finding some little areas where standards were detected is a big gain for this sector. It is important to validate the results and to include them in prospective analysis to go from strength to strength.

References

1. Nihon Tenmon Gakkai: Publications of the Astronomical Society of Japan 52, Tokyo (2000), 551–1206.
2. M. Leeuwinga: The Impact of Global High-Speed Networks on Radio Astronomy. Springer Science+Business Media, Netherland (2011).

3. R. Durbin, S.R. Eddy, A. Krogh, G. Mitchison: Biological Sequence Analysis: Probabilistic Models of Proteins and Nucleic Acids. Cambridge University Press, Cambridge (1998).
4. F. Eichinger, D.D. Nauck, F. Klawonn, *Sequence Mining for Customer Behaviour Predictions in Telecommunications.* In: ECML/PKDD 2006 Workshop on Practical Data Mining: Applications, Experiences and Challenges, Berlin (2006), 3–10.
5. J.A. Harding, M.Shahbaz, Srinivas, A. Kusiak, *Data Mining in Manufacturing A Review. Journal of Manufacturing Science and Engineering*, 128 (2006), 969.
6. K. Plantenberg, Introduction to CATIA V5 Release 19. Schroff Development Corporation, Kansas (2009).
7. R. Toogood, Pro/Engineer Wildfire 5.0 Advanced Tutorial. Schroff Development Corperation, Kansas (2009).
8. P. Smid, CNC programming handbook: A comprehensive guide to practical CNC programming, Library of Congress Cataloging-in-Publication Data, New York (2003).
9. MTS, Teachware Student's Book: CNC Basics, Berlin (1999), 158–169.
10. M.R. Berthold, C. Borgelt, F. Höppner, F. Klawonn: Guide to Intelligent Data Analysis: How to Intelligently Make Sense of Real Data. Springer, London (2010).
11. W. van der Aalst, A. Weijters, L. Maruster: Workflow Mining: Discovering Process Models from Event Logs. IEEE Transactions on Knowledge and Data Engineering 16 (2004), 1128–1142.
12. A. Datta: Automating the Discovery of As-Is Business Process Models: Probabilistic and Algorithmic Approaches. Information Systems Research, 9 (1998), 275–301.
13. A. Weijters, W. van der Aalst: Rediscovering Workflow Models from Event-Based Data using Little Thumb. Integrated Computer-Aided Engineering 10 (2003), 151–162.
14. Höppner, F. Klawonn: Finding Informative Rules in Interval Sequences. Intelligent Data Analysis 6 (2002), 237–256.
15. O. Cappé, E. Moulines, T. Ryden: Inference in Hidden Markov Models (2nd ed.). Springer, New York (2010).

Acronyms

CAM	Computer Aided Manufacturing
VLBI	Very Long Baseline Interferometry
KDD	Knowledge Discovery in Databases
CAD	Computer Aided Design
CATIA	Computer Aided Three-Dimensional Interactive Application
Pro/E	Pro/ENGINEER
NC	Numerical Control
CNC	Computerized Numerical Control
DNC	Direct Numerical Control
DNA	Deoxyribonucleic Acid

Chapter 21

REASONING-BASED ARTIFICIAL AGENTS IN AGENT-BASED COMPUTATIONAL ECONOMICS

Shu-Heng Chen
AI-ECON Research Center
Department of Economics
National Chengchi University
Taipei, Taiwan
chen.shuheng@gmail.com

In this chapter, we compare the development of the artificial agents in two popular kinds of agent-based computational economic models. One is the agent-based models guided by game experiments. The other is the agent-based financial markets. While the conversation between these two classes of agent-based models is rare, the two can be connected through the idea of the generalized reinforcement learning. This is because that the artificial agents in game experiments have been well developed into a hierarchical framework such that the cognitive capacity can be incrementally added to the artificial agents from a low-level one, such as zero-intelligence agents, to a high-level one, such as belief learning agents or level-k reasoning agents. However, this hierarchy has not been found in agent-based financial markets. Therefore, bridging the two classes of agent-based models through artificial agents can help build financial agents with different level of cognitive capacity. This step is crucial for the cognitive foundation of agent-based financial markets and is related to the recent piling-up empirical studies of cognitive finance.

21.1. Introduction

In this chapter, we compare the development of the artificial agents in two popular kinds of agent-based computational economic models. One is the agent-based models guided by *game experiments*. Artificial agents built in these models normally follow *reinforcement learning* or *generalized reinforcement learning* (Camerer and Ho, 1999; Camerer, Ho and Chong, 2002) to make decisions. The other is the *agent-based financial markets*. As generally reviewed by Chen, Chang and Du (2011), there are two different kinds of agent-based financial markets, namely, the H-type model and the SFI (Santa Fe Institute) model. Artificial agents built in these two models are different. The one extensively used in the H-type model, the *regime-switching agents* (Brock and Hommes, 1998), is also a kind of reinforcement learning agent.

Despite its possible connections through the idea of generalized reinforcement learning, the conversation between these two classes of agent-based model is rather

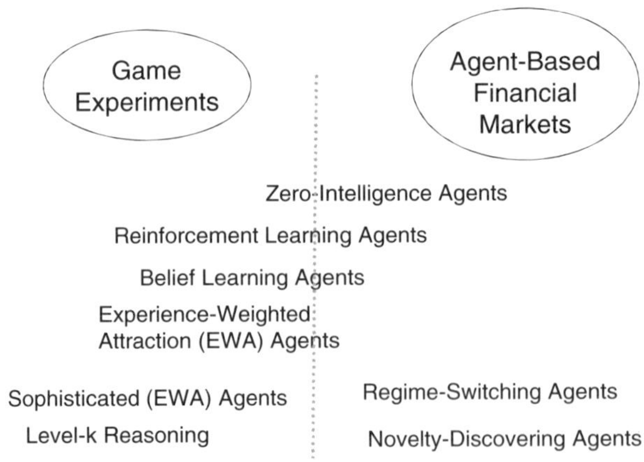

Fig. 21.1. Artificial Agents in Game Experiments and Financial Markets.

infrequent. Figure 21.1 lists all the main artificial agents used in these two classes of agent-based model, and shows how well each of them has been shared by the two camps. The artificial agents mentioned in this figure are zero-intelligence agents, reinforcement learning agents, belief learning agents, experience-weighted attraction (EWA) learning agents, level-k reasoning agents, regime-switching agents and novelties-discovering agents. From the position where they stand, one can see that a gap exists between the two. Few of these are simultaneously shared by both camps. Reinforcement learning agents seem to be the only ones that can connect the two, but, even for this one, the share is not equal between the two. They are much more inclined toward game experiments than financial simulations.[a] Further down the figure, we can see that some of the agents are exclusively employed in only one camp: level-k reasoning agents have only been applied in the game experiment; on the other hand, regime-switching agents and novelties-discovering agents have been only active in financial modeling.

One has to wonder if the gap is a natural consequence of the limited applicability of the agents developed in each of the camps, or whether there is a potential to explore them in different camps so that the gap can be possibly narrowed. In any case, a conversation between the two holds the key for the answer. In this chapter, we attempt to make the two talk and this talk is relevant for the theme of this handbook, *Reasoning-Based Intelligent Systems*. This is because the artificial agents in game experiments have been well developed into a *hierarchical framework* such that the cognitive capacity can be incrementally added to the artificial agents from a low-level one, such as zero-intelligence agents (Gode and Sunder, 1993), to a high-level one, such as belief learning agents (Fudenberg and Levine, 1998) or level-k reasoning agents (Camerer, Ho and Chong, 2004; Crawford and Iriberri, 2007a,b), as we demonstrate in Figure 21.2 (left panel).

[a]In agent-based financial markets, we see very few formal applications of generalized reinforcement learning. One exception is Pouget (2007).

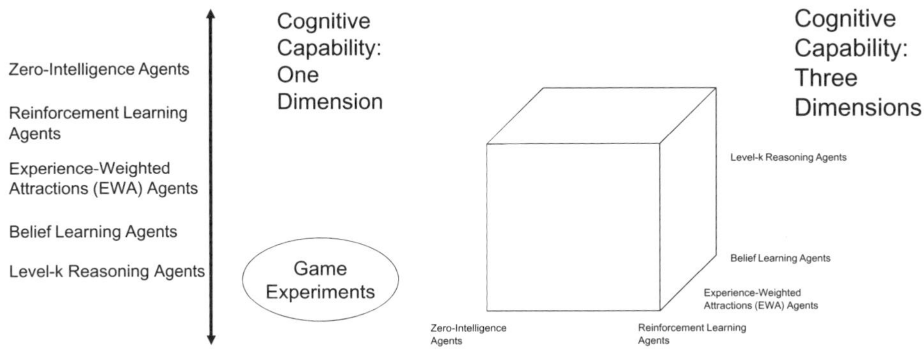

Fig. 21.2. Cognitive Capacity of Artificial Agents.

This conversation is beneficial for reflecting upon the artificial financial agents, since a similar hierarchy as shown in Figure 21.2 has not been found in agent-based financial markets. Therefore, bridging the two classes of the agent-based model through artificial agents can help us to sort out the cognitive hierarchy underlying various financial agents. This step is crucial for the cognitive foundation of agent-based financial markets and is related to the recently piling-up empirical studies of *cognitive finance*.[b]

In this chapter, we shall argue that the generalized reinforcement learning already incorporates three elements related to cognitive capacity, namely, *memory*, *consciousness* and *reasoning*, which may give a different cognitive map of the artificial agents as shown in Figure 21.2 (right panel). With these three attributes, we indicate how the artificial agents of SFI can be considered as an extension of the artificial agents of the H-type model in one or more of these three elements. A usual criticism of the application of genetic programming to artificial agents is its lack of a cognitive foundation. With this bridge and the related extension, one can see that this criticism may underestimate the significance of genetic programming in the development of cognitive financial agents. Cognitive financial agents can, therefore, be firmly developed using computational intelligence.

The remainder of the chapter is organized as follows. Section 21.2 presents agents with no capability of reasoning, and the most important class of these agents is the *zero-intelligence agent*. We then gradually make our agents *smarter* in the sense that they are able to learn and to reason. Section 21.3 introduces an important class of such agents, known as the *generalized reinforcement learning* agents. While generalized reinforcement learning agents have the cognitive capacity to memorize, to do counterfactual thinking, and to be conscious of surroundings, they are generally poor in terms of their capability to reason what their opponents are reasoning. Section 21.4, therefore, introduces the third important class of agents, known as the

[b]There is growing evidence that cognitive ability is an important predictor of financial outcomes. See, for example, Agarwal and Mazumder (2010); Finke (2010).

level-k reasoning agents, with a specific application to augment the generalized reinforcement learning agents. Sections 21.5 and 21.6 introduce artificial agents rarely used in game experiments, but which are dominant in financial markets, namely, *regime-switching agents* and *novelties-discovering agents*. There are followed by the concluding remarks which are given in Section 21.7.

21.2. Zero-Intelligence Agents

The idea of randomly-behaving agents seems to be very natural to start with when one has no clue as to how agents actually behave or want to have an easy start that does not involve too many complications or want to have a benchmark as a basis of comparison for other more developed models. None of the reasons above has anything to do with cognitive salience; however, this design has become popular simply because of the term *zero intelligence*.

The zero-intelligence agent was first initiated by Gode and Sunder (1993) in a double-auction market environment. Technically, the zero-intelligence agent was characterized as a *randomly-behaving agent*. This agent, when making the bid or ask decision, simply randomly picks any price from all those that will not impose a loss on the agent. This constraint is to ensure that *obvious stupidity* can be avoided, but these agents are obviously not able to learn. Even so, Gode and Sunder have shown that this kind of zero-intelligence agent can perform as well as human agents in the double auction experiments.

> Adam Smith's invisible hand may be more powerful than some may have thought; it can generate aggregate rationality not only from individual rationality but also from individual irrationality. (Ibid, p. 119)

This setting essentially binds agents' behavior only by their *budget constraints*. In other words, given the budget constraints and institutional arrangements, the rationality of the individual may not matter. Hence, zero-intelligence, as pointed out by Gode and Sunder (1993), can be long traced back to Gary Becker (Becker, 1962), who showed that a budget constraint is sufficient to guarantee the proper slope of the supply and demand curves.

In addition to double auction markets, the ZI agents have also been applied to other ACE models, including those for general equilibrium and financial markets. For the former, the applications are extended from the original partial equilibrium models to general equilibrium models; for the latter the simple continuous double-auction market is extended to order-book-driven markets and other related market mechanisms. While the concrete implementations of the idea of ZI need to be tailored to different applications, the basic idea is almost the same.[c] Recently, it has

[c]The zero-intelligence agent has been very influential in agent-based computational economics and finance. A general review can be found in Duffy (2006) and Ladley (2011). In addition, see Barr *et al.* (2008) for quite an extensive discussion on the use of zero-intelligence agents in the future of ACE.

been further applied to other ACE models, such as prediction markets (Othman, 2008; Klingert and Meyer, 2010; Othman and Sandholm, 2010), macroeconomics (Ussher, 2009a,b), and social networks (Tseng *et al.*, 2008, 2009). Randomly behaving agents are also extensively used in other models where the term zero-intelligence agents may be replaced by others, such as noise traders, irrational traders, etc.

21.3. Generalized Reinforcement Learning

A generalized reinforcement learning model was proposed by Camerer and Ho (1999) as a hybridization of the two competing classes of learning models in games, namely, *reinforcement learning* and *belief learning* (Fudenberg and Levine, 1998).[d] Camerer and Ho (1999) called this generalization *experience-weighted attraction* (EWA) learning. We shall use the term generalized reinforcement learning because it generalizes what is reinforced in learning, but still keeps the fundamental formulation of reinforcement learning unchanged. More specifically, it generalizes reinforcement learning by giving a more general version of attractions (propensities) and their dynamics (updating schemes), which broadens our understanding of the original reinforcement learning.

Consider a multiple-choice environment where agent i is presented with J possible strategies. A classic example is the *multi-armed bandit problem* (see Appendix A). Let $A_i^j(t)$ be the *attraction* of the jth strategy for individual i. Given $A_i^j(t)$ $(j = 1, 2, \ldots, J)$, agent i will make his choice in a stochastic manner by following a logit distribution:

$$P_i^j(t+1) = \frac{e^{\lambda A_i^j(t)}}{\sum_{k=1}^{J} e^{\lambda A_i^k(t)}}, \tag{21.1}$$

where $P_i^j(t)$ is agent i's probability of choosing j at time t. The equation above basically maps the attraction of each strategy monotonically to its choice probability.[e] This specific choice behavior is the fundamental of what is known as *reinforcement learning* initially studied in psychology (Bush and Mosteller, 1955), then later extended to computer science by Marvin Minsky,[f] games by Arthur Samuel (1901–1990), economics by John Cross (Cross, 1973, 1983) and neuroscience by Wolfram Schultz[g]. The parameter λ appearing in the logit Equation (21.1) measures the

[d] The contrasting studies of these two classes of learning even becomes a subject of neuroeconomics in the context of the *dual process model* (Charness and Levine, 2005; Fudenberg and Levine, 2006).
[e] The transfer function used in Equation (21.1) is a logit function; other alternatives include the power function and probit function.
[f] Marvin Minsky, one of the founders of artificial intelligence, wrote his PhD thesis on reinforcement learning in 1954.
[g] The recent progress in neuroscience indicates that humans or mammals are naturally endowed with a reinforcement learning mechanism in their brain. In fact, one of the most impressive recent results in neuroscience is the discovery of the relationship between the dopamine neural system and reinforcement learning. See Montague (2006), Chapter 4, for a vivid historical review of the research on the dopamine system and reinforcement learning.

sensitivity of agent i to attractions. The usual interpretation of λ is the intensity of agent i's motivation. It can measure the uncertainty of the attractions which agent i is facing.

The next issue concerns the constituents of attractions and their dynamics.

$$A_i^j(t) = \phi A_i^j(t-1) + (\delta + (1-\delta)I(s_i^j, s_i(t))\pi(s_i^j, s_{-i}(t)) \tag{21.2}$$

$I(s_i^j, s_i(t))$ is an indicator function. It is one if $s_i^j = s_i(t)$; in this case, the strategy j was the chosen strategy at time t. The use of this indicator function enables agent i to update the attraction of the chosen strategy with the received payoff $\pi(s_i^j, s_{-i}(t))$, which is a function of agent i's choice (s_i^j) and his opponents' choice ($s_{-i}(t)$).

The simple reinforcement learning does not consider the *foregone payoff* (the *hypothetical payoff*), i.e., the possible payoff had agent i chosen any other strategies at t. This corresponds to a special case of Equation (21.2) when $\delta = 0$. Camerer and Ho (1999) consider δ as the most important parameter in the generalized reinforcement learning, because it distinguishes two laws of effect, namely, the law of (*actual*) effect and the law of (*simulated*) effect. The latter may involve agent i's imagination capability.

ϕ is the discount factor, which is normally interpreted as the *memory decay*. However, in addition to memory decay, agent i may *deliberately* discount the old experience when the environment is changing. Hence, to take both conscious and unconscious "forgetting" into account, Camerer and Ho (1999) separate the physical time and *effective experience time*. For the latter, a depreciation of time is introduced as follows:

$$N(t) = \rho N(t-1) + 1 \tag{21.3}$$

They then use this experience time $N(t)$ to normalize the sum of the depreciating, experience-weighted previous attraction $A_i^j(t-1)$ plus the weighted payoff from t.

$$A_i^j(t) = \frac{\phi N(t-1)A_i^j(t-1) + (\delta + (1-\delta)I(s_i^j, s_i(t))\pi(s_i^j, s_{-i}(t))}{\rho N(t-1) + 1} \tag{21.4}$$

Equation (21.4) is the Camerer-Ho *generalized reinforcement learning*. We shall now see how this generalized reinforcement learning model can lead to two familiar and competing learning models in game experiments, namely, reinforcement learning (Section 21.3.1) and belief learning (Section 21.3.2).

21.3.1. *Reinforcement Learning*

There are two kinds of reinforcement learning models frequently used in the game-theoretic literature. They differ in how payoffs contribute to the level of reinforcement: one assumes that previous payoffs are accumulated (Harley, 1981; Roth and Erev, 1995), whereas the other assumes that previous payoffs are averaged (McAllister, 1991; Monkerjee and Sopher, 1994). Both of them can be framed as a special case of the Camerer-Ho generalized reinforcement learning.

The updating scheme of the *cumulative* reinforcement learning is:

$$R_i^j(t) = \phi R_i^j(t-1) + I(s_i^j, s_i(t))\pi(s_i^j, s_{-i}(t)) \tag{21.5}$$

Equation (21.5) is a special case of Equation (21.4) when $\delta = \rho = 0$ and $N(0) = 1$. The updating scheme of the *average reinforcement learning* is:

$$R_i^j(t) = \phi R_i^j(t-1) + (1-\phi)I(s_i^j, s_i(t))\pi(s_i^j, s_{-i}(t)) \tag{21.6}$$

Equation (21.6) is also a special case of Equation (21.4) when $\delta = 0$, $\phi = \rho$, and $N(0) = \frac{1}{1-\rho}$.

In both of the above two reinforcement learning models, $\delta = 0$, which assumes that people ignore *foregone payoffs*. This assumption has been well discussed in the literature. It is certainly easier to defend this ignorance when agents are placed in a poor-information environment so that they cannot possibly know what they might have lost or gained.[h] Nonetheless, ignoring possible foregone payoffs in all environments entirely does indicate that agent i's consciousness of the environment is very weak. Essentially, the agent with $\delta = 0$ acts as if he is placed in an independent individual choice situation, rather than a game-like interactive environment. What he did is basically very *reflexive*, just based on the memory that he had from the past received payoffs, and has no imagination of what he might have gained or lost. Neither makes any attempt to discover or learn about these unrealized payoffs.

21.3.2. *Belief Learning*

As we mentioned earlier, reinforcement learning was introduced to modeling learning in the multi-armed bandit experiment in the 1950s (see Appendix A). It was introduced to modeling learning in game experiments in the 1990s. The multi-armed bandit experiments and the normal-form game experiments are not necessarily fundamentally different, if one can express the former in the form of the latter. In this case, the machine can be considered to be the opponent who uses the mixed strategy to play and assign a fixed choice probability to each of the strategies. Furthermore, agent i is informed about this robot and hence also *believe* that it uses a mixed strategy. Nonetheless, agent i is not given the exact details on the opponent's choice probabilities that he has to learn from experience.

In the real normal-form game, the situation is, however, a little different. Agent i is now told explicitly that his opponent is a human, but has not been given any information on the strategies that he or she will play, be they mixed or not. In this situation, agent i needs to form the belief about how his opponent might play, and react based on this belief. This is the basic idea of belief learning.

[h]However, Ho, Wang and Camerer (2008) have recently extended the generalized reinforcement learning into the case where information regarding foregone payoffs is incomplete and has to be estimated.

There are many ways of forming beliefs, and some may involve a high degree of reasoning. Nonetheless, those sophisticated ways of forming beliefs are not included in the generalized reinforcement learning developed in Camerer and Ho (1999). In fact, what is modeled is just the belief formed by using a kind of average, taken over the observed repertoire (time series) of the opponent's chosen strategies, $\{s_{-i}(l)\}_{l=1}^{t}$. From there he can count, all the way up to t, how frequently each possible strategy has been played by the opponent, i.e., $\{N_{-i}^{h}(t)\}_{h=1}^{J}$ and $N(t) = \sum_{h=1}^{J} N_{-i}^{h}(t)$. Agent i's estimation (belief) of the probability that his opponent will play strategy k in the next period (period $t+1$), $B_{-i}^{k}(t)$, can then be represented as follows:

$$B_{-i}^{k}(t) = \frac{\rho N_{-i}^{k}(t-1) + I(s_{-i}^{k}, s_{-i}(t))}{\sum_{h=1}^{J}(\rho N_{-i}^{h}(t-1) + I(s_{-i}^{h}, s_{-i}(t))}, \quad k = 1, \ldots, J. \tag{21.7}$$

There is a recursive version of Equation (21.7), which updates the belief from period $t-1$ to t explicitly:

$$B_{-i}^{k}(t) = \frac{\rho N(t-1) B_{-i}^{k}(t-1) + I(s_{-i}^{k}, s_{-i}(t))}{\rho N(t-1) + 1}. \tag{21.8}$$

This form of updating the weights of observations involves multiplying those from one period ago ρ times as much as the most recent observations. When $\rho = 0$, only the most recent observation counts and one has the naive, Cournot-type learning (Cournot, 1838):[i]

$$B_{-i}^{k}(t) = I(s_{-i}^{k}, s_{-i}(t)). \tag{21.9}$$

Equation (21.9) says that agent i has a very static view of his opponents, namely, their strategies will remain unchanged. Hence, it could be naive in the sense of this passive expectation. When $\rho = 1$, all observations equally count and we have the familiar *fictitious play* (Brown, 1951):

$$B_{-i}^{k}(t) = \frac{N(t-1) B_{-i}^{k}(t-1) + I(s_{-i}^{k}, s_{-i}(t))}{N(t-1) + 1}. \tag{21.10}$$

Based on the estimated choice probabilities of his opponent (21.10), agent i can figure out his expected payoffs for each of his possible choices, j:

$$E_{i}^{j}(t) = \sum_{k=1}^{J} B_{-i}^{k}(t) \pi(s_{i}^{j}, s_{-i}^{k}) \tag{21.11}$$

[i] Antoine Augustin Cournot (1801–1877), a French economist and mathematician, introduced the first explicit model of learning in games. He assumed that players choose a best response to what they most recently observed.

Through the belief updating Equation (21.8), one can also find a recursive version of Equation (21.11):

$$E_i^j(t) = \frac{\rho N(t-1)E_i^j(t-1) + \pi(s_i^j, s_{-i}(t))}{\rho N(t-1) + 1}. \tag{21.12}$$

Belief learning is a special kind of generalized reinforcement learning if one treats the expected payoffs as attractions, i.e., $A_i^j(t) = E_i^j(t)$. Then with the three additional conditions: $\delta = 1$, $\phi = \rho$ and $A_i^j(0) = E_i^j(0)$, Equation (21.12) is identical to Equation (21.4).

21.3.3. *Cognitive Capability of Generalized Reinforcement Learning*

> While the super-rational theories require unrealistic strong assumptions, the naive reinforcement models suffer from the opposite sin of assuming players are too limited in their reasoning powers. (Stahl (2000), p. 106)

Reinforcement learning agents are often considered to be agents with minimal intellect because they demonstrate no strategic behavior. They act as if they were in the multi-armed bandit environment, and are ignorant or unconscious of the payoffs that they might have received from the slot machines had they tried them. *Imagination* is the word used in the literature to indicate the required cognitive efforts involved in this situation:

> The parameter δ measures the relative weight given to foregone payoffs, compared to the actual payoffs, in updating attractions. It can be interpreted as kind of "*imagination*" of foregone payoffs, in updating attractions, "*simulation*" of outcomes under alternative competitive scenarios ("*counterfactual thinking*" to psychologists), or responsiveness to foregone payoffs. (Camerer, Ho and Chong (2002), p. 141)

In a more general social interaction (not necessarily in a well-articulated normal-form game) the reinforcement-learning artificial agent may not even be aware of the presence of the opponents and mentally still position himself in a multi-armed bandit game, a game that just has himself and the machine.

Strategic thinking is developed in belief learning. Agents with belief learning form beliefs regarding their opponents. In addition to the choice history of himself, $\{s_i(l)\}_{l=1}^t$, the agent also observes the choice of history of his opponents $\{s_{-i}(l)\}_{l=1}^t$. The latter is not paid attention to by the reinforcement-learning agents. Despite this advance, what the belief-learning agent can do for his memory of the opponents' choice history $\{s_{-i}(l)\}_{l=1}^t$ is simply to memorize it, count it and update the count. Sophisticated analysis of the data, such as examining the statistical dependence of the time series of $\{s_i(l), s_{-i}(l)\}_{l=1}^t$ is not seen in the usual weighted fictitious play. Serious reasoning of what opponents might do is not available in belief learning. However, by using a kind of level-k reasoning, Camerer, Ho and Chong (2002) add the reasoning capability to generalized reinforcement learning, which is referred to as the *sophisticated* EWA model.

21.4. Level-k Reasoning and Sophisticated EWA

21.4.1. *Beauty Contest Games*

John Maynard Keynes in his *General Theory* (Keynes, 1936), Chapter 12, proposed a way to think of stock market investment in the context of newspaper beauty contests. In this newspaper competition, readers were asked to choose the six prettiest faces from 100 photographs. The winner was the person whose preferences were closest to the average preferences of all participants. Keynes assumed that contest participants do not choose faces that they personally find the most attractive; instead, some make their choices based on their expectations of others' expectations, and some, going one step further, make their choices based on their expectations of others' expectations of others' expectations, as well put in the following quotation:

> It is not the case of choosing those which, to the best of one's judgment, are really the prettiest, nor even those which average opinion genuinely thinks the prettiest. We have reached the third degree, where we devote our intelligences to anticipating what average opinion expects the average opinion to be. And there are some, I believe, who practice the fourth, fifth, and higher degrees. (Keynes (1936), p. 156.)

Investors in financial markets have much in common with these beauty contests. Consider the market-timing decision as an example. People holding stocks in a bull market consider when it is the best time to sell stocks and to exit the market. They certainly want to do it before a majority of people start doing it. So, they have to guess when this majority movement time will happen, and sell a few days ahead of this big maneuver. Nevertheless, other investors are gauging this timing in the same way, so some smarter investors have to, being one level higher, think what others think.

Partially because of this natural analogy between beauty contests and financial investment, in the middle of the 1990s, economists started to address this issue in an experimental setup, and developed the well-known *beauty contest experiments* (Nagel, 1995, 1998). It is called *beauty contest game* in honor of Keynes. In this game, contestants are told to guess a number from 0 to 100, with the goal of making their guess as close as possible to p times the average guess. In a world where all the players are known to be fully rational, in the sense that they will form expectations about the guesses of others and can carry out as many levels of deduction as necessary, the equilibrium in this game is zero (if $0 < p < 1$) and is 100 (if $p > 1$). Nonetheless, this theoretical result is not well supported by human-subject experiments.

For example, Richard Thaler ran this experiment with the *Financial Times* (Thaler, 1997). He found that although many contestants did guess zero or one, the most popular guesses were 33 and 22. The former is the right guess if everyone else chooses a number at random, and 22 is the right guess if everyone else picks 33. If we consider random guessing (the behavior of zero-intelligence agents) the level-0 reasoning, then 33 corresponds to level-1 reasoning and 22 corresponds to level-2 reasoning. With these findings, Thaler (Thaler, 2000) suggests to incorporated heterogeneity in reasoning into artificial agents:

> Although modeling how this game is actually played is not easy, some lessons are clear enough. An appropriate model would have to allow for two kinds of heterogeneity in sophistication. First, agents differ in how many levels of processing they engage in (33 is one level, 22 is two levels, and so on). Second, there is heterogeneity in how much agents think about the behavior of other agents. (Ibid, p. 135)

21.4.2. *Level-k Reasoning*

Since the middle of the 1990s, a series of efforts have been made to develop artificial agents who are heterogeneous in capability to reason other agents' reasoning (Stahl and Wilson, 1994, 1995; Nagel, 1995; Ho, Camerer and Weigelt, 1998; Costa-Gomes, Crawford, and Broseta, 2001; Bosch-Domenech, Montalvo, Nagel and Satorra, 2002; Crawford, 2003; Camerer, Ho and Chong, 2004; Costa-Gomes and Crawford, 2006; Ohtsubo and Rapoport, 2006; Crawford and Iriberri, 2007a,b). This literature is collectively known as *level-k reasoning*. There are several different versions of level-k reasoning, such as *theory-of-mind* (Stahl and Wilson, 1995), level-k reasoning (Crawford and Iriberri, 2007b), *cognitive hierarchy* (Camerer, Ho and Chong, 2004), and *Machiavellian Intelligence* (Ohtsubo and Rapoport, 2006), but the basic structure is, roughly, that higher-level agents are able to reason what low-level agents do, and they then best respond to it.

The basic idea behind level-k reasoning is that agents are placed in a *cognitive ladder*. Along this ladder, those who are placed at higher levels of this ladder are able to simulate the minds of those placed at lower levels, but not vice versa. In spirit, level-k reasoning is analogous to the *Chomsky hierarchy* in computer science. The Chomsky hierarchy is a hierarchy of formal grammars and the formal languages generated by the grammars. The hierarchy arranges four classes of grammars and the four generated languages in monotonically increasing order, so that any class of grammars (languages) appearing in this order is a *proper subset* of the one coming after it. It also implies a hierarchy of the automata which can recognize the grammars. By the same monotonic order, the automata at a higher level can simulate the automata at a lower level, but not vice versa.

21.4.3. *Sophisticated EWA Learning*

The sophisticated EWA learning proposed by Camerer, Ho and Chong (2002) can be regarded as an application of level-k reasoning to EWA learning. Camerer, Ho and Chong (2002) consider two types of agents, Level-0 and Level-I, with a given distribution of each type, namely, $1-\alpha'$ and α'. Level-0 agents do not know the existence of Level-1 agents, and behave as the EWA learning model describes. Level-I agents know the existence of Level-0 agents and know that they behave by following the EWA learning models. In addition, Level-I agents also "know" that the percentages of Level-1 and Level-0 agents are α' and $1-\alpha'$, respectively.[j] To make their own decisions, Level-I agents form their beliefs by taking into account both level-0 and other level-I agents' learning. Their attractions are accordingly modified as follows:

$$\begin{cases} A^j_{i,I}(t) = \sum_{k=1}^{J}[\alpha' P^k_{-i,I}(t+1) + (1-\alpha')P^k_{-i.0}(t+1)]\pi_i(s^j_i, s^k_{-i}). \\ P^j_{i,I}(t+1) = \dfrac{\exp^{A^j_{i,I}(t)}}{\sum_{j=1}^{J}\exp^{A^j_{i,I}(t)}} \end{cases} \tag{21.13}$$

The subscripts 0 and I appearing above refer to agents of Level 0 and I, respectively. Equation (21.13) can be read as a generalized and sophisticated version of Equation (21.11). The main difference lies in agents' belief formation. In the latter case it is formed exclusively based on the historical observations only, whereas in the former case it is formed by taking into account the expectations of Level-0 agents' expectations. With this sophistication, Level-I agents are able to replace the the *estimated* choice probability $B^k_{-i}(t)$ by the *true* choice probability $P^k_{-i.0}(t+1)$. In addition, Level-I agents also know the existence of other Level-I agents, like themselves, and know that they form the expectations in the same way as they did. Hence, the true choice probability of this group of agents, $P^k_{-i.0}(t+1)$ is also incorporated to replace the estimated one, $B^k_{-i}(t)$. Since for the EWA agents, the expected payoffs are taken as the attractions, we therefore use $A^j_{i,I}(t)$ directly instead of $E^j_{i,I}(t)$. Putting these together, we obtain the expected payoffs for the Level-I agents as described in Equation (21.13).

21.4.4. *Agents with Incremental Cognitive Capacity*

Table 21.1 proposes a summary of how the cognitive capacity can be incrementally added from the zero-intelligence agent all the way up to the sophisticated experienced-weighted attraction (EWA) agents (Camerer, Ho and Chong, 2002). The key cognitive elements used to distinguish these learning models are *memory*, *consciousness* and *reasoning*. Zero-intelligence agents have none of them. They have no memory of the past and even no memory of the present. They basically make decisions unconsciously. Hence, when making decisions, all human factors such as cognitive capacity, mental states, neurophysiological elements (dopamine), and whatsoever play no role. Reinforcement learning agents are endowed with some power of memory, which can help them to recall their past experiences of feeling in each of the same situations. However, they are not given a cognitive capacity that is sufficient for them to be aware of the entire environment with which their decisions

[j]Camerer, Ho and Chong (2002) actually considered the case when agents do not know the parameter α' and have to learn from the data.

Table 21.1. Artificial Agents with Incremental Cognitive Capacity.

Models	Memory	Consciousness	Reasoning
Zero-Intelligence	None	None	None
Reinforcement Learning	Short to Long	None	None
Belief Learning	Short to Long	Strong	Weak
EWA Learning	Short to Long	Weak to Strong	Weak
Sophisticated EWA	Short to Long	Weak to Strong	Weak to Strong
Regime Switching	Short to Long	Weak	None
Novelties-Discovering Agents (Autonomous Agents)	Short to Long	Weak to Strong	Weak to Strong

are embedded. Essentially, they are not aware of what their players have done and would not try to outguess what they will do. Without enough consciousness and reasoning, decisions made with reinforcement learning can be very *reflexive*, as if being automatically driven by the dopamine system.[k]

With the given memory, be it short or long, the belief learning agents are given additional cognitive capacity to be aware of the presence of their opponents. They are conscious of the effect of the opponents' actions on their payoffs. Their decisions are, therefore, based on such reasoning by taking this feedback into account. In particular, they are able to figure out, in an ex post manner, what would happen had they chosen different actions, and to take these so-called *foregone payoffs* into account. They, however, are not advanced enough to develop a sophisticated model to predict the behavior of their opponents' actions, apart from counting the frequencies of their past actions. The EWA (experience-weighted attraction) agents are a hybridization of the reinforcement learning agents and belief learning agents in the sense that their cognitive capacity to "imagine" all foregone payoffs is only partial. Their consciousness of the environment and the mental capability to engage in counterfactual thinking lies between those of the reinforcement learning agents and belief learning agents. The sophisticated EWA involves the incorporation of level-k reasoning into EWA learning. Therefore, agents with this design are clothed with the additional power to reason what other agents, agents with a lower level of cognitive hierarchy, might think.

21.4.5. *Cognitive Heterogeneity of Agents*

Cognitive hierarchy implies heterogeneity; it refers to different degrees of sophistication of agents. Evidence of the heterogeneity of human subjects in game experiments has been first found in Cheung and Friedman (1997). In Cheung and Friedman (1997), the artificial players are presented with a binary choice in the context of normal-form games. Their binary choice is stochastically determined by the familiar logit model (Equation (21.1)). The main input of the logit model is related to the

[k]For more details, see Montague (2006) and Glimcher (2010).

differences in the expected payoffs of the associated choices, which are derived with respect to the player's belief (subjective probability function). Cheung and Friedman (1997) then parameterized the decision rule (the logit model) and the learning rule (the belief updating scheme) of agents with two and one parameters, respectively. The three-parameter belief learning agent is then estimated based on four kinds of normal-form game experiments, involving a total of 393 human subjects. Cheung and Friedman (1997) then proposed a log-likelihood ratio test to test the null of a homogeneous agent (representative agent) with the alternative of complete heterogeneous agents (all agents are different). The null was rejected, indicating that heterogeneity across subjects in terms of learning and decision parameters is an important feature of their observations. In particular, the observed heterogeneity in the learning parameter allows us to further classify the majority of players, 257 out of 393, into three groups of agents with distinct memory lengths, from the shortest one to the longest one, corresponding to the *Cournot agents* and the *fictitious-playing agents*, respectively (Section 21.3.2).

Evidence on the heterogeneity of human subjects was also found in other studies (Camerer and Ho, 1998; Stahl, 2000; Ho, Wang and Camerer, 2008; Chen and Hsieh, 2011).[1] However, few have ever addressed what may cause the observed heterogeneities of these human agents. In light of the recent trend in cognitive and psychological economic experiments, for example, one may inquire whether these observed heterogeneities are related to the idiosyncratic differences in subjects' cognitive or psychological attributes. As an empirical foundation of ACE, this query can be quite crucial because the learning models developed in game experiments enable us to think of the heterogeneities of agents from a bounded-rationality perspective, and hence provide a justification for the development of heterogeneous agents in ACE. What is particularly intriguing is that these models may possibly be assembled into a hierarchical framework, from the ones with a lower cognitive capacity to the ones with a higher cognitive capacity. It demonstrates a way to model artificial agents with *incremental* cognitive capacity.

21.5. Artificial Financial Agents

In this section, we move to the agent-based financial markets, and examine the cognitive capacity of the artificial financial agents built in these markets in light of what we have learned from Sections 21.3.3 and 21.4.4. Before proceeding further, we would like to point out that generalized reinforcement learning has been applied to agent-based economic modeling, including electricity markets and financial markets. Therefore, they are not just limited to behavioral games. Nevertheless, these applications are relatively rare compared to others. Therefore, in the following, we

[1]It is not always the case that the null of homogeneous agents can be rejected by the laboratory data. For example, Camerer and Ho (1999) tested the null of homogeneous agents by assuming two types of agents as the alternative, and they found that the null cannot be rejected.

shall focus more on the two dominant classes of artificial financial agents, namely, the *regime-switching agents* and the *novelties-discovering agents*. The former is very standard in the H-type agent-based financial markets, and the latter is widely used in the SFI (Santa Fe Institute)-like agent-based financial markets.[m]

21.5.1. *Regime-Switching Agents*

Regime-switching agents are agents who follow a set of behavioral regimes, and switch over them. The switching mechanism is stochastic and is used to characterize the learning behavior of agents. Among many possible regime-switching agents, the most standard one is the fundamentalist/chartist model, a two-type agent-based financial model (Kirman, 1991; Lux, 1998; Hommes, 2002).

More generally, in the H-type agent-based financial model, agents have a set of H behavioral regimes. To decide which regime to activate in each period of time, the agents choose one of these regimes in a stochastic manner which is very similar to the generalized reinforcement learning as discussed in Section 21.3. For example, in the models initiated by Brock and Hommes (1998) and Hommes (2002), known as the *adaptive belief systems* (ABS), the logit model is applied for modeling the agents' stochastic choice. Another kind, the *heading-based agent-based financial models*, initiated by Kirman (1991) and Lux (1995), employ the idea of the Polya urn process as a basis for the stochastic choice, but this difference may be trivial since financial agents can be reinforced by various kinds of payoffs. In addition to monetary payoffs, they can also be reinforced by peer pressure (herding).[n]

21.5.2. *Cognitive Capacity of Regime-Switching Agents*

With this similarity, the cognitive capacity of calibrated financial agents can be examined using Table 21.1. First of all, most calibrated financial agents do take into account the foregone payoffs. For example, in the Brock-Hommes model (Brock and Hommes, 1998), the agents who chose to behave like fundamentalists (chartists) would also update the payoffs which they might have received had they behaved like a chartist (fundamentalist). Hence, as with the usual distinction made between reinforcement learning and belief learning, i.e., the law of actual effect and the law of simulated effect, they may be positioned closer to the latter. However, belief learning agents will form beliefs on what other agents may tend to do and react upon these expectations. This degree of consciousness is simply absent in these regime-switching agents. For example, they will not form expectations of the possible microstructure

[m]Based on the taxonomy given in Chen, Chang and Du (2011), the H-type agent-based financial markets and the SFI (Santa Fe Institute)-like agent-based financial markets are the two major classes of agent-based financial markets. A survey for the former can be found in Hommes (2006), whereas a survey for the latter is available in LeBaron (2006).

[n]Having said that, we notice that they can also be reinforced by risk aversion, and forecasting accuracy, etc. The idea of payoffs used in agent-based financial models is, therefore, broader than what has been commonly used in game experiments.

(percentages of fundamentalists or chartists) in the next period, and make their choice accordingly. This missing awareness is probably because of the overwhelming reliance on the use of simple programmed agents and the pursuit of simplicity in this type of agent-based financial model.

Second, without sufficient consciousness, reasoning capability as characterized by various cognitive hierarchies is also not presented in these calibrated financial agents. While level-k reasoning was partially motivated by Keynes's beauty contest as an influential metaphor of stock markets, its applications to agent-based financial models has not yet been realized. Accordingly, they are not heterogeneous in their cognitive capacity. As a matter of fact, while frequently termed as models of heterogeneous agents, most regime-switching agents in financial markets are *homogeneous ex-ante*, i.e., their heterogeneous behavior is just a different realization drawn from the same choice probabilities. The only source of agent heterogeneity is this random component.

The features of the cognitive capacity of these regime switching agents are summarized in Table 21.1 as a comparison with those agents developed in game experiments. The artificial agents built in game experiments (Sections 21.3 and 21.4) have two characteristics, namely, *incremental cognitive capacity* and *cognitive heterogeneity*. Neither of these two exists in the H-type agent-based financial models. Hence, the question is whether agent-based financial markets need artificial agents with cognitive hierarchy and cognitive heterogeneity. What are the additional insights one may gain when we allow our financial agents to be heterogeneous in their reasoning capability?

21.5.3. *Intelligence Quotients of Intelligent Algorithms*

We believe that these questions have not been seriously taken into account in current agent-based financial literature, while some initial attempts exist. These attempts usually tried to identify different intelligent algorithms with different levels of intelligence (Yeh, 2007, 2008). In other words, a cognitive hierarchy as suggested in Figure 21.2 is assumed implicitly. More generally, this doing is equivalent to assuming that all intelligence algorithms, such as genetic algorithms, neural networks, decision trees, nearest neighbors, support vector machines, etc., can be placed in a cognitive ladder like Figure 21.2.

However, the difference is that all artificial agents in Figure 21.2 are constructed with incremental cognitive capacity, whereas these intelligence algorithms, each motivated by different disciplines, are not constructed incrementally in this way. Therefore, one may abuse much power to assign their position in the hierarchy, not to mention the psychological ground to support this assignment. While this does not mean that these algorithms are equally smart and cannot be comparable, a theoretical foundation to make these tools have a psychologically-equivalent intelligence quotient has to be laid first. Of course, in principle, it is not impossible to do so. In fact, from the perspective of computational neuroscience, if brains with different cognitive capacities can be shown computationally, then all intelligent algorithms may also be examined in this vein.

21.6. Novelties-Discovering Agents

Novelties-discovering agents are the agents who, *with minimal external supervision*, are able to search for, to discover, and to take full advantage of hidden patterns, chances, and opportunities. With this capability, they of course can adapt to a changing environment. Due to their minimal dependence on the external supervision, they are also called *autonomous agents* in agent-based computational economics. The essence of economics is the existence of these agents and their behavior. Various versions of new economics can be developed without the conventional device of fully rational agents or Homo Economicus, but can hardly be convincing if agents are deprived of the capability to discover novelties.

It would probably be fair to say that the tools available for economists to build novelties-discovering agents or autonomous agents with a high degree of autonomy were rather limited before the early 1990s. In their pioneering paper introducing novelties-discovering agents to economics (Holland and Miller, 1991), John Holland and John Miller begin with the following:

> Economic analysis has largely avoided questions about the way economic agents make choices when confronted by a perpetually novel and evolving world. ... This is so,..., because standard tools and formal models are ill-tuned for answering such questions. However, recent advances... in the subdiscipline of artificial intelligence called machine learning, offer new possibilities. (Ibid, p. 365)

21.6.1. *Origin: Tournament Automation*

The idea of novelties-discovering agents was introduced, first, to game theory, and then to economics, through the attempt of *tournament automation*. The attempt was pioneered by Robert Axelrod in his automated *iterated prisoner's dilemma* (IPD) tournament (Axelrod, 1987), and was later continued by Martin Andrews and Richard Prager (Andrews and Prager, 1994), two Cambridge engineers, in their automated double auction (DA) tournament, originally held by the Santa Fe Institute (Rust, Miller, and Palmer, 1993, 1994).

Tournament automation is to replace man-made submissions to the tournament with machine-made (robot-made) submissions to the tournament. It is preferable to automate tournaments because normally tournaments have been designed as a way of tackling with open-ended questions. For example, in the aforementioned tournaments, the objective is to know the characterization of the viable strategies in the associated competing environments. However, there is no closed final answer to these questions; hence, running these tournaments in an open *evolutionary* or *co-evolutionary* manner seems to be the only legitimate way to do it.[o] This approach to studying social sciences and making social science computational has been well expounded by Robert Axelrod (Axelrod, 1997).

To support this way of doing social sciences, Robert Axelrod (Axelrod, 1987) used genetic algorithms to automate his early organized iterated prisoner's tournament (Axelrod, 1984). Since then the genetic algorithm has been used as a tool to involve novelties-discovering agents in many agent-based computational economic models, including agent-based financial markets (see LeBaron (2006) for a survey). Later on, Andrews and Prager continued Axelrod's idea to automate double auction tournaments using genetic programming. Genetic programming was then used by Shu-Heng Chen and his colleagues as a tool to build agent-based financial markets with novelties-discovering agents (Chen and Yeh, 2001, 2002; Chen and Liao, 2005; Chen, Liao and Chou, 2008).

21.6.2. *Outsmarting Opponents*

The reason for giving such a historical review is to point out that tournament automation is the origin of the novelties discovering agents, which are currently extensively used in agent-based computational economics. Through this origin, we can answer where the reasoning capabilities of financial agents are, if we do not find them in the H-type agent-based financial models. Our answer is that novelties-discovering agents are the reasoning agents missing in the H-type agent-based financial models. That is because they demonstrate the essence of level-k reasoning, i.e., *iterative reasoning*, even in a broader sense and are not limited to iterated dominance games.[p]

To see the novelties-discovering agents in relation to level-k reasoning agents, let us cite the findings from Axelrod's IPD open evolutionary tournament. In his evolutionary tournament, it was found, *with a little surprise*, that TIT-FOR-TAT was no longer the most effective strategy with respect to a fixed set of eight representative

[o]Robert Axelrod considered two conceptually distinct styles of open IPD tournaments, which later on become the stereotypes of automated open tournaments. The first style is based on matching autonomous agents with a fixed set of human-written programmed agents, whereas the second style is based on leaving autonomous agents to play against themselves. To distinguish between the two, we shall call the former the *evolutionary tournament*, and the latter the *co-evolutionary tournament*. In the evolutionary tournament, autonomous agents are adapted to a fixed environment, and the most effective strategy can be considered to be the optimal strategy, *optimal with respect to the fixed set of opponents*. The use of a genetic algorithm in this case is more like the application of a numerical optimization algorithm that searches for the optimal solution. In the co-evolutionary tournament, each agent adapts to the adaptation of other agents. The process may not even converge, and the effective strategy can change over time.

[p]Iterated dominance games, such as the beauty contest (Section 21.4.1), allow us to develop a step-by-step computation in the acquisition of a dominant strategy, and hence develop a complexity measure in the spirit of computational theory. The number of steps required to eliminate all dominated strategies is also called the depth of reasoning. See Camerer (2003) for details.

human-written programs which were retrieved out of the 62 entries in his early IPD tournament.

> Because TIT FOR TAT had done best in the computer tournament itself, I did not think that it would be possible to do much better with an evolutionary process. But as noted earlier, in about a quarter of the simulation runs with sexual reproduction, the population did indeed evolve substantially better strategies–strategies that were quite different from TIT FOR TAT. (Axelrod (1997), p. 26)

He then went further to discuss what the novelties-discovering agent was reasoning about his other opponents, who were all human-written programs, including the TIT-FOR-TAT artificial player. It was found that the novelties-discovering agent was able to find the more effective strategy by *distinguishing* the *exploitable* representatives from other *less exploitable* representatives, and then exploiting the former at a cost of having less cooperation with the latter, but still netting the two with a positive gain. In this case, the novelties-discovering agent was able to outsmart the TIT-FOR-TAT agent by being able to simulate what he, and other agents too, may do. In other words, this novelties-discovering agent stood at a higher level on the cognitive ladder which we mention in Section 21.4.2. Notice that he did not stand at a high level at the outset, but learned to be there through repetition.

A conjecture which is very similar to what Axelrod found was made by Rust, Miller, and Palmer (1993).

> The open question is whether strategies exist that are capable of dominating Kaplan's "wait in the background" strategy over a nontrivial range of environments. If we were to run another DA tournament, it seems likely that entrants would attempt to beat Kaplan by developing more sophisticated delay and "endgame" strategies rather than reverting to truthtelling mode after a fixed amount of time. (Ibid, p. 192)

In the quotation above, Rust, Miller, and Palmer (1993) was concerned with the robustness of Kaplan's "wait in the background" strategy, the best-performing strategy in the Santa Fe DA tournament, just like what had surprised Axelrod on the beatableness of the TIT-FOR-TAT strategy. John Rust and his colleagues, however, did not run another DA tournament. However, almost two decades later Chen and Tai (2010) re-ran the tournament in an automated fashion. They found that the Kaplan agent could indeed be beaten by the novelty-discovering agent driven by genetic programming.

These two examples, the beatableness of TIT-FOR-TAT in the IPD tournament and that of Kaplan in the DA tournament indicate that from the very beginning our designed novelties-discovering agents were there to learn who their opponents were, and tried to find their Achilles' heel and reacted to it. What they did is essentially the same as what the level-k reasoning agents would have done in a similar environment. Hence, unlike regime-switching agents, novelties-discovering agents are equipped with the reasoning capability (Table 21.4.4).

21.6.3. *Cognitive Capacity Hypothesis*

Empirical studies of level-k reasoning based on human-subject experiments have constantly shown that there is a distribution for k. Most frequently, k is one or two. That is, for human subjects, few of them are level-0, and also few of them are beyond level-3 (Stahl and Wilson, 1994; Duffy and Nagel, 1997; Ho, Camerer and Weigelt, 1998; Bosch-Domenech, Montalvo, Nagel and Satorra, 2002). The *cognitive economic experiments* which began in the early 2000s attempt to see whether there is a link between *cognitive capacity* and this *cognitive hierarchy* (level-k) (Devetag and Warglien, 2003, 2008; Burnham *et al.*, 2009; Georganas, Healy and Weber, 2010). There are two possibilities which make this research intriguing, namely, *nature* or *nurture*. In the former case, k is exogenously given, for example, possibly from genetic inheritance; in the latter case, k is endogenous evolving with experience and learning.

The *cognitive capacity hypothesis* basically says that k is exogenously given and can be positively related to cognitive capacity (Burnham *et al.*, 2009). Therefore, in every novel situation, human subjects tend to perform at the same cognitive hierarchical level. On the other hand, the *literacy hypothesis* predicts the opposite. The more the human subjects experience, the higher their cognitive hierarchies are. Hence, if we have an experiment by pooling differently experienced subjects, we tend to find that k is positively related to experience. While cognitive economic experiments seem to lend some support to the cognitive capacity hypothesis, the research is still too limited and premature to draw any strong conclusion.[q]

What, however, interests us is their possible implications for the design of artificial agents. Earlier, we mentioned that the agent community had already tried to identify different intelligence algorithms with their IQ, while in a somewhat causal manner. What we would like to suggest here is that this can be done more rigorously if we can ground this identification within cognitive psychology or cognitive neuroscience. Saying so is not too far-fetched considering that many intelligence algorithms are already naturally inspired, and also now computational neuroscientists are working on genuine brain networks (Sporns, 2010).

21.6.4. *Novelties-Discovering Agents with Cognitive Capacity*

Casari (2004) and Chen, Zeng and Yu (2008) are two pioneering studies in this area which give the designed novelties-discovering agents different cognitive capacities on psychological grounds. In both studies, they used the *population size* of genetic algorithms or genetic programming to parameterize the cognitive capacity of artificial agents. This idea is entirely motivated by cognitive psychology. In cognitive psychology, the human's working memory capacity is frequently tested based on the number

[q]The fundamental question is whether we can possibly examine the possible influence of cognitive capacity using laboratory games with human subjects. This question, as a whole, is what cognitive economic experiments have to answer.

of the cognitive tasks which humans can simultaneously process (Cappelletti, Guth and Ploner, 2008). Dual tasks have been used in hundreds of psychological experiments to measure the attentional demands of different mental activities (Pashler, 1998). This gives the inspiration to model *working memory capacity* directly through *population-based algorithms*, which can implement *parallel processing*, and genetic algorithms (GA) and genetic programming (GP) are such kinds of algorithms. The population size of GA or GP will directly determine the capability of parallel processing. Hence, the population size seems to be an appropriate choice with regard to mimicking the working memory capacity of human agents. A smaller population size, therefore, corresponds to a smaller working memory capacity, whereas a larger population size corresponds to a larger working memory capacity. In this way, an agent-based model composed of agents with different working memory capacities can be developed.

Casari (2004) literally treated the population size used in the genetic algorithm as being equivalent to the number of chunks that a human can process at a time. According to the famous "7 ± 2" rule proposed by George Miller (Miller, 1956), the capacity lies between five and nine. Casari (2004) then set the population size of genetic algorithms to 6, "which implies that decision-makers have a hardwired limitation in processing information at six strategies at a time." (Ibid, p. 261) It is demonstrated that the agent-based simulation with these artificial agents with the given *artificial* working memory capacity can simulate many patterns observed in common property resource experiments. This is probably the earliest article which connects the *population size* used in genetic algorithms to *working memory capacity* in cognitive psychology. Casari (2004), however, did not show the effect of his chosen working memory capacity (population size). Throughout his simulations, only the results with a population size of six are presented. It is, therefore, not clear whether human subject results are sensitive to working memory capacity (population size).

Later on, Casari's idea was extended by Chen, Zeng and Yu (2008) to genetic programming. In the context of the agent-based double auction market, they used genetic programming to model agents' adaptive behavior. This way of modeling agents is not new; however, they no longer assumed that agents are equally smart; instead, following the series of experiments which provide evidence of heterogeneity in subjects' working memory capacity, they used the population size of GP as the parameter to differentiate the agents' working memory capacity. They then simulated this agent-based double auction market, and examined the emergent properties at both the macro level (the market's performance) and the micro level (the individual's performance). It is found that working memory capacity has a positive and significant impact on the market efficiency, which is measured by the percentage of the realized consumer's and producer's surpluses. A more interesting part of their work is to examine the co-evolutionary dynamics when competing agents become equally smarter. Chen, Zeng and Yu (2008) show that, even in a competing situation like the double auction game, pairs of smarter agents can figure out a way to cooperate (collude) so as to create a win-win situation, whereas this collaboration is not shared by pairs of less smart agents.

In addition to the aggregate outcome, Chen and Yu (2011) also compared the strategies learned from agents with different working memory capacity. They analyzed the relationship between the complexity of profitable strategies learned by the agents and their associated working memory capacity. They found that some strategies, which are more complex but also more profitable, had never been found by agents with a smaller capacity, but could quite frequently be found by agents with a larger capacity. Further analysis of these strategies shows that additional capacity facilitates the combinatoric operation which agents need to cook up with more complex and profitable strategies.

Notice that artificial agents in both Casari's and Chen *et al.*'s designs are still novelties-discovering agents. They are able to learn, to discover, and to adapt. They just perform these tasks with different cognitive capacities (population sizes). Therefore, this design enables us to answer the questions regarding the nature or nurture under environments with different complexities.

21.7. Concluding Remarks

In a *Handbook on Reasoning-Based Intelligent Systems*, this chapter provides a review of the development of reasoning-based artificial agents in agent-based computational economics. We focus on two major domains of ACE: one is behavioral game experiments with human subjects, and the second is financial markets. The artificial agents reviewed in this chapter are distributed over these two different domains, although not evenly (Figure 21.1). Some artificial agents are quite extensively used in one domain, but not the other. Nevertheless, in this chapter we try to align them in such a way that the cognitive capacity of these artificial agents can be sorted in monotonic order. By adding the zero-intelligence agent to the list of artificial agents in game experiments, we are able to construct a cognitive spectrum for the artificial agents (Figure 21.2). The salient feature of using artificial agents in laboratory games to construct this spectrum is that this spectrum may have real counterparts for human beings. Hence, the term "systems" actually covers both artificial automata as well as natural automata. Our perspective for the reasoning-based intelligent system is, therefore, not just purely from an engineering viewpoint, but is also inspired and guided by the recent progress in cognitive psychology, computational neuroscience, and economics experiments.

Acknowledgements

The chapter is written based on the invited speech given at the *Conference on Quantitative Behavioral Finance*, University of Nice Sophia Antipolis, Nice, France, December 8-10, 2010, and the invited speech given at the *RIEM Workshop*, Southwestern University of Finance and Economics, Chengdu, China, December 24, 2010.

The author is grateful to Jorgen Vitting Andersen and Tong Zhang for their generous invitation and arrangement, and is also grateful for the comments received from Jorgen Vitting Andersen, Tong Zhang, Hepin Pan, Peter Ayton, Pietro Terna, Youxin Hu, Yu Wu, Han Li, Maxwell Pak as well as an anonymous referee. The NSC grant 98-2410-H-004-045-MY3 is also gratefully acknowledged.

Appendix A. Multi-Armed Bandit Problem

A mathematical formulation of the reinforcement learning model can best be presented with the familiar multi-armed bandit problem.[r] In the multi-armed bandit problem, the agent is provided with a set of J alternatives from which he has to choose one. The consequence of choosing the jth alternative is a payoff r_j which is random, $j = 1, 2, \ldots, J$. The stochastic structure of r_j is unknown to the agent. In a simple setting r_j is independent and identically distributed (i.i.d) with a density function $f_j(r)$ and an expectation μ_j. Suppose that the agent repeatedly played the multi-armed bandit problem N times with a sequence of decisions (choices) $d_1, d_2, \ldots, d_N$, where $d_i \in \{1, 2, \ldots, J\}$.

$$\pi_H = \sum_{i=1}^{N} r_{d_i}, \tag{A.1}$$

where r_{d_i} is the payoff of the decision made in the ith round. If the agent is further assumed to be risk neutral, then his goal is to maximize the expected value of the accumulated payoffs over the N rounds, $E(\pi_H)$.

The multi-armed bandit problem is interesting because to make a sensible decision the agent has to learn the underlying stochastic nature of the payoffs associated with each choice by simply trying these alternatives many times. However, the conundrum of this problem is that he may not be able to afford trying all these alternatives sufficiently long enough before he can have a good approximation (estimation) of the underlying distribution. Because for each time he embarks on an adventure (tries something new or less familiar), he may give up a gain which he feels quite confident of acquiring based on past experience. This conundrum is also known as the trade-off between *exploration* and *exploitation* in the machine learning literature. On the one hand, the agent should try out as many alternatives as possible and should not become satisfied too soon, and, on the other hand, agent should also be practically satisfied with the best alternative he has experienced.

Given the description above, the decision algorithm involved in this multi-armed bandit problem is a design of a *sampling plan*, which allows one to find the best alternative with as few samples as possible by not making many unnecessary observations at non-promising argument values. While the multi-armed bandit problem

[r]It is called the multi-armed bandit problem owing to the fact that it is comparable to finding the best slot machine out of a finite number of such machines.

has been intensively studied mathematically, its uncertain nature colors this problem with a heavy psychological consideration. Uneasiness, fear, and bewilderment can all arise; in particular, in the real world, there is no guarantee that the problem is repeated. When the i.i.d assumption is dropped and the payoff distribution changes over time, the more uncertain and vague environment may leave more room for emotion rather than reasoning.

The simplest version of the multi-armed bandit problem is the two-armed bandit, on which psychologists have conducted many human-subject experiments. Some early two-armed bandit experiments conducted by Jacqueline Goodnow and Laval Robillard at Harvard were reported in Bush and Mosteller (1955). The data used in Robillard's experiments have been further analyzed by Arthur (1993).

Appendix B. A list of Acronyms

To facilitate reading, a list of acronyms is provided in Table B.1.

Table B.1. A List of Acronyms.

Full Name	Acronyms
Agent-Based Computational Economics	ACE
Double Auction	DA
Experience-Weighted Attraction	EWA
Genetic Algorithms	GA
Genetic Programming	GP
Iterated Prisoner's Dilemma	IPD
Santa Fe Institute	SFI
Zero-Intelligence	ZI

References

Agarwal S, Mazumder B (2010) Cognitive abilities and household financial decision making. WP-2010-16, Federal Reserve Bank of Chicago.

Andrews M, Prager R (1994) Genetic programming for the acquisition of double auction market strategies. In: Kinnear K Jr. (ed.), Advances in Genetic Programming. MIT Press, Cambridge, MA, 355–368.

Arthur B (1993) On designing economic agents that behave like human agents. *Journal of Evolutionary Economics* 3(1): 1–22.

Axelrod R (1984) The Evolution of Cooperation. New York, NY: Basic Books.

Axelrod R (1987) The evolution of strategies in the iterated prisoner's dilemma. In: Davis L (ed.), Genetic Algorithms and Simulated Annealing. London: Pitman, and Los Altos, CA: Morgan Kaufman, pp. 32–41.

Axelrod R (1997) The Complexity of Cooperation. Princeton University Press.

Barr J. Tassier T, Ussher L, LeBaron B, Chen S.-H., Sunder S (2008), The future of agent-based research in economics. *Eastern Economic Journal* 34: 550–565.

Becker G (1962) Irrational behaviour and economic theory. *Journal of Political Economy* 70: 1–13.

Bosch-Domenech A, Montalvo J, Nagel R, Satorra A (2002) One, two, (three), infinity,...: Newspaper and lab beauty-contest experiments. *American Economic Review* 92: 1687–1701.
Brock W, Hommes C (1998) Heterogeneous beliefs and routes to chaos in a simple asset pricing model. *Journal of Economic Dynamics and Control* 22(8–9): 1235–1274.
Brown G (1951) Iterated solution of games by fictitious play. In: Koopmans T (ed.), Activity Analysis of Production and Allocation, pp. 377–380. New York: Wiley.
Burnham T, Cesarini D, Johannesson M, Lichtenstein P, Wallace B (2009) Higher cognitive ability is associated with lower entries in a p-beauty contest. *Journal of Economic Behavior and Organization* 72: 171–175.
Bush R.R, Mosteller F (1955) Stochastic Models for Learning. New York: John Wiley & Sons.
Camerer, C (2003) Behavioral Game Theory: Experiments in Strategic Interaction. Princeton, NJ: Princeton University Press.
Camerer C, Ho T-H (1998) EWA Learning in coordination gmes: Probability rules, heterogeneity, and time variation. *Journal of Mathematical Psychology* 42: 305–326.
Camerer C, Ho T-H (1999) Experienced-weighted attraction learning in normal form games. *Econometrica* 67(4): 827–874.
Camerer C, Ho T-H, Chong J.-K (2002) Sophisticated learning and strategic teaching. *Journal of Economic Theory* 104: 137–188.
Camerer C, Ho T-H, Chong J.-K (2004) A cognitive hierarchy model of games. *Quarterly Journal of Economics* 861–898.
Cappelletti D, Guth W, Ploner M (2008) Being of two minds: An ultimatum experiment investigating affective processes. Jena Economic Research Papers 2008-048, The Friedrich Schiller University and the Max Planck Institute of Economics, Jena, Germany.
Casari M (2004) Can genetic algorithms explain experimental anomalies? An application to common property resources. *Computational Economics* 24: 257–275.
Charness G, Levine D (2005) When optimal choices feel wrong: A laboratory study of Bayesian updating, complexity, and affect. *American Economic Review* 95(4): 1300–1309.
Chen S-H, Liao C-C (2005) Agent-based computational modeling of the stock price-volume relation. *Information Sciences* 170: 75–100.
Chen S-H, Hsieh Y-L (2011) Reinforcement learning in experimental asset markets. *Eastern Economic Journal* 37(1): 109–133.
Chen S-H, Tai C-C (2010) The agent-based double auction markets: 15 years on. In Takadama K, Cioffi-Revilla C, Deffuant G (eds), Simulating Interacting Agents and Social Phenomena, Springer, 119–136.
Chen S-H, Yeh C-H (2001) Evolving traders and the business school with genetic programming: A new architecture of the agent-based artificial stock market. *Journal of Economic Dynamics and Control* 25: 363–394
Chen S-H, Yeh C-H (2002) On the emergent properties of artificial stock markets: The efficient market hypothesis and the rational expectations hypothesis. *Journal of Economic Behavior and Organization* 49(2): 217–239.
Chen S-H, Yu T (2011) Agents learned, but do we? Knowledge discovery using the agent-based double auction markets. Frontiers of Electrical and Electronic Engineering (FEE) in China, forthcoming.
Chen S-H, Chang C-L, Du, Y-R (2011) Agent-based economic models and econometrics. Knowledge Economic Review, forthcoming.

Chen S-H, Liao C-C, Chou P-J. (2008) On the plausibility of sunspot equilibria: Simulations based on agent-based artificial stock markets. *Journal of Economic Interaction and Coordination* 3(1): 25–41

Chen S-H, Zeng R-J, Yu T (2008) Co-evolving trading strategies to analyze bounded rationality in double auction markets. In: Riolo R (ed.), Genetic Programming Theory and Practice VI, Springer, 195–213.

Cheung Y, Friedman D (1997). Individual learning in normal form games: Some laboratory results. *Games and Economic Behavior* 19(1): 46–76.

Costa-Gomes M, Crawford V (2006) Cognition and behavior in two-person guessing games: An experimental study. *American Economic Review* 96: 1737–1768.

Costa-Gomes M, Crawford V, Broseta B (2001) Cognition and behavior in normal-form games: An experimental study. *Econometrica* 69: 1193–1235.

Cournot A (1838) Researches sur les Principes Mathematiques de la Theorie des Richesses. Bacon N (ed.) English ed. Researches into the Mathematical Principles of the Theory of Wealth. New York: Macmillan, 1897.

Crawford V (2003) Lying for strategic advantage: Rational and boundedly rational misrepresentation of intentions. *American Economic Review* 93: 133–149.

Crawford V, Iriberri N (2007a) Fatal attraction: Salience, naivete, and sophistication in experimental hide-and-seek games. *American Economic Review* 97: 1731–1750.

Crawford V, Iriberri N (2007b) Level-k auctions: Can boundedly rational strategic thinking explain the winner's curse and overbidding in private-value auctions? *Econometrica* 75: 1721–1770.

Cross J (1973) A stochastic learning model of economic behavior. *Quarterly Journal of Economics* 87(2): 239–266.

Cross J (1983) A Theory of Adaptive Economic Behavior. Cambridge University Press.

Devetag G, Warglien M (2003) Games and phone numbers: Do short-term memory bounds affect strategic behavior? *Journal of Economic Psychology* 24: 189–202.

Devetag G, Warglien M (2008) Playing the wrong game: An experimental analysis of relational complexity and strategic misrepresentation. *Games and Economic Behavior* 62: 364–382.

Duffy J (2006) Agent-based models and human subject experiments. In: Tesfatsion L, Judd K (eds), Handbook of Computational Economics: Agent-based Computational Economics, Vol. 2. Elsevier, Oxford, UK, 949–1011.

Duffy J, Nagel R (1997). On the robustness of behavior in experimental 'beauty contest' games. *Economic Journal* 107: 1684–1700.

Finke M (2010) Cognitive ability and financial decision making. Working paper, Texas Tech University.

Fudenberg D, Levine D (1998) The Theory of Learning in Games. MIT Press.

Fudenberg D, Levine D (2006) A dual-self model of impulse control. *American Economic Review* 96(5): 1449–1476.

Georganas S, Healy P, Weber R (2010) On the persistence of strategic sophistication. http://healy.econ.ohio-state.edu/papers/Georganas_Healy_Weber-LevelKTests.pdf (last download: Feb 1, 2011)

Glimcher P (2010) Foundations of Neuroeconomic Analysis. Oxford University Press.

Gode D, Sunder S (1993) Allocative efficiency of markets with zero intelligence traders: Market as a partial substitute for individual rationality. *Journal of Political Economy* 101: 119–137.

Harley C (1981) Learning the evolutionary stable strategies. *Journal of Theoretical Biology* 89: 611–633.

Holland J, Miller J (1991) Artificial adaptive agents in economic theory. *American Economic Review* 81(2): 365–370.
Hommes C (2002) Modeling the stylized facts in finance through simple nonlinear adaptive systems. *Proceedings of the National Academy of Sciences of the United States of America* 99: 7221–28.
Hommes C (2006) Heterogeneous agent models in economics and finance. In: Tesfatsion L. Kenneth J (eds.) Handbook of Computational Economics, 2–23: 1109–1186, Elsevier.
Ho T-H, Camerer C, Weigelt K (1998) Iterated dominance and iterated best response in experimental p-Beauty Contests. *American Economic Review* 88: 947–969.
Ho T-H, Wang X, Camerer C (2008) Individual differences in EWA learning with partial payoff information. *The Economic Journal* 118(525): 37–59.
Keynes, J (1936) The General Theory of Employment, Interest and Money. BN Publishing.
Kirman A (1991) Epidemics of opinion and speculative bubbles in financial markets. In: Taylor M (ed.), Money and Financial Markets, 354–368. Blackwell, Cambridge.
Klingert F, Meyer M (2010) Multi-agent-simulation of prediction markets: Does manipulation matter using zero-intelligence traders? 3rd World Congress on Social Simulation (WCSS 2010), September 6-9, 2010, Kassel University, Kassel, Germany.
Ladley D (2011) Zero intelligence in economics and finance. *Knowledge Engineering Review*, forthcoming.
LeBaron B (2006) Agent-based computational finance. In: Tesfatsion L. Judd K (eds.) Handbook of Computational Economics: Agent-based Computational Economics, 2–24: 1187–1233, Elsevier.
Lux T (1995) Herd behavior, bubbles and crashes. *Economic Journal* 105: 881–896.
Lux T (1998) The socio-economic dynamics of speculative markets: Interacting agents, chaos, and the fat tails of return distribution. *Journal of Economic Behavior and Organization*, 33: 143–165.
McAllister P (1991) Adaptive approaches to stochastic programming. *Annals of Operations Research* 30: 45–62.
Miller G (1956) The magical number seven, plus or minus two: Some limits on our capacity for processing information. *Psychological Review*, 101(2): 343–352.
Monkerjee D, Sopher B (1994) Learning behavior in an experimental matching pennies game. *Games and Economic Behavior* 19: 97–132.
Montague R (2006) *Why Choose This Book?* (New York: Dutton).
Nagel R (1995) Unraveling in guessing games: An experimental study. *American Economic Review* 85(5): 1313–26.
Nagel R (1998) A survey of experimental guessing games: A study of bounded rationality and learning. In D. Budescu, I. Erev and R. Zwick (eds.), Games and Human Behavior: Essays in Honor of Amnon Rapoport (pp. 105–142). Hillsdale, NJ: Lawrence Erlbaum Associates.
Ohtsubo Y, Rapoport A (2006) Depth of reasoning in strategic form games. The Journal of Socio-Economics 35: 31–47
Othman A (2008) Zero-intelligence agents in prediction markets. In: Padgham L, Parkes D, Muller J, Parsons S (eds.), Proceedings of 7th International Conference on Autonomous Agents and Multiagent Systems (AAMAS 2008), May, 12–16., 2008, Estoril, Portugal, pp. 879–886.
Othman A, Sandholm T (2010) When do markets with simple agents fail? In: van der Hoek W, Kaminka G, Lesperance Y, Luck M, Sen S (eds.) Proceedings of 9th International Conference on Autonomous Agents and Multiagent Systems (AAMAS 2010), May, 10–14, 2010, Toronto, pp. 865–872.

Pashler H (1998). The Psychology of Attention. Cambridge, MA: MIT Press.

Pouget S (2007) Adaptive traders and the design of financial markets. *Journal of Finance* LXII(6): 2835–2863.

Roth A, Erev I (1995) Learning in extensive-form games: Experimental data and simple dynamic models in the intermediate term. Games and Economic Behavior, Special Issue: Nobel Symposium, 8: 164–212.

Rust J, Miller J, Palmer R (1993) Behavior of trading automata in a computerized double auction market, in Friedman D, Rust J (eds) Double Auction Markets: Theory, Institutions, and Laboratory Evidence, Redwood City, CA: Addison Wesley.

Rust J, Miller J, Palmer R (1994) Characterizing effective trading strategies: Insights from a computerized double auction tournament. *Journal of Economic Dynamics and Control* 18: 61–96.

Sporns O (2010) Networks of the Brain. MIT Press.

Stahl D (2000) Rule learning in symmetric normal-form games: Theory and evidence. *Games and Economic Behavior* 32(1): 105–138.

Stahl D, Wilson P (1994) Experimental evidence on players' models of other players. *Journal of Economic Behavior and Organization* 25: 309–327.

Stahl D, Wilson P (1995) On players' models of other players: Theory and experimental evidence. *Games and Economic Behavior* 10(1): 218–254.

Thaler R (1997) Giving markets a human dimension. Financial Times. 6, June 16, 1997.

Thaler R (2000) From Homo economicus to Homo sapiens. *Journal of Economic Perspectives* 14(1): 133–141.

Tseng J-J, Chen S-H, Wang S-C, Li S-P (2008) Scale-free network emerged in the markets: Human traders versus zero-intelligence traders. In: Terano T, Kita H, Takahashi S, Deguchi H (eds.), Agent-Based Approaches in Economic and Social Complex Systems V, Agent-Based Social Systems, Volume 6, 245–254, Springer.

Tseng J-J, Chen S-H, Wang S-C, Li S-P (2009) Emergence of scale-free networks in markets. *Advances in Complex Systems*, 12(1): 87–97.

Ussher L (2009a) Monetary policy reform with zero intelligent agents: An agent based model. Conference talk at the 15th International Conference on Computing in Economics and Finance, University of Technolgy at Sydney, Sydney, Australia, July 15–17, 2009.

Ussher L (2009b) Agent based models with zero intelligence and balance sheet constraints. Invited speech at First International Workshop on Managing Financial Instability in Capitalistic Economies (MAFIN), University of Reykjavik, Iceland, Sep 3–5, 2009.

Yeh C.-H. (2007) The role of intelligence in time series properties. *Computational Economics*, 30(2): 95–123.

Yeh C.-H. (2008) The effects of intelligence on price discovery and market efficiency. *Journal of Economic Behavior and Organization*, 68(3–4): 613–625.

Chapter 22

REASONING AND KNOWLEDGE ACQUISITION FROM MEDICAL DATABASE USING LATTICE SOM AND TREE STRUCTURE SOM

Takumi Ichimura*,‡, Takashi Yamaguchi† and Kenneth James Mackin†

**Faculty of Management and Information Systems,*
Prefectural University of Hiroshima
1-1-71, Ujina-Higashi, Minami-ku,
Hiroshima, 734-8559, Japan
ichimura@pu-hiroshima.ac.jp

†Department of Information Systems,
Tokyo University of Information Sciences
4-1 Onaridai, Wakaba-ku, Chiba, 265-8501 Japan
{tyamagu,mackin}@rsch.this.ac.jp

This chapter models the conceptual process of the classification of data for reasoning by using Tree Structure SOM and Lattice Structure SOM. We propose a new variant of structure growing SOM that puts appropriate neurons in the region on the map that has a high gradient in probability density estimation. A knowledge database is constructed automatically through data classification and rule extraction. In order to verify the validity of the proposed method, some experimental results using 2 different benchmark tests in the UCI Machine Learning Repository medical databases were described.

22.1. Introduction

Intelligent systems have made great advancements in computational reasoning by using knowledge extraction from databases which consist of large number of data samples. With the wide-spread use of open-source database management systems such as PostgreSQL and mysql, and the availability of a large number of real-life data samples through the Internet, interest has naturally increased in extracting useful relationships or knowledge existing in the raw data. Human intelligence attracts the extraction of relationship or regularity from such large data, and concepts are constructed from them. In this chapter, we focus on the following functions related to extraction of concepts: (1) classification of data sets, (2) explanation of the classification, and (3) reasoning by using the explicit structure of knowledge.

‡Corresponding author.

Classification of data is a prerequisite to construct the domain concept necessary for intelligent systems with reasoning functionality. For instance, "understanding" can be realized in intelligent systems by classifying cases into specific categories and by interpreting the meaning of these categories. Moreover, the relations among the data can be explicitly represented, and the concept and its structure become the basis of the reasoning mechanism.

This chapter will model the conceptual process of the classification of data for reasoning by using Self Organizing Map (SOM) [Kohonen (2000); Kohonen *et al.* (2000)]. Traditional SOM consists of two layers, which the first is an n dimensional input layer denoted by $x(t)$ and the second, called the "competitive layer" is a 2 dimensional map to visualize the output. However, the classification results by traditional SOM heavily depend on the initial value of the weight vector and the number of nodes in the map.

The first approach in this chapter is an adaptive learning method of SOM with lattice structure. The method implements a node generation/elimination algorithm on the map, which monitors the error between weight vector and input vector according to the given data set. If the error does not decrease enough during learning, the method assumes that the classification ability of the map is insufficient to find the optimal weight vector, and a new node is added to the appropriate location. On the other hand, even if the learning is sufficient to classify the data, if a given node is not activated, the neuron is regarded redundant, and the neuron is removed from the map. The proposed method can classify a given data set automatically with the optimal number of nodes in the competitive layer.

SOM shows powerful classification capability for a given data set under the condition that the map consists of optimal number of units, and this condition is met using the described node generation/elimination algorithm. However, the number of achieved categories is not always same, compared to solutions using other clustering methods. One of the reasons is that the initial weight vector is random and the classification performance using SOM depends on the initial weight vector. In order to improve the classification capability, we propose a classification method using Tree Structure SOM(TS-SOM) [Yamaguchi, Ichimura & Mackin (2010)]. Other growing type SOMs which could add and delete neurons to automatically find the optimal size of competitive layer have been previously reported [Fritzke (1994, 1995); Dittenbach, Merkel, & Rauber (2000); Rauber, Merkl & Dittenbach (2002); Forti, Foresti (2006)]. In TS-SOM, the SOM classifies the data set into 2 or more subcategories based on decision trees, and separate SOMs classify each subcategory into further subcategories independently. In this method, multiple SOMs classify the data set recursively until the SOM can divide the data no further, and the final category is found. The proposed method can use the feature of competitive learning in SOM, and the relationships in the data set are shown as the resultant tree structure.

Moreover, we proposed a new variant of structure growing SOM that puts appropriate neurons in the region on the map that has a high gradient in probability density estimation, in order to adjust the network structure to improve clustering [Yamaguchi, Ichimura & Mackin (2010)]. The requirement of pruning of nodes or the creation of layers in a tree structure of TS-SOM is determined by measuring the difference of error of weight vectors of the winner node and its neighbor nodes.

Furthermore, the knowledge extracted from the obtained tree based structure will represent the data structure explicitly. Especially in the case of large-scale data sets, Adaptive Tree Structured Clustering (ATSC) is an effective method for divisive hierarchical clustering with a function of sequential processing [Yamaguchi, Ichimura, & Mackin (2008a,b); Yamaguchi, Noguchi, Ichimura & Mackin (2009)]. We can infer the implicit relations in the data set by following the final constructed structure, and then make decisions according to the extracted knowledge or rules.

Using the proposed methods, a knowledge database is constructed automatically through data classification and rule extraction. To verify the validity of the proposed method, experimental results using 2 different benchmark tests in the UCI Machine Learning Repository [Fisher (1988); Forina *et al.* (1991)] and real medical databases [Suka *et al.* (2004)] are shown. The medical database contains some uncertainty in the information in the database such as missing data, or contradictory cases in which the inputs are same but outputs are different. We explain this using the proposed method, and the reasoning in intelligent systems using explicit knowledge database is described.

The remainder of this chapter is organized as follows. Section 22.2 explains the Planar Lattice NN and describes the neuron generation/elimination algorithm in PLNN. The tree search algorithm in Tree Structure SOM(TS-SOM) are explained in Section 22.3. The basic idea of neuron generation/elimination realized in TS-SOM is described in Section 22.4. For classification of large-scale data sets, Adaptive Tree Structured Clustering is proposed in Section 22.5. Section 22.6 describes the medical data set related to Coronary Heart Disease(CHD) database. Experimental results for classification of benchmark tests and CHD database is reported in Section 22.7. Section 22.8 summarizes the results with some discussions to conclude the chapter.

22.2. Planar Lattice Neural Networks

22.2.1. *An overview of Planar Lattice Neural Network*

In this section, we describe about the behavior in Planar Lattice Neural Networks (PLNNs) [1] developed as an extension of self-organizing maps (SOMs) [2] in the hidden layer of neural networks(NNs) as shown in Figure 22.1. This neural network(NN) can classify the training data set similar to SOM. The PLNN is a type of a 3 layered NN, which consists of input layer, interconnected hidden layer, and output layer. The interconnected hidden neurons adjusts the connection weights between input neurons and hidden neurons according to the signals to the network and the neighborhood $\mathcal{N}_i$ of neuron i. $\mathcal{N}_i$ is a set containing a group of neurons around i in the lattice. If $\mathcal{S}$ is the set of neurons and $\mathcal{N} = \{\mathcal{N}_i | i \in \mathcal{S}\}$ is the set of neuron neighborhoods, then the pair $\{\mathcal{S}, \mathcal{N}\}$ forms a graph in the usual sense. The neighborhoods system as shown in Figure 22.2 belongs to the class of Homogeneous Neighborhood System (HNS) defined in the following.

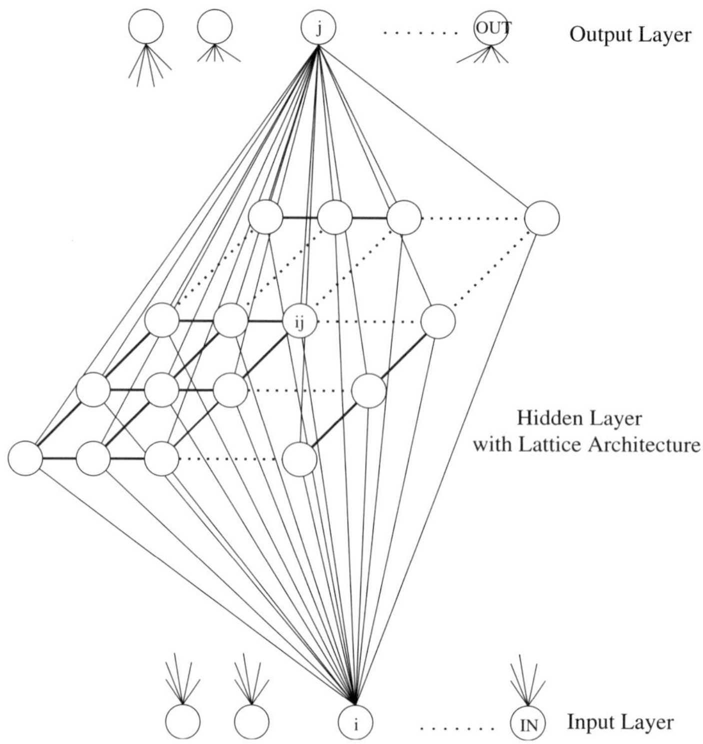

Fig. 22.1. An Overview of Planar Lattice Neural Network.

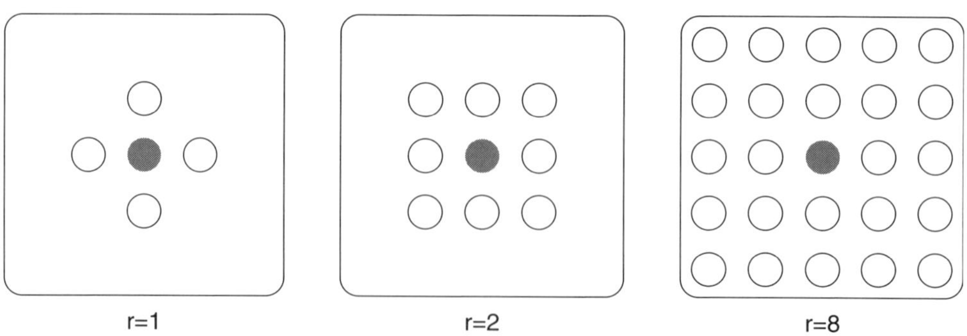

Fig. 22.2. Homogeneous Neighborhood System.

Definition 1: Homogeneous Neighborhood System is of the form:

$$\mathcal{H} = \{\mathcal{H}_i | i \in \mathcal{S}\}, \quad \mathcal{H}_i = \{j| \parallel \ell_j - \ell_i \parallel^2 \leq r\}, \quad i \in \mathcal{S}, \tag{22.1}$$

where $\mathcal{S}$ is the set of neurons, ℓ_i and ℓ_j are the lattice positions for neuron i and j, respectively.

If the network has IN input neurons and OUT output neurons, there are $IN \times OUT$ hidden neurons.

There are two function level variables (z_i, y_i) for each neuron in the lattice. z_i is pre-sigmoidal activation level and y_i is post-sigmoidal activation level. The adjustment of the pre-sigmoidal activation level is:

$$z_i[t] = a_1 \mathrm{F}(\mathrm{Met}(\mathbf{x}[t], \mathbf{w}_i^{IH})) + a_2 \sum_{j \in \mathcal{N}_i} y_j[t-1]\mu_{ij}, \tag{22.2}$$

where $z_i[t]$ is the pre-sigmoidal activation level for neuron i at time index t, the input vector of the network is $\mathbf{x} = (x_1, x_2, \ldots, x_h, \ldots, x_{IN})$, the input weight vector is $\mathbf{W}_i^{IH} = (w_{i1}^{IH}, w_{i2}^{IH}, \ldots, w_{ih}^{IH}, \ldots, w_{iIN}^{IH})$. w_{ih}^{IH} is the weight associating neuron i with input signal $x_h (i \le h \le IN)$.

F() is a positive monotonic decreasing function and Met() is a metric value that measures the distance between two vectors in a metric space. $\mathcal{N}_i$ is the neighborhood set of neuron i. y_j is the output activity of neuron j. The lateral interconnection weights between neuron i and j are the weight associating neuron i with j. The μ_{ij} depends only on the relative position in the lattice, $\mu_{ij} = \eta(\ell_i - \ell_j)$. $\eta()$ is the spatial impulse response function of the network as shown in Figure 22.3. The a_1 and a_2 in Equation (22.2) are two constants which are applied to the weights of network inputs and the outputs from other neurons in lattice, respectively. In this paper, we used a Euclidean metric for Met(), then F() is the following.

$$F(\mathbf{x}, \mathbf{w}_i^{IH}) = f\left(\sum_{h=1}^{IN} g(w_{ih}^{IH}, x_h[n])\right), \tag{22.3}$$

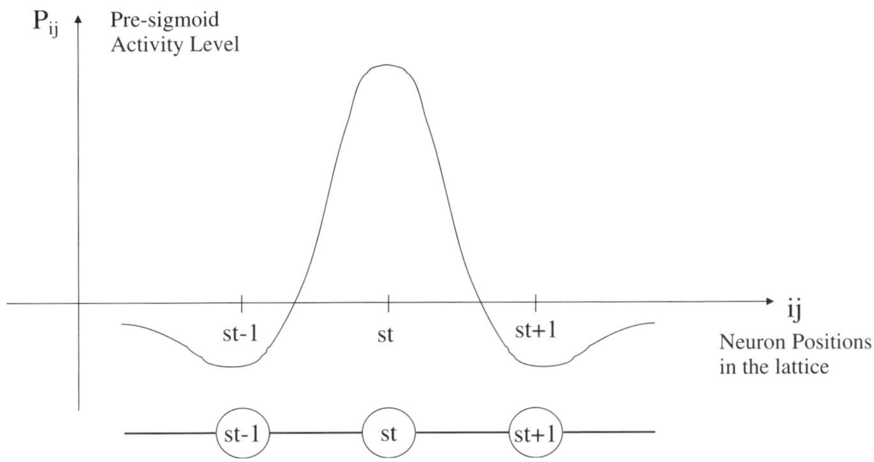

Fig. 22.3. Spatial Impulse Response Function.

where

$$g(w_{ih}^{IH}, x_h[t]) = (w_{ih}^{IH} - x_h[t])^2$$
$$f(u) = e^{-\lambda u}$$

The output signal for neuron is

$$y_j[t] = \sigma_i(z_i[t]), \tag{22.4}$$

where $\sigma_i()$ is the sigmoid function. The weight adjustment follows the Kohonen learning algorithm[2] as shown in Figure 22.4:

Algorithm:
for (each time index n) **do**
  fetch source signal $\mathbf{x}[t]$
  select i:$\| \mathbf{x}[t] - \mathbf{w}_i[t-1] \| \leq \| \mathbf{x}[t] - \mathbf{w}_k[t-1] \|, \forall k \in \mathcal{S}$
  for $(\forall j \in \mathcal{N}_i)$ **do**
    $\mathbf{w}_j[t] = \mathbf{w}_j[t-1] + \alpha\phi[t-1](\| \ell_j - \ell_i \|)(\mathbf{x}[t] - \mathbf{w}_j[t-1])$
  end
end

Fig. 22.4. Kohonen's Learning Algorithm.

As shown in Figure 22.4, $\mathbf{x}[t]$ is the input signal of the network at time index t. $\mathbf{w}_j[t]$ is the input weight vector of neuron j. $\mathcal{S}$ is the set of neurons. The learning rate α is a constant used to control the rate of convergence in the learning process. ℓ_i is the position of neuron i in the lattice. $\phi[t](d)$ is a time varying function used to control the range of influence of a neuron to other neurons in the lattice as follows.

$$\phi[n](d) = \begin{cases} 1 & \text{if } d \leq R[n] \\ 0 & \text{otherwise} \end{cases}, \tag{22.5}$$

where $R[t]$ is a positive definite monotonic decreasing function and the minimum value for $R[t]$ is 1 for the network to retain organizing capability.

The weights will be organized such that neighboring neurons on the network tend to have similar input weight vectors that the weight vectors represent neighboring regions in the input pattern space. In particular, it can be shown that the asymptotic values of the weight vectors will tend towards the weighted center of their influence regions. The influence region for neuron i is the region Ω_i in the input pattern space. During the training phase, whenever an input pattern falls in Ω_i, the $\mathbf{w}_i$ will be modified. That is, $\Omega_i = \cup_{j \in \mathcal{N}_i} \mathbf{v}_j$, where $\mathbf{v}_j$ is the Voronoi region for neuron j.

We follow the back propagation algorithm to adjust the weights between hidden neurons and output neurons as follows.

$$\mathbf{E} = \frac{1}{2}\sum_{k=1}^{OT}(t_k(x) - y_k(x))^2 \tag{22.6}$$

$$\Delta w_{ki}^{HO} = -\eta\frac{\partial \mathbf{E}}{\partial w_{ki}^{HO}}, \tag{22.7}$$

where η is the learning rate. HO and OT are the number of hidden neurons and output neurons, respectively.

22.2.2. *Neuron generation/elimination*

For a normal NN, we can observe the following behaviors during learning phase.

Case 1: If an NN does not have enough neurons in order to satisfactorily infer the result, then the input weight vector will tend to fluctuate greatly even after a certain period of the learning process.

Case 2: If an NN has enough neurons to infer and if the input weight vector of each neuron converges to a certain value, it is possible to remove unnecessary neurons from the network without affecting the inference ability.

In the first case, the network needs to generate a new neuron, as its parent's attribute is inherited. In the second case, it is necessary to delete a redundant neuron in the calculation. Based on such behaviors, we can determine the conditions for the neuron generation/elimination during the learning process.

22.2.2.1. *Neuron generation*

The situation for neuron generation appears when the representation capability of the network is insufficient. In general, given enough neurons in the hidden layer, an NN can form any mapping with any desired precision. Therefore, we can use the stabilized error as an index to determine whether the network needs to generate a new neuron. If a stabilized error after a specified period is larger than the desired value, then a new neuron may be generated.

The next problem is how to determine the position of the new neuron in the hidden layer. The desirable position can be found through monitoring the behavior of the neurons in the hidden layer. If a neuron strongly affects the final system error through the fluctuation in its input weight vector, then this neuron requires a higher processing capability. If the input weight vector of a neuron still fluctuates greatly even after an extended period of the learning process, this means that the neuron does not have enough processing capability to learn the sub problem the neuron must solve. Therefore, this neuron can be split into two, i.e. we can add another neuron to the exact same interconnection of the network, inheriting the original neurons attributes.

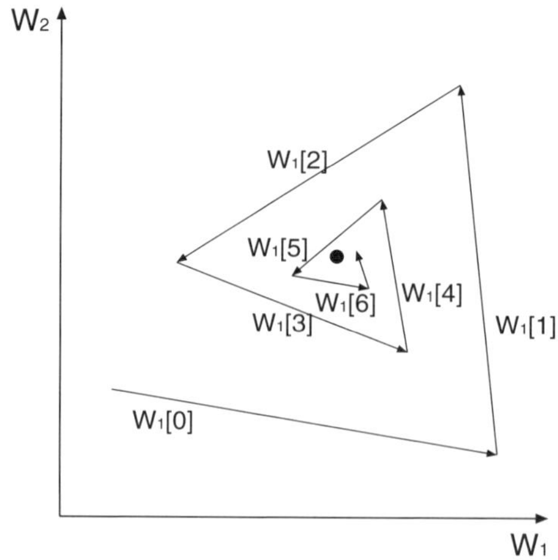

Fig. 22.5. Learning Weights of a Neuron.

Then we can give the following equation for generating a new neuron. Figure 22.5 shows an idea of learning the weights of a neuron by optimally searching the spaces of weights $\mathbf{W}_1$ and $\mathbf{W}_2$.

During the convergence process, the average length of movement between adjacent temporal weights gradually decreases. Based on this observation, we can define the following measure, called the walking distance(WD) for neurons:

$$WD_j[n] = \sum_{m=1}^{n} \mathrm{Met}(\mathbf{W}_j[m], \mathbf{W}_j[m-1]), \tag{22.8}$$

where $WD_j[n]$ is the WD for neuron j in the hidden layer at iteration n, $\mathbf{W}_j[m]$ is the input weight to neuron j at an arbitrary iteration m, and Met is some metric value that measures the distance between vectors in a metric space. In this subsection, Met() is given as an Euclidean metric:

$$\mathrm{EuMet}(\mathbf{X}, \mathbf{Y}) = \sqrt{\sum_{i=1}^{N} (x_i - y_i)^2} \tag{22.9}$$

where

$$N = \max(\mathrm{Dim}(\mathbf{X}), \mathrm{Dim}(\mathbf{Y})) \tag{22.10}$$

In Equation (22.8), WD_j is a measure for the time variance of the stored information for neuron j, and can be considered the activity of neuron j in the parameter space. WD that is too high for a particular neuron indicates that the processing capability of the neuron is insufficient, and a new neuron is needed to share its load.

Equation (22.8) can be approximated by the following operational measure:

$$WD_j[n] = \gamma WD_j[n-1] + (1-\gamma_w)Met(\mathbf{W}_j[m], \mathbf{W}_j[m-1]), \quad 0 < \gamma_w < 1 \tag{22.11}$$

A neuron j should generate another neuron if

$$\Delta\epsilon_j = \frac{\partial\epsilon}{\partial WD_j} \times WD_j > \theta_G, \tag{22.12}$$

where ϵ is the overall system error, WD_j is the walking distance for j, ϵ_j represents the contribution of neuron j to the overall system error, and θ_G is a threshold value.

22.2.2.2. *Neuron elimination*

In order to achieve a good overall system performance, we must remove unnecessary elements from the network. A network can gradually form an optimal structure in the development process by applying a mechanism with selection rules, which keeps effective neurons in the network and removes misplaced ones. To be more specific, we have based the selection mechanism on the following two observations:

(a) If a neuron does not form appropriate interconnections among other neurons, then it will become ineffective early in the development process.
(b) The neurons compete with each other for resources, because each neuron tries to inhibit other neurons from taking exactly the same functional role that it plays in the network.

The above observations suggest the following guidelines for the neuron elimination process. That is, we can remove a corresponding neuron when:

(a) It is not a functional element in the network.
(b) It is a redundant element of the network.

The criterion (a) can be checked by monitoring the output signal of a neuron. As a measure for this criterion, we can use the variance of the output signal of a neuron j:

$$VA_j = (\bar{O}_j - O_j)^2, \tag{22.13}$$

where O_j is the variance of the output signal level for neuron j, and $\bar{O}_j$ is an average of O_j. We can define the operational measure of VA_j as:

$$VA_j[n] = \gamma_v VA_j[n-1] + (1-\gamma_v)(O_j - Act_j[n])^2, \quad 0 < \gamma_v < 1 \tag{22.14}$$

where

$$Act_j[n] = \gamma_a Act_j[n-1] + (1-\gamma_a)O_j[n], \quad 0 < \gamma_a < 1 \tag{22.15}$$

$Act_j[n]$ is the operational measure of the average output signal for neuron j.

If the VA_j is zero for a given neuron j, then this neuron does not generate any additional information. In other words, this neuron does not perform any signal processing function. Therefore, we can eliminate such a neuron, when VA_j is smaller than a certain threshold value in a practical sense.

The situation (b) can be identified by watching the correlation between the output values or neurons in the network. If two neurons are completely correlated, that is, given the output of one neuron, the output of the other neuron can be deduced with probability equal to 1, then the neuron can be eliminated without affecting the performance of the network.

22.2.2.3. *Neuron generation/elimination in PLNN*

Figure 22.6 shows implementation of neuron generation process in the lattice of PLNN. In the subsection, we define $WD_{\max 1}$ as a neuron with the largest WD in the neighborhood $\mathcal{N}_i$, and $WD_{\max 2}$ as a neuron with the second largest WD. A new

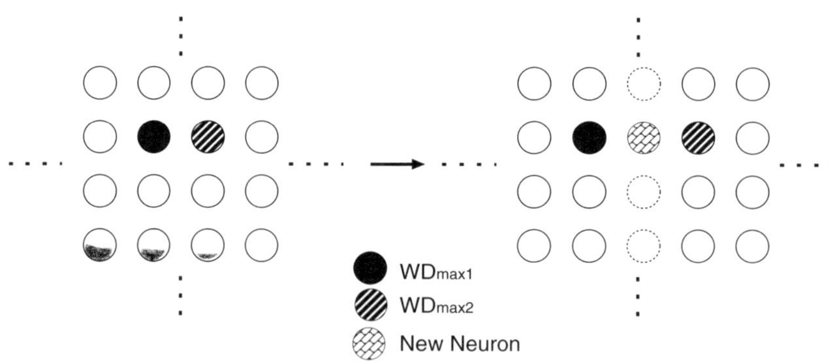

(a) generation case 1

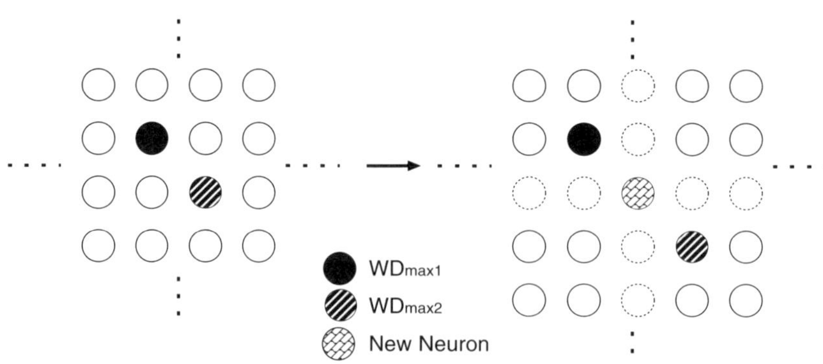

(b) generation case 2

Fig. 22.6. Neuron Generation in PLNN.

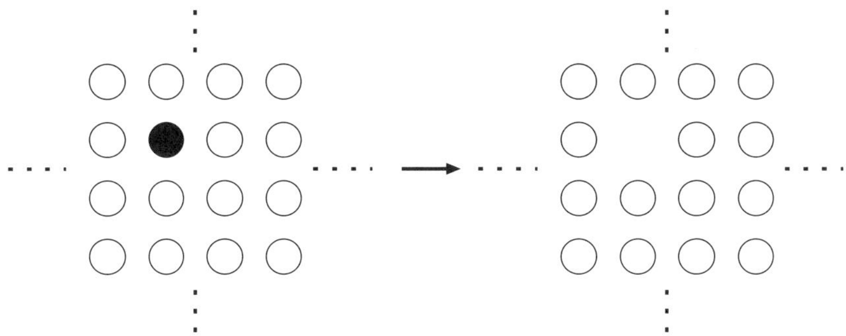

Fig. 22.7. Neuron Elimination in PLNN.

neuron is added in the middle of two neurons $WD_{\max 1}$ and $WD_{\max 2}$. A new neuron should set into the lattice according to the prior neuron arrangement, because neurons are arranged in a plane.

The situation of neuron generation is shown in Figure 22.6. Figure 22.6(a) shows that the $WD_{\max 1}$ neuron is placed in a row or a column to the $WD_{\max 2}$ neuron. In this case, the new neuron is added as shown on the right hand side of Figure 22.6(a). Figure 22.6(b) shows that the $WD_{\max 1}$ neuron and the $WD_{\max 2}$ neuron are on a diagonal line. In this case, the new neuron is added as shown on the right hand side of Figure 22.6(b).

On the other hand, if the neuron elimination rules are satisfied with the conditions described in Subsection 22.2.2, the specified neuron is eliminated as shown in Figure 22.7.

22.3. Tree Structured SOM

Tree Structured SOM (TS-SOM) is a faster SOM method applying tree search algorithm proposed by Koikkalainen [Koikkalaien *et al.* (1990)]. TS-SOM uses a hierarchical NN structure which consists multiple competitive layers arranged from small sized layer to large sized layer as shown in Figure 22.8. Each competitive layer is 2 dimensional grid map similar to that of basic SOM. In this figure, the circle filled with white color represents a node, the bold line represents the connection between parent and its child node. The parent node connects to the child node in the lower layer.

For intuitive interpretation, Figure 22.8 shows that the parent node in Layer 0 connects to 4 child nodes in Layer 1. The region of parent in the map is divided into subregions by 4 child nodes. The connections from Layer 1 to Layer 2 divides the region into 4 parts of subregions in the similar way. Therefore, TS-SOM is a hierarchical neural network structure which enables rough classification in the upper layer and precise classification in the lower layer.

The winner node is selected from 4 child nodes in Layer 2 and their parent node in Layer 1 is fired. Successively the node in Layer 0 receives the signals from their

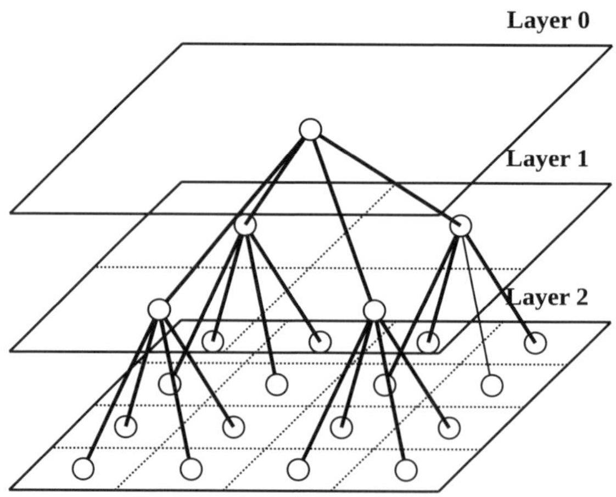

Fig. 22.8. Dimensional Neural Network Structure of TS-SOM.

nodes in Layer 1. That is, the connected path of fired nodes means an overview of structure to complex data samples.

Each layer is disciplined by using the training method of basic SOM independently from the top layer to bottom layer. The nodes in the lower layer are initialized by using the weight vector of parent node and its neighbor nodes. TS-SOM can build the hierarchical structure of map with precise classification capability and the obtained structure is the implicit topology of data sample from the upper layer to the lower layer.

The detailed algorithm of TS-SOM is described as follows. Let W_l be a weight vector in Layer l, where $W_l = \{\mathbf{w}_{j_l}\}$, $j_l = 1, 2, \ldots, m$ at Layer $l(l = 0, 1, 2, \ldots, l_{\max -1})$. Layer 0 is a root or top layer that has only a node. The j_lth node in Layer l connected to some child nodes in the lower $(l+1)$th layer as shown in Figure 22.8. Let C_{j_l} be a set of child nodes for the parent node j_l. In the case of 2 dimensional grid type map, the number of child nodes $|C_{j_l}|$ is 4 and number of nodes in lth layer $|W_l|$ is 4^l.

The initialization and update of weight vectors, and the determination of winner node are implemented by the TS-SOM training method according to the variant of layer structure. The winner node in TS-SOM employs a tree search algorithm, which the winner node is recursively determined from a root to child nodes in next lower layer. The winner node b_{i_l} in Layer l is selected from the child nodes of winner node $b_{i_{l-1}}$ in the previous layer by using Equation (22.16).

$$b_{i_l} = \mathop{\arg\min}_{j_l \in C_{b_{i_{l-1}}}, l \neq 0} \parallel \mathbf{x}_i(t) - \mathbf{w}_{j_l}(t) \parallel \tag{22.16}$$

As TS-SOM training process is finished to search the optimal weights in the layer, the weight vector in the successive layer is updated independently. The update

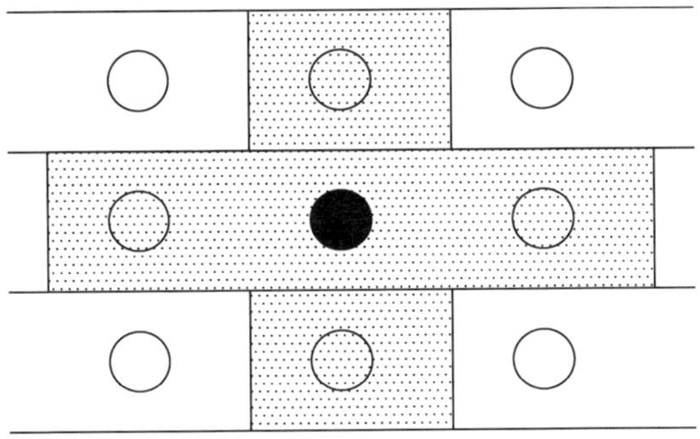

Fig. 22.9. Fixed Grid Type Neighborhood Function of TS-SOM.

of weight vector is followed Equation (22.17) which is similar to update of weights in the basic SOM.

$$\mathbf{w}_{j_l}(t+1) = \mathbf{w}_{j_l}(t) + \alpha(t) h_{b_{i_l} j_l} [\mathbf{x}_i(t) - \mathbf{w}_{j_l}(t)], \tag{22.17}$$

where $\alpha(t)$ is a learning rate which is a decreasing function according to training step t and $h_{b_{il} j_l}$ is the neighborhood function. Note that the fixed grid type neighborhood function can be used in TS-SOM as shown in Figure 22.9. The gray colored area is the neighborhood and a black colored point is a winner node in Figure 22.9. The Gaussian type neighborhood function that the width is decreasing with training step t is commonly used in basic SOM in order to find the fine tuned map by the adjustment in latter training steps. However it is not necessary in TS-SOM to apply this adjustment because the detailed classification works to train the weight vector in the lower layer [Koikkalaien *et al.* (1990)].

Before the training in each layer, the weight vectors are initialized with inherited some parameters from the parent node and the neighbor nodes of target node. The weight vector is initialized by 2 steps of following Equations (22.18) and (22.19)

$$\mathbf{W}_{j_l} \leftarrow \mathbf{W}_{k_{l-1}} \tag{22.18}$$

$$\mathbf{W}_{j_l} \leftarrow \mathbf{W}_{j_l} + \beta \bar{\mathbf{W}}_{j_l}, \tag{22.19}$$

where $l \neq 0, k_{l-1}$ is the parent node for jth node in Layer l, $\bar{\mathbf{W}}_{j_l}$ is a centroid of the set of nodes in the neighborhood for jth node in Layer l that denoted by $\mathcal{N}_{j_l}$ and is gray colored points in Figure 22.9, and $\beta (0 < \beta < 1)$ is a parameter that defines the modification rate for the nodes in neighborhood.

TS-SOM continues the training in the lower layer by using weight initialization method mentioned above, if the training is finished in parent layer. However, the number of nodes is greatly increased whenever a new layer is added since the number of nodes in lth layer is 4^l. For the preventing such a situation, we propose the adaptive learning method that optimizes the hierarchical NN structure by adding and deleting the units during the training process based on the node generation/elimination described in the Subsection 22.2.2. In addition, the proposed method for the updating weight vectors were different from original TS-SOM, because the updating method in TS-SOM may not be efficient if the NN structure dynamically changed.

22.4. Adaptive Learning Algorithm in TS-SOM

22.4.1. *NN structure adaptation*

We proposed a new variant of growing structure of SOM that puts appropriate nodes in the region on the map that has high gradient in probability density estimation in order to fit the network structure to good clustering [Yamaguchi, Ichimura & Mackin (2010)]. The pruning of nodes and the layer creation to a tree structure of TS-SOM is required to determine by measuring the mean squared error of weight vectors related to the winner node and its neighbor nodes, and the selected ratio of winner node.

In the TS-SOM, a new layer is added to bottom layer if the previous layer is trained to become a smaller error than the expected value. The proposed method can add partially a pyramidal unit with a parent node and 4 child nodes as the new layer. Figure 22.10 shows the process of growing NN structure. The child nodes can classify in precise by the arrangement of positions which divides their parent region into quarters in similar way of the behavior of original TS-SOM. Figure 22.11 shows the coordinate mapping for each partial layer in the case of a NN structure as shown in Figure 22.10. The hierarchical competitive layer can be mapped to a layer at leaf nodes in tree structure, such that a competitive layer can have partially

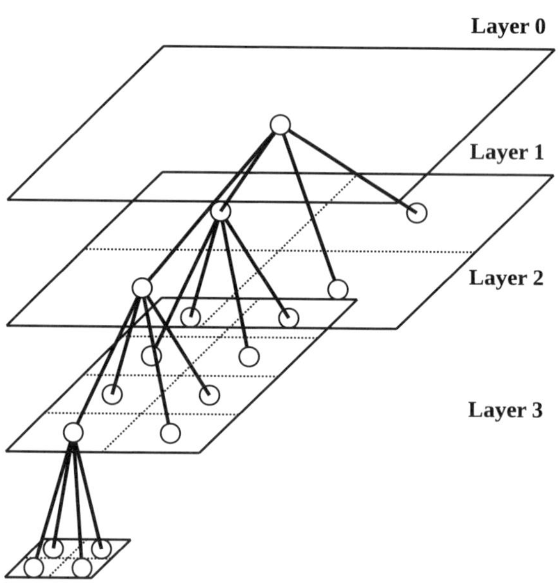

Fig. 22.10. NN Structure of Proposed Growing Type SOM.

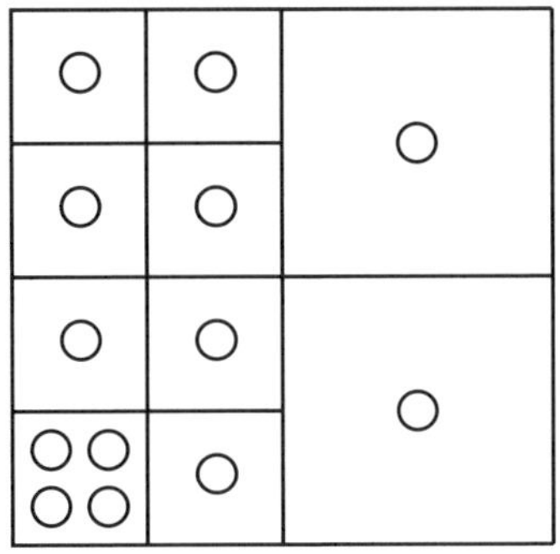

Fig. 22.11. The Coordinate Mapping for Multiple Competitive Layers.

high resolution region. Unlike TS-SOM, because the progress of training is different in the region on each layer, the adaptation of weight updating method should be considered for proper training. The weight updating methods are discussed in following Section 22.4.2. The detailed algorithm for NN structure is described as follows.

In the discovery of optimal NN structure, the procedure of addition or deletion node is performed if the criterion for target is satisfied during each training phase, τ. τ is the pre-determined parameter in the range [500, 1000], which means the training steps to make rough fitting weight vectors to input vectors as shown in basic SOM.

The proposed method can add an pyramidal unit for the partial space of its corresponding input vectors in early stage and can delete the unnecessary pyramidal units for clustering in latter stage. The elimination of redundant nodes occurs after the certain period of training, because it is not possible to learn appropriately if the node is deleted in spite of the fact that the weight vectors are not yet fitting to input vectors. The condition for distinction of switch node generation/elimination is defined by the overall training steps, T. If $1 < t < \gamma T$, then it is early stage. The elimination algorithms are implemented till the end of training after γT.

The initial structure of NN starts with the minimum tree structure, which consists of a root node where there are 1 parent node and 4 child nodes. The weight vectors in a root node are initialized by random values.

In the early stage of training, new child nodes are added to follow the node in the leaf node, when the error of weight vectors between target node and its neighbor is larger than the pre-determined threshold. Let $\mathcal{L}$ be the set of child nodes in all leaf nodes. The child nodes are added to follow the j_lth node such that $j_l \in \mathcal{L}$, if Equations (22.20) and (22.21) are satisfied.

$$\max_{k_l \in \mathcal{N}_{j_l}} \| \mathbf{w}_{j_l} - \mathbf{w}_{k_l} \| > \epsilon \tag{22.20}$$

$$\frac{p_{j_l}(\mathcal{S})}{|\mathcal{S}|} > \rho, \tag{22.21}$$

where $j_l \neq k_l$, k_l is a node in neighborhood $\mathcal{N}_{j_l}$ for j_lth node, $\epsilon(0 < \epsilon < 1)$ is a parameter which is defined the allowable error in the training, $p_{j_l}(\mathcal{S})$ is the frequency for the firing winner node when input data set $\mathcal{S}$ is given, and $\rho(0 < \rho < 1)$ is a parameter which is defined the minimum value for its frequency. The definition of neighborhood $\mathcal{N}_{j_l}$ is same as the neighborhood of TS-SOM as shown in Figure 22.9. A lot of redundant nodes will be generated because the nodes are added regardless of distribution in input data set, if Equation (22.21) is not considered. In order to avoid redundant nodes, Equation (22.21) is efficient, so that the nodes are not added to the target node that the frequency is low. The weight vectors of new child nodes are initialized by using Equations (22.18) and (22.19).

In the latter stage of training, the redundant child node in a leaf is deleted if the variance of frequency for firing winner node is smaller than an expected threshold. Consider that the deleting procedure of child nodes in a leaf has k_{l-1}th node as parent except the case of $l \neq 0$. 4 redundant child nodes that $j_l \in C_{k_{l-1}}$ and $j_l \in \mathcal{L}$ are deleted if the following equations are satisfied.

$$\frac{v_{k_{l-1}}(\mathcal{S})}{|\mathcal{S}|} > \theta \tag{22.22}$$

$$\frac{p_{k_{l-1}}(\mathcal{S})}{|\mathcal{S}|} < \rho, \tag{22.23}$$

where $v_{k_{l-1}}$ is a variance of frequency for firing winner node in the set of child nodes $c_{k_{l-1}}$, and $\theta(0 < \theta < 1)$ and ρ are pre-defined parameter. When 2 dimensional grid type competitive layer is used, the variance of frequency $v_{k_{l-1}}$ can be calculated by using Equation (22.24).

$$v_{k_{l-1}}(\mathcal{S}) = \frac{1}{4} \sum_{j_l \in C_{k_{l-1}}} \left(\frac{p_{k_{l-1}}(\mathcal{S})}{4} - p_{j_l}(\mathcal{S}) \right)^2 \tag{22.24}$$

The variance of frequency $v_{k^{l-1}}(\mathcal{S})$ is the estimator for the gradient of probability density. Thus the proposed method keeps to make the adaptive node arrangement according to the change of probability density. Equation (22.23) is for deleting the node that the firing frequency is extremely low as well as Equation (22.21).

22.4.2. *Gaussian type neighborhood learning model*

The weight updating method of TS-SOM can not be used in dynamic model and in partial NN structure, because the training in upper layer should be completed before the training in lower layer for the preserving topological relation between each layer. For solution of this problem, we propose a grid type neighborhood leaning model that recursively updates the weight vector from root to leaf in each winner node and its neighbor [Yamaguchi, Ichimura & Mackin (2010)]. Unfortunately, the method became slightly worse classification accuracy than TS-SOM from the observation of previous simulation results. The not better classification capability was caused by the error in the fixed grid type neighborhood, because the classification capability of TS-SOM also calculated a not good accuracy than that of basic SOM. In order to improve such a weak performance, we propose Gaussian type neighborhood leaning model. The proposed Gaussian type neighborhood leaning model updates the weight vector using Gaussian kernel without recursive procedure. The position of nodes is calculated on a global coordinate system corresponding to map in similar way as shown in Figure 22.11.

The determination of winner node works in the same way as TS-SOM. The winner node is recursively determined from a root to the child node in the following layer. The tree search algorithm determines the winner node b_{i_l} in each layer by Equation (22.16) and eventually determines the global winner node b_{i_g}, $b_{i_g} \in \mathcal{L}$ in layer g. After the global winner node b_{i_g} is determined, all weight vector in all layer is updated by using Equation (22.25).

$$\mathbf{W}_{j^l}(t+1) = \mathbf{W}_{j^l}(t) + \alpha(t) h_{b_i^g j^l}(g)[\mathbf{x}_i(t) - \mathbf{w}_{j^l}(t)], \tag{22.25}$$

where $h_{b_{i_g} j_l}(g)$ is Gaussian type neighborhood function that the width is defined by the depth of global winner node g. When 2 dimensional grid type competitive layer is used, $h_{b_{i_g} j_l}(g)$ can be defined by Equation (22.26).

$$h_{b_{i_g} j_l}(g) = \exp\left(-\frac{\| \mathbf{r}_{b_{i_g}} - \mathbf{r}_{j_l} \|^2}{2\sigma^2(g)}\right), \tag{22.26}$$

where $\| \mathbf{r}_{b_{i_g}} - \mathbf{r}_{j_l} \|$ is Euclidean distance between b_{i_g}th node and j_lth node on the map of competitive layer, and $\sigma^2(g)$ is a parameter that defines the width of Gaussian distribution. The position of j_lth node $\mathbf{r}_{j_l} = r_{j_l q}, q = \{1, 2\}$ is calculated by Equation (22.27).

$$r_{j_l q} = r_{k_{l-1} q} + 2^{-(l+1)}, l \neq 0, \tag{22.27}$$

where $r_{k_{l-1} q}$ shows the position of parent nodes. The distance of nodes can be calculated when the node is newly added. For fitting the width of Gaussian distribution to grid size in gth layer, $\sigma^2(g)$ is defined by Equation (22.28).

$$\sigma^2(g) = \frac{\delta^2(0)}{2^g}, \tag{22.28}$$

where $\delta^2(g)$ is an initial width of Gaussian distribution.

22.5. Adaptive Tree Structured Clustering

For a classification of large scale of dataset, sequential learning algorithms such as SOM have an advantage compared with batch learning algorithms in computational cost [Kohonen (2000); Kohonen *et al.* (2000); Vesanto (2000)]. Adaptive Tree Structured Clustering (ATSC) [Yamaguchi, Ichimura, & Mackin (2008a,b)] is a framework of Divisive Hierarchical Clustering Algorithm (DHCA) with a function

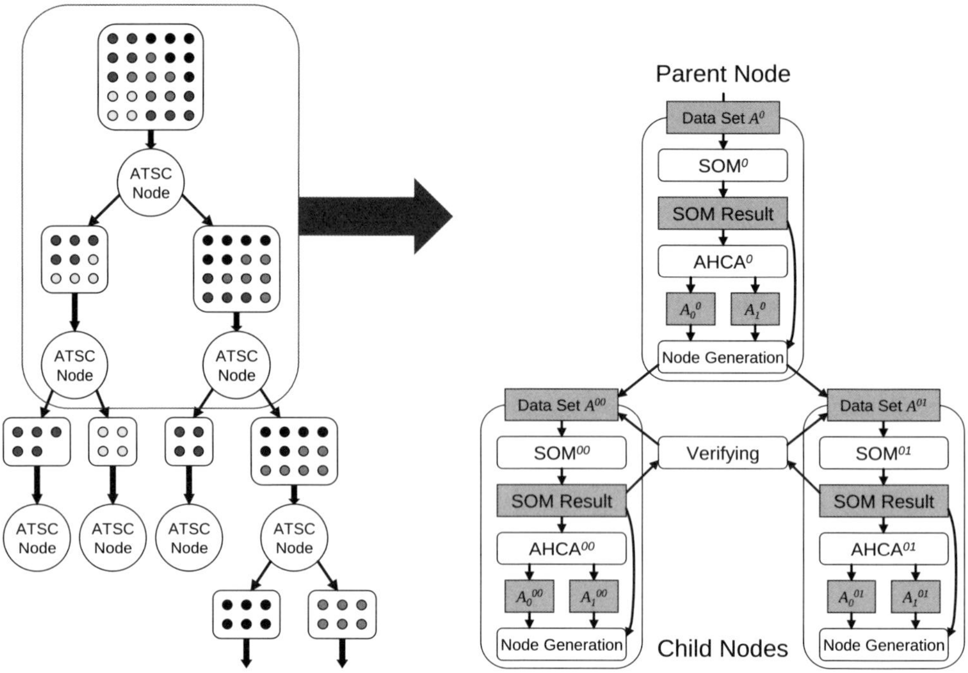

(a) ATSC data division process (b) ATSC node processes

Fig. 22.12. Model for ATSC Tree Structure and ATSC Node.

of sequential processing. Figure 22.12(a) shows a model of recursive data division process. Let $\mathbf{A}$ be the input data set, where $\mathbf{A} = \{\mathbf{x}_i | \mathbf{x}_i \in \mathcal{R}^n\}, i = 1, 2, \ldots, N$. $\mathbf{x}_i$ is ith sample of n dimensional real vector, and N is the number of samples. ATSC recursively divides the dataset $\mathbf{A}$ that is given in an ATSC node into K disjoint clusters A_k, $\{A_k | A_k \in \mathbf{A},\ A_k \neq \phi,\ \cup A_k = \mathbf{A},\ A_k \cap A_{k'} = \phi\}$, $k = \{1, \ldots, K\}, k' = \{1, \ldots, K\}, k \neq k'$. In this paper, we assume that $K = 2$. The criterion for selecting the divided cluster is defined by the change in variance of $\mathbf{A}$ and A_k. When each cluster $\mathbf{A}$ does not satisfy the criterion to select the divided cluster, recursive procedure is terminated. These processes can be considered to the K-ary tree generation process. In other word, the proposed method starts with a root node and recursively create K nodes. The node creation depends on the clustering result in each node. The each node is given a data subset from partitioned data in the parent node. As a result, the number of clusters is obtained as the number of leaves from a generated tree structure.

Figure 22.12(b) shows data division process of ATSC node. Each ATSC node consists of following 4 steps: (1) SOM, (2) Agglomerative Hierarchical Clustering Algorithm (AHCA), (3) Node Generation, and (4) Re-clustering. The criterion for the SOM training termination and re-clustering criterion are determined from the sum squared error between input vector and weight vector. We will discuss about the criteria of SOM training termination and re-clustering, because the criteria are interdependent on both the error of parent node and the divided child nodes.

22.5.1. *AHCA using SOM*

After the training by using SOM, a dataset $\mathbf{A}$ is divided into 2 subsets A_1 and A_2 based on the training results. When the training converges such that the input vector $\mathbf{S}$ is approximated by weight vectors $\mathbf{W}$, the winner unit c and units that have the value of closed weight vector with that of c, form a Voronoi cell [Kohonen (2000); Heskes (1999)]. The relative location of the winner units on the map shows the relationship between input vectors. Therefore clustering using SOM can be decided by the value of weight vector and neighbor of each winner unit.

When the set of winner units is $C = \{w_{c_i}\}, c_i = \{1, \ldots, c_{i_{\max}}\}$ and the set of disjoint subset of C is $B = \{B_k | B_k \in C,\ B_k \neq \phi, B_1 \cup \cdots \cup B_B = C,\ B_k \cap B_{k'} = \phi\}$, $k = \{1, \ldots, |B|\}, k' = \{1, \ldots, |B|\}, k \neq k'$; the clusters B_k is recursively merged using AHCA until the number of elements $|B|$ is equal to K from which the cluster B_k starts with each element as separate cluster. Final 2 subsets A_1 and A_2 are obtained from Equation (22.16) and B where $|B| = 2$. In the process of merging winner units, when the set of winner units in the neighbor of winner unit c_i is $\mathcal{N}_{c_i}$, the c_ith and the c_jth merged winner unit satisfies that $c_j \in \mathcal{N}_{c_i}, c_i \neq c_j, c_i = \{1, \ldots, c_{i_{\max}}\}, c_j = \{1, \ldots, c_{i_{\max}}\}, c_i \neq c_j$.

For distance function $d(B_p, B_q)$, we use following equation, Equations (22.29)–(22.32); single linkage method, complete linkage method, group average method, and Ward's method, respectively.

$$d(B_p, B_q) = \min_{\mathbf{w}_{c_i} \in B_p, \mathbf{w}_{c_j} \in B_q} \| \mathbf{w}_{c_i} - \mathbf{w}_{c_j} \| \tag{22.29}$$

$$d(B_p, B_q) = \max_{\mathbf{w}_{c_i} \in B_p, \mathbf{w}_{c_j} \in B_q} \| \mathbf{w}_{c_i} - \mathbf{w}_{c_j} \| \tag{22.30}$$

$$d(B_p, B_q) = \frac{1}{|B_p||B_q|} \sum_{\mathbf{w}_{c_i} \in B_p} \sum_{\mathbf{w}_{c_j} \in B_q} \| \mathbf{w}_{c_i} - \mathbf{w}_{c_j} \| \tag{22.31}$$

$$\begin{aligned} d(B_p, B_q) &= QE(B_p \cup B_q) - QE(B_p) - QE(B_q) \\ QE(B_k) &= \sum_{\mathbf{w}_{c_i} \in B_k} \| \bar{\mathbf{w}} - \mathbf{w}_{c_i} \|, \end{aligned} \tag{22.32}$$

where $\bar{\mathbf{w}}$ is a centroid of B_k. For updating distance $d(B_p, B_q)$, Lance-Williams update formula [Lance & Williams (1967)] is used. Lance-Williams update formula is defined as follow:

$$\begin{aligned} d(B_k, B_p \cup B_q) = {} & \alpha_1 d(B_k, B_p) + \alpha_2 d(B_k, B_q) + \beta d(B_p, B_q) \\ & + \gamma (d(B_k, B_p) - d(B_k, B_q)), \end{aligned} \tag{22.33}$$

Table 22.1. Parameter of Lance Williams update function.

Method	α_1	α_2	β	γ
Single linkage	0.5	0.5	0	−0.5
Complete linkage	0.5	0.5	0	0.5
Group average	$\frac{\|B_p\|}{\|B_p\|+\|B_q\|}$	$\frac{\|B_p\|}{\|B_p\|+\|B_q\|}$	0	0
Ward	$\frac{\|B_k\|+\|B_p\|}{\|B_k\|+\|B_p\|+\|B_q\|}$	$\frac{\|B_k\|+\|B_p\|}{\|B_k\|+\|B_p\|+\|B_q\|}$	$-\frac{\|B_k\|+\|B_p\|}{\|B_k\|+\|B_p\|+\|B_q\|}$	0

where $k \neq p, q$, and $\alpha_1, \alpha_2, \beta$ and γ are coefficients of each AHCA method defines as shown in Table 22.1.

22.5.2. *Node generation*

For automatically deciding the number of clusters, a node generation criterion is verified after the steps of SOM and AHCA. When the decreasing error ΔE is larger than a threshold θ, 2 new child nodes are created.

$$\Delta E(\mathbf{A}) > \theta E(\mathcal{S}), \tag{22.34}$$

where $\mathcal{S}$ is a dataset input into root ATSC node and $\mathbf{A}$ is a dataset input into current ATSC node. For the error of cluster E, the quantization error is used. Let $E(A_1 \cup A_2)$ be the quantization error of cluster before division, and $E(A_1)$ and $E(A_2)$ be the quantization error of clusters after division. ΔE and the quantization error are defined as follows:

$$\Delta E(A) = E(A_1 \cup A_2) - E(A_1) - E(A_2) \tag{22.35}$$

$$E(A_k) = \sum_{\mathbf{w}_{c_i} \in B_k} \| \bar{\mathbf{w}} - \mathbf{w}_{c_i} \|, \tag{22.36}$$

where $\bar{\mathbf{w}}$ is a centroid of B_k.

The child nodes have separate SOMs that trained with the corresponding subset. Let $Node_b$ be the bth parent ATSC node, $\{Node_k^b\}$, $(k = 1, 2)$ be child ATSC nodes, SOM_b be SOM map in parent node,$\{SOM_k^b\}$, $(k = 1, 2)$ be SOM map in child nodes. The parent node $Node_b$ divides the data set $\mathcal{S}$ into the subsets A_k from the result of SOM_b. The new child nodes $Node_k^b$ divide the data subsets A_k from the result of SOM_k^b.

22.5.3. *Re-clustering*

Let $\{A_k^{(t)}\}$ be K clusters at the training step t. At the training step T when the training termination criteria are satisfied, it is necessary to satisfy Equation (22.37)

$$\lim_{t \to T} A_k^{(t)} = \hat{A}_k \tag{22.37}$$

However, we may meet that the situation where the members of each cluster are not universally yielded by using data sets in which the data exist near the class boundary. This problem expected to be caused by SOM initialization problem that SOM training result is different in each separate runs by the randomly weight initialization and selecting order of training input.

For solving this problem, data set $\mathcal{S}$ is re-clustered by checking the error of each child nodes SOM. The error e_i^k between input vector $\mathbf{x}_i$ in data set $\mathcal{S}$ and weight vector $\mathbf{w}_{c_i}^k$ of corresponding winner node c_i in child node SOM_k^b is expressed as Equation (22.38).

$$e_i^k = \| \mathbf{x}_i - \mathbf{w}_{c_i}^k \| \tag{22.38}$$

Also the error $e_i^{\neg k}$ between input vector $\mathbf{x}_i$ and weight vector $\mathbf{w}_{c_i}^{\neg k}$ of corresponding winner unit c_i can be calculated by Equation (22.38).

If Equation (22.39) is satisfied then $\mathbf{x}_i$ is removed from the current subset A_k and added to subset $A^{\neg k}$:

$$e_i^k > \xi e_i^k > e_i^{\neg k}, \tag{22.39}$$

where ξ is the threshold for re-clustering. After $\mathcal{S}$ is re-clustered, SOM_k^b is retrained.

22.6. Coronary Heart Disease Database [Suka *et al.* (2004)]

22.6.1. *An overview of Framingham Heart Study*

The Framingham Heart Study was the first prospective study of cardiovascular disease. The study began in 1948 under the direction of the National Heart Institute (now known as the National Heart, Lung, and Blood Institute) in the United States. Participants were randomly sampled from the town of Framingham, Massachusetts. Examination of participants has taken place every two years and the cohort has been followed for morbidity and mortality over that time period.

Over the five decades, the Framingham Heart Study has provided valuable insights into the epidemiology and pathophysiology of Coronary heart disease (CHD). The study identified constitutional and environmental factors associated with the development of CHD and established the concept of risk factors and their joint effects.

22.6.2. *Six-year follow-up experience*

The following factors are known as a major risk factor for CHD and used for estimating the individual risk of developing CHD within a ten-year time period [Lucas and AbuHanna (1999)]: old age, smoking, high blood pressure, high cholesterol, and diabetes mellitus. Most of the factors were mentioned first in the reports on six-year follow-up in the Framingham Heart Study [Penny and Frost (1996); Wilson *et al.* (1998)].

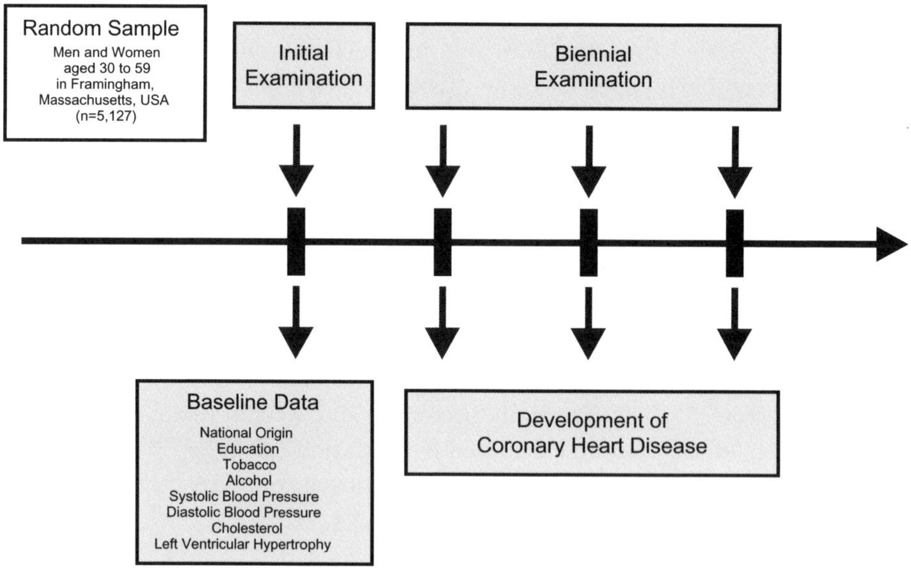

Fig. 22.13. Six-year Follow-up in the Framingham Heart Study.

Figure 22.13 shows the six-year follow-up in the Framingham Heart Study. The cohort consisted of 5,127 men and women aged 30 to 59 who were initially free of CHD. At the study inception in 1948, they received a physical examination, laboratory tests, and a lifestyle interview. The cohort has been followed by means of biennial examinations including a detailed medical history, a physical examination, and laboratory tests. There were 186 persons who developed CHD during the follow-up period. The six-year incidence of CHD in the age group of 45 or older was 9.1% in men and 4.5% in women.

22.6.3. *Database design*

The Coronary heart disease Database(CHD_DB) is designed to reproduce the original data of the Framingham Heart Study. Requisite information is derived from the reports on six-year follow-up in the Framingham Heart Study [Dawber *et al.* (1959); Kannel *et al.* (1961)].

Table 22.2 shows the data items of the CHD_DB. Each of the data sets consists of ten data items: ID, development of CHD, and eight items that were collected from the initial examination (i.e. baseline data). The eight items; ORIGIN, EDUCATE, TOBACCO, ALCHOL, SBP, DBP, TC, and LVH, were examined whether it was associated with the development of CHD over six-year follow-up in the Framingham Heart Study. Using the CHD_DB, researchers will develop a prognostic system that will discriminate between those who developed CHD (CHD cases) and those who did not (Non-CHD cases) on the basis of the eight data items as shown in Figure 22.14.

Table 22.2. Data items of Coronary heart disease database.

Data Item	Name	Value
ID	ID	Sequential Value
Development of CHD	CHD	0 = No; 1 = Yes
National Origin	ORIGIN	0 = Native-born; 1 = Foreign-born
Education	EDUCATE	0 = Grade School; 1 = High School, not graduate; 2 = High School, graduate; 3 = College
Tobacco	TOBACCO	0 = Never; 1 = Stopped; 2 = Cigars or Pipes; 3 = Cigarettes(<20/day); 4 = Cigarettes(≤20/day)
Alcohol	ALCOHOL	Continuous Value(oz/mo)
Systolic Blood Pressure	SBP	Continuous Value (mmHg)
Diastolic Blood Pressure	DBP	Continuous Value (mmHg)
Cholesterol	TC	Continuous Value (mg/dl)
Left Ventricular Hypertrophy	LVH	0 = None; 1 = Definite or Possible

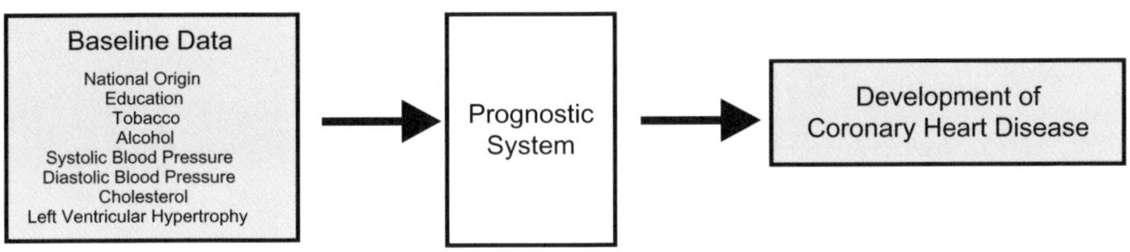

Fig. 22.14. Prognostic System for the Development of CHD.

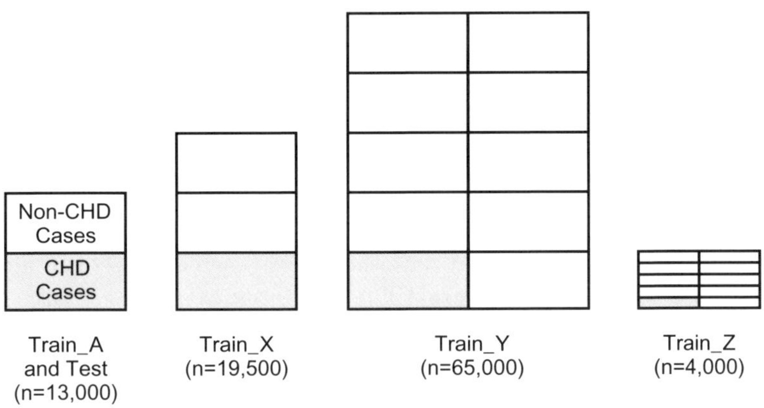

Fig. 22.15. Training and Test Data Sets.

The CHD_DB consists of four training data sets (Train_A, X, Y, and Z) and one testing data set (Test) as shown in Figure 22.15. We previously reported that the developed prognostic system highly depended on the quality of training data set [Suka *et al.* (2004)]. The small proportion of cases to non-cases in the training data set might contribute to a poor learning about the cases, and consequently, the developed prognostic system might have lack of ability to identify potential cases from the population at risk. Therefore, the four training data sets are designed to have different proportions of CHD cases to Non-CHD cases. The number of combination of the eight data items is expected to be about six and a half thousand: 2 (ORIGIN) ×4 (EDUCATE) ×5 (TOBACCO) ×3 (ALCOHOL) ×3 (SBP) ×3 (DBP) ×3 (TC) ×2 (LVH) = 6,480, where the numerical data are counted as three. In the original data of the Framingham Heart Study, the number of complete records was about four thousand and the proportion of CHD cases to Non-CHD cases in men aged 45 or older was about one to nine (i.e. the six-year incidence of CHD was 9.1%). Therefore, Train_A, Train_X, and Train_Y include 6,500 CHD cases and 6,500 (×1), 13,000 (×2), and 585,000 (×9) Non-CHD cases, respectively, while Train_Z approximates to the original data both for the number of total records and the proportion of CHD cases to Non-CHD cases. On the other hand, for the testing dataset, we prepared 6,500 CHD cases and 6,500 Non-CHD cases separately from the training datasets.

22.7. Experimental Results

22.7.1. *Benchmark tests*

First, we describe some experimental results for benchmark tests in the subsection to compare the classification performance before we report the computation results to the medical database in the Section 22.6.

For the evaluation of classification performance, we applied our proposed Tree based SOM to Iris [Fisher (1988)] and Wine [Forina *et al.* (1991)] data set, which are most popular benchmark data set in UCI Machine Learning Repository.

22.7.1.1. *Iris data set*

The Iris data set [Fisher (1988)] is the best known database to be found in the pattern recognition literature. The data set contains 3 kinds of classes with same ratio of class distribution (33.3%); "Iris Setosa", "Iris Versicolour", and "Iris Virginica". One class is linearly separable from the other 2 classes. The latter are not linearly separable from each other. There are 4 kinds of input attributes; "sepal length in cm", "sepal width in cm", "petal length in cm", and "petal width in cm". The data set consists of 50 instances in each, where each class refers to a type of iris plant. Table 22.3 explains the summary of statistics for Iris data set.

Table 22.3. Summary statistics for iris data set.

Class	Min	Max	Mean	SD	Class Correlation
sepal length	4.3	7.9	5.84	0.83	0.7826
sepal width	2.0	4.4	3.05	0.43	−0.4194
petal length	1.0	6.9	3.76	1.76	0.9490
petal width	0.1	2.5	1.20	0.76	0.9565

22.7.1.2. *Wine data set*

The Wine data sets [Forina *et al.* (1991)] are the results of a chemical analysis of wines grown in the same region in Italy but derived from three different cultivates. The analysis determined the quantities of 13 constituents found in each of the three types of wines.

The original data set has around 30 variables, but only the 13 dimensional version data set, in general, are used for benchmark experiments. There are 12 input attributes; "Malic acid", "Ash", "Alcalinity of ash", "Magnesium", "Total phenols", "Flavanoids", "Nonflavanoid phenols", "Proanthocyanins", "Proanthocyanins", "Color intensity", "Hue", "OD280/OD315 of diluted wines", and "Proline". There are 3 classes with different class ratio; 59 samples for class 1, 71 samples for class 2, and 48 samples for class 3. All attributes including class are continuous values.

Table 22.4 shows the summary of data sets descried in the section to compare Iris, Wine, and CHD_DB. The differences of the number of classes, the number of samples, ratio of samples in each class, and attribute information.

22.7.2. *Experimental results for benchmark test*

The proposed SOM with Gaussian neighborhood was compared in Iris data set with following 3 methods; basic SOM with Gaussian neighborhood, TS-SOM with fixed grid neighborhood, and ATSC. For distance functions of ATSC, Single Linkage, Complete Linkage, Group Average, and Ward method are used [Yamaguchi, Noguchi, Ichimura & Mackin (2009)].

The parameters for proposed methods are follows; $\tau = 500$, $T = 50,000$, $\gamma = 0.5$, $\epsilon = 0.01$, $\sigma = 0.01$, $\theta = 0,01$, $\alpha(t) = \alpha_0 \cdot \alpha_{dec}^t$, $\alpha_0 = 0.5$, $\alpha_{dec} = 0.99995$. The size of competitive layer in basic SOM and TS-SOM were determined by prior experiments [Yamaguchi, Ichimura & Mackin (2010)]. These methods are evaluated by the correctly ratio of classification, the quantization error, and the number of nodes in competitive layer. For the correct ratio of classification, each node was labeled by actual class label that has maximum frequency. Table 22.5 shows the average of classification result and of quantization error at 100 runs by comparing with TS-SOM and basic SOM. As a result, the performance of classification by TS-SOM is almost same as that of basic SOM, but the size of TS-SOM was decreased, because the TS-SOM method can delete the redundant nodes.

Table 22.4. Summary for data sets.

Data set	Class	Sample	Ratio	Attributes (Continuous/Discrete)
Iris	3	150	1:1:1	4/0
Wine	3	178	59:71:48	13/0
CHD_DB	2	13,000	1:1	4/4

Table 22.5. Classification performance of benchmark test(TS-SOM).

Method	Classification Accuracy	Quantization Error	Size
Iris Data Set			
Basic SOM	0.959	0.022	25.0
TS-SOM	0.901	0.033	21.0
Wine Data Set			
Basic SOM	0.961	0.204	25.0
TS-SOM	0.952	0.267	18.6

Table 22.6. Classification performance of benchmark test(ATSC).

	Classification Accuracy			Clusters	
Method	Ave.	Var.	Best	Ave.	Var.
Iris Data Set					
Single Linkage	0.600	0.054983	0.853	4.8	15.07
Complete Linkage	0.872	0.001983	0.940	9.9	1.43
Group Average	0.877	0.002866	0.953	9.4	2.48
Ward	0.880	0.002331	0.940	10.8	1.96
Wine Data Set					
Single Linkage	0.583	0.046892	0.938	2.9	3.88
Complete Linkage	0.929	0.000704	0.960	14.6	2.71
Group Average	0.926	0.001043	0.960	14.4	4.27
Ward	0.908	0.000552	0.940	13.0	2.00

On the other hand, for the classification performance of ATSC, classification accuracy and the number of cluster were investigated. The classification accuracy was derived by correct ratio of classification using class labeling. By class labeling, each cluster was labeled by teaching class label with the maximum frequency. Table 22.6 shows the classification result and quantization error by comparing with 4 kinds of distances in Table 22.1. For evaluating the accuracy and universality, the average and the variance are computed from the result of 10 times runs. The number of cluster could vary the universality of given a cluster and a tree structure.

Tables 22.7 and 22.8 show results on CHD_DB that has larger number of samples than previous two data sets. The run time by TS-SOM increased exponentially with the instance size of CHD_DB. The best classification performance in some trials was about 57% and 87 clusters were organized. On the contrary, the result by ATSC shows that classification performance as shown in Table 22.8 was improved similar to previous data sets, compared with the result by Basic SOM as shown in Table 22.7. However the improvement still remains smaller than for the previous two data sets. This problem will be occurred by the fundamental problem of SOM clustering. In ATSC, the boundary of each data division is defined by a set of Voronoi edges where Voroni vertex are weight vectors in the competitive layer. When the number

Table 22.7. Classification performance of CHD_DB(basic SOM).

	Classification Accuracy			Clusters	
Map Size	Ave.	Var.	Best	Ave.	Var.
8 × 5	0.583	0.000005	0.587	40.0	0.0
12 × 8	0.609	0.000024	0.614	96.0	0.0
16 × 10	0.628	0.000027	0.638	160.0	0.0

Table 22.8. Classification performance of CHD_DB(ATSC).

	Classification Accuracy			Clusters	
Method	Ave.	Var.	Best	Ave.	Var.
Single Linkage	0.582	0.003159	0.667	9.7	46.67
Complete Linkage	0.650	0.000116	0.667	29.6	11.37
Group Average	0.650	0.000088	0.665	31.0	14.44
Ward	0.651	0.000102	0.667	28.0	16.89

of weight vectors is insufficient, the boundary cannot be correctly approximated to divide the space of weight vector. Due to this problem, incorrectly classified samples are divided into small clusters in the child nodes. This problem occurs several times even if complex data sets such as CHD_DB are used. For solving such a problem, a method that adaptively changes the number of weight vector is necessary. The idea of solution methods will appear in [Yamaguchi, Noguchi, Ichimura & Mackin (2011)] and some experimental results using same data sets are discussed.

22.7.3. *Classification and knowledge in TS-SOM*

TS-SOM has a hierarchical structure due to the relations in a data set. Therefore, we can see an explicit and an implicit knowledge acquired NN structure from data set. Figure 22.16 shows an example of obtained structure by TS-SOM for CHD_DB data set. We can understand the knowledge in CHD_DB from structure of TS-SOM. The structure is depicted one of experimental results in different configuration of parameters.

However, the classification performance of CHD_DB by TS-SOM was not higher than other proposed methods in spite of long learning time. TS-SOM is mainly working to extract knowledge in the tree structure representation. That is, the larger the number of layer, the worse the classification in the bottom map, because of the less sample cases divided the map. Therefore, in case of the fine tuned structure, the classification performance of SOM does not become good.

Moreover, we suffers from the trade-off between the depth of TS-SOM and the size of map. A rare appearance in a small regions will discover scarce rules, but TS-SOM results in a deeper hierarchical structure of small maps focusing on the hierarchical structure. On the other hand, TS-SOM tends to be a lager maps in

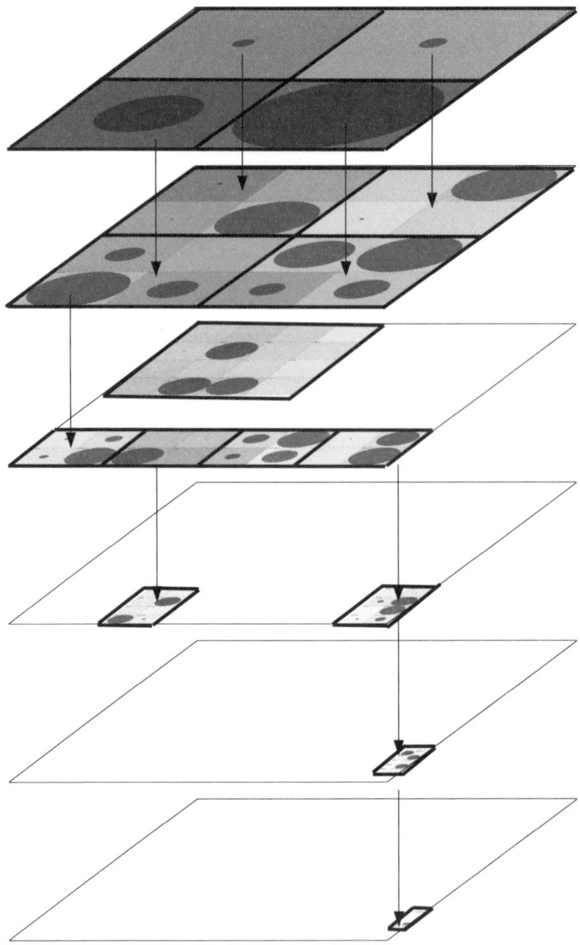

Fig. 22.16. The Knowledge Structure for CHD_DB.

each layer to explain the larger portions of the data set in their flat representation, but providing less hierarchical structure. We developed an interactive growing hierarchical SOM, where the parameters, depth of layer and the size of maps, are interactively determined by human intuition [Ichimura *et al.* (2010)]. The model treats both 2 parameters of depth of TS-SOM and size of map simultaneously. The detailed of the work will be represented in near future.

22.8. Conclusive Discussion

A variant of growing type TS-SOM that put some nodes in the regions that has high gradient in probability density estimation was proposed, in order to adapt the size of competitive layer and the structure for input data space. From the computer simulation, the proposed SOM can provide very close accuracy by training basic

SOM with the sufficient size in competitive layer. In the training of proposed SOM, the computational time tends to increase in the comparison with basic SOM and TS-SOM. The obtained structure by the TS-SOM suggests the implicit or explicit knowledge structure in the data set.

As for the result of CHD_DB, the improvement of classification accuracy is smaller than Iris and wine data sets. It is very important for solution of SOM problem, if the number of weight vectors is insufficient for partitioning complex data sets. The methods that adaptively add weight vectors should be investigated to lead correctly distinguish with the boundary of clusters for complex data sets.

However, our recently developed method, an ensemble ATSC [Yamaguchi, Noguchi, Ichimura & Mackin (2011)] is expected to yield novel clustering results and the generated optimal tree structure. We describe an ensemble ATSC method in which cluster ensemble is applied to each ATSC node for resolving non-universality of ATSC in [Yamaguchi, Noguchi, Ichimura & Mackin (2011)]. The proposed ensemble ATSC method provides the best accuracy and universality in all data sets. Specially, it is important that accuracy becomes good though general ensemble technique can improve only universality. The reason of this improvement causes the reduction of error in earlier partition. The universality in earlier partition affects the performance of ATSC or other DHCA. In the case of other DHCA, ensemble technique is expected to improve universality similar to the case of ATSC.

References

T.R. Dawber, W.B. Kannel, N. Revotskie, J.3rd. Stokes, *et al.* (1959). Some factors associated with the development of coronary heart disease: six year's follow-up experience in the Framingham Study, *Am. J. Public Health*, Vol. **49**, pp. 1349–1356.

W.B. Kannel, T.R. Dawber, A. Kagan, N. Revotskie, *et al.* (1961). Factors of risk in the development of coronary heart disease: six-year follow-up experience, *The Framingham Study. Ann. Intern. Med.*, Vol. **55**, pp. 33–50.

G.N. Lance and W.T. Williams. (1967). A general theory of classificatory sorting strategies, *Computer journal*, Vol. **9** pp. 373–380.

P. Koikkalaien, E. Oja (1990). Self-Organizing Hierarchical Feature Maps, *Proc. of International Joint Conference on Neural Networks*, Vol. **2**, pp. 279–284.

W. Penny and D. Frost (1996) Neural networks in clinical medicine, *Med. Decis. Making*, Vol. **16**, pp. 386–398.

P.W.F. Wilson, R.B. D'Agostino, D. Levy, A.M. Belanger, *et al.* (1998). Prediction of coronary heart disease using risk factor categories, *Circulation*, Vol. **97**, pp. 1837–1847.

T. Heskes (1999). Energy functions for Self-Organizing Maps, *Proc.of Workshop on Self-Organizing Maps (WSOM99)*, pp. 303–315.

P.J. Lucas and A. AbuHanna (1999). Prognostic methods in medicine, *Artif. Intell. Med.*, Vol. **15**, pp. 105–119.

T. Ichimura, S. Oeda, T. Yamashita, and E. Tazaki (1999). A learning method of neural network with lattice architecture, *Journal of Japan society Fuzzy Theory*, Vol. **14**, No. **1**, pp. 28–42.

J. Vesanto and E. Alhoniemi (2000). Clustering of the Self-Organizing Map, *IEEE Transactions on Neural Networks*, Vol. **11**, No. **3**, pp. 586–600.

T. Kohonen (2000). *Self-organizing maps 3rd edition*, Springer series in Information Sciences, Vol. 30, Springer.

T. Kohonen, S. Kaski, K. Lagus, J. Salojarvi, J. Honkela, V. Paatero, and A. Saarela (2000). Self organization of a massive document collection, *IEEE Trans.on Neural Networks*, Vol. **11**, No. **3**, pp. 574–585.

M. Suka, T. Ichimura, and K. Yoshida (2004). Development of coronary heart disease databases, *Proc. of the 8th International Conference on Knowledge-Based Intelligent Information & Engineering Systems (KES2004)*, Vol. **2**, pp. 1081–1088.

M. Suka, S. Oeda, T. Ichimura, K. Yoshida, *et al.* (2004). Comparison of proportional hazard model and neural network models in a real data set of intensive care unit patients, *Proc.of MEDINFO2004(the World Congress on Medical Informatics)*, pp. 741–745.

B. Fritzke. (1994). Growing cell structures — A self-organizing network for unsupervised and supervised learning, *Neural Networks*, Vol. **7**, No. **9**, pp. 1441–1460.

B. Fritzke. (1995). Growing Grid — a self-organizing network with constant neighborhood range and adaptation strength, *Neural Processing Letters*, Vol. **2**, No. **5**, pp. 9–13.

M. Dittenbach, D. Merkl, and A. Rauber. (2000). The Growing Hierarchical Self-Organizing Map, *Proc. of Int. Joint Conf. on Neural Networks*, Vol. **6**, pp. 15–19.

A. Rauber, D. Merkl, and M. Dittenbach. (2002). The Growing Hierarchical Self-Organizing Map: Exploratory Analysis of High-Dimensional Data, *IEEE Trans. on Neural Networks*, Vol. **13**, No. **6**, pp. 1331–1341.

A. Forti and G.L. Foresti. (2006). Growing Hierarchical Tree SOM : An Unsupervised Neural Network with Dynamic Topology, *Neural Networks*, Vol. **19**, No. **10**, pp. 1568–1580.

T. Yamaguchi, T. Ichimura, K.J. Mackin. (2008). Adaptive Tree Structured Clustering Method using Self-Organizing Map, *Proc. of Joint 4th International Conference on Soft Computing and Intelligent Systems and 9th International Symposium on advanced Intelligent Systems*, pp. 1407–1411.

T. Yamaguchi, T. Ichimura, K.J. Mackin. (2008). Analysis using adaptive tree structured clustering method for medical data of patients with coronary heart disease, *Proc. of the 4th International Workshop on Computational Intelligence and Applications*, pp. 139–144.

T. Yamaguchi, Y. Noguchi, T. Ichimura, and K.J. Mackin (2009). Applying Cluster Ensemble to Adaptive Tree Structured Clustering *Proc. of 5th International Workshop on Computational Intelligence & Applications Proceedings : IWCIA 2009*, IEEE SMC Hiroshima Chapter, ISSN:1883-3977, pp. 186–191.

T. Yamaguchi, T. Ichimura, and K.J. Mackin (2010). Adaptive Learning Algorithm in Tree-Structured Self-Organizing Feature Map, *Proc. of SCIS & ISIS 2010, Joint 5th International Conference on Soft Computing and Intelligent Systems and 11th International Symposium on Advanced Intelligent Systems*, pp. 1429–1434.

T. Ichimura, M. Kanda, A. Hara, T. Takahama (2010). A Classification Method by Control of Hierarchy in Growing Hierarchical SOM. *Proc. of the 35th Fuzzy Workshop*, pp. 1–4 (in Japanese).

R. A. Fisher (1988). Iris Data Set, *UCI Machine Learning Repository*, http://archive.ics.uci.edu/ml/datasets/Iris.

M. Forina, *et al.* (1991). Wine Data Set, *UCI Machine Learning Repository*, http://archive.ics.uci.edu/ml/datasets/Wine.

T. Yamaguchi, Y. Noguchi, T. Ichimura, and K.J. Mackin (2011). Cluster ensemble in adaptive tree structured clustering, *International Journal of Knowledge Engineering and Soft Data Paradigms (IJKESDP)*, Special Issue of *Proc. of 5th International Workshop on Computational Intelligence & Applications*, Vol. **3**, No. **1**, pp. 69–84.

Chapter 23

APPROXIMATE PROCESSING IN MEDICAL DIAGNOSIS BY MEANS OF DEDUCTIVE AGENTS

G. Fenza, D. Furno, V. Loia and S. Senatore
Dipartimento di Matematica e Informatica
Università degli Studi di Salerno
gfenza@unisa.it, loia@unisa.it, ssenatore@unisa.it,
domenico.furno@alice.it

This paper presents an integrated environment aimed at supporting the physician in the non-invasive data analysis and medical diagnosis. Goal is to define an architectural model, in order to build an open, distributed medical diagnosis system whose quality of diagnostic feedback relies on the acquisition of medical knowledge by means of a robust theoretical approach.The model combines both fuzzy techniques for the patients' data processing and software agent-driven diagnosis, for providing a support in the interpretation of analyzed data. Moreover, the system exploits service-oriented techniques as the ground infrastructure for managing the medical knowledge and identifying clinical diagnosis and pathologies. The design of the framework constitutes the final result obtained by applying the set of guidelines which provides the worthwhile requirements in the modeling of the diagnostic process.

23.1. Introduction

Traditional medical diagnosis in clinical study relies mainly on the physicians' expertise. During the medical practice, physicians use knowledge acquired from their experience in previous patients' cure. Some shared knowledge can be taken into account to lead the final diagnosis, although it may become difficult for physicians to keep up with all the information gained in the description of pathologies, symptoms and diseases. Computer-assisted technology has been employing as a valid support for physicians in making medical diagnosis or in supporting complex decisions. Knowledge-based systems have been developed exploiting approaches which combine soft computing techniques with the insights of expert physicians to produce tools for supporting health care and diagnosis.

This paper presents a general approach to the modeling of a medical diagnosis framework, describing the main phases and the correlated activities. In particular, this model combines fuzzy techniques and the agent-based paradigm for retrieving the patients' information, assisting the medical evaluation and providing the final diagnosis. The paper introduces general guidelines to profile the main activities in the modeling of medical context and, at the same time, to alleviate the complex design of medical diagnosis systems, by starting from the identification of the system requirements to yield a complete architecture model. Indeed, final result is the medical diagnosis framework, which is demonstrative of the integration of components built on embed the methodologies and technological approaches, presented in the modeling.

The paper is organized as follows. Section 23.2 presents an overview of works in Computational Intelligence domain which deal with the medical diagnosis. Section 23.3 introduces the modeling process through all its phases. Specifically, for each phase, the process model is applied on our case study. Then, Section 23.4 presents the overall architecture system, through the complete working flow and the roles played by all the system components. Finally, Section 23.5 details the achieved experimentation. Conclusions close the paper.

23.2. Related Works

In last years, computational intelligence has been exploited to solve many complex problems in medicine, diagnosis of disease and therapeutic treatments. Thus, a number of medical diagnostic decision systems have been developed to assist physicians and medical professionals. On the other hand, medical knowledge on the symptomatology, diagnosis, and the same clinical experience are sources of imprecision and vagueness. The nature of these data reveals the requirement of treating the uncertainty by means of robust theoretical modeling. Fuzzy techniques guarantee an appropriate coding of this kind of knowledge. Many approaches to medical diagnosis have shown their effective in the computational intelligence domain. Specifically, some applications and works in the main sub-domains are listed as follows.

— *Fuzzy Set theory* has been used for medical expert systems[1,17] and for modeling the diagnostic process[2]; similarly, fuzzy number and compositional rule of inference have been employed for fuzzy decision making in medical diagnosis.[3] These systems exploit linguistic representation of knowledge, which is easy to interpret and understand by physicians and medical professionals. However, the development of an expert system for medical diagnosis is not a trivial task: it requires a labor-intensive and iterative process from medical experts who have to validate step by step the several phases which bring to the final diagnosis.

— *Artificial Neural Networks* have been employed to combine numerical data recorded from sensory measurements, laboratory tests and images with clinical parameters in order to produce prognostic models for specific pathologies.[19,20] The trained neural network functioned as a black box where the human user cannot access or interpret the knowledge stored. This knowledge is difficult for physicians or medical professionals to understand the underlying rationale. In order to alleviate the well-established weaknesses of a neural network fuzzy systems, can be used which make use of fuzzy sets and the associated IF-THEN fuzzy rules to emulate the decisions or reasoning of a human expert, can be used to alleviate the well-established weaknesses of a neural network.[8] In fact, the output values of artificial neural networks have been translated to symbolic/linguistic rules and exploited as a knowledge base for fuzzy expert systems to support the physicians in making decisions for medical diagnosis.[18]

— *Genetic Algorithms* are stochastic optimization methods which apply a transformation process (crossover and mutation operators) based on the concept of natural selection. In medical diagnosis approaches, genetic algorithms often have been exploited in the selection of an optimal number of features[23,24] and then to generate fuzzy rules.[9,21] In general, genetic algorithms are exploited to support pattern recognition problems[25] in various domains, e.g. corporate decision making, text and image recognition or, in our specific study, in medical diagnosis.[22]
— *Clustering Algorithms* are widely exploited in the classification of pathologies and in the automatic exploitation of knowledge from data: usually the collection of symptoms is given as input to fuzzy clustering.[26,27] On the other hand, fuzzy clustering in combination with other approaches such as GA-optimization,[10] fuzzy mining medical rules,[28] neural networks,[29] image segmentation,[4] etc.
— *Hybrid systems* become a good compromise among these methodologies, when the resulting knowledge base is incomplete and difficult for a normal human being to read and apply. To guarantee the accuracy in a decision making system especially in the medical diagnosis area, an integration of symbolic/linguistic processing and numerical computation leads to the definition of hybrid intelligent architectures.[5,6,7] Hybrid architectures tend to be more appropriate in medical applications, because they require accuracy and interpretability. In particular, an incremental learning capability[20] is considered an important attribute for diagnosis systems, in order to guarantee an efficient way to incorporate novel medical knowledge without spending tremendous time in the learning process.

This approach represents a hybrid system, because it has been thought as an integrated environment, which combines theoretical support and technologies in the Computational Intelligence domain. Indeed, in the next sections, the modelling of a diagnosis system is described: the main phases are presented at a high abstraction level, and then, for each one, the corresponding possible implemented approaches are provided too.

23.3. Software Development Model

This section presents a methodic approach for profiling a software development model, describing a variety of tasks or activities that take place during the model design and provides mathematical supports to solving software (and hardware) problems at the requirements, specification and design levels.

The use of formal methods for model design is motivated by the expectation that, performing appropriate mathematical analysis can contribute to the reliability and robustness of a design.

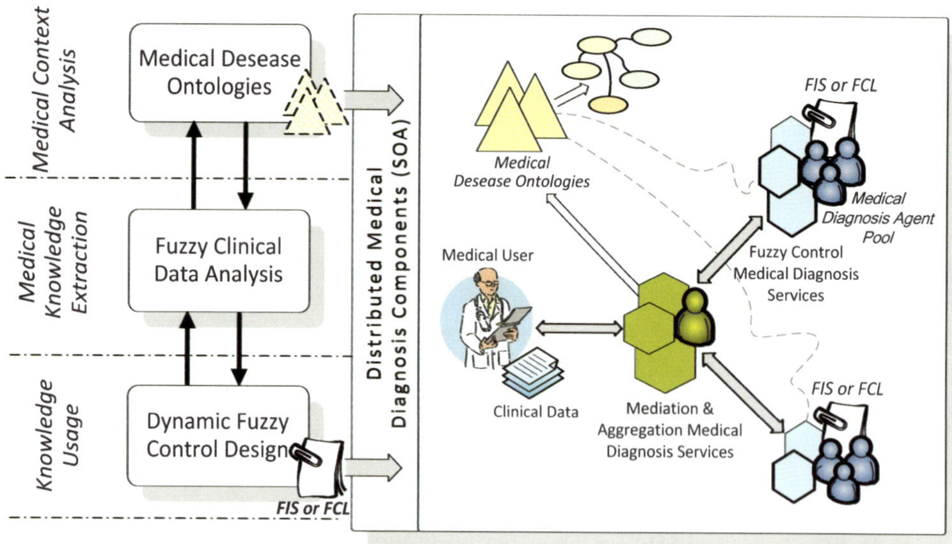

Fig. 23.1. Model design of medical diagnosis system

Figure 23.1 shows a sketched modeling which leads to design a straightforward medical diagnosis system. The working flow proceeds from one phase to another one in a purely sequential manner, with possible feedbacks when further specifications and requirements force to revise and/or readapt one or more phases in cascade. The output of each phase becomes the input of the successive one in order to incrementally reach the final system prototype. In fact, the final result is the development of the system architecture, which closely satisfies the requirements in the diagnostic process. Let us emphasize that this model represents a collection of general guidelines, which have been identified in our study and characterize the main activities required to produce a final acceptable system design. More specifically, as shown in Figure 23.1, our approach is essentially based on three main general activities: *Medical Context Analysis*, *Medical Knowledge Extraction* and *Knowledge Usage Design*.

The outputs of these three phases are analysis tasks, software components and algorithms. Some of them represent learning and reasoning capabilities, "injected" in enhanced components like software agents, or just are integrated together, to produce structured, services-oriented functionalities. In fact, Figure 23.1 shows also the final framework design, defined by the interactions of the *Distributed Medical Diagnosis Components*, coming from the entire modeling activity.

Next subsections introduce each phase: for each one, the main inherent problems and the individual goals are presented; then associated theoretical and methodological approaches are exploited too. Then, in correspondence of each general description of each phase of this model, an application instance is presented, applied to our case study, for an actual, specific medical context.

23.3.1. *Medical Context Analysis*

This phase accomplishes the analysis of the medical context in the reference domain. The pathologies of interest are studied, in order to elicit all the requirements and the inherent design constraints, necessary to get a complete overview of the medical field. The assistance of physicians or expert personnel plays a crucial role in the correct understanding of medical data and the right assessment of clinical measures; fair interpretations of the ranges of normality as well as the critical values are strongly required. Specifically, the clinical folders (describing the patient's clinical condition and history) must be analyzed; thanks to the medical expertise, the relevant data and processable information are elicited and interpreted.

One of main goals of this phase is to share a common language in the reference context which can guarantee a coherent understanding of the involved terms. Thus, a vocabulary composed of medical terminologies (which can be defined by means of semantic relationships) is taken into account.

This phase often needs the human skills (both the medical knowledge and the technical experience), especially when ad-hoc terminologies are required for describing a specific context.

Results of this activity are the harvesting of all the clinical data and terminologies, correctly formatted, according to the ontological definition. The requirements and the specifications are identified too.

23.3.1.1. *The Medical Diseases Ontologies*

Our case study deals with the anaemia, i.e. a well-known pathology, defined as a qualitative or quantitative deficiency of hemoglobin.

As said at the beginning of this section, in order to analysis this medical disease, the relative context and medical terminology have been taken into account.

Ontologies are exploited in this phase: they enable the semantic description of data and provide the inter-relationships among concepts and terms. One or more domain (specific) ontologies describe the medical domain and the pathology. The ontological description of medical concepts represents a way to get common machine-processable understanding, through a formal (OWL-based) language for encoding the set of concepts and the relationships among those concepts.

Last trends outline the ontology-based modeling of knowledge guarantees the immediate usability and future re-adaptability of a system (based on these criteria): potential changes in the system could be at zero efforts in terms of knowledge sharing and understanding.

Available ontologies often describe similar context, but exploit different notations and concepts naming, thus, when it is necessary, an alignment or a merging is required to make uniform the terminology.

Thus, the analyzed context sometimes needs an additional ontological support to describe the involved conceptualization. This situation could require the generation of an ad-hoc ontology. More specifically, an initial task requires the assembling

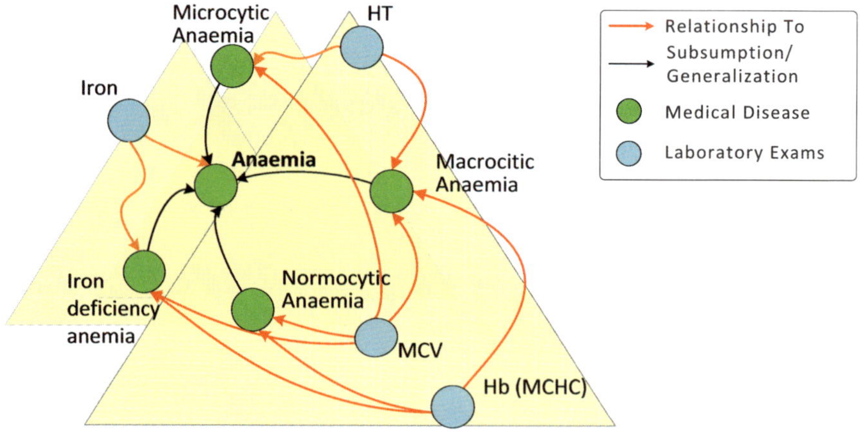

Fig. 23.2. Example of medical diagnosis ontology

of *information resources and expertise* which characterize consistently terms in the domain of interest, according to the OWL language. Then, the principal concepts, their properties, the relationships among the concepts, the linked reference ontologies are identified. Syntactic, logical, and semantic inconsistencies among the ontology elements are corrected too. Finally, the validation of the ontology by domain experts and its commitment to publishing are the conclusive activity of validation.

Figure 23.2 describes an example of conceptualization through the use of ontologies. Herein the anaemia pathology is described by more ontologies. As in the typical representation of an ontology, the concepts are depicted by circles, whereas all the relationships (generalization/specialization or other types) are depicted by arrows. Specifically, this ontological representation exploits two types of relationships: one describes the correlation with other similar or specialized pathologies (also coming from different ontologies); the other one emphasizes the links with clinical concepts (associated, for instance, to some laboratory examinations) which are related to the pathology or that semantically describe the pathology itself.

Let us note the ontologies allow the building of hierarchies of pathologies and emphasize the associated relationships. The ontologies become a valid tool to navigate through the semantic relationships and discover interesting correlations between concepts loosely correlated. The discovered information could be the result of services aggregation and composition activities, providing a more accurate reply to a service invocation with missing data.

23.3.2. *Medical Knowledge Extraction*

This phase consists of those activities that enable the retrieval of all the information useful for a complete and correct data analysis. More specifically, once the context is identified and analyzed, the real world is translated in corresponding mathematical model, in order to manage and process the intrinsic knowledge. This activity foresees a meticulous study of the patients' data, the interpretation of them and the actual mapping to a suitable model, based on a well-defined mathematical formalism. Homogenous data representation is often required to guarantee a right data processing.

Thus, given the mathematical model associated to an actual case study, this phase exploits techniques to collect and arrange the data according to similar characteristics, such as symptoms or patient's health assessments. In medical diagnosis approaches, several methodologies are exploited in the Computational Intelligence domain for the analysis and the elicitation of knowledge: genetic algorithms enable the extraction of relevant features[23,24] and clustering techniques[26,27] are widely employed in the extraction of knowledge automatically from clinical data. Moreover, combined approaches between fuzzy clustering and GA-optimization,[10] fuzzy mining medical rules[28] are demonstrated the responsiveness at this kind of activity.

23.3.2.1. *Fuzzy Clinical Data Analysis*

In our case study, this phase is accomplished by fuzzy clustering approach. More specifically, the main tasks foreseen in this phase are listed as follows:

- *Selection and Representation of clinical data*: data coming from the clinical folders (i.e. a sample of patients) have been exploited to extract features that characterize the medical context of the anaemia. Then, this collection of data enables the building of a relative medical store. Additional factors such as semeiotic elements and further medical evaluations could be considered in the mapping of the contextual model, according to the studied pathology.
- *Data Pre-processing*: collected data are processed through filtering and normalization procedures, in order to get an accurate and homogeneous representation of the information. Critical task in this step is the individuation of the values range associated to each characteristic and the relative mapping in the modeling. Let us note the pre-processing task is often strictly related to the technique of knowledge extraction exploited, because it has to be tailored according to the input specification, required for the appointed technique. For instance, in our case study, a data matrix is required as input to the fuzzy clustering. Specifically a matrix with patient's clinical folders as rows and value of laboratory exams as columns is built.
- *(Fuzzy) Clustering procedure*: once the patients' data are collected and arranged adequately, they are often processed by means of clustering techniques. The Fuzzy C-Means[31] (briefly FCM) algorithm represents a suitable way to partition the initial data collection. In fact, the generated clusters arrange data, according to similar symptoms or data characteristics. FCM is an unsupervised clustering algorithm, which considers a prior fixed number of clusters. In order to give more accurate partitions of data, many researchers have studied the cluster validity criteria.[32,33,34,35] Advanced approaches to (fuzzy) clustering methods

could be exploited, in accord with the nature of the data space and the inherent relationships.

Next subsection provides some technical details about the implementation in the given case study.

23.3.2.2. *Knowledge Extraction Implementation*

The medical context of our case study regards the anaemia pathology; in particular, the initial analysis converges to a selection of some specific characteristics that are representative for the study of this pathology. In fact, after a preliminary argument with physicians, four characteristics are identified: hematocrit (HT or HCT), haemoglobin (HB), red blood cells (RBC) and mean corpuscolar volume (MCV). They are the candidates for being the features of our clustering. In fact, the data matrix to input to the fuzzy clustering is composed of a set of rows, which represents the patients (in our experimentation, about 3000 patients) and a set of columns, which are the our selected features.

Let us note the values of the features did not required to be normalized, because the reference range of them exhibits values in comparable intervals and thus the preprocessing of data was not accomplished.

As said, we exploit the well-know FCM algorithm, particularly useful for flexible data organizationally. It takes as input a data matrix representing the given patients' measures and it tries to get an "optimal" partitioning of the feature space (composed by the data matrix). FCM aims at maximizing the homogeneity, grouping into the same cluster the patterns which are closer. Herein, each pattern is a row of matrix, i.e. some patient examination measures. FCM recognizes spherical "clouds of points" (clusters of patients' exams) in a multi dimensional data space (i.e. data matrix) and each cluster is represented by its center point (prototype or centroid). The function minimizes the weighted sum of the distances between data points x and the centroid v, according to this formula:

$$V(U) = \sum_{i=1}^{c}\sum_{j=1}^{n} uij^m \|x_j - v_i\|^2$$

where $c \geq 2$ is the number of clusters, $u_{i,j} \in [0, 1]$ is the membership degree of x_j in the i-th cluster and $m > 1$ controls the quantity of fuzziness in the classification process (usually in our application $m = 2$).

In this approach, each row of data matrix is a vector representing the examination measures (viz. four characteristics identified before) of each patient's folder $x = (x1, x2, x3, x4)$. After the FCM execution, data partitions are returned, in a prior fixed number c of clusters.

In order to give an intuitive idea of the clustering approach, let us set $c = 3$; applying the clustering algorithm, the final partition matrix shows three clusters.

Let it emerge that a cluster is composed of all the patients, whose parameters HB and HT are quite low; thus the presumable pathology could be normocitic anaemia (although the MCV is often tending to high); another cluster presents normal values for almost all the four features (sometimes the HT is low-medium, indeed the range is about [40, 42]): this cluster could be considered those one of the healthy patients.

The last cluster contains all the parameters out of the range of normality, usually the values are low; just sometimes the MCV parameter reports values closer to its normality range. This cluster could identify patients whose diagnosis is "iron-deficiency microcytic anaemia".

Additional details about this phase are provided in Section 23.5 in order to evidence the main problems seen in the fuzzy data analysis.

23.3.3. *Knowledge Usage*

This phase carries out a further processing on data, by means of intelligent reasoning and inference engine, aimed at producing possible pathology diagnosis. In computational intelligence literature, fuzzy rules have been often exploited in different approaches for decoding the knowledge in a way closer to the human understanding. In neural network[8] approach, *if-then* fuzzy rules enable the emulation of the decisions or reasoning of a human expert, whereas fuzzy number and compositional rules of inference have been employed for fuzzy decision making in medical diagnosis.[3,18]

23.3.3.1. *Dynamic Fuzzy Control Design*

The fuzzy if-then rules provide an interpretation of the clusters, closer to the human understanding and easier to evaluate, with respect to the values range of each feature.

Outputs of previous phase are prototypes and a partitioning matrix of the data patterns. This matrix contains the memberships of the elements of the given data set in each of the c clusters.

Generally, the linear interpolation of memberships is computed; in order to easily assign linguistic label, the membership function is computed by linear interpolation of the projected membership to each axis (according to the n-dimension space), for each cluster.

More specifically, exploiting the cylindrical extension technique[30] (a projection-based method of n-dimensional argument vector), the generic i-th fuzzy cluster can be described by a fuzzy rule, where each features is described through a fuzzy set. Each fuzzy cluster K_i with $i = 1, \dots, c$ can be represented through n functions $A_{i1}, A_{i2}, \dots, A_{\text{in}}$, obtained by the projected and interpolated memberships.

The generated rules are subject to a proper validation by expert physicians and then exploited in assistance of doctors' medical decision or patients' self caring

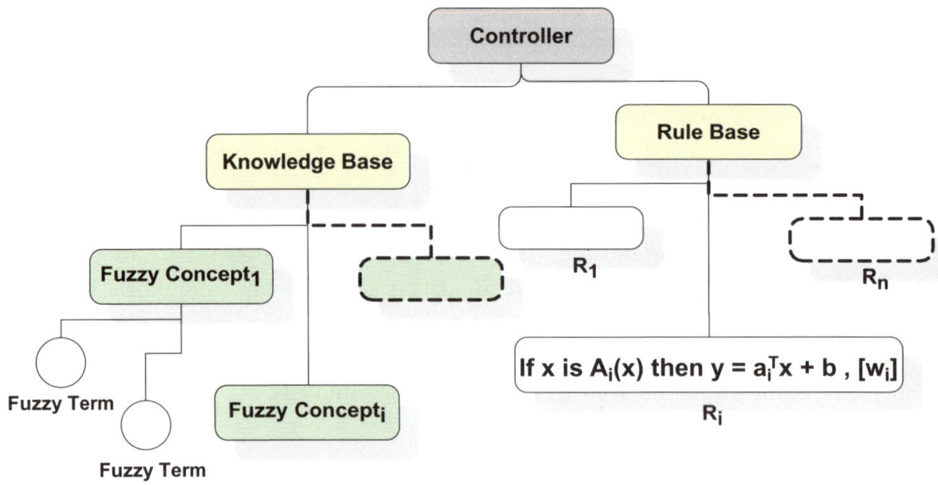

Fig. 23.3. Representation of a FIS/FCL files structure

demand. This way, the fuzzy rules are exploited to simulate the diagnosis process. In particular, these data are interpretable easily, because are described through linguistic (fuzzy) tools that make them more comfortable to human interpretation and understanding.

Last result of this phase is the serialization of the rules exploiting standard representations that enable the flexible translation and use of the interpreted data. In fact, these fuzzy rules are serialized by means of known languages, such as FIS employed by Matlab Simulink or FCL (Fuzzy Control Language).

In fact, a FCL/FIS file describes a fuzzy controller, with the input and output parameters, specific rules and declarations (see Figure 23.3). In particular, the controller contains a knowledge base, composed of fuzzy concepts and terms (defined through fuzzy sets) and a fuzzy rules base.

Due to the flexibility of this model, possible modifications of the knowledge base in the fuzzy controls correspond to a change of FIS/FCL files.

Let us note the use of fuzzy control in the design of the architectural model satisfies the requirements of a technical interoperability.

Just to provide an example, a fuzzy rule, supposed to be in the form of Takagi–Sugeno model[36] is described as follows:

$$\mathrm{R}_i: \textbf{If } x_1 \text{ is } A_{i1} \text{ AND } x_2 \text{ is } A_{i2} \text{ AND} \ldots \text{AND } x_n \text{ is } A_{\mathrm{in}} \textbf{ then } y\mathrm{i} = \mathrm{a}_{i1}\, x_1 + \mathrm{a}_{i2}\, x_{2+\cdots+}\mathrm{a}_{\mathrm{in}}\, x_n + \mathrm{b}_i, \text{ with weight } [\mathrm{w}_i]\ \mathrm{i} = 1, \ldots, c$$

where R_i is the i-th rule, $x_1, x_2, \ldots, x_n$ are the input variables, $\mathrm{A}_{i1}, \mathrm{A}_{i2}, \ldots, A_{\mathrm{in}}$ are the fuzzy sets assigned to corresponding input variables, variable y_i represents the value of the i-th rule output, and $a_i = [a_{i1}, a_{i2}, \ldots, a_{\mathrm{in}}]$ and b_i are parameters of the consequent function. As said before, the antecedent fuzzy sets are usually achieved by projecting the membership degrees in the fuzzy partitions matrix U onto the axes of individual antecedent variable x_j to obtain a point-wise defined antecedent fuzzy set A_{ij}. Then we approximate it by a normal bell-shaped membership function[37]; indeed the uniform structure of bell-shaped function is suitable for identification, analysis and optimisation. Hence, each fuzzy set A_{ij} is calculated from the sampled input data $x_j = [x_{1j}, \ldots, x_{nj}]^T$ and the fuzzy partition matrix $U = [u_{ij}]$ as follows:

$$A_{ij}(z) = \exp\left\{-\frac{1}{2}\left(\frac{z-\alpha_{iq}}{\beta_{iq}}\right)^2\right\}$$

where

$$\alpha_{iq}=\frac{\sum_{j=1}^{n} u_{ij}x_{qj}}{\sum_{j=1}^{n} u_{ij}} \quad \text{and} \quad \beta_{ij} = \sqrt{\frac{\sum_{j=1}^{n} u_{ij}(x_{qj}-\alpha_{iq})^2}{\sum_{j=1}^{n} u_{ij}}}$$

represent the mean and standard deviation of the bell-shaped membership function, respectively.

On the other hand, the computation of parameters $a_i = [a_{i1}, a_{i2}, \ldots, a_{\text{in}}]^T$ and b_i in the consequent part requires a deepened study, in order to evaluate the firing strength w_i of the i-th rule and the value of y_i.[38]

Goal is to evaluate the final output of the TS fuzzy model y_i for an arbitrary x_j input sample, which is calculated using the following formula:[38]

$$\hat{y}_k = \sum_{i=1}^{c}[w_i(x_k)(x_k a_i + b_i)]$$

with $k = 1, 2, 3, \ldots, n$ and where $w_i(x_k)$ represent the firing strength of i-th rule for the k-th pattern. Thus, in order to compute the parameters of consequent part of a rule, a regression model of the compact form $\hat{Y} = X'[a_i^T, b_i] + \varepsilon$ is evaluated: herein ε is the approximation error, X' assumes the form: $X' = [w_1, x_k^T, 1, w_2 x_k^T 1, \ldots, w_c x_k^T, 1]$. Vectors $[a_i^T, b_i]$ are determined using the least-squares method.[38]

Just to give a simple example of a fuzzy rule, in Figure 23.4, let us represent the third cluster, defined in the previous section. Besides the compact expression of the fuzzy rule (on the top of Figure 23.4), each feature is described by a fuzzy set, through a linguistic label which helps the interpretation of the relative fuzzy set. In this example all the values in the antecedent part of rule are out of the range of normality, consequently the identified cluster reveals a specific type of the anemia pathology.

23.4. Distributed Medical Diagnosis (SOA)

One of the goals of this approach is to design an open, distributed medical diagnosis system, able to provide diagnosis activities on heterogeneous medical diseases in the Web context. Existing diagnostic systems usually work as stand alone computer or

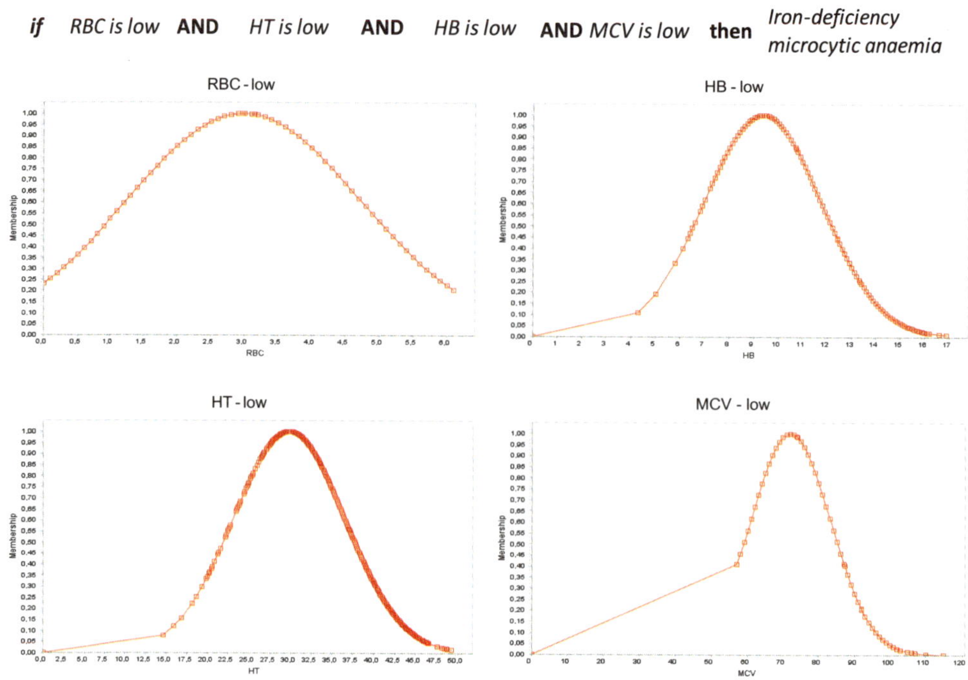

Fig. 23.4. Fuzzy rule generated by the third cluster

are available in a local network environment, without any interaction and accessibility from the Web network.[11]

This framework design integrates the studied techniques and methodologies, in order to yield a complete medical diagnosis model; moreover it provides a service-oriented entry point to enable medical request. To clarify the model, the following subsections provide additional details about the main components of the architecture. Finally, the whole flow of information and the main interactions among these components in the medical diagnosis system are described.

23.4.1. *Medical Diagnosis Services*

Web services represent collection of technologies that integrate business processes and services over the Web using standard XML-based protocols (SOAP, WSDL, UDDI); they can be discovered and joined dynamically over the Web to any Web-enabled client or device.

Our model adopts the web services as entry point to enable medical diagnosis query. In particular, basic unit of communication is a message, rather than an operation, thus the model defines a service-oriented architecture (SOA).

Agents support the web services activity, exploiting a standard mechanism as UDDI for their discovery. In fact, in this approach, the *medical diagnosis service register agent* provides a web service entry point to access to the medical diagnosis services. Theoretically, medical diagnosis services are based on fuzzy models

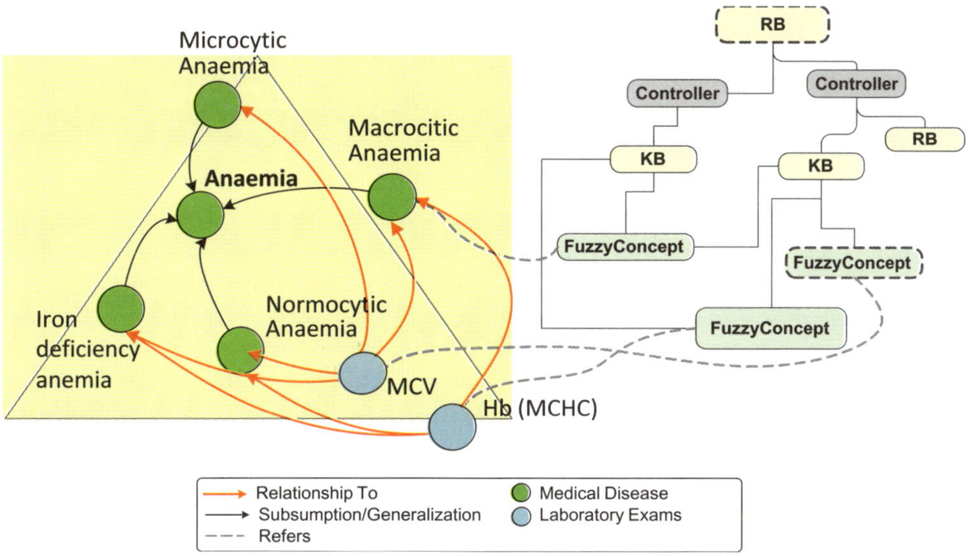

Fig. 23.5. Example of medical diagnosis ontology

which adopt inference procedures for diagnosis. Technically, the services offered over HTTP protocol represent fuzzy control services and SOAP messages contain the information about input and output controller.

Due to the nature of service-oriented architecture, in order to manage the set of medical diagnosis services, a model of mediation and aggregation should be designed too.

Essentially, two types of services could be defined as follows:

— *Fuzzy Control Medical Diagnosis Service*: this service is exposed by *medical diagnosis agents* which realize the inference task. Technically this kind of service represents a web oriented interface to the fuzzy controller.
— *Mediation & Aggregation Medical Diagnosis Service*: this service is exposed by *medical diagnosis service register agent*; this service guarantees the mediation among fuzzy control services in order to invoke and aggregate the computed results. This task is often accomplished by means of *medical diagnosis ontologies.*

Ontologies, beyond to allow the definition of medical concepts (i.e. examinations, observation, therapies, and possible relationships with other diseases), guarantee semantic interoperability among distributed medical diagnosis systems.

Using ontologies to support fuzzy controls specification enables the semantic definition of medical diagnosis services. In other words, in order to semantically support the fuzzy control activity, for each fuzzy concept in the knowledge base, an ontology concept specification exists in the model. This way, a shared general conceptualization reconciles different terminologies related to the same concept, although they are modeled differently in the distributed fuzzy knowledge bases. Thus, through ontology conceptualization these concepts share general definition.

Furthermore, using the ontological terms in the agent interactions, possible misunderstandings in the communication language are avoided and, at the same time, the right service is selected when two terms match. It can help the medical diagnosis service register agent to find the matching medical diagnosis services according to the user request.

In fact, the goal of dynamic publishing and discovering of web services, driven by agents and supported by ontologies, is reached by exploiting OWL-S languages [15] as a qualified OWL-based support for semantic web services. OWL-S can be mapped on UDDI [16] and enables the "injection" of semantic discovery capabilities in the *medical diagnosis service register agent* [13].

23.4.2. *Medical Diagnosis Agents*

The FIS or FCL serializations of fuzzy controls are injected in the *medical diagnosis agents*. Thus, for each set of fuzzy rules, there is a corresponding *medical diagnosis agent pool* which achieves a fuzzy inference service, according to the description of relative pathology in the rules.

More specifically, these agents are able to interpret fuzzy control serialization, to perform sensing (data acquisition and fuzzification), to work as actuators (defuzzification and final results) and to achieve fuzzy inference activities.

The fuzzy inference activities are published in the form of web services to the medical service register agent (*Fuzzy Control Medical Diagnosis Service*).

In particular, the design of agent behavior is based on the fuzzy control model serialized by means of standard formalism. This way, the agents reflect possible tuning of the controllers and the diagnosis model can be wholly sensitive to the environment change. More fuzzy controls can be associated to the same *medical diagnosis agent pool*.

23.4.3. *Medical Diagnosis Service Register Agent*

The *medical service register agent* supports the set of diagnosis services in order to enable a web service oriented interface between *user* (medical expert) and agent platform. In particular this agent achieves activities of mediation, indexing, brokerage, aggregation, discovery and matchmaking of semantic web services. A single instance of this agent class exists in the system.

The *medical service register agent* interprets the request coming from the user, then interacts with the *medical diagnosis agent pool* for discovering the web service (or an aggregation of them) which better replies to the service request (see Figure 23.6). When an aggregation of web service is required, the agent consults the ontologies, in order to get the matching of input/output conceptual terms. Let us consider the scenario in the Figure 23.6: the specified request is composed of patient data about *iron* and *MCV* values. In this case, more *fuzzy control medical diagnosis services* are associated to these values. In particular, some of these

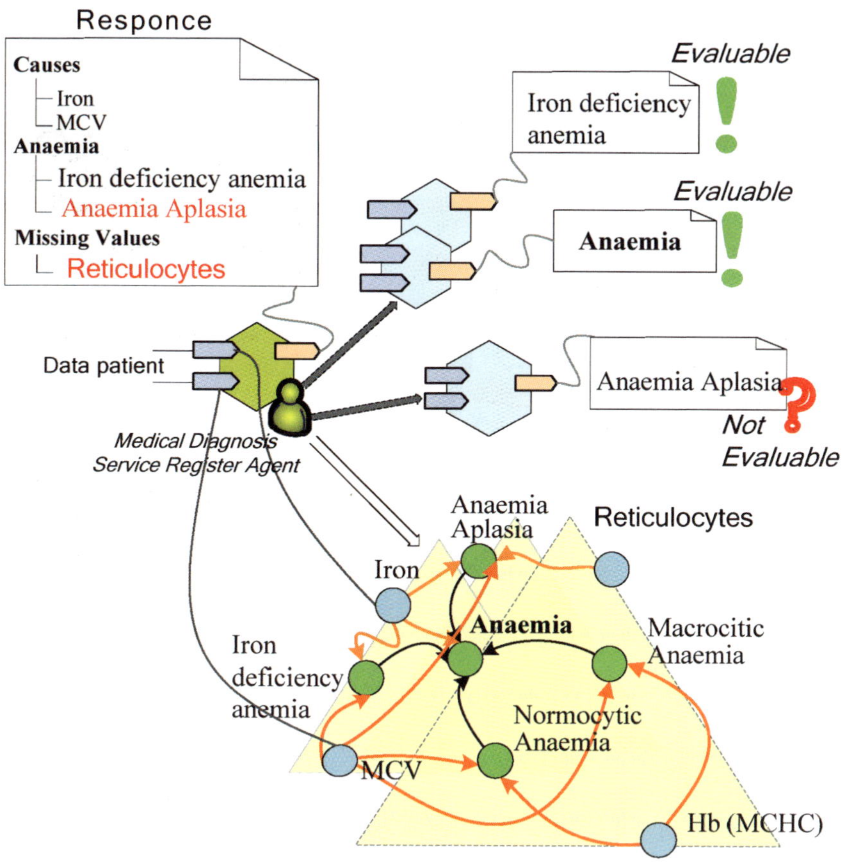

Fig. 23.6. Query/response activities of medical diagnosis service register agent

services can be directly invoked; some others missing values (in figure the anaemia aplasia cannot be evaluated because its associated input reticulocytes is missing in the query). Thanks to the conceptualization, the *medical diagnosis service register agent* could infer possible missing values: in this case, it can suggest to the user (i.e. physician) specific examination regarding the missing parameter (in the example, reticulocytes).

23.4.4. *Workflow of the system architecture*

Figure 23.7 shows the logical architecture depicting the whole working flow. The user request is filtered by the *medical service register agent.* This agent works as a mediator between the external environment (user, physician) and the agents platform. It interprets the query and starts an activity of discovery and selection of the services that are related to the incoming values. Then, it invokes fuzzy control services. In order to perform the fuzzy control service, the *medical diagnosis agents* read the FIS or FCL files and infer possible replies (in terms of pathology associated

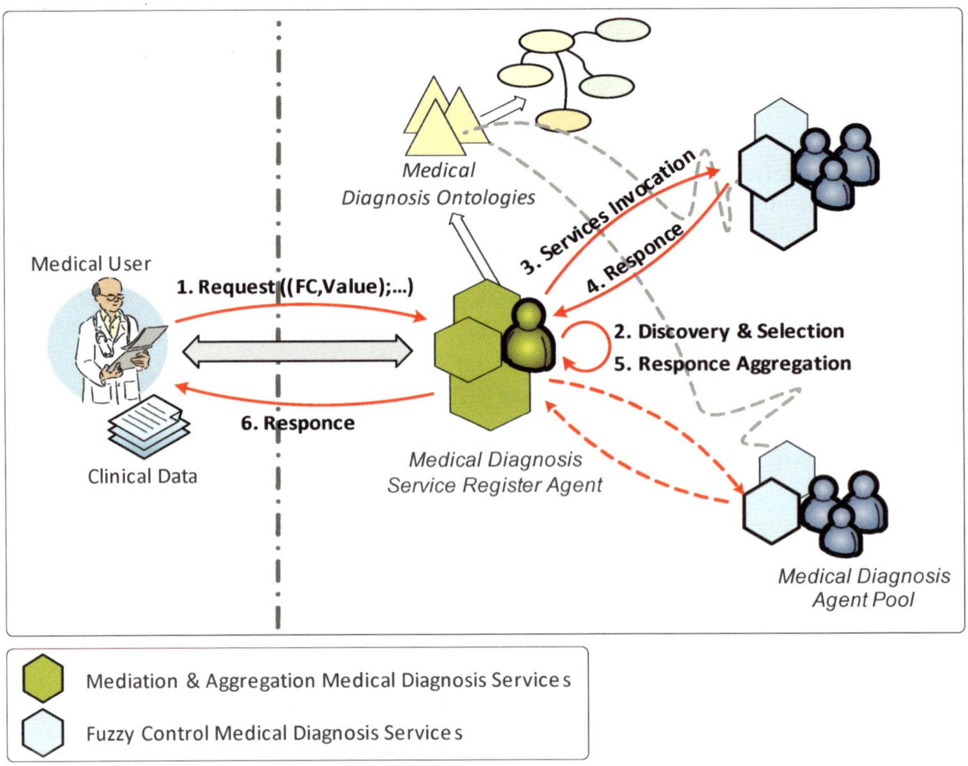

Fig. 23.7. Working flow of the medical diagnosis system

to the input patient data). Then, the *medical diagnosis agents* send the replies to the medical service register agent. This agent collects incoming answers and, when it is possible, achieves an aggregation activity. The final results are returned as output reply associated to the input request. *Medical diagnosis ontologies* enable semantic interoperability among distributed *medical diagnosis agents*, which are specialized for different pathologies. The ontologies assure the correctness of retrieval of agent services and support the aggregation and inference of incomplete diagnostic results. In fact, more than one resulting outputs could be collected by multiple *medical diagnosis agents*, thus final diagnosis is often not unique or only partial. Aggregation of results can help to discover further results (for instance, additional examinations, etc.) which could associate to a certain pathology.

23.5. Further Remarks on the Cases Study

The anaemia case study has been exploited to provide a sketched idea about the main activities of each phase of our model design. Further details during the actual experimentation of this concrete instance are described herein. Let us consider the main phases of our modeling.

Medical Context Analysis: the examined data sample comes from the pathologies related to the renal disorder, thus the patients are usually men. In our reference context, the range and the units of measurements for the four characteristics have been established, according to the medical knowledge, in the following manner:

— *haemoglobin (HB)*: haemoglobin concentration measurement is a common blood test, usually as part of a complete blood count. Results are reported in g/L, g/dL or mol/L. Normal levels (for men) are 13.5 to 16.5 g/dl (i.e. grams per deciliter);

— *hematocrit (HT or HTC)*: is the proportion of blood volume, occupied by red blood cells. It is normally about 46% for men.

— *mean corpuscular volume (MCV)*: is a measure of the average red blood cell volume. The normal range is typically 80–100 fL (femtolitres, viz. 10^{-18} m^3).

— *red blood cells (RBC)*: the number of red blood cells per volume of blood; it is reported in either millions in a microliter or millions in a liter of blood (uL). The ranges for a normal RBC count are: 4.7 to 6.1 million/uL for men.

Thanks to the physicians' support, a detailed values classification for each characteristic is given too. In other words, linguistic labels have been associated to specific range of each measurement.

Fuzzy Clinical Data Analysis: in the previous case study, we have executed the clustering algorithm with the number of cluster $c = 3$.

In order to estimate the number of clusters which better reflects the arrangement in the data space, validity criteria can be exploited to validate the resulting partitioning. Thus, we focus on the Xie-Beni index, which is defined as:

$$V_{xb}(U, V; X) = \frac{\sum_{i=1}^{c} \sum_{j=1}^{n} \mu_{ij}^2 \|x_j - v_i\|^2}{n \min_{i \neq j} \|v_i - v_j\|^2}$$

This index provides more reliable response when the number of clusters is wide and fuzzy weighting exponent allows many choices, but one of its drawback is that it monotonically decreases when the number of clusters gets very large and close to data points. Thus, sometimes the use of a punishing function is recommended on the Xie-Beni index. In our approach, we have exploited this method, obtaining the optimal, but trivial number of cluster $c = 2$. In other words, in this case, the Xie-Beni index allows us to elicit just two big clusters: the healthy people and sick people clusters.

In order to get a meaningful interpretation to the clustering activity, physicians support have allowed us to set the number of cluster $c = 10$. This choice identifies nine classes of the anaemia pathology plus the healthy people class.

After the clustering execution, the partitioning produces the expected clusters, interpreted as follows.

Cluster 0: the parameters HB, HT and MCV assume low values. This cluster could be identified as the class of iron-deficiency microcytic anaemia

Cluster 1: the parameters HB, HT assume low values whereas MCV are normal, tending to high. This cluster could correspond to the class of macrocitic anaemia

Cluster 2: the parameters HB, HT, MCV assume normal, whereas RCB is normal, tending to high. This cluster could correspond to the class of healthy people.

Cluster 3: the parameters HB and HT assume low-normal values, whereas MCV tends normal. This cluster is closer to the class of normocitic anaemia.

Cluster 4: the parameter HB is very low whereas HT is low and MCV and RCB assume normal values. This cluster could correspond to the class of normocitic anaemia.

Cluster 5: the parameter HB is very low-normal values whereas HT is low and MCV and RCB assume normal values. This cluster is similar to previous cluster, i.e. the class of normocitic anaemia.

Cluster 6: the parameters HB and HT are low whereas MCV and RCB assume normal values. This cluster is similar to the previous two clusters, i.e. the class of normocitic anaemia.

Cluster 7: the parameters HB, HT and MCV assume low values, whereas RCB assume normal values, tending to high. This cluster could be identified as the class of thalassemic anaemia.

Cluster 8: the parameter HB assumes low values, tending to the normality, HT is low whereas the parameters MCV assumes normal values. This cluster is closer to the class of normocitic anaemia.

Cluster 9: the parameters HB, HT, MCV and RCB assume normal. This cluster could correspond to the class of healthy people.

The interpretation of clustering reveals that there are some clusters which represent similar pathologies. By the analysis of our reference patients' data, it results that some pathologies defined by the physicians for these parameters, are not represented by the sample. For instance, patients with high value of the parameters HT and HB (i.e. erythrocytosis) do not appear.

In particular, in this case study the final rules generated by this clustering presents overlapping fuzzy sets that describe almost the same region in the domain of some model variable, as shown in Figure 23.8.

In such cases, the fuzzy model uses more fuzzy sets than necessary since these fuzzy sets represent more or less the same concept. It is desirable to avoid some degree of redundancy and thus unnecessary complexity in the acquiring of fuzzy models from data. A similarity measure for identifying similar fuzzy sets has been exploited in order to replace these ones by a common fuzzy set which is representative for the original ones. Thus, the common fuzzy set is created by merging the similar fuzzy sets.

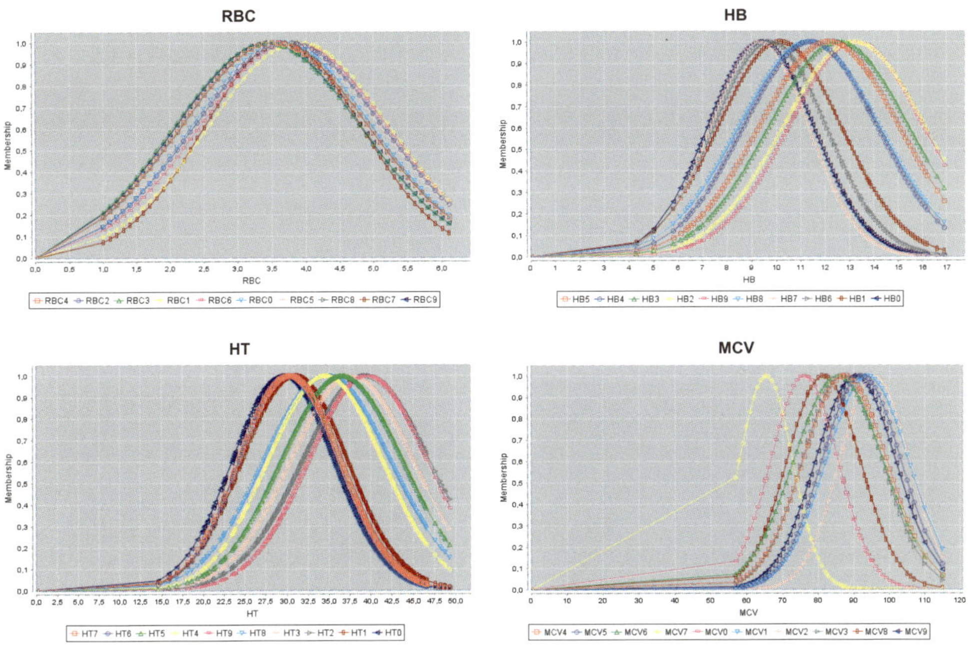

Fig. 23.8. Tendency of the features for the ten clusters

The definition of similarity between fuzzy sets[39] is the degree to which the fuzzy sets are equal. This definition is related to the concepts represented by the fuzzy sets. Overlapping fuzzy sets should have a similarity value greater than zero.

The similarity measure is based on the set-theoretic operations of intersection and union, to determine the similarity between fuzzy sets. Thus, if A and B are two fuzzy sets, the similarity between them S(A, B) is:

$$\mathrm{S(A,B)} = \frac{|A \cap B|}{|A \cup B|}$$

where $|\cdot|$ denotes the cardinality of a set, and the $\cap$ and $\cup$ operators represent the intersection and union respectively.

If the redundancy in the model is high, merging similar fuzzy sets might result in reduction of the number of rules from the rule base.

Rule base simplification algorithm[39] has been exploited to for merging fuzzy sets that are similar to one another. The algorithm iteratively merges similar fuzzy sets. In each iteration, the similarity between all pairs of fuzzy sets for each variable is considered. The pair of fuzzy sets having the highest similarity which is greater than a given threshold (i.e. $\mathrm{S(A,B)} > \lambda$) is merged to create a new fuzzy set. Then, the rule base is updated by substituting the new fuzzy set for the fuzzy sets merged to create it. The algorithm continues to evaluate the similarities in the updated rule base. The rule base is updated for rules with equal premise parts.

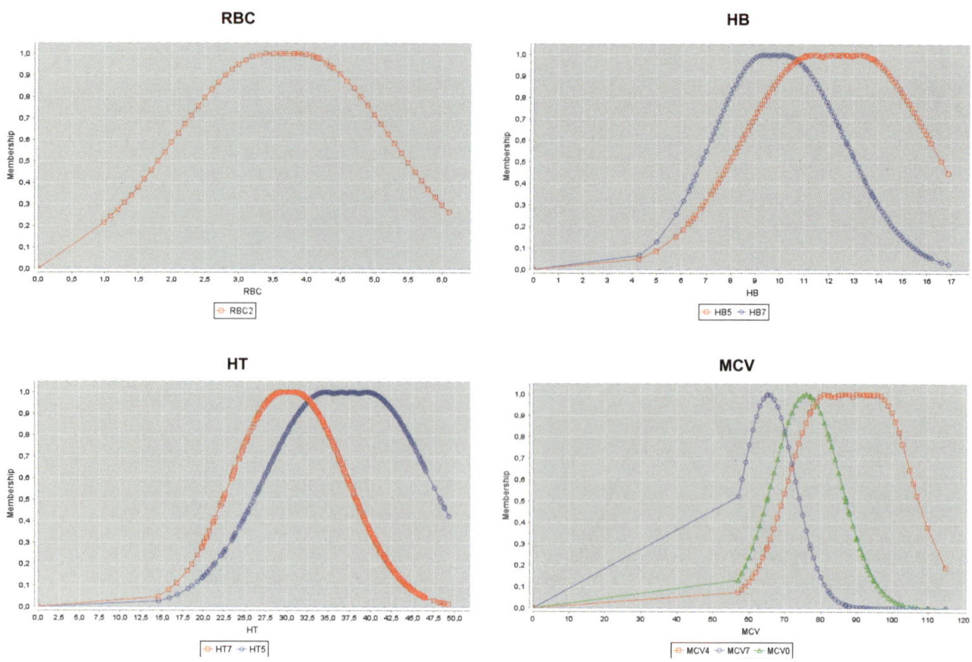

Fig. 23.9. Fuzzy sets relative to the Rule Base Simplification Algorithm

Applying this algorithm to our case study (with $\lambda = 0.7$), a set of only four rules are obtained.

Figure 23.9 shows the final fuzzy sets for all the features. As shown in figure, the merging of all the fuzzy sets relative to the parameter RCB generates just one final fuzzy set; thus it does not influence the characterization of the rules.

In summary the fuzzy rules are described as follows.

Rule 1: If HB is *very low* AND MCV is *low* ($\geq$ *normal)* AND HT is l*ow* then normocitic anaemia

Rule 2: If HB is *very low* AND MCV is (*normal* $\geq$) high AND HT is low then macrocitic anaemia

Rule 3: If HB is *very low* AND MCV is *very low* AND HT is *low* then iron-deficiency microcytic anaemia

Rule 4: If HB is *low* ($\geq$ *normal*) AND MCV is *normal* ($\geq$ *high*) AND HT is *low* ($\geq$ *normal*) then healthy people class

These rules identify four classes of pathologies, which reflect some clusters generated in our case study. Some linguistic labels in the rules have additional details in the parenthesis, just to emphasize the relative fuzzy sets present borderline values. For instance, MCV is *low* ($\geq$ *normal*) describes a fuzzy set for the parameter MCV, generally with “low” values, even though they are closer to normal range.

Because the parameter RCB does not revel additional information in the rules, the Cluster 7 (HB, HT and MCV assume low values and RCB is normal/high) in our experimentation does not appear. Besides, by studying the Cluster 7, the RCB is almost always normal, closer to the upper bound of the normality range.

23.5.1. *Additional Results*

In order to achieve a validation of our approach, we consider a known case study, the Wisconsin Breast Cancer benchmark, downloaded by the well-known UCI Repository of Machine Learning Databases (http://www.ics.uci.edu). Starting by this benchmark, let us analyze the phases describe in Section II. Each phase of our modeling process is reported and analyzed.

Medical Context Analysis: in this example, the context of pathology is well-delineated, i.e. breast cancer. The symptomatology, the clinical data and the related medical examination are not discussed in this example, because they are already given and correctly translated in the dataset provided. For the well-defined nature of the study case, the ontological relationships are not necessary for the characterization of the medical context.

Fuzzy Data Analysis: as said, the Wisconsin Breast Cancer, is a well-known benchmark, often used to test the effectiveness of classification and rule extraction algorithms [11]. This classification defines two classes of cancers: benign and malignant cancers, through a features set composed of nine measurements (with values in the range [1,10]): $\times 1$ clump thickness, $\times 2$ uniformity of cell size, $\times 3$ uniformity of cell shape, $\times 4$ marginal adhesion, $\times 5$ single epithelial cell size, $\times 6$ bare nuclei, $\times 7$ bland chromatin, $\times 8$ normal nuclei, and $\times 9$ mitosis. The reference data-set is composed of about 680 instances and the final class distribution is 65.5% benign and 34.5% malignant, respectively.

This benchmark has been given as input to the fuzzy clustering algorithm. Specifically, in this example, we have exploited the FCM algorithm. Figure 23.4 shows the data distribution of each feature on the population sample. In particular, last graph in figure, represents the expected "ideal" crisp classification of data in the two given class. Compared to our results, the classification error is not meaningful (the percentage of error is 3% and 4.7% in the benign and malignant cancers classes, respectively) and, although the nature "fuzzy" of our clustering, it represents correctly the two classes too. Additional techniques to validate the obtained results (validity criteria) or to refine the classification are not meaningful in this example, due to the well-defined benchmark considered herein.

Dynamic Fuzzy Control Design: The classification produced in the previous phase is translated into fuzzy rules. More specifically, exploiting the cylindrical extension technique (a projection-based method of n-dimensional argument vector: see [12]), the generic i-th fuzzy cluster can be described by a fuzzy rule, where each features is described through a fuzzy set.

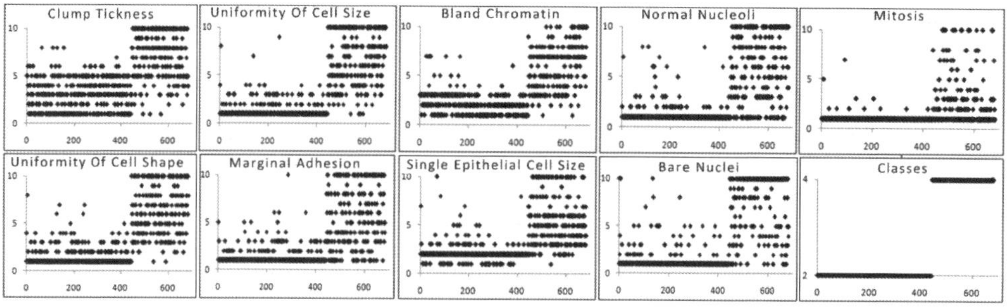

Fig. 23.10. Features representation and clustering distribution

This way, a collection of rules is generated. Just an example, a possible (simple) rule elicited by the given data sample could be:

R1 : **If** clump thickness is WIDE **then** "possible malignant cancers" (w=0.5)

where clump thickness is the feature, WIDE is the membership functions, interpreted as linguistic label. The expression "possible malignant cancers" represents one of clusters and w is the weight of the firing strength of the rule.

The generated fuzzy rules are collected and injected in the *medical diagnosis agent pool*, through a FIS or FCL file.

The following code is a sketched piece of code, in FCL language:

```
FUZZIFY Clump_Thickness
 TERM Clump_Thickness0 := (0.0, 0.0)... (1.0, 0.958)
                          (9.0, 0.24) (10.0, 0.223);
  TERM Clump_Thickness1 := (0.0, 0.0) ...(9.0, 0.751)
                          (10.0, 0.777);
END_FUZZIFY
```

Let us suppose Cluster_1 is the FCL term associated to "possible malignant cancers"

```
RULEBLOCK block_1
RULE 1 : IF (Clump_Thickness IS WIDE_0) THEN Cluster_1
IS w_05 ;
END_RULEBLOCK
```

where WIDE_0 and w_05 represent the linguistic variable and the firing strength of the rule, respectively.

Thus the agents pools are ready to interpret these files and infer new knowledge in order to process accurately the user request.

23.6. Conclusions

This paper presents a general approach for modeling a system which supports the medical diagnosis. Through a sequence of phases, general guidelines are given for dealing with medical profilation of pathologies and diseases. Then a specific instance of this model is described, adapted to a real case study. Final result is a flexible architecture for the diagnostic processing of clinical data. The nature of this approach makes it easily adaptable and scalable: a distributed architecture with a distributed knowledge is admissible, too. Furthermore the use of ontologies is fundamental for the usability of the system: they enable the semantic description of data and the inter-relationships among concepts.

A future extension of this approach foresees a building of a distributed model, that integrates different medical domains, in order to enable the access and consultation of different pathologies, thanks to adequate ontologies-based representation.

The case study carried forward in the whole modelling process, provides an idea of flexibility of this model and demonstrates its admissibility. It represents an effective research and validation strategy for investigating a real-life context.

Acknowledgment

This work has been supported by RS Group in the framework of the project "Phoenix", founded by the Minister of Productive Activities (MAP). We thank all the RS team for the fruitful discussion on the analysis of medical context, on the characterization of the pathology domain and for the identification of relevant data.

References

1. K. S. Park, Y.M. Chae, M. Park, Developing a knowledge-based system to automate the diagnosis of allergic rhinitis, *Beomed. Fuzzy Hum. Sci.* 2 (1996) 9–18.
2. K. P. Adlassnig, Fuzzy set theory in medical diagnosis, IEEE transactions on Systems, Man, and Cybernetics SMC, vol. 16, pp. 260–265, 1986.
3. J.F.-F. Yao, J.-S.Yao, Fuzzy decision making for medical diagnosis based on fuzzy number and compositional rule of inference, Fuzzy Sets and Systems, 120 (2001), pp. 351–366.
4. J.C. Bezdek, L.O. Hall, L.P. Clarke, Review of MR image segmentation techniques using pattern recognition, *Med. Phys.* 20(4) (1993) 1033–1048.
5. H. Song, S. Lee, D. Kim, G. Park, New methodology of Computer Aided Diagnostic System on Breast Cancer. *Proceeding of International Symposium on Neural Networks*, pp. 780–789, 2005.
6. J. Jang. ANFIS: Adaptive-Network based Fuzzy Inference System. *Proceedings of IEEE Transaction on System, Man and Cybernetics* 3 pp. 665–685, 1993.
7. C. Pena-Reyes, M. Sipper, Designing Breast Cancer Diagnostic System via a Hybrid Fuzzy-Genetic Methodology. *Proceedings of IEEE International Fuzzy Systems Conference* 1, pp. 135–139, 1999.
8. Nauck D. and Kruse R. Obtaining interpretable fuzzy classification rules from medical data. *Artificial Intelligence in Medicine*, 16: 149–169, 1999.
9. Ishibuchi H., Nakashima T., Murata T. (1999) Performance evaluation of fuzzy classifier systems for multidimensional pattern classification problems. IEEE Trans. SMC–B 29, 601–618.
10. Roubos J.A., Setnes M. (2000) Compact fuzzy models through complexity reduction and evolutionary optimization. In FUZZ-IEEE, pp. 762–767, May 7–10, San Antonio, USA.

11. Abonyi, J. and Szeifert, F. 2003. Supervised fuzzy clustering for the identification of fuzzy classifiers. *Pattern Recogn. Lett.* 24, 14 (Oct. 2003), 2195–2207.
12. F. Hoppner, F. Klawonn, R. Kruse and T. Runkler. Fuzzy Cluster Analysis-Methods for Image Recognition. J. Wiley, N. York, 1999.
13. Fenza G., Loia V., Senatore, S., "A Hybrid approach to Semantic Web Services Matchmaking", accepted on International Journal of Approximate Reasoning.
14. Available URL: http://ffll.sourceforge.net/fcl.htm
15. OWL Services Coalition, OWL-S: Semantic Markup for Web Services, 2004, Available: http://www.daml.org/services/owl-s/.
16. K. P. Sycara, M. Paolucci, A. Ankolekar, N. Srinivasan, Automated discovery, interaction and composition of Semantic Web services, *J. Web Sem.*, 1(1) pp. 27–46, 2003.
17. S. Zahan, "A fuzzy approach to computer-assisted myocardial ischemia diagnosis," *Artif. Intell. Med.*, vol. 21, no. 1–3, pp. 271–275, 2001.
18. S. Mitra,; Y. Hayashi,; Neuro-fuzzy rule generation: survey in soft computing framework. *IEEE Transactions on Neural Networks*, Volume 11, Issue 3, May 2000 Page(s):748–768.
19. M.E. Cohen, D.L. Hudson, Neural Network Modeling with Medical Data, ISCA Intelligent and Adaptive Systems in Software Engineering, 13:135–138, 2004.
20. J. Sim, W. L. Tung, C. Quek, "FCMAC-Yager: A Novel Yager-Inference-Scheme-Based Fuzzy CMAC," IEEE Transactions on Neural Networks, vol. 17, no. 6, pp. 1394–1410, Nov. 2006.
21. L. W. Man, L. Wai, S. L. Kwong, S. N. Po, and J. C. Y. Cheng, "Discovering knowledge from medical databases using evolutionary algorithms," *IEEE Eng. Med. Biol.*, vol. 19, pp. 45–55, July/Aug. 2000.
22. C. A. Peña-Reyes and M. Sipper, "Evolutionary computation in medicine: An overview," *Artif. Intell. Med.*, vol. 19, no. 1, pp. 1–23, 2000.
23. Meesad, P.; Yen, G.G., "Combined numerical and linguistic knowledge representation and its application to medical diagnosis," *IEEE Transactions on Systems, Man and Cybernetics, Part A*, vol. 33, no. 2, pp. 206–222, March 2003.
24. de Toro, F., Ros, E., Mota, S., Ortega, J., "Evolutionary algorithms for multiobjective and multimodal optimization of diagnostic schemes," *IEEE Transactions on Biomedical Engineering*, vol. 53, no. 2, pp. 178–189, Feb. 2006.
25. Lessmann, S.; Stahlbock, R.; Crone, S.F. Genetic Algorithms for Support Vector Machine Model Selection International Joint Conference on Neural Networks, 2006. IJCNN 06. Page(s):3063–3069.
26. G. Berks, D.G. Keyserlingk, J. Jantzen, M. Dotoli, H. Axer, Fuzzy clustering-a versatile mean to explore medical database, ESIT2000, Aachen, Germany.
27. Kiliç, K., Uncu, Ö., and Türksen, I. B. Comparison of different strategies of utilizing fuzzy clustering in structure identification. *Inf. Sci.* 177, 23, 2007.
28. G. Castellano, A. M. Fanelli, C. Menear, "A fuzzy clustering approach for mining diagnostic rules," Aoc. Of IEEE Coqferencr on Sysfrms, Man undCj∼bemelics Washington, USA, 2003.
29. Bagher-Ebadian, H., Soltanian-Zadeh, H., Setayeshi, S., Smith, S.T., "Neural network and fuzzy clustering approach for automatic diagnosis of coronary artery disease in nuclear medicine," *Nuclear Science, IEEE Transactions* on, vol. 51, no. 1, pp. 184–192, Feb. 2004.
30. Hoppner F., Klawonn F., Kruse R., Runkler T. Fuzzy Cluster Analysis — Methods for Image Recognition. J. Wiley, N. York, 1999.
31. J. C. Bezdek, Pattern Recognition with Fuzzy Objective Function Algorithms. Plenum Press, New York, 1981.

32. J.C. Bezdek, "Cluster validity with fuzzy sets," *J. Cybernet*, vol. 3, pp. 58–73, 1973.
33. M.P. Windham, "Cluster validity for fuzzy clustering algorithms," Fuzzy Sets and Syst., vol. 5, pp. 177–185, 1981.
34. B.L.M.R. Rezae and J. Reiber, "A new cluster validity index for the fuzzy c-means," Pattern Recognition Letters, vol. 19, pp. 237–246, 1998.
35. X. Xie and G. Beni, "A Validity Measure for Fuzzy Clustering," *IEEE Trans. Pattern Analysis and Machine Intelligence (PAMI)*, vol. 13, no. 8, pp. 841–847, 1991.
36. Takagi, T. and Sugeno, M.: 'Fuzzy identification of systems and its applications to modeling and control', *IEEE Trans. Syst., Man Cybern.*, 1985, 15, pp. 116–132.
37. C.C. Kung and J.Y. Su Affine Takagi-Sugeno fuzzy modelling algorithm by fuzzy c-regression models clustering with a novel cluster validity criterion, *IET Control Theory Appl.*, 2007, 1(5), pp. 1255–1265.
38. Dragan Kukolj Design of adaptive Takagi–Sugeno–Kang fuzzy models, *Applied Soft Computing* 2 (2002) 89–103.
39. M.Setnes, R.Babuska, U. Kaymak, and H. R. van Nauta Lemke Similarity Measures in Fuzzy Rule Base Simplification, IEEE Transactions on Systems, Man, and Cybernetics — Part B: Cybernetics, vol. 28, no. 3, June 1998.